Companion to Urban Design

Today the practice of urban design has forged a distinctive identity with applications at many different scales ranging from the block or street scale to the scale of metropolitan and regional landscapes. Urban design interfaces many aspects of contemporary public policy: multiculturalism, healthy cities, environmental justice, economic development, climate change, energy conservation, protection of natural environments, sustainable development, community liveability, and the like. The field now comprises a core body of knowledge that enfolds a rich history of ideas, paradigms, principles, tools, research, and applications, enriched by eclectic influences from the humanities, and social and natural sciences.

Companion to Urban Design includes more than 50 original contributions from internationally recognized authorities in the field. These contributions address the following questions: what are the important ideas that have shaped the field and the current practice of urban design? What are the major methods and processes that have influenced the practice of urban design at various scales? What are the current innovations relevant to the pedagogy of urban design? What are the lingering debates, conflicts and contradictions in the theory and practice of urban design? How could urban design respond to the contemporary challenges of climate change, sustainability, active living initiatives, globalization, and the like? What are the significant disciplinary influences on the theory, research, and practice of urban design in recent times?

There has never before been a more authoritative and comprehensive companion that includes core, foundational, and pioneering ideas, and concepts of urban design. This book serves as an invaluable guide for undergraduate and postgraduate students, future professionals, and practitioners interested in architecture, landscape architecture, and urban planning, and also interested in urban studies, urban affairs, geography, and related fields.

Tridib Banerjee holds the James Irvine Chair in Urban and Regional Planning at the University of Southern California School of Policy, Planning and Development. His research focuses on the design and planning of the built environment and the related human and social consequences. In particular, he is interested in the political economy of urban development, and the effects of globalization in the transformation of urban form and urbanism from a comparative international perspective.

Anastasia Loukaitou-Sideris is Professor of Urban Design at the University of California, Los Angeles, Department of Urban Planning. Her research focuses on the public environment of the city, its design, social meaning, and impact on urban residents. In particular, her work is characterized by a "user focus" in that she seeks to analyze and understand the built environment from the perspective of those who live and work there.

Companion to Urban Design

Edited by
Tridib Banerjee and
Anastasia Loukaitou-Sideris

Routledge
Taylor & Francis Group
LONDON AND NEW YORK

First published 2011
Paperback edition first published 2014
by Routledge
2 Park Square, Milton Park, Abingdon, Oxon OX14 4RN

and by Routledge
711 Third Avenue, New York, NY 10017

Routledge is an imprint of the Taylor & Francis Group, an informa business

British Library Cataloguing in Publication Data
A catalogue record for this book is available from the British Library

Library of Congress Cataloging in Publication Data
A catalog record for this book has been requested

ISBN: 978-0-415-55364-3 (hbk)
ISBN: 978-1-138-77654-8 (pbk)
ISBN: 978-0-203-84443-4 (ebk)

Typeset in Bembo
by Cenveo Publisher Services

Contents

List of figures x
List of tables xiii
List of contributors xiv
Acknowledgments xxi

Introduction - urban design 1

Part 1: Roots 7

Introduction

1 From CIAM to CNU: the roots and thinkers of modern urban design 9
Eugénie L. Birch

2 The open and the enclosed: shifting paradigms in modern urban design 30
Robert Fishman

3 Pedagogical traditions 41
Danilo Palazzo

Part 2: Theoretical perspectives 53

Introduction

4 Urban design: an incompletely theorized project 57
Niraj Verma

5 The two orders of cybernetics in urban form and design 70
M. Christine Boyer

6 Urban design and spatial political economy 84
Alexander Cuthbert

7 Critical urbanism: space, design, revolution 97
Kanishka Goonewardena

Part 3: Influences 109

Introduction

8 Urban design and the traditions of geography 113
Larry R. Ford

9 Influences of sociology on urban design 125
William Michelson

10 Influences of anthropology on urban design 137
Denise Lawrence-Zúñiga

11 Feminist approaches to urban design 150
Kristen Day

12 Environmental psychology and urban design 162
Jack L. Nasar

13 The law of urban design 175
Jerold S. Kayden

14 Political theory and urban design 186
Margaret Kohn

15 Interactions between public health and urban design 198
Marlon G. Boarnet and Lois M. Takahashi

16 Urban design and the cinematic arts 208
Rafael E. Pizarro

Part 4: Technologies and methods 219

Introduction

17 Design studios 223
Kathryn H. Anthony

18 Media tools for urban design 238
Martin H. Krieger

19 Visualizing change: simulation as a decision making tool 249
Peter Bosselmann

20 City design in the age of digital ubiquity 261
Eran Ben-Joseph

Part 5: Process 275

Introduction

21 Customs, norms, rules, regulations, and standards in design practice 277
William C. Baer

22 Decoding design guidance 288
Matthew Carmona

23 Urban design competitions 304
Ute Lehrer

24 The design charrette 317
Douglas S. Kelbaugh

25 Citizen design: participation and beyond 329
Jeffrey Hou

Part 6: Components 341

Introduction

26 Downtown urban design 345
Anastasia Loukaitou-Sideris and Tridib Banerjee

27 Suburbs: *rus in urbe*, the picturesque, and selfhood 356
John Archer

28 Planned communities and new towns 369
Ann Forsyth

29 Neighborhood spaces: design innovations and social themes 379
Ajay Garde

30 Spaces of consumption 391
Klaus R. Kunzmann

31 Cultural institutions: the role of urban design 405
Carl Grodach

32 Streets and the public realm: emerging designs 419
Elizabeth Macdonald

33 Mixed-life places 432
Mark Francis

34 Urban flux 446
Gary Hack

Part 7: Debates 463

Introduction

35 Compactness vs. sprawl 467
Reid Ewing, Keith Bartholomew, and Arthur C. Nelson

36 Living together or apart: social mixing, social exclusion, and gentrification 484
Ali Madanipour

37 Beyond placelessness: place identity and the global city 495
Michael Southworth and Deni Ruggeri

38 Old vs. new urbanism 510
Ivonne Audirac

39 Form-based codes vs. conventional zoning 526
Emily Talen

Part 8: Global trends 537

Introduction

40 City branding 541
Jon Lang

41 From metropolitan to regional urbanization 552
Edward W. Soja

42 Ethnoscapes 562
Clara Irazábal

43 Urban design for a planet of informal cities 574
Vinit Mukhija

Part 9: New directions 585

Introduction

44 Postmodern and integral urbanism 589
Nan Ellin

45 Ecological urbanism 600
Anne Whiston Spirn

46 Metropolitan form and landscape urbanism 611
Brenda Scheer

47 Intertwist and intertwine: sustainability, meet urban design 619
Randolph T. Hester and Marcia J. McNally

48 Smart growth: a critical review of the state of the art 632
Aseem Inam

49 Notes on transit-oriented development 644
Stefanos Polyzoides

50 Placemaking in urban design 654
Kathy Madden

51 Secure cities 663
Carolyn Whitzman

52 Design for resilient cities: reflections from a studio 674
Mahyar Arefi

Epilogue 686
Index 689

Figures

1.1	CIAM (1933) meeting and later publications	13
1.2	Michael Reese Hospital area before and after urban renewal	17
1.3	Patrick Geddes' and Team 10 drawings	23
8.1	Bangladeshi neighborhood in East London	117
8.2	Skyline in Jakarta	118
8.3	Stoops in Baltimore	119
8.4	Rapid change in West Los Angeles	120
10.1	New Chinatown, Los Angeles	141
10.2	Valley Blvd, City of San Gabriel, California	141
12.1	Basis for environmental response	163
12.2	Affordances for sitting	164
12.3	Dimensions of Environmental Appraisal	166
12.4	Behavior setting for outdoor eating	169
14.1	Political march, Avenida Juárez in Mexico City	193
14.2	Plaza de la República in Mexico City	195
17.1	East St Louis Action Research Project	225
17.2	Design studio at the Gulf Coast in the wake of Hurricane Katrina	228
18.1	Multiple views of Hiratsuka, a suburb of Tokyo	246
18.2	Paris Marville Google Map	247
19.1	Conceptional Representation of City Form, Venice Biennale	250
19.2	Simulated scenarios for San Francisco's skyline	255
19.3	San Francisco skyline with proposed Transit Tower	257
20.1	Real Time Rome	264
20.2	University College London 3-D virtual model	267
20.3	Illuminating Clay combines physical models and digital information	269
20.4	Wikitude by Mobilizy	269
21.1	Normative evaluation	281
21.2	Differences between criteria and standards	281
21.3	Eight permutations of rule forms	283
22.1	Example case study – Swindon	293
22.2	Example case study – Newhall	294
22.3	Coding and the development process	297

23.1 Potsdamer Platz and Leipziger Platz site 311
23.2 A+T buildings at Potsdamer Platz, Berlin 313
23.3 Sony Headquarters at Potsdamer Platz, Berlin 313
23.4 Part of the Sony Complex, Potsdamer Platz, Berlin 314
23.5 Debis at Potsdamer Platz, Berlin 314
24.1 Sites of University of Michigan charrettes 318
24.2 A Detroit charrette visit from Senator Levin 320
24.3 Four teams at a Detroit charrette in full swing 322
24.4 Several team leaders prepare for the public presentation 322
25.1 Open house in Seattle's Chinatown International District 335
25.2 Belltown P-Patch, citizen-initiated project in Seattle 336
25.3 Fruitvale Village, Oakland, California 338
26.1 Nicollet Transit Mall, Minneapolis 349
26.2 Lofts in downtown Los Angeles 351
27.1 Letchworth Garden City, England 359
27.2 Prairie Crossing, Grayslake, Illinois 364
28.1 Cumbernauld, Scotland 375
28.2 Tsukuba Science City on the outskirts of Tokyo 376
29.1 Neighborhood unit diagram 383
29.2 Conventional and neo-traditional suburban development 386
30.1 Das Schloss: Shopping Arcade in Berlin 394
30.2 Façade of war-demolished Schloss in Braunschweig 398
30.3 The Place: Chaoyang district, Beijing 400
30.4 Qianmen Street, Beijing 400
30.5 Venusfort, Tokyo 402
31.1 Centre Pompidou, Paris 410
31.2 Abandoibarra redevelopment area, Bilbao 412
31.3 Harley-Davidson Museum site plan 413
31.4 Lincoln Center Redevelopment, 65th Street Panorama 415
32.1 Octavia Boulevard, San Francisco 424
32.2 Street in Amsterdam 425
32.3 Rue des Petits Carreaux, Paris 425
32.4 Boulevard Magenta, Paris 426
33.1 Outdoor café in Oslo 433
33.2 The Pearl District in Portland, Oregon 436
33.3 Central Park, Davis, California 439
33.4 Central Park, Davis, California, was purposefully designed as a mixed-life place 442
34.1 Leon Krier's sketch for Dorset Village of Poundbury 448
34.2 Neighborhood in Poundbury, England 448
34.3 Street in Hong Kong 450
34.4 Shinjuku, Tokyo 451
34.5 Doge's Palace, Saint Mark's Square in Venice 452
34.6 Madeleine Church, Paris 452
34.7 Temporary scaffolding on the Washington Monument 453
34.8 The GreenPix Zero Energy Wall in Beijing 454
34.9 Times Square, New York 455
34.10 Times Square, New York 456

34.11 Murals in a North Philadelphia neighborhood 457
34.12 Mural in a North Philadelphia neighborhood 457
34.13 Village of Arts and Humanities, Philadelphia's Badlands 458
34.14 Melrose Avenue in Los Angeles 459
35.1 Endless Los Angeles 468
35.2 Satellite images of Portland and Raleigh 470
35.3 2003 Housing Supply vs. 2025 Housing Demand in the US 472
35.4 Public–Private Partnership at Baldwin Park, California 478
36.1 Times Square, New York 486
36.2 Urban generation project in Dublin, Ireland 488
36.3 Rappongi Hills complex in Tokyo 491
37.1 Pike Place Market, Seattle 497
37.2 Free Speech Movement site at UC Berkeley 498
37.3 Fourth of July celebration in Woodbridge, Irvine 503
37.4 Librino, Italy 505
37.5 Child's drawing of Librino 506
38.1 Aerial view of Seaside and Watercolor, Florida 512
38.2 Aerial view of Celebration, Florida 513
38.3 Gruen's Metrocore and Peter Calthorpe's Urban Network 515
38.4 Village Center One-Way-Couplets in San Elijo, California 517
39.1 Page from *SmartCode* 530
39.2 Page from *SmartCode* 531
40.1 Eiffel Tower and Paris in the background 544
40.2 La Rambla, Barcelona 545
42.1 Plaza Mexico in Lynwood, California 565
42.2 Plaza Mexico in Lynwood, California 565
43.1 Diagram of open space proposal in Kibera, Nairobi, Kenya 578
43.2 Usable open space in Kibera, Nairobi, Kenya 578
44.1 Palo Verde Library and Maryvale Community Center in Phoenix 592
44.2 The Grove at Arizona State University 593
46.1 Aerial Image of Texas Stadium 613
46.2 High Line Park in New York City 615
47.1 Authors' sketches for Augustus Hawkins Park in Los Angeles 625
47.2 Windmill at Augustus Hawkins Park 625
47.3 Children at Augustus Hawkins Park 626
48.1 Street in Portland 634
48.2 Rio Vista West transit-oriented development in San Diego 636
49.1 Diagram of transit station types by transect zone 646
49.2 Mission/Meridian Village, South Pasadena, CA 647
49.3 Del Mar Station, Pasadena, CA 648
49.4 Del Mar Station, Pasadena, CA 650
50.1 Los Angeles bus stop before improvements 655
50.2 Los Angeles bus stop after improvements 655
50.3 Project Driven Design Approach 658
50.4 Place/Community Driven Design Approach 659
50.5 Discovery Green, Houston 660
52.1 Rubik's Cube overlaid on study area 677
52.2 Schematic diagram from urban design studio 678
52.3 Elements of three concepts of resiliency 683

Tables

1.1	From CIAM to CNU: the roots of urban design	10
3.1	Graduate programs in Urban Design taught in English	45
4.1	Nature of problems	61
4.2	Goal and purpose	62
4.3	Institutional orientation	64
4.4	High and low theory	66
6.1	Theoretical foundations of three environmental disciplines	86
6.2	The design properties of cities within modernism and postmodern globalization	90
21.1	Types of rules, standards, and regulations related to urban development and the nature of their construction	284
22.1	Design guidance compared	291
22.2	The roles and motivations of key stakeholders within a typical coding process	298
22.3	Design codes, building on the site-based vision	301
23.1	Selection process in relation to certainty of outcome	307
33.1	Some principles of mixed-life places	438
33.2	A typology of mixed-life places and some movements that support them	440
35.1	Estimates of motor vehicle costs	473
35.2	Typical elasticities of travel with respect to four D variables	475
52.1	Concepts, themes, principles, and types of urban resiliency	680
52.2	Resiliency concepts by form, function, and flow	680

Contributors

Kathryn H. Anthony is Professor at the School of Architecture, University of Illinois at Urbana-Champaign. She holds the title of Distinguished Professor from the Association of Collegiate Schools of Architecture and is author of *Designing for Diversity: Gender, Race, and Ethnicity in the Architectural Profession, Design Juries on Trial: The Renaissance of the Design Studio* and over 100 publications.

John Archer is Chair of the Department of Cultural Studies and Comparative Literature at the University of Minnesota. His most recent book, *Architecture and Suburbia*, was the winner of the 2007 Alice Davis Hitchcock award from the Society of Architectural Historians.

Mahyar Arefi is Associate Professor at the University of Cincinnati, College of Design, Architecture, Art and Planning. He has practiced architecture and urban design for many years. His research interests include urban design and community development.

Ivonne Audirac is Associate Professor at Florida State University. Her current research includes the effectiveness of neo-traditional design and rural applications of sustainable development. She is editor of the book *Rural Sustainable Development in America.*

William C. Baer is Professor Emeritus at the University of Southern California. His has written on planning history, housing policy, planning standards, political theory and public policy. His publications include *Beyond the Neighborhood Unit: Residential Environments and Public Policy* (with T. Banerjee).

Tridib Banerjee is James Irvine Chair in Urban and Regional Planning and Director of Graduate Programs in Urban Planning at USC. He is the co-author of *Beyond the Neighborhood Unit* and *Urban Design Downtown: Poetics and Politics of Form*, and the co-editor of *City Sense and City Design: Writings and Projects of Kevin Lynch.*

Keith Bartholomew is Assistant Professor in the University of Utah's Department of City and Metropolitan Planning. He is the former director of the LUTRAQ project and co-author of *Growing Cooler: The Evidence on Urban Development and Climate Change.*

Eran Ben-Joseph is Associate Professor of Urban Planning and the Head of the City Design and Development group at the Massachusetts Institute of Technology. He is the author of *The Code of the City: Standards and the Hidden Language of Placemaking, Streets and the Shaping of Towns and Cities* (with M. Southworth), and *Regulating Place: Standards and the Shaping of Urban America* (with T. Szold).

Eugénie Birch, Lawrence C. Nussdorf Professor of Urban Research, University of Pennsylvania, is co-director of Penn Institute for Urban Research. Her most recent books are *Local Planning, Contemporary Principles and Practice* (with Gary Hack *et al.*) (2009), *Urban and Regional Planning Reader* (2009), *Growing Greener Cities: Urban Sustainability in the 21st Century* (2008) and *Rebuilding Urban Places after Disaster: Lessons from Katrina* (2006) (with Susan Wachter).

Marlon G. Boarnet is Professor in the Department of Planning, Policy, and Development at the University of California, Irvine. He has published extensively on the links between urban design and travel behavior. He is the co-author of the book *Travel by Design.*

Peter Bosselmann is Professor of Urban Design in Architecture, City and Regional Planning and in Landscape Architecture and Environmental Planning at the University of California, Berkeley. He is co-chair of Berkeley's Master of Urban Design program. His publications include *Urban Morphology – Understanding City Design* and *Representation of Places: Reality and Realism in City Design.*

M. Christine Boyer is William R. Kenan, Jr. Professor of Architecture at Princeton University. She is the author of *Dreaming the Rational City: The Myth of American City Planning 1890–1945, Manhattan Matters: Architecture and Style 1850–1900, The City of Collective Memory,* and *Cybercities.*

Matthew Carmona is Professor of Planning and Urban Design and Head of the Bartlett School of Planning, University College London. He has published in the areas of urban design, design policy and guidance, measuring quality in planning, and the management of public space.

Alexander Cuthbert is Professor of Planning and Urban Development at the University of New South Wales, Australia. His publications include *Designing Cities: Critical Readings in Urban Design* and *The Form of Cities: Political Economy and Urban Design.*

Kristen Day is Professor of Planning, Policy, and Design at the University of California, Irvine. Her research explores diversity and social justice in the design, use and meaning of urban environments.

Nan Ellin is Professor and Chair, Department of City and Metropolitan Planning, University of Utah. She is the author of *Postmodern Urbanism,* and *Integral Urbanism;* the editor of *Architecture of Fear,* and has collaborated with Edward Booth-Clibborn on *Phoenix: 21st-Century City.*

Reid Ewing is Professor, Department of City and Metropolitan Planning, University of Utah, Fellow of the Urban Land Institute, and author of *Best Development Practices, Transportation and Land Use Innovations*, and *Growing Cooler: The Evidence on Urban Development and Climate Change.*

Robert Fishman is Professor of Architecture and Urban Planning at the Taubman College of the University of Michigan. He is the author of *Bourgeois Utopias: The Rise and Fall of Suburbia* and *Urban Utopias in the Twentieth Century: Ebenezer Howard, Frank Lloyd Wright, and Le Corbusier.*

Larry R. Ford was Professor in the Department of Geography at San Diego State University and the author of the books *Southern California Extended* (with E. Griffin), *Cities and Buildings: Skyscrapers, Skidrows, and Suburbs*; *America's New Downtowns: Revitalization or Reinvention? Spaces Between Buildings,* and *Metropolitan San Diego.*

Ann Forsyth is Professor of City and Regional Planning at Cornell University. She is the author of *Constructing Suburbs: Competing Voices in a Debate Over Urban Growth, Reforming Suburbia: The Planned Communities of Irvine, Columbia, and The Woodlands*, and *Designing Small Parks: A Manual for Addressing Social and Ecological Concerns.*

Mark Francis, FASLA, is Professor Emeritus of Landscape Architecture and Environmental Design at the University of California, Davis. His design and research focuses on the use and meaning of public spaces. His books include *Urban Open Space, Public Space* and *Community Open Spaces.*

Ajay Garde is Associate Professor in the Department of Planning, Policy, and Design at the University of California, Irvine. His interests include sustainable growth and innovations in urban design, their impact on urban and suburban form, and their implications for public policy.

Kanishka Goonewardena is Associate Professor and Director, Program in Planning, Department of Geography at the University of Toronto, Canada. He is the co-editor of *Space, Difference, Everyday Life: Reading Henri Lefebvre* (2008) and the author of a chapter on Henri Lefebvre in *New Blackwell Companion to Major Social Theorists* (forthcoming).

Carl Grodach is Assistant Professor in the School of Urban and Public Affairs at the University of Texas at Arlington. His research focuses on the arts, culture and urban redevelopment.

Gary Hack is Professor and Dean Emeritus of the School of Design, University of Pennsylvania. He has prepared urban design plans for over 35 cities in North America and Asia and was a collaborator in the planning of the redevelopment of the World Trade Center site. He is the author of *Site Planning* (with Lynch), *Urban Design in the Global Perspective* (with Lin and Kuang) and *Global Regional Cities* (with Simmonds).

Randolph T. Hester is Professor and former chair of the University of California, Berkeley, Department of Landscape Architecture and Environmental Planning. He is the author of *Design for Ecological Democracy*, *Neighborhood Space*, *Planning Neighborhood Space with People*, and *Community Design Primer*.

Jeffrey Hou is Chair and Associate Professor in the Department of Landscape Architecture at the University of Washington. He is the editor of *Insurgent Public Space: Guerilla Urbanism and the Remaking of Contemporary Cities* and a co-author of *Greening Cities, Growing Communities: Learing from Seattle's Urban Community Gardens.*

Aseem Inam is Associate Professor of Urbanism at The New School in New York City. He is the author of *Planning for the Unplanned: Recovering from Crises in Megacities.*

Clara Irazábal is Assistant Professor of International Urban Planning at Columbia University's Graduate School of Architecture, Planning and Preservation in New York City. Her research focuses on processes and politics of placemaking, especially in Latin America and Latina/o US, and their impact on community development and socio-spatial justice.

Jerold S. Kayden is a lawyer, city planner, Professor of Urban Planning at the Harvard Graduate School of Design, and principal constitutional counsel to the National Trust for Historic Preservation. He is the author of *Privately Owned Public Space: The New York City Experience*; *Landmark Justice: The Influence of William J. Brennan on America's Communities*; and *Zoning and the American Dream: Promises Still To Keep*.

Douglas S. Kelbaugh is the former Dean and now Professor of Architecture and Urban Planning at the University of Michigan's Taubman College of Architecture and Urban Planning. He is the co-author of *The Pedestrian Pocket Book*, the author of *Common Place: Toward Neighborhood and Regional Design*, and *Repairing the American Metropolis: Common Place Revisited*, and the co-editor of *Writing Urbanism, a Design Reader.*

Margaret Kohn is Associate Professor of political theory at the University of Toronto. She is the author of *Radical Space: Building the House of the People* and *Brave New Neighborhoods: The Privatization of Public Space.*

Martin H. Krieger is Professor of Planning at the University of Southern California. His publications include: *Advice and Planning*, *Tools for the Crafts of Knowledge and Decision*, *What's Wrong with Plastic Trees? Artifice, and Authenticity in Design*. He is currently writing a book on *Taking Pictures in the City*.

Klaus R. Kunzmann has been Professor and Director of the Institut für Raumplanung at the Technical University of Dortmund. He is the author of *Venice, Venice, and Venice: Three Realities of the European Cities,* and *The Future of European City: Qingdao Celebration, or Las Vegas?* Recently he co-edited the book *China and Europe: The Implications of the Rise of China for European Space.*

Jon Lang is Professor of Architecture at the University of New South Wales in Sydney, Australia, and Director for urban design of the Environmental Research Group in Philadelphia. He has authored books on urban design, the relationship between people and the built environment, and modern architecture in India.

Denise Lawrence-Zúñiga is Professor of Architecture at California State Polytechnic University, Pomona. She is the co-editor of *House Life: Space, Place and Family in Europe* (1999) with Donna Birdwell-Pheasant, and *The Anthropology of Space and Place* (2003) with Setha Low.

Ute Lehrer is Associate Professor in the Environmental Studies Program of York University, Toronto, Canada. She has written on cities and globalization, economic restructuring and urban form, the political economy of the built environment, and immigrant landscapes.

Anastasia Loukaitou-Sideris is Professor and former Chair of the Department of Urban Planning at UCLA. She is the co-author of *Urban Design: Poetics and Politics of Form* and *Sidewalks: Conflict and Negotiation over Public Space* and the co-editor of *Jobs and Economic Development in Minority Communities.*

Elizabeth Macdonald is a Principal of *Cityworks*, Associate Professor in the Department of City and Regional Planning at the University of California, Berkeley, and co-author of *The Boulevard Book: History, Evolution, Design of Multiway Boulevards* and the *Urban Design Reader.*

Marcia McNally is an Associate Adjunct Professor at the University of California, Berkeley, Department of Landscape Architecture and Environmental Planning. Her work addresses the form of the ecological city, actions needed for sustainable outcomes, and the tools that inform decisions.

Ali Madanipour is Professor in Urban Design at the University of Newcastle upon Tyne. His books include *Design of Urban Space, Public and Private Spaces of the City*; *Social Exclusion in European Cities*; *Tehran: The Making of a Metropolis*; and *Whose Public Space?*

Kathy Madden is a Senior Vice President at Project for Public Spaces, Inc. She has been involved in evaluating public spaces and working with communities across the globe.

William Michelson is S.D. Clark Professor of Sociology Emeritus at the University of Toronto. His books include *Man and his Urban Environment: A Sociological Approach*; *Methods in Environmental and Behavioral Research*; *Environmental Choice, Human Behavior and Residential Satisfaction*; the *Handbook of Environmental Sociology*; and *Time Use: Expanding Explanation in the Social Sciences.*

Vinit Mukhija is an Associate Professor of Urban Planning at the University of California, Los Angeles. His research focuses on housing and the built environment in developing countries, and Third World-like housing conditions in the United States. He is the author of *Squatters as Developers? Slum Redevelopment in Mumbai.*

Jack L. Nasar is a Professor of City and Regional Planning at the Ohio State University and Editor of the *Journal of Planning Literature*. His books include *The Evaluative Image of the City*, *Design by Competition*, and *Designing for Designers.*

Arthur C. Nelson is Presidential Professor of City and Metropolitan Planning at the University of Utah where he also directs the Metropolitan Research Center and the Master of Real Estate Development program. His work focuses on market responses to planning and development policy.

Danilo Palazzo is an architect and Associate Professor of spatial planning at the Politecnico di Milano, Italy. He is the author of *Sulle spalle dei giganti* (1997); *Paesaggio e Territorio* (2001 with Canevari); *Transforming the Places of Production* (2002 with Fossa, Lane, Pirani); *Margini* (2006 with Treu); and *Urban Design* (2008).

Rafael E. Pizarro is Lecturer of Sustainable Urban Planning in the Faculty of Architecture, Design and Planning at University of Sydney and Visiting Lecturer (*Gastprofessor*) at the Technical University of Berlin. His book-in-progress is titled *Suburbanizing the Mind: Hollywood and the Globalization of American Suburbia.*

Stefanos Polyzoides is a Principal of Moule & Polyzoides, Architects and Urbanists of Pasadena, California. He is a co-founder of the Congress for the New Urbanism and member of its Board of Directors. He is the co-author of *Los Angeles Courtyard Housing: A Typological Analysis and The plazas of New Mexico* and author of *R.M. Schindler, Architect.*

Deni Ruggeri is Assistant Professor in the Department of Landscape Architecture at the University of Oregon. His experience includes years of practice as a landscape designer and urban designer. His research focuses on place identity. He is currently researching strategies for sustainable landscape design practices.

Brenda Case Scheer, AIA, AICP, is the Dean of the College of Architecture and Planning at the University of Utah. She is the co-author of *Suburban Form: an International Perspective*; *Design Review: Challenging Urban Aesthetic Control*; and *The Culture of Aesthetic Poverty.*

Edward W. Soja is Distinguished Professor of Urban Planning at UCLA. He is the author of *Postmodern Geographies: The Reassertion of Space in Critical Social Theory, Thirdspace: Journeys to Los Angeles and Other Real-and-Imagined Places, Postmetropolis: Critical Studies of Cities and Regions (2000)*, and the co-editor of *The City: Los Angeles and Urban Theory at the End of the Twentieth Century.*

Michael Southworth is Professor of City and Regional Planning and Landscape Architecture and Environmental Planning at the University of California at Berkeley. He is the author of *Streets and the Shaping of Towns and Cities* (with Eran Ben-Joseph), *The AIA Guide to Boston; Maps: A Visual Survey and Design Guide*, and *Ornamental Ironwork: An Illustrated Guide to Its History, Design, and Use in American Architecture* (with S. Southworth), and co-editor and contributor to *City Sense and City Design* (with T. Banerjee) and *Wasting Away* (by K. Lynch).

Lois M. Takahashi is Professor at the Department of Urban Planning at UCLA. She is the author of *Homelessness, AIDS, and Stigmatization: The NIMBY Syndrome at the End of* *the Twentieth Century,* and the co-author of *Rethinking Environmental Management in the Pacific Rim: Exploring Local Participation in Bangkok, Thailand.*

Emily Talen is Professor in the School of Geographical Sciences and Urban Planning at Arizona State University. She is the author of *New Urbanism and American Planning: The Conflict of Cultures*; *Design for Diversity: Exploring Socially Mixed Neighborhoods*; and *Urban Design Reclaimed: Tools, Techniques and Strategies for Planners.*

Niraj Verma is Professor and Director of the Wilder School of Government and Public Affairs at the Virginia Commonwealth University. He is the author of *Similarities, Connections, Systems: The Search for a New Rationality for Planning and Management* and editor of *Institutions and Planning.*

Anne Whiston Spirn (www.annewhistonspirn.com) is Professor of Landscape Architecture and Planning at MIT and director of the West Philadelphia Landscape Project, an action research program integrating research, teaching and community service. Her books include *The Granite Garden: Urban Nature and Human Design*, *The Language of Landscape*, and *Daring to Look.*

Carolyn Whitzman is Associate Professor of Urban Planning, Faculty of Architecture, Building and Planning at the University of Melbourne. She is the author of *Suburb, Slum, Urban Village,* and *The Handbook of Community Safety, Gender, and Violence Prevention,* and is the co-author of *Safe Cities: Guidelines for Planning, Design, and Management.*

Acknowledgments

Aside from the enthusiasm of Andrew Mould, our acquisition editor, who convinced us about the merit of this endeavor, we also benefited from the enthusiasm, encouragement, and interest shown by many of our colleagues and students toward this project. We take this opportunity to thank them all collectively, and Andrew Mould more specifically.

We thank our contributors, some 56 of them for their willingness, commitment, and original contributions that make this collection what it is – a companion to urban design. In writing their respective pieces they had agreed to take on the task of defining the contours of what we claim to be a field, and not just a profession or practice.

Of course we thank our respective academic homes – the School of Policy, Planning, and Development at USC, and the Department of Urban Planning in the School of Public Affairs at UCLA – for their institutional support and resources, especially in the form of graduate student assistance. We also acknowledge some support from the James Irvine Chair of Urban and Regional Planning fund at USC.

Finally, we thank some of our graduate students – Meredith Drake-Reitan and Felicity Hwee-Hwa Chan, two doctoral students at USC and Raabia Budhwani, a master's student in Planning, and Public Policy at USC for their invaluable help in editing and formatting a significant number of the submissions.

Introduction – urban design

Roots, influences, and trends

Today the field of urban design has emerged as an important area of intellectual pursuit, involving theory, research, and pedagogy, all intended to inform and improve practice. In the early stages of its modern professional identity, the field of urban design was defined by the interstices of the more established fields of architecture, landscape architecture, and urban planning with each claiming some proprietary rights and expecting their respective influence on practice. Today the practice of urban design, while still comprising participation from architecture, landscape architecture, and planning, has long eschewed its interstitial legitimacy. It has forged a distinctive identity with applications at many different scales – ranging from the block or street scale to the scale of metropolitan and regional landscapes, with such intermediate scales of applications as planned new communities, or conservation and design of urban neighborhoods. Because of its multiple scales of application, the practice of urban design now interfaces, if not engages, many aspects of contemporary public policy – multiculturalism, healthy cities, environmental justice, economic development, climate change, energy conservation, protection of natural environments, sustainable development, community livability, and the like. For students of the built environment and urban design, the field now comprises a core body of knowledge that includes a rich history of ideas, paradigms, principles, tools, research, and applications furthering aspirations of a good city form, and anticipating the consequences of the built environment on human activities and experiences. In its remarkable evolution, the field has become increasingly eclectic and interdisciplinary, enriched by influences from the humanities, and social and natural sciences.

Courses on urban design are increasingly a requirement not just for graduate or undergraduate professional studies in architecture, landscape architecture, and urban planning, but also undergraduate studies in urban studies, urban affairs, geography, and the like. General interest in these areas as courses of post-secondary studies is also growing as the world population and developing economies are undergoing an unprecedented urban transformation. The growing awareness of the importance of the quality of life and livability of the built environment extends much beyond the traditional design disciplines.

In recent years we have seen a bumper crop of readers and text books in urban design which are intended to meet the growing demand for introductory materials on the field. But the availability of these texts and readers, the latter mainly a collection of what the editors consider significant readings in the field, also point to the need for a more authoritative and comprehensive companion to these readers that includes core, foundational, and pioneering ideas and concepts. Such a volume will serve not only the students and future professionals, but also the teachers and practitioners of urban design.

Accordingly, we have sought to compile this *Companion to Urban Design*, which is composed of new writings and materials that are not necessarily addressed, critically or at all, in the introductory textbooks and readers on urban design. In inviting these contributions we expected the authors to be interpretive, reflective, and integrative. We did not require any definitive answers from them, nor did we seek any particular dogma or ideology. Indeed depending on their specific assignments our contributors have been critical, introspective, speculative, reflective, but not deterministic or dogmatic. This to be expected, for this collection represents a companion to a field which is still evolving, changing, and expanding its horizons. If anything, the collection presents a provisional view of the field, denying any smug claim that one has the definitive answer or paradigm to the complex nature of the emerging challenges to urban design. This collection also establishes quite effectively that unlike much of architecture and allied arts where single designers with specific clients is the norm, urban design experience is typically a collective, collaborative, and increasingly interactive effort. The clients of urban designers are usually the community or the public. Gone are the days of the Popes and Medicis who designed cities in an earlier time. Today urban design is a negotiated and mediated process that involves not just institutions but also the media public-at-large.

The essays in this volume are organized in nine distinct sections designed to address specific themes and questions. At the end of each chapter there is a list of suggested further readings selected and annotated by the respective authors.

Part 1 explores the intellectual roots of urban design. Eugenie Birch introduces the important thinkers, while Robert Fishman elaborates on the essential ideas and paradigms that have shaped the field and practice of urban design in the contemporary era. Danilo Palazzo explores the pedagogical traditions, principles, and philosophy of urban design education and how they have changed over the years.

Part 2 discusses the major debates, conflicts, and contradictions in our understanding of the production and consumption of urban space that must necessarily affect the theory and practice of urban design. Niraj Verma starts this part by describing Urban Design as "an incompletely theorized project" facing the normative versus the positivist tensions between its theory and its practice, and uncertainties about its institutional standing. The need to better theorize the discipline of Urban Design is also picked up in the next chapter by Alexander Cuthbert, who finds the current theories of urban design "wanting" and looks into spatial political economy as a method to understand better the essence of design challenges. Kanishka Goonewardena follows with a chapter that explores the nexus between urbanism and capitalism, pondering on urban design possibilities to disarticulate the two through "radical transformations of space." The last chapter in this part of the volume is by Christine M. Boyer who ponders the efforts and challenges faced by cybernetics in understanding and documenting the internal dynamics of contemporary urban systems, which are in a state of flow, continuously changing and re-assembling themselves.

A significant body of knowledge that informs the field of Urban Design is generated from other disciplines. Part 3 examines the various exogenous disciplinary influences on the theory, research, and practice of Urban Design in recent times. Thus, Larry Ford explains the contributions of Geography; William Michelson presents the essential ideas from the field of Sociology; Denise Lawrence investigates how anthropological studies have informed designers; Kristen Day discusses how Feminist Studies help Urban Design produce cities that are more equitable to women, while Jack Nasar evaluates the influences of Environmental Psychology on Urban Design. Marlon Boarnet and Lois Takahashi discuss the renewed interest in the links between Public Health and Urban Design. Jerold Kayden explains how the field of Law has influenced and continues to influence built form and the practice of design. Margaret Kohn argues how Political Theory contributes to the creation of vibrant public spaces and a democratic public realm. Finally, Rafael Pizzaro traces the connection between the Cinematic Arts and Urban Design.

In Part 4 we address the technologies and methods that have influenced or even transformed the practice of urban design at various scales. In recent decades, such methods have received a boost from new digital technologies that help urban designers better document existing urban contexts, envision alternative urban forms, and better communicate the impacts of different design scenarios. Like in all other design disciplines, the studio experience still remains the core of the design process. Kathryn Anthony starts this part detailing one of the oldest methods of design – the design studio – and describes its evolution over time and its current place in urban design education and practice. Martin Krieger explores how media tools from digital cameras to digital video devices, from cellular phones to GPS systems allow designers to "patch together" "multiple slices" of urban life and better document, understand, and represent the urban experience. Along the same lines, Ben Joseph details in the next chapter new digital tools such as Human–Computer Interactions (HCI), Augmented Reality (AR) and bottom-up, Internet delivery models, explaining how they can contribute to collaborative design processes, better understanding of spatial contexts, and even enhanced creativity. The last chapter in this part is by Peter Bosselmann, who examines the role of simulations in urban design and as a decision making tool.

Part 5 explores different processes utilized by urban designers in their search for a good city form, which can only be obtained through an incremental and additive process. The process through which these increments are designed and produced is all too critical. Thus, Ute Lehrer discusses and evaluates the process of urban design competitions, using the competition at Berlin's Postdamer Platz to draw some tangible conclusions. Similarly, Doug Kelbaugh draws from his significant experience with organizing and participating in design charrettes to examine the contribution of this process to better urban design. Urban design interventions affect multiple publics. Jeff Hou details and evaluates the challenges of participatory and bottom-up design processes that allow greater public participation in the design of neighborhoods and cities. In contrast, the next two chapters focus more on top-down rule systems designed by professional expertise that are set to ensure adequate (if not optimal) design forms. William Baer examines the rules, regulations, and professional standards that guide the process of design, while Matthew Carmona elaborates on the use of design guidance as a tool in the design process, focusing on one particular form of design guidance: the design code.

Part 6 casts its look at the spatial context of urban design, the different components that constitute the urban environment, seeking to distil socio-cultural and economic trends that influence built form. This part starts with a chapter on downtowns and central cities, written by the editors. It discusses the evolution in their urban form and context giving particular emphasis on downtown design trends in contemporary times. This is followed by a chapter by John Archer that offers a succinct account of the design principles and strategies that have influenced the shape and form of suburbs from the bucolic landscape of the late eighteenth century, to the picturesque designs of the nineteenth century, to the mass subdivisions of the post-war era, to the more recent New Urbanist reinvention of suburbs with a small-town feel. Planned communities and new towns have appeared mostly in suburban and exurban locations, and are dealt separately in the chapter by Ann Forsyth. This chapter outlines the different design traditions that have influenced new town planning and design as well as the issues and concerns that have accompanied the design of such large-scale urban developments. The neighborhood scale is an important focus of urban design. In the next chapter, Ajay Garde discusses the innovations, social themes, and current practices and innovations in the design of neighborhood spaces. Streets define and organize urban space, and represent a ubiquitous urban landscape and important component of a city's public realm. Elizabeth Macdonald discusses a number of innovative approaches for street design in the ensuing chapter. But streets are not the only public spaces that draw the attention of urban designers. The following chapter by Mark Francis concentrates on a greater variety of public open spaces (parks, plazas, waterfronts, sidewalks, urban gardens) and suggests some alternative concepts for their design. An important aspect of the public realm in contemporary times involves spaces of consumption. Indeed, many developments in the late twentieth and early twenty-first century city are designed to include spaces to precisely serve that purpose. The next chapter by Klaus Kunzmann elaborates on the spatial configuration of such spaces in cities and examines the urban design implications of consumption. Cultural institutions have always been important trademarks of cities but recent decades have witnessed unprecedented municipal efforts to develop flagship cultural complexes as a way of boosting a city's image and associated "buzz." The next chapter by Carl Grodach captures this phenomenon and also details the emerging trends in the design and planning of cultural complexes. While the previous chapters mostly focus on the permanent or largely durable components of urban environments, Gary Hack in the next chapter directs urban designers' attention to the "urban flux" – the more ephemeral, ad hoc, unpredictable and changing urban artifacts that can be found in city environments.

Each of the chapters in Part 7 critically presents and assesses an important debate in the field of Urban Design. Thus, equipped with new data regarding the cost of sprawl, Reid Ewing, Keith Bartholomew and Arthur C. Nelson revisit the old debate regarding the desirability of compact cities as an alternative to sprawl. Ali Madanipour revisits the debate regarding gentrification versus displacement and social exclusion asking whether urban design should facilitate social groups to live together or apart. In a rapidly globalizing world, the built environment of cities often becomes increasingly homogeneous, dominated by ubiquitous high-rise structures, big box retail establishments, and megamalls, lacking a sense or identity of place. In the next chapter, Michael Southworth and Denni Ruggeri take on the question of place identity in a global society investigating the role of designers in shaping place identity and local specificity. Another tension that has engaged urban designers in the last three decades is the emergence and proliferation of New Urbanism in the

United States, and the ensuing debate in urban design circles over the merit of its doctrines. Ivonne Audirac addresses this tension by contrasting the Old versus the New Urbanism, and discussing the contributions of the movement but also controversies surrounding it. The tension has now percolated to the level of zoning these days. While conventional zoning codes have received significant criticism in the urban design literature, the New Urbanist proposed alternative of form-based codes (FBCs) have also been controversial. Emily Talen compares and contrasts FBCs to conventional zoning codes and offers a response and a rebuttal to the various criticisms directed against the FBCs.

Part 8 focuses on the urban design challenges and opportunities presented by the new global economic order and the forces of globalization in the urban development of both developed and developing countries. For one, globalization has forced cities to compete in a global rather than a regional or national scale. Urban design then emerges as a tool that many cities use to boost their image and identity and ensure a certain type of branding. Municipal processes and efforts for city branding and marketing are detailed in a chapter authored by Jon Lang. Globalization has also brought about what Edward Soja deems as a profound urban restructuring and the emergence of a new form of development, which he coins "regional urbanization." His chapter delineates the attributes of this process, often characterized by the emergence of city-regions, and a blurring of the urban and suburban landscapes. Another outcome of the new global socioeconomic order is the emergence of ethnoscapes in cities. In the next chapter, Clara Irazábal discusses the characteristics of ethnospaces as new sociospatial typologies in global cities and investigates the role and opportunities for urban design in such settings. A phenomenon pre-dating globalization but one that is certainly intensified by it is the proliferation of informal settlements in the form of slums, spontaneous, ad hoc and non-regulated settlements in both developed and developing parts of the world. In the last chapter of this section Vinit Mukhija details the role that urban design can play and the positive interventions it can bring to informal settlements.

Part 9 focuses on some important new trends and directions related to the shape of urban form and the practice of urban design. These have come about in response to a multitude of contemporary challenges such as population growth, climate change, depletion of energy resources, natural and human-made disasters, and subsequent desires to create urban forms characterized by energy efficiency, sustainability, resiliency, and the potential for active living. In the first chapter of this section, Nan Ellin argues that we are increasingly witnessing urban design approaches characterized by an "integrative urbanism." Such designs seek to respond to the needs of different social groups and at the same time integrate the urban with the suburban, buildings with nature, and the local with the global. In the next chapter, Brenda Scheer discusses the movement of landscape urbanism which advocates looking holistically at the city as an evolving urban landscape. She explores the role of urban design in recovering and nurturing the natural systems of the metropolis, and the challenges and opportunities of design at the metropolitan scale.

In a similar vein, in the chapter that follows Anne Whiston Spirn discusses the related concept of ecological urbanism, exploring its historic roots and current trends. She offers a normative agenda for urban designers for treating the city as a habitat, an ecosystem, and a part of the natural world. Randy Hester and Marcia McNally follow with an investigation of the design principles that have contributed in the 1960s and 1970s to the formation of the concept of sustainability, and how in turn in recent times the concept of sustainability has influenced the practice of urban design. In the next chapter, Aseem Inam reviews the concept of smart growth and its relationship to urban design, pointing to successful efforts but also challenges in its implementation. Stefanos Polyzoides follows with his observations on Transit Oriented Development, which in his view is at the core of sustainable urbanism. The next chapter by Kathy Madden explains the concept of placemaking, a place- and community-driven approach that focuses on the design of the "ground floor" of the city, the public spaces of everyday life, where people congregate and socialize. Issues of safety and security are concerns for many urban residents. Carolyn Whitzman details the trend of designing secure cities comparing and contrasting two streams of thought about urban design, one associated with crime prevention through environmental design, and a more recent one which privileges the resident as an expert in building safe cities. Lastly, Mahyar Arefi discusses urban design strategies for the building of resilient cities, arguing that they should entail the identification of a city's liabilities, the transformation of such liabilities into assets, and the building of adaptable and flexible forms.

Part 1

Roots

Introduction

Many architects are fond of saying, in jest of course, that theirs is the second oldest profession in history. There might be some truth to this since building shelters is one of the earliest forms of organized human endeavors. Since urban design is functionally, if not etymologically, linked to architecture, and as urban design endeavors can be traced even in ancient times – from the Vastu Shilpa principles that dictated the design of ideal cities in India to the Chinese efforts over a millenium to tweak the ideal city form – it poses a problem for us to define the appropriate time frame for tracing the "roots" of urban design. Clearly the scope of this endeavor did not call for a comprehensive history of city and urban design. After all, there are already many authoritative works on this subject. In consultation with our contributors, then, we chose to focus on the most recent history of what can be considered contemporary urban design, beginning in the earlier half of the previous century. The chapters included in this section reflect this time period.

The opening chapter by Eugenie Birch includes a comprehensive review of the important projects, protagonists, and promoters of contemporary urban design, including its organized movements and institutional patronage. In framing this review the author chooses to bookend the two important movements of our time – CIAM (the French acronym of the International Congress of Modern Architecture) and CNU (Congress for New Urbanism) – which have shaped much of the thinking and practice of large scale urban design. This intriguing history is both about idealism and pragmatism, about individual and organized efforts, and about important projects and paradigms of urban design, much of it in the context of the socio-economic and political history of the US urban development during this period.

The following chapter by Robert Fishman frames the history of ideas in urban design in a similar time frame. The author defines the history of ideas as one of two competing paradigms – which he calls "the open" and "the enclosed," referring to the earlier modernist paradigm of city design shaped by Le Corbusier's idea of "tower in the park" on the one hand, and the yearning for a more compact urbanism of spaces defined by building façades and enclosed squares, characteristic of earlier cities. The author suggests that there might be a shift in the thinking about these two paradigms, as the concerns for sustainability, walkability, and mixed use urbanism continue to dominate public policy.

Finally, the third chapter in this section by Danilo Palazzo focuses on the history of the pedagogical tradition in contemporary urban design. The author carefully traces the institutional settings, innovations, and disciplinary influences that have shaped the training of urban designers in the US. In reporting this history, the author identifies key figures and programs which were influential in defining the professional training of urban designers.

These first three chapters of the *Companion* set the chronological stage and historical background of the chapters appearing in the following sections.

1

From CIAM to CNU

The roots and thinkers of modern urban design

Eugénie L. Birch

In the late nineteenth and early twentieth centuries, rapid urbanization in the western world stimulated dramatic responses in many disciplines, ranging from the social sciences to the design professions. Independently, they sought to understand and address urban issues. Sociologists, for example, wrote about deep differences in behavior among city and country residents, identified urban alienation, and offered remedies to promote community in cities (Tonnies 1887; Park and Burgess 1925; Perry 1929; Wirth 1938). Architects, city planners and landscape architects framed their professional practice around solving problems caused by population congestion (Le Corbusier 1924; Birch and Silver 2009). Their debates surrounded architectural style (neo-classical vs. modern), ideal settlement patterns (centralization vs. decentralization) and means to improve the internal organization of cities through better open space, land use, housing and circulation.

Among designers, these varied concerns led to the formation of a new field, variously labeled "der stadtebau (German)" "urbanisme (French)," "civic art," "civic design,""city design," and "urban design."[1] Now commonly known as urban design, its development spanned more than a hundred years, fed by synergistic relationships among five leadership groups (Precursors, Founders, Pioneers, Developers and Later Evolvers), who shaped its content and influence (see Table 1.1) In tracing the roots of modern urban design, this chapter will focus on the Founders, Pioneers and Developers, who were active from the 1920s through the 1970s, cognizant that the field's history is longer and broader but arguing that work in these years was formative.

Modern urban design emerged in the late 1920s as a loose organization of European and American architects and city planners, or Founders, who declared that they could solve ever-worsening urban problems (defined as unhealthy housing, inefficient land use and inadequate transportation) through enlightened city-building. Their highly conceptual work, mainly took the form of writing and unrealized projects. The Pioneers (architects, landscape architects, city planners and urban-focused writers/scholars) expanded the field in the 1930s and 1940s with contributions encompassing writing, a few projects and educational experiments. Developers emerged in the 1940s as European reconstruction and US urban renewal programs gave impetus to the growing movement. Drawn from the same

Table 1.1 From CIAM to CNU: The roots of urban design

Precursors	Founders	Pioneers	Developers	Later evolvers
Mid to late nineteenth and early twentieth centuries	1920s to early 1930s	Mid 1930s to late 1940s	Late 1940s through 1970s	Post 1970s
Designers	**Architects**	**Architects**	**Architects**	**Architects**
Georges-Eugene Haussmann (1809–1891)	Le Corbusier (1887–1965)	Joseph Hudnut (1886–1968)	Josep Lluis Sert (1902–1983)	Andres Duany (1949) and Elizabeth Plater-Zyberk (1950)
Ildefons Cerda (1815–1876)	Frank Lloyd Wright (1867–1959)	Clarence Stein (1882–1975)	David Crane (1917–2005)	Jonathan Barnett (1937)
Frederick Law Olmsted (1822–1903)	Hugh Ferriss (1889–1962)	CIAM members (1928–1956)	Edmund Bacon (1910–2005)	Peter Calthorpe (1943)
Daniel Burnham (1846–1912)	Eliel Saarinen (1873–1950)	MARS members (1933–1957)	Victor Gruen (1903–1980)	Alexander Cooper (1935)
Tony Garnier (1869–1948)		Walter Gropius (1883–1969)	Jacqueline Trywhitt (1905–1983)	Leon Krier (1946)
Paul Philippe Cret (1876–1945)			I M Pei (1917)	Alex Krieger (1951)
			Peter (1923–2003) and Alison Smithson (1928–1993)/Team 10	David Lewis (1922)
			Denise Scott Brown (1931) and Robert Venturi (1925)	Rem Koolhaas (1944)
			Ben Thompson (1918–2002)	
			Kenzo Tange (1913–2005)	
			Oscar Niemeyer (1907)/ Lucio Costa (1902–1998)	
			Urban Design Group (1967–1974)	

	City planners	**City planners**	**City planners**	**City planners**
	John Nolen (1869–1937)	Reginald Isaacs (1911–1986)	G. Holmes Perkins (1903–2004)	Gary Hack (1942)
	Frederick Law Olmsted Jr. (1870–1957)	Martin Meyerson (1922–2007)	Kevin Lynch (1918–1984)	Donald Appleyard (1928–1982)
	Henry Vincent Hubbard (1875–1947)			Allan Jacobs (1930)
				Alexander Garvin (1940)
	Landscape architects	**Landscape architects**	**Landscape architects**	**Landscape architects**
		Hideo Sasaki (1919)	Ian McHarg (1920–2001)	James Corner (1961)
			Laurie Olin (1938)	
Authors	**Authors**	**Authors**	**Authors**	**Authors**
Camillo Sitte (1843–1903)	Patrick Geddes (1854–1932)	Clarence Perry (1872–1944)	Jane Jacobs (1916–2006)	Jan Gehl (1936)
Ebenezer Howard (1850–1928)	Raymond Unwin (1863–1940)	Lewis Mumford (1895–1990)	William H. Whyte (1917–1999)	Clare Cooper Marcus (1934)
	Werner Hegemann (1881–1936) and Elbert Peets (1886–1968)	Sigfried Giedion (1888–1968)	Christopher Alexander (1936)	Anne Vernez Moudon (1947)
		Catherine Bauer Wurster (1905–1964)	Gordon Cullen (1914–1994)	Anne Whiston Spirn (1947)
		Christopher Tunnard (1910–1979)	Colin Rowe (1920–1999)	Joel Garreau (1948)
			Oscar Newman (1935–2004)	
			Paul Zucker (1888–1971)	
			Steen Eiler Rasmussen (1898–1990)	

Note: The dates proffered are rough guidelines as many participants were active in more than one era.

design communities but armed with new social science research, they wrote for scholarly and popular audiences, built projects and created advanced degree programs, thus promulgating the field in its solid theoretical and practical aspects.

The Founders

The idea of building a modern, rationally ordered city captured the imaginations of many designers, mainly architects in the early twentieth century; Swiss architect Le Corbusier is one example; the German-born architect Walter Gropius is another. Their fascination with the use of simple, mass-produced industrial products for housing soon led to an infatuation with the skyscraper, viewed as an emblem of its times, much like the cathedral in the Middle Ages. They sited low-lot coverage high-rises (often labeled "towers in the park") arguing that this building form could improve urban life by capturing the density required for city vitality while relieving ground-level congestion. It was a short step from designing such buildings to arranging them in geometric patterns in whole cites. Here, the designers called for replacing the obsolete industrial city with a modern one marked by land uses separated by function, superblocks of high-density districts (residential, downtown) and ample recreational areas knit together by modern highways. To display these ideas, Le Corbusier offered a succession of unrealized projects (*La ville contemporaine* [1922], *Voisin Plan* [1925] and *La ville radieuse* [1935]) and writing (*Urbanisme* [1924]).

In 1928, Le Corbusier and other like-minded designers founded the influential *Congrès Internationaux d'Architecture Moderne* (CIAM) that held annual congresses from 1928 to 1956 (with some interruption for World War II) in Europe and publicized their beliefs and work widely. Starting as a small group of two dozen at its initial meeting in La Sarraz, Switzerland, CIAM would grow rapidly, attracting more than 3,000 attendees to its annual congresses by the early 1950s (Pedret 2001: 151). Left-leaning and inspired by social justice as well as modern architecture, CIAM first focused on slums but soon took on the city. Their fourth congress, entitled "The Functional City," demonstrated this shift. Originally scheduled for Moscow to display the possibilities of modern architecture in a socialist setting, the organizers moved it elsewhere when the Soviets rejected Le Corbusier's entry to the Palace of the Soviets competition, concluding that "the *avantgarde* (sic) had no place in Stalin's Russia" (Giedion 1966: 698).

CIAM IV took place in summer 1933, on board the *SS Patris II* sailing from Marseilles to Athens and back and in a hotel in Athens. For fourteen days, the participants engaged in a comparative urban planning exercise, looking at thirty-three cities according to standards based on Patrick Geddes-recommended analytical surveys and codified by the Dutch urban planner Cornilis van Eesteren. They also had non-stop committee meetings to distill their work into a brand of modern city planning that encompassed four simple functional areas: housing, work, recreation and transportation (Geddes 1915; Somer 2007; Mumford 2000). Within this rubric, they promoted relieving congestion through slum clearance and rebuilding along the lines of Corbusier's "towers-in-the park" concept. Their doctrine contrasted directly with the competing British-based garden city vision that advocated decentralized, low-density satellite cities as a means of improving urban life.

Due to internal dissension about certain details, CIAM did not formally publish the CIAM IV proceedings as planned, but in 1941 Le Corbusier boiled down the results into a manifesto and issued it under the CIAM name as *La Charte d'Athènes* (1943). He captured the pre-World War II modernist urban design principles in 95 bullets, e.g.

> 1 The city is only a part of the economic, social and political entity which constitutes the region ... 9. The population density is too great in the historic, central districts of cities ... 30. Open spaces are generally insufficient ... 69. The demolition of slums surrounding historic monuments provides an opportunity to create new open spaces ... town planning is a science based on three dimensions, not two. This introduces the element of height which offers the possibility of freeing spaces for modern traffic circulation and for recreational purposes. .... (Tyrwhitt 1933)

At about the same time, Spanish architect and CIAM-member Josep Lluís Sert produced a longer English version, *Can Our Cities Survive? An ABC of Urban Problems, Their Analysis Their Solutions Based on the Problems Formulated by the CIAM* (1942). Together, these works defined CIAM-led urban design.

Although CIAM was dominant in promoting city-building ideas in the twentieth century, it was not alone. Visionary illustrator Hugh Ferris shared Le Corbusier's love of the skyscraper, producing dramatic images of its possibilities in *Metropolis of Tomorrow* (1929), while German architect Werner Hegemann and his American associate, Elbert Peets, documented the strength of urban cores, especially civic centers, in *The American Vitruvius, The*

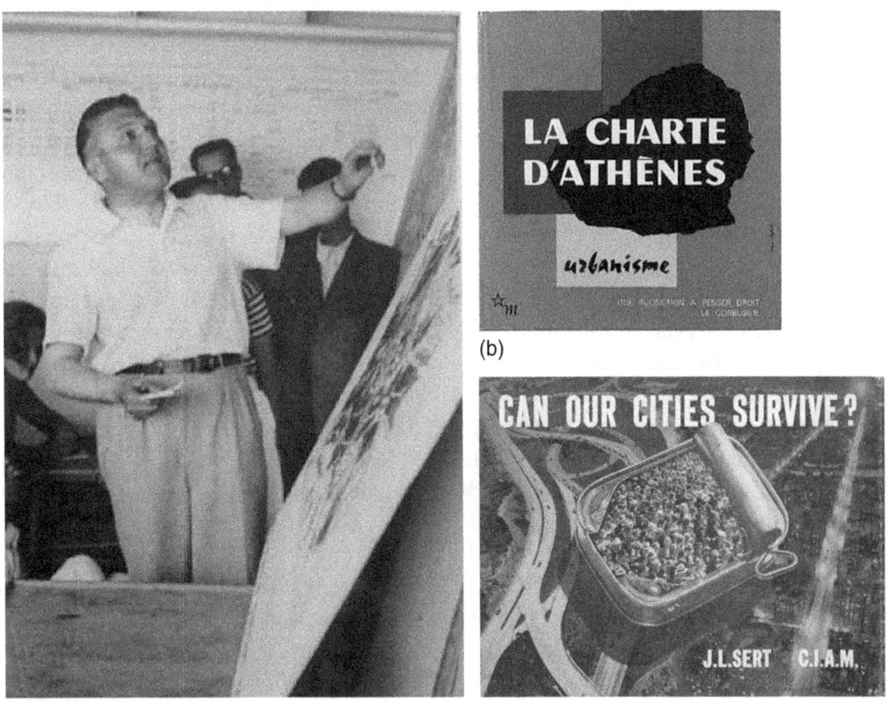

(a) (b) (c)

Figure 1.1 CIAM (1933) meeting and later publications.
Note: CIAM IV met in 1933 to discuss the "Functional City," through comparative urban planning (a) whose proceedings were issued as the *Athens Charter*, which was not published until 1943 and then again in a second edition 1957. The cover (b) is from the second edition (Paris: Editions des Minuit). (c) A year earlier, Josep Lluis Sert had published through Harvard Press a longer version of the congress and its city planning views in *Can Our Cities Survive?*

Architect's Handbook of Civic Art (1922). Notably, Hegemann argued that civic art required gleaning knowledge from the social sciences, humanities *and* design (architecture, city planning, fine arts and landscape architecture).

Prior to publishing *American Vitruvius*, Hegemann had practiced in the United States, hiring a Beaux-Arts-trained architect, Joseph Hudnut, as an assistant in 1917. Hegemann returned to Germany in the 1920s but came back to the US in the 1930s, meeting up with Hudnut again, who, by this time, was Dean of Columbia University's School of Architecture. As Dean, Hudnut sought to insert civic design into the curriculum and hired Hegemann and a rising landscape architect, Henry Wright, to teach the subject. In 1936, Harvard recruited Hudnut to be the founding Dean of the Graduate School of Design (GSD). There, he would aggressively pursue the idea of synthesizing the professions through civic design, hiring faculty who shared the vision and requiring all students to take a common introductory course (Pearlman 2008).

Meanwhile, decentralists were also active. Garden city proponents including Raymond Unwin, author of *Nothing Gained by Overcrowding* (1912) and *Town Planning in Practice* (seventh edition 1920) and designer (Letchworth Garden Suburb 1903–1920s), offered an alternative vision of civic design, one quickly adopted by leading American urbanist Lewis Mumford. He called for decanting the crowded metropolis into peripheral self-contained cities.[2] But it was American planning practitioners, Frederick Law Olmsted Jr. and his student, John Nolen, both steeped in landscape principles at Harvard, who executed these theories in the United States, drawing on European precedents, especially English and German efforts in housing reform and zoning. For example, Nolen's plans for model suburbs like Charlotte's Myers Park and for complete towns like Mariemont, Ohio, Kistler, Pennsylvania and Venice, Florida were notable examples of the translation of garden city principles in the United States. He articulated the "American brand," showing how to plan for places for relatively small populations (25,000 to 100,000) in self-contained satellite cities bounded by greenbelts, containing town centers, a range of housing choices, neighborhood-to-region park systems, buffered industrial sectors and hierarchical street arrangements (Nolen 1927). While practitioners like Nolen and Olmsted, Jr. were not directly involved with CIAM, their ideas fed the knowledge base of urban design, and in fact, would provide reference material for later expressions of urban design, especially new urbanism.

The Pioneers

While divisions among the Founders were present, opposing philosophies related to density and decentralization emerged more prominently among the many Pioneers who followed. While they built on the Founders' thinking, especially the CIAM pronouncements, they also added new ideas. Further, some Pioneers worked independently but were conscious of the others and drew inspiration according to their inclination and needs. Author Clarence Perry, for example, devised the "neighborhood unit," a physical/social arrangement centered on the grade school and its surrounding catchment area as a basic city-building block. Clarence Stein and Henry Wright adapted this idea at their Radburn New Jersey garden city experiment (1929).[3] Some CIAM followers were cognizant but critical of the neighborhood unit (and of garden cities), considering them contrary to their brand of urbanism. Architectural historian Sigfried Giedion (1966) falls in this category, writing that these concepts were "doomed to failure" because they often dissolved into low-density, incomplete suburbs (859). And he added, "No partial solution is possible; only preconceived and integrated planning on a scale embracing the whole structure of modern life in all its ramifications can accomplish the task..." (785).

But others like Sert absorbed (and modified) the various strains, including the neighborhood unit. Sert is an important transitional figure. Placed on Table 1.1 among the Developers, he had strong connections to the Founders and Pioneers through his early involvement with CIAM, his publication, *Can Our Cities Survive?* (1942), his practice and, finally, as a university dean. *Can Our Cities Survive?* was the first book-length account of CIAM views in English. It had the support of influential architects and public intellectuals, and launched Sert as an educator. Exiled from Spain in 1939 after designing the Republican government's pavilion for the 1937 Paris Exhibition, noted for its prominent placement of Picasso's *Guernica*, Sert used his first years in the US to write the chronicle. Supporters at Harvard, including CIAM member Walter Gropius, then Chair, Department of Architecture, Sigfried Giedion, CIAM secretary-general, then delivering the Charles Eliot Norton Lectures (that would form the basis of *Space, Time and Architecture* [1941]), and Dean Joseph Hudnut encouraged him. As an admirer of Lewis Mumford's *Technics and Civilization* (1934) and *Culture of Cities* (1938), he enlisted Mumford to read early drafts (Bacon 2008). This latter alliance was critical to the success of the book as Mumford would review it favorably in the *New Republic* (February 8, 1943).[4]

By 1941, Sert was well-established in the United States as a spokesman for CIAM urbanism but with added twists. In "The Human Scale in City Planning," a paper delivered at the pace-setting "New Architecture and City Planning" symposium organized by the New School and Cooper Union in 1944, he put his full support behind the neighborhood unit as a basic city-building block (Sert 1945). As he worked on Latin American commissions through his firm, Town Planning Associates (founded in 1941), he further adapted CIAM principles to focus on pedestrianized city cores (Hyde 2008). As CIAM vice president and president from 1944–1952, he promoted these changes, emphasizing them in CIAM meetings and publications. His second book, *The Heart of the City* (1952) (edited with Jacqueline Tyrwhitt and Ernesto N. Rogers), represents this work and informed US urban renewal, American suburban shopping centers and European New Town design.[5]

In the 1940s, Sert began lecturing at universities throughout the United States and by 1952 held a visiting professorship at Yale. When Harvard President James Conant named him GSD dean and chair, Department of Architecture a year later, the scene was set for the emergence of today's urban design in the educational arena. At his appointment, Sert had a clear idea of urban design as a three dimensional field, synthesizing architecture, city planning, landscape architecture and fine arts to improve the urban environment (Mumford 2009: 196). His emphasis on "three dimensional" underlined his belief that architecture would be the lead discipline.

The Harvard that Sert encountered was fertile ground for his ideas. His predecessor, Joseph Hudnut, had laid substantial groundwork, appointing G. Holmes Perkins, a strong believer in Sert's brand of urbanism in city planning and Bauhaus-founder, CIAM-member Walter Gropius as chair in architecture. While Perkins and Gropius were gone by the time Sert arrived, they had left their imprint.[6] For example, Gropius had supported the appointment of his former student, Reginald Isaacs, as chair of a combined department of city planning and landscape architecture (Pearlman 2008: 127; Mumford 2009: 36).

Gropius was familiar with Isaacs because between 1946–1948, he, along with landscape architect Hideo Sasaki, also Harvard-trained, served as consultants for the master plan for Chicago's Michael Reese Hospital, a seventy-year-old primary health-care facility located on the city's South Side. Headed by Isaacs and involving GSD-trained city planner Martin Meyerson, this project aimed to preserve the facility's $8 million investment and find room for needed new facilities while dealing with surrounding blight caused by White flight and Black in-migration. Isaacs and Meyerson and the planning team looked at the entire seven-square-mile community (and, in fact, were the force behind the formation of the influential Southside Planning Board) but focused on the immediate two-square-mile area adjacent to the hospital (Isaacs Collection, Wirth Collection). The resulting recommendations to clear and rebuild were "pure" CIAM urbanism (Mumford 2009: 42). As executed with help from Illinois's Blighted Areas Redevelopment Act (1947), the plan refashioned streets to create more than a dozen superblocks for several new hospital buildings and two neighboring residential complexes, Lake Meadows and Prairie Shores (designed by SOM and financed by New York Life) (see Figure 1.2). This project received a good deal of national attention as one of the first to condemn, clear and rebuild blighted land, according to CIAM ideals setting a pattern for future urban renewal (*Architectural Forum* 1952).

The Developers

Postwar urban design received input from others, notably Philadelphians Oskar Stonorov, Edmund Bacon, Robert B. Mitchell and Louis Kahn. Influenced not only by Le Corbusier and the CIAM teachings but also by Eliel Saarinen who headed an urban design program at Cranbrook Academy, where Bacon studied and by Mitchell's experience as director of the Federal Home Loan Bank's experiment with rehabilitation in Baltimore's Waverly neighborhood. This group first displayed its thinking in the Better Philadelphia Exhibition. Seen by almost 400,000 visitors in Fall 1947, the designers outlined an urban message that differed from the CIAM vision: first, the plans for the future city were to be a "logical outgrowth of past trends and a projection of established traditions...," second, "planning can be woven into the existing fabric of the city by a series of projects individually executed but planned together to produce a satisfactory and desirable end result without wholesale demolition," third, the center of the city, the "show window of Philadelphia" deserves immense attention and fourth, residential neighborhoods need new open spaces, new housing and relief from heavy traffic (Bacon 1948: 24, 26).

The Philadelphia Planning Commission, headed from 1943–1948 by Mitchell and from 1949–1971 by Bacon, would help execute these ideas. Both had a strong appreciation of the city's historic fabric as well as of modernism and would craft and implement a vision that blended elements of both. Bacon, for example, employed Kahn and Stonorov for several master plan projects, whose spirit was captured by the *Architectural Forum* in "The Philadelphia Cure: Clearing Slums with Penicillin Not Surgery," an article praising the work (1952). As time passed, the Philadelphia urban revitalization work would receive even wider publicity. Bacon appeared on the cover of *Time Magazine* as exemplifying the best of American urban design (*Time* 1964). The accompanying article noted that the city's seventy-five public and private urban renewal projects were overseen by a leader who "cherishes the old and adapts it to the new," preserves "[William]Penn's axis and provides new anchors," and aims to restore the inner city

(a)

(b)

Figure 1.2 Michael Reese Hospital urban renewal area before clearance (top) and after clearing and rebuilding (1953) (bottom).
Note: Overseen by Reginald Isaacs and planned by Walter Gropius and Hideo Sasaki, the hospital occupied part of the site (bottom half) and Lake Meadow housing development (designed by SOM) the remainder.

to the pedestrian, yet keep the car as an "honored guest." Offered as examples were Society Hill and Penn Center, both having strong pedestrian elements.

By the early 1960s, Martin Meyerson, Williams Professor of City Planning, Harvard (and soon to become Dean, College of Environmental Design, University of California, Berkeley, the school founded by his Harvard classmate William Wurster) and Jacqueline Tyrwhitt, Associate Professor of City Planning, Harvard, produced a compilation of model public- and private-sector projects, representative of the amalgamated urban design principles developed in the postwar period (Meyerson 1963). They focused on US examples but added some from Europe and Latin America, featuring works by I.M. Pei, Mies van der Rohe, Oskar Stonorov, Le Corbusier, Clarence Stein and Victor Gruen.

Notably missing in Meyerson's book was New York's Stuyvesant Town (1947), the 11,500 unit development that covered eighteen blocks of slum-cleared land in Lower Manhattan overseen by the city's redevelopment czar, Robert Moses. In some ways its absence signaled cracks in urban design theory that would break wide open in the 1960s and 1970s. Growing disenchantment would emanate from the social sciences and spill into the physical area with critical publications by Jane Jacobs (1961), Kevin Lynch (1960), Gordon Cullen (1961), Christopher Tunnard and Boris Pushkarev (1963), Robert Venturi, Denise Scott Brown and Steven Izenour (1972), Colin Rowe and Fred Koetter (1978), and William H. Whyte (1980).

Pre-shadowing Jacobs *et al.* Catherine Bauer (1934) had already advocated modernist urban design, attacked "scientific" urbanism, and, in particular, the neighborhood unit. In a 1945 essay, "Good Neighborhoods," she declared: "We cannot plan neighborhoods without a broad and progressive civic philosophy as to what really constitutes a 'good neighborhood' " and argued that contemporary planning, and its accompanying urban design, had become too formulaic, alienating, antidemocratic and dull (104). Three years later, Reginald Isaacs chimed in, blaming the narrowness of the contemporary proponents of urban design whose "major recruitment from the purely technical backgrounds of architecture, landscape architecture and engineering" as fostering a "misconception of the nature of city growth" (Isaacs 1948: 16, 19).[7] Both authors were especially incensed that federal agencies were not only recommending use of the neighborhood unit but also calling for its implementation through restrictive covenants, provisions written into deeds that forbade house sales or rentals to Blacks, Asians, Jews, Catholics and others.[8] Although the US Supreme Court put an end to such practices (*Shelley v Kraemer* [344 US 1948]), other issues remained. For example according to the critics, the neighborhood unit promoted a tendency of residents to "think introvertly within the narrow confines of their *neighborhood* and not to the purpose and well-being of the town or metropolitan areas" (Isaacs 1948: 21) or prevented children from learning to "live in the real world and eventually become effective leaders able to know and work with a variety of people and situations" (Bauer 1945: 108).[9] Bauer and Isaacs were not the only observers to question the neighborhood unit. British sociologist Ruth Glass while engaged in preliminary studies for a 1940s redevelopment plan of Middlesborough, the first major industrial city bombed by the German *Luftwaffe,* came to similar conclusions (Glass 1948).

While Bauer and Isaacs rejected important aspects of modern urban design and its principles due to their unintended political and social implications, others, like Jacobs, Lynch, Cullen, Venturi/Scott-Brown/Izenour, Rowe and Whyte turned to the physical arena in their critiques. They developed scholarship and practices that rejected ahistorical, expert-driven, standardized approaches. They called for individualized, empirically-based design based on appreciating the complexity and history of contemporary urban fabric. Further, they shied away from planning whole cities and focused on particular aspects of the public realm like the street or public squares. They stressed learning via observation of people and their use or reactions to urban places. They developed ideas of place-making that celebrated organic, walkable, mixed-use development. Of particular importance was the attention they gave to understanding urban morphology and the natural and social factors that shaped cities. Informed by the earlier work of such architects as Giambattista Nolli whose *Pianta Grande di Roma* (1748) detailed eight square miles of Rome including streets, squares, public spaces and monuments within them and Camillo Sitte (1889), who intensively analyzed the public realm in several European cities, these writers (especially Rowe and Koetter, Cullen, Whyte and Jacobs) focused on such basic city-building elements as block size, street widths, and public plazas composition. They bid urban designers to think about the individual experience in the urban environment by addressing small scale projects at the neighborhood or the block levels. In contrast to the Founders, they promoted a more "bottom-up," incremental approach to city design. Ultimately, this work would evolve into several important theoretical areas, including new urbanism and space syntax. This scholarship would also find its way into later twentieth-century practice including the executed plans for Battery Park City (New York), Reston City Center (Virginia) and the St Lawrence (Toronto) neighborhood.

Some of the critics developed their ideas through participation in the numerous urban design activities in the 1950s and 1960s. These included a long string of conferences on the subject sponsored by Sert at the GSD from 1956 to 1970; the Penn/Rockefeller Conference on Urban Design Criticism (1958) organized by Penn Urban Design program director David Crane (a one-time research assistant for Kevin Lynch); the Rockefeller Urban Design studies program that funded these conferences and basic research, the Ford Foundation support for the Harvard-MIT Joint Center for Urban Studies headed by Martin Meyerson (Harvard) and Lloyd Rodwin (MIT), and the initiation of urban design graduate programs at several universities, including Penn (1958), Harvard (1960), Rensselaer Polytechnic Institute and Washington University (1962) and Tokyo University (1965) (Krieger and Saunders 2009; Laurence 2006; Harvard-MIT Joint Center Papers).

The Rockefeller Urban Design studies program was an extraordinarily important force for the field. By seeding important urban design scholarship, it shaped urban design theory and practice for decades to come. Among those funded were Kevin Lynch, E.A. Gutkind, Jane Jacobs, Edmund Bacon, Christopher Alexander and Christopher Tunnard. Interest in the area had come from many sources, including the passage and implementation of the 1949 Housing and Slum Clearance Act and succeeding urban renewal legislation that originally supported large-scale demolition and rebuilding, the 1956 Federal-Aid Highway Act that supported the interstate system that ripped through urban neighborhoods, and postwar suburbanization that fostered massive subdivision construction and concomitant loss of rural lands adjacent to cities. These changes attracted the attention of scholars, journalists and design professionals who took an intense interest in understanding urban and suburban dynamics, looked for models of ideal development and did not hesitate to criticize what they saw occurring in their own cities. For example, Kevin Lynch's *Image of the City* (1960) codified individuals' perceptions of urban environments to demonstrate that they used specific features (paths, edges, nodes, landmarks and districts) to "read" their cities. Edmund Bacon's *Design of Cities* (1967) used historical and contemporary examples to demonstrate the spatial organization of regions, cities and neighborhoods with a special emphasis on the public realm. Christopher Tunnard despaired of the ugly suburban landscape in *Manmade America: Chaos or Control* (1963). And the most potent of all, Jane Jacobs' *Death and Life of Great American Cities* (1961) analyzed successful urban places, concluding that mixed use, high density, walkable places contributed safe, vital, sociable neighborhoods. She rejected top-down expert-driven designs in favor of locally determined accretive growth. A best-seller, this book would become the "bible" of urban design. Jacobs was largely responding to work emanating from the passage of the urban renewal and freeway legislation. And as a contemporary, Roger Montgomery, head of the first urban design office at the US Urban Renewal Agency later observed,

> The *Death and Life* [was] a causal factor in the demise of public housing, urban redevelopment and big freeway projects, Without question, the book and its author added energy to forces already at work derailing these three programs, but by the time Jacobs began to write, the big federal projects had already begun to unravel (Montgomery 1998: 272).

Urban renewal, in particular, fueled the growth of urban design as a profession. The federal government and local redevelopment authorities employed legions of designers to help conceive and execute government-sponsored projects. Notable were Victor Gruen, Lawrence Halperin and I.M. Pei. Gruen was particularly active in downtown master planning. When his *Plan for Fort Worth* (1956) received national attention, he became a most-in-demand consultant for reviving downtowns. His advice, to create urban pedestrian malls, emulating the suburban shopping center whose prototypes he was developing at about the same time, would be translated to more than 200 cities by the early 1980s. While on the whole, the wholesale application of Gruen's pedestrian mall concept to downtowns resulted in many failures, a few were successful. They include Denver's Sixteenth Street Mall, Burlington's Church Street Marketplace, Charlottesville's Downtown Mall and Santa Monica's Third Street Promenade. Some malls had rocky beginnings. For example, Santa Monica jumped on the bandwagon with an ill-conceived mall in the 1960s that failed by the 1970s but redesign, rebranding and adoption of a focused management scheme transformed it to the award-wining Third Street Promenade by the late 1980s (Pojani 2008: 141–143; Garvin 2002: 184–187). Pei's work in Cleveland (Erieview 1960), Boston (Government Center Master Plan 1961) Philadelphia (Society Hill Towers 1963) focused on in-town residential and civic center regeneration. As with the malls, these efforts had mixed success.

In the urban renewal era that spanned 1949–1973, urban design became an important public sector activity in many large cities. Municipal governments, supported by generous federal funding for slum clearance, formed redevelopment authorities (RDAs) that hired a cadre of experts, including urban designers, to help select sites for clearance and frame their rebuilding.[10] Motivated by a desire to transform obsolete and/or congested nineteenth-century industrial lands into twentieth-century uses including office centers, housing, and retail, the RDAs gave designers free hand to reshape large swaths of cleared land, conditioned only by the need for resale for redevelopment. The designers, mainly trained in the modernist, CIAM traditions, first tended to produce superblock, towers-in-the-park designs, but over the years added new thinking. Boston serves as an example. After the city's mayor, John F. Collins appointed Edward Logue as head of the Boston Redevelopment Authority (BRA), Logue hired urban designer David Crane to form an urban design group in the agency. In the next few decades, this group would redesign the city's waterfront, the Government Center, the South End and Charlestown. In this work, Crane developed and employed the "capital web" concept, that is, mapping and augmenting the public investments that would stimulate private investments that together could transform an urban district. Later designers, including Alexander Garvin, would adapt this idea in their writing and work (Garvin 2002).

Through the 1960s and 1970s decision-makers and practitioners refined urban design theory in practice. Notable was New York's Urban Design Group, established in 1967 by Mayor John V. Lindsay, "to bring new stature, coherence and boldness to [the] city's urban planning... [to] think about old problems in new ways" (Lindsay in Barnett 1974, na). Included among the original Urban Design Group were Jonathan Barnett, Alexander Cooper, Jacqueline Robertson and Richard Weinstein, all to be important as Later Evolvers. Although a cohesive group, they held a variety of mayoral appointments. Cooper was a member of the Planning Commission. Barnett was housed in the main office of Department of City Planning, charged with running the Urban Design group. (Cooper would later succeed him.) Robertson and Weinstein worked from the Office of the Mayor as heads of the Office of Midtown Planning (Robertson) and the Office of Lower Manhattan Development (Weinstein).

The members of this group did think in new ways. They rejected the CIAM/modernists' solutions of wholesale clearance as wasteful and disrespectful of existing urban fabric and resources. They sneered at their colleagues' belief that design should not be "contaminated" by the realities of finance and politics and determined how to motivate public and private decision-makers to undertake urban design goals. In so doing, they dug into understanding and manipulating in positive ways the laws and practices that determined in the city's development: zoning, landmark preservation, neighborhood planning, citizen participation, transportation planning and design review (Barnett 1974).

The Urban Design Group invented two important and powerful zoning devices: incentive zoning and special districts to achieve desired design objectives. Incentive zoning simply meant giving additional square footage on a designated site to a developer provided he gave a defined public benefit in return. Among the defined public benefits was the creation of plazas, theaters, and features like pedestrian bridges, mid-block pedestrian passages or arcades to improve circulation through an area. The special district permitted the redesign of multi-block areas. (The concept descended from urban renewal design but differed because, in this case, the land was in the hands of multiple owners.) The first special district, Greenwich Street in Lower Manhattan, for example, called for enhanced pedestrian circulation in a congested area faced with rapid development. The designers developed a general plan that treated foot traffic as well as other aesthetic considerations to enhance the walkers' experience. They then studied each block in the area to determine what improvements it could accommodate to support the plan and finally tied increased square footage of floor area to their provision. Thus a developer who provided an underground concourse or arcade or plaza or pedestrian bridge could gain extra rentable or for-sale space if he built the specified amenity. The Urban Design Group created additional special districts for critical areas of the city like Times Square and Lincoln Center. The plaza bonus, for example, resulted in more open space at the base of skyscrapers while the special district either protected the unique character or fostered particular improvements in a specified area. The adjacent Clinton neighborhood protected low income housing (Huxtable 1970, Barnett 1974). Reconfigured zoning transformed the city – by the turn of the century the ordinance would incentivize more than 500 privately-owned public spaces and accommodate more than 125 special districts (Kayden 2000, Garvin 2002). Many cities, witnessing the success of these techniques adapted them to their circumstances (Loukaitou-Sideris and Banerjee 1998). And, in the 1970s, William H. Whyte improved New York's incentive zoning by filming and analyzing individual behavior in the plazas, in order to perfect design details later incorporated into the ordinance (Whyte 1980).

In Boston, architect Ben Thompson, hired by developer James Rouse to oversee the restoration of Faneuil Hall Marketplace (1976) scored a huge hit with the resulting three-block, 6.5-acre site that mixed retail with dining, open market stalls with push carts, modern lighting and banners with gently restored buildings and pedestrian-oriented circulation.Thompson and Rouse would replicate this type of downtown revitalization successfully in Baltimore (Harbor Place 1980) and less successfully in New York (South Street Seaport 1985).

Through trial and error, the Developers determined how to manipulate building scale, sidewalk widths, landscaping, land use functions and other features to make aesthetically pleasing, economically successful places, ideas that inform today's designers in many practical ways. Descending from this work are Alan Jacobs' *Great Streets* (1995), Alex Garvin's *The American City: What Works What Doesn't* (2002) and the City of New York Department of Transportation's more recent *Street Design Manual* (2009).

As these activities gathered steam in the United States, similar changes were occurring in Europe where a new generation began to rethink CIAM-based urban design principles.While through the 1940s Sert and his associates had been modifying Le Corbusier's initial vision, the changes were not enough to satisfy a younger generation of British and Dutch architects. Participating in postwar reconstruction efforts where an estimated 10 million new dwellings were being built, they witnessed a variety of local and national housing and redevelopment strategies, many based on CIAM principles. (Pedret 2001: 46). The British housing programs included the London County Council's construction of thousands of in-town and peripheral housing estates and eight New Towns outside London; the Dutch built massive public housing projects in their major cities. For the upstarts "the CIAM's four functions [housing, work, recreation and transportation] were inadequate" and addressed issues "no longer deemed important" (Pedret 2001: 154).

What was needed were new ways of thinking, ones that were less mechanical, more tailored to "human associations" and focused on designing for individual needs in the habitat (different geographic scales) in which people lived. The younger architects soon expressed these sentiments at CIAM congresses, becoming more and more outspoken. By 1954, a rump group led by British architects Peter and Alison Smithson and Dutch architect Jaap Bakema met outside the CIAM meetings in Doorn, Holland to produce a short list of their beliefs, an eight point document, labeled the Doorn Manifesto. They offered a new four-dimensional analytic for planning: the house, street, district and city, associating each with a particular type of housing (see Figure 1.3). The teachings of Scottish botanist and pioneer sociologist, Patrick Geddes, drove their ideas. Geddes, reputedly inspired by youthful hikes in the rugged Scottish countryside, widely employed an image, the "Valley section," to express the symbiotic relationship between man and his environment. It was a cut from river to mountain overlaid with the occupations practiced in different scales. As Team they adapted the scheme, showing housing forms according to levels of urban development, arguing that "we must therefore study the dwelling and the

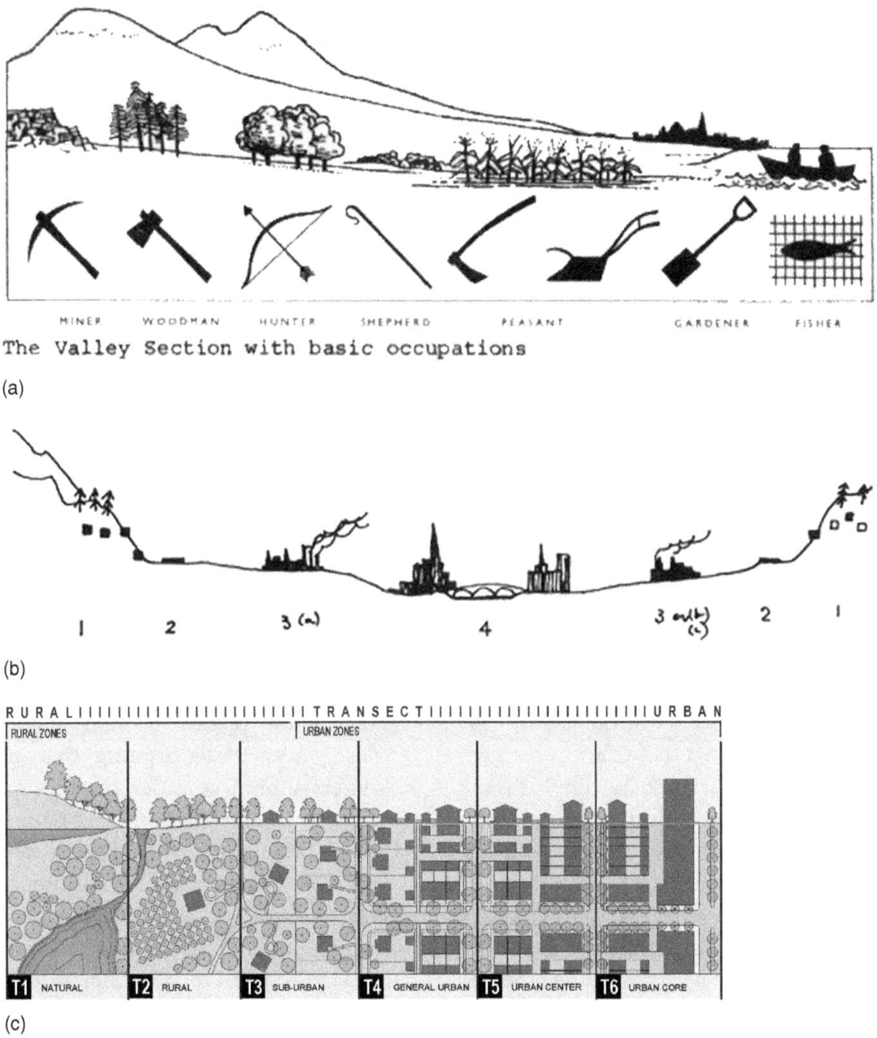

Figure 1.3 Patrick Geddes' (a), Team 10 (b), and New Urbanism (c) drawings.
Note: Team 10 drew heavily on Patrick Geddes' work, the Valley Section (1905) (top), to illustrate the different scales for design (1954) (middle). Five decades later, urban designer Andrés Duany composed a similar scale, the Transect, to illustrate the regional concepts of New Urbanism (bottom).

groupings that are necessary to produce convenient communities at various points on the valley section" (Smithson 1968: x). Their sketch bears a remarkable conceptual likeness to Andres Duany's Transect image that describes desirable settlement patterns advocated by the Congress for the New Urbanism today (see Figure 1.3).

Another follower of Geddes, Ian McHarg, was in Scotland in these formative years (1950–1954) but not associated with Team 10. He had studied at Harvard (1946–1950) and, recruited by his former teacher Dean Perkins, returned to the United States to build a landscape architecture department at Penn. Like Team 10 and his friend Lewis Mumford, he was quite disillusioned with the CIAM modernism that had dominated his education. In a later reflection, he noted his astonishment at the lack of appreciation of nature in the urban environment at Harvard (McHarg 1996: 82). He remedied the gap in the Penn curriculum and in his practice, Wallace McHarg, Roberts and Todd (now WRT) developed with Penn city planning professor, David Wallace, another colleague from Harvard. When he published his master work, *Design with Nature* (1969), he was repudiating the ahistorical, anti-nature training he had under received the Gropius/Perkins/Hudnut regime.

Designing for the human aspects of urban life became even more essential as sociologists Michael Young and Peter Wilmott published *Family and Kinship in East London* (1957), a study of public housing residents that revealed the intricate relationships, now known as social capital, hitherto unconsidered by the designers. Five years later, Penn's first city planning PhD graduate, Herbert Gans, offered the same type of insights about the US in *The Urban Villagers: Group and Class in the Life of Italian-Americans* (1962), a study of residents in Boston's West End, a slum cleared area, famous for the strenuous objections of its dislocated residents.

Urban designers attempted to address these issues. The Smithsons, for example, devised a "streets in the air" housing scheme, modernist in style but having open air hallways in high rise buildings meant to replicate street life in the slums they replaced.[11] They showed the idea in a competition entry (Golden Lane 1952) and executed it in Robin Hood Gardens (1967). While well-intentioned, the idea in execution failed in London (as much as for its harsh architecture, site planning and location as for the "streets" idea) and elsewhere, notably Chicago.

With disillusionment in the execution of modernist-urban-design-inspired redevelopment spreading throughout the US and Europe, new theories emerged, united by their common appreciation of the complexity of the city. Some theorists, like Christopher Alexander, shied away from actually specifying how to operationalize their ideas while others, like Gordon Cullen, Collin Rowe and Kevin Lynch, were more specific. In "A City is Not a Tree," Alexander argued that the human tendency to organize things in a linear fashion (like a tree) caused designers to create "artificial" cities, falsely compartmentalized (e.g. separation of land uses and vehicular and pedestrian traffic), when in reality, cities are complex systems of overlapping elements (or semilattices) that defy such simplification (Alexander 1965). Earlier, Cullen had produced *Townscape* (1961), a work that implicitly observed complexity and held that experiencing, observing and paying attention to detail in cities not only yielded appreciation of their organization and coherence but also gave clues to urban designers about how to produce better work. Collin Rowe and Fred Koetter picked up this theme in *Collage City* (1978) arguing that urban designers should consider the context of their sites. In *A Theory of Good City Form* (1981) Kevin Lynch offered more general "performance standards" (vitality, sense, fit, access, control) rather than precise prescriptions to guide urban design. By 1987, Alan Jacobs and Donald Appleyard, both professors of city planning, University of California, Berkeley, would capture these ideas in "Toward an Urban Design Manifesto," (1987), an attempt to produce a modern *La Charte d'Athènes.* With these encomiums, the field began to focus on smaller scale urban design concerns, often labeled the "public realm," those spaces (streets, plazas, waterfronts, parks) available to citizens bounded by the public and private buildings that surrounded them.

Other themes were emerging through the work of the Developers that would flourish after the 1970s: participatory decision-making, regionalism, growth management, and sustainability. Their fragile roots were present in many areas.

Jane Jacobs, for example, laid out the idea of taking into account local knowledge in urban design, an idea that stimulated "advocacy planning," first suggested by Penn city planning professor, Paul Davidoff in quite simple terms (Davidoff 1965) and elaborated by Roger Montgomery, the first urban design officer for the US Housing and Home Finance Administration (predecessor of the US Department of Housing and Urban Development) and later Dean of the College of Environmental Design, University of California, Berkeley (Montgomery 1966; also see Faga 2006).

With regard to regionalism/growth management, the original urban design manifesto, *La Charte d'Athènes*, began with a strong statement recognizing the importance of an urban region ("The city is only a part of the economic, social and political entity which constitutes the region..."). Team 10 adapted Geddes's Valley Section to illustrate the importance of thinking regionally and Edmund Bacon sketched how to apply regionalism with the vision of Philadelphia and its environs that graced *Design of Cities* (Smithson 1968, Bacon 1976). William H. Whyte proffered implementation strategies (conservation easements) as a way to protect the natural assets in rapidly suburbanizing regions, thus starting thinking about growth management (Whyte 1959). Later, planner Robert Yaro *et al.* would apply these concepts in their prize-winning *Dealing with Change in the Connecticut River Valley, A Design Manual for Conservation and Development* (1988). Supporting the larger scale regional visions would be social science research that studied contemporary settlement patterns and the merging of urban land uses from Boston to Washington (Gottman 1961). Peter Calthorpe and William Fulton (2001) extended this thinking to other parts of the United States. Among the techniques introduced by these authors was scenario building or visioning of alternative development patterns under varying design conditions. Finally in a recent study, Jonathon Barnett, Robert Yaro and others would investigate the power of natural systems and infrastructure in large scale (metropolitan to megapolitan) urban design (Barnett *et al.* 2007).

Ian McHarg pioneered today's sustainability movement with many of his ideas encompassed in modern landscape urbanism and green concepts. A new field of sustainable urban design is gaining ground around the world. One indicator of its attractions was the heavy attendance (over 400) at "Re-imagining Cities: Urban Design After the Age of Oil," a conference funded by the Rockefeller Foundation, in Fall 2008 at the University of Pennsylvania that explored the topic in its international implications and yielded suggestions for future urban design. Its organizers issued "Educating Urban Designers for Post Carbon Cities" a manifesto calling for designers to integrate sustainability in their work so that in the words of the Brundtland Commission, the field can meet the needs of today without compromising those of the future (Birch 2009).

Conclusion

While the roots of urban design emanated from responses to rapid urbanization in the nineteenth and twentieth centuries, they developed in several directions as three classes of theorists and practitioners, the Founders, Pioneers and Developers, elaborated and tested various approaches to city-building from the 1920s through the 1970s. They originally constructed models for entirely new cities: either dense centralized places often achieved through clearance and rebuilding as suggested by CIAM or low density, decentralized places as suggested by Garden City advocates. (Both groups aimed to codify the "good life" through techniques that would be widely employed in the immediate postwar period in US urban renewal projects and European reconstruction.) They were particularly enamored with the neighborhood unit and pedestrianized urban centers. The neighborhood unit is an example of a device that would later be discredited by those disillusioned by or unconvinced of its efficacy but later came back into favor through the advocacy of today's Congress for the New Urbanism. Walkable downtowns have experienced the same fate.

Standardized city-building schemes lost credibility when urban designers, assisted by scholars from the social sciences as well as proponents of earlier traditions in urban morphology, began to look more closely at how cities are organized. They discovered and acknowledged their complexity, offered goals for overall performance and learned of successes from the already built environment. Many moved from designing whole cities to designing places – streets and sidewalks, plazas and other elements of the public realm. In addition to undertaking these efforts, they refashioned the definition of professionalism, departing from elite, top-down processes to participatory models as they began to engage the people for whom they were designing in conversations about their needs and desires, integrating local knowledge into their plans.

At the same time, others began to address larger scale issues, those related to regional geographies and environmental issues. They would focus on natural systems as a determinant in urban design. And it is in this area, that today's sustainability concerns would emerge.

Urban design has deep roots that have nurtured a field through the twentieth and early twenty-first centuries. Its proponents have been responsive to current urban issues, developed experimental approaches and self-corrected them over time. Remarkable continuities and communications developed among the urban designers, especially the Founders, Pioneers and Developers, who often worked together, shared their ideas through conferences, writing and teaching. Unifying them is their dedication and engagement in developing integrated thinking to build places of long term value.[12]

Notes

1 Camillo Sitte, Austrian architect and author, called it "der stadtebau," (Sitte 1889, Collins 2006), Le Corbusier used the term "urbanisme" (Le Corbusier 1924) and authors/architects Werner Hegemann and Elbert Peets called it "civic art." (Hegemann and Peets 1922). The term "civic design" originated with David Crane, architect and head of the educational program at the University of Pennsylvania. "City design" was MIT's Kevin Lynch's term and various Harvard professors called the field "urban design." These differences of terminology in educational institutions also were reflected in their instructional programs: in 1958, Penn was the first to offer a formal urban design program and used a dual degree structure (a student received a first professional degree e.g. M.Arch, MCP) and a certificate in urban design. Harvard followed and created a separate track with the urban design degree appended to the first discipline. MIT offered a concentration and a certificate associated with a "professional degree". The terms also had nuances. Urban design was very much in the tradition of the Bauhaus and CIAM inspired the idea that there was a continuum of design from objects to interiors to buildings to urban areas. Civic design advocated the idea that there was a civic realm that needed designing, and that it was a different task from designing buildings or objects. City design espoused the view that designing cities was a task that involved more than physical considerations, including the economics and arrangements for development, sociological aspects, etc. (Hack 2009).

2 A decade later, Frank Lloyd Wright would go further with his concepts for very low density settlements as imagined in his Broadacre City (1932).

3 In a later transatlantic exchange, the British would insert it in the postwar new towns (Birch 1980).

4 *Can Our Cities Survive* went into two reprints (1944, 1947) and was also favorably reviewed in *Town Planning Review* (1943) and *Town and Country Planning* (1945).

5 Architect/planner Victor Gruen would deepen these ideas in *The Heart of Our Cities, the Urban Crisis: Diagnosis and Cure* (New York: Simon

& Schuster. 1964) as well as in his plans for numerous shopping centers including Southdale Shopping Center, Minneapolis, MN, featured in Martin Meyerson's *Face of the Metropolis* in 1963.

6 Perkins, appointed in 1945, had decamped to Penn in 1951 as the founding dean of the Graduate School of Fine Arts (Alofsin 2002) while Gropius, who differed strongly with Hudnut over the nature of the common introductory course, had retired in 1952.

7 Isaacs at this time was involved with the new city planning program at the University of Chicago, headed by Rexford Tugwell and numbering among its faculty political scientist Edward Banfield, economist Harvey Perloff and Martin Meyerson. This program was the first to insert a good dose of social science training in city planning.

8 This is what had happened, in effect, at Stuyvesant Town when its landlords, the New York Life Insurance Company, barred applications from Blacks. This may be why Meyerson neglected this otherwise "model" project of modern urbanism.

9 The Congress for New Urbanism that advocates the use of Perry's neighborhood unit avoids the homogeneity arguments by calling for mixed income dwellings within the unit. The CNU advocates seem to be unconcerned about the nimbyism potential of the device called out by Isaacs. But then, neighborhood planning in general has not addressed this issue.

10 Under the 1949 Housing and Slum Clearance Act and subsequent legislation, the federal government would cover two-thirds of the cost of purchasing, clearing and preparing a site for resale. The local government contributed the remaining one-third, a figure that could include the costs of providing new streets and public spaces and facilities.

11 Although the Smithsons and their colleagues hailed this idea as new, it was not, having been employed in Brooklyn, New York at Riverside Houses (1890) by William Field & Son.

12 The phrase developing "integrated thinking to build places of long-term value" comes from Marilyn Taylor, former chief of urban design, SOM and currently Dean and Paley Professor, PennDesign, September 8, 2009.

References

Alexander, C. (1965). "A City is not a Tree." *Architectural Forum*. 122: 1–2 (May, June 1965): 58–61 (Part 1), 58–62 (Part II).

Alofsin, A. (2002). *The Struggle for Modernism: Architecture, Landscape Architecture and City Planning at Harvard*. New York: W.W. Norton & Co.

Architectural Forum. (1952). "The Philadelphia Cure: Clearing Slums with Penicillin, Not Surgery," 96(2) (April): 112–119.

Bacon, E. (1948). "Are Exhibitions Useful?" *Journal of the American Institute of Planners* 14:2 (Spring): 23–28.

Bacon, E. (1967). *Design of Cities*. New York: Viking Press.

Bacon, M. (2008). "Josep Lluís Sert's Evolving Concept of the Urban Core." In E. Mumford and H. Sarkis (Eds.). *Josep Lluís Sert The Architect of Urban Design, 1953–1969*. New Haven, CT: Yale University Press.

Bauer, C. (1945). "Good Neighborhoods." *Annals of the American Academy of Political and Social Sciences*. 242 (November): 104–115.

Barnett, J. (1974). *Urban Design as Public Policy*. New York: McGraw Hill.

Barnett, J., R. Benfield, P. Farmer, S. Poticha, R. Yaro and A. Carbonell. (2007). *Smart Growth in a Changing World*. Chicago: APA Planners Press.

Birch, E. (1983). "Radburn and the American Planning Movement: The Persistence of an Idea." In D. Krueckeberg (Ed.). *Introduction to Planning History in the United States*. New Brunswick, NJ: Center for Urban Policy Research.

Birch, E. (2009). "Urban Designers Issue a Call to Arms." *Planning*, May: 56.

Birch, E. and C. Silver (2009). "One Hundred Years of City Planning's Enduring and Evolving Connections." *Journal of the American Planning Association*, 75(2): 113–122.

Calthorpe, P. and W. Fulton (2001). *The Regional City, Planning for the End of Sprawl*. Washington, DC: Island Press.

CIAM-France. (1943). *La charte d'Athènes avec un discours liminaire de Jean Giraudoux*. Paris: Plons.

City of New York. (2009). Street Design Manual. New York: New York City Department of Transportation.

Collins, C. and G. Collins (2006). *Camillo Sitte: The Birth of Modern City Planning: With a Translation of the 1889 Austrian edition of his City Planning According to Artistic Principles*. New York: Dover Publications.

Cullen, G. (1961). *Townscape*. London: The Architectural Press.

Davidoff, P. (1965). "Advocacy and Pluralism in Planning." *Journal of the American Institute of Planners,* 31(4): 331–338.

Faga, B. (2006). *Designing Public Consensus: The Civic Theater of Community Participation for Architects, Landscape Architects, Planners and Urban Designers.* Hoboken, NJ: John Wiley & Sons.

Ferris, H. (1929). *Metropolis of Tomorrow.* New York: I. Washburn.

Gans, H. (1962). *The Urban Villagers: Group and Class in the Life of Italian-Americans.* New York: The Free Press.

Garvin, A. (2002). *The American City: What Works What Doesn't.* New York: McGraw Hill.

Geddes, P. (1915). *Cities in Evolution.* London: Williams & Norgate.

Giedion, S. ([1941], 1966). *Space, Time and Architecture, The Growth of a New Tradition.* 5th Edition. Cambridge, MA: Harvard University Press.

Glass, R. (Ed.) (1948). *The Social Background of a Plan: A Study of Middlesborough.* London: Routledge & Kegan Paul.

Gottmann, J. (1961). *Megalopolis: The Urbanized Northeastern Seaboard of the United States.* New York: The Twentieth Century Fund.

Hack, G. 2009. Personal Communication. August 19.

Hegemann, W. and E. Peets (1922). *The American Vitruvius: The Architect's Handbook of Civic Art.* New York: The Architectural Book Publishing Company.

Huxtable, A. (1970). "Concept Points to 'City of the Future," *New York Times.* December 6'. p. 8 at 1, col. 3.

Hyde, T. (2008). "Planos, Planes y Planificación, Josep Lluís Sert and the Idea of Planning." In E. Mumford and H. Sarkis (Eds.). *Josep Lluís Sert The Architect of Urban Design, 1953–1969.* New Haven, CT: Yale University 54–75.

Jacobs, A. (1995). *Great Streets.* Cambridge, MA: MIT Press.

Jacobs, J. (1961). *The Death and Life of Great American Cities.* New York: Random House.

Jacobs, A. and D. Appleyard (1987). "Toward an Urban Design Manifesto." *Journal of the American Planning Association.* 53(1): 112–120.

Kayden, J. (2000). *Privately-owned Public Spaces: The New York Experience.* Hoboken, NJ: John Wiley & Sons.

Krieger, A. and W. Saunders (Eds.). (2009). *Urban Design.* Minneapolis, MN: University of Minnesota Press.

Isaacs, R. (1948). "The Neighborhood Theory." *Journal of the American Institute of Planners.* 14(2): 15–23.

Laurence, P. (2006). "The Death and Life of Urban Design: Jane Jacobs, The Rockefeller Foundation and the New Research on Urbanism, 1955–65." *Journal of Urban Design,* 11:2 (June): 145–172.

Le Corbusier. (1924). *Urbanisme.* Paris: Cres.

Le Corbusier. (1935). *La Ville Radieuse.* Boulogne-Billancourt: Éditions de l'Architecture d'aujourd'hui.

Loukaitou-Sideris, A. and T. Banerjee (1998). *Urban Design Downtown: Poetics and Politics of Form.* Berkeley and Los Angeles: University of California Press.

Lynch, K. (1960). *Image of the City.* Cambridge, MA: MIT Press.

Lynch, K. (1981). *A Theory of Good City Form.* Cambridge, MA: MIT Press.

McHarg, I. (1969). *Design with Nature.* New York: Natural History Press.

McHarg, I. (1996). *A Quest for Life.* Hoboken, NJ. John Wiley & Sons.

Meyerson, M. (with J. Tyrwhitt, B. Falk and P. Sekler) (1963). *Face of the Metropolis, The Building Developments That are Reshaping Our Cities and Suburbs.* New York: Random House.

Montgomery, R. (1966). "Improving the Urban Design Process in Urban Renewal." In J.Q. Wilson (Ed.). *Urban Renewal: The Record and the Controversy.* Cambridge: MIT Press.

Montgomery, R. (1998). "Is There Still Life in the *Death and Life?" Journal of the American Planning Association,* 64(3): 269–274.

Mumford, E. (2000). *The CIAM Discourse on Urbanism 1928–1960.* Cambridge, MA: MIT Press.

Mumford, E. (2009). *Defining Urban Design, CIAM Architects and the Formation of a Discipline, 1937–1969.* New Haven, CT: Yale University Press.

Mumford, L. (1934). *Technics and Civilization.* New York: Harcourt Brace.

Mumford, L. (1938). *Culture of Cities.* New York: Harcourt Brace.

Nolen, J. (1927). *New Towns for Old: Achievements in Civic Improvements in Some American Small Towns and Neighborhoods.* Boston, MA: Marshall Jones.

Park, R. and E. Burgess (1925). *The City.* Chicago: University of Chicago Press.

Pearlman, J. (2008). "Joseph Hudnut and the Prehistory of Urban Design." In E. Mumford and

H. Sarkis (Eds.). *Josep Lluís Sert The Architect of Urban Design, 1953–1969*. New Haven, CT: Yale University Press: 116–129.

Pedret, A. (2001). "CIAM and the Emergence of Team 10 Thinking 1945–1959." Unpublished PhD dissertation. Massachusetts Institute of Technology.

Perry, C. (1929). "The Neighborhood Unit." In *Regional Plan of New York and its Environs*. New York: Committee on Regional Plan of New York and the Its Environs.

Pojani, D. (2008). "Santa Monica's Third Street Promenade: The Failure and Resurgence of a Downtown Pedestrian Mall." *Urban Design International*, 13 (2): 141–155.

Rowe, C. and F. Koetter (1978). *Collage City*. Cambridge, MA: MIT Press.

Sert, J. (1942). *Can Our Cities Survive? An ABC of Urban Problems, Their Analysis Their Solutions Based on the Problems Formulated by the CIAM*. Cambridge, MA: Harvard University Press.

Sert. J. (1945). "The Human Scale in City Planning." *The New American Architecture and City Planning: A Symposium*. New York: Philosophical Library. 392–412.

Sert, J., J. Tyrwhitt and E. Rogers (Eds.). (1952). *The Heart of the City*. London: Humphries.

Sitte, C. (1889). *Der Städtebau nach seinen künstlerischen Grundsätze: Vermehrt um Grossstadtgrün*. Auflage 1. Wien.

Smithson, A. (1968). *Team 10 Primer*. London: Studio Vista.

Somer, K. (2007). *The Functional City: CIAM and the Legacy of Van Eesteren*. Rotterdam: Nai Publishers.

Time Magazine. (1964). "The City: Under the Knife, or All for Their Own Good." November 6. http://www.time.com/time/magazine/article/0,9171,876419,00.html (accessed 29 August 2010).

Tonnies, F. ([1887] 1935). *Gemeinschaft und Gesellschaft Abhandlung des Communismus und des Socialismus als empirischer Culturformen*, ("Treatise on Communism and Socialism as Empirical Patterns of Culture") Leipzig: Reisland.

Tunnard, C. and B. Pushkarev (1963). *Manmade America: Chaos or Control*. New Haven, CT: Yale University Press.

Tyrwhitt, J. (1933). *The Athens Charter 1933*. Translated from *La Charte d'Athenes* Paris, 1943. Accessed at http://www.getty.edu/conservation/research_resources/charters/charter04.html (accessed 30 August 2010).

Unwin, R. (1912). *Nothing Gained by Overcrowding, How Garden City Type of Development may Benefit both Owner and Occupier*. London: P.S. King and Company.

Unwin, R. (1920). *Town Planning in Practice An Introduction to the Art of Designing Cities and Suburbs*. 7th edition. London: Longmans Green & Company.

Venturi, R., D. Scott-Brown and S. Izenour (1972). *Learning from Las Vegas: The Forgotten Symbolism of Architectural Form*. Cambridge, MA: MIT Press.

Whyte, W. (1980). *The Social Life of Small Urban Places*. New York: The Conservation Foundation.

Whyte, W. (1959). *Securing Open Space for Urban America: Conservation Easements*. Washington, D.C.: The Urban Land Institute.

Wirth, L. (1938). "Urbanism as a Way of Life." *American Journal of Sociology*, 44(1): 1–24.

Yaro, R., R.G. Arendt, H.L. Dodson, and E.A. Brabec (1988). *Dealing with Change in the Connecticut River Valley: A Design Manual for Conservation and Development*. Washington DC: Lincoln Institute of Land Policy.

Young, M. and Wilmott, P. (1957). *Family and Kinship in East London*. London: Routledge & Kegan Paul.

Further reading

Barnett, J. (2008). *Redesigning Cities, Principles, Practice and Implementation*. Chicago: American Planning Association Planners' Press. The role of design in changing patterns of suburban growth, renewal of inner cities, and redressing dysfunctional metropolitan areas.

Carmona, M., T. Heath, T. Oc, and S. Tiesdall (2003). *Public Places Urban Spaces*. New York: Architectural Press. A guide to many different aspects of urban design; it presents some of the essential dimensions of urban design theory and practice.

Garvin, A. 2002. *The American City: What Works What Doesn't*. New York: McGraw Hill. A comprehensive review of some 300 urban design projects in 150 cities.

Larice, M. and E. Macdonald (Eds.). (2006). *Urban Design Reader*. London: Routledge. A collection of relevant urban design articles by well-known authors.

2

The open and the enclosed

Shifting paradigms in modern urban design

Robert Fishman

Anyone seeking to identify the shifting paradigms of modern urban design needs to look no further than the 92 acres of landfill along the Hudson River in lower Manhattan known as Battery Park City. Formed by the massive excavations for the twin towers of the World Trade Center in the 1960s, this magnificent site between the river and financial district became the perfect *tabula rasa* on which the profound transformations that shook urban design would be inscribed. The first plan from 1963 called for three rows of widely-spaced high-rise towers in an open, landscaped setting, an archetypal realization of the dominant "tower-in-the-park" paradigm dating back to Le Corbusier's 1925 Plan Voisin for Paris. When the 1963 plan was scrapped in the financial turmoil of the late 1960s, it was replaced in 1969 by a plan for a grandiose, futuristic, mixed-use "megastructure" proposed to run the entire length of Battery Park City, its cavernous interior spaces connected by the then-inevitable monorail (Gordon 1997).

But when the futuristic megastructure plan was in turn scrapped in the financial turmoil of the early 1970s, the next – and ultimately successful – plan took a surprisingly radical turn toward the past. Designed by the firm Alexander Cooper Associates to reflect the most successful existing neighborhoods in Manhattan, the plan ran a typical Manhattan grid over the landfill. The plan stipulated that a mix of high-rise and low-rise buildings would all be built out to the sidewalks to form solid street walls enclosing pedestrian-friendly narrow streets (some with ground floor retail) and small, enclosed parks. A wide but well-defined pedestrian "Esplanade," perhaps the most successful single feature of the plan, provided a grand public space along the riverfront. In a significant contrast to the former "megastructure," which would have been a single vast unified project, the designers provided that Battery Park City would be built out block-by-block over time by a range of developers whose differing designs would provide something like the variety of existing Manhattan streetscapes. Begun in 1979, the Cooper/Eckstut plan is only now reaching completion amid the turmoil of the rebuilding of the neighboring World Trade Towers site (Love 2006).

One can make sense of these vastly different plans by arguing that modern urban design has been dominated by a profound conflict between two very different paradigms regarding the role of the urban designer, each with deep roots in the history of cities and each with important implications for their future. The first paradigm, embodied in the initial tower-in-the-park plan for Battery Park City, celebrates the capacity of the urban designer to *open up* the too-solid fabric of the traditional city; to use modern design to relieve the inhuman overcrowding of the old city, and to replace it with a green open cityscape that would also provide room for the light-filled towers, great highways, and rapid communication that defined the modern age.

The second paradigm as embodied in the "neo-traditional" plan actually built, sees the primary role of the urban designer to *enclose space* – to create the human-scale "outdoor rooms" that provide the settings for the complex and informal communication, trade, and sociability that are the essence of urbanism. This second paradigm is respectful of the traditional fabric of the city and privileges continuity, walkability, small-scale enterprise, and neighborhoods over modernist innovation, scale, and speed.

The postwar era began with the first paradigm in the ascendant, especially as represented by Le Corbusier's remarkable synthesis of aesthetics and engineering in the compelling image of the "radiant city" and the "tower-in-the-park." Whether in downtown skyscrapers or in the "neighborhood units" that replaced the slums, this dream of a city of towers rising above open plazas and great highways embodied for its many champions the power and beauty that the modern city could attain. But history took another route, and the real story of urban design over the last fifty years has been the displacement of the urban design paradigm that sought to open up the city by the paradigm that sought to enclose space and to preserve the older urban fabric. This history begins with the international "citizen's revolt" against tower-in-the-park and highway urbanism in the 1950s; continues through Jane Jacobs's devastating critique of high modernist urban design in the 1960s; and concludes most recently with the trend toward sustainable urbanism. Ironically, the traditional urban fabric is proving more "modern" in its energy efficiencies and social "connectivity" than the more open designs that once seemed destined to shape the urban future (Farr 2008).

This "paradigm shift," to use Thomas Kuhn's famous phrase (Kuhn 1996), reflects a passionate debate *within* urban design but its outcome has ultimately been determined by those larger forces (such as industrialization, mass immigration, and more recently the energy crisis) that have the real power to shape the modern city. The "open" paradigm found its heroic rationale during the era of feverish growth of the Western European and North American city – roughly from 1800 to 1950 when the great metropolitan centers – what H.G. Wells called "the whirlpool cities" (Wells 1902) – drew literally millions from farms and villages into the super-dense vortices of cities like London, Paris, Berlin, Vienna, New York and Chicago. In these whirlpool cities the overwhelming "urban crisis" appeared to be overcrowding and congestion. The mass migration to the metropolis filled up the courtyards and alleyways in the older cores of large cities at the same time that these cities expanded inexorably in dense blocks into the countryside at the edge. The result was cities that were choking on their own traffic (even if this traffic was still horse-drawn); their overcrowded residents drinking polluted water and breathing polluted air; cities where providing even the minimum of light, space, and air for most residents seemed a utopian dream. (Mumford 1961; Hall 1998).

In response to this urban crisis of overcrowding and congestion, the great task of urban design appeared to be to *open up* the city, and designing paradigms for such openness pre-occupied the most brilliant efforts of urban designers of that era. But by the mid-twentieth century the very technologies – the railroad, electric tram, and subway – that had concentrated people in the whirlpool cities now permitted the urban population to spread out inexorably from their crowded cores. The mass ownership of automobiles in the United States and its eventual spread to Europe permitted a radical decentralization to low-density suburbs. In this new context, low-density automobile-dependent development became the norm – the "default setting" for urbanism – while the older urban values of density, walkability, and enclosure became goals that required the intense efforts and creativity of urban designers. In Battery Park City, for example, density and enclosure were no longer associated with the former slum districts of the nearby Lower East Side but with the ideal – at once new and old – of walkable urbanity. Hence the emergence of the enclosure paradigm as the preferred format for urban design, at least in those regions of Europe and North America where urban overcrowding was no longer a problem. By contrast, for those regions of Asia, Africa, and Latin America that are still in the "whirlpool" phase of urban development, the "open" paradigm with its towers-in-the-park design framework retains much of its importance and credibility (Campanella 2008).

Even in Western Europe and North America, the open paradigm still plays a vital, if limited, role in urban design, but its twenty-first-century incarnations tend to be drawn not from twentieth-century modernism but from the best work of the nineteenth century. In that era, the sheer difficulty of breaking through the dense urban fabric of existing cities required designers to adopt an admirable complexity and discipline in their attempts to realize the open paradigm. By contrast, twentieth-century modernist urbanism with its far greater technological resources often fell victim to inhuman scale and megalomaniacal ambitions. The earlier nineteenth-century open paradigm might best be defined by the interconnection of three major forms: (1) the multi-lane, tree-shaded boulevard, terminating in a grand public space and monument; (2) the parkway, a specialized boulevard at the urban periphery designed to connect the city to parks or rural open spaces; and (3) the "monumental" urban park, carefully planned as an alternative "green" environment while surrounded by dense building. As we shall see, these forms continue to inspire urban designers today.

This nineteenth-century design language of openness and movement will always be associated with its greatest achievement, the most successful "urban renewal" project of all time: the re-building of Paris undertaken by Emperor Napoleon III and his deputy Baron Eugene Haussmann in the mid-nineteenth century (Van Zanten 1994). From Paris, the form spread over the world under such rubrics as Beaux-Arts (named for the school of fine arts and architecture in Paris where it was best taught) or "City Beautiful," as it was called in the United States (Peterson 2003), and reached its most elaborate (but mostly unrealized) expression in Daniel Burnham and Edward Bennett's 1909 Plan of Chicago (Smith 2006). At the heart of this achievement was the network of Parisian boulevards and public spaces that Napoleon III and Haussmann cut through the dense fabric of Paris to open communications in a city where rapid movement from district to district was becoming impossible.

This "Haussmannization" used the power and resources of an absolutist regime to push through the massive demolitions that the imposition of the open paradigm on a dense city necessarily required. Nevertheless, the grand boulevards that resulted did more to justify the human costs than any subsequent "urban renewal" project (Jordan 1995). The boulevards were brilliantly designed to achieve a genuine urban complexity that complemented the finer-grained traditional urban fabric through which they ran. A Parisian boulevard is at once a high-capacity transportation system, with multiple lanes for both fast and slow moving traffic (then horse-drawn carriages and buggies, but now cars, bicycles, and buses); a "linear park" formed by carefully-arrayed rows of street-trees; a vital public commercial space including wide sidewalks and ground-floor cafés and retail establishments; and even a below-grade "sanitation system" formed by the water-pipes and sewers that run underground. The boulevards were designed to be lined by solid walls of apartment houses built to a uniform cornice height, whose bulk complemented and "framed" the width of the streets, and whose many windows and narrow balconies opening on the boulevard gave it a continuing life and animation. And the boulevards generally terminated in a monumental structure (e.g. the Paris Opera or the Arc de Triomphe) carefully placed in an expanse of open space that provided a monumental emphasis to the commercial/residential bustle of Parisian street life. Compared to the single-use automobile expressways of our time that leave a permanent scar on the city, the boulevard is a model of multi-faceted urbanity, and for that reason is again becoming a model for designers wishing to maximize both traffic and urban vitality (Jacobs *et al.* 2002).

One special Parisian boulevard, the Avenue de l'Imperatrice (now Avenue Foch), attracted particular attention from an American visitor, Frederick Law Olmsted, when he visited Paris in 1869 (Rybczynski 1999). Olmsted and his partner Calvert Vaux had designed New York's Central Park in 1858, their first park and the masterpiece of the nineteenth-century parks movement. Olmsted believed that the dense modern city was so destructive to both physical and mental health that the survival of its people required the creation of an alternative within it: an open, green world carefully designed as the "lungs of the city" to restore both body and mind. Along with the boulevard, the large urban park became the showpiece of the open paradigm. What intrigued Olmsted about the Avenue de l'Imperatrice was that it was a kind of linear park lined with tree-shaded villas that connected Paris to its largest park to the west, the Bois de Boulogne. Not only was this "parkway" an excellent model for a new kind of boulevard that could run through the periphery of the city (and indeed helped guide its development); but a unified network of parks and parkways could provide what Olmsted later called an "Emerald Necklace" at the urban edge to ensure a healthy balance of urban fabric and open space. In his great park/parkway projects for Brooklyn, Buffalo, Chicago, Boston (the site of the "Emerald Necklace"), and other American cities, Olmsted thus took the open paradigm to a regional scale (Zaitzevsky 1982). That regional scale was picked up and magnified by Daniel Burnham and Edward Bennett in their grand and grandiose 1909 Plan of Chicago, most productively in the designs for a great line of parks and parkways along the city's lakefront. "The lakefront belongs to the people," Burnham proclaimed at a time when the lakefront in fact belonged to the railroads and other polluting uses (Smith 2006, 22). But the Plan inspired another great achievement of the open paradigm, the network of Chicago parks along Lake Michigan, a network recently completed in 2004 with the opening of Millennium Park in the heart of Chicago's Loop, perhaps the most impressive recent achievement of American urban design (Gilfoyle 2006).

If the open paradigm reached its most ambitious scale in the 1909 Plan of Chicago, that Plan also showed, especially in the megalomaniacally-scaled "Civic Center," the dangers of that paradigm when Burnham and Bennett were not restrained (as they were in the parks) by a sense of human scale. Perhaps even more damagingly, the grand open spaces conceived by this and other Beaux-Arts and "City Beautiful" plans in the early twentieth century were soon overwhelmed by a tidal wave of automobiles, which brought a new level of congestion to the urban core and turned the most expansive open spaces into motorized maelstroms. Suddenly the neo-classical design language of the open paradigm seemed as obsolete as the elaborate carriages that once paraded along its boulevards. But the open paradigm found a new life and importance through its radical re-imagining in the 1920s and 1930s by the Swiss-French modernist architect and urbanist Le Corbusier. It was Le Corbusier's great achievement to bring the open paradigm into the age of the automobile and the skyscraper and to envision a totally re-formed modernist city that very quickly dominated first the imaginations and then the practice of urban designers (Fishman 1977).

Le Corbusier's Contemporary City (as he called it in the 1920s) or Radiant City (the name he introduced in the 1930s) was not the first to portray the modern city as a City of Towers, but it was the first to grasp the radical possibilities of high-rise building for urbanism. For Le Corbusier, the skyscraper was essentially a whole neighborhood extending upward instead of spreading out on the ground, its elevator system a "street in the air" (Le Corbusier 1924). It was therefore irrational to crowd skyscrapers together, as in New York City. Instead, each tower should stand free on its own landscaped "superblock," covering no more than 15 percent of the land. In such a "city of towers" one could for the first time encounter unprecedented density with unprecedented openness. The towers would free up space at ground level not only for beautiful parks and gardens but they would open up wide spaces between the superblocks for massive superhighways that would speed the new multitude of motorists around the city. Within each superblock a specialized system of roads would eliminate the multi-function "corridor street" with its (for Le Corbusier, irrational) mix of functions in favor of a hierarchy of single-function pathways ranging from pedestrian walkways to shopping streets. Whether in the now-functionally zoned and separated business center, residential areas, or industrial parks, each worker or resident would enjoy unlimited light, air, views, and mobility, in a truly radiant city (Le Corbusier 1935). As John Summerson put it, the park is not in the city (Olmsted's model); the city is in the park (Summerson 1963: 81).

Le Corbusier demonstrated, moreover, that he did not shy away from the massive demolitions that his version of the open paradigm would require for existing cities. In his Plan Voisin for Paris, he surpassed Haussmann (at least in his imagination), proposing to knock down 600 acres of traditional urban fabric in the historic core of central Paris and to replace them with eighteen 60-story cruciform-shaped glass-and-steel towers looming above highways and landscaped superblocks (Le Corbusier 1924). The project, which was never built, nevertheless demonstrated Le Corbusier's resolve that to be truly modern, one must be ruthless with the "obsolete" urban past. And, as he had hoped, the very daring and beauty of his designs gave an aura of inevitability to his designs. Here finally was a city that appeared to embody the full logic of modernity: the scale and speed; the standardization and separation of functions; the industrial materials and mass-production methods. From the utopian dream of an obscure outsider, Le Corbusier's radical modernist version of the open paradigm became the architectural avant-garde's accepted model for the modern city in the "Athens Charter" of the International Congress of Modern Architecture (CIAM 1933). After the (unplanned) urban destruction of the Second World War, the tower-in-the-park model became the shared ideal of architects and planners, government bureaucrats, and even capitalist developers (Mumford 2000).

But despite the aesthetic grandeur and functional logic of Le Corbusier's re-imagining of the open paradigm, the great new age of modernist urbanism and the open paradigm somehow never dawned. Le Corbusier may have disdained the confusions and the inefficiencies of the enclosed "corridor street," but we have learned that the complex, pedestrian-oriented life of these bustling streets nevertheless provided the essence of the urban experience, what Jane Jacobs would famously call "close-grained diversity" (Jacobs 1961, 5). Even when the "towers-in-the-park" did not degenerate into "towers-in-the-parking-lot," the pedestrian's experience at street level in these districts was a dispiriting combination of meaningless open space and inhumanly-scaled towers. During the 1950s and 1960s, the towers tended to inflate in scale as they became the favored design form for housing bureaucracies seeking to mitigate the postwar shelter crisis by constructing the maximum number of units on a given site. The results justified architect Rem Koolhaas's critique of the Bijlmermeer housing project outside Amsterdam as "boredom on a heroic scale" (Koolhaas 1995, 871). At worst, the towers degenerated into a new form of high-rise slum; the massive Pruitt-Igoe housing development in St. Louis, completed in 1958, deteriorated so quickly that many of its towers had to be demolished by 1972 (Fishman 2004).

The failure of the towers-in-the-park paradigm highlighted the continuing vitality of the older "obsolete" urban fabric the towers were supposed to replace. Despite decades of neglect, this fabric often had a wonderful human scale; a lively mix of functions, especially ground-floor retail. Even when these districts lost their manufacturing base, the loft spaces that became available were surprisingly adaptable to the "new urban economy" that appreciated small-scale flexible spaces. Unfortunately, urban design theory was so wedded to the open paradigm, that it long ignored the manifest evidence of failure. The traditional fabric was preserved by a grass-roots mixture of individual renovators – the so-called "gentrifiers"; by small property managers and speculators who operated at the fringe of the profession; and even by anarchists and artists who, as in Amsterdam and London, stubbornly "squatted" in abandoned buildings to save them from demolition (Tung 2001, 211–247). When, for example, artists began moving into the semi-derelict nineteenth-century industrial lofts in the newly-named "Soho" neighborhood in New York, they often had to hide their occupancy from building inspectors seeking to enforce codes prohibiting the conversion of factory buildings to residential use. Today Soho ranks as among the most desirable neighborhoods in the world, and the conversion of factories to residential "lofts" ranks as one of the most successful overall strategies for urban regeneration (Zukin 1982).

By the mid-twentieth century the strongest of these districts were able to challenge successfully those who threatened them with urban renewal, most famously in the neighborhood coalition that saved Washington Square Park in New York from Robert Moses's plan to run a highway through it (Fishman 2007), and a similar anti-freeway coalition which stopped the ugly Embarcadero Freeway in San Franscisco literally in mid-air from cutting the city off from its waterfront. The great manifesto of this movement appeared in 1961, written by a hitherto-obscure architectural journalist named Jane Jacobs, who had been a leader of the Greenwich Village group opposing Moses. In *Death and Life of Great American Cities*, Jacobs provided a stunning critique of the open paradigm, especially in its radical "demolitionist" form. Jacobs identified the life of cities with their *street life*, what she called "the ballet of the city street" that continuously brought together a diverse mixture of people, who not only supported the diverse enterprises that were the heart of the urban economy but gave a city its twenty-four-hour vitality. For this "close-grained diversity" to prosper, Jacobs argued, one needed density, mixed-use, and the enclosure provided by well-defined streets and public spaces, precisely what the "open paradigm" sought to overcome with its widely-spaced towers and functional zoning (Jacobs 1961).

Jacobs called for an urban design that would express the "intricate order" of cities, their "manifestation of the freedom of countless people to make and carry out countless plans," (Jacobs 1961) but she offered no detailed designs embodying that "great wonder," only the general principles that would indeed inform urban design in the four decades since the publication of her book. But as designers struggled to adapt her ideas, they discovered that an alternative paradigm did exist within urban design that stretched back to such nineteenth-century figures as the Viennese architect Camilo Sitte and the early twentieth-century English town planner Raymond Unwin. This paradigm was given new vitality by the English "townscape" movement of the 1950s and 1960s and most recently by the Congress for the New Urbanism. I have called this the "enclosure" paradigm, with Sitte as its first and in many ways archetypal exponent.

Sitte's book *City Planning According to Artistic Principles* was written in 1889 as a passionate critique of one of the greatest "open" designs of the nineteenth century, the Vienna Ringstrasse [Ring Street] (see Collins and Collins 2006). In the 1850s Viennese authorities began demolishing the massive but obsolete defensive walls, which had surrounded the core of the city, thus opening a vast area for the monumental structures – the Opera, the Parliament, the National Museums, the National Theatre, the City Hall and the University – that represented liberal culture and enlightened government in the Austro-Hungarian Empire. Set back from the roadway in ornamental parks and gardens, these widely-spaced, lavishly-ornamented structures in various historicist styles gave the Ringstrasse a scale and grandeur to rival anything in Paris or the rest of the world, and "Ringstrasse Vienna" was hailed as the embodiment of the new, open city (Schorske 1980). Surprisingly, one prominent Austrian urbanist protested: Camilo Sitte, who critiqued the disorienting vastness of the Ringstrasse spaces, the tendency of the buildings to "float" in the huge spaces, and the privileging of rapid movement over enclosure. By contrast, he found the true "artistic principles" of urban design in the narrow streets and especially the many tiny plazas of the old city. These irregular but carefully-formed spaces, often fronting churches, "humanized" the city, in Sitte's view, and gave a far better setting for a wide range of urban activities than the open spaces and constant movement of the Ringstrasse. "The ideal street," he argued, and even more the ideal square, "must form a completely enclosed unit" (Collins and Collins 2006, 117).

Sitte's re-discovery of the art of enclosure at the urban core found an unexpected but powerful echo at the urban periphery in the work of Raymond Unwin, a leader of the English "Garden City movement" and designer of what he called "the garden suburb" (Swenarton 2008). The Garden City movement might appear to belong to the "open" school of urban design, for its founder, Ebenezer Howard, wished to decentralize the metropolises of Europe and the United States and to move most of their population out to a regional network of planned "garden cities" of about

30,000 people, which would supplant the overgrown and overcrowded central cities. But Howard understood that it was important that this decentralization not sprawl out over the countryside but be concentrated in carefully-planned, mixed-income and mixed-use "garden cities" which would achieve a small-scale urbanity, walkability, and economic vitality along with close contact with nature (Fishman 1977). Howard chose Raymond Unwin and his partner Barry Parker to design the first English garden city, Letchworth, in 1903. And, in 1907, Unwin accepted the more difficult challenge of applying garden city principles to a new suburban development just north of London, Hampstead Garden Suburb (Unwin 1920).

Unwin had long been concerned with reforming the conventional English suburb of the time which (especially at the edge of London) stretched out along endless straight streets lined with row-houses, which formed an interminable and perpetually expanding gray edge to the city. By contrast, he conceived Hampstead Garden Suburb as tied to central London by rapid transit but as a distinct place of its own, with a pedestrian scale and a clear center and edge. Like Sitte, Unwin was an admirer of medieval urbanism, and he brilliantly utilized the courtyards and cul-de-sacs of traditional English cities to create a "landscape of cooperation," where small, enclosed open spaces lined with picturesque houses defined a neighborly common ground. Hampstead Garden Suburb was "mixed-use" with institutions at its core and shops at the edge; explicitly mixed-income with "artisans' cottages" mixed among substantial middle-class dwellings; green enough to distinguish itself from the gray suburbs that surrounded it, but dense enough to maintain a sense of enclosure, to ensure walkability, and preserve the bulk of Hampstead Heath (the parkland it bordered) from development (Miller and Gray 1992).

Hampstead Garden Suburb represented an ideal-type for a suburb designed within the "enclosure" paradigm, but even within the Garden City movement its careful balance of enclosure and greenery was rarely attempted. By the 1920s the movement was distracted by the coming of the automobile, and the many subsequent "garden suburbs" and "New Towns" such as Radburn, New Jersey (built in 1928 and coined the "town for the motor age") now tended to sprawl out almost like conventional suburbs. Only in the 1950s was Unwin's ideal of enclosure revived in the English "townscape" movement led by Frederick Gibberd, Gordon Cullen and Ian Nairn. They believed that the ideal "townscape" should consist of the pedestrian's "serial vision" of a series of dense, intricate, and enclosed spaces (Cullen 1961). This message was strongly reinforced from the perspective of sustainability by the landscape architect Ian McHarg, whose 1969 book *Design With Nature* emphasized the importance of "clustering" development to preserve farmland and unique and fragile eco-systems (McHarg 1969).

By the 1980s, this suburban wing of the enclosure movement was mature enough to link up with the urban wing coming out of Sitte and Jane Jacobs to create a truly regional enclosure paradigm that could run from such projects as Battery Park City at the core to "New Urbanist" garden suburbs at the edge. Within the central city, the principal emphasis has been on preservation of the existing built fabric and the transportation network that supports it, including adaptive re-use of older structures. When new buildings are required, they should be "contextual," reflecting the traditional typologies of the neighborhood, and organized into solid perimeter blocks fronting pedestrian-scale streets lined with ground-floor retail establishments. In addition to this mixed-use, the new residential stock should be mixed-income to promote true neighborhoods instead of single-class enclaves. The solid blocks and narrow streets that form the bulk of the neighborhood should be varied and relieved by carefully-enclosed small open spaces to serve as the defining public spaces of the neighborhood. More extensive open space for sociability and exercise might best be found in the spaces left behind by deindustrialization, most notably derelict waterfront sites that could be converted to scenic parks. For transportation, the enclosure paradigm favors a new incarnation of the nineteenth-century boulevards, multi-laned, multi-use streets for buses and trolleys as well as automobiles, tree-shaded and lined with housing to tie the boulevard back into the city.

At the periphery, the Unwin tradition of the "garden suburb" has been most strongly taken up in the United States by the Congress for the New Urbanism (CNU), a design and social advocacy movement founded in 1993. Reacting against the total automobile dependency of the typical cul-de-sac subdivision of the 1980s, the CNU had advocated in true Unwin fashion what two of its founders, Andres Duany and Elizabeth Plater-Zyberk, have called "traditional neighborhood design" (Duany and Plater-Zyberk 1992). First demonstrated in the Florida resort town of Seaside (1982), such neighborhoods achieve walkability and their own form of "urbanity" by adhering to the Unwin garden suburb principles of a clear center and edge; sufficient density to encourage walkability with houses on relatively small lots oriented toward the narrow streets; mixed-use and mixed-income, and well-defined and enclosed public spaces. Another CNU leader, Peter Calthorpe, has taken up Unwin's concern with transit, and his ideal of "Transit Oriented Development" (TOD) means building new suburbs around light-rail transit stops, both to give a walkable center to the development and limit sprawl, but also to provide rapid access to the regional downtown. The surprising re-birth of light-rail systems in the United States has given a renewed plausibility to the TOD (see also chapter by Polyzoides). Calthorpe himself has worked extensively in the metropolitan area that best embodies the ideal, Portland, Oregon (Calthorpe and Fulton 2001).

If the enclosure paradigm has the intellectual resources to design whole regions, the reality is that this paradigm (or anything like it) now accounts for only a small part of the built environment that has been created either in the United States or world-wide since 1945. The intentional and inadvertent destruction of traditional urban fabric continues unabated; Anthony Tung estimates that 50 percent of that fabric was destroyed in the course of the twentieth century (Tung 2001, 414). In the United States and Western Europe, the frantic over-production of low-density sprawl that resulted from the great international real estate "bubble" of the 2000s was perhaps the "last hurrah" of conventional sprawl development. Nevertheless, it added to the vast areas in our urban peripheries that (despite the efforts of New Urbanists and other reformers) are completely automobile-dependent. Even more disturbingly, the avid consumption of personal automobiles in the developing world has given rise to low-density, automobile-dependent "global suburbs" throughout the world.

Nevertheless, there has been a clear global trend, especially among younger people, to seek out dense, transit-oriented cities as the environment most congenial to contemporary life (Fishman 2005). Perhaps most importantly, the enclosure paradigm has been shown to have the best potential to produce energy efficient and sustainable cities just when we need them most. Where the dense, pedestrian-centered city was once a symbol of ecological and social crisis, the situation is now exactly reversed. It is the sprawling open paradigm that stands for unsustainable energy use, whereas the largest, densest cities like New York, Toronto, and Tokyo all exhibit energy consumption per capita at only a third of the average for their societies. As early as the 1960s, Lewis Mumford provocatively labeled the open paradigm with its towers and highways as "yesterday's city of tomorrow" (Mumford 1968, 116). The true twenty-first-century "city of tomorrow" is likely to be a complex blend of old and new, a synthesis of the open and enclosed paradigms into new forms never envisioned by their creators. But this new city which we are striving to design today will surely be a place where the human-scaled, traditional design-language of the street and the square will remain vital and enduring.

References

Calthorpe, P. and Fulton, W. (2001). *The Regional City, Planning for the End of Sprawl.* Washington, DC: Island Press.

Campanella, T. (2008). *The Concrete Dragon: China's Urban Revolution and What it Means for the World.* New York: Princeton Architecture Press.

Collins, C. and Collins, G. (2006). *Camillo Sitte: The Birth of Modern City Planning: With a Translation of the 1889 Austrian Edition of his City Planning According to Artistic Principles.* New York: Dover Publications.

Cullen, G. (1961). *Townscape.* London: The Architectural Press.

Duany, A. and Plater-Zyberk, E. (1992). *Towns and Town-making Principles*, 2nd ed. New York: Rizzoli.

Farr, Douglas. (2008). *Sustainable Urbanism: Urban Design with Nature.* Hoboken, NJ: Wiley.

Fishman, R. (1977). *Urban Utopias in the Twentieth Century: Ebenezer Howard, Frank Lloyd Wright, and Le Corbusier.* New York: Basic Books.

—— (2004). Re-thinking Public Housing. *Places* 16, #2 (Spring): 26–33.

—— (2005). "The Fifth Migration." *Journal of the American Planning Association*, 71, #4: 357–367.

—— (2007). "Revolt of the Urbs: Robert Moses and his critics." In Ballon, H. and Jackson, K. (Eds.) *Robert Moses and the Modern City: The Transformation of New York.* New York: Norton; 122–130.

Gilfoyle, T. (2006). *Millennium Park: Creating a Chicago Landmark.* Chicago: University of Chicago Press.

Gordon, D. (1997). *Battery Park City: Politics and Planning on the New York Waterfront.* Amsterdam: Gordon and Breach.

Hall, P. (1998). *Cities in Civilization.* New York: Norton.

Jacobs, A., Macdonald, E. and Rofe, Y. (2002). *The Boulevard Book: History, Evolution, and Design of Multiway Boulevards.* Cambridge, MA: MIT Press.

Jacobs, J. (1961). *The Death and Life of Great American Cities.* New York: Vintage Books.

Jordan, D. (1995). *Transforming Paris: The Life and Labors of Baron Haussmann.* New York: Free Press.

Koolhaas, R. and Mau, B. (1995). *Small, Medium, Large, Extra-Large: Office of Metropolitan Architecture.* (edited by J. Sigler) New York: Monacelli Press.

Kuhn, T. (1996). *The Structure of Scientific Revolutions*, 3rd ed. Chicago: University of Chicago Press.

Le Corbusier. (1924). *Urbanisme.* Paris: Cres. Translated as *The City of Tomorrow.*

—— (1935). *La Ville Radieuse.* Boulogne-Billancourt: Éditions de l'Architecture d'aujourd'hui. Translated as *The Radiant City.*

Love, T. (2006). "Urban Design After Battery Park City: Opportunities for Variety and Vitality in Large-scale Urban Real Estate Development." *Harvard Design Magazine*, 25 (Fall/Winter 2006–7): 60–70.

McHarg, I. (1969). *Design with Nature.* New York: Natural History Press.

Miller, M. and Gray, A. (1992). *Hampstead Garden Suburb.* Chichester: Phillimore.

Mumford, E. (2000). *The CIAM Discourse on Urbanism 1928–1960.* Cambridge, MA: MIT Press.

Mumford, L. (1961). *The City in History: Its Origins, its Transformations, and its Prospects.* New York: Harcourt.

—— (1968). *The Urban Prospect.* New York: Harcourt.

Peterson, J. (2003). *The Birth of City Planning in the United States.* Baltimore, MD: Johns Hopkins Press.

Rybczynski, W. (1999). *A Clearing in the Distance: Frederick Law Olmsted and America in the Nineteenth Century.* New York: Scribner.

Schorske, C. (1980). *Fin-de-siècle Vienna.* New York: Knopf.

Smith, C. (2006). *Plan of Chicago: Daniel Burnham and the Remaking of the American City.* Chicago: University of Chicago Press.

Summerson, J. (1963). Architecture, Painting, and Le Corbusier, in *Heavenly Mansions.* New York: Norton.

Swenarton, M. (2008). *Building the New Jerusalem.* Bracknell: IHS BRE Press.

Tung, A. (2001). *Preserving the World's Great Cities.* New York: Clarkson Potter.

Unwin, R. (1920). *Town Planning in Practice An Introduction to the Art of Designing Cities and Suburbs.* 7th edition. London:Longmans Green & Company.

Van Zanten, D. (1994). *Building Paris: Architectural Institutions and the Transformation of the French Capital, 1830–1870.* Cambridge: Cambridge University Press.

Wells, H. (1902). *Anticipations of the Reaction of Scientific and Mechanical Progress upon Human Life and Thought.* New York: Harper's.

Zaitzevsky, C. (1982). *Frederick Law Olmsted and the Boston Park System.* Cambridge, MA: Belknap Press.

Zukin, S. (1982). *Loft Living: Culture and Capital in Urban Change.* Baltimore, MD: Johns Hopkins University Press.

Further reading

Norma Evenson. (1979). *Paris: A Century of Change, 1878–1978.* New Haven, CT: Yale University Press. The best account of the meaning and impact of "Haussmannnization" (and later modernization) on Paris.

Le Corbusier. (1925). *The City of To-morrow and its Planning.* Translation from 1929 of his *Urbanisme*, (original edition Paris: G. Cres et cie, 1925). The brilliant and stunningly influential manifesto of the modernist version of the open paradigm.

Jane Jacobs. (1961). *The Death and Life of Great American Cities.* New York: Vintage Books. The key document for the mid-century transition from the open to the enclosed paradigm, still surprisingly rich and challenging especially for readers who get beyond the familiar opening chapters.

Raymond Unwin. (1994). *Town Planning in Practice: An Introduction to the Art of Designing Cities and Suburbs* (original edition London: T.F. Unwin, 1909; reprint with a new preface by Andres Duany and a new introduction by Walter Creese, New York: Princeton Architectural Press, 1994). The best embodiment of the enclosure paradigm in its classic "Garden City" phase and perhaps the most humane statement in twentieth-century urban design.

3

Pedagogical traditions

Danilo Palazzo

Writings on the pedagogy of urban design are relatively sparse in the extensive and burgeoning literature on the field. Of the seventy-eight selected excerpts from books and journals included in two recent "urban design readers" (Carmona and Tiesdell 2007; Larice and Macdonald 2007), none addresses this topic. A recent chronological anthology by David Gosling (2003) is an exception. Occasionally reflections on this topic have appeared in proceedings of conferences or seminars, some printed (Pittas and Ferebee 1982), some only in a typewritten form (Washington University 1962). Even such specialized journals as *Urban Design International* and *Journal of Urban Design,* have included very few papers dedicated to the teaching of urban design (Bartholomew 1980; Cuthbert 2001; Bakker *et al.* 2003; Radović 2004; Savage 2005). Other work on the matter has appeared in journals closer to the areas of architecture and urban planning (e.g. Tyrwhitt 1962; Kreditor 1980; Vernez-Moudon 1992) or has remained embedded in papers dedicated to the roles, challenges, and competencies of urban design, or concerned with teaching and training in the more generic realm of "design."

As an integrative profession and discipline "traditionally [...] allied with architecture and city planning" (Lang 2007: 464), urban design remains uncertain as a field. In Europe and elsewhere, both architecture and urban planning continue to compete for urban design tasks and activities, including its pedagogy (see Cuthbert in this volume). In North America, aside from architecture and urban planning, conflicts of competencies and claims on urban design also come from other professions, such as landscape architecture or, even landscape urbanism, a more recent subject of study (Waldheim 2006), which is gaining an increasing interest despite its redundancies with more mature related fields.

Apart from being a field that can be considered "an ambiguous amalgam of several disciplines" (Inam 2002) or a "no man's land" (Cuthbert 2001), urban design is also considered "largely fragmented in its practices, theories and methodologies" (Cuthbert 2007: 178). In addition to this, according to Anne Vernez–Moudon, "theories' guiding practice have remained at a paradigmatic level, based on different exemplary solutions" (Vernez-Moudon 1992: 331). Moreover, urban design educators, who came to the discipline from a variety of different origins, have used such theories in "somewhat eclectic ways" (Cuthbert 2001: 303). Nevertheless, the value of urban design lies in its role as a social practice, and urban design education needs to recognize that it is "an interdisciplinary approach to designing our built environment" (Vernez-Moudon 1992: 331) or, as Madanipour better clarifies, urban design can be defined as a "multidisciplinary activity of shaping and managing urban environments, interested in both the process of this shaping and the space it helps shape. [...] Urban design is part of the process of the production of space" (1996: 117).

Based on this preamble the chapter will investigate the topic from two points of view. The first part illustrates the five-decade history of urban design education, examining the thoughts of those who first introduced urban design into American or British universities. It will also look at the considerations of the reflective educators who tested the implications for education programs of matters like role, values, and competencies of urban designers, or topics such as social participation, significance of places, or emerging challenges such as globalization and the use of new technologies. This part ends with an attempt to depict the currently uncertain situation of world urban design graduate programs taught in English, followed by some further considerations.

The second part will scan some of the teaching techniques employed and developed by teachers in order to train urban design students for their future profession, looking into topics such as the field's interdependency with other professions, its responsibilities toward the social fabric, and the specific value of sites. The chapter ends with some final remarks describing the directions and topics that urban design education should consider in the future.

Urban design education

The term "urban design" was coined in the mid-1950s (Lang 2005) almost coincidentally with its first appearance in academic curricula in the United States. The first academic program was the University of Pennsylvania's Civic Design Program, started in 1956 (Barnett 1982; Strong 1990), followed by Harvard's Urban Design Program in 1960. Thereafter the term was imported into the UK, even though it is in the UK where the first course and the first department of "Civic Design" at Liverpool University began in 1909. The Liverpool University course was intended to train planners (Cullingworth and Nadin 2006), with a "close connotation to municipal government and functions such as 'Civic Centre'" (Cuthbert 2007: 180) and town planning (The Builder 1908) and hence cannot claim the progenitorship of today's urban design curriculum.

A few years before the creation of the Civic Design program at the University of Pennsylvania, G. Holmes Perkins, then Dean of the School of Fine Arts, set out a common program in architecture, city planning, and landscape architecture based on the argument that: "the work of the first three years of the [three] professional courses [...] is, except in rare cases, identical in content, reflecting the fact that all are parts of a common field whose processes and objectives are the same" (Perkins Holmes 1952 in Strong 1990:135). Clarence Stein, a former member and secretary of the Regional Planning Association of America, co-designer, with Henry Wright, of Radburn, NJ and author of *Toward New Towns for America* (1951), was asked by Perkins to draft a proposal for the program at the University of Pennsylvania. Stein established a definition of city design that still retains its clarity today: "CITY DESIGN is the art of relating: STRUCTURES to one another and to their NATURAL SETTING to serve contemporary living" (Stein 1955 in Strong 1990: 141).

Although the University of Pennsylvania's joint program did not survive, the principle of cooperative activity between architecture, city planning, and landscape architecture, and the definition of city design lived on, acting as keystones not only in the Civic Design program, which began a few years later at the same university, but in almost all graduate and postgraduate programs in urban design offered since the 1960s in universities worldwide. For example, at Harvard, the curriculum in "urban design" in the early 1960s suggested "in a quite limited and specific sense [...] an area of interaction between the three professions of architecture, landscape architecture and city planning" (Tyrwhitt 1962: 100). Thus conceptualized, urban design became a specialty for master degrees in architecture and city planning and not a degree in its own right.

In a 1979 urban design colloquium at the University of California at Berkeley, Kevin Lynch discussed the training of urban designers in American universities (Lynch 1980). His words provided personal, but significant, insights on the subject some twenty years after the first urban design course was offered in the US: "City design [the term he preferred to "urban design," although he began his academic career by using "civic design" (Lynch 1954)] is not a well-developed skill, and I know no school where it is adequately taught." (Lynch 1980: 655). Lynch proposed a "two-year graduate professional program" with three central, elementary skills that seemed to him indispensable. The first is "a sharp and sympathetic eye for the interaction between people, places, place events, and the institutions that manage them." The second skill to be developed is an understanding of the theory, technique, and values of city design. Lynch rejected the prevalent idea that design was non-analytical, socially irresponsible, concerned with images and representations, and reserved for the gifted few. The third skill to be acquired by a city designer is in communication. City designers must be prepared to understand and use the four social languages: written words, spoken words, mathematics, and graphic images.

This legacy notwithstanding, today, in the early years of this new century, reflections on urban design education are more dispersed in the literature and are more often centered on common concepts, such as globalization, sustainability of development at the macro and micro scale, and digital technology, that can address the contemporary and future forms of teaching urban design. Susan Savage (2005), underlining the role of school, calls for a pedagogic orientation that emphasizes practical knowledge, real-time learning, problem-driven and interdisciplinary approaches, ideas which are supported by other authors (Inam 2002; Bakker *et al.* 2003; Lang 2005).

Today urban design education is facing new challenges brought about especially by the driving forces of globalization and mobility of students. International student numbers grow every year in successful universities, raising "in each situation [...] implications for urban design education [...suggesting] that in the information age, universities and their constituent faculties are compelled to address globalization in their own programmes" (Cuthbert 2001: 300–301). Globalization and the attendant student flows, especially from East to West, raise questions about the applicability of Western analysis and design methods to the East (Radović 2004) or as Banerjee (1990: 175) has suggested, "environmental design education currently offered in the US or other Western universities may not be relevant or sufficient for students from developing countries."

Another relevant topic in urban design education is the interface with the reality of cities, societies, and places hence promoting a dialectic process with communities and sites, "a crucial aspect of environment" (Lynch and Hack 1984: 29), to stress the production of the public realm (Banerjee 2001; Hanson and Younés 2001, Arefi 2004; Arefi and Triantafillou 2005). Teaching the value of place and how to discover it through investigation and surveys remains central to every program, especially when there is a cultural hiatus between the nationalities of students and places and communities (extremely simplified: West and East, or on reverse). These situations are more and more frequent because of the mobility of students and teachers and the broad diffusion of international design studios where Western educated students encounter "other" places and "other" meanings. As Darko Radović pointed out, in the exploration of the place "it is necessary to broaden the views of participants, to be able to accept, at the very least, otherness, and even embrace, at the other extreme, the totally alien, a *tout autre*" (2004: 184). Fortunately "urban design provides an excellent field to encounter, experience and address the totality of the other" (Radović 2004: 178).

The dialectic process with sites and communities also allows urban design students to perform the so-called "work-integrated learning" that is the practice knowledge that they deserve to face, in their professional future, real problems affecting real people. Similar issues have surfaced in parallel literatures on design, planning, architecture, and landscape architecture education. In the 1960s and the 1970s social movements required that urban designers and architects "become the anonymous servant of the masses. The architect's ostrich-like fixation with imagery and aesthetics is challenged in the face of social need and participatory democracy" (McSheffrey 1978). Community Design Centers emerged in the 1970s in response to this new reality, drawing volunteer groups of architects, teachers and students "to exchange ideas concerning the provision of urban design services for moderate-to low-income communities" (Gosling 2003: 143, also chapter by Anthony in this volume). Some of them were strongly connected with universities. According to David Gosling, in the 1970s and 1980s there was a tangible transition "from education to practice," as highlighted in Jonathan Barnett's, *An Introduction to Urban Design* (1982). Thus claims for the contextualization of student work in the community, for education as reflection-in-action (Schön 1984; Shannon 1990), on the role of design (and its teaching) in the production of built environment and its effects on human health (Rodiek 2005), or about the ability of designers to assume the environmental, ethic, and cultural responsibilities of their acts (Levy 1990), appear as basic aspects of design education that are also intrinsic to the urban pedagogy.

A survey of urban design programs in universities

A study on the extent of urban design courses in different countries is yet to be written. Freeman (in Pittas and Ferebee 1982) compiled a "Directory of Graduate Programs in Urban Design in North America" dated 1981. Few scholarly articles are dedicated to reviewing specific aspects of undergraduate and graduate course syllabi, such as global urban topics (Ali and Doan 2006), land use planning (Miller and Westerlund 1990), physical planning (Pivo 1989) in the North American planning schools, or discuss architectural education in the US (Lyndon 1978). This chapter cannot fill this temporal and geographic void but few points can be made using data collected from various sources (see Table 3.1), which show the universe of graduate programs in urban design taught in English.

According to these data – which have been collected mainly from the web, selecting only those graduate level programs taught in English with "Urban Design" in the title – there are more than fifty graduate programs all over the world mostly concentrated in the US, UK, and Australia. Continental Europe and Asia have only a few programs. No programs, at

Table 3.1 Graduate programs in Urban Design taught in English

Country	*Institution*	*Course title*
India	Centre for Environmental Planning and Technology, Ahmedabad	Master of Urban Design[1]
	School of Planning and Architecture, Delhi	Master of Urban Design[1]
Singapore	National University of Singapore	Master of Arts (Urban Design)[1]
		Master of Architecture (Urban Design)[1]
New Zealand	University of Auckland	Masters of Urban Design (MUrbDes)[1]
Australia	University of New South Wales (Sydney)	Master of Urban Development and Design (MUDD)[1,2]
	University of Adelaide	Master of Urban Design[2]
	University of Melbourne	Master/PGDip in Urban Design[1,2]
	University of Sydney	Master of Urban Design[1,2]
		Graduate Certificate in Urban Design[2]
	Curtin University of Technology	Master of Urban Design[2]
	The University of Western Australia	Master of Urban Design[2]
South Africa	University of Cape Town	Master of Urban Design and City Planning[3]
		Master of Architecture (Urban Design)[3]
Canada	University of Toronto	Master of Urban Design Studies[1,4]
US	Arizona State University	Master of Urban and Environmental Design[4]
	City College of New York	Master of Urban Design[5]
	Cleveland State University	Master of Urban Planning, Design and Development[4,5]
	Harvard School of Design	Master of Architecture in Urban Design: MAUD[1,5]
		Master of Landscape Architecture in Urban Design: MLAUD[1,5]
	Kent State University	Graduate Certificate/Master in Urban Design[1]
	New York Institute of Technology	Master of Architecture in Urban and Regional Design[5]
	Pratt Institute	MSc in Architecture and Urban Design (Post-professional)[5]
	Savannah College of Art and Design	Master of Urban Design[1]
	University of California, Berkeley	Master of Urban Design Degree[5]
	University of Michigan	Master of Urban Design[1]
	University of Texas, Austin	Master in Urban Design[1]
	Washington University in Saint Louis	Master of Urban Design[5]
Sweden	Lund University	Master of Sustainable Urban Design[6]
	Royal Institute of Technology, Stockholm	Master of Urban Planning and Design[1,6]
Germany and China	Technische Universitaet Berlin and Tongji University Shanghai	Dual Master Program Urban Design (Berlin and Shanghai)[1]

Continued

Table 3.1 (*Continued*)

Country	*Institution*	*Course title*
Italy	Politecnico di Milano	MSc in Urban Planning and Policy Design[1]
Ireland	University College Dublin	MSc in Urban Design[1]
UK	Anglia Ruskin University	MPhil/PhD in Urban Design[8]
	Bartlett School of Planning, UCL	MSc in Urban Design[8]
		MSc in Building & Urban Design in Development[8]
	Birmingham City University	MA/PGDip/PGCert in Urban Design[8]
	Cardiff University	MA in Urban Design[7, 8]
	Edinburgh College of Art	PGDip/MSc in Architecture and Urban Design[8]
		PGDip/MSc in Landscape Architecture and Urban Design[8]
	Heriot Watt University	MSc/PGDip in Urban Design[8]
	Lincoln University	MSc/PGDip/PGCert in Urban Design[8]
	Liverpool John Moores University	MA in Architecture and Urban Design[8]
	London South Bank University	MA in Urban Design[7]
	Newcastle University	MA/PGDip in Urban Design[8]
	Oxford Brookes University	MA/PGDip/PGCert in Urban Design[7, 8]
	Queen's University, Belfast	MSc in Urban and Rural Design[7]
	University of Birmingham	MA in Urban Design[7]
	University of Dundee	MSc Spatial Planning *with* Sustainable Urban Design[7]
	University of Greenwich	MA in Urban Design[8]
	University of Liverpool	MA in Civic Design[7, 8]
	University of Nottingham	MA/PGDip in Architecture and Urban Design[8]
	University of Sheffield	MA in Urban Design[8]
	University of Strathclyde, Glasgow	MSc in Urban Design[7]
	University of the West of England	MA/PGDip in Urban Design[8]
	University of Westminster	MA/PGCert/PGDip in Urban Design[7, 8]

Note:
1 RUDI (2009a).
2 www.studyinaustralia.gov.au.
3 www.urbandesigninstitute.co.za.
4 ACSP (2007).
5 www.gradschools.com.
6 www.studyinsweden.se.
7 RTPI (2008).
8 RUDI (2009b).
Legend: PGDip – Postgraduate Diploma; PGCert – Postgraduate Certificate; MA – Master of Art; MSc – Master of Science.

least according to the sources investigated, are offered in the African continent (except for South Africa) or in South America. Urban design programs can also be found at Ph.D. level and at the undergraduate level but the graduate level seems to be offered the most.

Table 3.1 shows that, with very few exceptions, the graduate teaching of urban design is concentrated in the most developed countries. Additionally, teaching of urban design is mainly done in English-speaking countries. These issues lead to two considerations: first, Western universities have a massive responsibility to teach and disseminate urban design, through their international students, all over the world; second, urban design's language, literature, and terminology is mainly in English which means that there is also a risk of globalization due to the hegemony of one language over others.

Pedagogic techniques

Beyond the substantive developments that influence university curricula, there are pedagogic techniques that are employed worldwide in the active teaching of urban design. Most of these techniques are shared with the parent fields of architecture, landscape architecture, and planning. As discussed in detail by Kathryn Anthony in this volume, the Design Studio is the most popular and widespread method for teaching and training students of every level to work together, to accept a dialectic exchange with instructors and classmates, and to acculturate students to the "real-world" environment with all of the noises, intrusions, and nuisances that are typical to sharing work-space. "Studios are active sites where students are engaged intellectually and socially, shifting between analytic, synthetic, and evaluative modes of thinking in different sets of activities (drawing, conversing, model-making)" (Dutton 1987: 16). Donald Schön (1984) considers the studio a special form of reflection-in-action where design review plays an important pedagogical role both for students and teachers.

The studio, considered "a tradition of education for artistry" (Schön 1984) and "the heart and head of architectural education" (Dutton 1987), was subject to ongoing critiques and evolutions over the years on both sides of the Atlantic Ocean (Nicol and Pilling 2000; Salama and Wilkinson 2007). The most relevant innovations involving studio activities in universities are first, establishing links with the profession (Boyer and Mitgang 1996); second, increasing internationalization; and third, the use of new technologies.

Complaints about the relationship between schools and practicing professionals have been around "at least since the time of the École de Beaux-Arts" (Gutman 1984) and still remain on the agenda of the academia. Approaches for linking education to a professional aptitude of problem-solving include design seminars (Miller 1982) and workshop-like activities involving students, practicing professionals, government officials, local experts, and faculty. Such practices, devoted to problem-solving, also become part of "service-learning" as previously mentioned (Forsyth *et al.* 2000), or "communities of practice" (Schweitzer *et al.* 2008), which is work-based learning in collaboration with real communities with authentic needs acting as a model for learning. Also useful, as a didactic tool, is the use of juries composed, not only of academic and professionals, but also of public officials, stakeholders, and community members potentially affected by the design outcomes.

The internationalization of universities (Goldstein *et al.* 2006) has also affected studio practice as mentioned before. On the one hand there is the increasing activity of international field trips which "are fundamentally undertaken because of their educational merit, which is unsurpassed, if for no other reason than the sheer complexity of the experience" (Cuthbert 2001: 302). Sometimes field trips are associated with studio-format collaboration or a common participation to a competition between students from different countries who speak different languages but who benefit from the idea that "the language of spatial design is naturally a more communicable medium than speech or writing in circumstances when collaborators do not share a common language" (Abramson 2005).

On the other hand there is an increased internationalization of universities and hence of curricula in urban design and related fields. Due to the multicultural origins of students, an environment has been created in which "students now learn (and staff teach) in a global domain, working as teams across continents in ways made possible by digital technologies, international travel and the expectation that much practice is and will be international" (Bull 2004). At the same time, the growing internationalization has not assuaged the old doubts that "conventional design training in the US [or elsewhere in Western countries] often amounts to a socialization to professional world views and values of the Western world. [...Which generates, in their home-countries...] an imported vocabulary of architecture and urban design that incongruously mimics Western environmental forms, or worse, creates caricatures of traditional architectural and urban design" (Banerjee 1985: 28; 1990; also see Bakker *et al.* 2003; Parin 2004; Chettiparamb 2006).

A further means to enhance the skill of students to face international urban problems is through their participation in design competition. In almost the entire world, urban design competitions are progressively becoming a way to orientate urban transformations (OCCE 1998; Gospodini 2002; Beriatos and Gospodini 2004; Punter 2007). Competitions are launched by municipalities, governments, private owners, or even organized by groups of citizens. For professionals, participation in national or international urban design competitions is becoming a "must-do" studio activity, essential for getting themselves known and securing new work (see also chapter by Lehrer in this volume). As explained by Carmona (2006) however, participation in international competitions is full of risks – especially associated with overwhelming architectural emphasis and misunderstanding of place values and social behaviors. Simulating participation in a competition – or genuinely participating in national or international competitions for students (Palmer 1982) – can work as a didactic tool for creating multidisciplinary teams of students, obliging participants to deal with the competition program, and organizing a process and a timetable. An authentic or a simulated participation in international competition results in "confirm[ing] that urban design norms and principles are culturally specific, [... oriented to] capture a sense of local distinctiveness through their response to site context and the resulting urban form" (Carmona 2006: 123).

New technologies are introduced in the pedagogy of urban design, such as the use of GIS for analysis, powerful rendering software, and 3D modeling, satellite and aerial images, and intercontinental communication tools (some of them free of charge and widely available). Their effects on urban design pedagogy are still not fully evaluated. New technologies also allow distance education (Godshalk and Lacey 2001), which is becoming a flourishing field in the US and in Europe. However, teaching design at a distance presents problems (Alomyan 2004) resulting from individual differences and preparations, which are more controllable in face-to-face interaction or in-person group-evaluations. A more effective approach may involve the use of web-based-communication to link groups of students working on the same task but in different places. Studio practice also can be affected by the use of technology. Apart from the use of games like "SimCity" as a teaching tool (Gaber 2007) to build potential scenarios, innovative technologies are mainly used for design analysis and representations (Fraser and Bjornsson 2004, and also Bosselmann and Ben Joseph in this volume).

Conclusion

Contemporary urban problems (Rodiek 2005; United Nations 2004) will require urban designers to be prepared differently from the past. Urban design education could be the means for preparing professionals who see urban design not as "exercise in beautification of public spaces," but rather as an activity that will "reshape urban spaces [...] in the overall transformation of cities [...] to accommodate the new urban conditions" (Madanipour 2006:174, 191), and to participate in the production and reproduction of urban form as social space (Cuthbert 2007).

Urban designers, who "have the aptitude to give expression to creative intentionality matched with scientific knowledge and the capacity to manage processes" (Palazzo 2008: 268), need to improve, starting from the universities, their recognition and concern for contextual specificities, avoiding generic solutions and concepts valid elsewhere, and finding a balance between technical knowledge and creative expression. The classic argument, since the time of Plato and Aristotle through to Heidegger, of distinguishing between *téchne* and *poiesis*, where "*téchne* was the dimension of revelatory knowledge about the world, and *poiesis* was the dimension of creative, symbolic representation" (Corner 2002: 20), needs to be reconsidered in urban design teaching in order to give correct proportions to both sides of this apparent dualism. Awareness of both topics will induce urban design students to be soundly prepared to face urban problems, to understand places, social needs, and community roles, and to deal with decision-makers with solid technical skills; thus using *poiesis* and creativity to express their personal view on the matter. The correct balance between *téchne* and poeisis is also relevant in the *praxis* where urban designers have to apply their skills and knowledge in different places, within different cultures. A possible way to reach this outcome is to strengthen the teaching of processes and methodologies for approaching various issues and sites (Palazzo 2008; Steiner and Butler 2007; Lang 2005; Moughtin *et al.* 2003; Roberts and Greed 2001). Another way is to confirm the role of the studio as the place where the final product, with its aesthetic dimension, is assembled by means of a dialectic exchange with teachers and peers.

Finally, to avert the risk of one culture's dominance over others and of a diffusion of world-wide standardized solutions to urban problems, there is a need that urban design is taught in the universities of different countries, where local versions of teaching methods, languages, and applications are created. The aim is to banish the idea that urban design only applies to rich countries and, on the contrary, to reinforce the role of the urban designer as an honest broker and a promoter of design processes, and the active agent of social creativity for the realization of the public realm.

References

Abramson, D.B. (2005). "The 'Studio Abroad' as a Mode of Transcultural Engagement in Urban Planning Education," *Journal of Planning Education and Research* 25: 89–102.

ACSP – Associate of Collegiate School of Planning (2007). *Guide to Undergraduate and Graduate Education in Urban and Regional Planning*. 13th edition. Available <http://www.acsp.org> (accessed 25 June 2009).

Ali, A.K. and Doan, P.L. (2006). "A Survey of undergraduate Course Syllabi and a Hybrid Course on Global Urban Topics," *Journal of Planning Education and Research* 26: 222–236.

Alomyan, H. (2004). "Individual Differences: Implications for Web-based Learning Design," *International Education Journal* 4(4): 188–196.

Arefi, M. (2004). "The Pedagogy of the American City," *Urban Design International* 9: 103–117.

Arefi, M. and Triantafillou, M. (2005). "Reflections on the Pedagogy of Place in Planning and Urban Design," *Journal of Planning Education and Research* 25: 75–88.

Bakker, K.A., Le Roux, S.W. and Young, G.A. (2003). "Urban Design Education as Integral to 'Real-Time' Urban Revitalization

Processes: Salvokop, Pretoria," *Urban Design International* 8(3): 161–178.

Banerjee, T. (1985). "Environmental Design in the Developing World," *Journal of Environmental Planning and Research* 5: 28–38.

—— (1990). "Third World City Design" in Sanyal, B. (Ed.) *Breaking the Boundaries*, New York: Plenum Press.

—— (2001). "The Future of Public Space," *American Planning Association Journal* 67(1): 9–24.

Barnett, J. (1982). *An Introduction to Urban Design*, New York: Harper & Row.

Bartholomew, R.W. (1980). "Urban Design Education," *Urban Design International* 1(2): 47–54.

Beriatos, E. and Gospodini, A. (2004). "'Glocalising' Urban Landscapes: Athens and the 2004 olympics," *Cities* 21(3): 187–202.

Boyer, E.L. and Mitgang, L.D. (1996). *Building Community: A New Future for Architecture Education and Practice*, Princeton, NJ: Carnegie Foundation for the Advancement of Teaching.

Bull, C. (2004). "Editorial," *Urban Design International* 9: 173–174.

Carmona, M., (2006). "Designing Mega-projects in Hong Kong: Reflections from an Academic Accomplice," *Journal of Urban Design* 11(1): 105–124.

Carmona, M. and Tiesdell, S. (Eds.) (2007). *Urban Design Reader*, Oxford: Architectural Press.

Chettiparamb, A. (2006). "Bottom-Up Planning and the Future of Planning Education in India," *Journal of Planning Education and Research* 26: 185–194.

Corner, J. (2002). "The Origins of Theory," [1990] in Swaffield, S. (Ed.) *Theory in Landscape Architecture. A Reader*, Philadelphia: University of Pennsylvania Press.

Cullingworth, B. and Nadin, V. (2006). *Town and Country Planning in the UK*, London: Routledge.

Cuthbert, A. (2001). "Going Global: Reflexivity and Contextualism in Urban Design Education," *Journal of Urban Design* 6(3): 297–316.

—— (2007). "Urban Design: Requiem for an Era," *Urban Design International* 12: 177–223.

Dutton, T. (1987). "Design and Studio Pedagogy," *Journal of Architectural Education* 41(1): 16–25.

Forsyth, A., Lu, H. and McGirr, P. (2000). "Service Learning in an Urban Context," *Journal of Architectural and Planning Research* 17(3): 236–259.

Fraser, M. and Bjornsson, H. (2004). "Real-Time Digital Modeling in Design Education and Practice," *Urban Design International* 9: 187–196.

Gaber, J. (2007). "Simulating Planning," *Journal of Planning Education and Research* 27: 113–121.

Godshalk, D. and Lacey L. (2001). "Learning at Distance," *Journal of Planning Education and Research* 20: 476–489.

Goldstein, H.A., Bollens, S., Feser, E. and Silver, C. (2006). "An Experiment in the Internationalization of Planning Education," *Journal of Planning Education and Research* 25: 349–363.

Gosling, D. (2003). *The Evolution of American Urban Design*, Chichester: Wiley-Academy.

Gospodini, A. (2002). "European Cities in Competition and the New 'Uses' of Urban Design," *Journal of Urban Design* 7(1): 59–73.

Gutman, R. (1984). "Education and the World of Practice," *Journal of Architectural Education* 40(2): 24–25.

Hanson, B. and Younés, S. (2001). "Reuniting Urban Form and Urban Process," *Journal of Urban Design* 6(2): 185–209.

Inam, A. (2002). "Meaningful Urban Design," *Journal of Urban Design* 7–1: 35–58.

Kreditor, A. (1980). "The Neglect of Urban Design in the American Academic Succession," *Journal of Planning Education and Research* 9(3): 155–163.

Lang, J. (2005). *Urban Design. A Typology of Procedures and Products*, Amsterdam: Architectural Press.

—— (2007). "Urban Design as a Discipline in a Profession." In Carmona, M. and Tiesdell, S. (Eds.) *Urban Design Reader*, Oxford: Architectural Press.

Larice, M. and Macdonald, E. (2007). *The Urban Design Reader*, New York: Routledge.

Levy, R. (1990). "Design Education: Time to Reflect," *Design Issues* 7(1): 42–52.

Lyndon, D. (1978). "Architectural Education Here," *Journal of Architectural Education* 31(3): 2–7.

Lynch, K. (1954). "A New Look at Civic Design," *Journal of Architectural Education* 10(1): 31–33.

—— (1980). "City Design: What it is and How it Might Be Taught," *Urban Design International* 1–2: 48–53. Reprinted in Banerjee, T. and

Southworth, M. (Eds.) (1990). *City Sense and City Design*, Cambridge: MIT Press.

—— (1981). *A Theory of Good City Form*, Cambridge, MA: MIT Press.

Lynch, K. and Hack, G. (1984). *Site Planning* (3rd edition), Cambridge, MA: MIT Press.

Madanipour, A. (1996). *Design of Urban Space. An Inquiry into a Socio-Spatial Process*, New York: John Wiley & Sons.

—— (2006). "Roles and Challenges of Urban Design," *Journal of Urban Design* 11(2): 173–193.

McSheffrey, G. (1978). "Urban Design and Civil Strife," *Journal of Architectural Education* 32(2): 28–31.

Miller, I. (1982). "Design Seminar: An Urban Site," *Journal of Architectural Education* 35(4): 27–31.

Miller, D. and Westerlund, F. (1990). "Specialized Land Use Curricula in Urban Planning Graduate Programs," *Journal of Planning Education and Research* 9: 203–206.

Moughtin, C., Cuesta, R., Sarris, C., Signoretta, P. (2003). *Urban Design. Method and Techniques* (2nd edition), Amsterdam: Architectural Press.

Nicol, D. and Pilling, S. (2000). *Changing Architectural Education*, London: Spon Press.

OCCE (1998). *Thessaloniki 2000: On the Map of the European Metropolitan Cities. A Complete Catalogue of the Works, Projects and Architectural Competitions*, Thessaloniki: OCCE Publications.

Palazzo, D. (2008). *Urban Design*, Milano: Mondadori Università.

Palmer, E. (1982). "Student Design Competitions," *Journal of Architectural Education* 35(4): 17–21.

Parin, C. (2004). "The Recognition of Local Specificities in Cross-Cultural Design," *Urban Design International* 9: 197–207.

Perkins Holmes, G. (1952). "The School of Fine Arts," *University of Pennsylvania Bulletin*, LII: 24.

Pittas, M. and Ferebee, A. (Eds.) (1982). *Education for Urban Design*, Boston: Institute for Urban Design.

Pivo, G. (1989). "Specializations, Faculty Interest, and Courses in Physical Planning Subjects at Graduate Planning Schools," *Journal of Planning Education and Research* 9: 19–27.

Punter, J. (2007). "Developing Urban Design as Public Policy: Best Practice Principles for Design Review and Development Management," *Journal of Urban Design* 12(2): 167–202.

Radović, D. (2004). "Towards Culturally Responsive and Responsible Teaching of Urban Design," *Urban Design International* 9(4): 175–186.

Roberts, M. and Greed, C. (Eds.) (2001). *Approaching Urban Design. The Design Process*, Harlow: Pearson Education Limited.

Rodiek, J. (2005). "Human Habitats: a Focus for Design Education in the 21st Century," *Landscape and Urban Planning* 73(2/3): 81–85.

RTPI – Royal Town Planning Institute (2008). "RTPI Accredited Degree Programmes 2008–2009 Academic Year." Available HTTP <www.rtpi.org.uk> (accessed 25 June 2009).

RUDI – Resource for Urban Design Information (2009a). "A List of Courses in Urban Design and Related Subject Overseas." Available HTTP <www.rudi.net> (accessed 25 June 2009).

RUDI (2009b). "University Courses in Urban Design and Related subjects in the United Kingdom." Available HTTP <www.rudi.net> (accessed 25 June 2009).

Salama, A.M. and Wilkinson, N. (Eds.) (2007). *Design Studio Pedagogy*, Gateshead: Urban International Press.

Savage, S. (2005). "Urban Design Education," *Urban Design International* 10: 3–10.

Schön, D.A. (1984). "The Architectural Studio as an Exemplar of Education for Reflection-in-Action," *Journal of Architectural Education* 38(1): 2–9.

Schweitzer, L.A., Howard, E.J. and Doran, I. (2008). "Planners Learning and Creating Power," *Journal of Planning Education and Research* 28: 50–60.

Shannon, M.J. (1990). "Toward a Rationale for Public Design Education," *Design Issues* 7(1): 29–41.

Stein, C.S. (1951). *Toward New Towns for America*, Chicago: Public Administration Service.

—— (1955). "Proposed Department of City Design in the School of Fine Arts, University of Pennsylvania," manuscript, Graduate School of Fine Arts, University of Pennsylvania.

Steiner, F. and Butler, K. (Eds.) (2007). *Planning and Urban Design Standards*. Students' Edition, New York: John Wiley & Sons, Inc.

Strong, A.L. (1990). "G. Holmes Perkins: Architect of the School's Renaissance" in A.L. Strong, G.E. Thomas (Eds.) *The Book of the School. 100 Years*, Philadelphia: The Graduate

School of Fine Arts of the University of Pennsylvania.

The Builder (1908). "The Systematic Study of Town Planning" *The Builder.* Available HTTP <www.library.cornell.edu/Reps/DOCS/liverpool.htm> (accessed 15 March 2009).

Tyrwhitt, J. (1962). "Education for Urban Design," *Journal for Architectural Education* 17(3): 100–101.

United Nations (2004). *World Urbanization Prospects: The 2003 Revision*, New York: United Nations.

Vernez-Moudon, A. (1992). "A Catholic Approach to Organizing What Urban Designers Should Know," *Journal of Planning Literature* 6(4): 331–349.

Waldheim, C. (Ed.) (2006). *The Landscape Reader*, New York: Princeton Architectural Press.

Washington University (1962). *Education for Urban Design*, Proceedings of a conference, held at Washington University, School of Architecture, Jan. 8, 9, 10, 1962, Typescript.

Further reading

Cuthbert, A. (2007). "Urban Design: Requiem for an Era," *Urban Design International* 12: 177–223. A reflection on the role and value of the theoretical basis of urban design.

Goslin, D. (2003). *The Evolution of American Urban Design*, Chichester: Wiley-Academy. Discussion of five decades of urban design projects and theories and their relation to pedagogy.

Pittas, M. and Ferebee, A. (Eds.) (1982). *Education for Urban Design*, Boston: Institute for Urban Design. The proceedings of a meeting in Puerto Rico involving leading figures in urban design. A cornerstone of urban design pedagogy and practice.

Vernez-Moudon, A. (1992). "A Catholic Approach to Organizing What Urban Designers Should Know," *Journal of Planning Literature* 6(4): 331–349. A serious attempt to outline an epistemological map for urban design. It tracks the history of urban design theories and approaches.

Part 2

Theoretical perspectives

Introduction

In this section we consider the question: Do the practice and scholarship of urban design have adequate theoretical underpinnings? Is the practice of urban design sufficiently informed by theory? Has the scholarship of urban design made significant theoretical overtures? Or, is its theoretical terrain likely to be highly eclectic as is the case in urban planning, which draws from various disciplines in social sciences and political philosophy? Before we introduce the four contributions in this section which critically address the theoretical constructs that have shaped the major debates, conflicts and contradictions in our understanding of the production and consumption of urban space, we must submit that there are at least two different ways in which we might consider the theoretical discourse relevant to urban design. First, and as in the case of planning, theories can be either in, or of, urban design. The former involves theories that guide the practice of urban design. These are essentially "theories in action," as Schön (1983) has argued. The second category of theories addresses the social, economic, and political circumstances which must necessarily affect the conduct and practice of urban design. A second distinction of theories is made by Kevin Lynch who categorized theories to "explain the city as a spatial phenomenon" into three types: theories of making planning decisions or what he called "planning (or decision) theory;" theories that explain urban processes and outcomes, or what he called "functional theory" – obtained mainly from social sciences; and finally what he called "normative theory," that "deals with generalizable connections between human values and settlement form, or how to know a good city when you see one" (1981:37). We might note here that Lynch's distinction mainly applied to theories *in*, rather than *of* planning, his own contribution toward developing performance characteristics of good city form is a case in point.

The four essays included in this section reflect these two dichotomies. The essays by Niraj Verma and Christine Boyer fall in the category of theory in design, while the essays by Alexander Cuthbert and Kanishka Goonewardena are largely about design.

Verma argues that Urban Design remains "an incompletely theorized project" with considerable uncertainties about its institutional standing. Furthermore it continues to face tensions between its theory and its practice and between its inherent normative orientation and various positivist influences. Like Lynch, he emphasizes the normative dimension of design theory, and argues that while planning theory has been totally colonized, and indeed co-opted by positivist social sciences, urban design is still open to normative theorizing. In the manner that Habermas had considered modernity as incompletely theorized, Verma does not necessarily see the "incompletely theorized" aspect of urban design as a problem, but rather as promising and with potential to evolve. In making this point he argues that the relevant theories could be of two distinct genres – the high and the low theory. He does not see them as hierarchical, but rather the former as being more formal, axiomatic, and definitive. The latter on the other hand, while retaining the properties of the formal, may be open to the realm of the provisional and to propositions that may not be supported by axiomatic protocol. It is precisely these properties of low theory that keep them open to creativity, imagination, and normative possibilities.

The essay by M. Christine Boyer is also about theories *in* design, and she places them in the context of the two orders of cybernetics that have influenced our understanding and documentation of the internal dynamics of contemporary urban systems and the resulting urban form and spatial organization of cities. The first order of cybernetics, which is more "machinic" and which assumes that order can be obtained even in the face of the entropic tendencies of the urban system, according to Boyer, influenced earlier city design as characterized by the works of Gyorgy Kepes and his colleague Kevin Lynch, and such other well known designers as Philip Thiel, Christopher Alexander, Marvin Manheim, Donald Appleyard and the like. While this might seem a bit surprising to readers who may not think of Lynch's work as necessarily dogmatic or "machinic" because of his emphasis on user control and participation in design, the arguments she presents are quite intriguing. The second order of cybernetics focuses more on the organic, self-organizing or autopoeitic nature of complex systems, whose inspirations come from the work on life sciences. Boyer refers to the theoretical work of Rem Koolhaas, who in turn was influenced by Stefano Boeri's writings, as a case in point, even though ironically Koolhas' architecture does not resemble anything close to being self-organizing or organic.

The need to better theorize the discipline of Urban Design is also picked up in the chapter by Alexander Cuthbert, who finds the current theories of urban design "wanting" and looks into spatial political economy as a method to better understand the design challenges. Unlike Schön, who advanced the notion of "theories-in-action," Cuthbert is particularly dismissive of anything that can be claimed as theories *in* design. Instead he returns to the critical Marxian framework for examining production of space in the context of globalization and the new global economic order that continues to produce, on the one hand, urban forms that can be best described as "hyper-reality" (see Eco 1986) – "billion dollar theme parks and plasticulture urbanism" – and growing income inequality and the swelling ranks of slums and squatter settlements, on the other. He argues that these are the essential realities of the current transformations of the built environments that the future theories *of* urban design must address.

Finally, it is against this argument by Cuthbert that we may consider the fourth essay in this section by Kanishka Goonewardena on "critical urbanism." In developing a critical perspective, he explores the nexus between urbanism and capitalism, pondering on urban design possibilities to disarticulate the two through "radical transformations of space." Arguing that such concepts as social capital, multiculturalism, sustainability, democracy and human rights – commonly invoked by urban designers and planners in their concepts of improving urbanism – are co-opted by the capitalist liberal democracy, Goonewardena presents an alternative, "non-conforming" perspective on contemporary urbanism. This perspective of "critical urbanism" begins by drawing from the work of Guy Debord, especially his treatise on space, followed by the writings of Henri Lefebvre on the "social production of space." Using these two sets of writings as anchor points of this alternative perspective, Goonewardena embellishes his arguments with reference to the works of Kevin Lynch – especially his arguments for a normative theory of good city form – and also such notable authors as Adorno, Horkheimer, Jameson, Sadler and the like, as well as various other movements inspired by Marxian critical thoughts. The essence of this alternative perspective is that the urban phenomenon as the "intensely mediated site...at once social, spatial, and historical" should remain the critical focus of urban design as a "revolutionary struggle."

References

Eco, U. (1986) *Travels in Hyperreality*. New York: Harcourt, Brace, Jovanovich.

Lynch, K. (1981) *A Theory of Good City Form*. Cambridge: The MIT Press.

Schön, D. (1983) *The Reflective Practitioner: How Professionals Think in Action*. New York: Basic Books.

4

Urban design

An incompletely theorized project[1]

Niraj Verma

In his memoirs *Adventures of a Bystander* first published in 1978 and republished many times since, the management guru Peter Drucker described himself as a bystander who is on the stage but is not part of the action. Unlike audience or actors who affect what happens on stage, Drucker tells us that bystanders influence only themselves. But, by their very role bystanders see things that go unnoticed by others or at least see things in ways that are different from others. Fortunately, there is no requirement for bystanders to be gurus and from the perspective of a sympathetic bystander on the urban design stage I want to articulate a key sensibility about urban design and explore some of its implications.

This sensibility sees urban design as an incompletely theorized project with an ethos that goes to the heart of planning. The idea of incomplete theorization is an extension of the constitutional theorist Cass Sunstein's (2001: 9) notion of incompletely theorized agreements in decision-making where these agreements describe "a process by which people agree on practices or outcomes despite disagreement or uncertainty about fundamental issues." An incompletely theorized project results when such agreements arise out of fundamental issues that are constitutive of the field and its core mission.

In painting such a picture of urban design I do not suggest that the field is theoretically innocent.[2] Rather, the theoretical modesty of urban design is necessitated by its subject-matter and by the nature of its problems. Like many other fields within urban planning, urban design deals with what Rittel and Webber (1973) described as "wicked problems" that defy solution and that are constantly transformed into other problems. On one hand, urban design derives its identity by its similarities to design but it is also deeply aware of the risks of being "just design." On the other hand, it is as much a field of inquiry as it is a practice and while it relies on social science it is not a social science. Drawing on some classical and contemporary work from a variety of perspectives, including urban design (Lynch 1960, Banerjee and Loukaitou-Sideris 1990), design theory (Rowe 1987; Broadbent 2005), environmental psychology (Craik 1990), and the philosophy of science (Kuhn 1970), I will trace the theoretical precariousness to multiple enduring and fundamental tensions that affect the epistemology and the practice of urban design.

My essay is divided into three parts. In the first part I will develop an understanding of urban design as caught between two competing influences of the social sciences and design. This "essential tension" helps us to get our arms around an otherwise difficult to conceptualize terrain. The second part will dissect this further to show how this tension is derived and in turn influences the subject-matter, method, and institutional location of urban design. Finally, in part three by drawing on the philosophy of American pragmatism and particularly the works of William James, I will argue that the dialectical positioning of urban design within science and design suggests a preference for "low theory" over "high theory."[3] Low theory differs from high theory in its ethos. It is contingent, nuanced, and incomplete and has a precarious relation with its subject-matter. High theory, on the other hand, covets certitude and law-like propositions and ends up reducing its subject-matter so that it is devoid of emotive content.[4] Arguing that the recognition of its incompleteness yields a stance and an ethos that is necessary in urban planning, I want to suggest that in doing so urban design may have internalized some of the most important lessons of planning theory.

The essential tension of urban design

Urban designers occupy a unique if tenuous position in the academy. Although the majority of urban designers have a background in architecture their self-image is often at odds with that of architectural design. Like academics in other professional fields such as social work, planning, public health, and management, urban design scholars are scholars first and design professionals second. Although not unusual within planning, this is noteworthy in its contrast to architecture, where professional identity and success are prized even among academics.

Two significant developments are challenging this self-image of urban design. The first comes from research oriented architecture departments which are increasingly investing in pedagogy at the urban scale.[5] Were they to succeed, architecture's professional identity would likely rub-off on urban design thus reducing the professional–academic divide in urban design. The counter-influence comes from the social science disciplines that are increasingly coveting the urban realm as part of their inquiry and in some cases, e.g. Krugman *et al.* (1999) or Scott (1999), are making impressive contributions towards understanding the city.

These developments are not simply about opportunity and timing. Rather, they represent a fundamental tension in the field between the professional identity of design and the scholarly identity of the social sciences. The tension arises because although there are stark differences between the two identities both bring essential influences to urban design. Thomas Kuhn taught us that paradigm changes accompany different phases of science and that aside from occasional revolutions, "normal" science follows well-established rules and operates within clearly identifiable paradigms. "Normal" design, however, exhibits little such regularity. Subscribing to the culture of genius, where being touched by a muse is more important than fidelity to a paradigm, the genres of design are often incommensurable, each idiosyncratically different from its predecessors. This gives the field notoriety and currency and may even account for some amount of faddishness in its practice. But when something is a fad or when it is a genuinely creative leap is hard to separate.

Unlike fields like architecture, urban design derives additional complexity from its sheer scale. The sanitized story-book picture of the creative designer serving a client with seemingly endless resources – the proverbial Renaissance Prince whose good taste matches his pocketbook – is replaced by one with warring publics, coalitions, and protests on one hand and an assembly of differing preferences, values, and interests on the other. Perhaps for this reason while architecture saw its links to be primarily with the physical sciences – the physics of lighting and acoustics or the structural engineering of buildings and bridges are examples – urban design went to the social sciences. Its partners became psychology and sociology, a choice that is probative in explaining its contemporary stance.

While the physical sciences have little difficulty in embracing a crafts and design image of their work, social scientists are more conservatively tied to their science roots. In physics and medicine, for example, it is sometimes hard to tell science apart from engineering. As an example, the American Nobel laureate in physics, Percy Bridgman, developed a new device that enabled him to reach pressures more than 30 times what was previously possible. Without this bit of "design" or "engineering" several of his contributions to states of matter would be impossible. Similarly, biomedical engineering is an example of a profession where boundaries are eroding.

The social sciences, by contrast, are rigid about the divisions between application (policy and planning) and the basic science. Many sociologists steer clear of application and see it as a sign of contamination of their scholarship. Indeed, there is a hierarchy from theory to practice to actual application. If sociologists stay clear of policy, political scientists stop short of the messy world of practice, leaving that domain to public management. Consequently, the tension between knowledge and its application in the social sciences is not simply one of specialization and not even one of preference. Rather it is a case of scientists' self-image of their work – their own psychology – coming in the way of any reconciliation of the two tasks. In an interesting recent work, Richard Sennett (2008: 11) laments this division of theory and practice while celebrating what he calls the "craftsman's way of working."

> History has drawn fault lines dividing practice and theory, technique and expression, craftsman and artist, maker and user; modern society suffers from this historical inheritance. But the past life of craft and craftsmen also suggest ways of using tools, organizing bodily movements, thinking about materials that remain alternative, viable proposals about how to conduct life with skill.

Sennett's early training was as a musician and that has influenced his image of the craft. But, while the craft metaphor is relevant to urban design, it does not capture the breadth of the influences on it or its scale of application. The example of architecture suggests that a rigid separation of science (in this case primarily physical science) and design is unhealthy for many reasons, but most of all the danger is that it leads to a form of scientific determinism on one hand and an enigmatic, impossible to assess design on the other. Peter Rowe (1987) traces this determinism to the rigidity of the behaviorist influence on design and argues that this leads scholars to analyze and then describe the *overt* activities of design alone. Indeed, this is hardly new. Using the example of eighteenth- and nineteenth-century design, Rowe explains that during the heydays of the *Beaux Arts* and the *École Polytechnique*, a restrictive notion of science brought about a separation from design and restricted scientific contribution to design in the form of impoverished generalities rather than to a more meaningful science of design.

Given this context, neither the Kuhnian account of periods of "normal" science that is transformed in revolutionary ways nor the incommensurable genres of design capture the essence of urban design. Rather it is more useful to see changes in urban design as responses to its "essential tension" of science and design, where neither must be allowed to dominate.[6] Urban design may have multiple essential tensions and at a particular time some may be more salient than others. For our purposes the tension between science and design helps to understand some fundamental issues in an otherwise amorphous territory of urban design.

Epistemological considerations

By placing urban design in intellectual relief between social science and design, I do not mean to suggest that it is a hybrid while science and design are pure. As far as the practices of science and of design are concerned the story-book pictures of a pure, uncontaminated, non-political science have been roundly debunked (Mitroff 1983, Kuhn 1970, Proctor 1991) as have the glamorous pictures of design as a ceaseless string of creative, enigmatic actions (Cuff 1992). When I refer to science or design I am using the terms to denote ideal types, their meanings restricted to what philosophers call "the context of justification" rather than the "context of application." In turn this provides an epistemological contrast that can be used to illuminate many characteristics of urban design. The three characteristics of most relevance are: the nature of its problems, i.e. the subject-matter of urban design, the purpose and structure of its inquiry, and its institutional location within the academy.

The nature of urban problems

In a paper presented to the *American Association for the Advancement of Science* in 1969, Berkeley Professors Horst Rittel and Melvin Webber laid out a set of properties that showed the essential dilemma of planning. Characterizing planning problems as "wicked problems" they claimed that science had developed to deal with complex but "tame" problems. The paper highlighted several characteristics of wicked problems. Wicked problems are not solvable, i.e. each solution raises other wicked problems. There is no knowing when you have done enough towards a solution because there is always more. Wicked problems are symptoms of other wicked problems. Rittel and Webber (1973) contrasted this description of wicked problems with complex but tame problems. An example of a tame problem is the process of solving a mathematical equation. Here, once solved, solutions appear algorithmic in nature and an expert can with some confidence reach a solution that brings closure. Although published in a policy journal and addressed to planners, the origins of this thinking came from Horst Rittel's early work in design theory (1972) and it would be fair to say that at an epistemological level Rittel saw great similarities between design and planning.

For the most part urban problems share the characteristics of wicked problems. For example consider the issue of walkability. Is this a problem of lack of safe paths for pedestrians or is it about the easy availability of automobiles? Depending on how we construct the problem the solutions are dramatically different. Typical of wicked problems, methods to increase walkability have good or bad solutions while tame problems – even complex ones – can have correct or incorrect solutions. Further, each solution to a wicked problem is a symptom of other wicked problems. Problems in walkability may be a symptom of city governments not investing in infrastructure. Or they may be a specific case of the bigger issue of development patterns, including suburbanization.

Contrast this with possible research into walkability from a social science perspective. An early demand is to define just what walkability means. Is it about day trips to school or work or is it about evening strolls in one's neighborhood? And just how far is a neighborhood boundary? And what constitutes a trip? These are no parlor game questions. In transportation for instance, a rigid meaning of a "commute trip" led an entire generation of transportation planners to gloss over the complex nature of driving patterns that are typical of women drivers. On their way to work it is common for several women and some men to add tasks that are typically non-commute related. A driver may drop the kids at school or day-care. She may pick up groceries or dry-cleaning. Today, transportation planners deal with these complexities by way of "trip chaining," a term that illustrates the impoverished distinction between commute and non-commute trips that dominated the research landscape for a long-time.

If social sciences demand rigid and uncompromising "operational definitions" that increase internal validity, and if design forsakes definition for intuitions, urban design must somehow reconcile both demands. On one hand there is a cost to defining, and for wicked problems early definitions can lead to wrong solutions. On the other hand, despite its place-based focus, urban design must define its problems if it is to influence public policy. In other words, it must simultaneously recognize that it is dealing with wicked problems while seeking more broad-based negotiated formulations and solutions. Table 4.1 below shows the contrast while illustrating the urban design position.

Public participation, a recurrent mantra of the field of planning, becomes the mode by which urban design negotiates the competing demands of science and design. Negotiated formulations of problems become the norm when neither the self-assuredness of social science nor the bravado of design will do. In other words public participation serves not just a democratic function; it also results from an epistemological recognition of the nature of urban problems and the need to come up with solutions.

The nature of inquiry

In addition to problem definition the tension between science and design also affects other aspects of inquiry. Recounting a little bit of history will help to put this in perspective. The formal entry of social science into design can perhaps be marked by the start in 1981 of the journal, *The Journal of Environmental Psychology.* The new journal saw its task as recognizing and analyzing practice and also influencing it. Its first editors, David Canter and Kenneth Craik wrote about the field in one of the first issues:

> Its current vigorous state is held to be a product both of the way its practitioners have met the challenges of application and of the benefits accruing from the cumulative impact of several scientific research traditions.

Over the years this mandate changed a bit and although still deeply anchored in the psychological tradition, the journal has become even more steeped in matters of the built environment. Its list of topics now extends to "cognitive mapping," "spatial cognition and wayfinding," "design of, and experiences related to, the physical aspects of workplaces, schools, residences, public buildings and public spaces," "meaning of built forms," "theories of place, place

Table 4.1 Nature of problems

Social science	*Design/architecture*	*Urban design*
Fits within paradigm Tame problems	Hard to define and solve Wicked problems	Seeks negotiated solutions for wicked problems

attachment, and place identity," and "social use of space: crowding, privacy, territoriality, personal space." In other words, urban design may not be in the title of the journal but its subject-matter certainly permeates the journal's contents.

Yet environmental psychology is not urban design.[7] And the broader lesson here is that subject-matter by itself may be insufficient to capture the essence or ethos of a field. More than anything else as a science, environmental psychology's goal is explanation while design aims to create. To be sure, in the literature there is a meaning of science that sees it as a normative enterprise or as a science of values (Churchman 1982). Even in urban design, the UNESCO program "Paths of Thought" saw one of the authors (Solinis 2006: 79) ask, "Can urban design be the science of the ideal city?" But, regardless of these heroic calls for a value-laden science, once an experimental design is complete all that remains is a matter of rigorous data collection, hypotheses testing, and internally valid conclusions.

Whether or not social scientists recognize the difficulties of being neutral, many scientists will agree that their aspiration is in some ways to find the truth about the world. Immanuel Kant, the German philosopher, called truth the "agreement of knowledge with its object." So, if I am holding Kant's *Critique of Pure Reason* in my hand and I know that what I am holding is the *Critique* then I have the truth. This so-called correspondence theory of truth has inspired scientific methods to gain the truth or to come close to it. It forms the basis, for example, of the ideas of controlled experiment, measurement, external validity, objectivity and such concepts which attempt to increase our proximity to the truth. Since observation is mediated by the observer and all of the baggage that s/he carries, removal of that baggage is a way to the truth. Bacon called this the removal of "idols of the cave," an allusion to the non-scientific way of life that preceded modern living. Others call it the search for objectivity or ridding observation of bias.

Whatever the explanation, it belongs to the context of justification, where the search for neutrality and truth becomes a prime mover for the institution of science. Indeed, as the historian of science Robert Proctor (1991) has shown, the neutrality of science can be a device to protect it from religious meddling or, seen differently, as a way of retaining power. Similarly, counter-claims that all science is political are also driven by institutional forces and needs, such as the need to level the playing field of science for all or the need to reinstitute priorities in the allocation of public support.

If science is characterized by neutrality and the search for truth, design is avowedly normative (Banerjee and Loukaitou-Sideris 1990). It is not content to find the truth unless that truth is also utopian or ideal. In this sense with few exceptions, designers are motivated by Platonic virtues. From ideal forms to the ideal city, while the construction may differ, a utopian sensibility of perfection is shared. Even in scientific applications to design, such as in operations research, the recognition of constraints does not exclude the search for perfection, albeit within well-defined boundaries. Christopher Alexander's much celebrated "pattern language" (1977) and his decomposition algorithms are iconic examples but there are others as well (Broadbent 2005).

By contrast to the utopian idealism of pure design and the truth seeking idealism

Table 4.2 Goal and purpose

Social science	*Design / architecture*	*Urban design*
Neutral: seeks the truth	Utopian: seeks ideal forms	Pluralist: seeks shared or consensual meaning

of social science, realism is essential to contemporary urban design. Utopian purity takes away from a variety of objectives, many of which are shared with other fields within urban planning. These include the recognition of the diversity and legitimacy of human desires, the need to break barriers, the advocacy for under-represented interests, and the integration of communities of work and communities of home. As Isaiah Berlin (1990) has so pointedly described, by draining legitimacy from the present, utopian ideals legitimate any means for their fulfillment. Paraphrasing Berlin, "to make a utopian omelet there is no limit to the number of eggs that can be broken."

Quite different from the meaning of truth as correspondence between ideas and their objects, urban design creates meaning that is shared across constituents. If architecture is focused on designer and client and the relation between them, and social science is oriented towards the relationship of people, urban design addresses these sentiments by investing in place. Place, however, is not just a physical attribute. It is a socially constructed agora. Ultimately, place is about the diversity of people who inhabit it and there is recognition that diversity contributes to well-being and fosters an environment for the development and testing of identity and learning about oneself.[8, 9] Consider these early words from urban designers Stephen Carr and Kevin Lynch (1968: 1278) who presaged these sentiments almost 40 years ago in arguing that barriers in the city prevent genuine learning:

> Too often the city fences us away from other kinds of people. By the scale, impersonality, and even hostility of its places and institutions, the city tends to discourage independence of action and to encourage fear and feelings of powerlessness. The white mother and child in the suburb are kept from new experiences about as effectively as their black counterparts in a ghetto housing project.

Carr and Lynch (1968: 1279) explicitly connect this to identity: "the urban environment ... is a medium for transmitting the form and content of contemporary society, a territory to be explored, and a setting for the testing of identity."

Institutional context

If subject-matter and method contribute to the science-design tension, urban design's institutional position within the academy further cements the tension. As a professional field urban design must come to terms with its professional status and the academic need to justify itself on-par with other fields. Although institutional location can affect virtually every action within a field, I will limit myself to two aspects of institutional influence: how does the field grow and renew itself and how does it retain its knowledge?

The contrast between science and design is apparent in their rules of membership. While the scientific enterprise regulates its membership through some rather rigid criteria, design has few clearly established markers. So, for instance, enrollment in a graduate program and the earning of a PhD are taken for granted prerequisites for a social scientist. Design, on the other hand, has schools and colleges but it is possible to bypass these paths for one of individual experimentation and design. Certainly, the profession has rites of passage but this may be as much about the need to regulate supply into the professions rather than to protect the customer.

Consistent with closed membership criteria, the institution of science is nourished by a well-defined system of peer review; and while norms may be socially constructed they are well-known, well-regarded, and generally resistant to change.

The norms of design may be discernible, and as the history of design methods movement shows, there are quite a few similarities in the reasoning of designers, no matter what their specific craft. Yet, there are more iconoclasts within design, and clients – not peers – typically determine success. For example, architects in the professoriate are recognized by their buildings. Junior members suffer if they do not have clients to underwrite their work. For example, in the difficult economic climate of today, while many historians and anthropologists worry about the publication prospects of their dissertations given the demands on publishing houses to manage their accounts, architects lament the loss of their Renaissance client and the loss of building commissions.

Table 4.3 shows that the dialectic of peer and client gives urban design its public orientation and allows it to assert what Peter Katz (1994) has called "the primacy of the public realm over the private." We see this, for instance, even in cases where development is developer supported, as for example in much of the new urbanism. Here charrettes aim to include various stakeholders, including city and county officials and the community, into decision-making.[10]

Around the same time as the publication of Katz's book, Alan Kreditor (1990) raised the concern that urban design was neglected in the American academic succession because it had not kept up with the dramatic growth in urbanization. Similar sentiments about the inadequacy of urban design were expressed from across the Atlantic by John Punter and Matthew Carmona (1997: 9).

Table 4.3 Institutional orientation

Social science	*Design/ architecture*	*Urban design*
Peer assessment	Caters to clients	Public orientation

The design dimension of British planning has been much neglected as a subject of both academic enquiry and professional development. In academic terms, British urban design has been slow to develop a substantive body of thought that could underpin enlightened practice, and has rarely undertaken investigations of design control in action.

There is a paradox suggested by these critiques. On the one hand there is a yearning for the institutional independence of urban design. On the other hand there is hope for the kind of hierarchical knowledge where theories undergird practice. But, this is not the model of knowledge acquisition in urban design. There are canons in urban design but, even if we agree on them, they do not lead to the kind of theory as, for example, in Coleman's *Foundations of Social Theory* (1990).

Urban design has been conscious of its practical mission and responsibility to usher change even as it has aimed for new knowledge. In his introduction to *The Image of the City,* Kevin Lynch (1960: 1) outlined the purpose of urban design's emphasis on the visual environment: "...the function of a good visual environment may not be to simply facilitate routine trips, nor to support meanings and feelings already possessed. Quite as important may be its role as a guide and a stimulus for new exploration." This spirit would be echoed in a particularly interesting issue of *Daedalus* that brought leading city planners and others to write about *The Conscience of the City.* In a manner similar to Dyckman's (1961) ideas of "the educative city," Stephen Carr and Lynch (1968: 1277) would describe the role of the city as a generator of surprise and learning:

> The best learning happens by surprise; it is very different from the normal

> process of deliberate education...The routine business of life demands some regularity and enforces it through selective attention to what supports our efforts. But, often, when we have "nothing better to do," when we are waiting, in transit, on vacation, just hanging around...cities surprise us.

This kind of learning by surprise, as opposed to a structured pedagogy, puts it at odds with a scientific model that emphasizes rigor, repeatability, goals, and instruments. Although the city has been seen as a laboratory for learning, an idea that was popularized in no small way by the social scientists of the Chicago school, the "learning by surprise" vision of urban design has little in common with the text-book idea of the scientists' laboratory. The former is almost accidental; its design unclear and organic in the truest sense of the word. By contrast, the scientific enterprise be it in the city or in the laboratory is focused on controlled experiments, internal validity, and operational definitions.

What does this say about urban design? Are urban design developments cumulative – do they build on themselves, perhaps in a sort of autopoesis as Boyer argues in her essay in this volume? I am less sanguine. While we might find agreement on some aspects of urban design, it is unlikely that we will find an analog for the painstaking research where a scientist makes progress element by element, carefully testing hypothesis after hypothesis, and becoming doubly or triply sure before going public with any findings. Urban design has to balance scientific carefulness with the design tradition, where an intellectual iconoclast is more revered than a careful exponent. In design better the creative mind than the meticulously organized one. Indeed, public attention and eventually peer recognition in design result from not conforming, i.e. being different from everyone else.

The pragmatism of theoretical precariousness

The discussion suggests that urban design is not atheoretical as critics such as Cuthbert (in this collection) and in previous work (2006) have charged. Rather the nature of theory in urban design is different, and this results from its location within an epistemic tension between science and design. Cuthbert claims, with some dramatic flourish, that urban design has been colonized by the professions of architecture and planning and that this has impoverished it. Moreover, he claims that it can only recover by first locating itself within a politico-economic framework. While such a framework may be useful there are several assumptions in Cuthbert's critique that paint a quite specific picture of theorization. These assumptions are systematic – not idiosyncratic – and they add up to a picture that I have characterized as "high theory" and its converse, "low theory."

High theory is formal or at least lends itself to formal treatment. That is, it consists of a set of propositions that are axiomatic in nature. Low theory may share some characteristics of axiomatic theory but it will also admit other propositions that might fail the test of axiomatic rigor. These propositions may be self-referential or they might be contingent. They may follow or lead practice and yet, they are meaningful and useful in understanding the nature of the craft of urban design. Table 4.4 gives a brief contrast of high and low theory.

Meaning

Consider the interpretation of meaning in the two traditions. In an axiomatic system meaning is typically assigned by definition. Logicians call such assignments tautologies which do not have empirical meaning. The power of the formal system, however, is in the ability to manipulate the axioms

Table 4.4 High and low theory

	High theory	*Low theory*
1	Meaning is defined	Meaning is created
2	Search for Truth	Make a Difference
3	Axiomatic logic	Logic can be self-referential
4	Covets certitude	Recognizes contingency
5	Rigor has primacy	Relevance is prized
6	Theory precedes practice	Theory interwoven into practice
7	Rational	Rational and emotive

and observe the effects on the system under consideration. Take the case of walkability. The scientific method is to first define the parameters of walkability in a rigorous way – gender, age, distance walked, outdoor weather, etc. – and to then tabulate the various attributes on a walkability scale, aggregating them to arrive at a measure of walkability. We know, however, that safety and the perception of safety from accidents, crime, etc., plays into whether or not people walk (Loukaitou-Sideris 2006). So, from a position of high theory a neighborhood that is walkable may not be safe and hence not used for walking. This gives rise to the so-called implementation problem, where the design of the program is fine but somehow it doesn't come to fruition because of an implementation issue.

Low theory is different. It follows a pragmatic idea of meaning that looks towards consequences, not causes. Charles Peirce (1958, vol. 5, para 9), the American philosopher and mathematician put it thus: "In order to ascertain the meaning of an intellectual conception, one should consider what practical consequences result *by necessity* from the truth of that conception." So, for instance, a neighborhood is walkable not just because it satisfies a certain minimum walkability threshold, but because its residents actually walk in their daily activities. The difference between the two approaches – high or low theory – may not be that severe unless we also factor in that low theory drains legitimacy from a neighborhood that claims walkability but isn't so. In other words, it endogenizes implementation into the problem of walkability.

Anti-foundationalism

Although grand sounding, anti-foundationalism is actually a rather simple idea. Simply put, it means that if A depends on B it is permissible for B to depend on A for its meaning. That is to say, their foundations are relative or interdependent rather than absolute. Suppose we consider theory to be more fundamental than or logically prior to practice. Such a position is consistent with high theory. With low theory there is no such restriction but then how is circularity justified? Again, the pragmatic insight is that neither theory nor practices are pure categories to begin with and so at some level the circularity is inevitable. Take the example of a dictionary. Endless is circular and circular is endless but it still communicates meaning. That is because we never start with a blank slate. In the real world there is always prior knowledge, expectation, and anticipation, not the vapid blankness of an axiomatic system.

Truth

The final distinction between high and low theory is related to truth. Earlier we encountered Kant's meaning of truth as the agreement of knowledge with its object. This was the kind of logical definition of truth that applied to high theory. But, suppose we ask "What does agreement mean? How can we secure it?" In addition to the notion of objective truth the preparedness to agree becomes part of the truth. Agreement is as much psychological and social as it is logical. The pragmatic philosopher William James called this the "willingness to believe." Truth is not only the correspondence between two things; even before matching for correspondence is the willingness to believe. So, low theory emphasizes the strengthening of believability and not just the correspondence of knowledge with its object. I have written about this extensively elsewhere (Verma 1998). Here suffice to say that low theory's pragmatic truth implies that truth is as emotive and psychological as it is logical and objective.

High theory and low theory

The terms high and low theory are not meant to convey a hierarchy of any kind. The terms are convenient and less polemical than say positive theory and normative theory or such other variants. If we return to the tension of science and design, the biggest difference is that while high theory moves towards the sciences and sees the success of urban design as its proximity to the sciences, low theory resists science as well as design while learning from both of them. This is the meaning of precariousness. To borrow Meg Holden's (2008) metaphor, it is simultaneously "tough minded" and "tender minded." In other words, from the perspective of low theory, urban design has a contingent and nuanced relationship with its subject-matter. It is not a complete theory and knows that it cannot be a complete theory. It acts even in the face of theoretical deficit because waiting for the best can become the enemy of the good, something that Herbert Simon (1972) had called "bounded rationality." Even though such a tenuous and savvy urban design is incompletely theorized, it is able to take theoretically conscious action and to make agreements even while being theoretically modest. Indeed, the very tensions that give urban design its fundamental uncertainties may also be its biggest asset in helping it to understand its own incompleteness. This gives it suppleness and the freedom to compete and survive, while staying relevant to academia and practice.

Notes

1 Material from this paper was presented at a Colloquium at the GSD (Harvard) and at the ACSP Conference in Crystal City (2009). I am grateful to participants in these events, to editors of this handbook, and to my colleagues at University at Buffalo: Brian Carter, Mehrdad Hadighi, Lynda Schneekloth, and Ernest Sternberg for useful comments and conversations on a previous draft.

2 Moudon (1992) argues that the field lacks theoretical finesse. Since 1992, this situation may have been alleviated but not decisively changed. As an insider in the field, however, Moudon draws the boundaries of urban design to include only those who identify explicitly with it. By contrast, I include those whose writings substantially inform the core knowledge in the field.

3 At this point the terms "low" and "high" represent nominal categories and no gradation is implied.

4 See Sternberg (2000) on commodification in urban design.

5 This is still new among architecture programs. Berkeley's program is one of the earlier ones with such a focus.

6 The idea of an "essential tension" is inspired by Kuhn's book with that title. Simply put it stands for a ruling idea around which some of the most important arguments of a field can be organized.

7 Jack Nasar's essay in this volume presents a different view of the link with environmental psychology. But, in part this derives from his rather expansive view of environmental psychology.

8 See Arefi and Triantafillou (2005) for some insights on multiple uses of place in the pedagogy of urban design.

9 Through a set of case studies, Schneekloth and Shibley (1995) show the range of issues in place-making.

10 Indeed, there is even a market for charrettes. The National Charrette Institute located in Portland, OR, (http://www.charretteinstitute.org/charrette.html) has developed a three-day certificate program for those interested in using charrettes.

References

Alexander, C., Sara I., and Silverstein, M. (1977). *A Pattern Language:Towns, Buildings, Construction*, Oxford: Oxford University Press.

Arefi, M. and Triantafillou, M. (2005). "Reflections on the Pedagogy of Place in Planning and Urban Design" *Journal of Planning Education and Research*, 25: 75–88.

Banerjee, T. and Loukaitou Sideris, A. (1990). "Competitions as a Design Method: An Inquiry" *Journal of Architecture and Planning Research*, Vol. 7(2): 114–131.

Berlin, I. (1990). *The Crooked Timber of Humanity: Chapters in the History of Ideas*, New York: Knopf.

Broadbent, G. (2005). *Emerging Concepts in Space Design*, London: Routledge.

Canter, D. and Craik, K. H. (1981). "Environmental psychology" *Journal of Environmental Psychology*, 1: 1–11.

Carr, S. and Lynch, K. (1968). "Where Learning Happens," *Deadalus*. 97(4): 1277–1291.

Churchman, C. W. (1982). *Prediction and Optimal Decision: Philosophical Issues of a Science of Values*, Mahwah, NJ: Prentice Hall/Greenwood Press Reprint.

Coleman, J.S. (1990) *Foundations of Social Theory*, Cambridge, MA: Harvard University Press.

Craik, K. H. (1990). "Environmental and personality psychology: Two collective narratives and four individual story lines" in I. Altman, and K. Christensen (Eds.), *Environment and Behavior Studies: Emergence of Intellectual Traditions*, New York: Plenum Press, 141–186.

Cuff, D. (1992). *Architecture: The Story of Practice*, Cambridge, MA: MIT Press.

Cuthbert, A. (2006). *The Form of Cities: Political Economy and Urban Design*, Cambridge, MA: Blackwell.

Dyckman, J. (1961). "The Changing Uses of the City" *Daedalus*, 90(1): 111–131.

Holden, M. (2008). "The Tough Minded and the Tender Minded: A Pragmatic Turn for Sustainable Development Planning and Policy," *Planning Theory and Practice*, 9(4): 475–496.

James, W. (1947). *Pragmatism: A New Name for Some Old Ways of Thinking*, New York: Longman's Green. (Originally published 1907)

Katz, P. (1994). *The New Urbanism: Toward an Architecture of Community*, New York: McGraw-Hill.

Kreditor, A. (1990). "The Neglect of Urban Design in the American Academic Succession" *Journal of Planning Education and Research*, Vol. 9(3): 155–163.

Krugman, P., Fujita, M., and Venables, A. J. (1999). *Cities, Regions, and International Trade*, Cambridge, MA: MIT Press.

Kuhn, T. S. (1970). *The Structure of Scientific Revolutions*, Chicago: University of Chicago Press.

—— (1977). *The Essential Tension: Selected Studies in Scientific Tradition and Change*, Chicago: The University of Chicago Press.

Loukaitou-Sideris, A. (2006). "Is it safe to walk? Neighborhood Safety and Security Considerations and their Effect on Walking" *Journal of Planning Literature*, 20(3): 219–232.

Lynch, K. (1960). *The Image of the City*, Cambridge, MA: MIT Press.

Mitroff, I. (1983) *The Subjective Side of Science*, Seaside, CA: Intersystems Press (originally published 1974).

Moudon, A. V. (1992). "A Catholic Approach to Organizing What Urban Designers Should Know" *Journal of Planning Literature*, 6(4): 331–349.

Peirce, C. S. (1958). *Collected Papers* (Vol. I-VI). C. Hartshorne and P. Weiss (eds.) Cambridge: Harvard University Press.

Proctor, R. N. (1991). *Value-Free Science? Purity and Power in Modern Knowledge*, Cambridge, MA: Harvard University Press.

Punter, J. and Carmona, M. (1997). *The Design Dimension of Planning*. New York: Van Nostrand Reinhold.

Rittel, H. (1972). "On the Planning Crisis: Systems Analysis of the First and Generations" *Bedriftsokonomen*, 8: 390–396.

Rittel, H. and Webber, M. M. (1973). "Dilemmas in a General Theory of Planning" *Policy Sciences*, 4: 155–169.

Rowe, P. (1987). *Design Thinking*, Cambridge: MIT Press.

Schneekloth, L. and Shibley, R. (1995). *Placemaking: The Art and Practice of Building Community*, New York: John Wiley.

Scott, J. C. (1999). *Seeing Like a State: How Certain Schemes to Improve the Human Condition Have Failed*, New Haven, CT: Yale University Press.

Sennett, R. (2008). *The Craftsman*. New Haven, CT: Yale University Press.

Simon, H. (1972). "Theories of Bounded Rationality." In McGuire, C. and Radner, R. (Eds.) *Decision and Organisation*, Amsterdam: North Holland.

Solinis, G. (2006). "Utopia, the Origins and Invention of Western Urban Design" *Diogene,* 209: 79–87.

Sternberg, E. (2000). "An Integrative Theory of Urban Design" *Journal of the American Planning Association*, 66(3): 265–278.

Sunstein, C. R. (2001). *Designing Democracy: What Constitutions Do*, Oxford: Oxford University Press.

Verma, N. (1998). *Similarities, Connections, and Systems: The Search for a New Rationality for Planning and Management*. Lanham, MD: Lexington Books.

Further reading

Churchman, C. West. (1971). *The Design of Inquiring Systems*. New York: Basic Books. A fascinating philosophical journey where philosophers are viewed as designers of learning systems, exploring how the idea of design pervaded the thinking of Descartes, Spinoza, Hume, Locke, and Kant.

Lynch, K. (1981) *A Theory of Good City Form*. Cambridge, MA: MIT Press. A classic treatise on the performance characteristics of good city form that remains a major example of the normative theory of design.

Schön, D. (1983). *The Reflective Practitioner*. New York: Basic Books. A thoughtful description of the practice of design and planning, this elegantly shows the tensions faced by practitioners and their ways of dealing with them.

5

The two orders of cybernetics in urban form and design

M. Christine Boyer

The computer, a product of World War II, left no sector of Western society untouched in the wake of its postwar development. It produced a "machinic" assemblage revising both the terms of communication and the conceptualizations of life. Since nature has a tendency to organize itself, all physical processes were assumed to be rule-governed, simply or complexly computational, open to modeling by machine processes (Johnston 2008).

Two different cybernetic theories developed around this complex of ideas. First order cybernetics, or control theories, was introduced by Norbert Weiner (Weiner 1948). In this case order was achieved by taking information from the environment and feeding it back to a mechanism, thus correcting and regulating its trajectory towards a stated goal. The natural tendency to degrade the organized and destroy the meaningful in communication, i.e. entropy, was offset via this feedback of information. Second order cybernetics, the second computational theory promoted by Heinz von Foerster in the 1950s and developed further in the late 1960s by Humberto Maturana and Francisco Varela (Dupuy 2000), studied self-regulating or self-organizing autopoietic systems. Order in this case emerged from the ground up through highly distributed parallel interactions between layers of subsystems which achieved greater complexity and maintained equilibrium over time.

The first theory conceptualized the mind as a machine and equated thinking with computing. It led to the analogy in Artificial Intelligence that all cognitive processes could be simulated by a computer. Given a symbol system and a set of compositional rules or syntax, the computational machine could generate complex operations; it could think, learn, and even play chess. Logic controlled the precise steps of information management and defined a computer algorithm or program: an algorithm being a system of symbols connected according to a given set of rules (von Bertalanffy 1968). As these symbols were moved about – or processed – they emerged into patterns or information.[1]

The second cybernetic theory took a network approach and conceptualized machinic life as a heterogeneous collection of interacting processes and transforming behaviors. It modeled how dynamic systems achieved equilibrium, adaptability, and reproduction as self-directed actions, how they maintained constancy or stayed within a limited range of values over time, while simultaneously realizing the network of processes that produced them. This theory eventually mutated into Artificial Life.

Not always distinct and overlapping in time, these theories – one focused on countering entropy, the other on managing complexity; one more machinic, the other biological; one involved with thought processes, the other life processes – influenced how urban theorists conceptualized and perceived the form of the city. While the lines of influence are often implicit, the first theory can be said to have affected the work of Gyorgy Kepes and Kevin Lynch at MIT[2]; the second is reflected in more contemporary theories such as those promoted in Europe by Stefano Boeri with Rem Koolhaas and his associates (2000).

First order cybernetics and the language of vision

In the 1950s MIT was a hotbed of investigations into machinic processes. Norbert Weiner continued work on the transmission of messages that controlled machinery, society, linguistics, and communication.

> The commands through which we exercise the control over our environment [Wiener wrote in 1950] are a kind of information which we impart to it. Like any form of information these commands are subject to deformation in transit. ... In control and communication we are always fighting nature's tendency...for entropy to increase. (Wiener 1950: 26)

During the same years, Noam Chomsky (1957) developed his syntactic structure of language. Fusing symbolic logic with natural language, he studied how thought could be encoded in forms that could be manipulated purely by logical means. Meanwhile, artificial intelligence experts began to envision the mind as an information-processing machine, a manipulator of symbols and signs.

It is not surprising to find the visual artist Gyorgy Kepes, seeking to bridge the gap between artists and scientists at MIT after World War II, was also interested in how information theories, cybernetic controls, and symbol manipulating processes might be applied to the manner in which the new world of science and the industrialized cityscape were envisioned.[3] Kepes was especially enthusiastic about Wiener's 1950 book *The Human Use of Human Being*, and developed active interest in such concepts as feedback, noise, entropy, and information (Finch 2005).

Frank (1966) has argued that Norbert Wiener viewed the world as a multitude of "To whom it may concern" messages. Man has a particularly wide range of symbol recognition. He "learns to recognize and utilize signs as 'feedbacks' for orienting and directing much of his patterned conduct, evoking these signs as guidance." Man designs or articulates patterns of these signs to mirror external reality, and he invests these patterns with meaning to which he responds with purposeful goal-seeking conduct.

Commenting on his own 1976 painting "To Whom it May Concern," Kepes retrospectively acknowledged that he borrowed Wiener's expression because he believed that nature all around us was full of hidden messages "to whom it may concern" but only a person with an inner sensitivity could register the echoes of such (Goodyear 2004). It was Kepes' pedagogical aim to train the artist and the scientist to become sensitive decoders of messages sent and received from a variety of sources in the modern world; art could aid an individual to achieve a new equilibrium with the surrounding urban environment, to discover the invariant harmony beneath the constant flux and transformations of life (Kepes 1972).

Drawing from Moholy-Nagy's (1947) observation that "(T)he key to our age – seeing everything in relationship" Kepes continued this collaborative effort making "vision in relationship" the analogical basis for his many photographic works, exhibitions, and subsequent visual primers beginning with *The Language of Vision* (1944), then *The New Landscape in Art and Science* (1956), followed by *The Visual Arts Today* (1960), published as series of essays in *Daedalus* (winter 1960), and finally his *Vision + Value* series (1965–1972).

Called to teach at MIT's architectural department a few days after the atomic bomb brought an end to World War II, Kepes found himself aligned with concerns that science and technology had produced a world gone wrong, one in which chaos and disunity abounded. As a utopian humanist, he firmly believed the artist was a seismograph of current forces and must use his tools of perception to solve social ills. If artists were equipped with new visual perception, able to see forces previously invisible, then intractable world problems would be solvable. If man was to survive in a complex constantly transforming environment, he needed to learn how to visually master this world. In other words, man needed to be endowed with a specified system of representation which an artist could outline. It was the artist who must create a new set of visual representations or symbols of dynamic processes and organize these into new wholes, patterns, and maps. Also he must learn how to deploy this new vision of dynamic iconography toward positive social ends (Kepes 1944).

In his later work Kepes (1956) noted that in the twentieth century, we have become lost in an alien, menacing world. This is a "new landscape" in which the appearance of things no longer revealed their true nature, instead images faked forms, forms cheated functions, and functions were robbed of their natural sources. Man had to maneuver in a world of incomplete information, in which invisible processes not apparent to the human eye remained pervasive. In order to make sense of these invisible things, man had to become a symbol-maker. Because our distorted everyday environments rob man of the power to make experience coherent, proclaimed Kepes, we need new symbols to bring this new technical landscape into balance with the human environment:

> It is not with tools only that we domesticate our world. Sensed forms, images and symbols are as essential to us as palpable reality in exploring nature for human ends.... We make a map of our experience patterns, an inner model of the outer world, and we use this to organize our lives.

And further,

> ...[T]he essence of symbol making lies in the transformation of the ceaseless flow of sense data into clearly defined pictures, words and concepts. Symbol making is based on transformations, on the changing of substances or the changing of forms. (Kepes 1956: 18–22)

Thus the key to creative work lay in symbolic transformation: "the translation of direct experience into symbols which sum up experience in communicable form" (Kepes 1956:229). This means that the traditional concept of an image as a mirror held in front of nature is obsolete. Instead, the new patterns of information are pictures of processes (231).

His images and texts paid tribute to scientific and technological inventions that operated as control appliances. From automatic doors to radars and computers, all the instruments reacted, and after a fashion even thought, because they were connected to complicated electrical circuits that carried out the appropriate processes according to some rule-based machine. In this new environment, science and art could no longer be considered to be separate activities, but must be re-conceptualized as "ordering activities of the human mind" (Kepes 1956: 19–22). He wrote further:

> Images are the starting point of all thinking and feeling. .... Through images we participate in the world, responding emotionally to its sensible qualities and rhythms. .... Through images we become aware of the world's forms and structures. We mobilize ourselves to develop ideas and concepts...(22)
>
> Patterns are the meeting-points of actions. Noun and verb must be seen as one: process in pattern, *pattern* in *process*....We do not give up objective nature, but, where we formerly saw only things, we are now mobilized to see action patterns. (205)

Kepes found most visual patterns – be they cities, houses, objects of use, printed images, clothing, even facial gestures – no longer evoked the image or pattern of primary natural events. Such confusion, feeding perceptual life, dulled man's sensibilities (Kepes 1956: 207). Being focused on objects not processes, these forms misguide man, creating a serious blindness to the sensible world.

Man fails to develop a visual vocabulary of change and transmutation, of distortions and condensations, even though new scientific knowledge makes transformation a vital visual experience. Kepes (1956) suggested "...the essence of symbol-making lies in the transformation of the ceaseless flow of sense data into clearly defined pictures, words and concepts. Symbol-making is based on transformations, on the changing of substances or the changing of forms" (229). These processes establish "[a] new vocabulary of visual thinking" (230) focused on the fundamental significance of change.

First order cybernetics maintained that reason was abstract; embodied in a machine or a human. Abstract categories or symbols were the basis on which humans made sense of experience; they achieved meaning in the manner that they corresponded to or mirrored external reality. A collection of symbols was said to be a representation of the world – it spelled out a cognitive map. Thus thinking was embodied in abstract structures, with their own syntax and logical coherence. If the computer metaphor could extend to thinking, it could also be adapted to vision. The eye, as Kepes (1956) argued, operates essentially on externalized symbols, which are manipulated by an abstract syntax. The map they engender is merely a mirror of the external world (Golec 2002). And this eye had prosthetic devices – the X-ray, microscope, radar – enabling it to see into interior processes.

> The obvious world that we know on gross levels of sight, sound, taste and touch, can be connected with the subtle world revealed by our scientific instruments and devices. Seen together, aerial maps of river estuaries and road systems, feathers, fern leaves, branching blood vessels, nerve ganglia, electron micrographs of crystals and the tree-like patterns of electrical discharge-figures are connected, although they are vastly different in place, origin and scale ... Their similarity of form is by no means accidental. As patterns of energy-gathering and energy-distribution, they are similar graphs generated by similar processes. (Kepes 1956: 260)

Kepes first outlined his theory of vision in his primer *Language of Vision* written between 1939 and 1942 and published in 1944. In the opening sentence Kepes (1944: 176) declared that man is "torn by the shattered fragments of his formless world, incapable of organizing his physical and psychological needs." The "haphazard accumulation of scientific discoveries and a planless technological expansion," leaves man in need of "a new vital structure-order, a new form on a social plane, in which all present knowledge and technological possessions may function unhindered as a whole" (12).

As the title of his book conveys, Kepes claimed there was a syntactical construction to vision and made an explicit comparison between the letters of the alphabet and optical measures. Since letters can be combined to form innumerable words conveying endless meanings, so Kepes (1944) argued, optical entities through a similar syntactical combinatory generate limitless sensations of space. Because the mind was able to compute a series of symbols and ascribe new meaning to them, it could as well combine simple optical entities into new image patterns or comprehensible wholes.

This latent syntactical definition of language-like vision placed Kepes in a receptive position to absorb the cybernetic discourse on cognition being defined at MIT as computation performed on symbols and codes. In the preface of his 1956 book *The New Landscape in Art and Science*, Kepes (1956: 17) advances this theory of cognition further by declaring there is much evidence "that vision is itself a mode of thinking." Kepes is hinting, if not explicitly stating, that there is a fixed relationship between symbols and their meaning and that a priori representations mirror the world or surrounding environment in a fixed and determined manner. While not applying the term "cognitive" to his use of "map," Kepes assumed that information from the environment acts as input to the mind, which subsequently computes it in order to visually orient the body, stabilize it, and make it feel secure in its new surroundings.

Cognitive mapping and urban form

Already in 1944, Kepes had written: "To grasp spatial relationships and orient oneself in the metropolis of today, among the intricate dimensions of streets, subways, elevated trains, and skyscrapers, requires a new way of seeing" (67). Not only has the world of science transformed everyday vision, Kepes continued in *The New Landscape*, it has also affected the industrial landscape. In particular, the metropolis is a "giant focus of our unsettled world, [it] spreads out upon the land in widening rings of visual disorder." The cores of giant cities are

> bludgeoning us with their vulgar images, massive structures blot out open space; industrial areas beyond are dumped with factory buildings and the dingy barracks where we house our poor; the residential fringes are dotted with characterless cottages repeated endlessly. Everywhere, smoke and dirt screen out the sun; and our containers, advertisements, commercial entertainment, films, our home furnishings and clothes, our gestures and facial expressions mount up to grotesque, formless aggregates lacking sincerity, scale and cleanliness. (Kepes 1956: 69)

This chaotic environment, Kepes argued, shapes our vision and influences our imagination. Because it is without order, it destroys out self-confidence and has "robbed us of the power to make our experience coherent. When visual responses are warped, visual creativeness is impaired" (69). Consequently man must create a unified vision built up from a vocabulary of images augmented by optical devices that science and technology have provided: the telescope, microscope, X-rays, ultraviolet and infra red film, stroboscopic lighting, and the electron microscope, to name but a few. "We must feel the gaps in the structure and try to bridge them, map out a visually coherent panorama, a basis for equilibrium in our new world" (105).

Working collaboratively at MIT between 1954/5 and 1959, Kevin Lynch and Kepes conducted a Rockefeller-funded research project on the "Perceptual Form of the City." One of the outcomes was Kevin Lynch's famous book *The Image of the City* (1960). Lynch was interested in finding his way around the city and not getting lost in this human-made environment, for disorientation was to be avoided. Images of the city offered a sense of identity, well being, and belonging. They formed the basis of memory systems: they attracted attention and made a place memorable, storable in the mind. Thus the mental image, or "cognitive map," which spectators create as their image of the city could be used to guide subsequent design interventions. If a city was weak in imageability then its urban reformers should address points where its weaknesses lay.

Both investigators, Kepes and Lynch, were dealing implicitly or explicitly with symbolic logic and computation procedures. Thinking about city form as a logical manipulation of symbols, attention focused on the formal properties of these symbols and the rules by which they could be put together or pulled apart in order to generate good city form – well formed statements following syntactical rules. Thus cognitive mapping established an instrumental control over urban space and assumed that there were feedback loops from the environment to perceiving man. It implied there was a universal language of normative space allowing comparisons and contrasts to determine how far the image had deviated from good city form. Was it imageable or not? Alienating or supportive? Stability, equilibrium, control were the goals sought by a cognitive map which was none other than a mental representation of an external environment. It was a predetermined plan of action that reflected the principles of rational thought.

The cybernetic project is now well recognized to have been a failure. The brain is not like a computer, and the mind is not an information processing machine. Giving priority to top-down cognitive functions in language and reasoning led to a dead end. Eventually, dynamic complexity would raise its head to thwart humans' desire to control the physical environment, and order would be re-conceptualized from the bottom up, emerging out of chaotic situations, not the top-down imposition of representational form. In addition there were entropic systems, dissipating structures and disturbances of coherence that challenged the will to order.

Lynch was well aware of the limitations of this first cybernetic model of computation. For most of his career he never looked back at *The Image of the City* and his subsequent writings became more normative and holistic (thus an acknowledgment of the limits of the first model). He advocated user control and participatory design: a process that must be bottom up not one that imposed order from the top.

Kepes as well acknowledged the model's limitations. Concerned with ecological tragedies and disasters of potent technology, in another volume entitled *Arts of the Environment* published in 1972, Kepes began to consider the complex interactivity of biological systems.

> The increasing magnitude and complexity of interacting lives must make us realize that our future depends upon an understanding and control of our common system – a self-regulating, interdependent, dynamic pattern that moves from yesterday into today and from today into tomorrow. (5)
>
> We have begun to see that our extended body, our social and man-transformed environment, must develop its own self-regulating mechanisms to eliminate the poisons injected into it and to recycle useful matter. Environmental homeostasis

> on a global scale is now necessary to survival. (6)
>
> Scientists recognize that in the most precise ranges of observation, the observer and the observed interact. When observed and measured with maximum precision, the environment in both its largest and its smallest realism cannot be considered an independent objective world anymore. (7)

While not entirely abandoning his belief in control mechanisms, nevertheless Kepes was rethinking the relationship that formerly uncoupled organisms from interactivity with their environment. He was beginning to consider some of the features that constitute the second cybernetic revolution.

Second order cybernetics and urban form

The central fallacy of the urban theory of Kepes and Lynch and many others of those years, was the belief that a city expressed itself in its physical form and that analysis of form would provide sufficient information to establish procedures of intervention for its improvement[4]. This clearly utopian and humanist perspective appeared to be an anachronism by the end of the twentieth century. In place of outmoded concepts, a new language was needed to grasp shifting urban complexity and constant mutations of form. Now it is argued

> ... the city has lost its place, it tends to be everywhere and nowhere: it is an intangible space, a common, de-signified body which no longer forms an organism, an over-invested, exploded space, analyzed and surveyed, doubled and overwhelmed, while we struggle to catch up with keywords and periphrases – complexity, control, chaos, vectoriality, fractality; the generic, diffuse, oligoptic, or pandemonic city. (Tazi in Koolhaas *et al.* 2000: 43)

Stressing the need for new non-interventionary urban theories and language, Koolhaas *et al.* (1995) claimed

> [i]f there is to be a new urbanism it will not be based on the twin fantasies of order and omnipotence, it will be the staging of uncertainty, it will no longer be concerned with the arrangement of more or less permanent objects but with the irrigation of territories with potential; it will no longer aim for stable configurations but for the creation of enabling fields that accommodate processes that refuse to be crystallized into definitive form; it will no longer be about meticulous definition, the imposition of limits, but about expanding notions, denying boundaries, not about separating and identifying entities, but about discovering unnamable hybrids; it will no longer be obsessed with the city but with the manipulation of infrastructure for endless intensifications and diversifications, shortcuts and redistributions – the reinvention of psychological space. (969)

This contemporary city speaks of discontinuity and rupture. Its "...dynamics appear chaotic, unpredictable in their trajectory, and therefore...charged with uncertainty" (Simeoforidis 2000: 418). The use of the term "complexity" measures the amount of information contemporary urbanists do not posses but would need to make a complete structural and operational description of an urban system. However, these urbanists never attain nor aspire to such closure, therefore remaining in pursuit of adequate descriptions of the city's complexity, failing to define a methodology or strategy for urban design or planned intervention. Koolhaas explains: "The generic city's most dangerous and most exhilarating discovery is that planning made no difference whatsoever" (Koolhaas *et al.* 1995: 1255).

Drawing from the exhibition catalogue *Mutations* (Koolhaas *et al.* 2000), the following evaluation of urban complexity can be paraphrased. There is no solution to its bigness, no beginning or end to its myriad problems, no cause and effect relationships to unravel. The city has lost its face, its identity and thus comprehending the multiple, the nonlinear, and the interconnected become problematic for there is no totalizing overview. The contemporary city is nothing but information, everything is computerized. Its realism is its data sets; it constitutes a dynamic archive, a massive depository of local and global knowledge. In fact, the big city's "infrastructure concerns regimes of technical calculation of any and all kinds" (Kwinter and Fabricius in Koolhaas *et al.* 2000: 497). Thus information – as it was with Kepes and Lynch – is the basic element that characterizes the contemporary city – plain, raw, unadulterated information. But now information must be considered in all its dynamic aspects for it is transformative, it communicates between levels, and it flows around the world.

In all of these proclamations, the city is re-conceptualized as a self-organizing autopoietic system.[5] Hence urban patterns reflect different interactions at work in the construction of territorial form; its various subsystems "act as microcosms of autopoiesis...(extended families, ethnic and professional classes, cultural communities, leisure and consumer associations)..." (Boeri in Koolhaas *et al.* 2000: 371). Mutations in the urban terrain reveal "...autopoietic innovation[s] of inhabited space. Places and territories that seem able to adapt in original terms to the great global energies; limits within which the local...begins to fully manifest its staying power and long duration" (Boeri in Koolhaas *et al.* 2000: 369).

"Mutation," "complexity," "emergent," "information," "self-organization," and "autopoietic systems" are the keywords extracted from this discourse on the new urban terrain borrowed from second order cybernetics. While focused on biological systems, in particular immune systems, one of the authors of autopoietic systems theory, Francisco Varela, has allowed that his theory lends itself to generalization. Both society and culture, he notes, have a "unity that is living-like;" they are systems in which a higher order system emerges out of the interlocking of political, economic, cultural, communicative, legal, scientific, and religious systems (Varela 1979: 565).

Varela claims that an autopoietic system is a complex organization in which information transmittal, reception, and interruption proceed recursively through continual feedback and feed forward across multiple levels within the system. In turn, these interactions and the subsequent transformations to the organizational structure of the system they enact regenerate and realize a network of processes that re-constitute the system by itself. Hence constant communication – or interaction, the relay of information – is essential to every living system; it is what guarantees its self-maintenance over time (Varela 1979: 13, 56).

Unlike the open-ended goal-oriented system of Norbert Wiener, feeding back information from its external environment to the system itself, autopoietic systems are closed or self-referential systems. Such systems react to "environmental perturbations" that trigger compensating operations within the system itself (such as an immune system, or nervous system). They are nonlinear systems, with self-organizing abilities that achieve higher order complexity by reacting to disturbances and subsequently readapting their internal structure and modifying their behavior. In this manner self-maintenance or equilibrium emerges from within the interlocking layers and subcomponents of the system itself.

This is a theory that conceives of a system in terms of the processes that realize it – and this is the point of interest for urban theorists. How can they portray and investigate the urban system in terms of its internal dynamics, a structure that is continuously changing and re-assembling itself? The environment becomes the source of perturbations that are independent from the organization of the system itself. They can trigger but not determine the course of adjustments and transformations – hence they are not instructions or information to the system that causes the system to change its behavior as Wiener might have suggested. The definition of what is information to the system has been redefined. For the control model, information was input that was processed and produced an output, hence was applicable to the modernist concept of planning the city. For the "autopoietic" model, information refers to the complexity and consistency required to maintain its organization and leads to the viability of the system's functioning over time. Such theories when applied to urban form disavow any intention of a predetermined blueprint, plan of action, or rational organization to impose on the city; uncritically advocating that cities are determined by their own improvisations and experimental mutations. Without a plan of action laid out in advance, order or greater complexity is expected simply to emerge over time.

Thus concepts borrowed from autopoietic systems theory have little to offer urban design or any other interventionary urban strategy. Used to describe complexity and how information flows maintain internal stability, self-organizing theories leave a host of critical problems unexplored. The prevention of entropic dissipation is no longer the aim as it was with first order cybernetics; indeed chaos, the result of multiple interactions across autonomous layers and subcomponents of the city may reveal adaptive alternatives. At least that is how Rem Koolhaas describes Lagos: a city in constant flux, ingenuously self-organizing itself out of the depths of chaos. Markets spring up during hour long traffic jams as itinerant venders arrive to offer passengers in stalled vehicles all kinds of wares, while other markets beneath cloverleaf exit ramps spontaneously organize into cooking pots, metal wares all neatly arrayed (Koolhaas *et al.* 2000). No mention is made of Lagos' toxic waste dumps, the embezzlement of Nigerian oil revenues, forced migrations, near serfdom labor practices, informal settlements, to name a few intractable urban dilemmas.

Another theorist applying autopoietic systems theory is Stefano Boeri and the research network Multiplicity (2003) in their book *Uncertain States of Europe* (USE). Instead of the space of places described by geographers, these urban theorists focus on the space of flows in a network society (Multiplicity 2003). Rather than divide the map of Europe into states or spatial containers of people and political authority, each clearly delineated by linear boundary lines and blocked out from adjacent areas by separate colors, they see a network configuring Europe as a flat field of circulatory movement, lines drawn across the land or through the air, flows of messages, people, or ideas across points in space (Mattelart 1996). Focus lies on communication flows that hold the system together, not the disparity of uneven economic development, shrinking older industrial cities, the capital investment in gated communities, and the politics of land ownership and control, all of which have dramatic spatial effects. The exchange of information, immaterial and mobile, is hypostasized as the only force determining the unplanned and barely regulated form of the city.

Uncertainty over European national identity coincides with territory that can no longer be read topographically or mapped from a vertical perspective, where once the production of space and identity were joined in the compact European city of the twentieth century (Multiplicity 2003). Instead Europe has become a diffused city with differences erupting here and there in fits and starts across its territory, displaying unity and difference, as well as chaos and organization. Diffusion suppresses national borders, giving witness to a territory crossed with analogous processes and uncharted expansion. In other words European urban space is being transformed by accumulations and superimpositions of a finite set of standardized elements manipulated by a limited set of rules, but ones with which a multitude of individuals improvise and innovate creating in their wake unpredictable trajectories and a diffused spatial form. (Boeri in Koolhaas *et al.* 2000; Multiplicity 2003).

Older conventions of geography do more harm than good not allowing the cartographer to assemble the unlikes together, or to add one layer of information upon another, or to trace dynamic processes across a terrain, or to envision potentials as they begin to emerge. Nor do they allow the interpreter to see the arbitrary construction of geographical entities (such as single family homes juxtaposed to shopping centers, or an industrial park next to a car wash) or find expressions of uncertainty and indeterminacy in the overall urban pattern, yet stability and organization within its individual parts. Boeri believes that if the concept of European space is to acquire visibility and shape, although its boundaries may remain blurry and diffused, its space must be envisioned as a field shot throughout with autonomous subsystems having their own metabolisms of material and immaterial energies. The uncertain nature of European states, he claims, is the direct effect of action or interaction of a multitude of individual actors within these subsystems and hence why "European space is seen as an open, available context: a surface composed of heterogeneous, continually changing geographical environments, acted on by multiple of energies" (Boeri in Koolhaas *et al.* 2000: 360). Surface sprawl, or diffusion, defines a tangled multilinear ensemble composed of lines that follow certain directions, break, or bifurcate, before changing directions, becoming self-reflective or drifting about (Multiplicity 2003). A principle of variation – an infinity of adaptations and improvised solutions – is the sole regulator of the emerging urban composition.

Boeri's problem, however, is not merely how to model and interpret the surface mappings of dynamic processes that pulse across the face of Europe, it is also – as any geographer must note – a matter of language and the codes and concepts which enable subjects and spaces to form their identity. Like Kepes and Lynch, Boeri also turns to a linguistic paradigm for his urban theory not in order to exploit a syntactical metaphor of spatial organization, but to argue instead that new vocabulary is needed to interpret autonomous forces determining the impetuous expansion of inhabited space. He notes:

> European space, which is a palimpsest of projects sedimented in time, is also today the field of action for an indeterminate and changing number of subjects, many of whom maintain a temporary relationship with the territory. A battle of codes and interpretations ceaselessly unfolds upon this field, which is continually being rewritten, where almost nothing is ever erased, where the long-term structures are temporarily hidden by others which are less powerful and enduring, but currently more visible. (Boeri in Koolhaas *et al.* 2000: 375)

He offers a new vocabulary to interpret and describe this borderless nebula of European space, terms such as "linear attractor" (different buildings drawn to locate along major transportation corridors), "bowling pins" (introduction of autonomous elements), "islands" (introverted places of similar lifestyles and identical objects), "cloning zones" (the repetition of identical units), and "grafts" (insertions of replacement). A limited number of building prototypes such as single family homes, terraced housing, office-residences, and multi-use commercial structures are the elements of this urban language, which a myriad of independent users play with and interpret in open-ended fashion. The result is a horizontal expansion of the urban condition not recognizable as syntax and structure but as a crossing of dynamic energies flowing between society and space in fits and starts (Boeri in Mutations 2003: 424–451).

Unholy mixtures, variations, and anarchic statements proliferate, conjoined at a higher level without any perceptible plan. Focus is drawn to the awesome materiality and pragmatics of this urban language – its redundancies [noise], variations, and subversions grouped together by some force of attraction dictated by the codes and syntax that may control the assemblage at some higher unknown or unconscious level and that may at the same time be flouted and transformed. The grammar of this new urban language produces no articulate propositions – the rules are missing – but engenders a triumph of unsynchronized constructions defying synthesis or combinatorial logic. It is an impoverished language that infinitely repeats only a small sector of its very rich inherited architectural and spatial alphabet. The urban question then focuses on how these elements and their groupings are produced over time, how they work as utterances, in what assemblage they are inserted. Boeri remarks:

> Urban space in Europe today means, maybe more than anything else, this intermediate sphere that, like a real "phrase" between words and a discourse, absorbs the unpredictable variations of the world of life and identifies them according to a code inscribed in the materiality of the urban condition. The Forum, the block, the courtyard, the suburb on the public periphery, are inventions – but we should say reinventions – of this transformational device. The point is to ask ourselves if, how and where this device is still operating. (Boeri in Mutations 2003; 24)

Mutations in space require not only new vocabularies of decipherment but also new strategies of embodied observation. Hence the development of "Eclectic Atlases" offering a multitude of visual thinking and lateral modes of representation: research reports, photographic surveys, geographic descriptions, qualitative analyses, literary probes, collections of plans and projects. These atlases seek new logical connections between spatial elements, words that name the elements we see, and the mental images we project onto space. They reveal that "behind the apparent chaos, there is in fact an excess of organization, of regularity, an excess of evolutionary patterns" (Boeri in Koolhaas *et al.* 2000; 368). They are eclectic because they seek to represent the dynamics of inhabited space that are multidimensional, spurious and, experimental (Boeri in Multiplicity 2003; 104).

By creating a multiplicity of entangled lines that cut across questions of European identity, cartographic procedures, linguistic analogies, and theories of biological evolution, Stefano Boeri and his research network represent European space as a multilateral, multi-noded, multi-entry construction. In the end, however, it is just as abstract and detached from reality as any other map used as a metaphor for spatial ordering, a surface for notating field observations, or a medium of rationality. It may be supposed that the informational code within organisms produces their characteristic structure and behavior. But can this biological analogy be applied to the states of Europe and the diffused city? Can the invention of limited codes describe their properties and how they operate and interact with each other? And if a biological organism just is, neither progressing towards nor regressing from a state of perfection, then in adopting this model there is not only epistemological failure to describe in words the evolutionary processes we merely visualize, but a failure as well to design any operational procedures that might achieve an enhanced condition or better environment. Squeezed between homogenizing forces from above and fragmenting and fracturing energies from below, deploying an autopoietic analogy that assumes an organism always achieves stability or a strong sense of identity, may offer an inadequate perspective on the uncertain state of European identity and the geopolitical questions that trouble its space.

Conclusion

Norbert Wiener was concerned with the tendency for entropy to increase – how this imposed limitations on communication within and among individuals.

> Information is a name for the content of what is exchanged with the outer world as we adjust to it, and make our adjustment felt upon it. The process of receiving and of using information is the process of our adjusting to the contingencies of the outer environment, and of our living effectively within that environment. (Wiener 1950, 26–27)

As interpreted by Kepes and Lynch "living effectively within the environment" involved manipulating a symbolic representation of changing urban form; interpreting its organization as structured, as language-like, hence capable of change and renewal. Urban form had a structure – syntax and a set of transformational rules – defining a functional city of clearly separated zones and elements that could be willingly manipulated towards a more harmonious whole.

Heinz von Foerester countered this first order cybernetics by noting that self-organizing systems are ones whose internal order increases over time and that they find on their plate not only order but also noise. These systems and their subsystems become increasingly adaptive not only to themselves but to the conditions they bring about. Hence the human role of agency shifts from being actively and intentionally involved in directing the system's dynamics to being a passive operator, part of the system but not a director of the whole (Johnston 2008). As interpreted by Rem Koolhaas and Stefano Boeri, self-organizing urban complexity is an adaptive system operating through a multiplicity of interactive agencies and immaterial flows enabling complexity to build upon complexity. Urban form displays a type of coherence – or survival – despite diversity, change, and lack of central command and control devices. It presents a type of assemblage bringing the urban environment and human actors into interaction with energetic flows of historical sedimentation and evolutionary information.

The problem of agency remains to be addressed by these interpretive models: is the city a machinic assemblage operating without human intervention or is urban form capable of being directed and refigured to prevent or minimize "entropic disintegration?" If the problems of cities in the twenty-first century are to be addressed, those of megacities bursting their contours, of flows of migrants and displaced persons, of havoc wrought on urban terrain by extreme weather, military tactics, famine, neglect, ignorance, and so on; then we must change our models of urban form to facilitate more humane and adaptive response.

Notes

1 To consider an example of an algorithm, think of the Arabic notation of units: 1st column from right = 1s, 2nd = 10s, 3rd = 100s – figures written in columns makes arithmetic operations child's play.

2 The groups of urban designers influenced by first order cybernetics would include: (1) Philip Thiel, a student of Kepes, who developed a graphic notation system for spatial experience; (2) Chris Alexander and Marvin Manheim (who did a chapter for the Vision + Values series); (3) Donald Appleyard; (4) Ian Mcharg and his discussion of the ecological (natural) order. The work of Design Methods group is also noteworthy.

3 Working collaboratively at MIT, Kevin Lynch and Gyorgy Kepes undertook a Rockefeller funded study entitled "Perceptual Forms of the City." One of the outcomes was Kevin Lynch's book entitled, *The Image of the City* (Cambridge: MIT Press, 1960).

4 Rem Koolhaas, Harvard Project on the City, Stefano Boeri, Sanford Kwinter, Nadia Tazi, Hans Ulrich Obrist, (eds.) *Mutations* (Barcelona: ACTAR, 2000): 495.

5 "Organization" refers to a set of relations that must exist among the components of a system if that system is said to exist, and "self-organization" identifies the system as a self-producing or autopoietic organization (Auto: = stems from the Greek ἀύτος meaning self and poietic: = from the Greek ποιειν meaning to produce.)

References

Boeri, S., Kwinter, S., Tazi, N., and Obrist, H.U. (2000) (eds) *Mutations*, Barcelona: ACTAR.

Chomsky, N. (1957). *Syntactic Structures*, The Hague: Mouton.

Dupuy, J-P. (2000). *The Mechanization of the Mind: On the Origin of Cognitive Science*. Princeton, NJ: Princeton University Press.

Finch, E. (2005). Languages of Vision: Gyorgy Kepes and the "New Landscape" of Art and Science. Unpublished doctoral dissertation. CUNY: The Graduate Center.

Frank, L. (1966). "The World As a Communication Network" in Kepes, G. (ed.) *Sign, Image, Symbol*, New York: George Braziller.

Golec, M. (2002). "A Natural History of a Disembodied Eye: The Structure of Gyorgy Kepes's *Language*," *Design Issues*, 18(2): 3–16.

Goodyear, A.C. (2004). "György Kepes, Billy Klüver, and American Art of the 1960s: Defining Attitudes Toward Science and Technology," *Science in Context*, 17 (4): 611–615.

Johnston, J. (2008). *The Allure of Machinic Life: Cybernetics, Artificial Life, and the New Artificial Intelligence*, Cambridge, MA: MIT Press.

Kepes, G. (1944). *The Language of Vision*, Chicago: Paul Theobald.

—— (1956). *The New Landscape in Art and Science*. Chicago: Paul Theobald.

—— (1960). *The Visual Arts Today*, Middletown, CT: Wesleyean University Press.

—— (1972). (ed.) *Arts of the Environment*, New York: George Braziller.

Koolhaas, R., (2000) "Harvard Project on the City," Boeri, S., Kwinter, S., Tazi, N., and Obrist, H.U. (eds.) *Mutations*, Barcelona: ACTAR.

Koolhaas, R., and Mau, B. (1995). *Small, Medium, Large, Extra-Large: Office of Metropolitan Architecture* (edited by J. Sigler), New York: Monchelli Press.

Lynch, Kevin (1960) *The Image of the City*, Cambridge, MA: MIT Press.

Mattelart, A. (1996). *The Invention of Communication*. Minneapolis, MN: University of Minnesota Press.

Moholy-Nagy, L. (1947). *Vision in Motion*, Chicago: Theobald.

Multiplicity (2003). *Uncertain States of Europe*, Milan: Skira.

Simeoforidis, Y. (2000). *The West Arc of Thessaloniki: New Collective Spaces in the Contemporary City*, Athens: Untimely Books.

Varela, F. (1979). *Principles of Biological Autonomy*, New York: North Holland.

Von Bertalanffy, L. (1968). *General System Theory: Foundations, Development, Applications*, New York: George Braziller.

Wiener, N. (1948). *Cybernetics, or Control and Communication in the Animal and the Machine*,

Paris, France: Librairie Hermann & Cie, and Cambridge, MA: MIT Press.

—— (1950 reprint 1967). *The Human Use of Human Beings: Cybernetics and Society*, New York: Avon Books, 26–27.

Further reading

Dupuy, J-P. (2000). *The Mechanization of the Mind: On the Origin of Cognitive Science*. Princeton, NJ: Princeton University Press. A comprehensive intellectual history of the early cybernetic meetings (1946–1953), and the conceptual development of cybernetics and autopoietic systems as discussed in these postwar meetings.

Johnston, J. (2008). *The Allure of Machinic Life: Cybernetics, Artificial Life, and the New Artificial Intelligence*, Cambridge: MIT Press. A comprehensive look at the inheritance of postwar cybernetics as it has evolved into new concepts of machinic assemblages and machinic life.

Koolhaas, R., "Harvard Project on the City," Boeri, S., Kwinter, S., Tazi, N., and Obrist, H.U. (2000). (Eds.) *Mutations*, Barcelona: ACTAR. This well illustrated book exemplifies contemporary urban thought deploying images as a central component in the description of contemporary urban dilemmas.

Wiener, N. (1950 reprint 1967). *The Human Use of Human Beings: Cybernetics and Society*, New York: Avon Books, 26–27. This book deals with the impact of cybernetics on society – politics, culture, economics, and ethics.

Multiplicity (2003). *Uncertain States of Europe*, Milan: Skira. The book contains the work of Stefano Boeri and his associates that demonstrate how contemporary space is transformed by accumulations of information, layered on top of each other, open-ended, without overall direction or purpose.

6

Urban design and spatial political economy

Alexander Cuthbert

> If at first a theory is not absurd, then there is no hope for it.
>
> Albert Einstein

This chapter emerges from a deeply felt conviction that urban design should exist as an independent field within society and as an extended educational program within universities. But such legitimization will require serious intellectual engagement with the globalizing world we now inhabit. Academics and practitioners alike can no longer assume what their missions are. With a few exceptions, urban design has been content with theorizing itself unencumbered by the economic and political realities of the global capitalist system, and has been self-referential within a singularly contained ideology. I have referred to this as mainstream urban design. This would include everything written about it up until the new millennium (Cuthbert 2007). But what has been accomplished within this paradigm is an incoherent picture of the organization, production, transformation, and meaning of the built environment, a task some scholars have addressed with singular insight (e.g. Dickens 1979, 1980; Knesl 1984; King 1984; Clarke 1989; King 1996; Sklair 2005, 2006). In the process and because of the absence of its own synthesis with this knowledge, urban design has come perilously close to being a social technology devoid of any substantial understanding of the society it serves and affects.

In the process of redefining the discipline on the basis of substantial theory, many of our deeply held beliefs and images of ourselves as "urban designers" may have to be discarded, rethought, or reordered. For example, our concept of *urban* requires to be made distinct, while the narrow definition of *design* imported from architecture needs serious examination. In addition, one searches in vain for a definition of urban design that is not axiomatic or depthless (Cuthbert 2007:180–188; see also Gosling 1984; Rowley 1994). Since the inception of the architecture and planning professions around the beginning of the twentieth century, urban design has been colonized by both. As in all colonies, it has remained retarded in its possibilities and truncated in its development. In Australia, for example, the Australian Institute of Planners (AIP) now offers professional membership to all urban design graduates of whatever background. This effectively completed the process of colonization in Australia, as well as an intellectual division of labor, which enhanced both the AIP and the RAIA. If indeed urban design constitutes an independent "field" rather than a "profession," it remains so for academics. For others it is business as usual.

While the preceding example is not ubiquitous, annexation overall has fallen somewhat neatly into two major parcels. Architecture has retained a dominant interest in design guidelines and briefing, as well as the creative design of projects (witness the New Urbanism) – while Urban Planning regulates land use and development control. Due to the inherent complexity of urban design problems, a host of other professions are involved, including Landscape Architecture, Engineering, Law, Project Management, and associated disciplines. Most urban designers see this diversity of interest as an advantage, enriching the subject through cross breeding and a more robust gene pool, and this is not in question. But in ignoring the fact that most if not all professions have similar admixtures of knowledge, urban design remains colonized on the basis that its interdisciplinary nature demands the submergence of its own independence, and continuation of its subaltern status. So paradoxically, the very interdisciplinarity that supposedly enhances its relevance simultaneously diminishes its integrity. I maintain that such an ideology of denial has compromised urban design theory, and largely explains why urban design has failed to generate any significant interpretations beyond its own myopia. Hence the subject has neither an institutional presence as a profession, nor a coherent theoretical base as a discipline. It remains a colony, and like all colonies, is subject to serious underdevelopment.

Theoretical and real objects?

My position is that inter-disciplinarity and independence can co-exist in a profession of urban design as it does in other professions. Indeed I have tried to demonstrate that urban design has at least as much claim to independence and legitimacy as either architecture or urban planning by suggesting both the theoretical object as well as the real object of all three disciplines (Table 6.1). But such independence may come at some cost, demanding a significant departure from derived "theory" abstracted from its colonizers. Given the burgeoning interest of urban designers in political, economic, and social theory, a narrow referencing from within the professions of Architecture and Planning now seems inadequate and reactionary. I have elaborated this position in two prior books, *Designing Cities* (2003) and *The Form of Cities – Political Economy and Urban Design* (2006). Also involved is a forthcoming book on method currently nearing completion. Using exactly the same structure, these texts form a trilogy that reviews and critiques traditional theory in the discipline and redefines its content. The argument is also restated in a somewhat compressed manner in a special issue of *Urban Design International* under the title "Urban design: requiem for an era – review and critique of the last 50 years" (2007). Taken collectively, this corpus of work focuses on a new role for urban design by disengaging the subject both theoretically and politically from Architecture and Urban Planning while retaining necessary associations with both. I use the term *The New Urban Design* to distinguish it from the mainstream as it presently exists.

Following the Popperian principle that science advances not by proof but by disproof, mainstream urban design "theory" clearly has to undergo a Copernican shift in emphasis, if not in substance. In order to initiate this process, it is easier to reject the entire corpus of mainstream urban design as fundamentally *atheoretical* rather than attempting to reorganize the fragments from which it has been assembled. The reason for such rejection has several dimensions. First, it is easier to see the problem when past associations are set to one side, if only temporarily. Second, mainstream theory makes little or no distinction between the discipline and the environment that governs it. Third, most

Table 6.1 Theoretical foundations of three environmental disciplines

	Architecture	*Urban design*	*Urban planning*
Theoretical object	?	Civil society	The whole point of the diagram is for readers to answer these dilemmas themselves to illustrate their confusions. Public Interest? Efficiency? Equity? Social Justice?
Real object	The building	The public realm	The physical city? Settlements? Neighborhoods? Etc.? I would argue that one cannot list a variety of theoretical objects to one's heart's content – so is it the city or is it not?

Source: Cuthbert 2007:211.

urban design theory is self referential and legitimated on the basis of personalities (e.g. Christopher Alexander, Kevin Lynch, Rob Krier, Bill Hillier) and movements rather than any consistent integrity of its own, some of it bordering on mysticism. Fourth, and most importantly, mainstream urban design is atheoretical in a fundamental sense, that there are few if any substantial connections to primate disciplines in social sciences, arts, and humanities. The necessary shift in perspective should also accommodate the idea that all cities are *designed* by human action. While their aesthetics may fail to deliver high art in most cases, this happens to be the reality of social life. Only relatively recently has there been a small but significant movement to accommodate these ideas, accepting that the New Urban Design theory must locate itself first and foremost within the economic and political environment from which it emanates (Tafuri 1979; Knesl 1984; Sklair 2006; Kumic 2008).

In order to bring coherence to the somewhat anarchistic intellectual environment described above, there must also be some clear distinction between theory *of* urban design and theory *in* urban design. Both are necessary, but the former is largely absent. While this may seem obvious, only the latter seems to have been pursued over the course of the twentieth century. Paradoxically, and despite such criticism, there is no intention to suggest that all prior urban design knowledge has to be discarded. The inference is that if we add up the pieces from which urban design is currently composed, they do not provide us with a coherent picture of the discipline (or profession) based in substantive theory. Moreover, such coherence is both possible and necessary. The pieces of the jig-saw may all be present, but the image that allows us to see their relationship as well as the totality they represent is absent.

Spatial political economy

In order to overcome this problem, I maintain that the theoretical framework of *spatial political economy* can provide the intellectual base from which urban design can erect a theoretical scaffolding of its own. By this means, urban design processes and practices may be contextualized within significant discourses that continue to emerge from disciplines such as urban geography, sociology, and economics, as well as important associations within cultural studies, art history, and anthropology. Urban design is influenced by the economy, social relations, and politics of civil society. These constitute the rule systems that underwrite its formation and the origins of its material and intellectual existence. Hence the discipline of a new urban design becomes legitimated, not by its lateral connections to architecture and planning, but vertically in the first instance from society and space. Its internal dynamics may then be theorized as a product of social life, including the design process itself. But in order to grapple with new theory we also need new tools.

Spatial political economy has its point of origin in the political economy of Adam Smith within the Scottish Enlightenment of the mid-eighteenth century, specifically his treatise on *The Wealth of Nations* (Herman 2002). In so doing, modern economic theory came into existence. Smith opened up the possibility of civil society being isolated from the state, thus allowing the homeostatic properties of the market full sway, a principle that in recent times is elaborated within state neo-corporatism, guiding both the Thatcher and Reagan governments (Harvey 2006). The next great advance came in Marx's three volumes of *Capital* published in 1894, the greatest critique of the ravages of capitalism ever written. It was subtitled *A Critique of Political Economy*, where he railed against Smith's narrow vision of economics. Since that time Historical Materialism, which was the basis of Marxist thought, has undergone enduring evolutionary change within the sphere of political economy. Over time it has morphed into a political economy of the left in order to distinguish it from bourgeois economic theory or so-called neo-classical economics. Marx's intellectual construct was so immense that most thinkers within the social sciences have had to come to terms (willingly or otherwise) with his philosophy, economic theory, and methods of analysis, as well as their evolution over one and a half centuries.

Most of the great social scientists that came after Marx, such as Durkheim, Simmel, Weber, and others, were not concerned with space, a task that had to wait until the twentieth century and the Chicago School of Human Ecology (Coser 1977). They considered that the material conditions of existence sprang forth from immaterial forces, primarily the abstraction of nature and human labor into capital and its circulation within the world of finance and commodity production. What mattered was how wealth was created, transferred, stored, and distributed. The form adopted by the built environment was of no concern since its production lay in the economic and political circumstances of society, and these were what needed to be changed (Harvey 1985; Lyotard 1985; Castells 1989). More recently postmodernism criticized political economy for its failure to accommodate difference – issues involving the issue of space, as well as feminism; language and meaning; race and subjectivity. Over the last thirty years however, political economy has overcome these problems and has been generally adopted within the social sciences.

Interpretations in human geography, planning, and urban design

Three necessarily brief examples of spatial political economy and its use in human geography, urban planning, and urban sociology offer insights as to its application.

> We use the term "political economy" to encompass a whole range of perspectives which sometimes differ from one another yet share common

> concerns and similar viewpoints. The term does not imply geography as a type of economics. Rather economy is understood in its broad sense as social economy, or way of life founded in production. In turn, social production is viewed not as a neutral act by neutral agents, but as a political act carried out by members of classes and other social groupings ... political economic geographers practice their discipline as part of a general, critical theory emphasizing the social production of existence. (Peet and Thrift 1991:1)

The second example is taken from Brian McLoughlin's last paper *Centre or Periphery: Town Planning and Spatial Political Economy* (1994). McLoughlin was unique in the world of planning since he was the key proponent of the two theoretical frameworks of any substance to grace planning in the last fifty years, namely General System Theory (1970) and Spatial Political Economy (McLoughlin 1985, 1992, 1994; Huxley 1997). Using spatial political economy, he delivered an incisive analysis of the form and function of planning practice and education, to the point where its incoherence as a discipline became apparent – a pastiche of practices, "that has devolved into a ritualised choreography of routines [and] will survive purely as a ritualized technocracy" (Dear 1986: 379). For McLoughlin, spatial political economy represented a way of seeing the world as it *was*. In contrast, urban planning looks as the world as it *should be*. Hence his rejection of the entire idea of "planning" based in traditional historiography and practice:

> ... because under current social, economic and political conditions, there are no mechanisms for producing large-scale, long-term desired outcomes other than those that accord with dominant values and interests. Design, urban reform, modeling, systems, public policy, rational/procedural planning, and equity advocacy have all fallen short of their ideals, largely because, under capitalism, the very objects of urban/environmental practice – investment, development and the use of land – are all beyond democratic social control, whether expressed as state policy or as direct citizen participation. (Huxley 1997:742)

McLoughlin therefore considered planning to be a chimera, it was unreal; it was impossible; and it was fake. Planning education also fell under the same axe. Following from this, the sanction of the profession over planning programs suggested that tertiary education was also intimately connected to "planning the ideologies of planning" (Harvey 1985; see also Cuthbert 2006: 243–245). While all of this might seem surreal, McLoughlin's arguments are perceptive and balanced. He sought to remove the façade that "planning" had erected as a neutral and impartial agent in the development process (one which indeed he himself had helped to create). In fact it was yet another ideological construct that reinforced the class basis of capitalism through land regulation and development control (Scott and Roweis 1977).

Third, Manuel Castells uses spatial political economy from the perspective of a social scientist, and his search for a specifically *urban* sociology dominated much of the period from 1975–1985 (Pickvance 1976; Paris 1983), a debate which still echoes today. His text *The Urban Question* (1977) represents a critical threshold in the development of this project, one that owes much to Henri Lefebvre (1970). In contrast to most definitions of urban design, which are largely content free or so self-evident that any concept of refutability becomes impossible, Castells' encompassing statement is challenging:

> We define urban meaning as the structural performance assigned as a goal to cities in general (and to a particular city in the inter-urban division of labour) by the conflictive process between historical actors in a given society.
>
> We define urban functions as the articulated system of organizational means aimed at performing the goals assigned to each city by its historically defined urban meaning.
>
> We therefore define urban form as the symbolic expression of urban meaning, and of the historical superimposition of urban meanings (and their forms), always determined by a conflictive process between historical actors.
>
> We call urban social change the redefinition of urban meaning. We call urban planning the negotiated adaptation of urban functions to a shared urban meaning. We call Urban Design the symbolic attempt to express an accepted urban meaning in certain urban forms. (Castells 1983: 303–304)

From these necessarily brief examples, it should be clear that spatial political economy is a radically different method of scrutinizing social processes than the liberal views that permeate professions due to their dependence on the private sector and the state for projects, as well as their role as firms within capitalism. From the above, we can also deduce that spatial political economy *is not a theory* but a coalition of theoretical discourses. As in urban design, it is also a pastiche of propositions but one with real content on the basis of its connections to primate theory. The difference is that spatial political economy is first and foremost rooted to the fact that space, imagination, and design are all social products. As such, it provides the foundation for explanations of the world in which we live and the emergence of significant theory on the basis of critical thinking. As we have seen, the masters of sociological thought were unconcerned with space. But on the basis of this genius, Castells was able to transport social science into the urban dimension of the political and material role of space within society. The task we face as urban designers is to push this equation into its final state, from aspatial social process, to urban process, to the production of form. This story is just beginning and it offers us immense possibilities for reconstructing mainstream urban design into a political economy of urban design that is social, critical, informed, and connected to the world. In order to do this we could begin by rewriting the history of urban design from the perspective of political economy, one already attempted by Manfredo Tafuri (1979, 1987). But for the moment we can only take a brief look at how spatial political economy allows us insight into urban design in a globalizing world at the beginning of the third millennium.

Globalization and development

Existentially, it is possible to argue that globalization is all that mankind has ever experienced, since the limits of the known world were always "global" for its inhabitants. Today the term is used specifically to denote the progressive socio-economic and political integration of world finance, nation states, and populations. Part and parcel of this context is the shifting structure of trans-national capital in its search for ever-cheaper sources of labor, resulting in a new international division of labor (NIDL). At the time of this writing, the collapse of the world economy is currently challenging the Great Depression of the

1930s for primacy, and the global financial system is being restructured.

It is clear from the above that the conditions within which urban designers operate are being massively restructured on the basis of diminished trade imbalances, financial resources, the lack of liquidity in bank lending etc., to changes in land use, transportation, and urban administration the actual configuration of professional urban design practices (many for example going bankrupt or having to fire most of their employees). For urban designers, the significance of the built environment is that it represents a unique form of capital which is fixed in space rather than fluid as in Castells' *space of flows* – electronic communication and the internet (Castells 1989, 1996). As such, capital accumulation from space demands a process of creative destruction, where the built form of cities is endlessly destroyed, transformed, and rebuilt in order to recreate new capital. Urban design is an integral part of the accumulation process and not merely an aesthetic sideshow for the creation of symbolic capital. We can also acknowledge that the built environment within which urban design projects are embedded is not simply a stage to be dressed with aesthetically satisfying schemes, it is a structural part of the economy, as is the role of the urban designer within it. But as indicated above, the global economy is undergoing rapid and perhaps irreversible change, and with it the nature of the built environment and urban design. I have tried to indicate the spatial changes involved in the movement from modernism to globalized postmodern environments in Table 6.2.

A major difference between what I have called *the New Urban Design* and *Mainstream Urban Design* is that the former recognizes that its theoretical underpinning must explain the emergence of urban form from these circumstances, and theorize this transition appropriately. The New Urban design begins with the assumption that all environments are a product of design processes embedded in social action, and do not spontaneously arise from the software of architectural and planning practices or the frequently brilliant but ultimately fragmented discourses of the mainstream (Cuthbert 2007). Urban forms do not spring fully formed from the ground their own volition. They emanate in the first instance from the political economy of the time and are materially produced by it. Neither are forms of consciousness independent factors in the creative process. They too are socially produced (Harvey 1979; Knox 1982). The legitimation and retheorization of urban design is dependent on these conditions. In former eras, modes of production such as slavery, feudalism, and merchant and industrial capitalism generated economic and political structures that demanded specific urban forms.

Table 6.2 The design properties of cities within modernism and postmodern globalization

	Industrialism	*Post industrialism*	*Modernism*	*Post modernism*
Spatial effects	massification concentration centralisation	demassification diffusion dispersal	urban functions state symbols architectural 'styles'	urban landscape corporate symbols architectural rhetoric
Social implications	community base zoning suburban focus	locality based complex integration urban focus	paradigmatic syntactic design	eclectic metaphoric codification

For an expanded version see Cuthbert 2006:19.

The Roman spectacle demanded the Coliseum; Greek colonization in Asia Minor required the creation of the gridiron plan; merchant capitalism produced a new form, *Uffizi*, to reflect economic change dependent on a new type of administration based in "the office" and so on. Rather than eulogizing Art, the Florentine Uffizi Gallery actually symbolizes the ascent of bureaucracy (Mumford 1961). In turn, globalization and informational capitalism is demanding its own forms of environment derived from principles many times removed from these historical examples.

Luxury consumption, branding, image, and sign

Globalization represents a new deepening in capitalist social relations and the forms of consciousness necessary to sustain them. At the epicenter of these changes exists the concept of commodity. Whereas in the past commodity production was centered on fulfilling basic needs through the provision of use values, in developing countries basic needs have been met. While the need to provision daily life made limited demands on commodity *production*, desire is limitless and is rooted fundamentally to *consumption*. In transcending its material function, commodification has also been transformed from production to *consciousness*. What we are and what we desire fuse together, and the consciousness of the individual gradually becomes integral with that of the commodity, a phenomenon which is not a thing but a material and symbolic construct at the center of capitalism. As a result, "the dialectic of design movements is intimately connected to the development of capitalist markets" (Jenkins 2006:195).

The gradual absorption of consciousness into commodity fetishism has also been reinforced by the absorption of culture into the realm of production. Today, globalization is itself a generator of culture based on the universalization of products, informational capital, and the mass media (particularly "the People's Republic of Television," see Adorno 1991b: 136–153). In addition, Baudrillard maintains that culture itself has more to do with the production and consumption of signs than it has with the material world of objects. In the past, political economy saw culture as superstructural and ideological, a phenomenon that had no relation to the generation of wealth. Today culture has been absorbed into the realm of production, and denoted the *Culture Industry* that now forms part of many economies, developed and developing alike (Adorno 1991a; Scott 2000).

Architecture and urban design are an integral part of this process. Cultural and historical processes, structures, and events are all packaged in the interests of wealth generation, and the development of cities is to a large degree dependent on their capacity to commodify themselves through the branding and image generation associated with urban design and the public realm (Zukin 1996). Furthermore, the success of the urban brand to provide spectacles, artistic venues, "cappuccino environments," and overall amenity in the form of improved urban design becomes a magnet for the creative class, and hence the economic success of cities (Florida 2003). The corollary is that those that are unsuccessful will perish in a sea of decay (Harvey 1985). Thus, rather than architecture and planning defining urban design, urban design has become the central focus of concern. However, along with the need to provide a spectacular public realm comes the threat of ownership and/or colonization by neocorporate interests, and the threat this poses to civil society (Cuthbert 1995; Cuff 2003).

The form of cities

Such overall effects on the built environment are ubiquitous, reflecting the move from socially necessary production to socially unnecessary consumption, as well as the penetration of state and society by neocorporatism. The overall power of the commodity expressed in the image of the city and its promotion through neocorporatist ideology results in designed environments that mirror production. This begins with what Kumic (2008) refers to as "*the master brand*," branding the city within which branded architecture and urban design co-exist. Reflecting commodity fetishism, the brand represents a galaxy of desires compressed into physical space. Hence the sign language of brand and image progressively colonizes all public places from Times Square in New York to the Ginza in Tokyo to local shopping centers and public buildings and spaces, deepening in its compass from year to year (Chmielewska 2005). The concept of the brand and the accompanying logo (think Sydney Opera House) has been extended from shoes and perfume to cities, and *the city brand* is sought after and promoted by all urban administrations.

Much of this focuses on urban design as the vehicle for spectacles such as the Beijing Olympic Games, and the Shanghai World Expo in 2010, as well as international conventions, world fairs and expositions, summit conferences, premier league football, grand prix races, and international design competitions. In this process, the promotion of the brand image of the city coincides with the brand image of products. For example, the City of Sydney recently granted the right to turn its Olympic Park into a racetrack for the premier grand auto event of the year, thus enhancing the urban brand "Sydney" while simultaneously promoting the automobile industry, oil companies, and a host of others. Much public dissatisfaction with this idea was simply ignored in the interests of promoting the brand *Sydney*. Hence the brand becomes synonymous with the political appropriation of urban space as a general rule. Ownership of the image, branding, and urban design become welded together in the interest of commodity production and the consciousness that supports this coalition (Kumic 2008). Thus Marshall McLuhan's concept that *the medium is the message* becomes redundant as the medium and the message fuse together (Baudrillard 1981, 1997).

Several methods then nest within this idea. Closely related is the idea of theming to promote consumption, challenging traditional concepts of reality and authenticity. Building on the existing brand, even the "old style" Las Vegas has been recently rethemed to a fake version of its fake original, and the urban designs of themed environments now occupy an increasing proportion of the public realm and commodity circulation (Chaplin 2000). This overall process occurs in two major dimensions: first the theming of space, second the theming of its individual components. The theming of space through urban design projects is not new, and it could be argued that the concept is integral with the discipline. Any large-scale design is "thematic" in that L'Enfant's plan for Washington or Walter Burley Griffin's plan for Canberra imposed *themes* on nature. The difference is that neither was designed as an integral strategy to promote luxury consumption and commodity fetishism. Second, branding is now strategically accomplished through the process of iconic architecture and franchising. In a landmark article, Leslie Sklair (an economist) discusses the role of iconic architecture to transnationalism as follows:

> Iconic architecture is defined as buildings and spaces that are (1) famous for professional architects and/or the public at large and (2) have

> special symbolic/aesthetic significance attached to them. Architects can also be iconic in these senses. Also introduced in that article are distinctions between professional and public icons; local, national and global icons; and historical as contrasted with contemporary icons. The argument is located within a diachronic thesis suggesting that in the pre-global era (roughly the period before the 1950s) most iconic architecture was driven by the interests of the state and/or religion, while in the era of capitalist globalization the dominant force driving iconic architecture is the transnational capitalist class. (Sklair 2005:485; Sklair 2006)

Branded iconic architecture is also invested with the idea that its stimulus to the city brand has major economic benefits as claimed, for example, in Frank Gehry's Guggenheim Museum in Bilbao, and more recently in the Bird's Nest at Beijing's Olympic Park, Bahrain's World Trade Centre or Foster's Gherkin in London. The second aspect of branded architecture and urban design is the effect of multinational franchise architecture e.g. McDonald's, Starbucks, Burger King, and global commodity corporations such as Aldi and Ikea, as well as up-market brand stores such as those of famous fashion houses and culture outlets. This process is also enhanced by transnational architectural practices and their supporting retinue of firms – engineering, accounting, surveying, building services, etc. Using the example of branded chain hotels, Yahkleef promotes the idea that the detachment of global brands from cultural settings results in generic spaces that are place-constrained and have no referents except the commodity:

> Generic spaces are deterritorialized, disembedded, and lifted out from their context. Once cut loose from the joints of time and space, they take on features that are associated with the logic of flows (such as money, airports, hotels, information, etc.), which turns them into a direction rather than a reference that anchors them into a specific organizational culture or a specific nation.... Brands as generic spaces do not refer to any particular place (Casey 1997) or context. For Lash (2002), generic spaces can be seen as prototypes of natural, physical spaces that are contextless and identity-less. (Yahkleef 2004: 239)

Issues connected to branding, themed urban design, iconic architecture, the creative class, and the economic success of cities are thus all deeply interconnected to the new urban form, and the forces from which it is produced. In turn, this situation is generating a new consciousness of urban design that requires a quantum jump in current theory as a method of relocating and restructuring the discipline.

Conclusion

In compressing such a huge subject as globalization and urban form into one chapter, there is a clear temptation to somewhat overstate the case, and I am guilty as charged. Nonetheless, hyperbole has its place, and it is clear that there is an increasing gulf opening up between developed and developing countries, and a similar chasm between the class divisions in each. International monopoly capitalism, state neocorporatism, the exhaustion of nature and the allocation of sustainable strategies to the back burner imply one world of wealth and privilege and another of poverty and despair. Hence we see huge migrations of the poor towards sources of employment while the wealthy populate their countries in search of alternative tourist Meccas. All of these developments generate new urban forms from squatter camps and floating communities to billion dollar theme parks and plasticulture urbanism (Easterling 2005). As indicated above, this process is neither accidental nor unintentional. It reflects a new phase in the exploitation of labor power and nature in the interests of world capitalism. On the basis of global change, I have suggested both here and elsewhere in a more carefully argued case, that spatial political economy provides urban design with an opportunity to morph out of the current anarchy that pervades the discipline. In this overall context, and setting my own convictions to one side, we need at least to consider whether mainstream theory is up to the task of theorizing the New Urban Design and the place of the designer within the global economy, and if not, where we should go from here.

References

Adorno, T. (1991a). "The culture industry reconsidered." In Adorno, T. *The Culture Industry*, London: Routledge, 85–92.

—— (1991b). "How to look at television." In T. Adorno, *The Culture Industry*, London: Routledge, 136–153.

Baudrillard, J. (1981). *For a Critique of the Political Economy of the Sign*, St. Louis, MO: Telos Press.

—— (1997). *Fragments*, London: Verso.

Castells, M. (1977). *The Urban Question – A Marxist Approach*, London: Edwin Arnold.

—— (1983). *The City and the Grassroots, A Cross-Cultural Theory of Urban Social Movements*, Berkeley, CA: University of California Press.

—— (1989). *The Informational City*, Oxford: Blackwell.

—— (1996). *The Rise of the Network Society*, Oxford: Blackwell.

Chaplin, S. (2000). "Heterotopia deserta: Las Vegas and other spaces." In Cuthbert, A.R. *Designing Cities*, Oxford: Blackwell, 340–354.

Chmielewska, E. (2005). "Logos or the resonance of branding: A close reading of the iconosphere of Warsaw," *Space and Culture*, 8: 349.

Clarke, P.W. (1989). "The economic currency of architectural aesthetics." In Diani, M. and Ingraham, C. (Eds.), *Restructuring Architectural Theory*, Evanston, IL: Northwestern University Press, 48–59.

Coser, L. (1977). *Masters of Sociological Thought*, New York: Harcourt, Brace, Jovanovitch.

Cuff, D. (2003). "Immanent domain – pervasive computing and the public realm," *Journal of Architectural Education*, 57(1): 43–49.

Cuthbert, A.R. (Ed.) (2003). *Designing Cities*, Oxford: Blackwell.

Cuthbert, A.R. (1995). "The right to the city-surveillance, private interest and the public domain in Hong Kong," *Cities*, 12(5): 293–310.

—— (2006). *The Form of Cities*, Oxford: Blackwell.

—— (2007). "Urban Design: requiem for an era – review and critique of the last fifty years," *Urban Design International*, 12: 177–223.

Dear, M. (1986). "Postmodernism and planning," *Environment and Planning D: Society and Space*, 4: 367–384.

Dickens, P. (1979). "Marxism and architectural theory: a critique," *Environment and Planning B: Society and Space*, 6: 105–116.

—— (1980). "Social science and design theory," *Environment and Planning B*, 17:353–360.

Easterling, K. (2005). *Enduring Innocence: Global Architecture and its Political Masquerades*, Cambridge, MA: MIT Press.

Florida, R. (2003). *The Rise of The Creative Class*, Melbourne: Pluto.

Gosling, D. (1984). "Definitions of urban design," *Architectural Design*, 54(1/2): 16–25.

Harvey, D. (1979). *Social Justice and the City*, London: Edwin Arnold.

—— (1985). *The Urbanisation of Capital*, Oxford: Blackwell.

—— (2006). *The Spaces of Global Capitalism*, London: Verso.

Herman, A. (2002). *The Scottish Enlightenment – The Scots' Invention of the Modern World*, London: Fourth Estate.

Huxley, M. (1997). "'Necessary but by no means sufficient...' spatial political economy, town planning and the possibility of better cities: a commentary on Brian McLoughlin's last paper," *European Planning Studies*, 5(6): 741–751.

Jenkins, B. (2006). "The dialectics of design," *Space and Culture*, 9: 195.

King, A.D. (1984). "The social production of building form: theory and practice," *Environment and Planning D: Society and Space*, 6: 129–446.

King, R. (1996). *Emancipating Space – Geography, Architecture and Urban Design*, New York: Guilford Press.

Knesl, J.A. (1984). "The Powers of Architecture," *Environment and Planning D: Society and Space*, 1: 3–22.

Knox, P.L. (1982). "The social production of the built environment," *Ekistics*, 49 (295): 291–297.

Kumic, I. (2008). "Revealing the competitive city,' spatial political economy and city brands," Doctoral Thesis, Sydney University.

Lefebvre, H. (1970). *La Révolution Urbaine*, Paris: Gallimard.

Lyotard, J.P. (1985). *The Postmodern Condition*, Minneapolis, MN: University of Minnesota Press.

McLoughlin, J.B. (1970). *Urban and Regional Planning: A Systems Approach*, London: Faber and Faber.

—— (1985). "The systems approach to planning: a critique," *Centre of Urban Studies and Urban Planning*, Working Paper No. 2, The University of Hong Kong.

—— (1992). *Shaping Melbourne's Future: Town Planning, the State and Civil Society*, Melbourne: Cambridge University Press.

—— (1994). "Centre or periphery? Town planning and spatial political economy," *Environment and Planning A: Society and Space*, 26: 1111–1122.

Mumford, L. (1961). *The City in History: Its Transformations and Its Prospects*. New York: Harcourt, Brace, and World.

Paris, C. (1983). "Whatever happened to urban sociology? Critical reflections on social theory and the urban question," *Environment and Planning D.*, 1: 217–239.

Pickvance, C. (Ed.) (1976). *Urban Sociology-Critical Essays*, London: Methuen.

Peet, R. and Thrift, N. (Eds.) (1991). *New Models in Geography*, London: Hyman.

Rowley, A. (1994). "Definitions of urban design," *Planning Practice and Research*, 93: 179–197.

Scott, A. J. (2000). *The Cultural Economy of Cities*, London: Sage.

Scott, A.J. and Roweis, S.T. (1977). "Urban planning in theory and practice – a reappraisal," *Environment and Planning A: Society and Space*, 9: 1097–1119.

Sklair, L. (2005). "The transnational capitalist class and contemporary architecture in globalizing cities," *International Journal of Urban and Regional Research*, 29(3): 485–500.

—— (2006). "Ikonic architecture and capitalist globalisation," *City*, 10(1): 21–47.

Tafuri, M. (1979). *Architecture and Utopia – Design and Capitalist Development*, Cambridge, MA: MIT Press.

—— (1987). *The Sphere and the Labyrinth*, Cambridge, MA: Cambridge University Press.

Yahkleef, A. (2004). "Global brands as embodied 'generic spaces': the example of branded chain hotels:" *Space and Culture*, 7: 237.

Zukin, S. (1996). *The Culture of Cities*, Cambridge, MA: Blackwell.

Further reading

Easterling, K. (2005). *Enduring Innocence: Global Architecture and its Political Masquerades*, Cambridge, MA: MIT Press. This book is a *tour de force* in the literature – it illustrates the new forms of place and space that result from capitalism and its excesses, and in the process urban design is redefined.

Harvey, D. (2000). *Spaces of Hope*, Edinburgh: Edinburgh University Press. Rather than resorting to by now arthritic professional descriptors (Architecture; Planning; Urban Design etc.), Harvey uses the human body to explain the relationship between capitalist production on the one hand, and the production of social space on the other.

Harvey, D. (2007). *Spaces of Global Capitalism: Towards a Theory of Uneven Geographical Developmen*, London: Verso. Harvey dissects contemporary capitalism and suggests an encompassing new theory of geographic space, a work that distills his writing on the subject over the last ten years.

Scott, A.J. and Roweis, S.T. (1977). "Urban planning in theory and practice – a reappraisal." *Environment and Planning D: Society and Space*, 9(4)1097–1119. This remains the classic article on urban planning and urban design in the last thirty years, one of the few that stands *outside* the discipline and looks in.

Tafuri, M. (1976). *Architecture and Utopia-Design and Capitalist Development*, Cambridge MA: MIT Press. Tafuri's first book was arguably the very first attempt to give both a historical and contemporary account of the economic and social relations behind the social production of architecture.

7

Critical urbanism

Space, design, revolution

Kanishka Goonewardena

> Stadtluft macht frei!
> [*City air makes one free!*]
>
> German medieval saying (sometimes attributed to Max Weber)

> But if the history of the city is the history of freedom, it is also the history of tyranny. ...The towns may have supplied the historical battleground for the struggle for freedom, but up to now, they have not taken possession of that freedom.
>
> (Guy Debord, *The Society of the Spectacle*, 1967 § 176)

A consensus is haunting urban studies: the consensus of liberalism, postmodernism, and other dominant ideologies of our time that conform to what Francis Fukuyama called the "end of history"; or to what Alain Badiou, from a radically different perspective, named "capitalist-parliamentarianism." We recognize this pervasive concord readily from the series of buzz-words that it has unleashed since the world-historic triumph of liberal democracy in both academic and popular discourses, within which urbanism now exists as theory and practice: civil society, social capital (the topic of Fukuyama's 1996 book *Trust*, which followed *End of History*), multiculturalism, sustainable development, and, above all, democracy and human rights. These seemingly benign terms constitute today an unanswerable monologue to the extent that only a fool would dare to oppose them (fascism or unsustainable development, anyone?). So while ably representing the ideas of the rulers of the world, who are militantly fond of democracy and human rights in particular, this hegemonic opinion also represents itself as the only conceivable opposition to the major evils of the contemporary world including racism, sexism, ecological destruction, totalitarianism, and, since the celebrated conclusion of "actually existing socialism," terrorism, but not, of course, to the very basis of modern urbanism: *capitalism*. Yet not all students of urbanism are at such perpetual peace with the ruling mode of production and so incapable of stepping beyond the conceptual prison-house of its regnant discursive forms. And it is largely this non-conformist minority in urban design, architecture, and planning – feminist, ecological ... *and* Marxist – who will appreciate Guy Debord's lapidary theses in the legendary "space" section of *The Society of the Spectacle* – on the relationship between urbanism and capitalism.

To begin with, a point on which the mainstream of urban studies observes symptomatic silence:

> It is true that all the capitalist economy's technical forces should be understood as effecting separations, but in the case of urbanism we are dealing with the fitting out of the general basis of those forces, with the readying of the ground in preparation for their deployment – in a word, with the technology of *separation itself.* (Debord 1967 § 171)

Spectacle, written exactly a hundred years after the publication of the first volume of *Capital* by Marx, extends the latter's critique of classical capitalism to postwar consumer society, anticipating Fredric Jameson's (1991) influential theorization of "postmodernism" as the "cultural logic of late capitalism." In so doing Debord appropriated especially the concepts of alienation and fetishism elaborated by Marx, closely following Georg Lukács's (1972) extension of them by way of a theory of *reification* in *History and Class Consciousness.* Debord's key-term *separation* fuses the meanings of alienation, fetishism, and reification in order to capture in spatial as well as social sense the contemporary actuality of commodity form. What is of special interest to architects, planners, urban designers, *and* critics of capitalism here is this. "Urbanism," by which Debord refers to urban planning in postwar France and similar contexts, is not merely one among the many "forces of production" in late capitalism; rather, as "the mode of appropriation of the natural and human environment by capitalism" (§ 169), it is the one that creates the condition of possibility for the rest. For if "a society that molds the entire surroundings has necessarily evolved its own techniques for working on the material basis of this set of tasks," then in the case of capitalism "that material basis is the society's actual territory" (§ 169). In this sense, for Debord and his one-time close friend Henri Lefebvre, urbanism forms the foundation of capitalism. Theirs is quite a radical claim, not only for urban studies, but also for Marxism. In the practice-oriented branches of architecture, planning, and urban design that constitute urban studies, the nexus of urbanism and capitalism is typically understood, if at all, the other way around. Capitalism (dis)appears in these pragmatic disciplines as the natural – unexamined – basis of urbanism. Marxism, which takes capitalism as that which must be explained and transcended, has for its part focused on a few matters ranging from class struggle, state, ideology, hegemony, and even the unconscious – with some help from Sigmund Freud and Jacques Lacan – but rarely settled on what Lefebvre (1974) famously called "the production of space." Hence two *critical* questions for both urban studies and Marxism. Can capitalism live without urbanism? Can urbanism live without capitalism?

I should like to think then of "critical urbanism" as that which addresses these questions concerning the articulation and possible – indeed desirable – disarticulation between capitalism and urbanism. Now, what urban design has to do with it depends on what one understands by these two variously conjoined words. For my part, I am not interested in attempts to demarcate a professional practice or an academic discipline called "urban design" as distinct from architecture, planning, or anything else. I am interested instead in what designers laboring at the urban scale have done and continue to do, necessarily in association with many others engaged in the "production of space" – including architects, planners, and above all *activists* (who can also be architects and planners, in the same way that drivers can also be pedestrians). If urban designers can be so understood as belonging to a collective of what Marx following Lefebvre would have called "associated producers of space," then it should be possible to do some justice to their vocation by examining the dialectic of capitalism and urbanism with special references to radical interventions in it – which are at once aesthetic, technological, and political. What they ought above all to accomplish, from a "critical" standpoint, Debord explains with customary precision: "The proletarian revolution is that *critique of human geography* whereby individuals and communities must construct places and events commensurate with the appropriation, no longer just of their labor, but of their total history" (Debord 1967: § 176). Two points about the "revolution" here must not be missed: first, it is inconceivable without a radical transformation of space; second, its objective is not merely the equitable redistribution of social surplus, but the liberation of human subjectivity and the disalienated making of history as such. Accordingly, for Debord:

> The most revolutionary idea concerning city planning derives neither from urbanism, nor from technology, nor from aesthetics. I refer to the decision to reconstruct the entire environment in accordance with the needs of the power of established workers' councils – the needs, in other words, of the anti-State dictatorship of the proletariat, the needs of *dialogue* invested with executive power. The power of workers' councils can be effective only if it transforms the totality of existing conditions, and it cannot assign itself any lesser a task if it aspires to be recognized – and *to recognize itself* – in a world of its own design (§ 179).

These pre-postmodern French words may seem a long way removed from the world of urban design in Anglo-American academia. Yet we may make sense of them with some help from Kevin Lynch, one of the greatest thinkers on urban design in the last century. For in *A Theory of Good City Form* (1981), he identified with characteristic lucidity three types of urban theory, each responding to a distinct question about the city. First, "functional theory" – how did the city get to be the way it is and how does it work? Second, "normative theory" – what is a good city? Third, "planning" or "decision" theory – how do we go from the city we have to the city we love? One does not have to be a philosopher to note the striking correspondence of these three kinds of theory with the three *Critiques* of Immanuel Kant whose Anglo-American readings institutionalized the division of modern thought into three relatively but increasingly autonomous branches: Truth (understanding), Goodness (reason), and Beauty (judgment). A prime virtue of *Good City Form* has been, though not always noted by readers attracted to its phenomenology of urban space, the basic question it poses for architects, planners, and urban designers, which remains fundamental also for radical politics in the conditions of modernity. How can a radical urban praxis mediate between our knowledge of the city we have and our ideas of the city we want? Or, to put it in the terms of Kant's appropriation in German Idealism: how can we now rearticulate the True, the Good, and the Beautiful? No radical critique of capitalist modernity can avoid lamenting one way or the other their separation and independent development, most ominously in the form an "instrumental reason" that burst beyond its initial bounds of the True to kill the Good and the Beautiful, as Theodor Adorno and Max Horkheimer (1947) suggested in *Dialectic of Enlightenment.* Neither can critical urbanism. But it must also do more than mourn the murder of Goodness and Beauty by means-end rationality, and take a cue from the famous "Oldest System Program of German Idealism" co-authored by Hegel, Hölderlin and Schelling, which pleaded in 1796 that "*truth* and *goodness* are brothers *only in beauty*." In retrospect, Guy Debord and the Situationist International (1957–1972) – whose revolutionary-urbanism is usefully recounted in Simon Sadler's *The Situationist City* (1998) – appears to have taken exactly this point to heart in a unique combination of critical, visionary, and activist interventions in the name of "unitary urbanism," which to them meant an urban experience produced neither by capital nor the state, but by radical-democratic politics organized by ordinary people in their everyday life.

The Paris Commune provided for them – and for Lefebvre, especially in his 1965 pamphlet *La proclamation de la commune* – the inspiration as well as the model for urban revolution. Lefebvre, in fact, wondered aloud why the Commune has always been considered to be a socialist revolution but not an urban revolution, making the point that the two had to be *one* or not at all. In "Theses on the Paris Commune" Debord, Attila Kotányi and Raoul Vaneigem (1962) argued that "the apparent successes" of the workers' movement "are its fundamental failures (reformism or ... state bureaucracy), while its failures (the Paris Commune or the Asturias revolt) are its most promising successes so far, for us and for the future" (§ 1). According to both Lefebvre and the Situationists, who understood revolution according to Marx's radical concept of "people making their own history just as they please" rather than as some equitable political-economic redistribution of surplus value, "the biggest festival of the nineteenth century" offers us an invaluable lesson in the way that it brought everyday life into contact with history – by virtue of their *mediation* by the level of social reality called urban. "Underlying the events of that spring of 1871," wrote Debord *et al.*(1962), "one can see the insurgents' feeling that they had become the masters of their own history, not so much on the level of 'governmental' politics as on the level of their everyday life" (§ 2). As such, they found themselves in full accord with Engels' famous words in his 1891 postscript to Marx's study of the Commune in *The Civil War in France* (1988): "Look at the Paris Commune – *that* was the dictatorship of the proletariat." For "the Commune represents" to the Situationists "*the only realization of a revolutionary urbanism* to date – attacking on the spot the petrified signs of the dominant organization of life, understanding social space in political terms, refusing to accept the innocence of any monument" (§ 7). And an urban strategy of revolutionary politics duly followed from such observations, as already anticipated in the *Elementary Program of the Bureau of Unitary Urbanism* by Kotányi and Vaneigem published in *International Situationniste* #6 (August 1961):

> All space is already occupied by the enemy, which has even reshaped its elementary laws, its geometry, to its own purposes. Authentic urbanism will appear when the absence of this occupation is created in certain zones. What we call [the] construction [of situations] starts there (§6).

It is to this end of unitary urbanism – "the theory of the combined use of arts and techniques for the integral construction of a milieu in dynamic relation with experiments in behaviour" ("Definitions" in *SI* #1, June 1958) – that the remarkable repertoire of Situationist tactics – ranging from psychogeography ("the study of the exact laws and specific effects of the action of the geographical environment, consciously organized or not, on the emotions and behaviour of individuals"), dérive (drift: "a mode of experimental behavior linked to the conditions of urban society" and "a technique of transient passage through varied ambiences"), detournement ("the integration of present or past artistic production into a superior construction of a milieu"), and of course "constructed situation" ("a moment of life concretely and deliberately constructed by the collective organization of a unitary ambience and a game of events") – was deployed in the course of what Debord called "conscious alterations in everyday life."

Design in the context of "the production of space" constituted an essential dimension of Situationist urban-revolutionary politics. This should not be surprising, given the avant-gardist origins of the Situationists in 1957, following the fusion of a small group of artists and activists drawn mainly from CoBrA (*Co*penhagen-*Br*ussels-*A*msterdam group of radical artists), the Lettrist International (LI), and the International Movement for an Imaginist Bauhaus (IMIB). "We are bored in the city," wrote Ivan Chtcheglov (1953) (aka Gilles Ivain)," advocating "the need to *play* with architecture, time and space ... " and so interpellating the *subject* of "another city for another life": *Homo Ludens*. The Lettrists were swayed by this proto-Situationist's hope for "rooms more conducive to dreams than any drug, and houses where one cannot help but love." At the urban scale, Chtcheglov's Lettrist vision here recalled Fourier and anticipated Constant's legendary utopian design New Babylon: "The districts of this city could correspond to the whole spectrum of diverse feelings that one encounters by chance in everyday life," including what he called the "Bizarre Quarter – Happy Quarter (especially reserved for habitation) – Noble and Tragic Quarter (for good children) – Historical Quarter (museums, schools) – Useful Quarter (hospital, tool shops) – Sinister Quarter, etc." The IMIB supplied the other major – complimentary – current of design into the Situationist International. Its objective, according to artist Asger Jorn (1957), involved an amalgamation of art and science facilitated by "experimental artists ... get[ting] hold of industrial means and subject[ing] them to their own non-utilitarian ends." The LI journal *Potlatch* #27 (2 November 1956) elaborated on this point, recalling Walter Benjamin's reflections on technology and utopia, with a report on the proceedings of the IMIB congress held in Alba, Italy (2–8 September 1956). Its resolutions declared the "necessity of an integral construction of the environment by a unitary urbanism that must utilize all the arts and modern techniques"; the "inevitable outmodedness of any renovation of art within its traditional limits"; and the "recognition of an essential interdependence between unitary urbanism and a future style of life." Quoted in the same issue was Gil Wolman's statement in Alba:

> Comrades, the parallel crises presently affecting all modes of artistic creation are determined by an overall interrelated movement that cannot be resolved outside a general framework. ... Whatever prestige the bourgeoisie may today be willing to grant to fragmentary or deliberately retrograde artistic tentatives, creation can now be nothing less than a synthesis aiming at an integral construction of an atmosphere, of a style of life. ... A unitary urbanism – the synthesis that we call for, incorporating arts and technology – must be created in accordance with new values of life.

The Situationists were very much in accord here with the aesthetic avant-gardist agenda of transcending the opposition between art and everyday life, and with the political *van*guard of the time that saw the end of the opposition between politics and everyday life in the properly Marxist concept of "the withering away of the state." And they contributed to both the point that the city is central for such a revolution in and of everyday life. In the "Report on the Construction of Situations and on the International Situationist Tendency's Conditions of Organization and Action," Debord (1957) insisted that "integral art, which has been talked about so much, can only be realized at the level of urbanism."

While Debord and the Situationists may best exemplify the spirit of "critical urbanism" in the postwar West from the perspective I have sketched here, especially because of their inimitable union of critique, utopia, and activism leading up to the events of 1968 in France, its theoretical basis owes most to the longstanding work of Lefebvre. In a colorful interview with Kristin Ross (1997) recorded in 1983 and published in *October* 79, he recalls vividly the long nights of heated discussions with Debord and his friends in the late 1950s and early 1960s – on everyday life, the Commune, theories of the "moment" and "situation" and much else – before their "love affair" ended "badly, very badly," for reasons more personal than political. To begin with, the Situationists were drawn to the charismatic Marxist professor and heterodox communist philosopher for his *Critique of Everyday Life*, the first volume of which was published in 1947. For in it he proposed a highly original theorization of the forms of alienation evolving in postwar capitalism, complementing parallel investigations undertaken within the tradition of Western Marxism by Lukács, Antonio Gramsci and the Frankfurt School, especially Adorno and Benjamin. "Marxism," Lefebvre memorably said there, "really is a critical knowledge of everyday life." The Situationists, who collaborated with him on this concept for a few years, could not agree more when a character in a popular cartoon critically altered by them exclaimed: "Yes, Marx's thought is really a critique of everyday lie." Everyday life, a concept influentially theorized by Martin Heidegger (1927), does not have a particularly Marxist ring about it in much contemporary theory dominated by postmodern "cultural studies." Yet it was the focus of intense debate and discussion in the Soviet Union immediately after the October Revolution, especially in the writings of Leon Trotsky (1994) in *Pravda*, subsequently collected in *Problems of Everyday Life*. Here everyday life figured as the ultimate testing ground of revolution, that is, the terrain on which socialism had to be built. Among the first and foremost to operationalize this "reconstruction of the way of life" (*perestroika byta*) in a programmatic way were Soviet architects and urban planners, as superbly surveyed in *Town and Revolution* by Anatole Kopp (1970) – a catalytic influence on Lefebvre's thinking on the significance of the urban. According to the definition in Lefebvre's *Critique*:

> Everyday life, in a sense residual, defined by "what is left over" after all distinct, superior, specialized, structured activities have been singled out for analysis, must be defined as a totality. Considered in their specialization and their technicality, superior activities leave a "technical vacuum" between one another which is filled by everyday life. Everyday life is profoundly related to all activities, and encompasses them with all their differences and their conflicts; it is their meeting place, their bond, their common ground. And it is in everyday life that the sum total of relations which make the human – and every human being – a whole takes its shape and its form. In it are

> expressed and fulfilled those relations which bring into play the totality of the real, albeit in a certain manner which is always partial and incomplete: friendship, comradeship, love, the need to communicate, play, etc. (97)

Everyday life was also the topic on which Lefebvre chose to lecture at a major conference on Marxism in the US in 1988, where he underscored the contested nature of *la vie quotidienne*: that is, the opposition between "the everyday" (*le quotidien*), the embattled yet actually-existing humanity that Marx (1844) spoke of in the *Paris Manuscripts*, on the one hand; and, on the other hand, "everydayness" (*la quotidienneté*), the homogeneous, repetitive and fragmentary forms of being-in-the-world of capitalism. As philosopher Peter Osborne (1995) puts it, in Lefebvre's concept of everyday life "[t]here is the 'good,' but unrealized universality of an historically produced species-being and the 'bad,' abstract but realized universality of its alienated forms (money, the commodity, the state, etc.)" (191). In his 1961 talk in Paris, "Perspectives for Conscious Alterations in Everyday Life," Debord followed Lefebvre to the letter in theorizing "everyday life as the frontier between the dominated and the undominated sectors of life," effectively rendering it "the measure of all things: of the fulfillment or rather the nonfulfillment of human relations; of the use of lived time; of artistic experimentation; of revolutionary politics."

Lefebvre's writings on cities from the mid-1960s onwards related closely to his second and third volumes of *Critique of Everyday Life*. In fact, the former appears in his oeuvre as a way of addressing the political and philosophical questions posed by the latter for revolutionary strategy, as evidenced, for example, by the concluding chapter of *Everyday Life in the Modern World* (1971), which recapitulates some of the key themes of *La droit à la ville* (1968: translated as *Writings on Cities*), *The Urban Revolution* (1970b) and his best known work in the English speaking world, *The Production of Space* (1974). In *The Urban Revolution*, the most representative and original volume of his urban theory, Lefebvre submits an audaciously foresighted thesis: urbanization has superseded industrialization as the leading force, spatial as much as social, shaping late capitalism. On this formulation, the level of the social totality called urban no longer merely expresses social relations; it produced and reproduces them as well. Urban space now becomes a "productive force, like science," not least by its burgeoning role as an essential condition of possibility of capitalist accumulation (15). "The city, or what remains of it or what it will become, is better suited than it has ever been for the accumulation of capital," he wrote, emphasizing the urbanization of "accumulation, realization and distribution of surplus value" (35). His lucidity on the "role played by urbanism and more generally real estate (speculation, construction) in neocapitalist society" (159) anticipated the enormous contributions of David Harvey (1973, 1982) to urban theory: "As the principal circuit [of capital] – current industrial production and the movable property that results – begins to slow down, capital shifts to the second sector, real estate." With proleptic instincts on certain post-1973 realities, he also observes how "[i]t can happen that real-estate speculation becomes the principal source for the formation of capital, that is, the realization of surplus value" (160). Yet it would be an error to understand the *intent* and *content* of Lefebvre's contribution to both urban studies and Marxism within the boundaries of political economy, as did Harvey in the 1970s, or as a structuralist sociology, as did Manuel Castells in *The Urban Question* (1977), both of whom found him quite intriguing but too vague to their own specific enterprises – which in their turn came under some understandable postmodern criticism for neglecting questions of difference. Castells responded by abandoning Marxism; Harvey (1989) by reinvigorating his Marxism with due respect to some radical tendencies of his critics, drawing more generously on the richness of Lefebvre's wide-ranging work, which includes a meditation on difference more sophisticated than that of postmodern critics of *The Condition of Postmodernity*. In *Le manifeste différentialiste* (1970a) and *The Production of Space* (1974), Lefebvre made a telling distinction between *induced* or *minimal* difference – concrete abstractions of capital and state – and *produced* or *maximal* difference – which would be worthy of a society freed from those abstractions. Along with Adorno and Benjamin, he revived in these writings a reading of Marx as a philosopher of difference – and a critic of identity – that has been sorely missed in political-economic as much as postmodern renditions of him in urban studies.

Lefebvre produced a theory of the production of space; not a political economy of space. The latter is a perspective that he exceeded quite explicitly, by aspiring towards what in *The Urban Revolution* is repeatedly called *totality*: the entire content of the social active engaged in the making of history. It makes more sense indeed to see his perspective as a *critique* of political economy, along with Debord who said in *Spectacle*: "The city is the *locus of history*, because it embodies at once a concentration of social power, which is what makes the historical enterprise possible, and a consciousness of the past" (§ 176). Soon after these words were published, Lefebvre also specified the historical role of the city, with reference to a new concept of totality developed in *The Urban Revolution*, by underlining its supreme formal feature: *centrality*.

> The essential aspect of the urban phenomenon is its centrality, but a *centrality* that is understood in conjunction with the dialectical movement that creates or destroys it. ... However, centrality is not indifferent to what it brings together, for it requires a content. And yet, the exact nature of this content is unimportant. Piles of objects and products in warehouses, mounds of fruit in the marketplace, crowds, pedestrians, goods of various kinds, juxtaposed, superimposed, accumulated – this is what makes the urban urban. ... What does the city create? Nothing. It centralizes creation. And yet it creates everything. Nothing exists without exchange, without union, without proximity, that is, without relationships. The city creates a situation, the urban situation, where different things occur one after another and do not exist separately but according to their differences. The urban, which is indifferent to each difference it contains, often seems to be as indifferent as nature, but with a cruelty all its own. However, the urban is not indifferent to all differences, precisely because it unites them. In this sense, the city constructs, identifies, and delivers the essence of social relationships: the reciprocal existence and manifestation of differences arising from or resulting in conflicts. Isn't this the justification and meaning of this rational delirium known as the city, the urban? (115–18)

What is the theory of totality that Lefebvre proposed to theorize the urban? In the answer to this question lies not only his foremost contribution to Marxism, but also the indispensible theoretical foundation of "critical urbanism." This is explained most concisely in the groundbreaking "Levels and Dimensions" chapter of *The Urban Revolution*, which also provides a fine introduction to the totality of

Lefebvre's work. In the more orthodox interpretations of Marxism, the concept of totality – "mode of production" or "social formation" – derives more or less directly from the famous passage on "base and superstructure" in Marx's 1857 Preface to *A Contribution to the Critique of Political Economy*. Lefebvre did not reject this dialectical view, even if the urban as such did not figure prominently in it. But he did propose an alternative theory of totality and an attendant view of mediation. With due attention to urban space, this theory rests on a novel conception of *levels* of social reality that is integral to Lefebvre's explication of "the urban revolution." What are these "levels" and what is their architecture? Lefebvre sees the social totality as a dialectical articulation of *three* levels. At the "top," he identifies the *global* level (G) – "the far order of society" – by which he means universal and abstract logics that dominate the other levels "below": namely, *neo-dirigisme* and *neo-liberalism*, that is to say, the logics of state and market. This represents a prime object of Lefebvre's research in the 1970s, especially in his four tomes of *De l'État*. At the "bottom" lies the level of everyday life (EL) – "the near order of society" – his most enduring interest that lasted nearly sixty years from the beginning of his intellectual career in the 1930s until his death in 1990. Lefebvre considered everyday life as a reservoir of revolutionary energy – of human subjectivity not fully colonized by the global level (G), and so capable of resisting and transcending its abstract logics. Between these two levels, he located the crucial *urban* level – a *mediating* level between the global and everyday life (U/M). It is "projected" by the global level and, while retaining the relative autonomy of its own "forms-functions-structures ... *in* the city and *of* the city," introjects the contested dynamics of the vital level of everyday life beneath it. On the political import of this conception of totality – wherein the urban plays a pivotal role – Lefebvre is explicit: "during the critical phase" of the urban revolution, "these levels and dimensions tend to blur" as "[t]he city explodes" and "[t]he urban arrives" (88–90). Thus the "specifically urban level," for example, does not coincide simply with the physical space of the *city*, which in his conception clearly makes room for all three levels to operate within it. In fact, it is in this dialectically articulated sense that "the urban phenomenon," as an overdetermination of the three levels, becomes the most intensely mediated site of revolutionary struggle – at once social, spatial and historical. An "urban strategy" therefore assumes for Lefebvre a central role in the struggle for socialism, one that would be waged against the dominant logics of the global level (G), primarily if not exclusively on the intermediary urban terrain (U/M), drawing nourishment from the critical and utopian energies released from the contradictions of everyday life (EL). Whereas the hegemonic forces in Lefebvre's totality run from the global level through the urban level to the level of everyday life, counter-hegemonic struggle seeks to reverse their direction. Indeed, the urban-social revolution for Lefebvre as much as for Debord predicates itself precisely upon the prospect of everyday life acting on the urban level, and the urban level acting on the global level: $EL \rightarrow U/M \rightarrow G$. A revolution becomes possible for them only when the level of the everyday and the level of history can interact by way of the urban – as was witnessed in the Paris Commune.

The radical implication for "critical urbanism" as well as Marxism spelled out by the *Critique of Everyday Life* and *The Urban Revolution* and *De l'État* – and by the less voluminous writings by Debord and the Situationists – should be clear: there can be no social(ist) revolution without an urban revolution, no urban revolution without a social(ist) revolution, and neither without a revolution in everyday life. Now, it would be unwise to expect such an insight to be of much interest to those planners, architects, or urban designers who have made their professional or academic peace with "capitalist-parliamentarianism" at the "end of history." Fortunately for cities and citizens, the prospects of urban-revolutionary change rely not so much on such experts, but on radical-popular political movements exemplified by the Paris Commune or Paris 1968. It is to the activists in them that Debord and especially Lefebvre still speak, not as models to follow, but as resources for critique – as evidenced by the recent formation of the Right to the City Alliance across several US cities and, to speak from my own experience in Toronto, activist groups such as Planning Action, Toronto School of Criticism and Innovation (TSCI), and Creative Class Struggle, all of which include avid students of what I have defined as critical urbanism. This is the backdrop against which Lefebvre's novel concept of *the* right to city must be understood – not as another addition to the self-contradictory liberal-democratic list of "human rights," but rather the right to a radically different *world*. Lefebvre's insights on the urban therefore offer an invaluable *starting* point for critical urban theory to focus its theoretical horizons and sharpen its political vision, as shown by Kristin Ross's (1996) exemplary engagement with the problematics of gender and colonization in her penetrating study of French postwar modernization: *Fast Cars, Clean Bodies*. The extension of urban theory in such directions assumes paramount import in the current imperial conjuncture, as the far-flung order of our global social totality appears to be at a moment of geo-political-economic reformatting if not crisis, to enquire into the possible roles assumed by cities and their subjects in a new world system. Leading radical political thinkers of the world now emphasize the need for such efforts. Alain Badiou (2008), for example, highlighted the urgency of the "fundamental problem" posed for radical politics today by the global urban condition. Badiou's one-time student Slavoj Zizek (2006) recently asked: "what if the new proletarian position is that of the inhabitants of the slums of the new megalopolises?" (268). His answer takes off from Mike Davis (2006): "while we should of course resist the easy temptation to elevate and idealize the slum-dwellers into a new revolutionary class, we should nonetheless, in Badiou's terms, perceive slums as one of the few authentic 'evental sites' in today's society." Likewise, Tony Negri (2003) has underlined the centrality of urban struggles to revolutionary politics today, arguing in an essay floating on the Internet (http://www.generation-online.org/t/metropolis.htm) called "The Multitude and the Metropolis" that "the metropolis is to the multitude what the factory used to be to the working class." More proper names may be added to this list of cutting-edge thinkers who have turned on the paramount import of the metropolis for radical praxis, vindicating in no uncertain terms the fundamental thesis of Lefebvre's *The Urban Revolution*. It remains for critical urban theory – and urban designers mindful of their radical heritage – to return the compliment. For the future of "critical urbanism" now rests on delivering not only the aesthetics but also the politics capable of doing justice to the emancipatory possibilities alive in our Age of Empire and Planet of Slums.

References

Note: All of the quotations from Guy Debord and the Situations in this text can be found in the "situationist international online" website: http://www.cddc.vt.edu/sionline (accessed 15 March 2009). Adequate references to other key sources are provided within the text.

Adorno, T. and Horkheimer, M. (1944). *Dialectic of Enlightenment*, trans. by J. Cumming (1990), New York: Continuum.

Badiou, A. (2008). "Interview," *Critical Inquiry* 34 (summer).

Castells, M. (1977). *The Urban Question*, Cambridge, MA: MIT Press.

Chtcheglov, I. (1953). *Formulary for a New Urbanism*, trans. K. Knabb (2006): <http://www.bopsecrets.org/SI/chtcheglov.htm> (accessed 15 March 2009).

Davis, M. (2006). *Planet of Slums*, New York: Verso.

Debord, G. (1957). "Report on the Construction of Situations and on the International Situationist Tendency's Conditions of Organization and Action," in *Situationist International Anthology*, ed. K. Knabb, Berkeley, CA: Bureau of Public Secrets.

—— (1961). "Perspectives for Conscious Alterations in Everyday Life," in *Situationist International Anthology*, ed. K. Knabb, Berkeley, CA: Bureau of Public Secrets.

—— (1967/1994). *The Society of the Spectacle*, New York: Zone Books.

Debord, G., Kotányi, A. and Vaneigem, R. (1962). "Theses on the Paris Commune," pamphlet. *Internationale Situationniste*, 12, September 1969.

Engels, F. (1891). "Postscript," in K. Marx *The Civil War in France*. New York: International Publishers.

Harvey, D. (1973). *Social Justice and the City*, Baltimore, MD: Johns Hopkins University Press.

—— (1982). *The Limits to Capital*, Chicago: University of Chicago Press.

—— (1989). *The Condition of Postmodernity: An Enquiry into the Origins of Cultural Change*. Oxford: Blackwell Publishing Ltd.

Hegel, G.W.F., Hölderlin, F. and Schelling, F.W.J (1796). "The oldest system-program of German idealism" in F. Hölderlin (1988) selections in English *Essays and Letters on Theory*, Albany, NY: State University of New York Press.

Heidegger, M. (1927). *Being and Time*, trans. J. Macquarrie and E. Robinson (1962), London: SCM Press.

Jameson, F. (1991). *Postmodernism, or The Cultural Logic of Late Capitalism*, Durham, NC: Duke University Press.

Jorn, A. (1957). "Notes on the Formation of an Imaginist Bauhaus," trans. K. Knabb (2006), *Situationist International Anthology*. Berkeley, CA: Bureau of Public Secrets.

Kopp, A. (1970). *Town and Revolution: Soviet Architecture and City Planning, 1917–1935*, translated by Thomas E. Burton, New York: George Braziller.

Kotányi, A. and Vaneigem, R. (1961). *Basic Program of the Bureau of Unitary Urbanism*, trans. K. Knabb. <http://www.bopsecrets.org/SI/6.unitaryurb.htm> (accessed 15 March 2009).

Lefebvre, H. (1947). *Critique of Everyday Life*, trans. J. Moore (1992), London: Verso.

—— (1965). *La Proclamation de la Commune*, Paris: Gallimard.

—— (1970a). *Le Manifeste Différentialiste*, Paris: Gallimard.

—— (1970b). *The Urban Revolution*, trans. Robert Bononno, foreword Neil Smith (2003), Minneapolis: University of Minnesota Press.

—— (1971). *Everyday Life in the Modern World*, trans. S. Rabinovitch (1988), New Brunswick, NJ: Transaction Books.

—— (1974). *The Production of Space*, trans. D.N. Smith (1992), Cambridge, MA: Wiley-Blackwell.

Lukács, G. (1972). *History and Class Consciousness*, trans. R. Livingstone, Cambridge, MA: MIT Press.

Lynch, K. (1981). *A Theory of Good City Form*, Cambridge, MA: MIT Press.

Marx, K. (1844). *Economic and Philosophical Manuscripts of 1844*, translated by Martin Milligan, Moscow: Progress Publishers.

—— (1857). "Preface," in *A Contribution to the Critique of Political Economy*, trans. N. I. Stone (1904), Chicago: Charles H. Kerr & Co.

Negri, A. (2003). "The Multitude and the Metropolis," Online. Available HTTP: <http://www.generation-online.org/t/metropolis.htm> (accessed on 15 March, 2009).

Osborne, P. (1995). *The Politics of Time*, New York: Verso.

Ross, K. (1996). *Fast Cars, Clean Bodies*, Cambridge, MA: MIT Press.

—— (1997). Interview with Lefebvre recorded in 1983, *October* 79(winter).

Sadler, S. (1998). *The Situationist City*, Cambridge, MA: MIT Press.

Trotsky, L. (1994). *Problems of Everyday Life: Creating the Foundations for a New Society in Revolutionary Russia*, New York: Pathfinder.

Zizek, S. (2006). *The Parallax View*. Cambridge: MIT Press.

Further reading

Benjamin, W. (1999). *The Arcades Project*, trans. Howard Eiland and Kevin McLaughlin, Cambridge, MA: Belknap/Harvard.

Buck-Morss, S. (1990). *The Dialectics of Seeing: Walter Benjamin and the Arcades Project*, Cambridge, MA: MIT Press.

Goonewardena, K., Kipfer. S., Milgrom, R. and Schmid, C. eds. (2008). *Space, Difference, Everyday Life: Reading Henri Lefebvre*, New York: Routledge.

Part 3

Influences

Introduction

Part 3 presents a variety of disciplinary influences that continue to inform urban designers. Traditionally, urban design has been considered to be at the intersections of the "city building professions" of architecture, landscape architecture, urban planning, and civil engineering (Lang 2005). The traditional linkages of these disciplines to urban design are well established and significant since a central focus of urban design is to configure, articulate, and link spatial elements of the city's urban form. This physical and applied nature of urban design, although conventional and well explored, does not cover its expanding scale of applications or scope of inquiry and interests. The focus of urban design is not only physical or aesthetic but also social, economic, cultural, and political, and these aspects of urban design are closely interrelated.

Urban designers address the needs of multiple and at times anonymous publics, which represent their "substantive clients" (Mera 1967). They need to know how different needs, interests, and values are expressed in the urban form; how people interact in various settings, and how design can better support, enhance or even inhibit (as in the case of crime) social activities. Therefore, knowledge and tools from the social sciences – geography, sociology, anthropology, environmental psychology, and feminism – help inform urban designers in their intellectual and professional pursuits.

Additionally, urban designers should have a keen understanding of how space is produced, occupied, restructured, manipulated, controlled, and regulated, if they wish to influence the decision making realm that affects development (George 1997). The fields of political theory and law clarify the roles and relative power of various stakeholders in the development and decision making processes as well as their tools.

The ultimate goal of urban design is to increase the quality of life in cities through what Kevin Lynch had once called "the imaginative creation of possible form" (Lynch in Banerjee and Southworth 1990: 611), but a prerequisite for a good quality of life is of course health. The recognition of the important role that the built environment plays in affecting health outcomes is at the core of the recent interest in the connection between the disciplines of public health and urban design. Finally, the role that design can play as a communicative tool which helps create a consensus vision of "the good city" is only marginally explored. Here design may have much to learn from the filmic techniques of the cinematic arts as designers seek to better comprehend and often convey to diverse audiences the effects of alternative design scenarios.

The essays of this section detail the influences and contributions of nine different fields on urban design. As research in urban design has flourished in the last decades, urban designers have found it imperative to borrow knowledge and methodology from the social sciences to better understand the objects of their inquiry. As Larry Ford explains in his chapter, the discipline of geography has contributed knowledge in multiple ways and scales: from theoretical models explaining the internal structure of cities and comparative analyses of urbanization, to analyses of cultural landscapes and micro-environments, to the experiences and the meanings they may convey to their users.

As William Michelson suggests in his chapter, urban designers need to pose questions to bridge the social and physical realms, gather data, and apply multiple methods to study the built environment. Sociological research comes in handy, and methodologies such as observation and surveys are now widely used by urban design scholars, while simulations are also part of some urban designers' repertoire (as the chapters by Bosselmann and Ben-Joseph in Part 4 indicate).

To design for multiple publics in increasingly multicultural urban contexts, an urban designer has to acquire an understanding of the needs and values of different cultural and social groups often quite different from one's own. A major contribution of the field of anthropology is the ethnographic research, which according to Denise Lawrence, is to "sensitize design professionals" to different cultural mores and practices. Furthermore anthropological studies often reveal institutionalized negative and exclusionary practices inherent in certain urban design paradigms.

While anthropology has sensitized urban designers to the various social complexities and values of different social groups in general, recent feminist studies have focused on women and their distinct socio-spatial needs. In her chapter, Kristen Day explains how feminist studies have contributed to better design by emphasizing a women's perspective on how the city structure, infrastructure, and amenities can be better configured to be sensitive to women's concerns and needs in the built environment.

To reach urban design decisions about the shape and form of different environmental settings, designers should know about the interaction between environment and behavior – how variable configurations of the built environment may influence preferences and behaviors. This is the focus and realm of environmental psychology. Jack Nasar details how this field offers a knowledge base for urban design by obtaining such insights as how the public perceives the environment or, how design may improve structure and legibility of the environment, and how the evaluative image of a place can be discovered.

Legal institutions and instruments shape the built environment and guide the production of the built environment through an invisible but highly effective web of rules and regulations embodied in municipal ordinances and statutes, as well as in judicial opinions. In his chapter, Jerold Kayden discusses the sensitive balance between public and private interests in urban design outcomes and explains the scope of laws that shape urban design practice.

The public realm and the public space are a key consideration in urban design, often defining its primary scope. But visions of public space and its accompanying important qualities differ among different stakeholders in the city. Margaret Kohn explains the contributions of political theory in clarifying the concepts of public and private, and discussing issues of constitutional rights and power relations. Discussing how they affect the construction of public space, she offers pertinent normative frameworks to evaluate planning and design policies.

With the growing awareness that the structure and organization of the physical environment may have negative and serious health consequences, the built environment has become a major focus of intervention for the public health professionals. Marlon Boarnet and Lois Takahashi identify the areas where knowledge and research from the field of public health have begun to define the scope of urban design interventions to create healthier neighborhoods. They argue, however, that for the link between public health and urban design to be more meaningful the functional and the aesthetic/aspirational aspects of design should be bridged.

The contribution of the cinematic arts to urban design is not well explored, or fully understood. Yet increasingly contemporary spaces of shopping and entertainment are beginning to emulate multi-media experiences and beginning to look like backlots of Hollywood studios. Furthermore, historically cinema has used urban spaces as outdoor settings, and by choice of specific sets and their dramatic effects, may have influenced the long-term values and preferences of the increasingly media-savvy public. Raphael Pizzaro's chapter in this section offers some intriguing perspectives of this relationship. He argues that such contributions come at three levels. First, films about cities become part of the repertoire of experiences and images which influence designers' ideas about space and form. Second, they also act as interpretive media that help them understand cities. Finally, and at the same time, cinematic techniques may offer new tools in the practice of design and in design pedagogy.

References

George, V.R. (1997). "A procedural explanation for contemporary urban design," *Journal of Urban Design* 2(2): 143–61.

Lang, J. (2005). *Urban design: A typology of procedures and products*. Oxford: Elsevier.

Lynch, K. "Urban design," in Banerjee, T. and Southworth, M. (1990). (Eds.) *City sense and city design: Writings and projects of Kevin Lynch* Cambridge, MA: MIT Press, 511–34.

Mera, K. (1967). "Consumer sovereignty in urban design." *Town Planning Review* 37(4): 305–12.

8

Urban design and the traditions of geography

Larry R. Ford

This chapter explores the interrelationship between the fields of geography and urban design. Although the discipline of geography does not consider urban design as its specialty, the subject actually plays an important role in a variety of its subfields. While geographers do not normally specify that they are writing purposefully about urban design issues and policies, much of their work is relevant to urban designers in improving their understanding of urban form and landscape. In this chapter, I discuss some themes in urban geography, cultural geography, and philosophy of geography that relate to urban design. I argue that studies on *comparative urbanization* over time, *models of city structure, urban cultural landscapes,* and *the meaning and representation of space and place* represent geography's contributions to the field of urban design.

There are additional geographic traditions that will emerge in the chapter as well as the ones listed above, such as the representation of cities in cartography, art, and the media, but the real focus is on the ways in which cities have evolved in a variety of world cultural contexts over time. The emphasis in geography varies from many other disciplines such as, say architectural history, in that the focus is more on gradual, indigenous accretion rather than purposeful artistic design, and so micro-landscapes involving houses, gardens, and urban props usually take precedence over grand architecture or planning.

In writing this chapter I have partly relied on a review of recent articles published in *Geographical Review*, a well-known geographical scholarly journal, which is also accessible to a variety of educated readers.

Comparative approaches to urban form and landscape

One of the major ways that geography has contributed to urban design is through its monitoring of changes in urban form in a wide variety of cities all across the globe. While mainstream urban design often concentrates on important places and contexts, such as nineteenth-century Paris or seventeenth-century Amsterdam, geographers more often examine aspects of urban design in less famous locales and less celebrated time periods. For example, recent articles in the *Geographical Review* have focused upon such topics as "Continuity and Change in African Capitals" (Christopher 1985), "Postmodern Phoenix" (Schmandt 1995), and "Revisiting Rio de Janeiro and São Paulo" (Godfrey 1999). Other articles have dealt with squatter settlements in Kuala Lumpur (Aiken 1981), Mexican murals (Arreola 1984), historic preservation in Spain (Ford 1985), working class suburbs in Toronto (Harris 1991), and current uses of Italian piazzas (Fusch 1994). Many of these topics seem to lie somewhere between urban design as studied in art, architecture, and planning (aesthetic emphasis) and urban design as approached in the social sciences (emphasis on socio-economic change related to urban form).

Articles in geography differ from most of those in history and the social sciences in offering an extremely visual and often cartographic presentation, with contemporary historical photos, sketches, and maps depicting all kinds of phenomena. Geography often tends to focus on micro elements in the landscape such as houses, fences, murals, trees, and ethnic and religious symbols. Roughly one-quarter of the articles in the *Geographical Review* examine cities in the North American context while the rest are international in scope. The towns and cities studied relate more to the authors' individual research interests than to places that have played a formative role in the evolution of urban design practice. Thus, research on Zanzibar or Tijuana appears nearly as often as research on Paris or Vienna. The slums of Nassau are given equal time with the boulevards of Paris and the residential squares of London. The emphasis is more on the bottom-up aspects of urban design rather than the top-down influences. The contribution of these geographic studies is that they explore the hidden corners of cities and places of the world and examine placemaking by those who are rarely given a voice. For example, vernacular housing and garden traditions may play as important a role in local urban design as grand boulevards or monumental plazas.

This is not to say that geographers never look at monumental urban landscapes or the roles of the rich and powerful in shaping cities, as indicated by articles on the origins of European tree-lined boulevards (Lawrence 1988) and current efforts to redesign Moscow (Argenbright 1999). But even in these studies the emphases tend to be different. Sometimes research crosses into the area of environmental sustainability such as in studies of the impacts of grass lawns and the emerging desert landscapes in Tucson and other cities of the American Southwest (McPherson and Haip 1989). So far, however, few geographers have developed a focused interest in green architecture but some of this emerges in studies of traditional urban form and landscapes and the ways people have gotten by with less energy using traditional design norms.

Models of city structure

Since geography covers an exceptionally wide variety of cities and urban contexts without the connection to specific architectural eras or designers that often exists in architectural history, there is the danger that the studies could seem scattered and devoid of an adequate frame of reference. The schematic models of city structure developed by geographers solve this problem. Although in the social sciences, models are often thought to be mathematical in nature, this is not usually the case for those models that are useful to urban design. Many of these models attempt to explain the internal structure and organization of cities. Derived from the Chicago School Models of the 1920s and 1930s, they use combinations of concentric rings, sectors, and rectangles in varying degrees of complexity. They are essentially cartographic with geometric shapes reinforced with patterns, arrows, and a minimum of verbiage. These models provide a setting or context for research on specific types of cities in particular regions. Ideally, they are complemented by cartographic representations, air photos (satellite imagery, these days), and pictures of specific places.

Geographers have developed models of the structure of the Latin American city, Southeast Asian city, Indonesian city, Northwest European city, Mediterranean city, Eastern European city, Middle Eastern city, Sub-Saharan African city, South Asian city, and Chinese city (both traditional and People's Republic). In addition, there are hybrid models that add colonial additions to older spatial contexts. More place-specific models such as those for Spain, Italy, or Argentina are also available. A handy introduction to many of these models appears in the edited volume *Cities of the World*, frequently used in courses on comparative urbanization (Brunn *et al.* 2008).

The purpose of a schematic model is not so much to display a predictable geometric form as to provide an understanding of the processes of city formation. The neat concentric zones and sectors used in models rarely, if ever, show up in real cities but they can help in the conceptualization of dynamic processes.

Consider, for example, one of the more widely used models of Latin American city structure.[1] The initial model was built on Gideon Sjoberg's conception of a classic preindustrial city albeit with considerable modifications. The model postulated a strong "downtown" or "centro" resulting from both a traditional attraction to a highly symbolic core and the continued reliance on core-centered public transportation. In addition to a strong core, this model postulated a spine of important economic and cultural activities leading outward from the core. The primary reason for this is the relative inability of Latin American cities to expand an adequate infrastructure in every direction. Therefore, the urban elite wanting paved streets, reliable electricity, water, and sewerage, good public transit, access to shops and entertainment and a government regulated land market (without the threat of squatters), locate outward along the spine rather than scattering to a wider variety of suburban locations. In contrast, in the US and other more "developed" countries, households moving to the periphery expect the infrastructure to appear quickly in every part of an expanding metropolis.

A sector of wealthy residential areas is found on either side of this spine since stereotypically, Latin Americans value access and "movimiento" or action over and above the rustic isolation and privacy sometimes sought by Anglo-Americans. Thus the model includes one dominant spine and sector of elite activity emanating from the core. Within the elite sector, the housing market operates more or less like that of the US with professionally built homes and available (though high-interest) mortgages. There is also a small filter-down mechanism such that older houses may be sold to lower status residents as the wealthy move further out in the sector. This stands in stark contrast to the market in the rest of the urban area.

The patterns described above have been both the result of and reinforced by more traditional interventions of urban design and planning. The Laws of the Indies in 1573 mandated an orderly grid focused on a monumental "plaza mayor" in the cities of Spanish America. This helped to define the symbolic as well as functional importance of the city center. Later, especially during the mid-nineteenth century, the French tradition of the grand boulevard was introduced. Not only in Mexico but also in cities such as Buenos Aires, it was a matter of voluntary emulation of Paris. These boulevards often became the dominant spines, like the Paseo de la Reforma in Mexico City.

Outside of the spine-sector, Latin American cities can be characterized as having a series of "reverse" concentric zones, as social status decreases outward from the center. This is in direct contrast to the zones postulated for the US by the

University of Chicago sociologist Ernest Burgess in 1925. In the Latin American rings, most housing is self-built and the infrastructure has been developed very slowly. The inner ring is the "zone of maturity" since it has been around long enough for streets to be paved and electricity to be widely available. The houses have been upgraded over time and are reasonably substantial. There are also neighborhood stores, schools, and a variety of services. The landscape is poor but "finished" in appearance. The second ring is a "zone of accretion" where everything looks as though it is under construction with piles of bricks, unfinished building skeletons, and rough paving. When people save a little money, they gradually upgrade their dwellings.

Finally, the outer ring is a "zone of peripheral squatter settlements," where houses are often little more than sheds put together with makeshift materials and the infrastructure is almost entirely absent. In time, these settlements are upgraded and become part of the zone of accretion just as that zone joins the ring of maturity. Over time, neighborhoods filter up rather than down. Two sectors of squatter settlements are also depicted in the model as extending inward to the center of the city. These represent the difficulty of overcoming steep hills, flood plains and other challenging physical obstacles to the process of gradual upgrading compared to the wealthier countries of the global north.

Since all cities change over time, by 1996 a new and improved model with "paste on" embellishments to the early model was developed (Ford 1996). The additions included middle class tract houses, an industrial zone, areas of central city gentrification related to a new awareness of historic preservation, and a peripheral highway with a shopping mall. Such elements were appearing in many Latin American cities but their locations were typically different from those in North American cities. In fact, the creation of a model of Latin American city structure responded to the fact that while North American and international urban forms such as skyscrapers, industrial parks, shopping malls, highways, and historic districts were becoming more common in Latin America, this did not mean that cities were evolving toward becoming like those in the North. Urban forms are always being adopted, adapted and hybridized.

Geography's schematic city models are today as relevant as ever. For example, there is a need for new models for the booming cities of East Asia and those in Eastern Europe where formerly socialist urban forms are being transformed by capitalist economies. The new developments in Dubai and elsewhere in the Persian Gulf are rapidly transforming the classic Islamic city into something never seen before, except perhaps in Disney World. The models emerging for these places help to contextualize disparate research and to place new or largely forgotten places into an ongoing urban design discussion. In all these places, culture, in its broadest sense (house form, domestic life, religion, landscape tastes, economic system etc.) must be considered in the attempts to make sense of evolving urban forms (Ford 1993).

Ethnic complexity has become increasingly important in urban design, and designers should frequently respond to the diverse needs of a heterogeneous public. While ethnic "quarters" have been around as long as cities themselves, globalization has amplified their presence and impacts on urban form (Figure 8.1). In Los Angeles, for example, the Latino areas in the southern and eastern parts of the metropolis occupy a space that is larger than the total area of most cities around the world. The colors, yards, fences, business signs, and land uses in these areas do not always conform to Anglo patterns of urban design. In addition, especially in North America, many ethnic areas are celebrated and

Figure 8.1 Bangladeshi neighborhood in East London. Source: Larry Ford.

enhanced for tourism and other economic purposes. The landscapes of Chinatowns, Little Italys and Little Saigons are often exaggerated and "zoned in" to attract attention and investment. Some of this investment may be international as in the case of money from Hong Kong flowing into Vancouver. These factors need to be included in generalizations about and models of urban form.

The urban cultural landscape: architecture and city structure

Geographers have long been interested in particular kinds of architecture and other types of landscape features which help them interpret a city's cultural landscape. During the early decades of the twentieth century, the focus was most often on housing types, barn types, traditional building materials, and other aspects of a mostly rural regional identity. By the 1960s, the focus had changed and studies of urban architectural elements became more popular. My own PhD dissertation in 1970, for example, looked at the role of the skyscraper in urban design in the US and Argentina. The skyscraper affords an excellent example of a building type that is not only highly monumental and symbolic but also one that can play an important role in guiding and shaping the location of urban activity (Figure 8.2). In Cleveland, Ohio, for example, the 742-foot-tall, 2.2 million square-foot Terminal Complex completely reorganized the form and structure of downtown during the late 1920s. This project provided a new visual and functional center for the city while contributing to the underutilization of structures only a few blocks away, which led to the formation of a skid row district. Even at the heart of New York City one major project can reorganize urban form. Rockefeller Center, for example, has focused and anchored Midtown Manhattan for the

Figure 8.2 Skyline in Jakarta. Source: Larry Ford.

past 70 years. The project's size attracts attention, and its famous public spaces, including a winter skating rink, help it to serve as a kind of "plaza mayor" for Midtown Manhattan.

In addition to its role in shaping city structure, the skyscraper also fits nicely into the traditional interest geographers have displayed in understanding landscape tastes, especially as they relate to the topic of cultural diffusion. Skyscraper construction is easily traceable around the world. The worldwide distribution of skyscrapers can tell us a lot about changing levels of technology and evolving cultural values. There are many cities, for example, which have long had the technical expertise to build towers but simply chose not to. European cities have been slow to accept the tall building but are only recently doing so with great enthusiasm from La Defense in Paris and Canary Wharf in London to the new landmark "Twisting Torso" tower in Malmo, Sweden.

For a variety of social, political, and economic reasons, East Asia is now the skyscraper capital of the world and the designers of these landscapes are usually internationally recognized "passport architects." Asian skyscrapers can tell a great deal about fluctuations in the global economy. Many American architectural teams, for example, move back and forth between the US and East Asia with economic booms and busts. When the California economy was weak during the early 1990s, many firms did much of their work in Asia. With the 1997 Asian crisis, most returned to a booming West Coast. In the process, a great deal of cultural hybridization takes place as projects around the globe learn and borrow from the experiences of similar projects in other socio-political contexts (Ford 1998).

Studies of the processes of development have become increasingly intertwined with the resulting landscapes. In China, for example, the combination of a command economy coupled with unbridled capitalism has meant that huge projects can be built quickly with little opposition or "nimbyism." The Pudong area across the river from old Shanghai, for example, was designed, planned, engineered, and fleshed out with some of the tallest skyscrapers in the world in little more than a decade. Experts from all over the world, including those who had worked on the La Defense towers of Paris, were involved in the effort. At the same time, the governmental agencies of China had to learn new ways of zoning, lending, and monitoring a new level of urban development.

Geographers have also examined less monumental vernacular aspects of the urban environment such as housing. Studies of housing styles, eras, and regional variations have been popular topics for decades as geographers have traced the evolution and locations of bungalows, shotgun houses, dingbats, camelbacks, and a wide variety of other residential structures as parts of the urban scene (Figure 8.3). Alley housing, townhouses, warehouse conversions, and high-rise condos have also attracted attention. Mapping urban housing types is one way of unraveling the social geography of the city. Spatial patterns of social trends such as gentrification and ghettoization can be related to the distributions of particular types of housing. Some landscapes have more staying power than others. The study of housing thus provides an opportunity for geographers interested primarily in urban design and the cultural landscape to interact with those whose concerns are more social and economic (Cybriwsky 1978; Datel 1985). Beginning in the 1970s, geographers looked at the interface between architecture, housing size and infrastructure, ethnic neighborhoods, and battles over territory and neighborhood identity in a wide variety of American cities. British geographers

Figure 8.3 Stoops in Baltimore. Source: Larry Ford.

followed suit with studies of London where the term gentrification was first coined in 1964.

Studies of architecture and gentrification have also involved the examination of converted industrial and commercial buildings in central cities. The geography literature has long focused on the location of manufacturing but this has changed in recent decades as spaces of consumption in a post-Fordist economy have replaced many spaces of production. The old mills and warehouses have been converted to condos, brew pubs, fancy shops and boutique hotels. This change from traditional studies of industrial complexes emphasizing location theory and production to studies of the design of places of spectacle and consumption is one of the interesting aspects of historical geography. As cities change functionally, their landscapes display new meanings in different ways (Figure 8.4). Image and symbolism become tools for conforming to and enhancing changes necessary in a global economy.

Geographers have also been turning a critical eye to the design of shopping malls, waterfront developments, festival marketplaces, and theme villages (Goss 1993). The topic of placelessness versus authenticity looms large in many of these studies. Led by geographers focusing on various aspects of social theory, writers have discussed the roles that shopping centers and their component parts (individual store design and displays) have played in creating a culture of consumption and the acceptance of make-believe geographies. The modern shopping mall, and its many variations, have gained attention as designers have failed to make interesting places in most other parts of the urban environment. In some cities, the mall has become the only vibrant social place, albeit for only some segments of the public.

Geographers have examined the spatial organization and symbolic design elements

Figure 8.4 Rapid change in West Los Angeles. Source: Larry Ford.

of shopping malls in order to uncover the ways artificial "places" are being created and experienced. Historic architecture often supplies a theme, especially in festival marketplaces like Quincy Market in Boston or Pier 39 in San Francisco, but even new malls can reference exotic and charming landscapes. Horton Plaza in San Diego, for example, is meant to reference the colorful confusion of a Tuscan hill town. In all of these malls, design is meant to enhance the shopping experience by providing settings that link the consumer to other times and places.

While malls in North America have been the focus of much of this research, indeed an entire issue of the *Canadian Geographer* was devoted to studies of the West Edmonton Mall, there is an increasing interest in everything from Victorian arcades in Britain to the immense complexes being built in Asia (*Canadian Geographer* 1991). In extreme cases, new histories are designed into the landscape of entire towns. Leavenworth, Washington has become entirely "Bavarian" in appearance, while Solvang, California has experienced a total "Danish" make-over (Frenkel and Walton 2000). Terms such as "Disneyfication" and "Rousesification" (after developer James Rouse) have been invented to describe this phenomenon. Even Tijuana, Mexico, the quintessential border town, was modeled after Olvera Street in Los Angeles, which was created as a Mexican-themed tourist district in the 1920s (where many of the trinkets initially sold in Tijuana were made).

The representation of cities in cartography and art

Geographers have long explored the interface between landscape and the depiction of landscapes in art. Beginning with articles such as "English Landscape Tastes" in the 1960s, geographers have looked at the ways in which places have inspired art and then, in turn, been redesigned to look more like famous artistic depictions (Lowenthal and Prince 1965). English landscapes portrayed by Constable and Turner for example, sometimes influenced planning decisions aimed at making valued landscapes more authentic (Rees 1982). The large number of influential Impressionist paintings of Haussmann's new boulevards in Paris during the late nineteenth century helped to diffuse Parisian design elements to cities all around the world. In turn, the new landscape elements of Paris have inspired new types of art.

Artistic depictions can also be used to recreate the historical geographies of places. Visitors to the east coast of Mexico, for example, often made drawings as well as maps of the places they visited during the early 1800s (Arreola 1982). These pictures, combined with verbal descriptions, provided insight into not only what the towns were like but also what the visitors of that era saw as worth recording. Historical geographies can also be painted into the urban landscape through the use of murals. Murals are most often associated with Latin America, especially Mexico, and with Latino communities in the US, but in recent decades, a wide variety of pictorial landscapes have been examined (Arreola 1984). In both Chicano Park in San Diego and in the Catholic neighborhoods of Belfast, murals have been used to record and celebrate the histories of places.

Cartography has also played an important role in the contribution geographers have made to the field of urban design. Indeed, art and cartography have overlapped for at least 8,000 years as they have helped people to see, shape, and reshape cities. Early maps were often pictorial with buildings, walls, towers, and rivers clearly drawn. The nature of particular maps tells us a lot about the purposes of their makers. Sometimes skylines were portrayed to provide landmarks for sailors, while at other times they were employed to show what was important symbolically. Of course, not all cartographers were geographers in the current academic sense but by the nineteenth century, the specialty was part of the academic field.

Today geographers have been keen on analyzing old maps to both understand the processes of city creation and to examine various aspects of place perception. Robert Churchill, for example, in his article "Urban Cartography and the Mapping of Chicago" examined the relationship between types of historical maps and the beginning of city planning ideas in Chicago (Churchill 2004). A combination of maps and pictures have been used to understand the development of very different types of suburbs – and indeed, the meaning of the concept of suburbs, in studies of such diverse settings as Toronto and Cape Town (Duncan 1973).

Today, geographers provide important tools of spatial analysis to urban designers through the use of GIS (geographic information science/systems). Computer cartography and various spatial technologies such as three-dimensional "fly throughs" of proposed and existing cityscapes and web-based mapping allow professionals to gather and present vast amounts of information at very high speed and complexity. This, however, is a different topic that is covered in another chapter of this book.

Pondering the meanings of sense of place

A final thread of geographic contributions to urban design is that of the connections of humans to place. Geographer Yi-Fu Tuan, with a series of books such as *Topophilia* (Tuan 1974), has asked us to think deeply about the experience of being in different settings. In topics such as "symbols of cosmos and urban forms" he roams the world examining the relationship between meaning and urban design. Although few other geographers have been quite as philosophical as Tuan, a significant literature has blossomed around the topic of the interpretation of ordinary landscapes and how to read the city through them (Meinig 1979). Much of this work involves non-western landscapes and the role of culture in designing ideal cities. A related issue is the meaning of place, especially place as designed and built as part of a cultural landscape. Edward Relph in his seminal work *Place and Placelessness* explores the issue of authenticity in urban design (Relph 1976). Amos Rapoport in *House Form and Culture* also examines the close relationship between people and the authentic landscapes they create to surround them (Rapoport 1969). For example, the design of cities such as Amsterdam with its big-window facades could not be created with courtyard-focused traditional Islamic house types where privacy is of the utmost importance.

Conclusion

In this chapter I demonstrated geography's many contributions to urban design. By making invisible and ordinary landscapes and corners of the world more visible, geographers have added to the inventory of spaces that urban designers should care for. Their comparative studies of urban form and their city model formations have provided urban designers with a better understanding of city structures in different contexts – their formation, evolution, and change. Geographers' in-depth studies of the evolution of different urban elements – from skyscrapers to residential backyards – have enriched urban designers' understandings of diverse cultural values. Geographers' documentation of places through cartography has opened windows into historic landscapes, while research into the sense and meaning of place has helped designers better understand their idiosyncrasies.

Perhaps the best way of concluding this chapter is to quote from a plaque in the lobby of one of Ohio's first skyscrapers: "We figure to ourselves the thing we like and then we build it up, each temple nobler than the last. So build we up the beings that we are." The disciplines of geography and urban design are certainly contributing to building "nobler temples."

Note

1 I was involved in the model's creation and subsequent revision. "Model of Latin American City Structure" (Griffin and Ford 1980) was reprinted in a wide variety of articles and textbooks. It also led to the publication of competing models and occasional critiques and suggested improvements. A later article, "A New and Improved Model of Latin American City Structure" (Ford 1996) summarizes an on-going debate on this issue.

References

Aiken, R.S. (1981). "Squatters and Squatter Settlements in Kuala Lumpur." *Geographical Review* 71(2): 158–175.

Argenbright, R. (1999). "Remaking Moscow: New Places, New Selves." *Geographical Review* 89(1): 1–22.

Arreola, D. (1982). "Nineteenth-Century Townscapes of Eastern Mexico." *Geographical Review* 72(1): 1–19.

—— (1984) "Mexican American Exterior Murals." *Geographical Review* 74(4): 409–424.

Brunn, S., Hays-Mitchell, M. and Zeigler, D. (2008). *Cities of the World: World Regional Urban Development.* Lanham, MD: Rowman, & Littlefield.

Canadian Geographer. (1991). Special Issue on the West Edmonton Mall. 35(3).

Christopher, A.J. (1985). "Continuity and Change of African Capitals." *Geographical Review* 75(1): 44–57.

Churchill, R.S. (2004). "Urban Cartography and the Mapping of Chicago." *Geographical Review* 94(1): 1–22.

Cybriwsky, R. (1978). "Social Aspects of Neighborhood Change." *Annals of the Association of American Geographers* 68: 17–33.

Datel, R.E. (1985). "Preservation and a Sense of Orientation for American Cities." *Geographical Review* 75(2): 125–141.

Duncan, J. (1973). "Landscape Taste as a Symbol of Group Identity." *Geographical Review* 63: 334–355.

Ford, L. (1985). "Urban Morphology and Preservation in Spain." *Geographical Review* 75(3): 265–299.

—— (1993). "A Model of Indonesian City Structure." *Geographical Review* 83(4): 374–396.

—— (1996). "A New and Improved Model of Latin American City Structure." *Geographical Review* 86(3): 437–440.

—— (1998). "Midtowns, Megastructures and World Cities." *Geographical Review* 88(4): 528–547.

Frenkel, S. and Walton, J. (2000). "Bavarian Leavenworth and the Symbolic Economy of a Theme Town." *Geographical Review* 90(4): 559–584.

Fusch, R. (1994). "The Piazza in Italian Urban Morphology." *Geographical Review* 84(4): 424–438.

Godfrey, B. (1999). "Revisiting Rio de Janeiro and São Paulo." *Geographical Review* 89(1): 94–121.

Goss, J. (1993). "The 'Magic of the Mall': An Analysis of Form, Function, and Meaning in the Contemporary Retail Built Economy," *Annals of the Association of American Geographers* 83: 18–47.

Griffin, E. and Ford, L. (1980). "A Model of Latin American City Structure." *Geographical Review* 70(4): 397–422.

Harris, R. (1991). "A Working-Class Suburb for Immigrants, Toronto 1909–13." *Geographical Review* 81(3): 318–332.

Lawrence, H. W. (1988). "Origins of the Tree-Lined Boulevard." *Geographical Review,* 78(4): 355–374.

Lowenthal, D. and Price, H. (1965). "English Landscape Tastes." *Geographical Review* 55: 188–222.

McPherson, E. G. and Haip R. A. (1989). "Emerging Desert Landscape in Tucson." *Geographical Review* 79(4): 435–449.

Meinig, D.W., ed. (1979). *The Interpretation of Ordinary Landscapes*. Oxford: Oxford University Press.

Rapoport, A. (1969). *House, Form and Culture.* Englewood Cliffs, NJ: Prentice-Hall.

Rees, D. (1982). "Constable, Turner and Views of Nature in the Nineteenth Century." *Geographical Review* 72(3): 253–269.

Relph, E. (1976). *Place and Placelessness.* New York: Pion.

Schmandt, M.J. (1995). "Postmodern Phoenix." *Geographical Review* 85(3): 349–363.

Tuan, Yi-Fu. (1974). *Topophilia*. Englewood Cliffs, NJ: Prentice-Hall.

Further reading

Ford, L. (1998). *Cities and Buildings: Skyscrapers, Skidrows, and Suburbs*. Baltimore, MD: John Hopkins Press. A compelling account of the evolution of different city segments, and an analysis of the link between the built environment and urban economies.

Harvey, D. (2000). *Spaces of Hope*. Berkeley and Los Angeles: University of California Press. Inspired by utopianism, the book casts emphasis on possible designs that can promote social justice and living with nature.

Scott, A.J. and Soja, E. (1996). *The City: Los Angeles and Urban Theory at the End of the Twentieth Century*. Berkeley and Los Angeles: University of California Press. A collection of essays examining the built environment and social dynamics that have resulted in a contemporary metropolis.

Tuan, Yi-Fu. (1974). *Topophilia*. Englewood Cliffs, NJ: Prentice-Hall. A thorough examination of environmental perceptions, experiences, and values, and how they contribute to the creation of place.

9

Influences of sociology on urban design

William Michelson

The discipline of sociology contributes to knowledge and practice in urban design in numerous ways through the application of sociologically-relevant questions and research methods. In this chapter, I will first discuss the nature and extent of diversity within sociology and how this interfaces with similar diversity within urban design. Second, I will examine some of the characteristic differences in approaches that sociologists have brought to the study of design. Then I will turn to some of the main methodological tools that have proven useful and some major applications in urban design.

Sociological range

Sociology focuses primarily on the implications of social groups on human life. While this contrasts with the nominal focus of other social sciences, the boundaries are often fuzzy and are often crossed.

Social groups vary greatly in scale and focus. Two people can make a group. At the micro end of the spectrum, sociologists devote considerable attention to the family. At the macro extreme are interests in world systems and globalization. In between are multiple layers of scale, and sociological research often addresses substantively different interests within the same level, such as health, education, work, religion, politics, and the economy. Sociologists are often concerned with the distribution of power and resources across groups, dealing with inequalities in such realms as gender, race, and ethnicity. Some sociologists concentrate on youth; others, on the elderly. There are urban sociologists, and there are rural sociologists – even suburban sociologists. The American Sociological Association had 46 substantively different special interest *sections* in 2008, while the International Sociological Association, not to be outdone, has 55 *research committees*.

The topics covered in this volume suggest much the same pattern within the field of urban design. While larger in scale than the architecture of specific buildings, urban design covers a range from neighborhood spaces to intra- and inter-regional differences.

Sociologists also vary on a continuum as to whether their thinking on a subject represents, at one extreme, interested speculation from a sociological perspective or, at the opposite extreme, the results of fully documented research approximating a scientific experimental design. If this chapter, within its space limitations, were to give explicit consideration to any and all sociological observations of even latent relevance to urban design, it would do injustice to more focused work carried out with the intention to inform the process of urban design. Hence, the following examination is restricted to the consideration of explicit, developed interests in both sides of the social science–urban design linkage. This excludes, for example, Louis Wirth's famous statement (1938) about urbanism as a way of life, which simply takes for granted that cities are large, dense, and socially heterogeneous – and indeed impermeable to human intervention. Excluded as well are the many eloquent, intriguing, and imaginative observations by Richard Sennett on urban life as driven by a fear of exposure and retreat from public activity (1970; 1977; 1990). Even the recent rediscovery of *place* by American sociologists, with few exceptions (e.g. Fitzpatrick and LaGory 2000), largely neglects all but the social, economic, and political characteristics of the local areas that they examine (e.g. Jargowsky 1996; Dreier *et al.* 2001).

To understand the contribution of sociology to urban design requires a differentiation of both the sociological and design sides of their linkage. Given the range and complexity of practice within each, the matrix of possible points of contact and contributions is large and complex even after restricting our focus to sociological contributions with well developed interest in one or another aspect of urban design.[1]

Approaches

Among the relatively few sociologists with developed environmental interests, there is a fundamental source of difference. Some study the impact of environments on people. Others study the impact of people on environments.

Environment and behavior

Some see environment as causing such outcomes as human behavior, health, crime, and the like. What happens *in consequence* of exposure to built environments? Is high density a cause of pathology? How are suburbs as a place to grow up? Under what spatial circumstances are women more (or less) likely at risk on public transportation? What outdoor configurations of space attract people to congregate?

Degrees of causality

At one extreme, causality can be *deterministic*. The environmental context determines what happens to people or what they do. There was a high degree of determinism in a paradigm called *human ecology* which was espoused in the first decades of the 1900s by members of the University of Chicago Department of Sociology (Park *et al.* 1925). They adapted the structure and processes of plant ecology to lend scientific credence to urban growth, development, and life. Their view was that the cost of land, which they saw as uncontrollable by individuals and authorities, determines both the intensity and land use of built environments, which, in sequence, would have a bearing on who lives there and what behaviors emerge among the particular groups of residents attracted to live there. Yet, research actually supporting such deterministic causality is rare, reserved largely to extreme situations such as highly dilapidated housing or disasters (cf. Wilner, *et al.* 1962).

More common now is a lesser degree of causation called *probabilism*. Under probabilism, spatial and other parameters of a physical context make it likely that a certain outcome might occur. For example, studies shortly after World War II informed by field theory (Lewin 1936) suggested that residential site planning could influence which neighbors would be likely to come into regular interaction with each other, all else equal (e.g. Festinger *et al.* 1950; Whyte 1956).

A third, yet milder form of causality is *possibilism*. In this regard, the environmental context creates the conditions under which certain kinds of behavior or interaction become physically possible. Although creating possibilities hardly offers a strong degree of predictive causality, it represents a basis for design criteria in non-traditional, democratic societies, in which designers seek to present opportunities rather than to reinforce traditional, lock step conventions. Much housing research in the environment-behavior tradition is possibilistic (cf. Michelson 1977).

The dividing line between probabilism and possibilism is not definitive. Both lend themselves to the creation of urban designs intended to provide the spatial foundations within which eventual residents or users will be able to accomplish their objectives. Gerald Suttles, for example, applied these considerations to the development of new residential enclaves intended to attract with their designs and then serve particular sub-sectors of the population. He called these *contrived communities* (1972, Chapter 4).

Is it ironic that the realities of person-environment relationships suggest that the most frequently found forms of causality are the least prescriptive for designers? I suggest that it is nonetheless challenging for designers to be aware of what to include and facilitate with their designs, lest their products obviate something crucial to personal or local life – as so often occurs.

It should not be surprising that the paradigms and theories that have been applied to how environments impact on people are so varied. The theoretical hegemony of the deterministic Chicago School has been diluted by other schools (e.g. Los Angeles, New York) that are more liberal in their explanation of characteristic urban life styles and behaviors (*City and Community* 2002). In addition, detailed analyses of environment and behavior at more micro levels by a wider variety of social scientists have added to the range of paradigms and methods devoted to environment and behavior (cf. Michelson and Van Vliet 2002).

Proliferation

A North American-based interdisciplinary organization of researchers taking this type of approach to environment and people called the Environmental Design Research Association (EDRA) met formally for the first time in 1970 and has stimulated research exchange ever since. Several small scale newsletters morphed into a continuous, respected journal, *Environment and Behavior*. This eventful decade saw the start of an outpouring of texts and readers on this approach. Many were by psychologists, in the name of "environmental psychology" (e.g. Proshansky *et al.* 1970; Toepfer *et al.* 1972; Ittelson *et al.* 1974; Heimstra and McFarling 1974; Gifford 1987; Bonnes and Secchiaroli 1995). Others represented writers of other disciplines, for example geography (Rapoport 1977) and sociology (Michelson 1970). While these and other such works are understandably not identical in coverage, the substance of these books represents more overlap of interests than discipline-related differences. EDRA, the organization, fostered a state of the art overview after more than a decade of activity (Moore *et al.* 1985), and two years later an encyclopedic, two-volume *Handbook of Environmental Psychology*, with disciplinarily-diverse authorship was issued (Stokols and Altman 1987). A *Handbook of Environmental Sociology*, parts of which addressed sociological contributions to urban design, followed in the new century (Dunlap and Michelson 2002).

Researchers in several European countries (particularly the Scandinavian nations and The Netherlands) simultaneously developed particularly pragmatic versions of this approach, with research institutes and laboratories devoted to the empirical exploration of how well particular built environments function. With the assistance of government funding, the equivalent of industrial research and development cycles were instituted, bringing research findings to bear on design practice and development – and hence improving housing, workplaces, institutions, and public spaces on an ongoing basis (cf. Thiberg 1985).

Social structure and environment

Other sociologists took a different tack, focusing on how the actions of people bring about the environments that we get. Stability and change are viewed not as inevitable but rather as functions of collective human action. This opposite approach is referred to as *structural*, a reflection of the social structure.

The first recognized criticisms of the Chicago School's deterministic explanation came in mid-twentieth century from sociologists demonstrating how residents of selected local neighborhoods were acting in an organized way to maintain residential areas that reflected shared backgrounds, sentiments, and symbolism. There was nothing subsocial about organized groups working to keep their areas stable and consistent with shared values (e.g. Firey 1947).

Some sociologists presented schemes to understand the presence, influence, and diversity of interest groups in local areas (e.g. Form 1954; Long 1958). What was inevitable, they said, was that any issue will bring forward a variety of interest groups, each working explicitly and rationally to accomplish some impact in what happens to the contextual issue du jour. This was a relatively nondeterministic approach, as it observed that winners and losers on any given issue could not be uniformly predicted – nor that there were always the same players on the fields of influence and decision-making.

Karl Marx and his theories were rediscovered in the years following the Vietnam War along with the political uncertainties and protests that emerged in the late 1960s and 1970s. A New Urban Sociology sought to replace the Chicago School as a main organizing way of thinking. This put the spotlight not on explaining behavior as a function of context, but on illustrating how such macro, structural factors as the ownership of capital and social class drive urban development, growth, and change (cf. Pahl 1970; Castells 1978; Tabb and Sawers 1978, 1984; Lefebvre 1991, 1996).

Structural research expanded greatly in the years after the Marxian thrust, as the explanatory factors researchers explored broadened beyond orthodox Marxian analysis. An influential book by John Logan and Harvey Molotch (1987) took up from the pre-Marxian writers who had focused on interest groups. Logan and Molotch addressed the trend for cities to renew themselves with massive projects focusing on sports stadiums, convention facilities, hotels, and other large buildings catering to leisure and tourism. They found similar pro-growth coalitions in many cities consisting not just of holders of capital but of the many groups in society expecting to gain from an influx of modern facilities in the leisure sector: people in real estate, banking, the hospitality industry, construction, unions, and, not least, urban politics (related to expected increases in tax base). The authors conceptualized this coalition as an *urban growth machine*. The concept of the urban growth machine indicates how the conscious efforts of various respected interest groups result in some urban outcomes but not others. In a retroactive look at this concept ten years later (and twenty years after Molotch [1976] first introduced it), Logan *et al.* (1997: p. 605) clarified "that the principal effect of growth machines is to bend the policy priorities of localities toward developmental, rather than redistributional, goals." Subsequently, Gottdiener and Hutchison (2000) describe an even newer "New Urban Sociology" built around a "socio-spatial model" which focuses even more broadly on the interplay of the structural processes in urban society and their spatial manifestations.

Emerging in this structural approach is a special focus on cultural influences on the urban structure. John Hannigan (1998) describes a trend in which the urban growth machine forces in many cities have focused on the creation of new or restored leisure districts, representing public theme parks: what Hannigan referred to as *Fantasy City*. Mark Gottdiener referred to this phenomenon as "*The Theming of America*" (1997). Sharon Zukin has emphasized in much of her work (e.g. 1995) the importance of support for the arts and its practitioners for the revitalization of urban districts and tax bases.

The tone of housing research has changed dramatically in the past forty years, as the bulk of sociological research and researchers has changed over from behavioral to structural approaches. Logan and Molotch (2007) note in retrospect the priority in urban development decision-making under the urban growth machine accorded to *exchange values* (i.e. monetary gain for specific subgroups) over *use values* (i.e. benefits to the ongoing welfare of the population as a whole).[2] This change of emphasis can be observed not only in the differential numbers of researchers following the two approaches, but also in the work of individuals. For example, Suttles analyzed the dynamic interrelationships of behavior and territory among members of different groups in Chicago in his early research publications (1968, 1972). Just over two decades later, his focus turned to the process by which decisions on proposed large downtown developments in Chicago got made, linking private sector exchange interests to public sector planning in what he called a "Land-Use Confidence Game" (Suttles 1990).

Methodologies and major applications

Modern cameras come with the technical sophistication necessary to capture images under highly varying conditions, either through pre-programmed scenes or freely allowed manual manipulation. In Sociology, research methodology represents a tool kit from which to choose in order to capture both the essence and detail of evidence necessary for a given topic or analysis. As with photography, there is no single best choice for all situations. The objective is to be able to choose a way to capture most appropriately the particular reality that is out there, rather than to represent only what is in the mind of the sociological equivalent of the expressive artist. Not surprisingly, the methods brought to urban design by sociologists (and kindred social scientists) are various.

Challenges

There are three challenges facing the connection of sociological methodology to urban design. The first is the need to pose questions that bridge the social and physical realms explicitly and appropriately. The second is to gather data that are not only appropriate for the task(s) at hand, but also sufficient to consider alternative explanations to the questions asked. And the third is to conceive of and then apply multiple methods, so as to enable *triangulation* – akin to having spotlights from two or more directions focus on a target, so as to be able to illuminate that target more fully and to provide more confidence in the results.

Guidance

Several books address the special considerations pertaining to research bridging social science and environmental concerns and elucidate useful methods for doing so (Zeisel 1975, 1981; Michelson 1975; Bechtel *et al.*(1987, 1990).

Prominent methods

Observation

Observation, simply put, can take two forms: participant and non-participant. In participant observation, the researcher puts himself/herself directly into the situation being studied, to experience more or less the same phenomena as those normally present. As it is usually evident that the researcher is not native to the scene, skills need be developed for acceptance, enough that the participant observer can get exposure to a relatively accurate view of ongoing behavior and collective life. This takes time and dedication.

For example, Herbert Gans (1962, 1967) carried out two major research studies of significant types of residential settings: an inner-city ethnic enclave of Italian-Americans in Boston about to be transformed by renewal (1962) and Levittown, a newly built, planned suburban town in New Jersey within commuting range of Philadelphia (1967). In the former study, Gans described convincingly the extent that the life style of the resident group was supported by the physical structure and land-use of their enclave, despite the area being designated as a slum by planning officials. Gans, as a result of his participant observation, made a strong case of such areas being viewed more as *urban villages* than as slums, though the results did not deter local officials from destroying the area and dispersing the population, replacing both the existing urban form and the residents in the process. In Levittown, Gans established that the priorities and salient activities of the residents of this newly-built suburb, focusing heavily on their children, were more of a function of their social class and stage in the life cycle than the physical design of the area. Although the fabric of the respective areas provided opportunities for how the residents chose to live, Gans interpreted his findings as critical of environmental determinism. His approach was later conceptualized as *compositional*, emphasizing the composition of the population (Fischer 1976; see also Gans 1968). A more recent study in this tradition was in Celebration, Florida, a town developed by the Disney Corporation on new urbanist principles (Ross 1999).

Keith Hampton, in association with Barry Wellman, did participant observation of note in a new "wired" subdivision north of Toronto (Hampton and Wellman 2003). Hampton lived for an extended period in this community, the infrastructure for which emphasized the advent of the eventual computer revolution. He combined participant observation with on-line survey techniques to examine the development of local community in a situation in which most people were accorded the tools to bypass in person communication with neighboring persons. He found that computer-assisted contact enabled strong contact patterns with *both* neighbors and those farther away.

Nonparticipant observation involves paying close visual attention to human activities and physical traces of them without the pretense of being an insider. There are varying degrees of nonparticipant observation, with one extreme requiring no interaction with those being observed.

Jan Gehl, a Danish architect, has used observational techniques throughout many decades, traveling the globe making careful observations and records of where and how people occupy and use external spaces, turning dead spaces into social spaces.

He counts how many people are there, what they are doing, and what factors help make their appropriation of this space possible. And he documents his observations with photographs and maps (Gehl 1987). In consequence, he provides consultation to urban designers and political leaders in many different cities. The well-known business journalist, William H. Whyte, Jr., utilized similar observational techniques while assessing public use of plazas created at the ground level of major corporate buildings in New York City (1980).

Surveys

Surveys are a major method within Sociology. While some argue in favor of the first hand view of behavior offered by observational techniques, observation is limited to areas that an observer can realistically cover. Surveys fill the gap for getting information that can not be observed. They enable the gathering of objective, subjective, and statistical information from potentially large and dispersed numbers of people. And depending on the situation, surveys, observation, and other methods can complement each other, as was the case in Hampton's research (Hampton and Wellman 2003). While surveys take many forms and are ubiquitous, some respond more directly than others to the content and needs of urban design.

Survey research can be *cross-sectional* or *longitudinal*. Cross-sectional surveys are done once. Longitudinal surveys involve repeated applications of the survey, ideally but not always with the same respondents, in order to document longer term stability or change in the lives and feelings of people subjected to the same conditions. Longitudinal surveys require a greater commitment over time by both researchers and respondents, as well as more substantial funding. But they are an improvement over the snapshot in time offered by cross-sectional surveys if data on longer term environmental impacts are pertinent.

A classic example of longitudinal survey research is Suzanne Keller's study of Twin Rivers, New Jersey (2003). Keller studied the extent and circumstances of community development in Twin Rivers, New Jersey, a large planned settlement of town houses and apartments for 10,000 residents that opened in 1970. Keller studied Twin Rivers for thirty years. Surveys in each of three succeeding decades provided her with information on people's background and expectations, how they regarded their new residential environments, and how social interaction developed in the particular setting and dynamics of Twin Rivers. In this study, the surveys were complemented by continuous monitoring of local issues and how they transpired, personal observations, and a host of archival data across thirty years.

Another type of survey has shown merit when used in conjunction with more conventional survey content in environmental studies. The time-use survey provides information on people's actual behavior during specified time periods: usually a day but sometimes as long as a week. People respond to interviews or diaries to provide information in serial order from the start of an immediate day in their lives, listing in a matrix-like log each activity in which they have participated from the time they arise, and, for each, what they did, from when to when, where this took place, and who else was with them during the episode. Sometimes people are also asked their subjective feelings about each such episode. This kind of survey provides behavioral information for great numbers of people which is more complete and accurate than their offhand estimates of what they spend time doing, and it includes what they do at home and away. It provides a data set that enables simultaneous, integrated analysis of several vital components of behavior: who, what, where, and when.

This survey is neither hypothetical nor value laden (Michelson 2005).

Planners find this useful for transportation planning. The classic application was by F. Stuart Chapin, Jr. (1974), who gathered such data from a large sample in Washington, D.C. for the purpose of assessing the typical daily rounds among different activities and land uses by various segments of the population, so as to have a rational behavioral data base for future transportation and land use planning. Transportation planners are longstanding advocates of trip logs; but the time-use survey is more precise, complete, and behaviorally grounded.

I have used this type of survey in several contextual analyses – of families living in different housing types and locations (Michelson 1977), on community structure in the lives of employed mothers and their families (1985), on emergent behaviors in Swedish experimental housing projects (1993), and on the behavioral dynamics of home-based work (1998, 2000). In each case, time-use analysis was part of a multi-method package.

Sociologists borrow special surveys from the other social sciences. Psychologists specialize in scales and indices measuring people's subjective orientations – some of them with regard to environmental contexts. Moos (1976), for example, assembled a variety of such scales regarding different types of environment. Geographers have expanded sociological work on cognitive mapping (also known as mental mapping), a survey technique in which respondents draw their own images of their city or nation, exposing in the process their perceptions of the environment in question – of what they are aware or unaware, and why.

Simulations

There have been major efforts to assess people's environmental priorities by the creation of simulation games, in which participants are forced to make tradeoffs in the choice of preferred environmental situations and designs. At one time, this was in the form of questionnaires and board games (Robinson 1987). However, the level of complexity and attraction of these exercises has grown exponentially with the home computer revolution. Sim City became a favorite if only for recreation purposes, and succeeding versions are increasingly detailed and demanding.

Much useful work has been done in Northern Europe's full-scale simulation laboratories. These are inside spaces into which mockups of planned built environments can be constructed, usually with modular blocks or panels, and then, once functionally furnished, made available for people's reactions, while researchers carefully observe them and keep records. In Amsterdam, for example, housing authorities have tested out proposed designs on people who are the likely residents, monitoring their reactions and suggestions. In Lund, Sweden, proposed new hospital rooms have been built and supplied with patients, nurses, doctors, and orderlies, whose hands on experiences help shed light on the time and motion components of design. There, too, the concept of apartments with flexible walls and room arrangements has been tested by potential tenants. The simulation lab concept crossed the ocean to the University of California's Irvine campus, where the Program in Social Ecology adapted it to the full-scale testing of office settings, in conjunction with major office suppliers.

Social scientists associated with design institutions have been at the forefront in developing full-scale simulations, but a more complete review of the topic can be found in Peter Bosselman's chapter in this volume.

Analysis of available data

The preceding research methods all reflect the need to obtain new data for analysis on questions pertaining to urban design. Nonetheless, the trend of recent decades is for social scientists to gain increasing access to large, costly data sets collected under governmental and corporate auspices. The average researcher now has access to quantities of data that were previously unimagined. The crucial question is whether available data address the secondary analyst's needs sufficiently well.

One of the most discussed research projects in the history of urban design research was made possible by available data on the locus of crime in housing projects, collected by police and housing authorities in New York. Oscar Newman's *Defensible Space* (1972) used these data in conjunction with his analyses of the spatial parameters of carefully selected housing projects to examine the extent and ways that project design had a bearing on the commission of crimes within these spaces.

Researchers pursuing structural approaches are likely to rely on both data and written archives, mixed with non-standardized, in-depth interviews of key informants – respondents who, far from randomly selected, are picked for their presumed knowledge of the situation. These methods accompany an emphasis on case study. As cases occur in actual places and in real time, the information needed to understand what happened is usually out there in file folders, libraries, archives, public records, and the minds and memories of particular people. Meyerson and Banfield's study (1955) of how many new housing projects in Chicago ended up being located in the wards of opposition aldermen was a dramatic early illustration of this methodology.

Final comments

The interests and methods that sociologists bring to the table have clearly been useful in the pursuit of urban design and its supporting research. However, this is not a tidy, uniform package, and, more importantly, it is not always conveyed by a practicing sociologist. Yet, surely the practice of urban design has been expanded and enlightened by the field of sociology and an interdisciplinary interchange of ideas and methods.

In the near absence of deterministic dynamics, it is inevitable that ultimate responsibility for urban design lies in the hands and heads of design professionals (not to speak of the formal decisionmakers with legal responsibility and the omnipresent, informal interests including investors, residents, and other users). Architect Christopher Alexander (1964) provided helpful leadership in how to make choices from among the many bits and pieces of design that eventually make up the total design. Choices from among the potentially many specific behaviorally-relevant solutions to parts of the total design should take into consideration which ones are compatible with each other as the overall design gets put together. Successful application of sociological evidence thus involves not just informed creativity but also careful integration. It surely helps if the needs and wants of the diverse stakeholders, active and latent, are recognized and reconciled.

Notes

1 The Ekistic Grid, developed, promulgated, and revised over time within the journal *Ekistics*, illustrates this situation accurately and helpfully. This grid categorizes research content within Ekistics, the science of human settlement, according to the respective scales of design and human group involved in any particular design concern. (cf. Doxiadis 1968)

2 This contrast is pointed out in an Australian on-line real estate and planning medium (*Scribbling on Bricks* 2008).

References

Alexander, C. (1964). *Notes on the Synthesis of Form*, Cambridge, MA: Harvard University Press.

Bechtel, R., Marans, R. and Michelson, W. (Eds.) (1987). *Methods in Environmental and Behavioral Research,* New York: Van Nostrand Reinhold. Reprinted in 1990 by Robert E. Krieger Publishing Company, Inc. Malabar, Florida.

Bonnes, M. and Secchiaroli, G. (1995). *Environmental Psychology: A Psycho-social Introduction*, London: Sage.

Castells, M., translated by Lebas, E. (1978). *City, Class and Power,* London: Macmillan.

Chapin, F. Stuart, Jr. (1974) *Human Activity Patterns in the City: Things People Do in Time and in Space*, New York: Wiley-Interscience.

City and Community (2002):1 (entire issue).

Doxiadis, K. (1968). *Ekistics: An Introduction to the Science of Human Settlements,* New York: Oxford University Press.

Dreier, P., Mollenkopf, J., and Swanstrom, T. (2001). *Place Matters*, Lawrence, KS: University Press of Kansas.

Dunlap, R. and Michelson, W. (Eds.) (2002). *Handbook of Environmental Sociology*, Westport, CT: Greenwood Press.

Festinger, L., Schachter, S., and Back, K. (1950). *Social Pressures in Informal Groups,* New York: Harper and Bros.

Firey, W. (1947). *Land Use in Central Boston,* Cambridge, MA: Harvard University Press.

Fischer, C. (1976). *The Urban Experience*, New York: Harcourt Brace Jovanovich.

Fitzpatrick, K. and LaGory, M. (2000). *Unhealthy Places: The Ecology of Risk in the Urban Landscape*, New York: Routledge.

Form, W. (1954). "The Place of Social Structure in the Determination of Land Use," *Social Forces,* 32: 317–323.

Gans, H. (1962). *The Urban Villagers*, New York: The Free Press of Glencoe.

—— (1967). *The Levittowners: Ways of Life and Politics in a New Suburban Community*, New York: Pantheon Books.

—— (1968). *People and Plans: Essays on Urban Problems and Solutions,* New York: Basic Books.

Gehl, J., translated by Koch, J. (1987). *Life Between Buildings: Using Public Space,* New York: Van Nostrand Reinhold Company.

Gifford, R. (1987). *Environmental Psychology: Principles and Practice*, Boston, MA: Allyn and Bacon.

Gottdiener, M. (1997). *The Theming of America*, Boulder, CO: Westview Press.

Gottdiener, M. and Hutchison, R. (2000). *The New Urban Sociology 2nd ed.*, Boston: McGraw Hill.

Hampton, K. and Wellman, B. (2003). "Neighboring in Netville: How the Internet Supports Community and Social Capital in a Wired Suburb," *City and Community* 2: 277–311.

Hannigan, J. (1998). *Fantasy City: Pleasure and Profit in the Postmodern Metropolis*, London: Routledge.

Heimstra, N. and McFarling, L. (1974). *Environmental Psychology,* Monterey, CA: Brooks/Cole.

Ittelson, W.H., Proshansky, H., Rivlin, L. and Winkel, G. (1974). *An Introduction to Environmental Psychology*, New York: Holt, Rinehart and Winston.

Jargowsky, P. (1996). *Poverty and Place*, New York: Russell Sage Foundation.

Keller, S. (2003). *Community: Pursuing the Dream, Living the Reality*, Princeton, NJ: Princeton University Press.

Lefebvre, H., translated by Donald Nicholson-Smith (1991). *The Production of Space*, Oxford: Blackwell.

Lefebvre, H., translated and introduced by Eleonore Kofman and Elizabeth Lebas (1996). *Writings on Cities*, Cambridge, MA: Blackwell.

Lewin, K. (1936). *Principles of Topological Psychology,* New York: McGraw-Hill.

Logan, J. and Molotch, H. (1987, 2007). *Urban Fortunes: The Political Economy of Place,* Berkeley, CA: University of California Press.

Logan, J. and Molotch, H., Whaley, Rachel B., and Crowder, Kyle (1997). "The Character and Consequences of Growth Regimes: An Assessment of Twenty Years of Research," *Urban Affairs Review* 32: 603–630.

Long, N. (1958). "The Local Community as an Ecology of Games," *American Journal of Sociology,* 64: 251–261.

Meyerson, M. and Banfield, E.C. (1955). *Politics, Planning, and the Public Interest: The Case of Public Housing in Chicago,* Glencoe, IL: Free Press.

Michelson, W. (1970). *Man and his Urban Environment: A Sociological Approach,* Reading, MA: Addison Wesley.

—— (Ed.) (1975). *Behavioral Research Methods in Environmental Design,* Stroudsburg, PA: Dowden, Hutchinson & Ross, Inc.

—— (1977). *Environmental Choice, Human Behavior, and Residential Satisfaction,* New York: Oxford University Press.

—— (1985). *From Sun to Sun: Daily Obligations and Community Structure in the Lives of Employed Women and Their Families,* Totowa, NJ: Rowman & Allanheld.

—— (1993). "Grounding Time-Use in Microspace: Empirical Results," *Social Indicators Research* 30: 121–137.

—— (1998). "Time Pressure and Human Agency in Home-based Employment," *Society and Leisure*. 21: 455–472.

—— (2000). "Home-based Employment and Quality of Life: A Time-Use Analysis," in Edward Diener (ed.) *Advances in Quality of Life Theory and Research,* Dordrecht: Kluwer.

—— (2005). *Time Use: Expanding Explanation in the Social Sciences,* Boulder, CO: Paradigm Publishers.

Michelson, W. and van Vliet, W. (2002). "Theory and the Sociological Study of the Built Environment" in Riley Dunlap and William Michelson (eds.) *Handbook of Environmental Sociology,* Westport, CT: Greenwood Press.

Molotch, H. (1976). "The City as a Growth Machine," *American Journal of Sociology* 82(2): 309–330.

Moore, G.T., Tuttle, D.P. and Howell, S.C. (1985). *Environmental Design Research Directions: Process and Prospects,* New York: Praeger.

Moos, R. (1976). *The Human Context: Environmental Determinants of Behavior,* New York: Wiley.

Newman, O. (1972). *Defensible Space: Crime Prevention through Urban Design,* New York: Macmillan.

Pahl, R.E. (1970). *Whose City? And Further Essays on Urban Society,* London: Longman.

Park, R.E., Burgess, E., and McKenzie, R.D. (Eds.) (1925). *The City,* Chicago: University of Chicago Press.

Proshansky, H., Ittelson, W. and Rivlin, L. (1970). *Environmental Psychology: Man and his Physical Setting,* New York: Holt, Rinehart and Winston.

Rapoport, A. (1977). *Human Aspects of Urban Form: Towards a Man-Environment Approach to Urban Form and Design,* Oxford: Pergamon Press.

Robinson, I. (1987). "Trade-off Games as a Research Tool for Environmental Design" in Bechtel, Marans, and Michelson (eds.), op. cit., pp. 120–161.

Ross, A. (1999). *The Celebration Chronicles: Life, Liberty, and the Pursuit of Property Value in Disney's New Town,* New York: Ballantine Books.

Scribbling on Bricks: Readings in Property, Planning and Economics (2008) http://bricks.civilpandemonium.com/index.php?groupid = 83, (retrieved August 26, 2008.)

Sennett, R. (1970). *The Uses of Disorder: Personal Identity and City Life,* New York: Vintage.

—— (1977). *The Fall of Public Man,* New York: Alfred A. Knopf.

—— (1990). *The Conscience of the Eye: The Design and Social Life of Cities,* New York: Alfred A. Knopf.

Stokols, D. and Altman, I. (Eds.) (1987). *Handbook of Environmental Psychology,* New York: Wiley Interscience, 2 vols.

Suttles, G. (1968). *The Social Order of the Slum: Ethnicity and Territory in the Inner City,* Chicago: University of Chicago Press.

—— (1972). *The Social Construction of Communities,* Chicago: University of Chicago Press.

—— (1990). *The Man-Made City: The Land-Use Confidence Game in Chicago,* Chicago: University of Chicago Press.

Tabb, W.K. and Sawers, L. (Eds.) (1978, 1984). *Marxism and the Metropolis,* New York: Oxford University Press.

Thiberg, S. (Ed.) (1985). *Bostadsboken (The Housing Book),* Stockholm: Byggforskningsrådet.

Toepfer, C.T., Bicknell, A.T., Fox, L., Kirk, W., and Sayre, R. (Eds.) (1972). *Environmental Psychology: Selected readings,* New York: MSS Information Corps.

Whyte, W.H., Jr. (1956). *The Organization Man,* Garden City, NY: Anchor Books.

—— (1980). *The Social Life of Small Urban Spaces,* New York: Project for Public Spaces.

Wilner, D.M., Price, Valkley, R., Pinkerton, T.C. and Tayback, M. (Eds.) (1962). *The Housing Environment and Family Life: A Longitudinal Study of the Effects of Housing on Morbidity and Mental Health,* Baltimore, MD: The Johns Hopkins Press.

Wirth, L. (1938). "Urbanism as a Way of Life," *American Journal of Sociology,* 44: 1–24.

Zeisel, J. (1975). *Sociology and Architectural Design,* New York: Russell Sage Foundation.

—— (1981). *Inquiry by Design: Tools for Environment-Behavior Research,* Monterey, CA: Brooks/Cole.

Zukin, S. (1995). *The Cultures of Cities,* Cambridge, MA: Blackwell Publishers.

Further reading

Lofland, L.H. (1998). *The Public Realm: Exploring the City's Quintessential Social Territory,* New York: Aldine de Gruyter. Socially relevant public spaces.

Macionis, J.J., and Parrillo, Vincent N. (2010). *Cities and Urban Life, 5th ed.,* Upper Saddle River, NJ: Prentice Hall. Widely-used text, with diverse coverage and documentation.

Palen, J.J. (2005). *The Urban World,* New York: McGraw-Hill. Excellent text, from the pen of a long-standing urban scholar.

Thorns, D.C. (2002). *The Transformation of Cities: Urban Theory and Urban Life,* New York: Palgrave Macmillan. A sophisticated historical and theoretical approach to urban life.

10

Influences of anthropology on urban design

Denise Lawrence-Zúñiga

Contributions from the discipline of anthropology to the field of urban design address questions about the needs of diverse social groups that design schemes are intended to accommodate. Anthropology's theoretical and methodological foundations in descriptive ethnography provide the means to investigate and form deep or "thick" understandings of the everyday lives of ordinary people. Ethnographies of urban populations provide rich documentation of behavior patterns, meanings people attach to their surroundings, and values and aspirations for the future that can inform and guide the urban designer. Anthropological studies often produce unexpected findings, revealing order and complexity where disorder is anticipated, power where marginality is assumed, contradictions embedded in design ideologies, and unintended consequences of design projects. Anthropology provides cultural information about people but also critiques urban design processes and practices, thereby contributing to the enlightenment of designers whose design vision may ignore the realities of urban populations. Anthropology's own self-critique as a postcolonial discipline suggests a reflective process for assessing urban realities that can improve responses to functional and aesthetic needs through urban design.

The discipline of anthropology, popularly recognized as the study of humankind, was born of nineteenth-century colonialism and took as its subject local native peoples, or the "Other." Socio-cultural anthropologists routinely moved to exotic places to live for extended periods among the people they studied and described them in ethnographies, initially conceptualized as holistic, ahistorical accounts of relatively self-contained local communities. The study of urban populations, however, challenged assumptions about cultural isolation and presumed rootedness to particular places. The techniques of participant observation and intensive interviewing yielded rich data on the complexities of everyday urban life that were often overlooked by researchers employing survey methods. While attentive to the larger socioeconomic and political contexts that impact local groups, anthropologists have sought to understand the cultural logic and patterns of activity specific groups of people create to enable them to live together in predictable and orderly ways, and to adapt and survive.

Ethnography not only documents cultural phenomena but interprets and represents those data to larger audiences (Marcus and Fischer 1986). Postcolonial anthropology has shifted away from attempts to write objectively about cultural groups as autonomous entities, seeking instead to understand and explain the interdependencies between local groups and macrostructural influences, while reflexively examining the theoretical constructions used in explanations. One critical dimension of this inquiry has been to reconsider space and place in cultural theory as more than invisible background for, or material evidence of, cultural patterns, and to view them as mutually constituting features of social life (Lawrence and Low 1990; Low and Lawrence-Zúñiga 2003). This shift in ethnography to include socio-spatial perspectives provides urban designers more useful cultural information for inclusion in design processes. Anthropological research, by challenging and illuminating formal design assumptions or procedural strategies, also contributes to achieving a better fit for these populations' needs. The following examines a selection of ethnographic and critical studies, loosely organized by chronology, in a variety of settings in the developed and developing world.

Of squatters, slum dwellers, and the urban working poor

The postwar period increasingly drew anthropologists to study the urban poor and rural migrants seeking a better life in rapidly industrializing cities of the developing world, especially in Latin America. Oscar Lewis (1959; 1966) famously argued that the adaptation to and perpetuation of urban poverty found in slums was due to a cultural phenomenon which produced and reproduced dysfunctional individual personality traits at home, and resulted in family isolation, and social and political marginality. Lewis's "culture of poverty" theory, highly criticized for "blaming the victims" of poverty and later discredited (Valentine 1968), stimulated active debate among social scientists about causes and remedies. Other anthropologists (Mangin 1970; Epstein 1972) focused on settlements built by the poor, new migrants to the city who, lacking sufficient resources to rent or buy property, challenged the state's authority by seizing or "squatting" on public or privately owned land. Squatter settlements, variously called informal, marginal, spontaneous, irregular, or illegal settlements, or "squatments," often lacked basic sanitation, infrastructure and public services. William Mangin, who collaborated with architect/planner John C. Turner in Lima, Peru (Mangin 1970; Turner and Fichter 1972), interpreted life in informal settlements positively, arguing that despite their poverty, residents had a sense of security and control over houses they had built themselves, and over their own lives. Indeed, researchers have found urban migrant settlements highly organized around informal mutual aid, social clubs and informal economic activity (Mangin 1970).

To government officials, the middle classes and professionals, however, slums and squatter settlements appeared chaotic and unsanitary, socially disorganized, criminally dangerous, and pathologically problematic as concentrations of extreme poverty. Epstein (1972; 1973) observes that Latin American officials and urban elites actively sought to eradicate, reform or simply hide from view local slums and squatments. In Brasilia, satellite towns were created to accommodate squatting migrant workers who came to build Brazil's capital city, but whose presence was never planned for in the original design (see also Madhu Sarin for Chandigarh 1982). Although settlements acted as reserves for labor willing to work at depressed wages in construction and domestic services, workers' needs were never a government priority and their self-built houses were only allowed to exist out of view of and contact with middle-class zones in the planned city (Epstein 1972: 56). Epstein (1972) and Perlman (1976) both argue that the social and institutional blindness to the poor, a "myth of marginality", has been supported and legitimized by notions such as the culture of poverty, which blame squatters' material conditions on their supposed personal characteristics, a view which anthropologists have endeavored to correct with ethnographic research.

Observations that squatter settlements do not disappear, despite eradication efforts, eventually prompted some planners and government policy makers to reconsider them as a potential solution to housing large numbers of urban migrants by relying on squatters' own efforts and desire to build a home. Lobo (1982), for example, details how the Peruvian government initiated a "remodeling" of a squatter settlement in 1973 that also included the possibility for residents to acquire secure title to their properties. Many Latin American governments initiated "sites and services" programs for new settlements, and programs to "upgrade" infrastructure in existing settlements, including title transfers, to more effectively accommodate poor populations (Low 1988). Despite innovations in some countries, other sites are less accommodating. Alan Smart (2001) reports that Hong Kong authorities persist in attempts to eradicate long established squatter settlements because land values and demands for development are high, but residents resist and engage in a thriving housing market even though they cannot hold legal title. Moreover, not all urban slums are squatter settlements. Pellow (2002) shows in the stable and successful stranger settlement of Sabon Zongo in Accra, Ghana, despite its ramshackle appearance, Hausa landlords and their tenants form positive, lasting attachments to each other and to place, effectively neutralizing the attraction of newer housing.

Anthropologist Lisa Peattie (1972; 1987), perhaps the best known critic of urban design practices, participated on an American planning team consulting with the Venezuelan government in the 1960s to design a new planned industrial city, Ciudad Guyana. Her ethnography of the poor working-class barrio where she lived for two-and-half years reveals how class differences undermined residents' efforts to have their legitimate needs considered by, or get answers from, Venezuelan government officials (Peattie 1972: 88). Peattie (1987) admits frustration in not being able to influence the planning team and in an extended critique identifies a series of obstacles. While planners aimed for "efficiency, amenity, social equity and community" in their design, she observes the new city lacked these (Peattie 1987: 15). Peattie argues that the designers' vision of a future modern and progressive city ignored the actual people who already lived there, imagining in their place totally different inhabitants. Moreover, urban designers' reliance on architectural models and representations to measure their professional design accomplishments caused them to misunderstand social realities and fail to institute processes that would guide or mobilize actors to a desirable outcome (Peattie 1987: 60–68). Peattie also observes that professional specialists on the planning team, especially the social scientists, were recruited by political elites, who represented corporate interests, in order to legitimize rather than improve the outcome (1987: 164). Although designers, transportation planners, economists, and engineers competed to control the design process, their professional practices were protected by political institutions as long as the urban design was consistent with Venezuela's national project.

The lessons for urban designers drawn by Peattie and other anthropologists revolve around two themes. First, ethnographic descriptions of lower income populations in developing countries provide critical data to inform and sensitize design professionals to the social complexities and sense of community present in many poor communities, and encourage designers to serve and support these populations in addition to their elite clients. Second, the ethnography of urban design practices reveals their social and political construction around the asymmetrical distribution of power. Designers' conceptual and representational tools may be ineffective counter forces to this power which limits understandings of social realities and obscures alternative design strategies and solutions.

Ethnography: urban sites and identities

The sheer size and heterogeneity of cities has stimulated debate among urban anthropologists about the appropriate unit of analysis: families, gangs, ethnic or tribal groups, neighborhoods, or ephemeral events. Of particular interest to urban designers are site-specific ethnographies that focus on the political-economic and cultural dimensions of single or multiple ethnic or racial groups or social classes (for reviews of this literature see Low 1999; Sanjek 1990; Smart and Smart 2003). Low (1999) describes two research directions in which ethnicity is used as a construction within the larger social context of the city: the "ethnic city" where, despite the presence of multiple ethnic groups, one ethnicity becomes identified with a city as a result of its self-contained cultural strategy for economic and political dominance; and "ethnic enclaves" which are distinct territorial units formed within the larger urban setting based on ethnic identity with reference to occupational position, immigrant status, marginality, or degree of discrimination. An example of an ethnic city is found in Miami (Portes and Stepick 1993) where the economic and political control once exercised by a white middle class has shifted to a Spanish-speaking Cuban-born immigrant population who have come to dominate the city commercially, politically, and culturally and transform its aesthetic and social character. Another is Monterey Park, California, which was transformed by Hong Kong and Taiwanese émigré professionals and entrepreneurs in the 1980s who now dominate cultural, economic and, increasingly, political institutions creating a distinctly Chinese suburban city (Fong 1994). The potential danger for urban designers in the ethnic city is that other cultural groups may be overlooked.

Ethnic enclaves such as American Chinatowns (Wong 1988; Kwong 1987; Loo 1992; Zhou 1992) (Figure 10.1) traditionally transform their physical environments to assert and express cultural and commercial self-sufficiency: distinct ceremonial gateways, businesses and community organizations, foreign language signs, and distinct architectural forms. Chen (1992) and Fong (1994) observe that as Chinese populations have moved from Chinatowns to the suburbs, the geographic spread of distinct businesses and signage and hybrid architectural designs follows (Figure 10.2). Because successive generations of ethnic populations tend to move away from the original enclave, the ethnic identities of particular sites are dynamic and complex, and cultural meanings and memories attached to place may be buried as new groups move in and make their mark. Low *et al.* (2002) argue that urban design and planning processes may also pose a danger by threatening to erase historical representations and cultural attachments of diverse ethnic groups to urban spaces unless the voices of all the groups are heard.

New York City provides exceptionally fertile ground for anthropological studies of ethnic enclaves, including West Indian (Foner 2001a), Brazilian (Margolis 1993), and Jewish (Kugelmass 1996) communities, and studies of multiethnic communities (Foner 2001b; Sanjek 1998). Relations between members of ethnic groups and government officials, planning professionals and elites, or between ethnic groups

Figure 10.1 New Chinatown, Los Angeles. Source: Denise Lawrence-Zúñiga.
Note: Opened in 1938 to accommodate residents and commercial activity; relocated from the original old Chinatown site.

Figure 10.2 Valley Blvd., City of San Gabriel, California. Source: Denise Lawrence-Zúñiga.
Note: San Gabriel is a contemporary suburb dominated by Chinese and other Asian immigrants.

themselves, are often fraught with miscommunication and conflict. Early ethnographies of African American urban social life (Leibow 1967; Hannerz 1969) adapted "culture of poverty" themes to describe family pathologies and "ghetto" life as different from the mainstream to explain persistent poverty, and conflated race, class and place in constructions of the "black ghetto" or "inner-city" community (Gregory 1998: 9–10). Gregory describes middle and lower income African-American and other residents of a multiethnic New York community whose community activism in the 1980–1990s brought them into contact with this essentialized rhetoric associated with black communities. When activists opposed a light railline through their community, multiple public agencies employed racialized discourse linking deviant and criminal behavior with place – that is, with the same neighborhoods in which the black homeowner activists lived – which divided community loyalties and subdued opposition. That urban designers' and planners' discourse praising the global economic benefits of the project was privileged over residents' complaints, Gregory interprets as exemplary of state hegemony and its effective capacity to "command the social processes through which meanings are publicly articulated, communicated, and invested with contextual authority and social legitimacy" (1998: 246).

Ritualized performances of protest, sentiment and identity in public settings such as streets, squares and parks serve as vehicles of expression for those who are politically disenfranchised or lack recognition as major stakeholders in public deliberations. The well known Pasadena Rose Parade acquired an evil twin in 1978, the satirical Doo Dah Parade, when resident artists and activists organized to contest urban redevelopment plans to gentrify Old Pasadena (Lawrence 1982; 1987). Brooklyn's West Indian Labor Day Parade, a multiethnic event attracting political elites (Kasinitz 1992) and the Greenwich Village Halloween Parade, said to "eroticize the city" (Kugelmass 1994: 165), originated as expressions of specific group identities but have grown in popularity to become lasting ritual symbols of urban places. Urban Native Americans, who do not typically form residential enclaves, express their tribal affiliations away from the reservation by ritually gathering to celebrate powwows at public parks and recreation centers (Weibel-Orlando 1999). Ritualized events such as parades, festivals and cultural performances are ephemeral and may be largely invisible to outsiders, yet they operate as important anchors for the local expression of identity and place attachment (Lawrence 1992), and deserve attention and consideration in urban design processes.

Global cities and cultural hybridity

The radical transformation of urban landscapes due to intensified global flows of people and material resources in the second half of the twentieth century challenges anthropology's prior understandings and produces new ways of writing about and explaining rapidly changing urban populations. Gupta and Ferguson (1997) argue that anthropology's practice of identifying a particular culture with a particular location (e.g. Ethiopian culture is identified with Ethiopia) has made space an unacknowledged organizing principle in the study of socio-cultural phenomena. The failure to recognize the tacit association of people and place creates problems for analyzing new migrant cultures, cultural variations in one locality, post colonial hybrid cultures, and the challenges that autonomous cultures pose to the hegemonic nation-state. With global flows and transnational communications pulverizing space in postmodern society, recent migrants "re-territorialize" urban spaces by inscribing remembered or imagined communities in new physical settings, producing new concepts of community, solidarity, identity and cultural difference (Gupta and Ferguson 1997: 37).

For Appadurai (1996) the spatial production of locality involves social groups in the practical and discursive construction of the "ethnoscape" through rituals of home building and place making, which in turn produces local subjects. That is, a particular immigrant group might engage in a pattern of home construction activities based on their cultural ideal, or ethnoscape; the material products of those creative efforts in turn influence both creators and their many neighbors. In the global deterritorialized, diasporic, and transnational world, the production of locality "as a structure of feeling, property of social life and ideology of a situated community" is a struggle (Appadurai 1996: 189). In building locality, or neighborhood, groups colonize and compete by deploying history, environment and imagination to produce new contexts for defining power relations. While nation-states conceive of localities as sites for incubating and reproducing citizens through national mythologies and celebrations, or through more "disciplinary" techniques such as garbage collection or requiring building permits, immigrants challenge the states' exclusive claims through transnational mobility and electronic communications which produce a more compelling and expansive "neighborhood" in which to experience social life (Appadurai 1996: 190–191).

An example of the production of new urban localities and their social, material and imaginary dimensions can be found in Bubinas (2005) who describes the development of an Asian Indian commercial center, Gandhi Marg, which was officially recognized by the city of Chicago in 1991 with a street sign. Gandhi Marg consists of 150 businesses that cater to a transnational immigrant population in constant contact with the Indian homeland, but also commodify Indian culture for broader popular consumption. Bubinas argues that Gandhi Marg re-territorializes Indian-American identity in an Indian place for commerce and political power thereby moving beyond the idea of ethnic enclave as a segregated immigrant residential zone (2005: 171–173). The theoretical shift in urban ethnography to acknowledge the autonomy of cultural groups in expressing their own identity provides urban designers with specific understandings of the spatial and aesthetic forms that meet group needs, but also suggests arenas in which cultural groups may resist or subvert urban designs.

Other new types of localities are produced in developing countries by transnational and transcultural flows of materials, knowledge and ideas, and mediatized images of the built environment. Distinctive localities in house construction and neighborhood building in suburban and peri-urban locations around developing city centers are produced for and consumed by new elites as a mark of class distinction (Miller 1995; Bourdieu 1984). Beal (2000) describes old- and new-money elites in Amman's outskirts who build outlandish villas symbolizing contrasting traditional and modern lifestyles, and provoking competition about taste and debates about which style better indicates authentic Jordanian citizenship. Pellow (2003) describes the construction of new villas in peri-urban areas in Accra, Ghana, by transnational migrants to the US who earn money and learn about American house styles that they translate into an African idiom for extended family living. In Beijing's new suburbs, the consumption of middle-class modern lifestyles through exclusive residential compounds is central to identity formation for "Chuppies" (Chinese urban professionals) and contributes to growing social disparities in Chinese society (Fleischer 2007).

Class distinction also characterizes historic preservation efforts which draw legitimacy from state regulations, market forces and neoliberal policies to aesthetically transform neighborhoods and privilege the interests of professionals and middle classes over lower income residents. Williams (1988) describes a multiethnic neighborhood in Washington, D.C. where newly arrived middle-class white home owners advocated historic designation to resist developer threats to their nostalgic construction of community, but they encountered indifference among their neighbors, both older non-white owners and renters. When upper income blacks displaced lower income blacks in an historic Chicago neighborhood, they justified their behavior by using "racial uplift" rhetoric, referring to a shared racial identity, to naturalize inequalities between renters' and owners' interests (Boyd 2005). State imposition of historic preservation regulations without consideration for their complexity or local consent can result in homeowners' subversive techniques, such as nocturnal construction, to upgrade antiquated homes, as Herzfeld (1993) describes in the historic center of Rethemnos, Crete. Historic preservation regulations and practices privilege and naturalize advocates' knowledge and taste, while tacitly legitimizing the exclusion of lower income residents (Lawrence-Zúñiga forthcoming). Their advocacy by urban designers as a means to "preserve history" risks the selective promotion of elite concepts of history not shared by everyone.

Other exclusionary localities based on fear appear in the international proliferation of gated communities and fortress architecture constructed to segregate residents from urban crime and danger (*City and Society* 2004). Low (2003) argues that Americans' encounters with crime, or fear of crime, has caused them to seek safety in gated housing developments, where they effectively avoid unwanted contact with the "others" they fear: workers, Mexicans, the poor, and newcomers. Caldeira (2001) argues that São Paulo's middle and upper classes have increasingly sought walls to separate and protect them from violence and danger on the streets. While new condominiums are advertised as a total and secure way of life, ironically, the "dangerous" but excluded lower classes must still enter to provide needed domestic services. According to Caldeira, the fortress mentality and private enclaves in São Paulo and Los Angeles threaten the vitality of the public sphere (1999: 125). Urban designers play critical roles in the evaluation and promotion of new urban forms, whether they are generated and financed privately, such as gated communities, or involve state legitimization of private investment in historically preserved neighborhoods. These new forms create potential conflicts between and within ethnic and class segments of urban society, and they may have unrecognized impacts on the larger public sphere which ethnography can clarify.

Public space ethnography and the critique of urban design practices

The anthropology of urban public spaces originated with proxemics, the study of human uses of space as a form of non verbal communication the patterns of which vary by culture (Hall 1959; 1969). The systematic observations of public behaviors revealed distinct patterns of body positions and distancing in specific settings, which contributed to an understanding of the social and normative orders of specific spaces (Goffman 1966). While these approaches provide standard conceptual and methodological tools for urban designers seeking to understand the ordinary behavioral patterns of everyday spaces, recent anthropological research reveals how particular urban design practices tend to include or exclude specific populations from these spaces. Low contrasts the social production of space, involving its physical creation conditioned by social, economic, ideological and technological factors with the social construction of space, referring to people's everyday experiences through social exchanges, memories, images and uses that give spaces meaning (Low 2000: 128). In a study in the capital city of San José, Costa Rica, Low analyzes how urban design professionals, and political elites, produce urban public spaces, encoding them with ideological meanings and practical aspirations that may contain contradictory goals. City officials created a design that would return the Parque Central to its historic glory after hearing complaints about its decline from middle-class professionals; the design displaced lower-class users in favor of elites who, ironically, do not use it (Low 2000: 188).

Similar cases that ignore or misunderstand segments of the population have been replicated by anthropological studies of urban design and redevelopment processes. Cooper (1993) and Sieber (1993) find waterfront redevelopment schemes in North America and Canada often privilege upscale users to the detriment of the working classes who once used them. Rutheiser (1996) criticizes Atlanta's "Imagineering" urban redevelopment scheme in preparation for the 1996 Olympic Games that packaged the city as a commodity, while doing little to resolve long term social problems. And McDonogh found planners' discourse exhorting residents in Barcelona's historic but notorious Ravel district to be "good citizens" by appreciating redevelopment design proposals, effectively silenced their opposition and reproduced the same inequalities the scheme sought to overcome (McDonogh 1999: 368). These studies illustrate how the larger social production of space in which urban designers play a pivotal professional role, creates outcomes that do not satisfy and may exclude local populations, either as a result of contradictions between goals or because of social realities that are ignored, just as Peattie (1987) observed.

Some anthropologists have examined formal design theories themselves as instruments of state power in urban social life. Rotenberg (1995) explores how the history of landscape design philosophies expressed in the production of a succession of Viennese gardens, from which the public was often excluded, showed that each new garden style contested the social veracities and power relations represented in the previous one. Following Foucault, Rabinow (1989) investigates the "emergence of modern urbanism" in French colonial Morocco under Governor-General Lyautey (1912–1925) who applied scientific norms and techno-social forms to create parallel but "superior" French modern urban plans, juxtaposed to traditional Moroccan cities, with which to control and regulate citizens' behavior (Rabinow 1989: 277).

Like previous scholars of Brasilia, Holston (1989) shows how a radical vision for a classless capital city resulted in residential segregation. His ethnography specifically investigates how modernist design and planning principles, derived from CIAM's 1933 Athens Charter and used to organize physical spaces, disrupt citizens' familiar perceptions and behavior. In Brazil's "traditional" colonial city of Ouro Preto, for example, buildings act as "figure" to enclose private behaviors, contrasting with streets and plazas that serve as "ground" for public activities. Brasilia's modernist design inverts and neutralizes this spatial order by eliminating the figural street and intersections, and makes all buildings public monuments surrounded by open public space. Holston observes that Brasilia's residents rejected and subverted modernist design intentions by converting the rear service access in commercial buildings to front entrances, reproducing the familiarity of the traditional colonial urban plan (1989: 139). In a later study, Holston (2008) considers workers' self-constructed houses "sites of insurgent citizenship" where everyday expressions of vitality and creativity contest state control and challenge modernism's grand planning theory. He argues that planning and design goals might be better achieved if schemes could incorporate possibilities for the multiple citizenships documented in the ethnographies of new urban residents who re-territorialize urban landscapes in rapidly changing global cities.

Along these same lines, Herzfeld (2006) proffers a critique of state agencies in his recent comparison of historic preservation-driven urban redevelopment in Italy, Greece and Thailand. Herzfeld argues that nations such as Thailand, which struggle to achieve legitimacy in the global arena, employ a "western-derived model of statehood" to inform their urban redevelopment practices (Herzfeld 2006: 145). Their imposition of aesthetically driven architectural "forms of order" on urban landscapes results in the categorization of local inhabitants as "matter out of place," as polluted, that must be sanitized, excluded or reformed. Both Holston and Herzfeld see urban design as a totalizing if not hegemonic aesthetic regime which is fundamentally exclusionary in its application and practice and which ethnography may help correct.

Conclusion

Urban anthropologists have long sought to understand and represent the cultural patterns, values and aspirations of the people they study within contexts of deep social complexity and asymmetrical power. For the urban designer, ethnographic research is critical for revealing and explaining the needs of overlooked populations in planning processes, to help avoid their outright exclusion or inappropriate solutions. Consistent with Peattie's insightful analysis in the 1980s, however, anthropologists' ethnographic research on urban design theories, techniques and processes has also provided a way to influence the social production of design. Encouraging urban designers to consider social, economic and political constraints on their professional roles, in addition to the cultural needs of specific urban populations, is important to anthropologists who endeavor to reveal why these populations are overlooked in the first place.

References

Appadurai, A. (1996). *Modernity at Large: Cultural Dimensions of Globalization,* Minneapolis: University of Minnesota Press.

Beal, E. (2000). "Real Jordanians Don't Decorate Like That! The Politics of Taste among Amman's Elites," *City & Society* 12(2): 65–94.

Bourdieu, P. (1984). *Distinction,* Cambridge, MA: Harvard University Press.

Boyd, M. (2005). "The Downside of Racial Uplift: The Meaning of Gentrification in an African American Neighborhood," *City & Society* 17(2): 256–288.

Bubinas, K. (2005). "Gandhi Marg: The Social Construction and Production of an Ethnic Economy in Chicago," *City & Society* 17(2): 161–179.

Caldeira, T. (1999). "Fortified Enclaves: The New Urban Segregation," in S. Low (ed.) *Theorizing the City,* New Brunswick, NJ: Rutgers University Press.

—— (2001). *City of Walls,* Berkeley, CA: University of California Press.

Chen, H. (1992). *Chinatown No More,* Ithaca, NY: Cornell University Press.

City and Society (2004). "Gated Communities and Other Forms of Urban Segregation." *City & Society* 16(2).

Cooper, M. (1993). "Access to the Waterfront: Transformations of Meaning on the Toronto Lakeshore," in R. Rotenberg and G. McDonogh (eds.) *The Cultural Meaning of Urban Space,* Westport, CT: Bergin & Garvey.

Epstein, D. (1972). "The Genesis and Function of Squatter Settlements in Brasilia," in T, Weaver and D. White (eds.) *The Anthropology of Urban*

Environments, Boulder, CO: Society for Applied Anthropology.

—— (1973). *Brasilia, Plan and Reality: A Study of Planned and Spontaneous Urban Development,* Berkeley, CA: University of California Press.

Fleischer, F. (2007). "To Choose a House Means to Choose a Lifestyle: The Consumption of Housing and Class-Structuration in Urban China," *City & Society* 19(2): 287–311.

Foner, N. (ed.) (2001a). *Islands in the City: West Indian Migration to New York,* Berkeley, CA: University of California Press.

—— (2001b). *New Immigrants in New York,* New York: Columbia University Press.

Fong, T. (1994). *The First Suburban Chinatown,* Philadelphia, PA: Temple University Press.

Goffman, E. (1966). *Behavior in Public Places,* New York: Simon & Schuster.

Gregory, S. (1998). *Black Corona: Race and the Politics of Place in an Urban Community,* Princeton, NJ: Princeton University Press.

Gupta, A. and J. Ferguson (1997). "Beyond 'Culture': Space, Identity and the Politics of Difference" in A. Gupta and J. Ferguson (eds.) *Culture, Power, Place: Explorations in Critical Anthropology,* Durham, NC: Duke University Press.

Hall, E. (1959). *The Silent Language,* Garden City, NY: Doubleday & Co.

—— (1969). *The Hidden Dimension,* Garden City, NY: Doubleday & Co.

Hannerz, U. (1969). *Soulside: Inquiries into Ghetto Culture and Community,* New York: Columbia University Press.

Herzfeld, M. (1993). *A Place in History: Social and Monumental Time in a Cretan Town,* Princeton, NJ: Princeton University Press.

—— (2006). "Spatial Cleansing: Monumental Vacuity and the Idea of the West," *Journal of Material Culture* 11(1/2): 127–149.

Holston, J. (1989). *The Modernist City: An Anthropological Critique of Brasilia,* Chicago: University of Chicago Press.

—— (2008). *Insurgent Citizenship: Disjunctions of Democracy and Modernity in Brazil,* Princeton, NJ: Princeton University Press.

Kasinitz, P. (1992). *Caribbean New York,* Ithaca, NY: Cornell University Press.

Kugelmass, J. (1994). *The Greenwich Village Halloween Parade,* New York: Columbia University Press.

—— (1996). *The Miracle on Intervale Avenue: The Story of a Jewish Congregation in the South Bronx,* New York: Columbia University Press.

Kwong, P. (1987). *The New Chinatown,* New York: Hill & Wang.

Lawrence, D. (1982). "Parades, Politics and Competing Urban Images: Doo Dah and Roses," *Urban Anthropology* 11: 155–176.

—— (1987). "Rules of Misrule: Notes on the Doo Dah Parade," in A. Falassi (ed.), *Time Out of Time: Essays on the Festival,* Albuquerque, NM: University of New Mexico Press.

—— (1992). "Transcendence of Place: The Role of La Placeta in Valencia's Las Fallas," in I. Altman and S. Low (eds.) *Place Attachment,* New York: Plenum.

Lawrence, D. and S. Low (1990). "The Built Environment and Spatial Form," *Annual Review of Anthropology,* 19: 453–505.

Lawrence-Zúñiga, D. (forthcoming) "Cosmologies of Bungalow Preservation: Identity, Lifestyle and Civic Virtue," *City & Society.*

Leibow, E. (1967) *Tally's Corner,* Boston, MA: Little, Brown & Co.

Lewis, O. (1959). *Five Families: Mexican Case Studies in the Culture of Poverty,* New York: Basic Books.

—— (1966). *La Vida: A Puerto Rican Family in the Culture of Poverty – San Juan and New York,* New York: Random House.

Lobo, S. (1982). *A House of My Own: Social Organization in the Squatter Settlements of Lima, Peru,* Tucson, AZ: University of Arizona Press.

Loo, C. (1992). *Chinatown: Most Time, Hard Time,* New York: Praeger.

Low, S. (1988). "Housing Organization and Social Change: A Comparison of Programs for Urban Reconstruction in Guatemala City," *Human Organization* 47: 15–24.

—— (1999). "Introduction: Theorizing the City," in S. Low (ed.) *Theorizing the City,* New Brunswick, NJ: Rutgers University Press.

—— (2000). *On the Plaza: The Politics of Public Space and Culture,* Austin, TX: University of Texas Press.

—— (2003). *Behind the Gates: Life, Security and the Pursuit of Happiness in Fortress America,* New York: Routledge.

Low, S. and D. Lawrence-Zúñiga (eds.) (2003). *The Anthropology of Space and Place,* Malden, MA: Blackwell.

Low, S.D. Taplin, S. Scheld, T. Fisher (2002). "Recapturing Erased Histories: Ethnicity, Design, and Cultural Representation – A case Study of Independence National Historical Park," *Journal of Architectural and Planning Research* 19(4): 282–299.

Marcus, G. and M. Fischer (1986). *Anthropology as Cultural Critique: An Experimental Moment in the Human Sciences,* Chicago: University of Chicago Press.

Margolis, M. (1993). *Little Brazil: An Ethnography of Brazilian Immigrants in New York City,* Princeton, NJ: Princeton University Press.

Mangin, W. (ed.) (1970). *Peasants in Cities.* Boston, MA: Houghton Mifflin.

McDonogh, G. (1999). "Discourses of the City: Policy and Response in Post-Transitional Barcelona," in S. Low (ed.) *Theorizing the City,* New Brunswick, NJ: Rutgers University Press.

Miller, D. (ed.) (1995). *Acknowledging Consumption,* London: Routledge.

Peattie, L. (1972). *A View Form the Barrio,* Ann Arbor, MI: University of Michigan Press.

—— (1987). *Planning: Rethinking Ciudad Guyana.* Ann Arbor, MI: University of Michigan Press.

Pellow, D. (2002). *Landlords and Lodgers,* Chicago: University of Chicago Press.

—— (2003). "New Spaces in Accra: Transnational Houses," *City & Society* 15(1): 59–86.

Perlman, J. (1976). *The Myth of Marginality,* Berkeley, CA: University of California Press.

Portes, A. and A. Stepick (1993). *City of the Edge: The Transformation of Miami,* Berkeley, CA: University of California Press.

Rabinow, P. (1989). *French Modern: Norms and Forms of the Social Environment,* Cambridge, MA: MIT Press.

Rotenberg, R. (1995). *Landscape and Power in Vienna,* Baltimore, MD: Johns Hopkins Press.

Rutheiser, C. (1996). *Imagineering Atlanta: Making Place in the Non-Place Urban Realm,* New York: Verso.

Sanjek, R. (1990). "Urban Anthropology in the 1980s: A World View," *Annual Review of Anthropology* 19: 151–186.

—— (1998). *The Future of Us All: Race and Neighborhood Politics in New York City,* Ithaca, NY: Cornell University Press.

Sarin, M. (1982). *Urban Planning in the Third World: The Chandigarh Experience,* London: Mansell.

Sieber, T. (1993). "Public Access on the Urban Waterfront: A Question of Vision," in R. Rotenberg and G. McDonogh (eds.) *The Cultural Meaning of Urban Space,* Westport, CT: Bergin & Garvey.

Smart, A. (2001). "Unruly Places: Urban Governance and the Persistence of Illegality in Hong Kong's Urban Squatter Areas," *American Anthropologist* 103(1): 30–44.

Smart A. and J. Smart (2003). "Urbanization and the Global Perspective," *Annual Review of Anthropology* 32: 263–285.

Turner, J. and R. Fichter (eds.) (1972). *Freedom to Build,* New York: Macmillan.

Valentine, C. (1968). *Culture and Poverty,* Chicago: University of Chicago Press.

Weibel-Orlando, J. (1999). *Indian Country, LA: Maintaining Ethnic Community in Complex Society,* Revised edition, Urbana-Champaign, IL: University of Illinois Press.

Williams, B. (1988). *Upscaling Downtown: Stalled Gentrification in Washington, D.C.,* Ithaca, NY: Cornell University Press.

Wong, B. (1988). *Patronage, Brokerage, Entrepreneurship and the Chinese Community of New York,* New York: AMS Press.

Zhou, M. (1992). *Chinatown: The Socioeconomic Potential of an Urban Enclave.* Philadelphia: Temple University Press.

Further reading

Hancock, Mary E. (2008). *The Politics of Heritage from Madras to Chennai.* Bloomington, IN: Indiana University Press. Historical and ethnographic account of the spatial dimensions of cultural memory focusing on its expression in the materiality of landscape and the built environment in postcolonial India.

Holston, James (2008). *Insurgent Citizenship: Disjunctions of Democracy and Modernity in Brazil.* Princeton, NJ: Princeton University Press. It investigates how working-class residents of two peripheral settlements in São Paulo, Brazil, have mobilized to assert their "rights" and claims to citizenship, both materially in the expression of self-built homes and by activating the public sphere.

Low, Setha, Dana Taplin, and Suzanne Scheld (2005). *Rethinking Urban Parks: Public Space and*

Cultural Diversity. Austin, TX: University of Texas Press. An excellent discussion of and handbook for incorporating and promoting cultural diversity in the design and planning of large scale public spaces.

Pellow, Deborah (2008). *Landlords and Lodgers: Socio-Spatial Organization in an Accra Community*. Chicago: University of Chicago Press. For urban designers who believe that slums should be razed because they house socially dysfunctional populations in overcrowded and insalubrious conditions, this ethnography of an Accra Zongo should provide a contemporary dissuasion. Pellow describes highly dense and sociable living conditions that cause many residents who can easily afford improved housing elsewhere to choose the slum because of emotional and social attachments.

Weiss, Brad (2009). *Street Dreams and Hip Hop Barbershops: Global Fantasy in Urban Tanzania*. Bloomington, IN: University of Indiana Press. This award-winning ethnographic account describes young African men, who gather in urban barbershops and streets, whose creative impulses find recognition in the global circulation of hip hop music, fashion and fame.

11

Feminist approaches to urban design

Kristen Day

The design of cities and suburbs in the second half of the twentieth century has often neglected women's needs and their lived experiences. Women are disadvantaged in settings that were not created with their views and experiences in mind (Greed 2006). Consider, for example, the creation of isolated suburbs where mothers strive to care for households and participate in paid employment without ready access to nearby stores, schools, and jobs; the design of transportation systems meant to accommodate single adult commuters on their journeys to work, rather than women with children running errands; and the layout of urban environments that does not ensure safe travel. Increasingly, we recognize that fundamental changes in urban design and form are needed to create cities that are more equitable for women.

In the last three decades, research and practice have begun to address this gap. Scholars in urban planning, geography, architecture, anthropology, environmental psychology, and other fields have explored women's relationships with built environments (cf. Ahrentzen 2003; Altman and Churchman 1989; Anthony 2001; Berkeley and McQuaid 1989; Dandakar 1993; Greed 1994; Miranne and Young 2000; Rendell, Penner, and Bordon 2000; Rose 1993; Rothschild 1999; Spain 1992; Sprague 1991; Weisman 1992; Wilson 1991). This chapter focuses specifically on the gaps and opportunities revealed by feminist approaches to urban design.

Feminist perspectives

While no single definition of "feminism" prevails, feminist perspectives share a belief that justice requires freedom and equality for women. These approaches argue that patriarchy – a social system that attaches power to masculine gender – disadvantages women. Patriarchy burdens women through the gendered division of labor and activities, gendered access to resources, and the construction of gendered identities (Law 1999).

Feminist perspectives emphasize the differences between women and men (Greed 2006; Sandercock and Forsyth 1992). If we assume that no differences exist, then we may create systems and spaces that reinforce the status quo (Rakodi 1991; Wallace and Milroy 1999; Weisman 1992). In considering difference, we must also consider differences among women themselves (Anthony 2001). Race/ethnicity, class, sexuality, religion, physical ability, age – all shape women's experiences and their relative privilege. Increasingly, feminist scholars recognize that the views and experiences of white, middle class women (whose voices dominated the US women's movement until the 1980s), do not represent the priorities and experiences of all women (Sandercock and Forsyth 1992).

The consideration of gender complicates and enriches urban design scholarship. Historically, urban design has emphasized the human experiences of place and the needs of users who will occupy the places created by designers and others. If, however, this focus on "residents" or "users" ignores gender and other identities, then it may mask differences in needs, perceptions, and experiences of the built environment (Rakodi 1991). Feminist approaches to urban design correct this oversight, by exploring how women's identities shape their use of urban environments, and how the design of cities and communities can better accommodate women's needs. Key groups of women to consider are those who are most disadvantaged by current design and planning practices, such as lower income workers, working mothers and single headed households, and older women (Rakodi 1991).

Many of the classic works on women and environments were written in the 1980s (see for example, Hayden 1980, 1984; Leavitt and Saegert 1989; Matrix 1984; Mazey and Lee 1983; McDowell 1983; Stimpson *et al.* 1981; Wekerle, Peterson, and Morley 1980). This classic literature focuses primarily on the experiences of white, middle class women (Miraftab 2007). More recently, empirical research has expanded to involve diverse groups of women in settings that vary by place type and geographic location. Scholars increasingly address the use of urban environments by women in developing countries (cf. Chhibber 2002; Dandekar 1993; Njoh 1999). The experiences of minority and low-income women in US and Western environments have received less systematic attention.

This chapter reviews research and theory tied to the experiences of women in different urban settings. In each setting, women's experiences can be understood as constrained, constraining, and/or as resisting (after Shaw 1994; Day 1999a). Constraints disadvantage women's use of environments. Constraints include housework and childcare responsibilities that limit women's ease of travel, and traditional gender norms for safety and modesty that hinder women's freedom in public spaces. Women's use of urban environments is potentially constraining when these experiences reinforce or reproduce oppressive gender relations. Examples include recreational spaces for women that encourage frivolous consumption (many shopping environments fit this description) or spaces that reinforce our preoccupation with women's physical appearance (such as nail and tanning salons). The use of urban environments can constitute resistance when women claim their own space and challenge restrictive gender norms about where they belong. Examples might include women's health centers and women's bookstores.

Feminist critiques of the separation of land uses

Women are fundamentally restricted by the separation of land uses and the distinction between public and private roles. In Western cities, this distinction has its roots in the Victorian "separation of spheres," which delineated separate economic and spatial realms for women and men (Franck and Paxson 1989; Hayden 1984; Rose 1993). Historically, private (domestic) spaces and virtues were associated with women, and public spaces and activities with men. The capitalist economy (dividing production and reproduction) and suburbanization further reinforced this dualism (Valentine 1992). For many low income women and women of color, however, restriction to home and domestic sphere was a "luxury" that was rarely achievable (Rose 1993). These women's daily routines necessitated significant time spent working in other women's homes and in public settings.

The rigid separation of land uses into public and private, urban and suburban, still disadvantages women in multiple ways. Dolores Hayden's landmark *Redesigning the American Dream* (1984), documents how traditional suburban environments encourage individual consumption and impede women in performing their multiple roles as workers and mothers. At the same time, in the US and elsewhere, low income, predominantly minority women remain isolated in urban environments with limited employment, housing, and educational opportunities (Massey and Denton 1993). The problems identified by Hayden and other feminist writers – the lack of public transportation to everyday destinations, the unwieldy distances between homes and places of employment, the absence of nearby shops – will sound familiar to today's urban designers and planners. These critiques resonate with the more recent New Urbanist, Smart Growth/sustainability, and Active Living movements. Arguably, these newer movements have had more impact on design and planning practice (Greed 2006). Feminist perspectives, however, have been notably absent from these recent movements, raising questions about how best to link feminist scholarship and urban design practice.

Feminist approaches to urban design have blurred rigid distinctions between public and private, bringing some "private" issues into public conversation (for example, sexual assault in public spaces, Day 2000a), and reframing some "public" issues as private decisions (for example, the legal definition of who can live in a household, Ritzdorf 1994). In women's lives, rigid boundaries between public and private may be meaningless and constraining.

Women's use of public spaces

Contemporary Western and especially US urban design scholarship reveals a nostalgia for a perceived loss of public life (Brill 1989) and a scorn for the increasing privatization of public spaces (cf. Huxtable 1997; Sorkin 1992). Critics advocate a return to the traditions of idealized, "truly public" spaces to overcome limitations on civil rights (free speech, assembly), increased exclusion, and a growing focus on consumption in public space.

From a feminist perspective, however, there is no such thing as "truly public" space that is experienced in the same way by all groups (Mozingo 1985; Ruddick 1996). Gender shapes women's experiences of public space. The oft-celebrated right to observe and mingle with strangers in public space, for example, is not shared equally among women and men. Women are less likely to approach strangers in public space and more likely to be approached by them, than are men (cf. Henley 1977). Experiences of objectification (of the male gaze) can shape some women's use of urban environments (Borlsoff and Hahn 1997; Gardiner 1989). Also, the characterization of an idealized public sphere where all come together in equal and free exchange of ideas, does not resonate with some women's experiences (Fraser 1992). In accounts of urban life, women are typically characterized as part of the "background," rather than as part of the "action" (Lofland 1975, in Sandercock and Forsyth 1992).

For many women, responsibility for home and children and fear for safety constrain their activities in public space (Franck 2002; Franck and Paxson 1989; Harrington *et al.* 1992). Gendered social norms further limit women's public space participation (Gardiner 1989; 1994), by encouraging women to curtail their behavior to keep up socially desirable self-presentations of femininity.

Women's bodily experiences of public spaces are also distinctive. For example, women may have smaller "personal space" bubbles than men. People tend to stand closer to women than to men, and women move out of the way for others more often than do men (Mozingo 1989). Women are touched more in public spaces than are men. Women often find crowding less stressful, compared to men, and may even find some crowded situations appealing (Mozingo 1989), assuming that crowding does not involve groping or sexual harassment.

Women's use and experience of public spaces differ significantly with race/ethnicity, culture, sexuality, age, and physical ability. Recent years have seen an increase in research on women's use of public spaces around the world and especially in developing countries (cf. Alizadeh 2007; Chhibber 2002; Mazumdar and Mazumdar 2001; Mills 2007; Sangwha 1999; Seedat *et al.* 2006). Much of this research involves case studies of women in one country or city. Still needed are comparative studies that integrate these cases and advance theories of women and public space.

There is danger in overstating women's constraints in public space. Certainly, women enjoy public spaces and traverse them freely under many circumstances (Lofland 1984; Wilson 1991). Indeed, women's use of public space can constitute resistance, when women define their own identities through participation in self-determined, meaningful activities. Consider, for example, women's use of lesbian bars (Wolfe 1992), or creation of feminist public art (Lacy 1995), or young Latinas' claims on dangerous urban street environments (Hymas 2003), and even homeless women's occupation of highly visible public spaces (Casey *et al.* 2008), as cases in point.

If the goal of urban design is to create accessible, diverse, and open public spaces, then we must recognize that no single setting will meet the needs of all groups at all times (Franck and Paxson 1989). Rather, it is more appropriate to think about a network of spaces that can accommodate the meaningful characteristics of specific social groups. Public spaces will be more useful for women if these spaces provide perceived and actual safety and facilitate women's multiple roles by allowing women to conveniently entertain children, complete work tasks, and/or accomplish household responsibilities such as shopping or other errands. Examples include airports that offer play spaces for children and fitness centers that provide child care.

Women and transportation

Since the late 1970s, feminist scholars have examined the role of gender in travel behavior and the implications of women's travel for the design of cities and transportation systems (cf. early work by Giuliano 1979; Rosenbloom 1978; 1980). This research is part of a broader recognition of the mobility needs of "transportation disadvantaged" groups including women, older adults, and others (Law 1999). Early studies characterized women as deprived in their access to cars, dependent on public transportation, and burdened in their travel by children and household responsibilities (Coleman 2000). Later studies have provided more nuanced descriptions of the travel experiences of diverse women. Research on women and transportation focuses predominantly on developed countries and especially emphasizes women's work trips (Law 1999).

Women's mobility continues to be constrained by factors that include gendered division of household and childrearing labor, gendered access to time and money, gendered attitudes about women and travel, and segregated patterns of urban land uses (Law 1999; Njoh 1999). The separation of land uses, discussed earlier, has important implications for women's mobility, making it more difficult for women to travel between different uses and increasing children's and others' dependency on women for transportation.

Women's travel and mobility are distinctive in many ways. Because of their greater responsibility for children and households, women's trips are more likely to be multipurpose and "trip chained" (multiple trips strung together), compared to men's travel (Blumenberg 2004; Hamilton 2000; Hu and Young 1999). Women generally make about the same number of trips as men, but women's trips are often shorter and more local in nature, making support for travel to nearby destinations especially important. Due to differential access to cars and the shorter nature of some women's work trips, women are more likely to travel on foot or public transportation than are men (Greed 2006), though Black and Latina women do not necessarily have shorter commutes to work (Law 1999).

At the same time, women may be less likely to cycle to work compared to men, due to safety concerns, a lack of changing facilities at work, and beliefs about women's proper appearance (Greed 2006).

Planning for public transportation has typically concentrated on work trips during prime commuting times (Blumenberg 2004). This is problematic, since women (who frequently work part time) are less likely to travel at rush hour than are men (Greed 2006; Njoh 1999; Rakodi 1991). Planners sometimes view women's nonwork trips as a nuisance that slows and interferes with public transportation planning (Greed 2006). Instead, we must recognize women's travel as essential activity and design transportation systems to serve the times when women – and men – need to travel. This may mean, for example, more investment in bus transportation during evenings and weekends, rather than the creation of additional park and ride facilities to serve workers during traditional commuting hours (Hamilton 2000).

Public transportation should consider the needs of women with children, who may face special burdens while traveling (strollers, need for restrooms, etc.); women conducting household errands that require carrying heavy or bulky loads; and older women, who are less likely to have driver's licenses (Coleman 2000; Pickup 1989; Rosenbloom and Winsten-Bartlett 2002). Such consideration would improve access to public transportation from different parts of the city, and lead to the design of systems with chairs for sitting and waiting, fewer steps, places for strollers and bags on board, and other accommodations.

The realities of women's travel may cause us to reconsider our prescriptions about what constitutes "good urban design and planning." Contemporary urban designers strongly advocate a shift away from cars to public transportation to promote sustainability and to increase physical activity. And yet many of women's car trips actually provide efficient transportation for others in the household (Greed 2006; Law 1999). Many such trips (chained together, involving children) would be difficult to accommodate by most public transportation systems, especially in suburban environments where public transportation is more limited. Car travel may be more necessary for women with young children than for other groups (Hillman *et al.* 1974, in Pickup 1984). In fact, in terms of increasing low income and single mothers' mobility and their access to more and better jobs, policies to increase auto ownership may actually be more helpful than focusing exclusively on increasing access to public transportation in urban environments (Blumenberg 2004).

Women and safety in urban environments

Extensive research examines women's experiences of fear and safety in the city. Women consistently report greater fear in urban environments than do men (Gordon and Riger 1989; Grabowsky 1995; Stanko 1987). Fear especially impacts those women with the fewest resources to ensure their safety. In the US and other Western countries, women who are most fearful include older women, women with limited education and lower incomes, and women of color (Gordon and Riger 1989; Pain 1997a; Thompson *et al.* 2002). These women are more likely to reside in high crime neighborhoods, which may explain their higher fear (Gordon and Riger 1989; Loukaitou-Sideris and Fink 2009; Pain 1997b).

Physical features associated with women's (and men's) fear of crime include the presence of hiding places, limited vistas, and low potential for escape (Fisher and Nasar 1992; Nasar and Fisher 1992); graffiti; poor maintenance; dense vegetation; and inadequate lighting (Cooper Marcus and Wischemann 1983; Day 2000a; Nasar and Fisher 1992; Wekerle and Whitzman 1995). Fearful places include pathways, alleys, bus and transit stops, parking lots, tunnels, and natural areas (Cooper Marcus and Wischemann 1983; Gordon and Riger 1989; Loukaitou-Sideris and Fink 2009; Loukaitou-Sideris *et al.* 2002). Women's fear is especially heightened at night time (Valentine 1992; Warr 1990). Social incivilities, such as public drinking, panhandlers, and rowdy crowds, are also tied to fear in urban environments (Day 2000a; Rohe and Burby 1988).

Women's fear in urban environments is attributed to many factors, including past victimization, women's sense of themselves as physically weak, warnings of women's vulnerability, and especially women's specific fear of sexual assault (Gordon and Riger 1989; Loukaitou-Sideris and Fink 2009). Women are victims of crime in both public and private places. Yet women's primary association of fear with public spaces belies the reality that women are more often victimized in private and domestic environments (Gordon and Riger 1989; Koskela and Pain 2000; Valentine 1992).

For women of color, the notion of safety in urban environments is broader than the absence of assault or disorder. Safety also involves feeling welcome and accepted in a setting (see Day 1999b). Walking alone in a neighborhood, hiking in an urban park, or participating in community events require reassurance that individuals will not "stand out" uncomfortably in terms of race or ethnicity, and will not be targeted by race harassment or violence.

Fear functions as a form of social control over women's use of urban environments, since women are persuaded to significantly curtail their travel and behavior in public spaces out of fear (Deegan 1987; Valentine 1989). Women have made considerable strides in reversing their exclusion from public spaces, and yet social rules for appropriate behavior for women still restrict their full and equal access. These social norms designate "unseemly" places where women should not go – especially not alone or at night, or else risk sexual assault or harassment and be blamed for any harm that may occur (Gardiner 1989; 1994). More recently, researchers have expanded the study of women and fear to also examine women's resistance to fear in urban environments (Hyams 2003; Koskela 1997). This research is important for helping us to understand women as bold and assertive users of urban environments and not only as victims.

The question of fear in urban environments is one of the few areas in urban design research where we also see research that addresses *men's* experiences from a gender (and typically a feminist) perspective. Such research is still in the early stages. For many men, fear in urban settings is intimately tied to their masculine identities. Settings can be judged fearful depending, in part, on whether they challenge men's masculine identities. Men's fear in urban environments may be tied to the need for control and to potential confrontation with other men (Day 2006; Day *et al.* 2003). Race and racism critically shape men's experience of fear and of being feared in urban settings (Brownlow 2004; Day 2006; hooks 1992).

Feminist urban designers and planners have undertaken several initiatives to enhance women's safety in cities (see also the chapter by Whitzman in this volume). One example is the groundbreaking work of METRAC in Toronto, where a special committee has implemented numerous planning projects to increase women's safety (Modlich 1986; Wekerle and Whitzman 1995). Similar efforts have also taken place in the Netherlands (Sandercock and Forsyth 1992). Feminist scholars warn us that we must exercise caution in turning to urban design as the (only) solution to enhancing women's safety in urban environments (Koskela and Pain 2000). Many of the underlying issues that cause women's fear and danger will not be resolved by better lighting and safer transit, as important as these issues are. Indeed, increasing women's safety will also require a fundamental rethinking of women's roles and place in the city.

Conclusions

Research on women and environments – in urban design and in other fields – has proliferated over the past three decades. Researchers have shifted their focus over time in accordance with changes in urban design and women's studies scholarship. As in other areas of feminist research, the emphasis is increasingly on the construction of gender identities in urban environments, and less on the identification of constraints to women's use of cities (Law 1999). This shift in focus has both costs and benefits. It encourages us to identify structural factors that disadvantage women in urban environments, but it may neglect practical issues that must be addressed to improve "conditions on the ground."

Despite many recommendations to improve gender equity in urban design and planning, actual impacts on design and planning practice have been limited (Greed 2006). Model programs do exist, such as the METRAC program in Toronto, discussed earlier (Wekerle and Whitzman 1995). In other examples, in Italy, recent legislation allows mayors to coordinate the hours of employment, retail, and other facilities, to allow women to balance employment with their substantial family responsibilities (Belloni 1998). In Oslo, Norway, municipal government officials undertook a comprehensive process to incorporate women's perspectives into local planning decision making (Skjerven 1993, in Greed 2006). These are isolated cases, however. We have yet to see a more widespread movement to enhance gender equity in city planning and design. This limited impact may reflect the fact that women still occupy peripheral positions in planning and design decision-making, despite their large numbers in schools of planning (Greed 2006; Sandercock and Forsyth 1992). We must continue to promote the advancement of women and men who support feminist agendas to positions of power in planning and design.

We must also recognize the numerous ways in which women play leadership roles in the shaping of cities and communities. Women are leaders in creating urban gardens; spearheading neighborhood improvements; grassroots organizing; supporting urban parks; establishing national women's policy think tanks; documenting public history; and in struggles around housing, childcare, and neighborhood preservation (cf. Bland 1989; Cranz 1981; Dubrow 2007; Feldman and Stall 1994; Hayden 1997; Rakodi 1991; Spain 2001). These efforts are often driven by a feminist "ethic of care" for places and for the people that occupy them (Day 2000b;

Krenichyn 2004). We must acknowledge that, in a time when cities are abandoning their public responsibilities, these activities can sometimes exploit women's free and unpaid work in the name of "women's empowerment" (Miraftab 2007). At the same time, however, women's leadership in these efforts represents a powerful force for advancing equity in urban design and planning. We should work to strategically link women's community work to formal planning and design processes and resources and to other planning movements (sustainability, active living, etc.) that share similar values.

Finally, we must work to reduce the constraints that shape women's use of urban environments (and especially those tied to caring for children and households), while at the same time challenging the restrictive gender roles that disadvantage women. Often, the most strategic solutions will not be design interventions. We must work with policy makers and others to address underlying issues tied to women's roles and status, while we continue to improve the quality of urban environments to support women's and men's lives.

References

Ahrentzen, S. (2003). "The space between the studs: feminism and architecture." *Signs,* 29(1), 179–206.

Alizadeh, H. (2007). "Changes conceptions of women's public space in the Kurdish city." *Cities,* 24(6): 410–421.

Altman. I. and Churchman, A. (Eds.) (1989). *Women and the environment.* New York: Plenum Press.

Anthony, K.H. (2001). *Designing for diversity. Gender, race, and ethnicity in the architectural profession.* Urbana and Chicago: University of Illinois Press.

Belloni, M.C. (1998). "Tempi delle città: Italy's urban time plans and policies." *Time & Society,* 7: 249–263. At: http://tas.sagepub.com/cgi/content/abstract/7/2–3/249.

Berkeley, E.P. and McQuaid, M. (Eds.) (1989). *Architecture: A place for women.* Washington, DC: Smithsonian Institution Press.

Bland, S. (1989). "'Miss Sue' of Charleston: Saving a neighborhood, influencing a nation." In Berkeley, E.P. and McQuaid, M. (Eds.) (1989). *Architecture: A place for women.* Washington, DC: Smithsonian Institution Press, 63–76.

Blumenberg, E. (2004). "En-gendering effective planning: Spatial mismatch, low-income women, and transportation policy." *Journal of the American Planning Association,* 70(3), 269–281.

Borlsoff, D. and Hahn, D.F. (1997). "The mirror in the window: Displaying our gender biases." In Drucker, S. J. and Gumpert, G. (Eds.), *Voices in the street. Explorations in gender. Media and public space.* Cresskill, NJ: Hampton Press, 101–117.

Brill, M. (1989). "Transformation, nostalgia, and illusion about public life and public environments." In Altman, I. and Zube, E. (Eds.), *Public places and spaces.* New York, NY: Plenum, 7–29.

Brownlow, A. (2004). "A geography of men's fear." *Geoforum,* 36(5): 581–592.

Casey, R., Goodie, R., and Reeve. K. (2008). "Homeless women in public spaces. Strategies of resistance." *Housing Studies,* 23(6): 899–916.

Chhibber, P. (2002). "Why are some women politically active? The household, public space, and political participation in India." *International Journal of Comparative Sociology,* 43: 409–429.

Coleman, C. (2000). "Women, transport, and cities: An overview and an agenda for research. In Darke, J., Ledwirth, S. and Woods, R. (Eds.), *Women and the city. Visibility and voice in urban space.* New York: Palgrave, 83–97.

Cooper Marcus, C. and Wischemann, T. (1983). *Campus open space: An underutilized potential.* Mimeo. Department of Landscape Architecture, University of California.

Cranz, G. (1981). "Women in urban parks." In Stimpson, C.R., Dixler, E., Nelson, M. and Yatrakis, K.B. (Eds.), *Women and the American city* (pp. 76–92). Chicago: University of Chicago Press.

Dandekar, H. (Ed.) (1993). *Shelter, women and development: First and third world perspectives.* Ann Arbor, MI: George Wahr.

Day, K. (1999a). "Introducing gender to the critique of privatized public space." *Journal of Urban Design,* 4(2): 155–178.

—— (1999b). "Embassies and sanctuaries: Race and women's fear and welcome in privatized public space." *Environment and Planning D: Society and Space,* 17(3): 307–328.

—— (2000a). "Strangers in the night? Women's fear of sexual assault on urban college campuses." *Journal of Architectural and Planning Research,* 16(4): 289–312.

—— (2000b). "The ethic of care and women's experiences of public space." *Journal of Environmental Psychology,* 20: 103–124.

—— (2006). "Being feared: Masculinity and race in public space." *Environment and Planning A,* 38: 569–586.

Day, K., Stump, C., and Carreon, D. (2003). "Confrontation and loss of control: Masculinity and men's fear in public space." *Journal of Environmental Psychology,* 23: 311–322.

Deegan, M.J. (1987). "The female pedestrian: The dramaturgy of structural and experiential barriers in the street." *Man-Environment Systems,* 17, pp. 79–86.

Dubrow, G. (2007). "Honoring the landmarks of feminist planning thought while embracing the future." *Journal of the American Planning Association,* 73(1): 114–115.

Feldman, R.M. and Stall, S. (1994). "The politics of space appropriation: A case study of women's struggles for homeplace in Chicago public housing." In Altman, I. and Churchman, A. (Eds.), *Women and the environment* New York: Plenum Press, 167–200.

Fisher, B. and Nasar, J.L. (1992). "Fear of crime in relation to three exterior site features. Prospect, refuge, and escape." *Environment and Behavior,* 24(1): 35–65.

Franck, K. (2002). "Women and environment." In Bechtel, R. and Churchman, A. (Eds.) *Handbook of environmental psychology*. New York: John Wiley & Sons, 347–362.

Franck, K. and Paxson, L. (1989). "Women and urban public space." In Altman, I. and Zube, E. (Eds.), *Public places and spaces*. New York: Plenum, 121–146.

Fraser, N. (1992). "Rethinking the public sphere: A contribution to the critique of actually existing democracy." In Calhoun, C. (Ed.), *Habermas and the public sphere*. Cambridge, MA: MIT Press, 109–142.

Gardiner, C.B. (1989). "Analyzing gender in public places: Rethinking Goffman's vision of everyday life." *American Sociologist,* 20(1): 42–156.

—— (1994). "Out of place: gender, public places, and situational disadvantage." In Friedland, R. and Boden, D. (Eds.), *Now Here. Space, time and modernity*. Berkeley, CA: University of California Press.

Giuliano, G. (1979). "Public transportation and the travel needs of women." *Traffic Quarterly,* 33: 607–616.

Gordon, M.T. and Riger, S. (1989). *The female fear*. New York: Free Press.

Grabowsky, P. (1995). "Fear of crime and fear reduction strategies." *Trends and Issues* Paper 44. Australian Institute of Criminology, Canberra.

Greed, C.H. (1994). *Women and planning: Creating gendered realities*. London: Routledge.

—— (2006). "Making the divided city whole: Mainstreaming gender into planning in the United Kingdom." *Tijdschrift voor Economische en Sociale Geografie,* 97(3): 267–280.

Hamilton, K. (2000). *Public transport audit London*. At: www.uel.ac.uk/womenandtransport.

Harrington, M., Dawson, D. and Bolla, P. (1992). "Objective and subjective constraints on women's enjoyment of leisure." *Society and Leisure,* 15(1): 203–221.

Hayden, D. (1980). "What would a non-sexist city be like? Speculations on housing, urban design, and human work." *Signs,* 5(3): S170–187.

—— (1984). *Redesigning the American dream: The future of housing, work, and family life*. New York: W.W. Norton.

—— (1997). *The power of place. Urban landscapes as public history*. Cambridge, MA: MIT Press.

Henley, N.M. (1977). *Body politics*. Englewood Cliffs, NJ: Prentice-Hall.

Hillman, M., Henderson, I., and Whaley, A. (1974). *Mobility and accessibility in the outer metropolitan area. Political and Economic Planning Report to the Department of the Environment*. London: Policy Studies Institute.

hooks, b. (1992). *Black looks. Race and representation*. Boston, MA: South End Press.

Hu, P.S. and Young, J.R. (1999). *Summary of travel trends: 1995 Nationwide Personal Transportation Survey*. Washington DC: U.S. Department of Transportation, Federal Highway Administration.

Huxtable, A.L. (1997). *The unreal America. Architecture and illusion*. New York: The New Press.

Hymas, M. (2003). "Adolescent Latinas bodyspaces: Making homegirls, homebodies, and homespaces." *Antipode,* 536–558.

Krenichyn, K. (2004). "Women and physical activity in an urban park: Enrichment and support through an ethic of care." *Journal of Environmental Psychology,* 24: 117–130.

Koskela, H. (1997). "'Bold walk and breakings': Women's spatial confidence versus fear of violence." *Gender, Place, and Culture,* 4: 301–319.

Koskela, H. and Pain, R. (2000). "Revisiting fear and place: Women's fear of attack and the built environment." *Geoforum,* 31: 269–280.

Law, R. (1999). "Beyond 'women and transport': Towards new geographies of gender and daily mobility." *Progress in Human Geography,* 23(4): 567–588.

Lacy, S. (1995). (Ed.). *Mapping the terrain. New genre public art.* Seattle, WA: Bay Press.

Leavitt, J. and Saegert, S. (1989). *From abandonment to hope: Community-households in Harlem.* New York: Columbia University Press.

Lofland, L. (1975). "The 'thereness' of women: A selective review of urban sociology." In Millman, M. and Kanter, R. M. (Eds.), *In another voice: Feminist perspectives on social life and social science.* New York: Anchor Books.

—— (1984). "Women and urban public space." *Women and Environments,* 6(2): 12–14.

Loukaitou-Sideris, A. and Fink, C. (2009). "Addressing women's fear of victimization in transportation settings. A survey of U.S. transit agencies." *Urban Affairs Review,* 44(4): 554–587.

Loukaitou-Sideris, A., Ligget, R. and Iseki, H. (2002). "The geography of transit crime. Documentation and evaluation of crime incidence on and around the green line stations in Los Angeles," *Journal of Planning Education and Research,* 22: 135–151.

Massey, D.S. and Denton, N.A. (1993). *American apartheid: Segregation and the making of the underclass.* Boston, MA: Harvard University Press.

Matrix (Eds.) (1984). *Making space: Women and the man made environment.* London: Pluto Press.

Mazey, M.E. and Lee, D.R. (1983). *Her space, her place.* Washington, DC: Association of American Geographers.

Mazumdar, S. and Mazumdar, S. (2001). "Rethinking public and private space: Religion and women in Muslim society." *Journal of Architectural and Planning Research,* 18(4): 302–324.

McDowell, L. (1983). "Towards an understanding of the gender division of urban space." *Environment and Planning D: Society and Space,* 1: 59–72.

Mills, A. (2007). "Gender and mahalle (neighborhood) space in Istanbul." *Gender, Place, and Culture,* 14(3): 335–354.

Miraftab, F. (2007). "Planning and gender as seen from the global South." *Journal of the American Planning Association,* 73(1): 115–116.

Miranne, K.B. and Young, A.H. (Eds.) (2000). *Gendering the city. Women, boundaries, and visions of urban life.* New York: Rowman & Littlefield.

Modlich, R. (1986). "Women Plan Toronto," *Women and Environments,* 8(1).

Mozingo, L. (1985). "Public space in the balance." *Landscape Architecture,* 2: 43–47.

—— (1989). "Women and downtown open spaces." *Places,* 6(1): 38–47.

Nasar, J.L. and Fisher, B. (1992). "Design for vulnerability: Cues and reactions to fear of crime." *Sociology and Social Research,* 76(2): 48–58.

Njoh, A.J. (1999). "Gender-biased transportation planning in sub-Saharan Africa with special reference to Cameroon." *Journal of African and Asian Studies,* 34(2): 216–233.

Pain, R. (1997a). "Old age and ageism in urban research: The case of fear of crime." *International Journal of Urban and Regional Research,* 21(1): 117–128.

—— (1997b). "Social geographies of women's fear of crime." *Transportation Institute of British Geographers,* 22: 231–244.

Pickup, L. (1984). "Women's gender-role and its influence on travel behaviour." *Built Environment,* 10: 61–68.

—— (1989). "Women's travel requirements: Employment, with domestic constraints." In Grieco, M., Pickup, L. and Whipp, R. (Eds.), *Gender, transport and employment: The impact of travel constraints.* Avebury: Aldershot.

Rakodi, C. (1991). "Cities and people: Towards a gender-aware urban planning process?" *Public Administration and Development,* 11: 541–559.

Rendell, J., Penner, B., and Bordon, I. (2000). *Gender space architecture.* London: Routledge.

Ritzdorf, M. (1994). A feminist analysis of gender and residential zoning in the United States. In I. Altman & A. Churchman (Eds.) *Women and the environment* (pp. 255–279). New York: Plenum Press.

Rohe, W.M. and Burby, R.J. (1988). "Fear of crime in public housing." *Environment and Behavior,* 20: 702–720.

Rose, G. (1993). *Feminism and geography: The limits of geographical knowledge.* Minneapolis, MN: University of Minnesota Press.

Rosenbloom, S. (1978). "Editorial: The need for study of women's travel issues." *Transportation,* 7: 347–350.

—— (Ed.) (1980). *Women's travel issues: Research priorities and needs.* Washington DC: Department of Transportation, Research and Special Programs Administration.

Rosenbloom, S. and Winsten-Bartlett, C. (2002). "Asking the right question. Understanding the travel needs of older women who do not drive." *Transportation Research Record,* 1818: 78–82.

Rothschild, J. (Ed.) (1999). *Design and feminism: Re-visioning spaces, places, and everyday things.* New Brunswick, NJ: Rutgers University Press.

Ruddick, S. (1996). "Constructing difference in public spaces: Race, class and gender as interlocking systems." *Urban Geography,* 17(2): 132–151.

Sandercock, L. and Forsyth, A. (1992). "A gender agenda. New directions for planning theory." *Journal of the American Planning Association,* 58(1): 49–59.

Sangwha, L. (1999). "The patriarchy in China: An investigation of public and private spheres." *Asian Journal of Women's Studies,* 5(1): 9–.

Seedat, M., MacKenzie, S, and Mohan, D. (2006). "The phenomenology of being a female pedestrian in an African and an Asian city: A qualitative investigation." *Transportation Research* Part F, 9: 139–153.

Shaw, S.M. (1994). "Gender, leisure, and constraint: Towards a framework for the analysis of women's leisure." *Journal of Leisure Research,* 26(1): 8–22.

Skjerven, R. (1993). *Manual of alternative municipal planning.* Oslo: Ministry of Environment.

Sorkin, M. (1992). "Introduction: Variation on a theme park." In M. Sorkin (Ed.), *Variation on a theme park: The new American city and the end of public space* (pp. xi–xv). New York: Hill and Wang.

Spain, D. (1992). *Gendered spaces.* Chapel Hill: University of North Carolina Press.

—— (2001). *How women saved the city.* Minneapolis, MN: University of Minnesota Press.

Sprague, J.F. (1991). *More than housing: Lifeboats for women and children.* Boston, MA: Butterworth Architecture.

Stanko, E.A. (1987). "Typical violence, normal precaution: Men, women, and interpersonal violence in England, Wales, Scotland, the US." In J. Hanmer and M. Maynard (Eds.), *Women, violence and social control* (pp. 122–134). London: Macmillan.

Stimpson, C. Dixler, E., Nelson, M., and Yatrakis, K. (Eds.). (1981). *Women and the American city.* Chicago: University of Chicago Press.

Thompson, J.L., Allen, P., Cunningham-Sabo, L., Yazzie, D., Curtis, M. and Davis, S.M. (2002). "Environmental, policy, and cultural factors related to physical activity in sedentary American Indian women." *Women and Health* 36(2): 59–74.

Valentine, G. (1989). "The geography of women's fear." *Area,* 21(4): 385–390.

—— (1992). "Images of danger: Women's sources of information about the spatial distribution of male violence." *Area,* 24(1): 22–29.

Wallace, M. and Milroy, B.M. (1999). Intersecting claims: Possibilities for planning in Canada's multicultural cities. In T. Fenster (Ed.), *Gender, planning and human rights.* London: Routledge.

Warr, M. (1990). "Dangerous situations: Social context and fear of victimization." *Social Forces,* 68(3): 891–907.

Weisman, L.K. (1992). *Discrimination by design: A feminist critique of the man-made environment.* Urbana, IL: University of Illinois Press.

Wekerle, G. and Whitzman, C. (1995). *Safe cities. Guidelines for planning, design and management.* New York: Van Nostrand Reinhold.

Wekerle, G.R., Peterson, R. and Morley, D. (Eds.) (1980). *New space for women.* Boulder, CO: Westview Press.

Wilson, E. (1991). *The sphinx in the city: Urban life, the control of disorder and women.* Berkeley, CA: University of California Press.

Wolfe, M. (1992). "Invisible women in invisible places. Lesbians, lesbian bars, and the social production of people." *Architecture and Behavior,* 8(2): 137–158.

Further reading

Dandakar, H. (Ed.) (1993). *Shelter, women and development: First and third world perspectives.* Ann Arbor, MI: George Wahr. Proceedings of a conference that sought to draw links between issues of shelter, women, and development, and to advocate for gender-sensitive housing policies.

Franck, K. and Paxson, L. (1989). "Women and urban public space." In I. Altman and E. Zube (Eds.), *Public places and spaces* (pp. 121–146). New York: Plenum. One of the first articles to give a historic overview of women's use of public spaces in the city.

Hayden, D. (1984). *Redesigning the American dream: The future of housing, work, and family life*. New York: W. W. Norton. Excellent analysis of the interplay between gender roles and housing design.

Wilson, E. (1991). *The sphinx in the city: Urban life, the control of disorder and women*. Berkeley, CA: University of California Press. A compelling critique of how planners and urban reformers have repeatedly sought to regulate women in cities.

12

Environmental psychology and urban design

Jack L. Nasar

Prior to the development of environmental psychology, architects and planners gave normative descriptions – such as city beautiful, city efficient, the radiant city, or Broadacre city – about the ways they thought the world should be (Lang 1987). Their definitions of the environment and human responses, and the causal links were vague. Psychologists had precise definitions and methods for testing, but they often neglected the physical environment and tested variables under unrealistic conditions with little relation to people's everyday life. Environmental psychology applies social science methods and theories to real world questions about human experience in everyday physical environments. Unlike the normative approach, it seeks to describe the world the way it is – how we use it and, in turn, how it affects our behavior – to build a knowledge base for urban design. Unlike psychology, it emphasizes large-scale physical environments in which people exist. It often takes a multi-level, multi-disciplinary, social ecological approach to examine relationships between characteristics of the physical environment, humans, context and human responses (King *et al.* 2002). It is evolving a knowledge base for urban design decisions about the context and characteristics of places.

The scientific approach carries values of "honesty, doubt, respect for evidence, openness, accountability and tolerance and indeed hunger for opposing points of view" (Overbye 2009). It uses feedback for continuous improvement (Petzinger 1999). For design this entails a cyclical process in which one gathers information for a plan, implements the plan, systematically evaluates it from the user's perspective (post-occupancy evaluation) afterwards, and uses the evaluation to improve it and future plans (Nasar 1999; Preiser *et al.* 1988).

Figure 12.1 shows a model of the relation between socio-physical characteristics of places and human responses. Socio-physical attributes of places interact with human characteristics to affect user evaluations and behavior. The *Individual* refers to characteristics, such as personality, affective state, socio-cultural experience, expectations and intentions of the person evaluating the setting. *Setting Attributes* refers to social and physical characteristics of the environment. The social characteristics include purpose, culture, age or gender of the individuals using the setting. The physical characteristics are characteristics, such as size, shape, order, or legibility, of the environment. *Perception* refers to direct responses of our senses to the structure of forms with little to no mental activity.

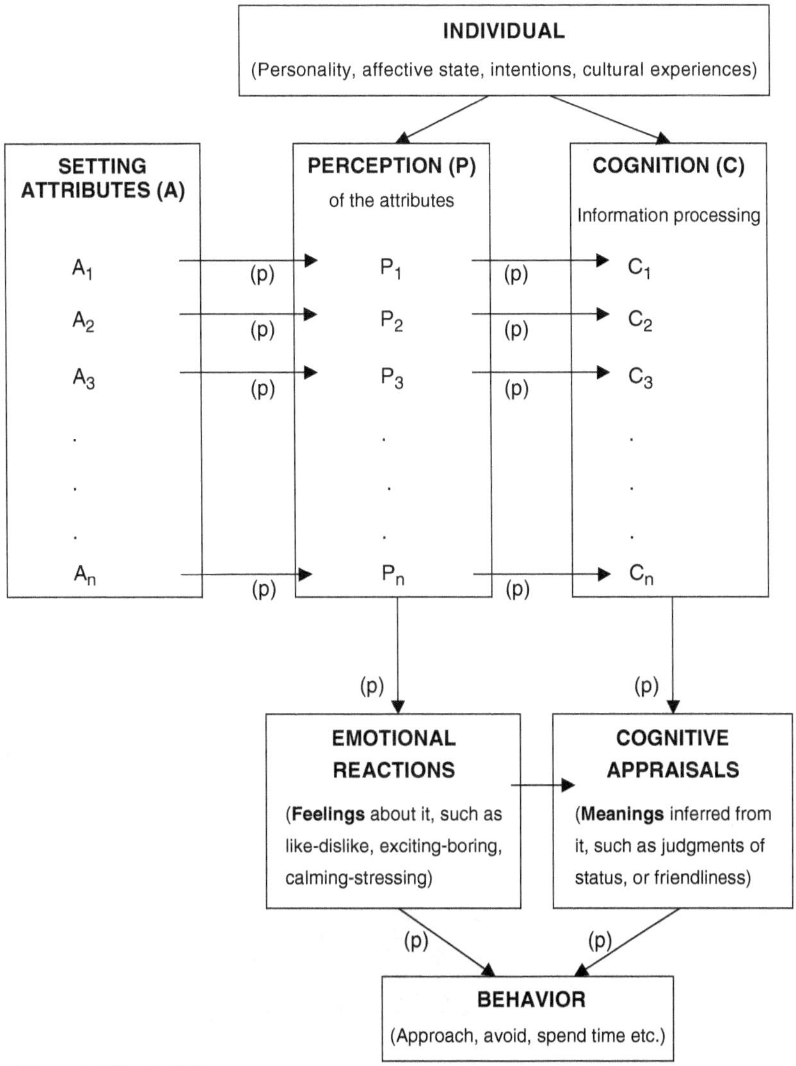

Figure 12.1 Basis for environmental response.

Our perception has limits. We cannot see, hear, smell or feel everything; and we notice some things more than others.

Perceptions influence our thoughts. *Cognition* involves the ways we categorize, remember, and represent our experiences of the environment. We identify or recognize environments (such as a plaza). We see a structure or pattern in environments, and infer meanings about them (such as judging it as a safe plaza) (Lynch 1960). Perceptions of characteristics of the setting and the population evoke *Emotional Reactions*, our affective responses to the place. *Cognition* and *Emotional Reactions* yield *Cognitive Appraisals*, connotative meanings, such as inferences about the overall safety or friendliness of the people. These reactions and appraisals can affect *Behavior*. Thus, human behavior in settings varies with visceral emotional reactions and reflective thought, which is affected by perceptions and the socio-physical characteristics of the setting. This chapter discusses each part of the model.

The chapter focuses on areas of agreement in response to the visual environment. Although individuals differ, there are substantial areas of agreement. In shaping places for use, urban designers need to know about the likely effects of their designs on the public who experiences it (Lynch 1960), as well as about the areas of consensus among most people. While other senses affect our experience, vision dominates. Most research focuses on the visual experience.

Environmental perception

Three theories of environmental perception offer ideas for urban design. One, *adaptation level theory*, holds that people adapt to the prevailing level of stimuli (Wohlwill and Kohn 1973). Individuals in a crowded place adapt to the crowding, but the adaptation has a cost: it involves stress, particularly if the stimulus is unpredictable and perceived as uncontrollable (Evans and Cohen 1987). Crowding, traffic congestion, traffic or airport noise, pollution, litter, fear of crime, and dislocation are stressors. Designs that reduce or make such stressors appear more predictable or controllable can reduce stress. Thus, providing nearby nature or community gardens is desirable (Kaplan and Kaplan 1989) in part because it gives people perceived control of and potential escape from everyday stress.

Another theory, the *ecological approach to perception*, sees the environment as made up of structured and meaningful stimulus information (Gibson 1979; Heft 2001). Active observers detect functionally significant environmental structures that support their activities and provide a ground for exploration. Of particular importance is the concept of *affordances*. We perceive substances and surfaces of settings and objects primarily in terms of their relational properties to us. For example, chairs at the appropriate height relative to one's leg length are experienced as affording sitting-on, but many other solid raised surfaces with the requisite body-scaled properties are also sittable (Whyte 1980) (Figure 12.2). While designers

Figure 12.2 Affordances for sitting. Source: Jack Nasar.

often focus on form, users experience the functional opportunities (affordances) of a place. The ecological approach suggests that for desired functions – such as sitting, shelter, or social interaction – urban designers should seek the appropriate affordances for users.

A third theory, *probabilistic functionalism*, argues that human evaluations of environments have probabilities associated with the person's perception of physical cues, which has probabilities associated with actual physical attributes in the environment (Brunswik 1956). The cues and probabilities derive from one's experience of functioning in environments. The model suggests that designers focus on salient attributes in human perception and evaluation, discussed later in this chapter.

Environmental cognition

Being lost is often frustrating, stressful and a potential threat to survival. Legibility – the ease with which one can comprehend and navigate environments – can lessen these threats. Places vary in legibility, which relates to imageability – identity, structure and meaning (Lynch 1960). We identify or recognize objects, we see a recognizable structure or pattern, and we see meaning in, or evaluate, them. Legibility involves identity and structure. Consideration of legibility led to two innovations for urban design (Lynch 1960). First, tallying people's responses to the environment can define shared responses to guide urban form. Second, one can use the shared elements of people's cognitive maps to enhance legibility. Five kinds of elements affect legibility (Lynch 1960):

Landmarks: visible points of orientation, such as St. Louis Arch

Paths: shared channels of travel, such as roads or highways

Nodes: gathering points, or concentration of activities, or convergence of paths, such as Times Square

Districts: areas, such as London's Soho, that people judge as having a consistent character, or distinctive urban form that differs from other areas

Edges: linear features, such as a river, rail line or highway that separate one area from another.

Research confirms that distinct landmarks, paths, districts and edges arranged in a coherent structure enhance legibility (Evans 1980; Golledge 1987). Sharp edges around a district, the co-occurrence of nodes and landmarks along major paths give a coherent and thus legible structure to cities and urban places (Appleyard 1976; Lynch 1960). Imageable elements tend to be distinctive, in that they stand out from their context. This distinctiveness arises from differentiation from the immediate context. Thus, typically landmarks have vertical differentiation. Street size, paving and organization can offer horizontal differentiation. Scale matters. Citywide elements would have a greater overall differentiation than would neighborhood elements, which would have a greater differentiation than street-level elements. Horizontal or vertical physical differentiation and simpler layouts enhance legibility (Cubukcu and Nasar 2005).

Landmark buildings have distinctive form, use significance, and visibility (Appleyard 1969). Landmarks should have a clear contour, complex shape, a unique style, movement of people and natural elements (trees and shrubs) around the base of the building, intense use, large relative size, centrality and proximity to a major orientation point, singular use, visibility to many people, and easy access (Evans *et al.* 1982).

Environmental evaluations

Recall that in addition to identity and structure, imageability includes meaning (Lynch 1960). Places convey an ambiance or meanings that we feel (Rapoport 1993).

For urban design, the meanings should be consonant with the functional goals of the place for the public experiencing it. Any design varies in the likelihood that it will evoke a specific meaning among people experiencing it, but some attributes will more likely evoke a meaning than others. Urban designers can use those shared meanings to craft designs compatible with purposes of settings for many users. Environmental psychology has established attributes associated with preference, or *likeability*. To plan for those substantial areas of agreement, urban designs should incorporate the public meanings, their *evaluative image* of places (Nasar 1998).

As adults in a particular culture we learn the non-verbal language of our "recognizable cultural landscape" (Rapoport 1993, p. 36). Our shared environmental meanings help us make sense of things. Studies confirm substantial agreement on likability (Kaplan and Kaplan 1989; Nasar 1998 1999). A meta-analysis covering 40 studies, 1,001 environments, 5,301 respondents from 432 samples, 21 countries, and 13 groups found agreement in preference ($r = 0.82$) for all of the groups by ethnicity, political affiliation, gender, culture, student versus non-student, expert versus non-expert (Stamps 1999). Likability relates more to characteristics of places than to characteristics of people (Nasar 1998; Stamps 2000) with one major exception.

Architects differ from the public in responses to "high-style" or atypical designs versus "popular" styles or more typical designs (Devlin and Nasar 1989; Nasar 1999). These differences yield designs incompatible with the users (Nasar 1999; Vischer and Cooper Marcus 1986). Environmental preferences of the public have stability over time and thus can accurately predict future public preferences (Nasar 1999; Stamps 1997).

For urban design, which involves public money, public property, or is visible to the public, the design should satisfy the public. It should consider the user public's evaluative responses to environments. *Evaluations of places* are personal judgments about their emotional quality – such as their attractiveness. *Feelings in places* are the person's internal emotions – such as feeling pleased. *Evaluations* of a place (it feels unsafe) might affect *feeling in it* (I feel unsafe), but they may also arise independently. Individuals might feel happy regardless of their location.

Recall that Brunswik's (1956) model of perception suggested that designers focus on salient attributes in human evaluation and perception. Human evaluation of (and feelings in) environments have three salient factors (Russell and Snodgrass 1987): pleasantness (pleasure), excitement (excited), and calmness (calm) (Figure 12.3). The vertical axis – arousing (arousal) – is independent of evaluation. The diagonal axes mix pleasantness and arousing. Exciting places are *more* pleasant and arousing than boring ones; and calming places are *more* pleasant but *less* arousing than distressing ones. Although research has focused on pleasantness, it suggests some attributes that affect excitement or calmness.

Thoughts about place can include evaluations that go beyond recognition of the place (its denotative meaning) to inferences

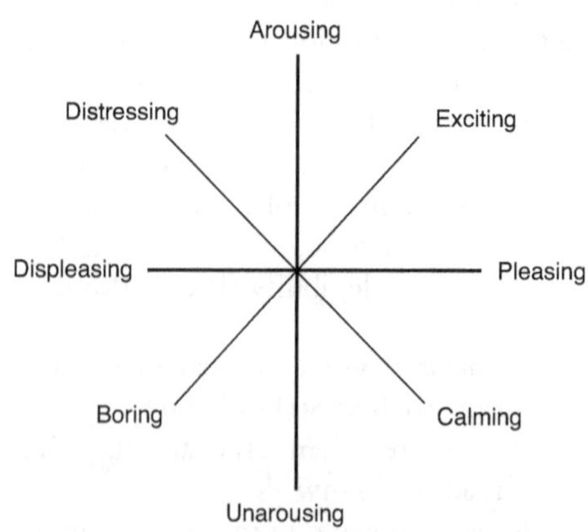

Figure 12.3 Dimensions of Environmental Appraisal.

about its qualities and the qualities of its occupants (connotative meanings). Connotative meanings include inferences about such things as safety, and the prestige, status, and friendliness of people inside (Nasar 1989; Nasar and Fisher 1993; Rapoport 1993).

As for perception, research using a variety of methods and respondents has identified six salient perceptual-cognitive aspects of environments that also affect evaluation: openness, naturalness, upkeep, historical significance, complexity and order.

Openness

Openness refers to prospect, visual scope, and related attributes (such as spaciousness, vista, and enclosure). The "visual scope" ("vistas and panoramas which increase depth of vision") and defined space ("a strong physical form") strengthen the memorability of nodes, and people prefer defined openness, "well managed panoramas" (Lynch 1960: 44, 76, 106). Open views allow people to see, predict, and more easily navigate; and people prefer moderate and defined openness (spatial definition) to wide-open or blocked vistas (Kaplan and Kaplan 1989; Nasar 1983, 1998). Another spatial variable – *mystery* – involves the promise of new information ahead. People judge curved paths (deflected vistas) as offering more new information ahead (higher in mystery) than straight ones. In situations perceived as safe, people prefer mystery (Kaplan and Kaplan 1989), but when people feel vulnerable, such as walking alone after dark, a deflected vista is ominous. It would afford a potential predator concealment from which he or she could see and surprise a passerby. When feeling vulnerable, people should feel safer in (and prefer) open prospect and absence of refuge (hiding places) ahead. Blocked prospect and places of refuge (concealment or hiding places) ahead increase fear of crime, avoidance, and actual crime (Nasar 1999; Nasar and Fisher 1993).

Naturalness

Naturalness refers to people's perception that a place is natural or has more "natural" elements (such as vegetation, water, or mountains) than artificial ones (such as buildings, signs, or sidewalks). Some places that people perceive as "natural" – such as a farm field or trees in an urban plaza – are not natural in that they depend on human intervention. Human preference increases with perceived naturalness; people prefer environments they perceive as natural over ones perceived as artificial; preference increases with the addition of natural elements; and the experience of nature can be restorative or calming (Kaplan 1995; Kaplan and Kaplan 1989; Nasar 1994, 1998; Ulrich 1991).

Upkeep

Upkeep refers to the perceived maintenance. More broadly in the negative form, poor upkeep is associated with physical incivilities – such as litter, boarded up or vacant buildings and lots, and graffiti – which convey cues of disorder (Perkins and Taylor 1996). According to the "broken windows theory," people perceive signs of decay as cues to a break-down in the social order and control, which increases fear of crime and crime (Wilson and Kelling 1982). Research confirms that incivilities lessen preference and sense of community, and increases fear of crime and crime (Keizer *et al.* 2008; Nasar 1983; Perkins and Taylor 1996; Wyant 2008). People have a more favorable image of well-kept places (Nasar 1998). By removing or buffering incivilities with desirable elements (such as trees), one can make settings more appealing and calming.

Historical significance

Historical significance also depends on perception. An environment can be authentically historic or simply look historical. People prefer perceived historical significance, historical styles and historical areas (Nasar 1998; Whitfield 1983). Studies across four US cities, three different sets of houses and responses converged on preference for vernacular, historical styles or familiar, typical styles over high and atypical styles (Nasar 1989; Devlin and Nasar 1989).

Complexity and order

Complexity refers to the number and variability of elements, such as height, shape and layout, in an environment. Order refers to the perceived structure or the degree to which an environment appears coherent, congruous, legible, or clear. In theory, *complexity* increases uncertainty, arousal and interest; and order reduces uncertainty and arousal (Wohlwill 1976). Too much complexity would create an overload of uncertainty and arousal. Too little would be boring. People should like moderate complexity or a mix of complexity with order. Studies confirm that people like order and that as order decreases, interest and excitement increase (Kaplan and Kaplan 1989; Nasar 1998). For complexity, studies suggest that people prefer either moderate complexity or a mix of order and moderate to high complexity (Nasar 1994).

Summary

Humans see meaning in the appearance of settings, and these meanings can support or interfere with the intended function. Successful urban designs will convey meanings compatible with functional goals. Urban designers can create such designs through understanding setting characteristics that evoke desired meanings. They need to determine the ambience needed for an environment's purposes – for example, the degree to which it should appear appealing, exciting, calming, friendly, or unsociable. They need to pick the attributes most likely to evoke that ambiance and specify the way to manipulate those attributes to achieve it. A post-occupancy evaluation after implementation can assess the result and improve the knowledge base.

Research findings provide some directions for achieving certain evaluative responses. *Pleasant* environments should have naturalness, good upkeep, order, moderate complexity, enclosure, low to moderate novelty and popular or historical styles; *calming* environments should have lots of nature, water, open vistas, and order; and *exciting* environments should have high complexity, low order, low naturalness, high novelty and possibly high (or unfamiliar) styles.

Other meanings, such as perceived sociability, or status, and projects with distinctive needs, may require the development of a plan for appearance related to the visual qualities needed for the particular situation (Nasar 1998; Nelessen 1994). For this, interviews with the relevant population can uncover the desired ambiance, and identify the relevant features and the way to manipulate them to achieve that ambiance. This consumer-oriented approach has the benefit of involving people in decisions that affect them.

Behavior in the environment

In the 1950s *ecological psychologists* at the Kansas Field Station, following ecological principles, sought to understand how humans behaved in everyday environments. They recognized an ecological interdependence between behavior and the environment; and through observation they learned that situations had a larger impact on behavior than did an individual's characteristics

Figure 12.4 Behavior setting for outdoor eating. Source: Jack Nasar.

(Barker 1968; 1987). This shift in emphasis from the person to the environment led to the concept of *behavior setting*. Behavior settings are real entities, with time and place boundaries, a fit between their physical components and the people behaving in them and program of events (Wicker 1979). People in a drug store exhibit "drug store behavior," but in a gymnasium, they exhibit "gymnasium behavior" (Figure 12.4). These standing patterns of behavior arise from the socio-physical and time characteristics of settings.

The ideas from ecological psychology contribute to useful research methods and concepts for urban design. To discover person-environment relationships, ecological psychologists observed behavior in natural situations via *specimen* and *setting records*. *Specimen records* involve following individuals and recording what they said, did, what people said or did with them, and where they were (Barker 1987). *Setting records* center on recording behavior in a behavior setting (Barker 1968; 1987).

For urban design, the unobtrusive observation of naturally occurring behavior in public settings led to several breakthroughs. It highlighted cultural differences in response (Hall 1966), suggesting that designers should attend to such differences and nonverbal cues. An interrelated set of observations suggests ways to build a community. First, functional distances, such as shared walking routes or natural gathering places, affect informal interpersonal interaction more than proximity (Festinger *et al.* 1950). Second, different distances between people support different kinds of interactions – intimate, personal, social and public. The distances vary with culture (Hall 1966), but a comfortable social distance for Americans is four to twelve feet (Sommer 1969). Third, territorial spaces and markers, such as a marked front yard in a townhouse, signal ownership and control to the individual and others. Combination of these concepts can create settings such as tot lots, dog walks, community gardens, front steps, shared mailboxes, laundry rooms, and some back-yards that support informal interaction by giving users a territory they can comfortably occupy and functional connections to others at a comfortable social distance (Hall 1966; Sommer 1969). When they fit a population, such settings can lead to informal interaction, friendship formation (if people perceive one another as similar), and sense of community.

Jane Jacobs' (1961) observations of her West Village neighborhood gave her an understanding of what made a vital city. Adopting her ideas, Newman (1972) examined crime in different housing projects and before and after modifications. He found that provision of natural surveillance, access control, territorial definition, and milieu/image reduced crime. Though sociocultural factors play a role (Newman and Franck 1980), research confirms reductions in crime associated with improved natural surveillance, territoriality, access control, and milieu/image – improving upkeep and removing incivilities (Cozens *et al.* 2005; Perkins and Taylor 1996; Wyant 2008).

Whyte (1980) used time lapse photography of public plazas to learn how people used them and the factors attracting users. Affordances that attracted use include sittable space, movable seating, connection to the street and people watching, food, deciduous trees (affording sun protection in the summer and sunlight in the winter), water, and triangulation (something that links strangers and leads them to interact). Researchers have done similar observational studies in other contexts. For example, observations of behavior, incremental changes, and evaluations helped transform Copenhagen into a pedestrian-oriented city (Gehl 1987).

The livable street project used self-reporting to understand effects of traffic on residents (Appleyard 1981). It evaluated three streets, with light traffic, moderate traffic, and heavy traffic. As the traffic level increased, residents reported less social interaction (neighboring, acquaintances on the street), a smaller definition of their home territory, and higher levels of noise and danger. Examination of twenty-one streets confirmed the findings (Appleyard 1981). These studies led a shift from engineering streets to maximize traffic flow to using traffic calming – median islands, speed humps, traffic circles, curb extensions and chokers (which narrow a street), chicanes (S-shaped curves often done with a pair of curb extensions), and *woonerfs* – to slow down traffic and create pedestrian-friendly streets. *Woonerfs* are streets designed for shared use by motorists and others. They may have a gateway, curves to slow vehicles, trees and play equipment, and no lanes, curbs, or long-term on-street parking (Appleyard 1981). Research is also considering how environments can encourage physical activity to reduce obesity and associated health risks such as cardiovascular disease, cancer, type two diabetes, osteoporosis, injurious falls, premature mortality, and mental disorders. Affordances for physical activity include: perceived aesthetics, safety from crime and from traffic, pedestrian activity, good upkeep, higher density, mixed use, shorter block length, sidewalks, connectivity and accessible destinations (Ewing and Cervero 2001; Handy *et al.* 2002).

Individual differences

Human responses to environments also vary (Rapoport 1993; Zube, Pitt, and Evans 1983). Children go through stages in spatial cognition and notice and use different attributes differently from adults (Evans 1980; Heft and Wohlwill 1987). They develop landmark knowledge, in which they know discrete objects, but do not integrate them into a configuration; they develop route knowledge in which they mentally connect points in space; and they develop survey (or Euclidean) knowledge, in which they know the interconnection between features and routes, such that they grasp the interrelationships of routes and objects in space (Hart and Moore 1976). Adults go through similar stages – landmark to route to survey knowledge – in developing mental images of environments (Evans 1980; McDonald and Pellegrino 1993). The importance of paths or landmarks varies with the characteristic of the environment (Appleyard 1970; Heft 1979; Evans *et al.* 1981). Older adults notice different attributes than younger ones and have declines in their spatial ability (Evans 1980; Passini *et al.* 1998). Men and women differ in their spatial abilities, probably related to differences in environmental experience (Evans 1980; Webley and Whalley 1987). Mental maps vary with culture (Gulick 1963) and have systematic distortions related to the environment and observer (Evans 1980).

Neighborhoods differ in socio-physical characteristics (Michelson and van Vliet 2002; Popenoe and Michelson 2002). Research suggests six neighborhood types by place, people, culture and meaning, each a good fit for its residents: Small Town, Center, Residential Partnership, Retreat, Residential Partnership/Small Town, and Residential Partnership/Center (Brower 2000). Centers are activity hubs, the place to meet people, with lots to see and do, world class facilities, good public transport, tourists, and a diversity of residents. Small Towns have their own local institutions and meeting places and a small town feeling, in which people know one another, take care of one another, and have long-term personal relationships. Residential Partnerships (considered good for raising children) are residential enclaves, separate from work and entertainment. Retreats allow residents, who are private, independent, and go their own way, to remove themselves from other people and activities. Residential Partnership/Small Town and Residential Partnership/Center merge residential use with either a small town feel or a central location. In sum, such individual differences suggest that there is no one best solution. Urban designers need to understand and work with people to find compatible solutions.

Conclusion

Environmental psychology has brought advances to inform urban design in the creation of humane places. Designers can use the findings or study the situation to derive situation-specific guidelines. Questions remain. A mix of controlled and naturalistic studies can enhance our understanding of causality and applicability for urban design. Treating the design/planning process as an applied science inquiry can contribute to an evolving knowledge base. This involves a cyclical process of planning, programming, design, construction, occupancy, post-occupancy evaluation and, if appropriate, back to planning (Preiser and Nasar 2008). This approach gives priority to and incorporates occupants into designs that are vital and desirable for them.

References

Appleyard, D. (1969). "Why Buildings are Known," *Environment and Behavior* 1: 131–156.

—— (1970). "Styles and Methods of Structuring a City," *Environment and Behavior* 2:101–117.

—— (1976). *Planning a Pluralistic City*, Cambridge, MA: MIT Press.

—— (1981). *Livable Streets*. Berkeley, CA: University of California Press.

Barker, R.G. (1968). *Ecological Psychology: Concepts and Methods for Studying the Environment of Human Behavior*. Stanford, CA: Stanford University Press.

—— (1987). "Prospecting in Environmental Psychology: Oskaloosa Revisited." In Stokols, D. and Altman, I. (Eds.) *Handbook of Environmental Psychology* (Vol. II), New York: Wiley.

Brower, S. (2000). *Good Neighborhoods: A Study of in-Town and Suburban Residential Environments*, New York: Praeger.

Brunswik, E. (1956). *Perception and the Representative Design of Psychological Experiments*, Berkeley: University of California Press.

Cozens, P.M. Saville, G. and Hillier, D. (2005). "Crime Prevention through Environmental Design (CPTED): A Review and Modern Bibliography," *Property Management* 23: 328–356.

Cubukcu, E. and Nasar, J.L. (2005). "Relation of Physical Form to Spatial Knowledge in Large-Scale Virtual Environments," *Environment and Behavior* 37: 397–417.

Devlin, K. and Nasar, J. (1989). "The Beauty and the Beast: Some Preliminary Comparisons of 'High' Versus 'Popular' Residential Architecture and Public Versus Architect Judgments of Same," *Journal of Environmental Psychology* 9: 333–344.

Evans, G.W. (1980). "Environmental Cognition," *Psychological Bulletin* 88: 259–287.

Evans, G.W. and Cohen, S. (1987). "Environmental Stress." In Stokols, D. and Altman, I. (Eds) *Handbook of Environmental Psychology* (Vol. 1), New York: Wiley.

Evans, G.W., Marero, D.G. and Butler, P.A. (1981). "Environmental Learning and Cognitive Mapping," *Environment and Behavior* 13: 83–104.

Evans, G.W., Smith, C. and Pezdak, K. (1982). "Cognitive Maps and Urban Form," *Journal of the American Planning Association* 48: 232–244.

Ewing, R. and Cervero, R. (2001). "Travel and the Built Environment: A Synthesis," *Transportation Research Record* 1780: 87–114.

Festinger, L.A., Schachter, S. and Back, K. (1950). *Social Pressures Informal Social Groups*, New York: Harper and Row.

Gehl, J. (1987). *Life between Buildings: Using Public Space* (translated by Jo Koch), New York: Van Nostrand Reinhold.

Gibson, J.J. (1979). *An Ecological Approach to Visual Perception*, Boston, MA: Houghton Mifflin.

Golledge, R.G. (1987). "Environmental Cognition." In Stokols, D. and Altman, I. (Eds.) *Handbook of Environmental Psychology*, New York, Wiley.

Gulick, J. (1963). "Images of an Arab City," *Journal of the American Planning Association* 29: 179–198.

Hall, E. T. (1966). *The Hidden Dimension*, New York: Doubleday.

Handy, S., Boarnet, M., Ewing, R. and Killingsworth, R. E. (2002). "How the Built Environment Affects Physical Activity: Views From Urban Planning," *American Journal of Preventative Medicine* 23: 64–73.

Hart, R.A., and Moore, G.T. (1976). "The Development of Spatial Cognition: A Review." In Downs, R.M. and Stea, D. (Eds.) *Image and Environment: Cognitive Mapping and Spatial Behavior*, Chicago, IL: Aldine.

Heft, H. (1979). "The Role of Environmental Features in Route-Learning: Two Exploratory Studies," *Environmental Psychology and Nonverbal Behavior* 3: 172–185.

—— (2001). *Ecological Psychology in Context: James Gibson, Roger Barker, and the Legacy of William James*, Mahwah, NJ: Lawrence Erlbaum.

Heft, H. and Wohlwill, J. F. (1987). "Environmental Cognition in Children." In Stokols, D. and Altman, I. (Eds.). *Handbook of Environmental Psychology*. New York: Wiley, 175–203.

Jacobs, J. (1961) *The Death and Life of Great American Cities*. New York: Vintage.

Kaplan, S. (1995) "The Restorative Benefits of Nature: Toward and Integrative Framework," *Journal of Environmental Psychology* 15: 169–182.

Kaplan, R. and Kaplan, S. (1989) *Experience of Nature*, New York: Cambridge.

Keizer, K., Lindenberg, S. and Steg, L. (2008) "The Spreading of Disorder," *Science* 332: 1681–1685.

King, A.C., Stokols, D., Talen, E., Brassington, G.S. and Killingsworth, R.E. (2002) "Theoretical Approaches to the Promotion of Physical Activity: Forging a Transdisciplinary Paradigm," *American Journal of Preventive Medicine* 23(2S): 15–25.

Lang, J. (1987). *Creating Architectural Theory: The Role of the Behavioral Sciences in Environmental Design*. New York: Van Nostrand Reinhold.

Lynch, K. (1960). *The Image of the City*, Cambridge: MIT Press.

McDonald, T.P. and Pellegrino, J.W. (1993). "Psychological Perspectives On Spatial Cognition." In Gärling, T. and Golledge. R.G. (Eds.) *Behavior and environment: Psychological and Geographical Approaches*, Amsterdam: Elsevier.

Michelson, W. and van Vliet, W. (2002). "Theory and the Sociological Study of the Built Environment." In Dunlap, R.E. and Michelson, W. (Eds.) *Handbook of Environmental Sociology*, Westport, CT: Greenwood Press.

Nasar, J.L. (1983). "Adult Viewer Preferences in Residential Scenes," *Environment and Behavior* 15: 589–614.

—— (1989). "Symbolic Meanings of House Styles," *Environment and Behavior* 21: 235–257.

—— (1994). "Urban Design Aesthetics: The Evaluative Qualities of Building Exteriors," *Environment and Behavior* 26: 377–401.

—— (1998). *The Evaluative Image of the City*, Thousand Oaks, CA: Sage.

—— (1999). *Design by Competition: Making Design Competitions Work*, New York: Cambridge.

Nasar, J.L. and Fisher, B. (1993). "'Hot Spots' of Fear of Crime: A Multiple-Method Investigation," *Journal of Environmental Psychology* 13: 187–206.

Nelessen, A. (1994). *Visions for a New American Dream: Process, Principles, and an Ordinance to Plan and Design Small Communities*, Chicago: American Planning Association.

Newman, O. (1972). *Defensible Space*, New York: Macmillan.

Newman, O. and Franck, K. (1980). *Community of Interest*. New York: Doubleday.

Overbye, D. (2009). "Elevating Science, Elevating Democracy," *The New York Times* Jan. 27 D1.

Passini, R., Rainville, C., Marchand and Yves, J. (1998). "Wayfinding and Dementia: Some Recent Research Findings and a New Look at Design," *Journal of Architectural and Planning Research* 15: 133–151.

Perkins, D.D. and Taylor, R.B. (1996). "Ecological Assessments of Community Disorder: Their Relationship to Fear of Crime and Theoretical Implications," *American Journal of Community Psychology* 24: 63–107.

Petzinger, T. (1999). "A New Model for the Nature of Business: It's Alive!" *The Wall Street Journal*, February 26.

Popenoe, D. and Michelson, W. (2002). "Macro-environments and People: Cities, Suburbs, and Metropolitan Areas." In Dunlap. R. E. and Michelson, W. (Eds.). *Handbook of Environmental Sociology*, Westport, CT: Greenwood Press.

Preiser, W.F.E. and Nasar, J.L. (2008). "Assessing Building Performance: Its Evolution From Post-Occupancy Evaluation," *Archnet, International Journal of Architectural Research* 1: 84–99.

Preiser, W.F.E., Rabinowitz, H.Z. and White, E.T. (1988). *Post-Occupancy Evaluation*, New York: Van Nostrand Reinhold.

Rapoport, A. (1993). *The Meaning of the Built Environment: A Nonverbal Communication Approach*, Tucson, AZ: University of Arizona Press.

Russell, J.A., and Snodgrass, J. (1989). "Emotion and Environment." In Stokols, D. and Altman, I. (Eds.), *Handbook of Environmental Psychology* (Vol. 1), New York: John Wiley.

Sommer, R. (1969). *Personal Space*, Englewood Cliffs, NJ: Prentice Hall.

Stamps, A (1997). "Of Time and Preference: Temporal Stability of Environmental Preferences," *Perceptual and Motor Skills* 85: 883–896.

——(1999). "Demographic Effects in Environmental Preferences: A Meta-Analysis," *Journal of Planning Literature* 14: 155–175.

—— (2000). *Psychology and the Aesthetics of the Built Environment*, Boston: Kluwser Academic.

Ulrich, R.S. (1991). "Stress Recovery during Exposure to Natural and Urban Environments," *Journal of Environmental Psychology* 11: 201–1991.

Vischer, J.C. and Cooper Marcus, C. (1986). "Evaluating Evaluation: Analysis of a Housing Design Awards Program," *Places* 3: 66–85.

Webley, P. and Whalley, A. (1987). "Sex Differences in Children's Environmental Cognition," *Journal of Social Psychology* 18: 192–213.

Whitfield, T.W.A. (1983). "Predicting Preference for Everyday Objects: An Experimental Confrontation between Two Theories of Aesthetic Behavior," *Journal of Environmental Psychology* 3: 221–237.

Whyte, W.H. (1980). *The Social Life of Small Urban Spaces*, New York. Conservation Foundation.

Wicker, A. (1979). *An Introduction to Ecological Psychology*, Monterey, CA: Brooks/Cole.

Wilson, J.Q., and Kelling, G. (1982). "Broken Windows," *Atlantic Monthly* 211: 29–38.

Wohlwill, J.F. (1976). "The Environment as a Source of Affect." In Altman, I. and Wohlwill, J.F. (Eds.) *Human Behaviour and Environment: Advances in Theory and Research* (Vol. 1), New York: Plenum.

Wolhwill, J.F. and Kohn, I. (1973). "The Environment as Experienced by the Migrant: An Adaptation-level Approach." *Representative Research in the Social Psychology* 4: 135–164.

Wyant, B.R. (2008). "Multilevel Impacts of Perceived Incivilities and Perceptions of Crime Risk On Fear of Crime," *Journal of Research in Crime and Delinquency* 45: 39–64.

Zube, E.H., Pitt, D.G. and Evans, G.W. (1983). "A Lifespan Developmental Study of Landscape Assessment," *Journal of Environmental Psychology* 3: 115–128.

Further reading

Altman, I. and Chemers, M. (1980). *Culture and Environment*. Los Angeles, CA: Brooks/Cole.

Covers personal space, territoriality, privacy and culture.

Gärling, T. and Evans, G. (1991). *Environmental Cognition and Action: An Integrative Multidisciplinary Approach*, New York: Oxford. Review of environmental cognition.

Geller, E.S., Winnett, R.A. and Everett, P.B. (1982). *Preserving the Environment: New Strategies for Behavior Change*. New York: Pergamon. How to change people's behavior to save the environment.

Goltsman, S.M. and Iacofano, D. (Eds.) (2008). *The Inclusive City: Design Solutions for Neighborhoods and Urban Space*, San Francisco: MIG Communications. Research based guidelines for inclusive designs.

Zeisel, J. (2006). *Inquiry by Design: Environment/Behavior/Neuroscience in Architecture, Interiors, Landscape and Planning*, New York: W.W. Norton. Methods, case studies and findings in environment behavior research.

13

The law of urban design

Jerold S. Kayden

The law of urban design in the United States finds definition through a heady brew of legislation, judicial opinions, constitutions, and private agreements that together guide the physical layout and appearance of the built environment. Urban design rules appear most commonly in locally enacted ordinances, usually expressly authorized by state statutes that restrict the conduct of private individuals. Government decisions taken under these rules are sometimes explained in written documents. Laws of urban design also reside in agreements entered into consensually by private parties, for example, in the form of residential community association by-laws. Federal and state constitutions authorize governments to do certain things and limit the doing of other things in ways that contribute to the law of urban design. Judicial opinions announcing whether government actions have exceeded the bounds of the ordinances or statutes or, more fundamentally, whether the ordinances, statutes, and accompanying decisions themselves have infringed impermissibly on the constitutional rights of individuals, add to the tapestry.

Consider, then, the following:

1 Is it legal for a city to require private property owners to build the city's recently developed urban design plan, in terms of use, shape, bulk, and public spaces?
2 Is it legal for a town to prohibit a homeowner from painting her house purple?
3 Is it legal for a city to prohibit flat roofs on office skyscrapers?
4 Is it legal for a local government to require use of brick as a façade material?
5 Is it legal for a conservation commission to prohibit a landowner from constructing anything on her property in order to preserve a view shed between the public road and the mountains?
6 Is it legal for a historic preservation commission to stop an owner from demolishing or even modifying the exterior of a historically significant building?
7 Is it legal for a design review commission to deny approval for development of a new building because a majority of the members do not "like" the architecture?
8 Is it legal for religious institutions to build a church or temple or mosque in an area zoned for single family residential uses?
9 Is it legal for government to take one person's property against the will of the owner, as long as compensation is paid, and give it to another party who agrees to provide a better urban design?

For better and worse, law makes a powerful imprint on the design of the built environment. Just as technology, market preferences, and artistic impulse influence urban form, law shapes cities and towns through rules and judgments embodied in statutes, implementing decisions, and judicial opinions. Zoning ordinances, subdivision controls, design review procedures and guidelines, historic preservation ordinances, and sign controls, among other laws, intentionally and unintentionally influence the look and feel of cities and towns in ways not fully appreciated by scholars, practitioners, or members of the public. Zoning's height and setback restrictions sculpt residential and commercial skyscrapers and define their relationship to the street and sidewalk, while lot area controls create patterns of scattering or clustering for homes in the suburbs. Design review commissions control colors, materials, and styles of architecture, in an attempt to make sure that new structures are compatible with the existing context of surrounding neighborhoods. Billboard laws may prohibit offsite billboards, while sign laws control the size and styling of on-premise identification signs. Historic preservation commissions designate individual landmarks and districts, thereby gaining the power to disallow even minute changes to façade and structure.

What happens when these government-enacted urban design laws infringe in a given case or across the board upon such constitutionally protected individual rights of private property, free speech, due process, equal protection, or religious practice? Judges enter the act to balance the community's legally implemented preference for specific urban design outcomes against the individual infringement, a balancing act made especially treacherous when the urban design outcome enters the subjective and vague arena of aesthetics. This chapter describes the scope of laws that individually and combined represent the basis for, expression of, and check upon government and collective private actions shaping urban design. The chapter furthermore explores administrative debates framing the formation and application of urban design law, including the tension between rule and discretion.

Government power

The current register of government-enacted urban design laws principally includes zoning, design review, historic preservation ordinances, and subdivision controls. At the heart of these laws are rules that, in furtherance of publicly determined urban design principles, affect what individuals may build on their privately owned property. The rules may be mandatory or voluntary. They may be clear-cut in application or administered with a healthy dose of discretion. They may be detailed or broadly framed.

From a legal point of view, the first question is whether government is empowered to adopt such rules. Government obtains its mandate to act from the consent of the governed, and that consent *ab initio* is set forth in federal and state constitutions that lay out the very nature of government. Thus, say constitutions, the legislative branches may do this, the executive branches may do that, and the judicial branches may review them both. At state levels and, derivatively at local levels, governments enjoy an inherent authority to act under their so-called "police power" to protect and promote the health, safety, morals, and general welfare of their citizens. So what is urban design in this context? To be sure, the accomplishment of urban design goals may always be classified as protecting and promoting health, safety, morals, and especially the general welfare. Good urban design produces a built environment that is productive, functional, equitable, sustainable, and inspirational. Nonetheless, the earliest understandings of the police power saw a difference between regulations that advanced traditionally understood outcomes. The former were within the contours of the police goals of health and safety, and regulations that sought to achieve aesthetically desirable power; the latter were not.

The strict prohibition against government efforts to seek aesthetically desirable outcomes slowly dissipated, especially as it became clear just how hard it was to distinguish between health and safety outcomes on the one hand and aesthetic ones on the other. Zoning provides a classic example. Introduced by New York City in 1916, and approved constitutionally by the US Supreme Court in 1926, comprehensive zoning laws were justified on the basis of achieving broader social and economic goals than those suggested by billboard controls, but their proponents still avoided explicit statements about aesthetics. That said, zoning's trio of use, shape, and density controls undeniably sculpted the profile of structures, as well as their relationships to lot, street, and precinct. New York City's ziggurat towers and Chicago's boxier buildings directly resulted from standards in their zoning codes.

Entering the mid-twentieth century, changes in attitudes, if not a growing recognition of the futility of distinguishing outcomes, accelerated the decline of the "no aesthetics" principle. The first opening became known as the "aesthetics plus" doctrine. Where the mere hint of aesthetics previously poisoned government action, now, as long as the goal of aesthetics was coupled with health, safety, morals, or general welfare goals, the law would pass muster. US Supreme Court Justice William O. Douglas opened the floodgates, at least rhetorically, with his ringing endorsement of beauty in his 1954 *Berman v. Parker* opinion:

> The concept of the public welfare is broad and inclusive. ... The values it represents are spiritual as well as physical, aesthetic as well as monetary. It is within the power of the legislature to determine that the community should be beautiful as well as healthy, spacious as well as clean, well-balanced as well as carefully patrolled.

A quarter century later, Justice William J. Brennan in his 1978 *Penn Central Transportation Company v. New York City* emphasized the importance of culture and history in upholding the city's landmarks preservation law. In the 1980s, the Court acknowledged the validity of aesthetics in considering the validity of billboard and poster prohibitions. State courts similarly have tolerated the promotion of aesthetics, and there are relatively few states today that would question government's inherent power to pursue aesthetically driven outcomes.

A further question involving government authorization to act in furtherance of urban design objectives is presented by the locus of most urban design laws. Control over the use of land has historically resided with local, rather than state or federal, governments. Interestingly, local governments are the least grounded in constitutional law. The federal constitution makes no mention of local governments whatsoever, and in most states, a mere majority vote of the legislature could dissolve the jurisdictional lines separating city from town or village. That would lead to one single state jurisdiction. Empirically, however, the idea of anything other than local control of land use is anathema to the polity. The absence of regional government is enduring testament to the power of locally based authority, even as the very distribution of land use power to local government structures has created a skewed urban design relatively unreflective of regional concerns.

Given the fact that local governments are in most cases legally creatures of the state, a legal question arises whether they can implement urban design law on their own authority, or whether they need express authorization from the higher level authority of the state. In legal jargon, the issue revolves around whether the local government is located in a "Dillon's Rule" state (named after a judge who wrote about the subject) or a "home rule" state (or has its own charter to operate on its own). Within a Dillon's Rule state, the local government must find within state legislation – often the state's zoning act – language that expressly empowers the city to enact the type of urban design law it seeks to adopt. Thus, for example, a section in the state statute will specify that local governments may adopt incentive zoning, and local governments that adopt an incentive zoning provision must follow that express provision to the letter. In home rule states, local governments may think up and enact zoning techniques on their own, even if there is no express state legislative language, as long as there is no state language expressly or impliedly forbidding what they want to do.

The distinction between Dillon's Rule and home rule is often clearer in law than reality. Dillon's Rule states do not restrict as much as commonly believed, and home rule states do not liberate as much as commonly hoped. Nonetheless, in theory, a planning department has more leeway to innovate within a home rule rather than within a Dillon's Rule state. In practice, advice given by in-house counsel or a city's legal department is often overly conservative, urging municipalities within either a Dillon's or home rule state to hew closely to state legislation and not do anything unless expressly authorized. Since such lawyers, especially those outside the planning department, are more concerned with law and less interested in urban design policy, they have little to gain and a lot to lose by going out on the legal innovation limb. From a purely legal point of view, it might be easier to say no, and the only thing that suffers is innovation, a salient reminder that legal advice may be overly restrictive with regard to a city's ability to innovate in the zoning area.

Individual rights

Authorization to act in furtherance of urban design objectives is necessary, but not sufficient. The political check of the voting booth may protect the interests of the majority, but this system axiomatically fails to protect the minority. It is left to federal and state constitutions, as interpreted by judges, to protect individuals against government action with regard to private property, free speech, due process, equal protection, and freedom of religion. In 1981, Supreme Court Justice Brennan penned the famed, to some notorious, phrase, "After all, if a policeman must know the Constitution, then why not a planner." Thus, while the fundamental authority of government to impose an urban design vision on everyone is no longer in doubt, government must nonetheless follow a constitutionally written script. Individual rights must be respected. The federal constitutional clauses of significance include the following:

- Fifth Amendment's "Just Compensation" Clause
  "... nor shall private property be taken for public use, without just compensation"
- Fourteenth Amendment's "Due Process" Clause
  "... nor shall any State deprive any person of life, liberty, or property, without due process of law"
- Fourteenth Amendment's "Equal Protection" Clause
  "... nor shall any State ... deny to any person within its jurisdiction the equal protection of the laws"

- First Amendment's "Free Speech" Clause
  "Congress shall make no law ... abridging the freedom of speech"
- First Amendment's "Free Exercise" Clause
  "Congress shall make no law ... prohibiting the free exercise [of religion]"

State constitutions have similar provisions, even if the wording may differ. As a rule, state constitutions may grant greater, but never lesser, constitutional protections to individuals than that granted by the federal constitution. More than any other constitutionally protected individual right, it is the right of private property that limits what government may pursue in terms of urban design outcomes. The federal Constitution's Just Compensation Clause and state constitutional corollaries command that private property not be taken for public use without paying just compensation. Federal and state due process clauses prevent government from depriving individuals of property without due process of law. Although the generality of these constitutional phrases makes it difficult to define a bright line rule separating acceptable government infringements from unacceptable ones, a treasure chest of federal and state judicial opinions provides a decent feel for how courts might react in a given fact pattern. One thing is clear: private owners do not have, and have never had, an unlimited right to use their property as they see fit. To begin with, the common law of nuisance, *sic utere tuo ut alienum non laedes*, stretching back centuries, restricted owners to uses of their property that did not unreasonably injure others' use of property.

The first decades of the twentieth century in the US introduced a fast-growing, rapidly industrializing nation with newly incompatible, cheek-by-jowl land uses and a general belief in scientific city planning solutions for escalating urban problems. When government regulatory approaches began to supplant case-by-case application of nuisance law as a more efficient, predictable check on private land use, the US Supreme Court emerged as arbiter of how much state intervention would be constitutionally acceptable. With no significant guiding precedent on land-use regulation from the nineteenth century upon which to rely, the Court made up its mind as it went along, and the bulk of its initial jurisprudence approved the state's exercise of regulatory authority.

The 1915 *Hadacheck v. Sebastian* opinion is one of the earliest examples. Hadacheck operated a brick yard in Los Angeles in violation of a local ordinance and was thrown in jail for doing so. He alleged that, used for brick-making, his eight-acre tract was worth $800,000, whereas for residential or any other purpose – and he said there were no other purposes to which it could be put – it was worth $60,000. Although Sebastian, the city's police chief, did not dispute the specific contention of value diminution, he did deny that the ordinance as applied would "entirely deprive Hadacheck of his property and the use thereof." Hadacheck claimed both a deprivation of property and a taking of property without compensation, thereby situating his claim under due process and just compensation clause labels.

In language so sweeping that it still catches constitutional land-use experts by surprise, the Court heartily endorsed the government's exercise of the police power:

> It is to be remembered that we are dealing with one of the most essential powers of government, one that is the least limitable. It may, indeed, seem harsh in its exercise, usually is on some individual, but the imperative necessity for its existence precludes any limitation upon it when not exerted arbitrarily. A vested interest cannot be asserted against it because

> of conditions once obtaining. … To so hold would preclude development and fix a city forever in its primitive conditions. There must be progress, and if in its march private interests are in the way they must yield to the good of the community. The logical result of petitioner's contentions would seem to be that a city could not be formed or enlarged against the resistance of an occupant of the ground and that if it grows at all it can only grow as the environment of the occupations that are usually banished to the purlieus.

Two Supreme Court opinions from the 1920s approved new methods of government restriction on private property while drawing the line on extreme deprivations. The 1922 *Pennsylvania Coal Co. v. Mahon* decision is best known for Justice Oliver Wendell Holmes' declaration, "if regulation goes too far it will be recognized as a taking." There, a Pennsylvania state statute known as the Kohler Act forbid coal companies from conducting subsurface mining in ways causing subsidence of houses on the surface, even in cases where the coal company had expressly retained the subsurface rights for itself when it sold the surface rights to the homeowner. The Act made it "commercially impractible" to mine the coal, leading the Court to conclude that the law had "very nearly the same effect for constitutional purposes as appropriating or destroying [the property right to mine the coal]." In such an extreme case in which the coal-mining property interest was effectively destroyed, the Court found a taking:

> Government hardly could go on if to some extent values incident to property could not be diminished without paying for every such change in the general law. As long recognized, some values are enjoyed under an implied limitation and must yield to the police power. But obviously the implied limitation must have its limits, or the contract and due process clauses are gone. One fact for consideration in determining such limits is the extent of the diminution. When it reaches a certain magnitude, in most if not in all cases there must be an exercise of eminent domain and compensation to sustain the act.

If *Pennsylvania Coal* bookended *Hadacheck*, *Village of Euclid v. Ambler Realty Co.* bookended *Pennsylvania Coal*. There, the Supreme Court decisively affirmed the constitutionality under due process and equal protection challenges of comprehensive zoning. Ambler owned 68 acres in Euclid and wanted to develop its tract for industrial uses which, it alleged, would yield a value of $10,000 per acre. Limited by Euclid's zoning to residential uses, the land would have a value of $2,500 or less per acre, Ambler claimed. In its general exposition, *Euclid* sounds like *Hadacheck,* except more so:

> Building zone laws are of modern origin. They began in this country about twenty-five years ago. Until recent years, urban life was comparatively simple; but with the great increase and concentration of population, problems have developed, and constantly are developing, which require, and will continue to require, additional restrictions in respect of the use and occupation of private lands in urban communities. Regulations, the wisdom, necessity and validity of which, as applied to existing conditions, are so apparent that they are now uniformly sustained, a century ago, or even half a century ago, probably would have been rejected as arbitrary and oppressive. Such regulations are sustained,

> under the complex conditions of our day, for reasons analogous to those which justify traffic regulations, which, before the advent of automobiles and rapid transit street railways, would have been condemned as fatally arbitrary and unreasonable. And in this there is no inconsistency, for while the meaning of constitutional guaranties never varies, the scope of their application must expand or contract to meet the new and different conditions which are constantly coming within the field of their operation. In a changing world, it is impossible that it should be otherwise. But although a degree of elasticity is thus imparted, not to the *meaning*, but to the *application* of constitutional principles, statutes and ordinances, which, after giving due weight to the new conditions, are found clearly not to conform to the Constitution, of course, must fall.

Reviewing the heart of the Euclid ordinance – its exclusion of business, industry, and, most controversially, apartment houses, from single-family residential districts – the Court accepted the proffered justifications as "sufficiently cogent to preclude us from saying, as it must be said before the ordinance can be declared unconstitutional, that such provisions are clearly arbitrary and unreasonable, having no substantial relation to the public health, safety, morals, or general welfare." The Court did not demand irrefutable arguments to buttress the regulatory action, and reminded all that, if the validity of the legislation is "fairly debatable," the legislative judgment should stand.

Today, the gold standard for understanding the balance between government's interest in urban design outcomes and the economic rights of private property owners is found in Justice Brennan's 1978 regulatory takings *magnum opus, Penn Central Transportation Company v. New York City*. In that case, New York City's landmarks preservation commission had designated the 1913 beaux arts Grand Central Terminal a landmark. Penn Central, its owner, wanted to build a skyscraper above or in place of the terminal, but was denied permission by the commission. That action prevented the company from realizing millions of dollars in annual lease revenue.

In its six-to-three decision favoring the city, the Court took *Pennsylvania Coal's* statement that a regulation could go too far and dressed it up with several factors to determine what too far would be. Judges should consider both the economic impact of the regulation, particularly with regard to its effect on the owner's distinct investment-backed expectations, and the character of the governmental action. It was the application of these factors to the facts of the case that would demonstrate, once again, that property rights, as economic rights, would rarely impede government efforts to achieve urban design and planning objectives. The fact that Penn Central indisputably would lose money derived from speculative development would not suffice for a finding of unconstitutionality. The fact that it had conceded to making a reasonable return on the existing terminal tenants would suffice for constitutional purposes. Though much trumpeted, cases following *Penn Central* have not upset its basic approach.

From time to time, government has chosen to exercise its power of eminent domain, rather than its police power, to achieve urban design objectives. Under the Just Compensation clause of the federal constitution, government may take land from private owners against their will as long as the taking is for a public use and just compensation is paid. In the 2005 *Kelo v. City of New London* case, the Court heard a constitutional challenge from several homeowners seeking to overturn the city's decision to take their properties and give them to a private developer willing to implement the city's idea of a superior urban design plan. In a five-to-four decision, the Court constitutionally sided with the city, finding a public use and declining to second-guess the plans prepared by the local urban officials. The public outcry that followed, however, led a majority of states to amend their legislation with regard to eminent domain exercises. Of course, individuals use property for more than economic purposes. A home is not only an investment; it is an expression of one's individuality regardless of profit or loss. Courts have recognized that the ability to use one's land may implicate rights of liberty, speech, religion, and privacy. Under a variety of conceptual approaches springing primarily from due process and equal protection provisions, judges have reviewed government actions that favor one design over another. They have yet to state categorically that design, as embodied in a person's home, *is* the owner's "speech," and thus worthy of the high constitutional protections afforded classically defined speech conveying political or ideological messages. They have also declined, generally, to find that the designer herself can claim a constitutionally protected right to her architectural expression. At the same time, perhaps sensing that there is something highly individual, as well as communal, about the built environment, judges have devised rules that confine government's attempt to advance the "communal" at the expense of the "individual." Most noteworthy are cases stating that design review laws must provide standards that reassure the reviewing judge that the law and decisions taken pursuant to it are neither arbitrary and capricious nor too vague for individuals to follow. Rule-based laws are least vulnerable to constitutional attack to the extent they state their standards in black and white (or purple). Design laws operated with a healthy dose of discretion run a higher risk. Subjective, beauty-in-the-eye-of-the-beholder decision-making is more suspect; objective, straightforward criteria are not. Predictability is prized; the average person, let alone the average judge, is not to be left in the dark about how an application for development approval will be handled.

Where do administrators and reviewing judges find objective, predictable standards for anchoring their decisions? The most popular repository is the surrounding neighborhood, where architectural style, construction material, massing, cornice lines, and other design elements may be seen, assimilated, and copied. Procedural safeguards of public notice and hearings, as well as written decisions by design review administrators, help convince reviewing judges that the gauntlet has been fair to participating runners. Design commissions composed of professionals, scholars, and representatives from such interested groups as property owners, unions, and relevant geographic areas contribute to a sense of fairness.

If constitutions protect individuals only against state (government) action, what happens in a world in which private governance regimes replace public governance regimes, in the mall, the gated community, or the privately owned public space? Should the "private" regulator of such places enjoy a regulatory *carte blanche* simply because it is nominally not a public government? Americans living in privately owned, privately managed communities find themselves subject to privately created, privately administered design codes. When it comes time to repaint the house or replace a window, the code tells them what they can and cannot do. Because these rules are implemented through privately created bodies, the provisions of federal and state constitutions generally do not apply. It is hard to be sympathetic to such individuals since they voluntarily purchased their home in a community expressly governed by such codes. The principal remedy for disgruntled individuals is to convince a voting majority of fellow residents to change the rule or decision, or to move out. At the same time, if the American built environment continues its tilt toward privately managed built environments, it is easy to imagine legislative interventions that may limit the authority of the private regulator. It is even easy to imagine, as has occurred from time to time in state courts, that judges will consider privately owned and managed spaces to be sufficiently similar to public spaces that the constitutional protections become relevant.

Administrative considerations: rule versus discretion

Law is about more than government authority and individual rights. The way it is administered equally affects its character. The most significant administrative debate with practical consequences for urban design is whether such laws should be grounded within a rule- or discretionary-based legal framework. Rule-based law, often given the label "matter-of-right" or "as-of-right," expressly specifies through text, map, and/or diagram what an owner can and cannot build on her property. To the extent they are necessary, approvals are ministerial in that the government staff reviewer is measuring not whether the project represents good urban design, but simply whether it meets the letter-of-the-law set forth in the text, map, and/or diagram. Discretionary-based urban design laws, on the other hand, vest case-by-case, subjective decision-making authority in the hands of city staff and officials, who determine proposal-by-proposal what the owner may do based on a substantive review.

For much of the twentieth century, the rule-versus-discretion debate was fairly clear-cut as laid out by the two dominant planning regulatory regimes in the world. The German/American zoning scheme was rule-based, while Great Britain's town planning scheme was discretion-based. Under the British *Town and Country Planning Act of 1947*, applicants needed to obtain "planning permission" for most development activities, whereas standard zoning states in advance what owners can and cannot do. Today, the British system has moved toward the rule-based model, while the American model has incorporated an enormous amount of discretion. They meet somewhere in the middle of the pond.

Indeed, in practice, urban design law today is an amalgam of rules and discretion. For some cities, smaller projects are exempt from discretionary review. Sometimes, an ordinance on its face may appear to be rule-based, but in fact no one can possibly build under the rules, so discretionary triggers are consistently pulled. Sometimes the variance-granting body gives out so many variances, often illegally if one takes seriously the legal standard of hardship for a variance, that they begin to subvert the basic plan suggested by the otherwise as-of-right zoning.

What are the generic arguments for and against rule and discretion? The principal argument for the rule-based approach is that it provides predictability, if not certainty, for developers and lenders who above all else prize predictability and certainty unless that predictability and certainty is that the developer predictably and certainly cannot develop anything. Indeed, a predictable and certain zoning district allowing *de minimus* development is not treasured highly by those owning property within it. A second argument is that it is easier and cheaper to administer a rule-based system, since high-level administrators with expertise are not needed. Third, a rule-based system is less susceptible to the corruption of politics, not in the sense of illegal bribery, but in the sense of allowing improper considerations to color a decision. Fourth, rules force planners to decide about planning in a more comprehensive, future-oriented way, rather than making things up as they go along. Fifth, planners too often are co-opted by developers in the discretionary system and give in more than they should.

The arguments for discretion revolve around a different take on design and planning. Discretion proponents might agree that rules produce certainty, but they disagree about the possibility of composing compelling rule-based criteria. First, discretion advocates see the impossibility of reducing to standard rules the qualities that make for well-designed urban environments. When rules are stated, they say, developers provide letter-of-the-law compliance or find loopholes that, in either case, produce mediocrity. Second, discretion allows for an engagement with developers encouraging a collaborative inventiveness absent from the rule-based approach. Third, discretion allows planners to get exactly what they want, even if the owner does not want to produce it. Fourth, discretion is not as disliked as rule-based proponents may claim. Developers and their servants (lawyers, expediters, architects, and planners) have invested much time honing the navigation skills ideal for discretionary approval and are not as ready, as developers' complimentary words about rules might suggest, to jettison those skills to the nasty winds of rule-based law.

There are combination approaches that attempt to marry the best of rules and the best of discretion. Under such approaches, the rules are, indeed, set forth clearly in advance, but the issue of determining whether the developer has met the rules is left to skilled planners and designers rather than less skilled, ministerial inspectors from the building permitting and licensing bureaucracies. In theory, the city planner must approve the project if it meets the rules, but she can in the process urge the developer to do better than just meet the rules during the review of the development proposal. Such a process has gone under the name "certification" in at least one jurisdiction, New York City, and the marriage has lasted for many years.

Conclusion

This chapter has reviewed the many elements of law that, together, may be deemed urban design law. The legal regime is constructed from legislation enacted primarily at local and state levels, actions taken by government pursuant to such legislation, constitutional provisions protecting individual rights, judicial opinions interpreting the application of constitutional provisions, and private agreements made between consenting individuals. The dynamic tension between government power, exercised on behalf of the collective, and individual rights protected by constitutions, has provided much of the excitement in the evolution of urban design law. The evolving challenge of a public realm increasingly provided and managed by private actors will require adaptation of prevailing legal norms and invention of new ones. Designers and planners, no less than lawyers and developers, should accept the invitation to build this new legal regime.

Author's principles of urban design law

1. The more express authorization by state legislation, the better.
2. The stronger the inherent or carefully documented evidence of aesthetic values or goals, the better.
3. The more detailed the standards guiding the exercise of discretion, the better.
4. The more an average person would understand the rules, the better.

5 The more the process – notice, hearing, record, written decision – the better.
6 The less "final" decision-making authority delegated to a non-legislative body, the better.
7 The less deprivation of all economically viable use of the entire property, the better.
8 The less a regulation places a disproportionate regulatory burden on one property owner, when such burden is more properly borne by the public as a whole, the better.
9 The less a regulation directly or indirectly limits freedom of expression, the better.

References

Berman v. Parker, 348 U.S. 26 (1954).

Hadacheck v. Sebastian, 239 U.S. 394 (1915).

Kelo v. City of New London, 545 U.S. 469 (2005).

Penn Central Transportation Company v. New York City, 438 US 104 (1978).

Pennsylvania Coal Company v. Mahon, 260 U.S. 393 (1922).

San Diego Gas & Electric Co. v. City of San Diego, 450 U.S. 621 (1981) (Brennan, J., dissenting).

Village of Euclid v. Ambler Realty Co., 272 U.S. 365 (1926).

Further reading

Barnett, J. (1974). *Urban Design as Public Policy.* New York: Architectural Record. Together with Marcus/Groves' *The New Zoning,* this classic is essential reading for those interested in how legislation can be used to direct the urban design of cities.

Blaesser, B.W. (2002). "Smart Growth: Legal Assumptions and Market Realities," and Kayden, Jerold S., "The Constitution neither Prohibits nor Requires Smart Growth," in Szold, T.S. and Carbonell, A. (Eds.). *Smart Growth: Form and Consequences.* Cambridge: Lincoln Institute of Land Policy. 128–157, 158–179. The two articles present different points of view in the property rights legal debate as it affects enactment of legislation designed to promote smart growth.

Duerksen, C.J. and Goebel, R.M. (2000). *Aesthetics, Community Character, and the Law.* Chicago: American Planning Association. This concise volume is an excellent primer on how individual rights such as free speech may limit government's ability to regulate urban design for aesthetic purposes.

Kayden, J. S. (2004). "Charting the Constitutional Course of Private Property: Learning from the 20th Century," in Harvey Jacobs, ed., *Private Property in the 21st Century: The Future of an American Ideal.* Cheltenham: Edward Elgar, 31–49. Review of how the U.S. Supreme Court has interpreted the federal constitution as it enables and circumscribes the ability of government to regulate the planning and design of American cities in the face of private property rights.

Kayden, J.S. (2005). "Using and Misusing Law to Design the Public Realm." In Ben-Joseph, E. and Szold, T.S. (Eds.) *Regulating Place: Standards and the Shaping of Urban America.* New York: Routledge 115–140. This article presents a case study of how law pro-actively sculpted the urban design of New York City to produce, for better and worse, a new category of public space for the public to use.

Marcus, N. and Groves, M.W. (Eds.). (1970). *The New Zoning: Legal, Administrative, and Economic Concepts and Techniques.* New York: Praeger. A classic book that reviews, among other things, the legal and administrative mechanism and limitations associated with energetic efforts to influence the urban design of cities.

14

Political theory and urban design

Margaret Kohn

There is a New Urbanist development called Town of Tioga just outside Gainesville, the sprawling, suburban town where I used to live. The Town of Tioga features broad sidewalks, leafy, tree-lined streets, beautifully landscaped parks, and a town center with a children's playground, meeting hall, and swimming pool. The relatively high price of real estate in the Town of Tioga reflects the fact that it is selling more than just "McMansions." It is selling community. The style is New Urbanist; the large, imposing houses typically have porches and sit on fairly modest lots; the scene is unmarred by automobiles or driveways and curb cuts because the two-car garages are located off alleys in the back. The communal spaces – the playground, walking trails, and parks – distinguish it from other housing developments. The appeal of these common spaces is reflected in housing prices, which are typically around 30 percent higher than similar sized houses in other new developments nearby. The proliferation of developments like the Town of Tioga suggests that the market has proved adept at providing common spaces, at least to those who can afford it. This is affluent enclavism with a twist; there are no gates at the entrance, and row houses and neo-craftsman cottages are situated practically adjacent to mansions (see Low 2003 on gated communities). The real estate literature promised "a return to what made classic communities great" and, in a way, it delivered. Begrudgingly, I was enchanted.

This chapter is the attempt to think through this spell from the perspective of political theory. Urban design is concerned with creating vibrant public spaces and political theory can contribute to this project by clarifying the meaning of the term public and its relationship to other values such as democracy and equality. New Urbanism is a good example of the way that design can be used to elide the distinction between communal space and public space; political theory can help clarify the difference between them and explain why citizens should appreciate the latter (for a slightly different version of this distinction, see Hénaff and Strong 2001). Community is so appealing because it is a seductive substitute for public life. Like an artificial sweetener, which offers all of the pleasure without the calories, communal space promises the pleasures of sociability without the discomforts of the unfamiliar. It offers the fellowship of a shared world without demanding the sacrifices of sharing with those who have less to offer. In a community, we share with others who are similar to ourselves. But in a pluralistic democracy, we must also share with people who are different. Democratic solidarity depends upon a public realm – a public good – that allows individuals to build sympathy with one another in spite of their ethnic, religious, and economic differences.

The commons and the public

In the past few years, some of the most thought-provoking critiques of privatization have come from scholars writing about "the commons." The core idea is that citizens collectively own an array of resources that should not be exploited for private gain. The term "commons," a somewhat archaic concept usually associated with pre-capitalist agriculture in England, is artfully redeployed by these scholars to suggest that there is a populist alternative to the Scylla of big government and the Charybdis of corporate control. David Bollier (2002: 4), for example, describes the commons as "the vast range of resources that the American people own." In his book *Silent Theft*, he specifies that the commons include "tangible assets such as public forests and minerals, intangible wealth such as copyrights and patents, critical infrastructure such as the Internet and government research and cultural resources such as the broadcast airwaves and public spaces" (Bollier 2002: 2–3). Lawrence Lessig (2002: 9) defines the commons more broadly as a resource "in joint use or possession to be held or enjoyed equally by a number of persons." The examples that he offers are (public) streets, parks, and beaches, Einstein's theory of relativity, and creative works that are in the public domain.

The reappropriation of the term commons is a recent response to a large body of scholarship that had discredited it. In the aftermath of Garrett Hardin's influential article, "The Tragedy of the Commons," (1998) the term commons became associated with the exploitation of natural resources. In the article, Hardin used the example of a common pasture to illustrate the problem of over-exploitation and the need for a private property regime. He claimed that each individual, pursuing his rational self-interest, would choose to graze the maximum number of cattle on the common pasture even if this would lead to overgrazing and the destruction of the pasture. Since the benefit of each additional cow went to the individual but the cost was shared by the group, there was no incentive to conserve. The metaphor of the pasture was taken up by economists and politicians who argued against any kind of public goods or public property, which, they felt, were doomed to be destroyed by self-interested, inefficient behavior. Only private ownership could ensure the proper incentives for responsible stewardship (see Blackmar 2005).

Political scientists such as Elinor Ostrom, however, have concluded that Hardin was too pessimistic. Ostrom (1990) has documented the way that informal norms or formalized practices can ensure the long-term viability of common property regimes (Hess and Ostrom 2007). Thus, after a long period of disrepute, the commons began to experience a comeback as an alternative to the bureaucratic inefficiency of public property on the one hand and the hyper-individualism of the market on the other. Whereas in the hands of Hardin's right-wing followers the rhetoric of the commons was used as an argument for private ownership, its new proponents (on the left) redeploy it as a solidaristic *alternative* to public (state) ownership, as I have discussed elsewhere (Kohn 2004).

There are good reasons for adopting the rhetoric of the commons. The term is etymologically related to community, a word with largely positive connotations whereas the alternative – public – is associated in many people's minds with bureaucratic red tape and inadequate government programs (public schools, public assistance, public transit). The rhetoric of the commons also lends itself to a powerful critique of privatization by way of historical analogy with the enclosure movement that transformed English agriculture in the seventeenth and eighteenth centuries. Just as English lords enclosed common lands in order to appropriate the resources for their personal enrichment, contemporary corporations today are privatizing common resources (scientific discoveries, natural resources, public spaces) for their exclusive benefit. The rhetoric of the commons also makes it possible to identify the similarities between otherwise dissimilar things noted above that are all part of our common wealth.

Despite these compelling features, I am hesitant to adopt the term commons and instead want to defend the more familiar (but unpopular) concept of the public. The main reason for my choice of terminology is that the term commons can legitimately be applied to forms of joint ownership that are still extremely elitist and exclusionary. According to Lessig (2002: 20) "The commons is a resource to which anyone *within the relevant community* has a right without obtaining the permission of anyone else" (my emphasis). Although this may initially seem inclusive, it can actually be very exclusive, at least in the cases where residential communities are extremely stratified and segregated. The crucial caveat is that one must be a member of the relevant community. Gated communities and other Common Interest Developments (CIDs) often provide extensive collective amenities for their residents: swimming pools, golf courses, play grounds, etc. These amenities are available to all residents without obtaining anyone's permission and therefore meet Lessig's definition of a commons. Yet, these types of commons do not provide an alternative to the balkanization produced by private interests or a solidaristic, egalitarian oasis within the market economy (see McKenzie 1994 for example).

In the Town of Tioga, for example, the children's playground and swimming pool are gated and accessible only to residents with an entry code. They provide an opportunity for residents to socialize with one another but simultaneously decrease their contact with the more diverse range of people who inhabit the broader polity (see Gordon 2004).[1] Moreover, residents of the Town of Tioga and other similar Common Interest Developments (CIDs) have no motive to support property taxes that pay for public recreational amenities such as parks and playgrounds. The short-term consequence is the increased segregation of leisure time and the long-term consequence may be the disappearance or deterioration of public places that are accessible to the poor (Young 1999).

The term commons is problematic because it erases the distinction between fundamentally different kinds of collective property. The commons of a gated community is not the same as the Boston Common. The latter is a public place (accessible to everyone) and the former is akin to a clubhouse, a place shared by members. We need a language that helps us distinguish between apparently similar forms of collective ownership that have very different social and political effects.

Roman law provides a useful starting point because it distinguished between several different forms of non-exclusive property. *Res nullis* was the term used to describe things belonging to no one such as abandoned property or uncultivated lands. It designated property that had not yet been appropriated for individual or shared use. *Res communes* referred to things that were open to all by their nature. Typical examples included the ocean or the air, things that could not be separated into proprietary parcels. The next two categories are particularly important for our purposes: *res publicae* and *res universitatis* (Rose 1986). According to Bouvier's law dictionary, *res publicae* are things belonging to the state, such as bridges, roads, and waterways. *Res universitatis* refers to things belonging to cities or other corporate entities, such as theaters, market houses, and the like. They differ from things that are public, inasmuch as the latter belong to a nation (Bouvier and Rawle 1984).

Initially the spaces that fell into the category of *res universitatis* were relatively inclusive. Theaters and stadiums were owned by municipalities and functioned as staging grounds for spectacles that unified the city by bringing residents together. Cicero, for example, thought that they were key political institutions because they fostered civic pride and civic identity. But after the break-up of the Roman empire and the emergence of feudalism in Europe, *res universitatis* came to describe the shared property of increasingly exclusive institutions such as universities, monasteries, and guilds. These were places jointly owned by a corporate body and accessible only to their members.

The early universities, guilds, and monasteries functioned as a commons on the inside but were perceived as private property from the outside (Rose 2003). If, following MacPherson, we define property as "an enforceable claim of a person to the use or benefit of a thing," then *res universitatis* is private property from the perspective of non-members (MacPherson 1978).[2] In keeping with this distinction from Roman law, I will use the term commons to refer to *res universitatis*: places that are owned collectively for the exclusive use of group members. The paradigmatic modern examples of this type of commons are the parks, playgrounds and pools that are owned by homeowners associations; these developments that have both common and private property (individual houses and shared amenities) are called Common Interest Developments (CIDs). By contrast, I will use the term *public* to refer to places that are generally accessible and reflect the diversity of the broader polity. Although such places are usually owned by the state, they can sometimes be owned by other entities and, by law or custom, be dedicated to use by all citizens.

Of course, public spaces are also regulated in order to resolve conflicts between uses that are perceived as incompatible; for example, in order to ensure safety, many parks separate off-leash areas for dogs from playgrounds designated for small children. It is difficult to distinguish between regulations that exist in order to coordinate different types of uses and those that are meant to exclude undesirable people through restrictions on conduct such as loitering (Ellickson 1996). Because of its visibility, public space has always been a site of contestation over collective identity and individual behavior. In the early years of the parks and playground movement there were struggles between groups who desired open fields for sporting competition and those who insisted that parks should be an aesthetic site of contemplation. As Lynn Staehli and Don Mitchell (2008) have shown, similar conflicts animate contemporary sites such as the plaza in downtown Santa Fe, where vendors, business leaders, historic preservationists, civic boosters, teenagers, indigenous people, and workers have different visions about how to govern the symbolic heart of the city. Nevertheless, there is still an important distinction between a public space and a more limited commons; both are regulated, but in public spaces, the principle of fairness requires that everyone's basic liberty be respected and everyone's voice counted equally in the process of determining the regulations (King 2004).

A critic might object that denouncing CIDs as elitist and exclusionary implies a wholesale and unsustainable assault on private property. When a residential community association or developer provides its own parks, pools, and playgrounds, it is no different from a family's decision to install play equipment or a pool in its backyard.

As long as individuals can use their property as they see fit, there is no reason that neighbors cannot get together to share such amenities. To argue otherwise would not only be inconsistent but would also have the perverse effect of excluding the middle classes – the groups most likely to buy into Common Interest Developments – from the amenities enjoyed on private property by the rich.

The objection is a convincing response to an argument for banning Common Interest Developments (CIDs), but this is not the intent of the argument. My goal is to show that the new rhetoric – including the architectural language – of the commons is problematic because it disguises the difference between public and collectively held private property. The rhetoric of the commons has become so popular that it is now a popular name for new shopping malls, even though malls are privately owned and allow access to people only as long as they behave as consumers not as citizens (Barber 2001). These new malls-qua-commons try to recreate the atmospherics of turn-of-the-century downtowns even as their retail practices tend to destroy these objects of nostalgia.

Many of the scholars working on sub/urbanism from the perspectives of human geography, cultural studies, and architectural criticism have pursued a strategy of de-mystification (Sorkin 1992). These studies show how popular reform projects such as urban renewal, gentrification, new urbanism, and festival marketplaces have negative consequences that are not initially apparent. The cultural critique emphasizes that the aesthetic appeal of historical allusions or artifacts often serves to disguise rather than transform the alienating features of modernity, whether in the form of homogeneous, chain-store retailers or automobile based suburbs. The socio-political critique draws attention to the way that a superficial diversity of styles masks an underlying racial and socio-economic homogeneity. This research has enriched our understanding of recent trends in the production of space but it is based on the problematic false consciousness model. Its target audience are those people who "naively" believe that Ye Olde Towne Center or the Sweet Home Plantation Commons are public places. But this critical theorizing itself naively assumes that revealing the exclusive, elitist, or consumerist nature of these places will motivate people to seek/create public alternatives. It assumes – but often does not explicitly defend – the value of public goods. What is missing in much of this literature is a positive defense of the value of public space (as an exception, see Young 1999). A number of political theorists have explored the meaning of the term public and its relationship to other concepts such as citizenship, rights and justice. The final sections of this chapter provide a brief overview of normative approaches to public space.

The bourgeois public sphere

In *The Structural Transformation of the Public Sphere*, Jurgen Habermas (1991) analyzed the concept of the public in the history of modern political thought as well as the emergence and decline of the bourgeois public sphere as a site of a distinctive political practice. In Habermas's socio-historical account, the café was the paradigmatic site of the bourgeois public sphere. The café was a political space with its own characteristic rules, informal norms, and scripted behaviors. It was the social milieu of the new liberal politics, a place that brought together artisans, intellectuals, the commercial middle classes, and even aristocrats. Habermas's analysis of the bourgeois public sphere was striking because of its attentiveness to the sites of the new form of power. The bourgeois public sphere was an arena of rational-critical discussion about the common good. Habermas defined the public sphere as "a realm of our social life in which something approaching public opinion can be formed" (Habermas 1989:136). After analyzing the coffeehouses, salons, clubs, and journals, Habermas concluded that the public sphere existed wherever private individuals engaged in critical debate that exerted influence over government. It provided a link between the established channels of political authority and private economic and domestic interests.

The concept of the bourgeois public sphere highlighted the political significance of civil society. It located liberal politics in a particular social milieu rather than simply an intellectual field or historical period. For Habermas, however, the public sphere was not a physical place. It was an analytic construct that could not be reduced to its constitutive sites and locations. It was an ideal type, abstracted from empirical regularities in order to highlight their salient features.

The notion of public sphere was universal in the sense that it was, in theory, accessible to everyone and oriented towards general rather than private, individual interests. Paradoxically, however, it was also bourgeois, not only because the public was made up of educated urban clerks, merchants, and professionals but also because the bourgeois era created the conditions that made this type of public sphere possible. One key condition was the rise of privacy; the separation of work from home and increase in leisure time created a sphere of inwardness and subjectivity, which was a necessary precondition of inter-subjectivity. According to Habermas, only individuals with distinctive judgments and views were capable of rational, critical debate.

The Structural Transformation of the Public Sphere is different from the more abstract theories of deliberative democracy that are advanced today. While a number of scholars have criticized Habermas for dismissing the popular public sphere or overlooking the elitist dimensions of the bourgeois public sphere, the method of his study actually opens up these questions for the reader. By locating deliberation in space and time, it allows us to see the power relations that determined who participated. The concept of the *bourgeois* public sphere draws attention to its class character and Habermas also emphasizes the structural conditions that made it possible for private people to develop independent judgment about matters of common concern.

Although the concept of the public sphere has continued to exert influence, Habermas himself was more ambivalent about its relevance for contemporary society. The second half of his book explains the reasons for the decline and disappearance of the public sphere. These include the rise of the mass media, which is so intrusive that it destroys the inwardness and subjectivity; the emergence of a politics of interest groups rather than ideas (especially class conflict); the influence of a performative rather than deliberative public sphere; and growth of a leisure industry that encourages spectatorship. Yet despite his supposition that these trends amount to a "refeudalization" of public life, Habermas concludes on an oddly optimistic note. He suggests that the conditions for a resurgent public sphere may be emerging in affluent societies where struggles over resources may be replaced by arguments about the public good.

The concept of the public sphere enriches contemporary debates about space and cities in a number of ways. It has inspired theorists to look at the sites and practices that anchor democratic citizenship. It emphasizes the political salience of the conversations and activities that bind people together in civil society. It also defends an ideal of rational-critical debate that can serve to criticize associations that are systematically exclusive, manipulative, deceptive, or oriented to the promotion of private interests. Some critics have faulted Habermas for presenting an idealized account of the bourgeois public sphere and there is undoubtedly some truth to this criticism, but he also painted a vivid picture of public life, one that reminds us that democratic citizenship is not practiced only at the ballot box and public reason is useless when practiced for citizens rather than by them.

Other contemporary theories of public space

It is beyond the scope of this chapter to summarize the ways that Habermas modified and developed these ideas in his subsequent books, but most commentators would probably agree that in his later work, Habermas focused more on the characteristics of communication rather than on space and place. Other theorists, however, have continued to explore the relationship between public space, citizenship, and rights. In "Roads to Freedom," Arthur Ripstein (2008) has developed a Kantian account of public space. He argued that roads are the paradigmatic public spaces (Figure 14.1); roads are legitimate government responsibilities because they are necessary to sustain a system of private freedom. Without pubic roads, each person would be effectively imprisoned on his or her own private property. The owners of adjacent property would be able to arbitrarily limit their neighbors' movement and therefore their freedom. Jeremy Waldron (1991) has taken this argument a step farther, noting that homeless individuals do not have any private property therefore their very existence depends on the accessibility of public space. The state cannot forbid individuals from performing basic life functions in public if they have no other place to perform them; to do so would be a violation of the basic rights to life and liberty.

Even neo-liberal theorists who are not convinced by the idea that there is a right to public space still recognize that there are cases when it makes sense for the government to provide certain types of public goods including public spaces. According to classical economic theory, public intervention is necessary in the cases of market failures. The paradigmatic examples of public goods are things such as national defense and clean air that are "non-severable;" in other words, you cannot provide them to one person without also providing them to others. Economists also use the phrase "transaction costs" to explain why it may be difficult for a private entrepreneur to provide amenities, even when people are willing to pay for them. For example, people may be willing to pay to maintain a park, but they will not pay the extra cost that would be necessary to build a gate and pay the salary of the gatekeeper. Although this has been an influential argument in favor of the public provision of roads, parks and plazas, it has an important weakness. Private companies have been quite creative at finding ways to circumvent these challenges by using technology to minimize transaction costs or creating high value amenities that are marketed to elites who can absorb the high transaction costs. Country clubs, electronically controlled toll roads, and Common Interest Developments are illustrations of the way that communal spaces can be commodified (see Foldvary 1994 for a detailed discussion).

Democratic theorists have provided a very different rationale for public space. They emphasize that the market has developed an ingenious variety of places that interpolate us as consumers but few that foster an identity as citizens. According to this line of critique, the market tends to create specialized landscapes that attract targeted socio-economic groups but has

Figure 14.1 Political march, Avenida Juárez in Mexico City. Source: Tridib Banerjee.
Note: Street as the setting for political action; supporters of defeated candidate Andrés Manuel López Obrador of the Democratic Revolution Party (PRD) march along Avenida Juárez in Mexico City to demand vote recount after the Mexican presidential election in July 2006.

no incentive to provide public spaces that are shared by rich and poor alike. In "Constructing Inequality," Susan Bickford (2000) argues that cities and suburbs have become hostile environments for democratic participation and imagination. This is partially due to the extreme segregation reinforced by the built environment. Gated communities are the most obvious manifestation of this logic of exclusion, but there are more subtle and flexible practices and architectural cues that also create zones of safety and zones of danger. Prickly plants and "bum proof" benches are designed to drive away homeless people; malls provide limited seating that does not accommodate groups of people who might socialize rather than shop; police and private security selectively enforce rules in order to force undesirables to move along (Davis 1992). According to Bickford (2000) these strategies hide the existence of inequalities and social problems while also obscuring the diversity of our polity. These sanitized environments also have the effect of stunting our imagination, making other people's lives seem completely alien and unconnected to the experiences of the more affluent.

In "Residential Segregation and Differentiated Citizenship" Iris Marion Young (1999) makes a similar argument;

she points out that segregation not only undermines our civic capacities but also is a source of injustice. She argues that more affluent areas typically have better municipal services, superior amenities, more convenient public transportation, better schools, and attractive physical environments. Residential segregation limits people's options and forces them to live in places that are costly in relation to the standard of living that they provide. Segregation is wrong because it undermines both freedom and equality while also mystifying the existence of inequality. Because of the design of cities and suburbs, affluent people may never even see the living conditions in the poorer parts of town. This in turn fosters a kind of cognitive and normative distortion, and a form of denial, whereby relatively privileged people perceive their own struggles to pay a hefty mortgage, private school tuition, or car payments as real adversity. According to Young (1999: 242), "Segregation thus makes privilege invisible to the privileged in a double way: by conveniently keeping the situation of the relatively disadvantaged out of sight, it thereby renders the situation of the privileged average."

Young also argues that segregation impedes political communication. It does this in two ways. Segregation diminishes the number of sites that might provide opportunities for discussions about identity, difference, and injustice. Segregation, however, is not just a matter of physical barriers. It has a psychological dimension with political implications. In segregated cities there are few opportunities for the types of informal interaction that dismantle stereotypes and build sympathy, therefore when encounters do occur, they frequently lead to misunderstandings or hostility. For Young, however, the solution isn't simply integration; there are legitimate reasons to seek out ethnic enclaves that might provide specialized services, businesses, and support, especially for people who are marginalized in the majority culture. It is important, however, that these neighborhoods do not command vastly different resources and that there are ample opportunities to blur the boundaries, diffuse tensions, and build coalitions between them.

Conclusion

For democratic theorists, public space is necessary because it fosters capacities for citizenship, but there are different ways of conceiving of our essential civic capacities. Some theorists emphasize the agonistic character of public space (see Villa 1992, Young 2002). They see it as a place to agitate, to demonstrate, to provoke, to perform, and to force even unwilling spectators to confront difference (Mitchell 2003). Others emphasize that public space is an important site of social integration (Figure 14.2). The City Beautiful Movement of the late nineteenth century, for example, promoted monumental building projects as a way to bolster civic pride and social order (Mattson 1998). Today, downtown boosters tend to promote urban infrastructure in terms of economic development. Parks, plazas, and markets are ways to attract the sought-after, mobile workers of the creative class or at least they can generate revenue by drawing tourists and their dollars (Florida 2002, 2005). There are other urbanists, including academics, planners, and designers who try to promote a different vision of public space, one that sees public space as an alternative to the privatism of the home and the commercialism of the shopping mall.

Political theorists have contributed to debates about urban design in at least three ways: conceptual clarification, normative analysis, and critical theorizing. Conceptual theory attempts to systematically define concepts such as public and private in order to clarify the multiple and ambiguous

Figure 14.2 Plaza de la República in Mexico City. Source: Tridib Banerjee.
Note: Public space as a stage for political theater; crowd gathers at Plaza de la República in Mexico City to demand recount of the votes for presidential election in July, 2006.

character of the subject matter and bring some order to the subject (see Weintraub and Kumar 1997). New Urbanist communities are by no means the only places that use design in order to blur the line between public and private. Some places are legally accessible to the public but are designed in order to discourage people from using them (Low 2000; Low and Smith 2006). Examples include a number of the privately owned public spaces built in New York City in exchange for density allowances; some of these plazas are sunken below grade, partially fenced, or constructed of materials that are dissimilar from the surroundings and these features suggest that people are not welcome (Kayden 2000).

Normative analysis applies theories of justice to evaluate planning and policies dealing with public housing, residential segregation, and recreational amenities. Some urban designs seek to create an atmosphere that is vibrant and welcoming to diverse users. Others have numerous features that exclude people who are not part of the target demographic groups; these might include guard houses, seating for customers only, uncomfortable seating, plants rather than lawns, marked perimeters, etc. Theories of justice and theories of rights can help explain what is wrong with the de facto segregation that these designs encourage.

Finally critical theory is an approach that reads the city itself as a text in order to reveal patterns of domination, exclusion, and power relations that are difficult to recognize because of the way that they are taken for granted in our experience of daily life. Using these tools can help us think more critically about urban life and demystify enchanting places such as the Town of Tioga.

Notes

1 A recent study using GIS mapping systems has documented the racial segregation of Common

Interest Developments (CIDs) in California. After analyzing hundreds of neighborhoods, it found a statistically significant and dramatic difference in the racial composition of traditional urban neighborhoods and CIDs. For example, the percentage non-Hispanic black in traditional neighborhoods was 11.3 percent, in CIDs 3.7 percent. Similarly, there were twice as many Hispanics in traditional neighborhoods compared to CIDs. The difference was even more dramatic in the suburbs (Gordon 2004).

2 CB MacPherson, Property: Mainstream and Critical Positions (Toronto: University of Toronto Press, 1978), 5.

> In the case of private property the right may, of course, be held by an artificial person, that is, by a corporation or an unincorporated grouping created or recognized by the state as having the same (or similar) property rights as a natural individual. The property which such a group has is the right to use and benefit, and the right to exclude non-members from the use and benefit, of the things to which the groups has a legal title. Corporate property is thus an extension of individual private property.

References

Barber, B. (2001). "Malled, Mauled, and Overhauled: Arresting Suburban Sprawl by Transforming Suburban Malls into Usable Civic Space." *Public Space and Democracy,* Minneapolis, MN: University of Minnesota Press.

Bickford, S. (2000). "Constructing Inequality: City Spaces and the Architecture of Citizenship," *Political Theory* 28(3): 355–376.

Blackmar, E. (2005). "Appropriating 'the Commons': The Tragedy of Property Rights Discourse." In Low, S. and Smith, N. (Eds.) *The Politics of Public Space,* New York: Routledge.

Bollier, D (2002). *Silent Theft: The Private Plunder of our Common Wealth,* New York: Routledge.

Bouvier, J. and Rawle, F (1984). *Bouvier's Law Dictionary and Concise Encyclopedia.* 3rd revision, Buffalo, NY: William S. Hein.

Davis, M. (1992). *City of Quartz: Excavating the Future in Los Angeles,* New York: Vintage Books.

Ellickson, R. (1996). "Controlling Chronic Misconduct in City Spaces: Of Panhandlers, Skid Rows, and Public-Space Zoning," *Yale Law Journal,* vol. 105: 1165–1248.

Florida, R.L. (2002). *The Rise of the Creative Class and How It's Transforming Work, Leisure, Community and Everyday Life,* New York: Basic Books.

—— (2005). *Cities and the Creative Class,* New York: Routledge.

Foldvary, F.E. (1994). *Public Goods and Private Communities: The Market Provision of Social Services,* Brookfield, VT: Aldershot.

Gordon, T. (2004). "Moving Up by Moving Out? Planned Developments and Residential Segregation in California," *Urban Studies,* 41(2): 441–461.

Habermas, J. (1989). "The Public Sphere: An Encyclopaedia Article." In Bronner, S.E. and Kellner, D.M. (Eds.) *Critical Theory and Society: A Reader,* New York: Routledge, 136–142.

—— (1991). *The Structural Transformation of the Public Sphere: An Inquiry into a Category of Bourgeois Society,* Cambridge, MA: MIT Press.

Hardin, G. (1998). "The Tragedy of the Commons." In Baden, J. and Noonan, D. (Eds.) *Managing the Commons,* Indianapolis, IN: University of Indiana Press.

Hénaff, M. and Strong, T.B (2001). *Public Space and Democracy,* Minneapolis, MN: University of Minnesota Press.

Hess, C. and Ostrom, E. (2007). *Understanding Knowledge as a Commons: From Theory to Practice,* Cambridge, MA: MIT Press.

King, L. (2004). "Democratic Hopes in the Polycentric City," *Journal of Politics* 66(1): 203–233.

Kohn, M. (2004). *Brave New Neighborhoods: The Privatization of Public Space,* New York: Routledge.

Lessig, L. (2002). *The Future of Ideas: The Fate of the Commons in a Connected World,* New York: Vintage.

Low, S. M. (2000). *Behind the Gates,* New York: Routledge.

MacPherson, C.B. (1978). *Property, Mainstream and Critical Positions,* Toronto: University of Toronto Press.

Mattson, K. (1998). *Creating a Democratic Public: The Struggle for Urban Participatory Democracy during the Progressive Era,* University Park, PA: Pennsylvania State University Press.

McKenzie, E. (1994). *Privatopia: Homeowner Associations and the Rise of Residential Private Government,* New Haven, CT:Yale University Press.

Mitchell, D. (2003). *The Right to the City: Social Justice and the Fight for Public Space,* New York: Guilford Press.

Ostrom, E. (1990). *Governing the Commons: The Evolution of Institutions for Collective Action,* Cambridge: Cambridge University Press.

Ripstein, A. (2008). "Roads to Freedom," Paper presented at the University of Toronto Centre for Ethics.

Rose, C. (1986). "The Comedy of the Commons: Custom, Commerce, and Inherently Public Property," *University of Chicago Law Review* 53(3): 711–781.

—— (2003). "Romans, Roads, and Romantic Creators: Traditions of Public Property in the Information Age," *Law and Contemporary Problems* 66(1–2): 89–110.

Sorkin, M. (1992). *Variations on a Theme Park: Scenes from the New American City and the End of Public Space,* New York: Hill and Wang.

Staeheli, L.A. and Mitchell, D. (2008). *The People's Property? Power, Politics, and the Public,* New York: Routledge.

Villa, D. (1992). "Postmodernism and the Public Sphere," *American Political Science Review* 86(3): 712–729.

Waldron, J. (1991). "Homelessness and the Issue of Freedom," *UCLA Law Review* 39: 295–324.

Weintraub, J.A. and Kumar, K. (1997). *Public and Private in Thought and Practice: Perspectives on a Grand Dichotomy,* Chicago: University of Chicago Press.

Young, I.M. (1999). "Residential Segregation and Differentiated Citizenship," *Citizenship* Studies 3(2): 237–252.

—— (2002). *Inclusion and Democracy,* Oxford: Oxford University Press.

Further reading

Davis, M. (1992). *City of Quartz: Excavating the Future in Los Angeles,* London and New York: Verso. This is an extremely influential book that examines how power and class have marked the built environment in Los Angeles.

Miller, K. (2007). *Designs on the Public: The Private Lives of New York's Public Spaces,* Minneapolis, MN: University of Minnesota Press. Written from a design perspective, this book chronicles the history of six iconic public places in New York City.

Mitchell, D. (2003). *The Right to the City: Social Justice and the Fight for the City,* New York and London: Guilford Press. This book explores the concept of public space and the role that it plays in democratic politics.

15

Interactions between public health and urban design

Marlon G. Boarnet and Lois M. Takahashi

Interest in the links between urban design and public health has exploded in recent years (e.g. Jackson 2003; Corburn 2004; Srinivasan *et al.* 2003). In this chapter, we review the historical interaction between public health and urban design and we summarize insights from the past decade's burst of research that seeks to build bridges between the two fields. We note that public health has typically conceptualized the urban design field in limited ways. We call for urban designers to take an active role in moving toward a more holistic view of what urban design is and can offer to the study and practice of public health, and we close with some observations on how a more sophisticated and robust urban design–public health link can be built.

There are two ways in which public health's interaction with urban design has been limited. First, we differentiate between urban design and the built environment. Urban design reflects human agency in managing, organizing, and ordering the physical environment with specific human purposes, while the built environment tends to be more of a descriptive framework of built form elements and their relationships. Urban design is the process and activity that leads to deliberate change in the built environment; the built environment is the outcome of human intervention. Public health research and practice has largely been directed toward the built environment, with comparatively little attention to the process of how the built environment is produced and how urban design and the built environment interact in an iterative process. While this focus on the built environment (the outcome of urban design) has not been exclusive, it is strong enough to reduce the focus of public health's attention to a relatively static view of the existing built form, abstracting from process and human agency and hence from urban design's long tradition of inquiry into the goals and methods for city building.

Second, we note that urban design has both an aesthetic and a functional tradition, and those two influences and goals have been evident, in varying degrees, throughout the modern history of the field. Public health, though, has allied more easily with the functional view of urban design, and so highlights part but not all of the urban design endeavor. We develop this idea further by reference to some history.

The birth of urban planning and the interaction of urban design and public health

In the late 1800s, both the planning and public health professions were concerned with the perceived (and very real) pathologies of urban dwelling in industrializing economies. Cities were viewed as congested, unsanitary breeding grounds for disease, filth, and (reflecting attitudes of the time) social decay. Epidemics of contagious illness were common, and were typically viewed as spreading out of slum neighborhoods and threatening the whole city. As a response, the sanitary movement of the post Civil War period in the US focused on cleaning cities, developing water and sewer infrastructure, and deconcentrating city populations by encouraging the development of lower density settlements (Peterson 1983; Sloane 2006). The task was to use infrastructure – primarily common sewer systems – and (to a lesser extent) land development to combat urban contagions (Corburn 2007; Peterson 1983).

This functional view was soon supplanted for a brief period by the City Beautiful movement. The City Beautiful movement elevated the role of aesthetics, reflecting the grand traditions of city building on a broader scale. Daniel Burnham's Chicago Plan of 1909 epitomized the peak of the aesthetic tradition reflected in the City Beautiful movement (Legates and Stout 1998), and also was one of the markers for the birth of planning as a field in the United States (Hall 1989). Sloane (2006, 12) cites Peter Hall's (1988) assessment of Burnham's Chicago plan, saying that in the plan beauty "clearly stood supreme," with health "almost nowhere."

In short order, the young field of urban planning had been influenced by both a modernist view of city building that was grounded in the use of scientific and technological advances intended to solve urban ills and a grand city-building strategy that reflected the long-standing tendency to link urban design to inspirational and even utopian visions of the city (e.g. Legates and Stout 1998). Those two viewpoints – the aspirational and aesthetically-focused "City Beautiful" and the narrower and instrumental "City Functional" (Hall 1989) – have long been evident in urban design thinking and practice. Public health, with its basis in scientific measurement and problem solving, allied more easily with the "City Functional." The issue was not so much that public health and planning were joined only in the early sanitary movement, but that the functional approach to urban design provided a more ready link for the public health community throughout the twentieth century. As a later example of links between public health and urban design, in 1948 the American Public Health Association's Committee on the Hygiene of Housing used the neighborhood unit as the basis for healthy neighborhoods (Corburn 2007), a focus with roots in the ideas of Clarence Perry (e.g. Banerjee and Baer 1984; Lawhon 2009).

In sum, public health's focus has been one of measurement of the built environment, linking more easily to the outcome of urban design than to the design process itself and incorporating a bias toward a functional rather than a more holistic view of city building. In the extreme, the built environment in this view is not the whole of a neighborhood or the context for communal interaction and inspiration, but instead a set of characteristics to be narrowly measured and manipulated toward specific health goals. A discussion of recent public health – urban design research illustrates these biases, starting with research on physical activity, which is possibly the most narrowly functional of the current body of public health research that incorporates concepts from urban design.

The modalities of urban design and health research

Physical activity

Research on physical activity and the built environment was almost nonexistent ten years ago. Since then, special issues on the topic have appeared in the *American Journal of Preventive Medicine* (2002), the *American Journal of Public Health* (2003), the *American Journal of Health Promotion* (2003), and the *Journal of the American Planning Association* (2006), among others. The major scholarly conference for planners in the United States added a "planning and human health" track in 2004, and popular ranking schemes for planning departments include "planning and health" or similar categories. In 2007 there were over 200 articles published on the topic "Built Environment and Policy – Physical Activity" (Active Living Research web site 2009).

This body of research is grounded in two motivating literatures. Public health scholars had for years focused on behavioral change, encouraging persons to lead less sedentary (more physically active) lives. The health benefits of physical activity had been well established by the late 1990s (e.g. US Department of Health and Human Services 1996; Paffenbarger *et al.* 1986; Leon *et al.* 1987; Ekelund *et al.* 1988; Blair *et al.* 1989; Morris *et al.* 1990; Sandvik *et al.* 1993). Yet behavioral change alone had proven insufficient to increase physical activity rates, and by the late 1990s, public health scholars were turning their attention to the built environment (Owen *et al.* 2004). Some of the popular reports from that time hinted at a certain naïve environmental determinism, suggesting in the extreme that environmental changes through urban design would be the fix for an increasingly sedentary society. The scholarship, especially as the research moved forward, adopted a more nuanced tone, viewing the built environment as the context within which behavior occurs, such that design interventions in the built environment might facilitate or hinder individual physical activity (e.g. Transportation Research Board/Institute of Medicine 2005). A second motivating literature, research on travel behavior, had moved to aggressively pursue individual level data on travel, paired with data from geographic information systems (GIS). These data innovations allowed detailed analyses that avoided ecological fallacies inherent in the use of aggregated data. (Previous examples of research on individual travel and urban design existed, but were more episodic and constituted precursors to the large explosion in such studies that occurred in the 1990s. For early work, see, e.g. Hanson and Hanson 1981; Vickerman 1972; and Kain and Fauth 1977).

After an initial period of somewhat separate research, the public health field added a greater focus on objective (as opposed to self-reported) measurement of walking and physical activity to the transportation data sources, which were typically travel surveys at that time, and planners brought enhanced methods to measure the built environment using GIS, which was less familiar to public health researchers (Boarnet 2004). After roughly a decade of research, the following conclusions can be drawn.

There is a clear association between built environment elements and walking (e.g. Frank 2000, Greenwald and Boarnet 2002; Handy *et al.* 1998; Handy *et al.* 2006; Rodriguez *et al.* 2006; Krizek and Johnson 2006; Boarnet *et al.* 2005; Boarnet *et al.* 2008; Ewing *et al.* 2003; Doyle *et al.* 2006), and public health researchers and policy makers have energetically embraced planning's condemnation of sprawl (Frumkin 2002). Inferring causality is more difficult, largely because of a lingering debate about whether persons who are predisposed to walk choose to live in walking oriented neighborhoods, or whether built environment elements directly influence walking propensity and the amount of walking. Recent studies, which focus more broadly on all travel behavior, suggest that it is some of both, but that the presence of built environment elements does exert some independent effect on travel behavior (Mokhtarian and Cao 2008).

Urban design in this literature is relatively absent. There is an implicit assumption that identifying built environment elements related to walking will then result in urban design interventions that support these research conclusions. Hence, to clarify urban design as the process and the built environment as the outcome, the public health focus has been on the built environment. Much of this research focuses explicitly on developing quantified measurements, or audit instruments, of built environment characteristics (e.g. Boarnet *et al.* 2006; Clifton *et al.* 2007; Cunningham *et al.* 2005; Day *et al.* 2006; Ewing *et al.* 2006; Hoehner *et al.* 2005; Hoehner *et al.* 2007; Lee and Moudon 2006; Saelens *et al.* 2006; Williams *et al.* 2005). The built environment is something to be measured, possibly on a block-by-block scale, dissected into its elements, and manipulated for purposes of human health.

The sanitary engineers of the late 1800s would find clear kinship in this viewpoint. An exceptionally ambitious sanitary survey in Memphis in 1879–1980 included an exhaustive house-by-house assessment of living conditions that filled 96 folio volumes (Peterson 1983: 25). That effort was motivated by a yellow fever epidemic that killed approximately one in ten residents of the city (Peterson 1983: 25). Today, urban design audit instruments (e.g. Clifton *et al.* 2007; Day *et al.* 2006) are similarly exhaustive, block-by-block inventories often used in areas with the highest obesity levels.

There is little if any room for concepts of aesthetics, inspiration, or grand city-building in the public health approach to physical activity and the built environment. The focus on measurement is due in part to the strong influence of social scientific, quantitative, and (in the form of transportation researchers) engineering traditions that are at the heart of much of the existing research on physical activity and urban design. The challenge, which urban designers are well positioned to address, is that the existing physical activity – urban design discussion and synergy must be broadened to include not just the built environment, but the human agency that creates that built environment, and to make room for urban design not simply as a functional practice in the service of health outcomes (important though that may be), but also as an inspirational endeavor that includes the grander traditions of city building. Some clues as to a more holistic view can be gleaned from reviewing the interaction between health and urban design in other contexts.

Accessibility and disability – regulating urban design for access

Access to health and social services is a fundamental concept for research on health disparities, inequities, and social determinants of health. Research since the 1990s has tended to focus on existing social conditions (socio-economic status, lack of available services), institutional settings, individual behavior, and logistical challenges (such as lack of transportation), arguing that low service use can be traced to particular combinations of these characteristics or factors for distinct populations (Crane and Takahashi 2008). Yet, in conceptualizing service use and access, little mention is made of specific built environment elements.

Despite the dearth of either public health or urban design research, one direct link between health and urban design has been forged by the disability rights movement (Johnson 1999). The Americans with Disabilities Act of 1990 (or ADA, most recently amended in 2008) highlights various obstacles to mobility, employment, housing, and civic engagement. This federal legislation explicitly identifies "architectural" barriers as a form of discrimination that excludes persons with disabilities from employment, housing, and services that able bodied individuals enjoy (Americans with Disabilities Act of 1990). Disability in the ADA is defined as having any mental or physical impairment that reduces people's abilities to care for themselves, and interferes with basic functioning (e.g. walking, eating, hearing, etc.). Examples of research that highlights legal implications of ADA for urban design include Mazumdar and Geis (2001; 2002).

Today ADA requires public agencies and private businesses to accommodate persons with disabilities in the use of public transportation (e.g. wheelchair accessibility on buses), and access to hotel rooms, restaurants, movie theaters, grocery stores, schools, and museums, in ways that are integrative (meaning that the accommodation should not be separate or different from other existing services or facilities). Specifically, the Act considers the lack of accommodation a form of discrimination (see ADA 1990: 11). Though implementation of the ADA has created more accessible built environments for persons with disabilities, the National Council on Disability (2004) indicated that physical obstacles still remain in many places.

Health disparities at the neighborhood level

As in the physical activity literature, public health researchers studying broad health issues such as health disparities (or concentrations of illness/morbidity or death/mortality in specific racial/ethnic, age, gender, or other social groups or medically underserved places) or the social-environmental factors that influence or cause illness or death (e.g. social determinants of health and disease) have increasingly seen the role of the built environment as an important factor needing clarification. In some ways, this body of research remains tied to the creation and testing of built environment inventories (as with the physical activity literature), but in other ways, public health researchers have begun to view the built environment as representative of more complex dynamics, that is, reflective of human action, but also constraining and enabling human agency (Corburn 2004; also see for example the *Journal of Urban Health* and the *International Society for Urban Health* – http://www.isuh.org/).

Robert Sampson and his colleagues have been especially influential in exploring local neighborhood attributes and health outcomes, albeit drawing from Chicago School of Human Ecology and social capital debates rather than urban design approaches per se (Sampson *et al.* 2002; Sampson *et al.* 1997). From a public health perspective, the focus on neighborhoods is a departure from the typical public health approach, which has tended to emphasize large population studies (to establish epidemiological trends), with neighborhood level analyses focusing instead on local factors affecting individual behavior (rather than socio-demographic, attitudinal, or knowledge factors alone). What this literature has highlighted is the important role of community-level factors, including the built environment, in creating and reinforcing structural, institutional, community, and individual barriers to health care and resources (leading to disadvantage and inequality), and how such factors might best be measured and assessed (Sampson *et al.* 2005). Of particular concern is the structural differentiation that leads to or causes health inequalities and inequities. Though sociological measures of neighborhood, such as residential stability and racial segregation, tend to predominate, the role of the built environment remains relatively unclear. Relatively few health researchers have assessed the role of the built environment and urban design on health behaviors (an exception is Grusky and Swanson 2004).

Housing and asthma

Public health researchers have argued for over a decade that low quality housing, especially the presence of mold, is associated with heightened risk and prevalence of asthma in children. Strachan (1988), for example, found that after controlling for housing tenure, household size, presence of smokers in the household, and cooking with gas appliances, the presence of mold made asthma three times more likely. Though housing design is more the purview of architecture than urban design, this body of research makes clear that there is a need for design and designers to understand how aging structures and poor materials directly influence health and well-being.

Heat islands and health

Poor quality housing and urbanization have led to illness and death for specific segments that are likely to become worse with the extreme weather patterns associated with climate change. Heat waves tend to have the highest impact in central cities because temperatures are higher and night cooling is lower than in less paved, and less built up areas (McMichael 2000). Such climate related impacts on health tend to be concentrated in less mobile (e.g. elderly persons) and lower-income populations. Public health and planning have begun to focus on such issues, but have not provided clear ways forward in terms of urban design interventions.

Searching for a synthesis

The gap between urban design and health is twofold: first, the difference between viewing the city as a set of functional instruments and seeing the city as an integrated whole with aspirational dimensions, and second, the difference between focusing on the built environment without attention to the process that produced that outcome versus examining both the design process and its outcome. Can these gaps be bridged? There are encouraging signs. Here we discuss two possible contributions of public health to urban design – one small, and one large.

In an instrumental, measurement-oriented way, the literature on physical activity has increased attention to sidewalk infrastructure, the street environment, parks and open spaces, and the physical elements of the non-motorized travel experience. Similarly, the research and practice on the ADA has raised the visibility of design treatments that increase accessibility for persons living with disabilities. Both efforts are important, and both have increased awareness of the critical importance of urban design as remediation. All in all, these design elements of urban living are important pieces of the whole, but can the field of public health do more than draw attention to the occasional overlooked design treatment? We suggest a possible path toward such a larger view.

Bridging the functional and the aesthetic/aspirational might begin with a shared focus on the neighborhood not as a collection of parts to be manipulated, but as a place for human living. The role of urban design as process and human agency must be restored – the built environment cannot be the whole of the focus. The recent attempts in the health disparities literature to examine neighborhoods in a more comprehensive way, to articulate the social determinants of health, and to link broadly to human health and society, are a start as they focus on the structural obstacles to addressing inequity and inequality. There are also methodological necessities for bridging this divide. Upscaling the unit of analysis to a meso scale – neighborhoods large enough to capture the lived-in built environment, but somewhat smaller than cities or metropolitan areas – is one possibility. This meso scale is still typically under-researched in the physical activity literature, though is being considered in research on heat islands. Physical activity researchers, drawing on their link to transportation planning, might broaden their focus beyond the block-level streetscape and aggregations thereof, to neighborhoods large enough to be centers of activity, living, shopping, and working, and shift the focus from dissected elements to these spaces of activity. Major metropolitan areas are pursuing growth visioning plans that seek to focus infill development in urban nodes. Examples, often called "blueprint planning", include the Sacramento Region Blueprint Plan (2009) and the Southern California Association of Governments (2009) COMPASS plan. Those plans inherently view neighborhoods as the lynchpin of metropolitan planning, yet theory and practice that can inform the details of urban neighborhood building lack specifics. Neighborhood building, both its functions and its aesthetics and aspirations, should be a vital core element of public health and urban design efforts.

The involvement of public health, if the focus is on the links between the built, natural, and social environment and impacts on human well being, can provide a framework for moving beyond seeing the city as a simple set of tools to be manipulated. Instead, public health can contribute to viewing cities as a place to live. Explicit links to modern blueprint planning efforts and more holistic concepts drawn from the New Urbanism and Smart Growth movements (which have embraced both aspiration and function, and both urban design process and outcome) can help build a knowledge base that moves beyond a purely functional approach to urban design and health. Such links will require that the city and its neighborhoods become the center of analysis – a shift that would be large for the physical activity literature, but somewhat smaller for researchers examining neighborhood effects on health disparities, housing, and climate impacts on health. We suggest that both health and design researchers examine ways to reconceptualize built environment elements not as a variable set to be manipulated, but as the fabric of communities that are the central object of thought and practice. Having said that, the focus should not be strictly on the aesthetics of the city, but on the role of neighborhoods in human well being and aspiration. Such a focus can combine public health and urban design in ways that can be deeper and longer lasting than the episodic and, at times, limited alliances of those two fields in years past.

References

Active Living Research web site. (2009). http://www.activelivingresearch.org/resourcesearch/referencelist, accessed March 19, 2009.

Americans with Disabilities Act of 1990. http://www.ada.gov/pubs/adastatute08.htm#12101, accessed 23 March 2009.

Banerjee, T. and Baer, W. C. (1984). *Beyond the Neighborhood Unit*. New York: Plenum.

Blair, S.N., Kohl, H.W., Paffenbarger, R.S., Clark, D.G., Cooper, K.H. and Gibbons, L.W. (1989). "Physical fitness and all cause mortality: A prospective study of healthy men and women." *Journal of the American Medical Association,* 262 (17): 2395–2401.

Boarnet, M.G. (2004). *The Built Environment and Physical Activity: Empirical Methods and Data Resources.* Background paper prepared for Transportation Research Board/Institute of Medicine committee on Physical Activity, Health, Transportation, and Land Use, July 18,

2004. http://trb.org/downloads/sr282papers/sr282Boarnet.pdf, accessed March 27, 2009.

Boarnet, M.G., Anderson, C., Day, K., McMillan, T., and Alfonzo, M. (2005). "Evaluation of the California safe routes to school legislation: urban form changes and children's active transportation to school." *American Journal of Preventive Medicine,* 28(2, suppl2): 134–140.

Boarnet, M.G., Day, K., Alfonzo, M., Forsyth, A., and Oakes, M. (2006). "The Irvine Minnesota Inventory to measure built environments: Reliability tests." *American Journal of Preventive Medicine,* 30: 153–159.

Boarnet, M., Greenwald, M. and McMillan, T. (2008). "Walking, urban design, and health: Toward a cost-benefit analysis framework." *Journal of Planning Education and Research,* 27(3): 341–358.

Clifton, K., Livi Smith, A., and Rodriguez, D. (2007). "The development and testing of an audit for the pedestrian environment," *Landscape and Urban Planning,* 80 (1–2): 95–110.

Corburn, J. (2004). "Confronting the challenges in reconnecting urban planning and public health." *American Journal of Public Health,* 94(4): 541–546.

—— (2007). "Reconnecting with our roots: American urban planning and public health in the twenty-first century." *Urban Affairs Review,* 42: 688–713.

Crane, R. and Takahashi, L.M. (2008). "Planning for accessibility." In Hack, G., Birch, E., Sedway, P.H. and Silver, M.J. (Eds.) *Local Planning: Contemporary Principles and Practice,* Washington, DC: ICMA Press, 359–363.

Cunningham, G.O., Michael, Y.L., Farquhar, S.A., and Lapidus J. (2005). "Developing a reliable senior walking environmental assessment tool," *American Journal of Preventive Medicine,* 29: 215–117.

Day, K., Boarnet, M.G., Alfonzo, M., and Forsyth, A. (2006). "The Irvine Minnesota Inventory to measure built environments: Development," *American Journal of Preventive Medicine,* 30: 144–152.

Doyle, S., Kelly-Schwartz, A., Schlossberg, M. and Stockard, J. (2006). "Active community environments and health: The relationship of walkable and safe communities to individuals' health." *Journal of the American Planning Association,* 72(1): 19–31.

Ekelund, L.G., Haskell, W.L., Johnson, J.L., Whaley, F.S., Criqui, M.H., and Sheps, D.S. (1988). "Physical fitness as a predictor of cardiovascular mortality in asymptomatic North American men: The Lipid Research Clinics mortality follow-up study." *New England Journal of Medicine,* 319: 1379–1384.

Ewing, R., Schmid, T., Killingsworth, R., Zlot, A., and Raudenbush, S. (2003). "Relationship between urban sprawl and physical activity, obesity, and morbidity." *American Journal of Health Promotion,* 18(1): 47–57.

Ewing, R., Handy, S., Brownson, R., Clemente, O., and Winston E. (2006). "Identifying and measuring urban design qualities related to walkability," *Journal of Physical Activity and Health,* 3(Suppl 1): S223–S240.

Frank, L. (2000). "Land use and transportation interaction: Implications on public health and quality of life." *Journal of Planning Education and Research* 20(1): 6–22.

Frumkin, H. (2002). "Urban sprawl and public health." *Public Health Reports,* 117(May–June): 201–217.

Greenwald, M. and Boarnet M.G. (2002). "Built environment as determinant of walking behavior: Analyzing nonwork pedestrian travel in Portland, Oregon." *Transportation Research Record,* number 1780: 33–42.

Grusky, O. and Swanson, A.N. (2004). "Signs of HIV." *Contexts,* 3(1): 52–59.

Hall, P. (1988). *Cities of tomorrow: An intellectual history of urban planning and design in the twentieth century.* New York: Blackwell.

—— (1989). "The turbulent eighth decade: Challenges to American city planning." *Journal of the American Planning Association,* 55(3): 275–282.

Handy, S., Clifton, K., and Fisher, J. (1998). *The effectiveness of land use policies as a strategy for reducing automobile dependence: A study of Austin neighborhoods.* Research Report SWUTC/98/467501-1. Austin, TX: University of Texas.

Handy, S., Cao, X., and Mokhtarian, P. (2006). "Does self-selection explain the relationship between built environment and walking behavior? Empirical evidence from Northern California." *Journal of the American Planning Association,* 72(1): 55–74.

Hanson, S. and Hanson, P. (1981). "The travel-activity patterns of urban residents: Dimensions and relationships to sociodemographic characteristics." *Economic Geography,* 57(4): 332–347.

Hoehner, C.M., Ramirez, L.K., Elliott, M.B., Handy, S.L., and Brownson R.C. (2005). "Perceived and objective environmental measures and physical activity among urban adults." *American Journal of Health Promotion*, 28(2S2): 105–116.

Hoehner, C.M., Ivy, A., Ramirez, L.K., Handy, S., and Brownson, R.C. (2007). "Active neighborhood checklist: A user-friendly and reliable tool for assessing activity friendliness," *American Journal of Health Promotion*, 21: 534–537.

Jackson, R.J. (2003). "The impact of the built environment on health: An emerging field." *American Journal of Public Health*; Sep 2003; 93(9): 1382–1383.

Johnson, R.A. (1999). "Mobilizing the disabled." In Freeman, J. and Johnson, V. (Eds.) *Waves of Protest: Social Movements since the Sixties*. Lanham, MD: Rowman & Littlefield, 25–46.

Kain, J.F. and Fauth, G.R. (1977). *The impact of urban development on auto ownership and transit use*. Discussion Paper D77–18. Cambridge: Department of City and Regional Planning, Harvard University (December).

Krizek, K. and Johnson, P.J. (2006). "Proximity to trails and retail: Effects on urban cycling and walking." *Journal of the American Planning Association*, 72(1): 33–42.

Lawhon, L.L. (2009). "The neighborhood unit: Physical design or physical determinism?" *Journal of Planning History*. Published online February 3, doi: 10.1177/1538513208327072.

Lee, C. and Moudon, A.V. (2006). " The 3Ds + R: Quantifying land use and urban form correlates of walking," *Transportation Research Part D*, 11: 204–215.

Legates, R.T. and Stout, F. (1998). "Modernism and early planning: 19870–1940." In Legates, R.T. and Stout, F. *Early Urban Planning, 1870–1940*, London: Routledge.

Leon, A.S., Connett, J., Jacobs, D.R., and Rauramaa, R. (1987). "Leisure-time physical activity levels and risk of coronary heart disease and death: the Multiple Risk Factor Intervention Trial." *Journal of the American Medical Association, 258*(17): 2388–2395.

Mazumdar, S. and Geis, G. (2001). "Interpreting accessibility standards: Experiences in U.S. courts", in Preiser, W.F.E. and Ostroff, E. (eds.) *Universal design handbook,* New York, NY: McGraw-Hill, 18.1–18.20.

—— (2002). "Accessible buildings, architects, and the ADA law: The MCI Center Case", *Journal of Architectural and Planning Research,* 19(3): 195–217.

McMichael, A.J. (2000). "The urban environment and health in a world of increasing globalization: issues for developing countries." *Bulletin of the World Health Organization,* 78(9): 1117–1126.

Mokhtarian, P.L. and Cao, X. (2008). "Examining the impacts of residential self-selection on travel behavior: A focus on methodologies." *Transportation Research B*, 42 (3): 204–228.

Morris, J.N., Clayton, D.G., Everitt, M.G., Semmence, A.M. and Burgess, E. H. (1990). "Exercise in leisure time: Coronary attack and death rates." *British Heart Journal* 63: 325–334.

National Council on Disability. (2004). *Livable communities for adults with disabilities*. Washington, DC: National Council on Disability.

Owen, N. Humpel, N., Leslie, E., Baumann, A., and Sallis, J.F. (2004). "Understanding environmental influences on walking: Review and research agenda." *American Journal of Preventive Medicine,* 27(1): 67–76.

Paffenbarger, R.S., Hyde, R.T., Wing, A.L., and Hsieh, C.C. (1986). "Physical activity, all-cause mortality, and longevity of college alumni." *New England Journal of Medicine,* 314: 605–613.

Peterson, J.A. (1983). "The impact of sanitary reform upon American urban planning, 1840–90." In Krueckeberg, D. A. (Ed.). *Introduction to Planning History in the United States,* New Brunswick, NJ: Rutgers, Center for Urban Policy Research.

Rodriguez, D.A., Khattak, A., and Evenson, K. (2006). "Can new urbanism encourage physical activity? Physical activity in a new urbanist and conventional suburban neighborhoods." *Journal of the American Planning Association,* 72(1): 43–54.

Sacramento Region Blueprint Plan. (2009). http://www.sacregionblueprint.org/sacregion blueprint/home.cfm (accessed March 27, 2009).

Saelens, B.E., Sallis, J.F., and Frank, L.D. (2003). "Environmental correlates of walking and cycling: Findings from the transportation, urban design, and planning literatures." *Annals of Behavioral Medicine,* 25: 80–91.

Saelens, B.E., Frank, L.D., Auffrey, C, Whitaker, R., Burdette, H., and Colabianchi, N. (2006). "Measuring physical environments of parks and playgrounds: EAPRS instrument development and inter-rater reliability." *Journal of Physical Activity & Health*. 3 (Sup. 1): 190–208.

Sampson, R.J., Raudenbush, S., and Earls, F. (1997). "Neighborhoods and violent crime:

multilevel study of collective efficacy." *Science* 277: 918–924.

Sampson, R.J., Morenoff, J.D., and Gannon-Rowley, T. (2002). "Assessing 'neighborhood effects': Social processes and new directions in research." *Annu. Rev. Sociol.* 28: 443–478.

Sampson, R.J., Morenoff, J.D., and Raudenbush, S. (2005). "Social anatomy of racial and ethnic disparities in violence." *Am J Public Health,* 95: 224–232.

Sandvik, L., Erikssen, J., Thaulow, E., Erikssen, G., Mundal, R., and Rodahl, K. (1993). "Physical fitness as a predictor of mortality among healthy, middle-aged Norwegian men." *New England Journal of Medicine,* 328: 533–537.

Sloane, D.C. (2006). "From congestion to sprawl." *Journal of the American Planning Association,* 72(1): 10–18.

Southern California Association of Governments. (2009). COMPASS Blueprint Plan. http://www.compassblueprint.org/, accessed March 27, 2009.

Srinivasan, S., O'Fallon, L.R., and Dearry, A. (2003). "Creating healthy communities, healthy homes, healthy people: Initiating a research agenda on public health and the built environment." *American Journal of Public Health,* 93(9): 1446–1450.

Strachan, D.P. (1988). "Damp housing and childhood asthma: validation of reporting of symptoms." *BMJ* (297): 1223–1226.

Transportation Research Board/Institute of Medicine. (2005). *Does the built environment influence physical activity? Examining the evidence.* TRB Special Report number 282. Washington, D.C.: National Academy of Sciences.

U.S. Department of Health and Human Services. (1996). *Physical activity and health: A report of the surgeon general.* Atlanta, GA: U.S. Department of Health and Human Services, Centers for Disease Control and Prevention, National Center for Chronic Disease Prevention and Health Promotion.

Vickerman, R.W. (1972). "The demand for non-work travel." *Journal of Transport Economics and Policy,* 6(2): 176–210.

Williams, J.E., Evans, M., and Kirtland, K.A., Canver, M.M., Sharpe, P.A., Neet, M.J. and Cook, A. (2005). "Development and use of a tool for assessing sidewalk maintenance as an environmental support of physical activity," *Health Promotion and Practice,* 6: 81–88.

Further reading

American Public Health Association (2008). *Reducing the Burden of Poor Health and Health Inequities Through Transportation and Land-Use Policies* http://www.apha.org/advocacy/policy/policysearch/default.htm?id = 1378 (accessed 23 August 2010). A policy statement released by APHA concerning the interactions of urban design and public health disparities.

http://www.cdc.gov/healthyplaces/ (accessed 23 August 2010). The U.S. Centers for Disease Control and Prevention, through its Healthy Places initiative, is a clearing house for research, tools and measurement, policy initiatives, and examples on urban design and health practices and programs, including Health Impact Assessment.

http://www.who.int/hia/en/ (accessed 23 August 2010). The World Health Organization (WHO) has information and tools for Health Impact Assessment, a policy oriented means of evaluating the effects of urban design on community health.

Handy, S., Boarnet, M.G., Ewing, R., and Killingsworth, R. (2002). "The Built Environment and Physical Activity: Contributions from the Field of Urban Planning." *American Journal of Preventive Medicine,* 23(2): 64–73. This article summarizes the classic transportation theory and methods for analyzing travel demand, and how those can be adapted to the study of walking and other active travel modes.

Transportation Research Board and Institute of Medicine. (2005). *Does the Built Environment Influence Physical Activity?* Special report 282. Washington, D.C., The National Academy of Sciences/National Research Council. This report summarizes the findings of a National Research Council panel which studied the evidence on physical activity and the built environment. The findings are summarized at these web sites: http://onlinepubs.trb.org/onlinepubs/trnews/trnews237activity.pdf http://onlinepubs.trb.org/onlinepubs/sr/sr282summary.pdf. (accessed 23 August 2010).

16

Urban design and the cinematic arts

Rafael E. Pizarro

This chapter explains the influence of the cinematic arts in urban design by showing the connections between city, urban design, and films. Following a general introduction to the topic, I elaborate on three specific ways in which these connections are evident: first, movies influencing designers' ideas about urban space and urban form, second, cinematic techniques as tools in the practice of design and in design pedagogy, and third, films as interpretive media to understand cities and urban societies. Concluding, I point out that the three areas are in great need of more scholarly attention and offer some personal reflections on ways to respond to this need. For those interested in pursuing further research on the "cinematic city," I include an additional bibliography (further reading) on the cinematic arts and the city at the end of the chapter.

The cinematic arts have a natural kinship with urban design. Film's immediacy in relating characters to urban space makes it undoubtedly an urban design-related media (Strickland 2006). Filmic montage, for example, the fragmentation of the visual field and its reassembly into a narrative, is not too different from the way people experience real cities (Russell 1992; Hight 2004). The appeal to use filmic techniques to aid in urban design processes dates back to the mid-twentieth century. In the 1960s, renowned urban scholars Gordon Cullen, Kevin Lynch, Donald Appleyard, and Philip Thiel started exploring "serial vision," a technique commonly used in movie scripting and storyboarding, as a way to "read" and design urban space. The connection between cinema and the city, however, dates even further back. At the end of the nineteenth century, the fortunes of cinema and the city became inextricably linked with Louis Le Prince positioning his experimental camera to "film" people on the Leeds Bridge in 1888 and the Skladanowsky Brothers shooting the first scenes of Berlin in 1892 (Barber 2003). Their animated photographs (literally "moving pictures") led to the Lumière brothers' invention of the *Cinematograph* in 1895 (ibid.) giving birth to a representational technique capable of capturing and reproducing two fundamental dimensions of the urban experience: time (motion) and sound. Since then, the cinematic arts have offered to the layperson an alternative way to experience urban reality, to urban designers a potential new tool to represent and design urban space, and to urban scholars a new medium to observe and understand urban phenomena.

The connection between cinema and urban growth is also tight. In Europe, the motion picture evolved during a period of tremendous urbanization. Its becoming one of the major arts was, largely, a function of late nineteenth-century European urban development. Seemingly, there was a natural interaction between the accelerating urban life and cinema, a medium capable of observing and commenting upon it (Uricchio 1988). Something similar occurred in the United States after World War II. In the 1950s, television became a key infrastructure apparatus for the suburbanization of America (Spigel 2001), emerging as the post-war suburban corollary to European cinema and the European twentieth-century metropolis (Hight 2004). Since then, in Europe and the United States, cinema has become both a product of the changing structure of cities and a technology for understanding those changes (Shiel and Fitzmaurice 2001). Later, towards the end of the twentieth century, cities and cinema became intertwined with each other with the identities of places inextricably bound up in their cinematic representations; the cinematic landmarks of Los Angeles, Paris, New York, London, for example, turned into iconic symbols of wealth, power, status, style and culture (AlSayyad 2006). According to Paul Virilio, with Lumière's first projections "the [cinema] screen abruptly became the city square [and] the crossroads of all mass media" (1997: 384). Virilio claims that we have learned as much about cities from their cinematic representations as we have from urban scholarship. As he put it, "more than Venturi's Las Vegas, it is Hollywood that merits urbanist scholarship" (1997: 384).

Today, there are three ways in which the cinematic arts, the city, and urban design are connected: first, the city in cinema influencing architects' and urban designers' design ideas, second, urban designers borrowing movie-making techniques as aids in the design process and in the teaching of urban design, and third, representations of cities in cinema as alternative interpretive media to understand social, economic, and cultural processes in cities.

The cinematic city influencing architectural and urban design ideas

The cinematic arts and spatial design started interacting when modernism and commercial cinema came into maturity in Europe in the 1920s and 1930s. During the 1920s, a shared agenda between the architectural and cinematic avant-garde seemed to exist when French and German directors started incorporating architectural modernist design features in their film productions (Penz and Thomas 1997). That symbiotic relationship between cinema and design migrated to the United States in the post-World War I years when, with the rise of fascism, some of Europe's great modernist architects fled to the US finding a new home in the booming Hollywood film industry. From Paramount Studios' "Bauhaus Modernism" to Metro Goldwyn Mayer's "Streamline Modernism," Tinseltown was an enthusiastic adopter of mega-set modernist designers (Albretch 1987; Kroiz 2006; Ramirez 2004). Conversely, movie set designs started influencing the design of real places. The set in the film *Robin Hood* (Dir. Douglas Fairbanks 1922), for example, inspired the design for an English medieval village purposively referred to as "Robin Hood Style," and the stage-design for *The Thief of Baghdad* (1924) inspired developer Glenn Curtiss in 1926 to design the Floridian community of Opalocka as a "Baghdad-themed" residential development (Kroiz 2006).

Today, the distinction between cinematic and real spaces is blurring as "real" places begin to look more and more like studio back lots (e.g. the case of the re-modeling of the Las Vegas strip) (Lukinbeal 1998; Penz 2003; Vale and Warner Bass 2001). This new *real*-city-imitating-*reel*-city phenomenon also points at a connection between the new "cities of spectacle" (Debord 1994) and landscapes, images, and symbols derived from movies (Knox 2005).

The intensification of theming, branding, and marketing of cities – Harvey's "degenerative utopias" of global capitalism (2000) – seems to have accelerated, fueled by the image of cities in the cinematic arts. According to Wards (2005), some films such as *The Truman Show*, for example, have even made architecture an integral part of the storytelling. In the film, New Urbanist's development Seaside in Florida "becomes the real/unreal set for the real/unreal life led by Jim Carrey's character. Immaculately manicured and picturesque, Seaside's townscape can only be imagined as something erected in the studio, yet it is a real place inhabited by real people" (online, no page number). Furthermore, in a case study of Hollywood urban imagery influencing urban development in Latin America (Pizarro 2005), interviews with architects and urban designers of upscale neighborhoods in Colombia revealed that the images of American suburbia in Hollywood movies and television series influenced those designers' decisions to design and build American-like suburbs.

But designers have not only been inspired by the imagery of cities in films. Some urban designers and urbanists have also used filmic techniques, such as storyboarding and sequential visions, for example, to describe, analyze, and design urban space, and even to teach urban design in design schools, as I explain in the following section.

Cinematic techniques as tools for urban designers and design educators

The use of filmic techniques in architectural and urban design, in urban space analysis, and in studio teaching also dates back to the beginnings of the twentieth century. Modernist architect Bruno Taut, for example, inspired by Russian cinematographer Sergei Eisenstein's montage method, argued that film could serve as an instrument of aesthetic education because "the mobile cinematographic record almost replace[s] the guided tour around and through [a] building," allowing for the first time a person to assess the building in its totality (Taut 1917 in Huber 2005, 88). Later in the 1960s, urban scholars and designers such as Donald Appleyard, Gordon Cullen, Kevin Lynch, and Philip Thiel used the notion of serial vision as a technique to understand, explain, and even design urban space (Appleyard *et al.* 1964; Cullen 1971; Lynch 1960; 1981; Thiel 1981). Appleyard, for example, built realistic scale models of urban and suburban environments and made sets of sequences of photos taken at ground level to simulate the real visual experience of moving through a city. Likewise, Gordon Cullen advocated depicting urban environments as an array of sequential views to simulate the moving through urban space. In his proposal for the new town of Lhmtrisant in Wales, Cullen drew superb sequences that included a filmstrip as a framing device to present the project to the public (Russell 1992).

Other examples of these early adopters include David Gosling (Gosling and Maitland 1984) who advocated using similar comic strip-like storyboards to analyze and design urban environments. His study of the Blackpool funfair, for example, "made explicit use of both sequential and simultaneous views of rollercoaster riders" to recreate the riders' viewing of the city (Russell 1992, 48). Even "cognitive mapping," a common methodological technique for urban analysis originally developed by Kevin Lynch, is, allegedly linked to the cinematic arts.

> Since the 1990s, cognitive mapping served American urban geographers to map a city's … spatial dimension and to incorporate the imaginary space of media into urban research. Empirical analysis of cities is itself defined by technology and implies that media (here most importantly,

> film) themselves determine the construction of cognitive maps. (Huber 2005, 83)

Furthermore, Huber claims that cinematic representations of cities "can be seen as instruments serving to produce and construct the urban, *affecting the practices of urban design* ..." (ibid. Italics are mine).

Despite these examples of filmic techniques as pedagogical, representational, and analytical tools in the field of urban design, the literature on their utility in the learning processes of design students is still limited (Strickland 2006; Leigh and Kenny 1996; Pizarro 2009; Webb 1987). Yet, it is easy to anticipate the flourishing of academic studies in this area as urban design educators become increasingly aware of the great amount of time students spend in cyber reality. Indeed, by virtue of their constant exposure to the digital, students already experience the world in a way similar to how cinema represents reality: simultaneously, non-linear, juxtaposed, fragmented, and by bits (Fraisse 1984; Pizarro 2009). The similitude between this form of experiencing reality and how film narratives are assembled may make filmic techniques ideal tools for training design students, at least in the visual aspects of design (Bridges and Charitos 1997). The fundamental question in this area remains whether design students can learn about urban spatial experience and designing cities from the way film directors portray urban space. The advent of the term "cinematic urbanism" (AlSayyad 2006), however, seems to herald a shift in the way we conceive some aspects of design education and in the pedagogical tools we use to train design students. Indeed, it is foreseeable that cinematic techniques such as storyboarding, montage, zooms, jump edits, pans, close-ups, framing, tracking shots, sequencing, and depth of field may become part and parcel of the future training of designers (Pizarro 2009; Robertson 2007; Stickells and Mosley 2009).

The third area of contact between the cinematic arts and urban design is in the use of filmic representations of cities to learn about urban societies. This use of films as interpretive media has arisen as an alternative to more conventional methods to study cities such as direct observation, surveys, census data analysis, examination of maps, GIS, and the like.

Cinematic arts as interpretive media to study cities

There is no doubt that our understanding of the European and American cities (and increasingly of the "Third World" city) would be incomplete without attending to their portrayals in the cinematic arts (Shiel and Fitzmaurice 2001). The history of using films to understand urban phenomena also dates back to the end of the nineteenth century, but because the urban imagery in those early films typically concentrated on monuments, famous streets, popular sights, and parades, such analyses usually overlooked some of the grave economic and social contradictions of the nineteenth-century city (Uricchio 1988). Nonetheless, the analyses made by thinkers such as Simmel, Ostwald, Behne, and Benjamin already show the significance of film (and of photography) as a visual representation and reproduction technology to interpret social spatial phenomena and to decipher the production of space (Huber 2005).

The use of film as an interpretive tool for the study of cities is undoubtedly the most common area of contact between the cinematic arts and the city, although this scholarship has developed largely outside the field of urban design. It is cultural theorists, visual and urban anthropologists, human and cultural geographers, sociologists, critical theorists, and, of course, film scholars who have produced most of the literature in this area (see for example the authors in the additional bibliography at the end of the chapter). The lack of studies in the part of urban design and planning academics is rather puzzling given that these scholars' object of study is the city (Russell 1992; Tewdwr-Jones 2005). As Beauregard put it, "urban designers argue about the forces causing the growth and decline of cities but very rarely, if at all, reflect on how rhetorical inventions influence their interpretations" (1993, xi). "After all," says another scholar, "our sense of what the urban "is" is inflected by a range of interpretations, atmospheres, inherited viewpoints, dialogues and scenarios derived from the cinematic arts" (Atkinson and Willis 2007, 820).

The usefulness of films in understanding urban realities is therefore undeniable. For one, films can provide a handy international survey of cities enabling comparison and contrast between urban forms and the use of space across different societies. For another, films can depict urban experiences at different scales of the city, from the metropolitan to that of the individual building, helping us to understand complex relationships among all aspects of urban design. As Strickland (2006, 51) argued, "with their cameras' depth of field, films can capture the city's multiple streams of activity, the forms and spaces containing them, and their hierarchy." Furthermore, urban spaces, places, and people portrayed in films can reflect prevailing ideologies, cultural norms, societal structures, and moral imperatives otherwise difficult to capture with more conventional forms of urban analysis. And, as Leigh and Kenny point out (1996), although there is no doubt that "the need for dramatic effect influences the scenarios and characterizations [of cities] constructed by filmmakers, the resulting images draw on our collective knowledge of urbanism [and], in doing so, they serve as an index in contemporary views of the city" (52).

Most interpretive works using cinematic representations of cities, however, focus on western societies – chiefly the American and the European. And despite the growing number of interpretive writings based on movies set in Asian, African, and Latin American contexts (Foster 2002; Mazumdar 2007; Podalsky 1998; 2004; Zhang 1996; Shiel and Fitzmaurice 2001), they still represent a small portion of this type of scholarship. No doubt, this is related to the meager number of films produced in the global south (India being the shining exception, of course, as it is recognized as the number one producer of movies in the world) compared to the sheer number of movies produced in Europe and North America. This is unfortunate because films can help us understand urban societies of the non-Western lineage better. The complexities of the enormous cities that grew together in the colonial era, which now have become world-cities in themselves, can defy conventional urban scholarship. Films would open windows not only on the consequences of colonialism but on these countries' emergence from those consequences into the global economy.

The above areas of connection between cities, urban design, and cinema no doubt represent fertile grounds for strengthening the field of urban design, but there is still ample room for future development. As the bibliography at the end of this chapter shows, the earliest writings about the city in film date back only to the 1970s (Gold's 1984). In the following concluding section, I suggest directions for further development in these areas.

Conclusion

As this chapter has shown, since the beginnings of the twentieth century the cinematic arts have, in various ways and at different degrees, influenced urban design practice, its pedagogy, and the learning about cities. Yet, these three areas of influence are still in need of greater attention from urban design scholars.

The first area, urban images in films influencing the practitioner's ideas about the morphological and the aesthetic qualities of urban space, is perhaps the most puzzling and challenging to unpack. Although some of the examples in this chapter may serve as evidence of this influence, it is difficult to gauge to what degree, and by way of what perceptual mechanism, urban landscapes portrayed in films inform practitioners about positive or negative qualities of urban environments. Of course, the few cases referred to in this chapter can not lead to any kind of generalization in this area. The challenge to generalize comes from the difficulty in establishing in a systematic way direct cause-effect relationships between cinematic urban images and a practitioner's design ideas. However, the case study this author conducted in Colombia (Pizarro 2005), yielded clear evidence that designers' prescriptions for American-type detached single-family home low-density suburban developments in three Colombian cities had been influenced by the images of American suburbs in Hollywood TV and cinema. In that work, I suggested that if we add some of the findings in audience research (Screen Theory) that have yielded strong evidence of spectator-behavior cause-effect relationships (Berquist and Greenwood 1977; Brown and Schulz 1990; Jenkins 1992; Taylor 1989) with Stuart Hall's "encoding/decoding" reception theory (a model of mass communication with elements of Gramsci's cultural hegemony theory), it is possible to develop a theoretical model to further study cause-effect relationships between urban images in the cinematic arts and real life urban design prescriptions. Hall's model tells us that audiences (e.g. the developer, the urban designer) "read," ideological discourses embedded in the products of mass communication (e.g. urban images in films or TV ads) and, by way of the workings of cultural hegemony, "translate" and transform those discourses into equivalent social practices (e.g. reproduction of like-urban places in reality). In turn, the re-appearance of the product in the market reifies its value for the cultural industries (e.g. Hollywood) and for the actual product "manufacturers" (e.g. the developer, the designer), hence creating an endless circle of like-products in the market, and, of course, in the media.

Even stronger evidence of spectator-behavior cause-effect relationships has been found in the case of video games, as the study made by Atkinson and Willis (2007) about game players of the video game *Grand Theft Auto 3: Liberty City* has shown.

Although the methodological and theoretical complexity of such studies may seem daunting, they do deserve further exploration as they will no doubt increase our knowledge of how design ideas are formed and how the image of cities in the cinematic arts contribute to such formation.

The second area, using filmic techniques as part of the design process in the professional practice and in the pedagogy of urban design, is perhaps the one most developed of the three areas. As pointed out in the chapter, already in the 1960s some urban designers were using serial vision to understand movement through space in their design projects, consultancy work, or even in the teaching of design. Today, the practice has gained currency in some design schools (e.g. the University of Cambridge, MIT, University of Michigan, Sydney University) with studio instructors encouraging students to apply the technique to help them visualize their projects (Penz 1994). Yet, this and other cinematic techniques are far from becoming part of the standardized methods used in the teaching of urban design. This is surprising as, after all, the 24 frames-per-second that makes an image appear as moving are all but a technical reproduction of the way the human eye perceives reality (an essential fact to consider in designing urban space) (Snickars 2000). I argue that if to these "moving pictures" we added sound, another characteristic of films (and also of the aural dimension of real space), we will have at least two of the sensorial elements of the urban experience (the others being gravity, touch, odors/taste, temperature, and humidity) and thus achieve a closer approximation to the real sensorial/phenomenological experience of being in cities (Pizarro 2009; Yang *et al.* 2007).

It is worth noting that many design schools already offer classes in computer animation and digital media giving students the opportunity to conceive spatial design in a way similar to how it is represented in cinematic works. The downside of these courses is that the students taking them are usually more interested in the technologies themselves than in actual urban design and also, paradoxically, the time the students spend in front of the computers tend to keep them further secluded in the digital world (Pallasmaa 2005). In addition, these classes rarely include the actual filming with cameras or recording sounds outside the classroom.

This is, nevertheless, a promising area in design education as the ready availability of handy-cams (and of video modes in digital photographic cameras and cellular phones) and the ever growing ease of using movie editing and computer animation programs to create virtual worlds will surely make cinematic techniques part of urban design pedagogy in the near future.

The third area, films as interpretive media to understand city life and urban societies, also holds great prospects for the field of urban design. As AlSayyad (2006) has suggested, the myriad representations of cities in films could well form the base of a cinematic epistemology of the city. This epistemology would not only contribute to establishing urban design as an academic field with its own knowledge base, but it will surely give urban scholars and designers a new perspective to study and design urban space. To be sure, furnishing a cinematic epistemology of the city with a theoretical framework constitutes a research project in itself, but such a framework may be grounded in the intellectual foundations of thinkers such as Walter Benjamin, Jean Baudrillard, David Harvey, Mikhail Bakhtin, Henri Lefebvre, Gilles Deleuze, Felix Guattari, Marshall Berman, Jean Baudrillard, Jacques Lacan, Michel Foucault, and Sergei Eisenstein, whose multifarious appreciations of space have made them the most cited in the works of the other disciplines that look at the city in the cinematic arts such as geography, sociology, anthropology, and cultural/critical studies.

In summary, the influence of the cinematic arts in urban design is undeniable, yet, their potential to enhance urban design practice, its pedagogy and to use them in the learning of cities can only be realized if such influence is readily acknowledged by, and willfully incorporated in the curricula of design schools, in the practice of the profession, and in urban design scholarship.

References

Albretch, D. (1987). *Designing Dreams: Modern Architecture in the Movies*. London: Thames and Hudson.

AlSayyad, N. (2006). *Cinematic Urbanism: A History of the Modern from Reel to Real*. New York & London: Routledge.

Appleyard, D., Lynch, K. and Myer, J.R. (1964). *The View From the Road*. Cambridge, MA: MIT Press.

Atkinson, R. and Willis, P. (2007). Charting the Ludodrome: The mediation of urban and simulated space and rise of the flaneur electronique. *Information, Communication and Society* 10(6): 818–845.

Barber, S. (2003). *Projected Cities: Cinema and Urban Space*. London: Reaktion Books.

Beauregard, R. (1993). *Voices of Decline: The Postwar Fate of Us Cities*. Cambridge, MA: Blackwell.

Berquist, G. and Greenwood, J. (1977). Protest against racism: The birth of a nation in Ohio. In *Slow Fade to Black: The Negro in American film, 1900–1942*, edited by T. Cripps. New York: Oxford University Press.

Bridges, A. and Charitos, D. (1997). On architectural design in virtual environments: Principles of architecture and cinema in the design of spatial entities. *Design Studies* 18 (April): 143–154.

Brown, J. and Schulze, L. (1990). The effects of race, gender, and fandom on audience interpretations of Madonna's music videos. *Journal of Communication* 40(2): 88–102.

Cullen, G. (1971). *The Concise Townscape*. New York: Van Nostrand Reinhold.

Debord, G. (1994 [1967]). *The Society of Spectacle*. (Trans. Donald Nicholson-Smith). New York: Zone Books.

Foster, D.W. (2002). *Mexico City in Contemporary Mexican Cinema* Austin, TX: University of Texas Press.

Fraisse, P. (1984) Perception and estimation of time, *Annual Review of Psychology* 35: 1–36.

Gold, J. (1984). *The City in Film: A Bibliography*. Department of Social Science – Oxford Polytechnic, UK.

Gosling, D. and Maitland, B. (1984). *Concepts of Urban Design*. New York: St. Martin's Press.

Harvey, D. (2000). *Spaces of Hope*. Edinburgh: Edinburgh University Press.

Hight, C. (2004). Inertia and interiority: 24 as a case study of the televisual metropolis. *The Journal of Architecture*, Volume 9, Issue 3, pp. 369–383.

Huber, N. (2005). Center or nexus: Berlin's "new" politics of belonging. *Journal of Urban History* 32(1): 82–103.

Jenkins, H. (1992). *Textual Poachers: Television Fans and Participatory Culture*. New York: Routledge.

Knox, P.L.(2005). Creating ordinary places: slow cities in a fast world, *Journal of Urban Design* 10(1): 1–11.

Kroiz, L. (2006). Stealing Baghdad: the city of Opa-locka, Florida and the Thief of Baghdad. *The Journal of Architecture* 11(5): 585–592.

Leigh, N. G., and Kenny, J. (1996). The city of cinema: Interpreting urban images on film. *Journal of Planning Education and Research* 16(1): 51–55.

Lukinbeal, C. (1998). Reel-to-real urban geographies: the top five cinematic cities in North America, *The California Geographer* 38: 64–78.

Lynch, K. (1960). *The Image of the City*. Cambridge MA: MIT Press

—— (1981). *A Theory of Good City Form*. Cambridge MA: MIT Press.

Mazumdar, R. (2007). *Bombay Cinema: An Archive of the City*. Minneapolis, MN: University of Minnesota Press.

Pallasmaa, J. (2005). *The Eyes of the Skin: Architecture and the Senses*. London Academy Editions.

Penz, F. (1994). Cinema and architecture: Overlaps and counterpoints, studio-made features in the film industry and studio-based experiments in architectural education. *Architectural Design* (London, England) 64 (November/December): 38–41.

—— (2003). Screen Cities: Introduction. *City*, 7(3) (November): 361–411.

Penz, F. and Thomas, M. (Eds.), (1997). *Cinema and Architecture*. London: British Film Institute.

Pizarro, R.E. (2005). Suburbanization of the mind: the Hollywood urban imaginarium and American suburbia in the Colombian Caribbean. Ph.D. Dissertation. The University of Southern California.

—— (2009). Teaching to understand the urban sensorium in the digital age: lessons from the studio. *Design Studies* 30(3): 272–286.

Podalsky, L. (1998). Cityscapes and alienation: Buenos Aires in the Argentine cinema, 1950–60. *Nuevo Texto Crítico* 11:21–22 (Jan–Dec), 77–92.

—— (2004). *Specular City: Transforming Culture, Consumption, and Space in Buenos Aires, 1955–1973*. Philadelphia, PA: Temple University Press.

Ramirez, J.A. (2004). (Trans. John F. Moffit) *Architecture for the Screen: A Critical Study of Set Design in Hollywood's Golden Age*. Jefferson, NC: McFarland.

Robertson, S. (2007). Visions of urban mobility: the Westway, London, England. *Cultural Geographies* 14(1): 74–91

Russell, F. (1992). New ways of city viewing. *Cities* 9(1): 43–48.

Shiel, M. and Fitzmaurice, T. (2001). *Cinema and the City: Film and Urban Societies in a Global Context*. Oxford: Blackwell.

Snickars, P. (2000). Architectonics of seeing: architecture as moving images. In Fullerton, J. and Soderbergh Widding, A. (Eds.) *Moving Images: From Edison to the Webcam*. Sydney, Australia: John Libbey & Co.

Spigel, L. (2001). *Welcome to the Dreamhouse: Popular Media and Postwar Suburbs*. Durham, NC: Duke University Press.

Stickells, L. and Mosley, J. (2009). *Film/Architecture/Narrative*. The University of Sydney. Faculty of Architecture, Design and Planning (unpublished).

Strickland, R. (2006). Background into foreground: film as a medium for teaching urban design. *Places* 18(2): 44–51.

Taut, B. (1917). "Mitteilung," *Der Städtebau* 14 (2/3): 32–33.

Taylor, H. (1989). *Scarlett's Women*: New Brunswick, NJ: Rutgers University Press.

Tewdwr-Jones, M. (2005). Film, space and identity: the use of the city as cinematic landscape. Paper presented at the 2005 Association of European Schools of Planning (AESOP) congress, July 13–17, Vienna.

Thiel, P. (1981). *Visual Awareness and Design: An Introductory Program in Perceptual Sensitivity, Conceptual Awareness, and Basic Design Skills*. Seattle, WA: The University of Washington Press.

Uricchio, W. (1988). The city reviewed: Berlin's film image on the occasion of its 750th anniversary. *Film and History* 18(1): 16–25.

Vale, L.J. and Warner Bass, S. Jr., (Eds.). (2001). *Imaging the City: Continuing Struggles and New Directions*. New Brunswick, NJ: Rutgers Center for Urban Policy Research (CUPR) Press.

Virilio, P. (1997 [1991]) The Overexposed City. In Leach, N. (Ed.) *Rethinking Architecture: A Reader in Cultural Theory*. London: Routledge.

Wards, S. (2005). Film and Architecture: what are the relationships between cinema and architecture? Following the architecture screenings at the Adelaide Film Festival, Steven Ward considers the possibilities. *Architecture Australia* 94.3 (May–June 2005): p.21(2). (online journal) http://www.archmedia.com.au/aa/aaissue.php?article = 2&issueid = 2005 05&typeon = 1. Accessed on July 2, 2009.

Webb, M. (1987). The city in film. *Design Quarterly* 136: 5.

Yang, Perry Pei-Ju, Simon Yanuar Putra, and Meutia Chaerani. (2007). Computing the sense of time in urban physical environment. *Urban Design International* 12(2–3): 100–115.

Zhang, Y. (1996). *The City in Modern Chinese Literature and Film: Configurations of Space, Time, and Gender*. Stanford, CA: Stanford University Press.

Further reading

AlSayyad, N. (2006). *Cinematic Urbanism: A History of the Modern from Reel to Real*. New York & London: Routledge. An alternative urban history of modernity/postmodernity through the lens of cinema arguing that urbanism cannot be viewed independently from the celluloid city.

Barber, S. (2003). *Projected Cities: Cinema and Urban Space*. New York: Reaktion Books. A survey of the connections between cinematic images and cities focusing on the urban cinema cultures of Europe and Japan and exploring urban film imagery at moments of urban turmoil.

Clarke, D.B. (Ed.) (1997). *The Cinematic City*. London: Routledge. An authoritative compendium combining cultural studies theory and fictional filmic narratives arguing that cities are shaped by the cinematic form and that cinema owes much of its nature to the historical development of the city.

Jousse, T. and Paquot, T. (Eds.) (2005). *La Ville au Cinema*. Paris: Cahiers du Cinema. An encyclopaedic volume exploring the representations of the city on the screen and including assertions by architects, town planners, landscape designers of how cinema has influenced their practice.

Konstantarakos, M. (Ed.) (2000). *Spaces in European Cinema*. Exeter, UK and Portland, OR: Intellect. Exploration of how space is constructed in European cinema, and the ideological and artistic aspects in European filmic narrations.

Lamster, M. (Ed.) (2000). *Architecture and Film*. Princeton, NJ: Princeton Architectural Press. Depicts how architecture and architects are treated on screen and how these depictions filter and shape the ways we understand the built environment.

Penz, F. and Thomas, M. (Eds.) (1997). *Cinema and Architecture*. London: British Film Institute. Traces the relationship between film-making, architecture, and urban planning through the twentieth century.

Shiel, M. and Fitzmaurice, T. (Eds.). (2003). *Screening the City*. London: Verso. A compelling examination of the relationship between cinema and the changing urban experience in Europe and the United States since the 1930s.

Shiel, M. and Fitzmaurice, T. (2001). *Cinema and the City: Film and Urban Societies in a Global Context*. Oxford: Blackwell. Highlights the changing structure and nature of cities in the global era and the ways in which cinema has become both a product of these changes and an interpretive medium for understanding them.

Part 4

Technologies and methods

Introduction

In this section we include a collection of essays that focus on the changing nature of the pedagogy of urban design and a deeper understanding of the tools and methods used in urban design pedagogy and practice, especially in the context of the information and communication technology revolution (Castells 1996). The following chapters address the following relevant topics: the evolving nature of the studio culture; media tools and the documentation of the built environment; the technology of simulation in imagining and visualizing change; and the emerging digital resources for solving urban design problems.

In her chapter Kathryn Anthony reviews the origin and evolution of design studios as a distinctive pedagogic format for the training of urban designers. The studio mode of urban design instruction reflects its lineage of architecture, which many practitioners consider, in jest, "the second oldest profession" in human history. But one major difference between the studio based learning in architecture and urban design is that while the former emphasizes individual "desk crits," the latter depends more on collaborative and team work and "pin up" reviews from multiple or a panel of instructors and often members of the public representing the community client. Collective brain storming and argumentative process often define the outcome of the urban design studio experience, rather than the more introspective and often zealously guarded private process of a studio in architecture. Furthermore, as Anthony discusses in her essay, increasingly urban design studios involve real clients not just as sounding boards, but also as active participants of the process. She talks about the rise of the community design centers and the involvement of urban design studios in such community oriented projects. She presents the experience of the East St. Louis Action Research Project, as well as post-Katrina workshops and other such studios as case examples of urban design involving community clients. While in architecture the client usually remains hypothetical and is often some imagined corporate or wealthy individual or family, in urban design studios the client is typically real and involves tangible entities like local community groups, business improvement districts, non-profit organizations, or local public agencies, for example.

Anthony's essay offers a rather informative review of the history of the studio process and indeed the culture associated with it, including some of its abuses and tolls on students, not unlike the residency experience of graduate medical students. She also talks about the Ecole de Beaux Arts origin of the contemporary charrette phenomenon – another vestige of the architectural ancestry – which is becoming increasingly common in urban design studios, and in professional practice (also covered by Doug Kelbaugh in Part V). In defining the evolving studio culture, Anthony, quite appropriately, draws from Schön's (1983) empirical work on understanding the creative nature of a desk-crit or the pin-up review, the kind of open-ended conversation between students and instructors that leads to creative learning and outcome. The essay thus defines the contemporary context of the urban design studio culture and sets the stage for the following essays which focus on the advances of media tools, simulation, and the ubiquitous digital world within which the future learning and practice of urban design will occur.

The chapter by Martin Krieger is about documentation of the urban phenomenon, especially the sensed experience of everyday urbanism that includes not only the immediate urban space or the larger urban form, but also the sound, smell, and perhaps even the touch and taste of the urban experience – that is, the experience of the urban sensorium (see Goonewardena 2005) at any given location. Visual documentation of existing urban space has always been a part of urban design methodology toward understanding the context, defining the base line conditions, analyzing the current misfits and anticipating possibilities for the future. The methodology also includes the techniques for representation of the place in time – its past, present, and future.

While the above are the implicit premises of Krieger's chapter, his treatment of the material is quite complex and diverse, weaving history, theory, philosophy, and technology into a compelling narrative about how urban designers may use new media tools as they become increasingly accessible to both the professionals and the public in a digital age. He begins with a historiography of visual – especially photographic – documentation of cities going back to photographs taken by Marville of the Hausmann's transformation of the nineteenth century of Paris, and thus emphasizing the very craft of documentation as well as the role of archival materials, especially comparing the present and past images of urban spaces. In another section of his essay Krieger focuses on the phenomenology of images, especially the process of patching together slices of images of the urban world taken over time in discrete intervals, on the process of making sense of the whole. His essay also includes practical advice for the urban designer, the inherent archival values of documentation, and what one should know about the archival mechanics and requirements. In concluding this essay, Krieger refers to his own on-going experiments with using the contemporary digital tools, involving digital cameras and phones, commercial software and Google street maps, and the emerging technology of surveillance allowing for the simultaneous documentation of the same event as in the opening scene of "The Bourne Identity," thus creating possibilities for new ways of "patching" together information about the urban world.

In somewhat of a similar vein and arguing that urban design is an "anticipatory activity," Peter Bosselmann proposes that design imagery is intrinsic to design practice. Arguing that much of this imaging involves "visualizing change," he focuses on the functions and techniques of simulation, and its role in design decision-making. He defines the scope of his essay as answering three basic questions related to the meaning, goodness, and possibilities of simulations in informing policy. He concludes by discussing case examples of simulating the magnitude, rate, and nature of change in the built environment.

The need for simulation existed many centuries ago, because abstract geometry of drawings and other presentations of technical data and analysis could never fully represent what the proposed environment would be like in reality. But it was always a function of available technology. In making this point, Bosselmann refers to the fifteenth and sixteenth century drawings of Brunelleschi and Leonardo da Vinci as examples.

The quality of simulation is judged by veridicality or truthfulness of the simulation. This raises the question of validity and the essay discusses ways in which the validity of simulations has been tested in empirical studies. Bosselmann also considers the unavoidable politics of simulation, which can be used to emphasize or "sell" a particular point of view, or a building or project, as commonly done by private developers to influence future clients or decision-makers. The essay concludes by discussing a recent example of use of simulation in making urban design decisions for the future development of downtown San Francisco.

In the final essay of this section, Ben-Joseph reviews the digital technology, much of which is now available on-line and downloadable for the general public, as also pointed out by Krieger, and the ubiquity of this digital world. He argues that the pedagogy and practice of urban design that requires collaboration, cognition, and creativity will increasingly depend on and draw from innovations in the ubiquitous digital world, and the inexorable developments in the hardware technology that no doubt will follow. Ben-Joseph sees these developments as promising, and indeed advancing productivity in collaborative ventures between partners separated by distance, located say on the other side of the planet, or by other situational differences. Thus these digital technologies could help forge ties between the planner and the lay citizen with different parochial interests, as they may share information and insights about common urban experiences, and engage in developing mutually agreed upon ideas about designs for change. Similarly, these digital technologies may expand the urban designers' comprehension of the urban world, a point also emphasized by Krieger, thereby making them increasingly savvy about their analysis and interpretation of the world. These digital tools no doubt will help the designers to simulate the environments of their design imaginations, which is an important part of the creative process, as Bosselmann has also emphasized in his essay.

Collectively, these essays capture the contemporary thinking and the trends in the pedagogy and practice of urban design, with exciting promises for creativity, collaboration, and community engagement, the features that distinguish urban design from its allied design arts.

References

Castells, M. (1996). *The Rise of the Network Society, The Information Age: Economy, Society and Culture Vol. I.* Cambridge, MA; Oxford, UK: Blackwell.

Goonewardena, K. (2005) "The Urban Sensorium: Space, Ideology and the Aestheticization of Politics" *Antipode* 37(1): 46–71.

Schön, D. (1983) *The Reflective Practitioner: How professionals think in action.* London: Temple Smith.

17

Design studios

Kathryn H. Anthony

Throughout the evolution of architectural education, from its origins in the nineteenth century to the early part of the twenty-first century, one teaching method has remained predominant above and beyond all others: the design studio. As the bedrock of architectural schools, studio culture has influenced generations of architects around the world. And it has long fascinated family and friends of budding young architects, along with others outside the profession. What is it about this mysterious form of education that keeps students slaving away for hours, days, and weeks on end, with little or no sleep?

This chapter discusses the evolution of the studio as the predominant method of teaching design, and it examines its place in urban design education and practice. It begins with a historical overview of design studios and design juries. Next it addresses how design studios have evolved over time, examining such questions as:

- How have academic urban design studios engaged real-world community clients and community members, and what has been the impact of their work?
- How has the Internet provided opportunities for more collaborative teaching models in design studios?

In exploring these questions we will draw upon examples that include design studios conducted as part of the East St. Louis Action Research Project (ESLARP) at the University of Illinois at Urbana-Champaign. Started in 1987, ESLARP has become one of the longest-running community service projects at the University of Illinois, engaging students from architecture, urban and regional planning, and landscape architecture and other disciplines to work collaboratively on urban design issues in one of the most economically distressed cities in the US. Drawing from these examples, we will critically assess how well or how poorly studio culture prepares young urban designers for the challenges of their profession, and how studios can be used more effectively for their training.

The evolution of studio culture

An atelier culture, much like the medieval guilds, helped form the basis for the design studio. Academic studio culture originated in the nineteenth century at Paris's École des Beaux-Arts (School of Fine Arts). There the design problem, requiring learning by doing, superseded the lecture as the primary method of teaching architecture.

Students began their study of design upon entering the architectural curriculum, where they belonged to different ateliers, or studios, led by a patron, or master through the use of the esquisse, an initial sketch solution to a problem to be developed further. Older students, or anciens, typically helped the younger ones. Practicing architects taught design. A jury of practitioners evaluated the students' projects behind closed doors. Students retrieved their work noting the jurors' marks, with little or no comment. Their fate ultimately rested "in the hands of the gods" – that is, jury members – who decided whether they passed or failed.

The École introduced the design charrette, a practice still used in many schools today (for more on this see chapter by Doug Kelbaugh). Many ateliers were located in neighborhoods distant from the school. When their projects were due, freshmen architecture students pulled a cart from one studio to another, collecting the older students' completed design projects and rushing them to the large gallery where the jury was to judge them soon afterwards. When they saw the cart approaching, other freshmen would stand outside the studio, shouting "La Charrette! La Charrette!" warning students to hurry up, complete their finishing touches, and prepare to submit their work. The term "charrette" has come to mean a design competition or exercise under tight time constraints, and the intense flurry of activity and sleepless nights just before the project is turned over to the jury.

In the mid-nineteenth century, architecture in the US began to develop into a full-fledged profession. The American Institute of Architects (AIA) was formed in 1857, the first architectural school in the US opened at MIT in 1865, and by the turn of the twentieth century, eleven schools had been established. The Association of Collegiate Schools of Architecture (ACSA), the umbrella organization that oversees all academic programs in North America, was formed in 1912 (Cuff 1991).

The design studio culture was introduced to North American schools around the turn of the twentieth century. In the early twentieth century, many architecture schools had at least one Paris-trained professor who brought this tradition with him. Over 500 Americans attended the École des Beaux-Arts between 1850 and 1968, when it closed its doors. The overwhelming majority were men. Among the notable exceptions was Julia Morgan. After completing her studies there in 1902, Morgan later became the first woman licensed to practice architecture in California, with a prolific career that included the design of Hearst Castle (1919–1947) in San Simeon along with hundreds of buildings. By the 1930s, the studio and design jury had become firmly entrenched in American architectural education, attaining the prominence that remains today.

The influence of the German Bauhaus school (1919–1933) and the teaching methods of its founder, Walter Gropius, who later headed the architecture department at Harvard (1938–1952), soon superseded that of the École des Beaux-Arts. Instead of neoclassical monuments, the machine, mass production, and modern technology served as inspirations for design. The Bauhaus Building in Dessau, Germany contained 28 live-in studios for students with baths and a basement gymnasium. Design studio culture became even more of a world of its own. And the jury system remained.

Sometime during the late 1940s through the 1960s, juries went public, switching from a closed to an open format. Students orally presented their work one by one before a jury, with their classmates, passersby, and total strangers listening in. Instead of a cryptic letter grade, students received detailed comments from the jurors. Design juries thus became marathon sessions – usually a minimum of three to four hours at a stretch – a practice that remains today in architecture schools around the world. In extreme case, design juries can last all day long and even into the evening. As design studio courses typically meet for several hours a week, and students are expected to work on their project for many hours outside class, students tend to remain somewhat cloistered in the studio culture. The intense time commitment required to complete a design project is such that many architecture students almost live in the studio space, with little time to make friends from other academic disciplines or to participate in the larger life of the university. In fact they can make excellent roommates since they are rarely home.

At most accredited architectural schools in the US, students are assigned a permanent spot in studio for the duration of the academic term, and only one student occupies a desk at a time. However, at a handful of American schools where space is at a premium, and at some architectural schools abroad, students occupy "hot" seats shared by others when they are not in class, as is customary in most university classrooms. At architecture studios in Milan Polytechnic, which admits a very large number of students, this is common practice. Hot seats fundamentally alter the studio culture, as students tend to see each other only in class, when the instructor is present. They are far less convenient for students who must carry their models and drawings with them from home rather than leave them in the studio. Whenever possible, architectural school administrators in the US far prefer the practice of permanent studio space in order to create a climate where students mix and mingle and have opportunities to view and critique each others' work after hours (Figure 17.1).

Studio culture and design juries have inspired several critiques over the years. One of the earliest was Donald Schön's chapter on "Design as a Reflective Conversation with the Situation," in *The Reflective Practitioner: How Professionals Think in Action* and several subsequent works (Schön 1983, 1985, 1987 and 1991). He describes the iterative nature of the desk critique, where designers make a representation of a plan, program or image of an

Figure 17.1 East St. Louis Action Research Project (ESLARP). Source: Kathryn H. Anthony.

artifact to be constructed by others. The complex marking process results in design moves that produce unintended consequences. By shaping the situation based on the designer's initial appreciation of it, the situation then "talks back" and the designer responds to the situation's back-talk (Schön 1983: 78–79). Schön describes a desk crit between a hypothetical instructor he calls Quist, teaching a student named Petra, how to learn about design, what he refers to as "reflection-in-action," i.e. the role of observation and reflection, balancing between "hard" and "soft" thinking. Schön argues, "Each move is a local experiment which contributes to the global experiment of reframing the problem" (Schön 1983: 94) ... "And if they are good designers, they will reflect-in-action on the situation's back-talk, shifting stance as they do so from 'what if?' to recognition of implications, from involvement in the unit to consideration of the total, and from exploration to commitment" (Schön 1983: 103). He views design training as an open-ended conversation between student and instructor, much like a piano or violin teacher showing a student how to play an instrument.

My research for *Design Juries on Trial: The Renaissance of the Design Studio*, based on systematic observations, videotape recordings of juries, diaries of design students, and interviews and surveys of hundreds of students, educators, and practitioners conducted over a seven-year period exposed both the positives and negatives of studio culture, calling for a shakeup in both design education and practice (Anthony 1991). In my book, I argued that at their best, design juries can be extremely valuable learning experiences for students, a chance to hear fresh opinions about their projects from design critics who see their work for the first time. Yet, I was also highly critical of a system that at its worst required students to stay up all night with little to eat and no sleep, where sexual harassment could run wild after hours, and where insensitive instructors or critics demolished students' egos with inappropriate harsh, destructive, and personal criticism in front of their peers. For example, one critic asked a senior design student, "Have you ever taken freshman design? Yes? Then you need to retake it!" More recently, at a final review of a year-long master's thesis design project, another critic exclaimed, "I'm convinced that you just chose the wrong site for that project. It's just the wrong site! How could you possibly pick such an awful site?" Students respond to such vicious critiques with defensive verbal and nonverbal behavior, often leaving the jury demoralized, bitter, confused, and in tears.

Such aspects of studio culture are hardly the basis for successful professional practice. Ironically, they could also prove especially intimidating to individuals who still remain vastly underrepresented as professionals in the field, notably women and students of color. Many students uncomfortable with these aspects of studio culture have quit the architecture major altogether. Several accomplished practitioners interviewed for *Design Juries on Trial* still recalled scars from searing design juries that they had experienced in school. I called for design studios and juries to become more responsive to the designers who compete, the clients who pay, and the public that lives and works in the spaces created. As I argued then and would argue today, "the increasingly complex nature of the professional world – reliance on design teams and joint development efforts, and larger and more complex design projects – has left the designer trained as a solo artist, engrossed in competitive, individual pursuits, out in the cold" (Anthony 1991: 167).

Since the publication of *Design Juries on Trial*, national architectural student leaders and educators have become increasingly vocal critics of studio culture. The American Institute of Architecture Students Studio Culture Task Force was formed in 2000 in the wake of a tragedy. Having fallen asleep at the wheel, an architecture student died in a car accident while driving home after spending two consecutive sleepless nights working on his final project. He collided head-on with a truck. In extreme cases at least a dozen other architecture students, too, met their death by sleep deprivation around design jury time. Thomas Fisher (1991) pointed out such atrocities in his article, "Patterns of Exploitation." Scores of other design students continue to be injured while building models or using hazardous laser-cutters with little or no sleep. As 2004–2005 AIAS President Jacob Day puts it,

> A story has been told of a student who lost his life in an automobile accident caused by sleep deprivation. A dozen stories have been told of similar instances. Thousands of stories have been told of cut fingers, damaged cars, life-changing critiques, friends lost and lives changed. All for an education in the art and science of architecture. (Kellogg 2005: 2)

In *The Redesign of Studio Culture: A Report of the AIAS Studio Culture Task Force,* AIAS called for significant improvements to studio culture (AIAS Studio Culture Task Force *et al.* 2002). In 2004, AIAS hosted a major studio culture summit at the University of Minnesota, bringing it to the forefront of their advocacy agenda (Kellogg 2005). In 2004 the National Architectural Accrediting Board (NAAB) adopted a thirteenth condition for Accreditation (Condition 3.5) requiring schools to have a written policy about the culture of their studio environments:

> The school is expected to demonstrate a positive and respectful learning environment through the encouragement of the fundamental values of optimism, respect, sharing, engagement, and innovation between and among the members of its faculty, student body, administration, and staff. The school should encourage students and faculty to appreciate these values as guiding principles of professional conduct throughout their careers.
>
> The *APR* must demonstrate that the school has adopted a written studio culture policy with a plan for its implementation and maintenance and provide evidence of abiding by that policy. The plan should specifically address issues of time management on the part of both the faculty and students. The document on studio culture policy should be incorporated in the *APR* as Section 4.2. (NAAB 2004:5)

In 2005, the AIAS established the Studio Culture Task Force to study effects of current architectural education practices on students and to consider alternatives, resulting in the AIAS (2008) publication, *Toward an Evolution of Studio Culture.* This document included the results of the 2007 Administrators Survey on Studio Culture, the 2008 AIA Council of Presidents Survey on Studio Culture, lessons learned from peer reviewed studio culture policies, along with a summary of best practices, guidelines, and recommendations for more effective studio culture.

These AIAS reports reveal that at its best, the design studio culture has immense pedagogic value. Among its greatest virtues: one-to-one communication between faculty and student, peer-to-peer learning, Socratic discourse, learning by doing, and rewarding visual literacy. Yet with its roots in nineteenth-century France, studio culture has elements that no longer meet the needs of twenty-first-century architecture. The changes called for in the debate included: conducting more rigorous research, defining best practices, and communicating about meaningful issues.

To increase students' skills and abilities to relate to an increasingly diverse population, I have called for all schools of architecture to require students to have at least one design studio experience well out of their individual comfort zone – to design for a population, setting, or issue that is highly unfamiliar to them, not a typical part of their life experience. For the white suburban student, this could involve designing for an ethnic or racial minority culture in an inner city (African American, Latino, Native American), an age group out of their normal range (young children or elderly), or persons with physical disabilities. And I call for schools to require students to have at least one design studio working with real – not imaginary – clients and people who would use the spaces they design (Figure 17.2). These clients and users should participate in the design studio throughout the project's inception, the interim reviews, and the final reviews.

Yet today, such experiences are still the exceptions rather than the rule in most architecture schools. More often than not, in design studios the client remains a fictitious person who never appears, and students are allowed to design in dreamland, accountable only to their design critics. It is as absurd as training surgeons without patients.

Studio culture as a vehicle for teaching urban design

In their landmark study of architectural education, *Building Community: A New Future for Architecture Education and Practice*, authors Ernest Boyer and Lee Mitgang (1996) stress the importance of what they term "service to the nation," encouraging schools to increase and make better known the storehouse of architectural knowledge to enrich communities, and to prepare all architects for lives of civic engagement and ethical practice. They recognize that

> schools of architecture deserve huge credit for performing, collectively, millions of dollars worth of pro bono work every year through their involvement in a variety of *pro bono* housing and community projects in some of America's most depressed urban and rural communities. (Boyer and Mitgang 1996: 130)

While the design studio lends itself to the study of virtually any building type, it presents both special challenges and opportunities for teaching students about

Figure 17.2 Design studio at the Gulf Coast in the wake of Hurricane Katrina. Source: Kathryn Anthony.

urban design. Instructors can draw upon the collaborative nature of studio to assign teams of students to collect vital prerequisite information about site analysis, design precedents, and the types of people likely to use the new buildings or spaces. Because studio courses are usually held all morning or all afternoon long, they lend themselves to field trips to visit the site and meet with key individuals who have a stake in the project. Yet because urban design projects inevitably present a far more complex set of issues than stand-alone buildings, issues that span well beyond the scope of one academic term, it is all too easy for students to receive only a superficial glimpse of what these projects would involve in real life. And unless design instructors go out of their way to invite representatives of several kinds of people affected by the proposed designs to participate as vital members of the design studio – in developing the design program, in critiquing student work, in participating in design juries – students will leave with an unrealistic view of what urban design is all about.

To address this dilemma, and to take advantage of the visionary ideas that architecture students can offer, several schools of architecture and planning have instituted urban design projects through community design centers whereby students, faculty and staff have opportunities to work directly with community leaders and to immerse themselves in issues that outlast one academic semester. As examples of service learning, community design studios are outgrowths of the progressive educational philosophy of John Dewey who advocated that for learning to occur, an interaction of knowledge and skills with experience is key.

Sparked in part by the civil rights movement, the first community design centers originated in the 1960s, and several sprouted up in subsequent decades. As of 2000, the Association of Collegiate Schools of Architecture's survey of community design programs at North American schools reported a total of 47 university-affiliated Design Centers, 24 university-based Community Research Centers, and 15 Design/Build Programs (Cary 2000). Over half of the university-based programs were established in the 1990s. Such centers are models of multidisciplinary teamwork, engaging planners, urban designers, architects, and landscape architects with scholars and professionals from related fields. They provide rich opportunities for scholarship and research. Today's architecture faculty and students around the country continue to engage in significant pro bono work.

Several such centers are described in a recent monograph (Hardin and Zeisel 2005). In its introduction, architectural educator Anthony Schuman discusses the pedagogy of engagement and the tensions between advocacy and activism. Schuman argues that while the field of "planning has demonstrated a continuous evolution towards a socially engaged practice, architectural education has not" (Schuman 2005:8).

Although the emergence of socially based architecture in response to the Civil Rights movement of the 1960s and 1970s sparked the establishment of several community design centers, both architectural education and the architectural profession were soon preoccupied with postmodernism, the deconstructionist movement, and more recently, digitally produced forms of sculptural design that were previously impossible to draft and to build. Instead of being rewarded for their involvement, some architectural faculty members affiliated with early service-learning programs faced denial of tenure and promotion, and some lost their jobs. Unfortunately, community design work does not fit neatly into typical categories of academic promotion such as funded research, scholarly publications, or creative artistry. The end results, which usually value process over product, rarely result in buildings that would be publishable in the leading design magazines. Many schools of architecture and their faculty members still face such quandaries today.

Nonetheless, the major professional organization of architectural educators, the Association of Collegiate Schools of Architecture (ACSA) recognizes the value of such community design centers. Since 1998 the ACSA has offered its Collaborative Practice Award to honor best practices in school-based community outreach programs. Award recipients demonstrate how faculty, students, and community and civic clients work to realize common objectives.

Community design centers affiliated with schools of architecture include the Chattanooga Planning and Design Studio, with partial funding by the University of Tennessee at Knoxville, TN; Community Design Center at University of Cincinnati, OH; and the University of Arkansas Community Design Center.

The Detroit Collaborative Design Center at the University of Detroit Mercy School of Architecture, established in 1995, is a year-round, fully operating facility that provides opportunities for staff, faculty, and students to work with non-profit community development organizations to promote quality design solutions and respond to local concerns ("Detroit Collaborative Design Center"). The staff teaches a community design studio every semester. Examples of projects include a Community/University Center (2001), a gymnasium for a range of abilities at Friends' School (2002), and a variety of design-build projects. The latter include a fascinating work entitled "FireBreak," where students collaborated with local artists and residents to fabricate and construct a series of installations in and around burned out houses on Detroit's east side.

In the wake of Hurricane Katrina in 2005, several architecture, planning, and landscape architecture faculty around the nation initiated urban design projects to address cities and communities devastated by the storm. At Mississippi State University's School of Architecture, the Gulf Coast Community Design Studio (GCCDS) worked with members of the East Biloxi, MS community to provide early damage assessment maps, planning assistance, and design services. As of 2008, over 80 homes in East Biloxi, both rehabilitations and new construction, were completed through the work of GCCDS ("Gulf Coast Community Design Studio").

At the University of Washington, "PK (Post Katrina) studio," the first collaborative studio including the college's three departments – architecture, landscape architecture, and planning – focused on Terrytown, an unincorporated bedroom community of 25,000 on the west bank of the Mississippi River, just five minutes drive from downtown New Orleans. While Terrytown suffered water and wind damage from the storm, the greatest loss resulted from looters who plundered a regional shopping mall and set it on fire. As one architecture student explained, this real-world studio experience pushed her and her classmates "to design and think at a different level, where the options we suggest have to be feasible and realistic." (Lewis 2007).

Harvard's Graduate School of Design partnered with Tulane University on several initiatives concerning the recovery and rebuilding of post-Katrina New Orleans. Harvard's faculty developed three new courses and a forum focusing on the role of universities in urban rebuilding and recovery following natural disasters. One such course was a design studio named "Cities in Crisis: Memory and Community in Architecture and Planning" ("GSD Partners with Tulane University in New Orleans Hurricane Recovery" 2005). Other schools with architectural design studios focusing on the aftermath of Hurricane Katrina included Arizona State University,

Columbia University, University of Kansas, and Washington University in St. Louis. In sum, community design work like the projects described here present an extremely valuable focus for studios. They are a win-win situation, forming a valuable experience for communities needing design services and a memorable part of students' education.

The East St. Louis Action Research Project

Design studios conducted as part of the East St. Louis Action Research Project (ESLARP) at the University of Illinois at Urbana-Champaign have applied the studio model to the study of urban design problems for over 20 years. Established in 1987, ESLARP is one of the longest running community design centers ("East St. Louis Action Research Project 2009"). Students and faculty in architecture, landscape architecture, and urban and regional planning have left their mark through various neighborhood revitalization efforts in East St. Louis, Illinois, one of the nation's most economically depressed cities. ESLARP has since evolved to include the Graduate School of Library and Information Science, the Department of Recreation, Sport and Tourism; and the College of Law. As of 2006, the city was 98 per cent African American, with 35 per cent living below the poverty level and a median household income of $21,324, about half the national median (Harwood 2006).

Some ESLARP projects have been undertaken as architectural studios, while others have involved a collaborative studio model combining students and faculty in two or more of these environmental design disciplines, either concurrently or during subsequent semesters. Critical components of ESLARP are two outreach weekends per semester, when student volunteers enrolled in these studios, along with others from across campus, travel to East St. Louis to participate in short-term "clean-up, fix-up, paint-up" projects. Although at first glance, one might question the educational value of such engagements, they enhance the pedagogy of urban design by immersing students in a problem setting for a prolonged period of time (including an overnight stay), placing them side-by-side with residents and community leaders. And they provide immediate rewards when participants see short-term projects through to completion.

Projects have produced tangible improvements in poor neighborhoods, enhanced the quality of life and increased the ability of neighborhood organizations to complete community development efforts. ESLARP projects have resulted in neighborhood beautification, housing improvement, job creation, and park development.

Early completed works include construction of a local farmers' market and the Illinois Avenue playground. UIUC students and faculty later succeeded in lobbying city officials and government agencies to route a light-rail line through and to provide a station stop in East St. Louis' Emerson Park neighborhood. This station, that opened in 2001, now offers easier access to job opportunities on both sides of the Mississippi River. UIUC volunteers "blitz built" four affordable single-family homes and helped persuade the Emerson Park Development Corporation to construct new mixed-income housing units resembling market-rate townhouses in architectural style and density. Students conveyed input from local residents into the design of 464 housing units at Parson's Place. The rehabilitation of homes in the city's South End neighborhood, the opening of a full-service supermarket and construction of the Jackie-Joyner Kersee Youth Development Center are other notable accomplishments.

During 2002–2006 ESLARP students, faculty and staff assisted renowned jazz artist

Eddie Fisher and his wife Christina Fisher with the interior architectural design, code reviews, site work, and interior and landscape construction of the Village Theater, home of Community Concepts in Centreville, Illinois. Community Concepts is a grassroots agency providing technical and skill-based training in computer literacy, academic tutoring, media and theater production, and leadership to at-risk youth in St. Clair County. It also serves as a community performance and entertainment venue, a 7,500 square-foot complex with a 250-seat stage, offices, dressing rooms, and labs for 22 computers.

ESLARP can claim to its credit the renovation of a 3.6-acre site containing a 17,000 square foot shelter and 2,600 square foot garage into The Joseph Center of Eagle's Nest of St. Clair County, a not-for-profit project providing transitional housing and support for homeless male veterans in the metropolitan St. Louis area. Here men can reside in a supportive environment and begin to get their lives back on track. Offering 26 units of housing, Eagle's Nest is one of only three sites in the US that offers around-the-clock, long term care and counseling for homeless military veterans. University of Illinois at Urbana-Champaign students assisted in grant writing, collaborative participatory design charrettes, and volunteer on-site labor. The dedication and home warming ceremony for The Joseph Center took place on November 13, 2009 (Dearborn forthcoming A; Eagle's Nest of St. Clair County 2009).

Impact of community design studios on future urban designers

How do community design studios help train future urban designers? What kinds of impacts can these studios have? According to faculty, the experience can be life changing. Twenty-five faculty engaged in these studios describe numerous educational benefits in Hardin and Zeisel's (2005) *From the Studio to the Streets: Service-Learning in Planning and Architecture*. Here are some of the common themes that run throughout their accounts.

Community design studios strengthen the discipline by fostering a sense of caring about others with greater needs, helping students to become more sensitive to the needs of disadvantaged communities, and exposing them to people with whom they are less familiar. They help students develop a wide range of professional skills such as land use and building condition surveys, demographic analysis, cost estimating, site planning, participatory design processes, public workshops, design charrettes, and zoning analysis. They provide models of multidisciplinary teamwork, involving planners, urban designers, landscape architects, architects, and scholars and researchers from related fields (Hardin and Zeisel 2005).

At Lawrence Technological University, Joongsub Kim and James Abernethy administered surveys to students, studio clients, community residents, and guest critics participating in their community outreach programs. Over 95 percent of the 45 respondents reported that the studio experience was positive. Benefits included gaining some real-life experience, learning from diverse perspectives, experiencing a sense of community, promoting community building, learning from a variety of disciplines, building relationships with stakeholders, and networking. As one respondent put it, "I learned that reality out there is messy" (Kim and Abernathy 2005: 152). Among the few negative drawbacks were disagreements and the inability to make decisions expeditiously.

How have ESLARP alumni applied what they learned from urban design studios in their subsequent professional experience? Lynne Dearborn of the University of Illinois sent an 80-question survey to

525 ESLARP alumni, of which 133 surveys were returned. Survey results formed the basis for subsequent 30 telephone interviews, 10 from each of the three disciplines. Of the respondents, 62 percent had taken at least one ESLARP–based course, and 38 percent had taken two to five such courses. Most (63 percent) were employed in traditional design firms and planning departments. Planners were mainly employed in government-related work, while architects and landscape architects were primarily employed in the private sector. Survey results showed that 58 percent had participated in community service during the past year, more than the 44 percent of American adults who volunteer annually.

At least three-quarters of ESLARP alumni reported that they were either much better or better than most people on a variety of characteristics: respecting the views of others, thinking critically, tolerant of people who were different from them, effective in accomplishing goals, more able to see the consequences of actions, and able to lead a group. Interviews revealed that the ESLARP experience stressed the importance of sensitivity to client needs, client communication, and better understanding the constituencies who are left out of the design and planning processes. Transactions between students and real clients in contexts with multiple-real-world complexities were important in providing experiences that expanded their professional horizons (Dearborn forthcoming B).

Several former ESLARP student-volunteers have moved into leadership positions with neighborhood development organizations in East St. Louis. Others now lead planning and community development projects at city and federal agencies in New York and Washington, DC (Kline 2007).

ESLARP's impact on the residents of East St. Louis has been positive. According to one resident, the visits from the University of Illinois students are like "a shot of penicillin ... [ESLARP] has had a great impact on our community. We can all say it's a sewer, it's a pit, it's worthless. They still bring that hope and that vision" (Kline 2007: 7).

Recalling Donald Schön's (1983, 1985, 1987, 1991) work, community design studios provide yet another source of iteration in the design process, a source far better grounded in reality than that provided solely by the design instructor in the typical one-to-one desk crit. Students must reflect upon their design ideas not only with their instructor but also with the community, underscoring the notion of urban design as a collaborative process. Students learn about the importance of listening and negotiating skills as well as the value of spending time on-site with the people whose lives will ultimately be affected by their designs. This helps them to reframe the design problem and view the consequences of their design moves in a more meaningful way.

Impact of new technology on studio teaching

How has the Internet provided opportunities for more collaborative teaching models in design studios? By the end of the twentieth century, the advent of the Internet and increased use of personal computing caused dramatic shifts in traditional studio culture. The Internet has made it far easier for students to study design precedents and gather relevant research needed for their design projects. Software programs like GoogleEarth have revolutionized the process of site analysis. With new software and projection equipment, students are now able to present digital images of their designs before a much wider audience than ever before. By zooming in and zooming out, they can highlight minute details of their project that would otherwise be impossible for a large audience to see. They can produce paperless projects that are more eco-friendly.

Yet students and faculty continue to grapple with new modes of technology, and their incorporation into design studios is still being perfected. When delivering digital presentations, students often face technical difficulties with large files that fail to open if not compressed properly. When relying exclusively on digital designs and 3D renderings, design critics often struggle to grasp the scope and scale of the project, especially if images are shown sequentially. What seems to work best is if two screens and two laptop computers are used simultaneously, one to show all the students' boards altogether, and the other to highlight a portion of the project in detail. This allows the audience to see both the big picture and the small picture at once. And despite the sophisticated computer renderings that students can now produce, critics and clients still tend to gravitate to 3D models that they can hold in their hands.

Several instructors have developed course web sites documenting the studio experience, where professors post the design program along with completed student projects. Design practitioners, clients, and citizen groups can view students' work and provide critiques from afar. No longer must they travel for miles in order to participate in academic design juries as jurors or critics. Yet because of the high value that continues to be placed on face-to-face interaction, in-person juries still remain the norm in most architectural schools.

Some examples from my teaching experiences shed light on how media technology is integrated into design courses and provide a useful model for documenting urban design studios. For several years I taught a series of health care design studios, all of which are documented on web sites (Anthony 2003, 2004, 2005, 2007). Architects based in the Chicago and St. Louis offices of Cannon Design, the firm that sponsored our studio, were able to view students' interim and final designs, as were client representatives such as the head nurse at a local outpatient surgi-center and the director of our university's psychological services center. The web sites provide a permanent archive of each studio that would otherwise have disappeared once the course was over. They also can serve as a valuable teaching tool for students, educators and practitioners elsewhere.

Videoconferencing provides similar opportunities for enhanced collaboration with students, faculty, practitioners, community groups, and others. My graduate course on design entrepreneurship included two videoconferences a week apart with two top representatives from Real Estate and Workplace Services at Google. The first videoconference session was held at Google headquarters in Chicago, while the second occurred on the University of Illinois at Urbana-Champaign campus. During the week in between, students were asked to redesign a twenty-first-century design studio inspired by what they had seen at Google, and we sent electronic versions of the students' designs for Google staff to review. During the second videoconference Google staff provided students with detailed criticism and feedback on their designs. This kind of interaction would have never been possible without new technology.

Using a variety of computer software programs, students can now design from the convenience of their homes, rather than being tied to drafting boards in studio. In fact, many students now prefer to work at home. As a result, some design studios may appear almost deserted at times – a situation that was hardly the case in years past. Students may now spend more time at home, in computer labs, and in facilities with state-of-the-art laser cutters to prepare complex models and drawings that would have been impossible to produce just a few years ago. The nature of visual design presentations becomes more and more sophisticated with each passing term, and the bar continues to be raised. Although they may spend less time in the studio than before, students still spend most of their time in studio-related activities, and still have little free time of their own.

Yet most architecture faculty still relies on the traditional studio model, encouraging students to continue to work in studio as much as possible. Faculty continues to place great value on the traditional studio setup that allows students to view each other's work and to collaborate with teams for site analysis and model building, and that provides opportunities for the work to be displayed for desk crits, pin-up critiques, and reviews.

Urban design studio projects are among the most important of all design projects addressed in architectural schools. Will the 2008 election of President Barack Obama, with his personal history of community organizing and crusade for community service provide an impetus for a reinvented, twenty-first-century generation of urban design studios and community design centers? One where all students of architecture, landscape architecture, and urban planning are required to take at least one such course? Or where their professional internship would require it or their professional licensure demand it? If this was to happen, the creative synergy of design students, faculty, and community members would have the potential to revitalize urban designs throughout the world.

Conclusion

Throughout their lengthy history, design studios, and their unique studio culture, have held enormous pedagogical potential to transform students and their urban environments. Although that potential has not always been realized, the formation of community design centers, along with student and faculty involvement in service learning projects are steps in the right direction. Compared to most other academic courses, these opportunities offer an education that extends beyond the classroom, and is far likely to be remembered above all others. Because they offer a myriad of pedagogical benefits, such courses should be mandatory in the education of urban designers. Yet they often remain on the margins at most architecture schools, falling outside the mainstream of design education. Even at schools offering community design centers and related urban design studios, student participation is not always required. As a result, it is possible for students to graduate and miss out altogether on this valuable real world experience.

Acknowledgments

The author thanks Wambaa Mathu for his valuable research assistance with this chapter and the editors for their helpful comments.

References

AIAS Studio Culture Task Force (Koch, A., Schwennsen, K., Dutton T.A. and Smith, D.) (2002) *The Redesign of Studio Culture: A Report of the AIAS Studio Culture Task Force*, Washington, DC: AIAS.

American Institute of Architecture Students (2008) *Toward an Evolution of Studio Culture: A Report of the Second AIAS Task Force on Studio Culture. Lessons Learned, Best Practices and Guidelines for an Effective Studio Culture Narrative*, Washington, DC: AIAS. Online. Available HTTP: <http://www.aias.org/studioculture/AIAS_Toward%20an%20Evolution%20of%20Studio%20Culture_2008.pdf> (accessed 25 February 2009).

Anthony, K.H. (1991) *Design Juries on Trial: The Renaissance of the Design Studio*, New York, NY: Van Nostrand Reinhold.

—— (2003) *Architecture 372, Senior Healthcare Design Studio*. Online. Available HTTP:

<http://www2.arch.uiuc.edu/kanthony/arch372SP03/> (accessed 13 May 2009).

—— (2004) *Architecture 272, Junior Healthcare Design Studio*. Online. Available HTTP: <http://www2.arch.uiuc.edu/kanthony/arch272sp04/home.htm> (accessed 13 May 2009).

—— (2005) *Architecture 475/476/572-Spring 2005*. Online. Available HTTP: <http://www2.arch.uiuc.edu/kanthony/arch476_572SP05/> (accessed 13 May 2009).

—— (2007) *Graduate Healthcare Design Studio*. Online. Available HTTP: <http://www2.arch.uiuc.edu/kanthony/arch572sp07/homepage.html> (accessed 13 May 2009).

Boyer, E.L. and Mitgang, L.D. (1996) *Building Community: A New Future for Architecture Education and Practice*, Princeton NJ: The Carnegie Foundation for the Advancement of Teaching.

Cary, J.M. Jr. (ed.) (2000) *The ACSA Sourcebook of Community Design Programs at Schools of Architecture in North America,* Washington, DC: ACSA Press.

Cuff, D. (1991) *Architecture: The Story of Practice*, Cambridge, MA: MIT Press.

Dearborn, L. (forthcoming A) 'Rehab, Rebuild, Renew: ESLARP's Work with Two Community Partners', in C.L. Wilkins (ed.) (forthcoming) *Activist Architecture: The Philosophy and Practice of Community Design Centers*, New York: Princeton Architectural Press.

—— (forthcoming B) 'Applying Service-Learning Experience in Professional Practice: What ESLARP Alumni Reveal about Action, Investigation, and Reflection', in C. Doble and P. Horrigan (eds.) (forthcoming) *Erasing Boundaries – Supporting Communities*, Washington, DC: Island Press.

"Eagle's Nest of St. Clair County: Serving veterans with excellence." Eagle's Nest of St. Clair County (2009). Available at: http://www.thejosephcenter.org/ (accessed 19 November 2009).

Fisher, T.R. (1991) 'Patterns of Exploitation', *Progressive Architecture,* (May): 9.

Hardin, M.C. and Zeisel, W. (eds.) (2005) *From the Studio to the Streets: Service-Learning in Planning and Architecture*, Washington, DC: American Association for Higher Education.

Harwood, S.A. (2006) 'East St. Louis and the East St. Louis Action Research Project', *Illinois Planning News, Official Bi-Monthly News of the Illinois Chapter of the American Planning Association*, Number 78 (April 8). http://www.ilapa.org/news/2006/Mar06/ILAPA_Mar-Apr.pdf (accessed 11/19/09)

Kellogg, C. (2005) The Studio Culture Summit. Organized by the American Institute of Architecture Students. Held October 8–10, 2004 at the University of Minnesota An Overview Report by Clark Kellogg. Washington, DC: AIAS. Online. Available HTTP: <http://www.aias.org/studioculture/summitreporthighres.pdf (accessed 12 November 2009).

Kim, J. and Abernathy J. (2005) 'Service-Learning as a Holistic Inquiry and Community Outreach Studios', in Mary C. Hardin and William Zeisel (eds.) (2005) *From the Studio to the Streets: Service-Learning in Planning and Architecture*, Washington, DC: American Association for Higher Education.

Kline, G. (2007) *Positive Partnership*. Online. Available HTTP: <http://news.illinois.edu/news/07/07f_pm_eslarp.html> (accessed 24 February 2009).

Lewis, P. (2007) *Post-Katrina: Design for Recovery*. Online. Available HTTP: <http://uwnews.washington.edu/ni/uweek/uweekarticle.asp?articleID = 31206> (accessed 26 February 2009).

National Architectural Accrediting Board (NAAB) (2004) "NAAB Conditions for Accreditation For Professional Degree Programs in Architecture. 2004 Edition." Washington, DC: The National Architectural Accrediting Board. See also 'Accreditation'. Online. Available HTTP: http://www.naab.org/accreditation/2004_Conditions_2.aspx> (PDF file of NAAB 2004 Conditions downloaded from this site accessed 12 November 2009).

Schön, D.A. (1983) *The Reflective Practitioner: How Professionals Think in Action*, London: Temple Smith.

—— (1985) *The design studio: an exploration of its traditions and potentials*, London: RIBA Publications for RIBA Building Industry Trust.

—— (1987) *Educating the Reflective Practitioner*, San Francisco: Jossey-Bass.

—— (1991) *The Reflective Turn: Case Studies In and On Educational Practice*, New York: Teachers Press Columbia University.

Schuman, A.W. (2005) 'Introduction: The Pedagogy of Engagement', in Mary C. Hardin and William Zeisel (eds.) *From the Studio to the Streets: Service-Learning in Planning and Architecture*, Washington, DC: American Association for Higher Education, 1–15.

Further reading

Anthony, K.H. (1991) *Design Juries on Trial: The Renaissance of the Design Studio*, New York: Van Nostrand Reinhold. Available HTTP: http://www2.arch.uiuc.edu/DesignJuriesOnTrial/ (accessed 12 December 2009). The book unlocks the door to the mysterious design jury system, exposing its hidden agendas, helping students overcome intimidation, confrontation, and frustration and offers educators strategies to create more effective, efficient ways to evaluate students' designs.

Bosworth, F.H. Jr. and R.C. Jones (1932) *A Study of Architectural Schools*. New York, NY: Scribner. This classic source provides a critical historical account of architectural education in the US and its intellectual origins in the French École des Beaux Arts; it is based on the authors' visits to almost 50 North American schools of architecture.

Harvard Graduate School of Design (2005) *GSD Partners with Tulane University in New Orleans Hurricane Recovery*. Online. Available HTTP: <http://www.gsd.harvard.edu/news/archive/katrina_gsd_tulane.html> (accessed 26 February 2009).

Mississippi State University's School of Architecture (2005) *Gulf Coast Community Design Studio*. Online. Available HTTP: <http://www.gccds.org> (accessed 25 February 2009).

Schön, D. (1985) *The Design Studio: An Exploration of its Traditions and Potentials*, London: RIBA Publications for RIBA Building Industry Trust. A detailed account and case study of the iterative "reflection-in-action" process that occurs between a student and instructor while developing a studio design project, with far-reaching implications for other fields of higher education.

Stevens, G. (1998) *The Favored Circle: The Social Foundations of Architectural Distinction*. Cambridge, MA: MIT Press. This book x-rays the architectural profession to uncover its underlying value system; Chapter 5 provides a cross-cultural critique of architectural education in Britain, France, Germany, and the US.

University of Detroit Mercy School of Architecture (1995) *Detroit Collaborative Design Center*. Online. Available HTTP: <http://architecture.udmercy.edu/DCDS07/dcdc.htm> (accessed 25 February 2009).

University of Illinois at Urbana-Champaign (2009) *East St. Louis Action Research Project*. Online. Available HTTP: <http://www.eslarp.uiuc.edu> (accessed 23 February 2009).

18

Media tools for urban design

Martin H. Krieger

Urban design focuses on the structure and experience of the built and social environment, usually at the scale of a neighborhood or a district. Contemporary media tools allow us to envision environments at different scales, to propose designs with the current environment in mind, and to present that work to an audience varying in their political interests and design sophistication. New media tools have opened up opportunities for sensing, documenting, understanding, and representing the urban experience, for exploration and communication. What is striking is how tools go from being steep-learning-curve technologies, to readily-employed applications. So, I will focus here on general principles rather than on any particular technology or application. What is also apparent is that these media technologies and applications allow for new modes of understanding, while closing off others. Skills fundamental to the training of their teachers, are displaced by students' new media methods. New forms of hand-eye/ear relationships, replace those of drawing, for example.

Material facts matter, and they may be documented visually and aurally. Just as physiological processes are produced by anatomy, chemistry, and electricity, city life is produced by material circumstances. If newspapers are to be sold, there will be vending boxes at certain corners, if there is to be worship, there must be places available for that worship. For understanding city life, what is everyday and ordinary and material is just as indicative and richly symbolic as is the extraordinary and unique and conceptual. Media documents can capture this material richness: the interrelated *choreographies* of where and how people worship and work, and the *industrial engineering* and the coordinated networks, systems and infrastructures that support industry and residents, and that allow them to live near each other.

I use the term *media* or perhaps *multimedia* to suggest that we shall be concerned not only with the visual, and not only with individual images, but with the full range of *cinematic* modes now possible with computation and user-friendly applications (see Daley 2003, on "multimedia literacy"). Storytelling and montage are crucial features of the cinematic arts, and they play a prominent role in what I describe here. Moreover, the visual and the aural inform each other, as do the other senses; we expect that the account we give of our experience is consistent across the senses.

I will begin with discussing the work of some model documenters of urban life, in particular those who have been concerned with multiple images. What makes urban life so peculiar is that we are in the middle, always, never at a vantage point. So a corpus of media allows one to adequately explore a city only if it is all round and multi-aspectival. Phenomenologists are the philosophers who have described this unity in multiplicity, even if the mechanism for "how we do it" is not so explicit in their work.

The big transformation in media has been the capabilities given by digital and computational resources, and the capacity to make many images or recordings, and to more readily manage a large corpus. I close with a description of the urban sensorium epitomized by this corpus and use the notion of storytelling to connect multiples to practice.

Our models will be Charles Marville (1816–1879), Denis Diderot (1713–1784) and August Sander (1876–1964), Eugène Atget (1857–1927) and Hilla and Bernd Becher (1934–, 1931–2007).

Marville was commissioned by Baron Haussmann to document Paris before and after it was eviscerated and pierced by grand boulevards, and Marville made something like 500 photographs of Paris's streets. For his *Encyclopédie* (1751–1772), Diderot went out and documented how artisans did their work (he interviewed them) and then used engravings to discuss those processes. His *Déscriptions des Arts et Métiers* (Diderot 1993) is a detailed account of arts and crafts and technological processes of the period. Diderot describes building construction, shipbuilding, woodworking, and scientific-instrument making, among other processes. These descriptions are accompanied by those detailed engravings. This excerpt from d'Alembert's preface to the *Encyclopédie* (d'Alembert 1995) gives us good insight into the work at hand:

> The section on the mechanical arts required no fewer details and no less care ... [E]verything impelled us to go directly to the workers.
>
> ...We took the trouble of going into their shops, of questioning them, of writing at their dictation, of developing their thoughts and of drawing therefrom the terms peculiar to their professions ... We have seen some workers who have worked for forty years without knowing anything about their machines. With them, it was necessary to exercise the function in which Socrates gloried, the painful and delicate function of being midwife of the mind, *obstetrix animorum*.
>
> ...But the general lack of experience, both in writing about the arts and in reading things written about them, makes it difficult to explain these things in an intelligible manner. From that problem is born the need for figures ... A glance at the object or at its picture tells more about it than a page of text.
>
> We have sent designers [draftsmen?] to the workshops. We have made sketches of the machines and of the tools, omitting nothing that could present them distinctly to the viewer...
>
> ...Moreover, it is workmanship that makes the artisan, and it is not in books at all that one can learn to work by hand. In our Encyclopedia the artisan will find only some views which he would not perhaps ever have had and some observations which he would have made only after several years of work. We will offer to the studious reader, for the satisfaction of his curiosity, what he would have learned by watching an artisan operate, and to the artisan we will offer what one might hope he would learn from the philosopher in order to advance toward perfection ...

This excerpt highlights three important lessons that can be drawn from the attempts at documenting the *métiers:*

- The very act of documentation forced the artisans to articulate the importance of, or to re-examine, a particular act in the creation of an artifact (e.g. a piece of woodwork) – *midwifery of the mind.*
- Words are insufficient to describe these processes, and *pictures* are essential for conveying the nuances of the craft.
- A single image or a description of one facet of a craft is insufficient. The historian attempts to put together a *story*, using multiple images, and it is that story that triggers re-examination of the artisan's craft.

Sander photographed a wide variety of workers in Germany (1927–1945). Atget devoted himself to typologies of Paris and its environments, and the Bechers did much the same for the Ruhr as well as other parts of Europe and for North America.

Characteristic in each of these cases is *series* and *multiples*, what I shall call *urban tomography*, in which *aspectival variation* – actual and imagined – is the crucial mode of inquiry.[1] We have varied experiences of things and people, and we imagine other such experiences, possible and much less possible. Those aspectival variations allow us to investigate the meaning of those situations and objects, and in the end those variations must all fit our general idea of what the world is like.

Many individual images are iconic and works of art. But, what shall concern us here is that they are part of larger *systematic documentary* endeavors. Just as tomography provides images of multiple two-dimensional slices of an object, so allowing for three-dimensional reconstruction of the body or the earth, so these multiples (now "slices of life") allow for seeing a situation in multiple ways – albeit the images are not so readily "combined" as they are in most tomography.

I should note that the analogy is not related to the reconstruction problem in computed tomography, which starts out with a series of projections of a slice along lines (say by angle) and then computing by Fourier transforms a reconstructed image of a two-dimensional slice. But, more analogously, medical tomographers may record overlapping slices to check their reconstructions since the slices need to coincide to some extent due to the overlap.

I shall not discuss the long tradition of rendering in city planning, architecture, and urban design, which will be addressed in another chapter (see Kostof 1991, for many examples).

Documenting in-the-middle: the urban sensorium

In practice, designers document the environment from a vantage point – aerial photographs, bird's-eye views, and street-level pictures are cases in point. But, in actuality, we are *surrounded* by ordinariness, complex and multi-focal, with no assurance that what we do not attend to is unimportant. We are always immersed in the surround, just what we might appreciate in a surround-sound recording of ordinary everyday life – with no distanced vantage point-of-view. Perhaps, "all" of sensory experiences should be recorded, in focus, and distinct. Such would be a time-capsule. But, of course, this is impossible. And so we make tomograms, systematically, multiple slices that in effect give us more than any arbitrary selection might provide.[2]

A useful archive allows for *inquiry* and for *inference*. Such inquiry and inference is *phenomenological*, patching the world we already know with newly-seen aspects (and in the process of so doing, revising our notion of that world we already know).

And, naturally, we'll discover that there are aspects missing in the archive, for again past documenters can never fully anticipate future inquiries. Still, we may infer what we cannot know – we already have done so, all along. And, by doing actual documentation, we discover unanticipated topics worthy of attention (such as *ritual* and *play* as in teenagers playing *Rock Band* at home, or as William H. Whyte [1988] found in the streets of Manhattan).

Rather than a story of what has disappeared, cities are also stories of persistence and inadvertent survival, vestiges, and reconstruction, and re-facing. We can see this in the Marville's *c.* 1870 images compared to 2009 images of the same sites in Paris. Maps will need to be multilayered, as streets are renamed, reconfigured, and created and destroyed. Places may have new street-names and addresses; buildings may not move even if the streets they are "on" are transmogrified.[3]

Such documentation must be archival – so it not only survives, but also can be found a century hence. Movies and sound and images need to be in robust formats, even as digital coding schemes keep changing and improving. And, again, places need to be located not only by street, but by latitude and longitude, by date and time, and by point-of-view and compass-direction of the camera's lens or the microphone.

Google Street View might be seen as a model. Systematic survey of a city, involving something like a dozen views from each point, and powerful search and display user-friendly software, would allow one to view a place. In fact, one might review Marville's *c.* 1870 scenes of Paris with 2009 images on Street View. Of course, Google is approximately close but not often close enough to Marville's camera position, quite insufficient to make an identical point-of-view re-photograph. That would require being at the same position, pointing in the same direction, with the correct angle of view, with perhaps the same time of day and time of the year – so that occlusions, proportions, and the extent of the view were roughly the same, and shadows were similar, too. By the way, one should attend to identifying details such as fenestration to be sure one is at the right position. Also, it would help to know details of the history of Paris's streets and their reconstruction as we try to make comparisons.[4] Now, if we were to actually go to Paris, and re-photograph Marville's scenes, we could do a much better job of rephotography. We can move around, check out places that Street View did not encompass, using our bodies get a better feel for the environment – something that is much harder to do with just our eyes and our imaginations with Street View.

Exploring a city using a corpus of media

We start out with an archive or library of videos or photographs or sound recordings (we'll call them *media*), indexed by space, time, and subject tags – in effect, a spreadsheet that you can search and filter. Ideally, one has also "logged" the videos, so one has internal tags indicating content within a video at particular times. One searches the library for, say, videos of interest, uploads them to your machine, perhaps rapidly goes through them (or "scrubs" through them), and searches their internal logs, to see if they really are of interest, and so settles on a selected corpus for current study. You are now free of internet delays since you have the videos of interest on your local machine.

Of course, to "start out with an archive" presumes on how the archive was developed. Some of the materials are likely to be chance recordings, or recordings for purposes other than yours. But it might be that part of the corpus was deliberately recorded, perhaps many points of view of a single event (as in instant replays in television football), perhaps multiple videos from a single place and time period with not too much literal overlap among them except they are from roughly the same location/time interval. In surveillance videos, the cameras are usually fixed (but think of *The Bourne Ultimatum* [2007], where the cameras are smartphones carried by people who can be ordered to be in certain positions).

A template (say three by three) becomes the organizing scheme for displaying the video. Each video clip is dragged into a frame within the template which also assigns its audio track to a particular speaker that is spatially or directionally appropriate.[5] Maps are available showing the location from which each video has been made. You might demand that the videos be played so that the clock time on each is the same – which means that some videos might well begin or end before others. Hence the temporal sequence of actions is preserved, and simultaneous multi-views of action are displayed simultaneously. For still photographs, Ruscha (1966) did this many years ago for Sunset Boulevard in Los Angeles. (Note that because of sound's speed of 1000 ft/sec, views that are simultaneous in clock-time, will have sounds from the same source appearing at noticeably different clock-times.)

Examining the videos might lead to further questions. Repeated replays will allow for more careful examination. And now relevant videos can be called up from the corpus for further comparisons. Rearrangement of the videos in the template to make for easier visualization in space is again simple to do, and the sound will follow the videos.

In effect, you will have views of many aspects, in time and space, time emphasizing cause, and space given by surround views and sound. The open question is how such a level of flexibility will allow researchers, community members, policy-makers, and investigators to see more, to study more carefully, and to make convincing arguments more quickly.

I should note that we are not trying to produce a montage or merged image/sound-field, as one might do with Microsoft *PhotoSynth* or a *QuickTime Virtual Reality* presentation. And, in general, one does not have the luxury and problem of a very dense surveillance array (fixed or moving cameras) – the problem being how to compare and contrast hundreds of videos (although this is just what the technical genius – Caltech graduate that she is – does in the TV show *Criminal Minds*). We need methods that allow for lots of missing pieces, and allow for trained but ordinary users.

A phenomenology of patching the urban world

How do you use multiple slices of life to get a better sense of where you are and your situation? Now, it is not so simple as piling two-dimensional slices on top of each other to get a three-dimensional image, as in conventional tomography. It is not so algorithmic as in X-ray crystallography, where you guess what the crystal structure might be, compute its diffraction pattern and compare that pattern with the experimental one, then refine the guess based on the mismatch between the guessed and the measured patterns, and iterate and modify. But it does matter that you have a sense of how to "pile" various slices of life "onto" each other, and it does matter that you start out with a good guess. Then you will be able to get a better sense of where you are.

Knowing where you are is a matter of understanding the connection of the local facts on the ground to a structure that can accommodate those facts. Also, places are locationally tagged by specific substantive facts, so that you know where you are by the smell, the sound, the street furniture, and by your notion of how they all fit together into a whole, that notion to be modified or replaced as you learn more about such specific substantive facts.

Technically, some key notions are: patching together and a covering of the world, local-global connections; mathematically: a presheaf glued to become a sheaf, co-homology, and index theorems; phenomenologically: identity in manifolds (unity in multiplicity), imaginative draftsmanship and fulfillment, an object is a set of presentations of itself.

In our everyday lives, we always have some sense of where we are and what we are up to. We have also local nearby information: what we see, hear, smell, etc. We move around, things change around us as well, and somehow we employ (just how?) that local information to have a more adequate global sense of where we are. We use all of our sensory capacities, so that if it makes sense visually but not aurally, we wonder if we understand what is going on. If it makes sense in several sensory dimensions, we feel more confident. And, we explore our tentative notions by trying out crucial "test cases."

What we learn about a few localities may not tell us how to revise that global picture, but at some point we can "figure it out" much more adequately. We revise our global picture, a picture based on our previous global pictures – and what we have learned now fits in better.

The cognitive process never begins from scratch (see Bruner *et al.* 1956). We start out with some rough guesses or past experiences or presuppositions that a set of patches have to accommodate. We then *collate* (although this word does not say how we do this) the various aspects/slices/videos/tomograms to get a revised sense of the whole, a whole about which we already have informed notions. So that collating is actually a filling-in of detail – a fulfillment, so to speak, of a new picture. Figuring-out is not a calculative process here, rather it is a matter of what might be called *imaginative draftsmanship*, so that once you have figured out what it is you are seeing, you can now account for how it appears in all the various slices.

One of the consequences of paying attention to local information is that much of the time we do not need GPS-type information to know where we are. The local context, as we know it, tells us where we are, especially if the objects or sounds are in a well-defined area such as an airport or a neighborhood, an area we may have surveyed ahead of time. In effect, every image or sound is already tagged (not formally, but experientially) by its location and even era or time period through the substantive facts of scene and sounds.

"Figuring out" and "patching together" start out with a suspected answer, not with a *tabula rasa*, and that answer is modified, or discarded to be replaced with another potential answer. Consider learning about a place from a series of photographs and/or a set of aural recordings. Can we imagine an adequate global picture of what is going on, so that we can actually fit those images and recordings into a more adequate global picture of what is going on, more adequate than the picture we already have? We will have to keep in mind that some sources of information, such as fragrance or odor, travel in peculiar ways, and that sound diffracts around objects, and images may be reflected by mirrors. Again, it is essential to have a rough idea of what the setting is – a house or an airport – before one begins.

Places are delineated neither by latitude and longitude, nor by direction and depth, but by substantive contextual facts that allow us to see a place as coherent and meaningful. For example, artists may create places that may not fit together to explore the notions we have of space and place. But what is distinctive here is the emphasis on substantive contextual facts and those presupposed guessed solutions. For example, it helps to know that a couch will terminate with an arm (or perhaps the lack of an arm), and that tables have tops and supports. In effect, the problem of the wholeness of the world is not merely a matter of filling-in, but also of knowing the substantive facts about how the world is constituted in general, and, again, having an idea about what the answer might be.[6]

We may face similar challenges when we are trying to understand the organization of an institution, whether it be a bureaucracy, a terrorist cell, or a discipline. We have lots of local information, perhaps from many places and times, about connections and linkages. Some of that information is reliable, some of it is less so or is inferential. Yet we want to figure out how the institution is structured, and so to find a way to pull together (or, collate) all of our information, taking into account the quality of that information. Pulling-together is actually a filling-in of a presumed or guessed structure, a guess based on just some of the information. Now, a potential structure might demand that we ignore parts of our supposed information, since those parts do not fit and we suspect they might not be so reliable. That is, we may propose possible structures, sacrificing some information and inferring aspects for which we have no information at all. To start with, we might presume some general structures (one person at the top, a network with several centers, a network of networks, and so on). Investigators in various fields do this "by hand" using their best intuitions and their experience of previous structures. Phenomenologists call this process of collation or pulling-together and the revision of global pictures "unity in multiplicity," "identity in manifolds" (Sokolowski 2000).

Put differently, how do we get hold of the experienced *objective* world? We make *dense* observations, ubiquitous in space and time; systematic and random but then well indexed (co-ordinated in space, time, subject): *tomograms*. So we may present the world back to ourselves. Examining these dense observations, we start out *already* with a preliminary idea of what the world is like and a supposed meaning of what we are seeing and hearing. And, then we *fill in* details, correcting preconceptions. So we *figure out* what is going on, as an act of *imaginative draftsmanship.* We re-discover the *tissue of negligible detail* that makes up the concrete particular world. Those details and those preconceptions are aspects of purposeful activities and actions, and so what we might be discovering is a *choreography* and a dance. Later, we can *theorize the world into abstraction.*

Digital and archival

The dominance of digital devices has made the technologies of photographic film and analog magnetic tape recording less available for everyday practice. So, let us assume one is using a digital camera, a digital video device (even a cellphone), or a digital audio recording device. Many of these devices will now add metadata of date and time and mode of recording and perhaps GPS-derived spatial coordinates.

After a century of practice, we know how to make archival film and even vinyl records, but for digital records our experience is much more limited. For archival purposes, still images should be bitmaps (as in tiff), although jpg seems to be here to stay. Video should be in .mp4 or .jpeg 2000 motion video or perhaps in mini-DV format but now made into a DVD-video, and sound should be pulse-code-modulation coded, as in .wav and .aiff files. And all these files need to be archived on longer-lasting discs ("gold") as well as on hard drives.

One must index one's corpus, by date, by time, by place, by GPS coordinates, and by topic. These indexes may be done as spreadsheets, but printing them out is a must so that we do not have to worry about what Excel 2029 will look like. And, these files need to be saved alongside the digital image/video/audio files.

Practical advice

Urban designers should document all round in the environments they are studying, not just pointing toward a particular place of interest. They might document from the point of view of that place looking outward, and of course record the surrounds, looking out, looking in, and looking sideways. Some of David Hockney's work is archetypal here, individual pictures made of hundreds of snapshots of parts of a scene, much as we have maps of the surface of Mars or of the Moon, but now the overlaps are rather more provisional.

The goal should be a way of interacting with the corpus of individual images, so being encouraged to explore a situation close-up, then from a more distant perspective, and then behind obstructions as well. In some cases stitched images might work well, but often there are multiple points of view that may encourage one to be an analytic cubist as in Duchamp's *Nude Descending a Staircase* (1912). Temporal comparisons may involve side-by-side presentation or even overlays with variable transparency. Only a well-indexed corpus can be accessed by any such system, and so documentation must include indexing, the latter often taking as much time as the fieldwork itself.

Examples

I have photographed more than 850 storefront houses of worship in Los Angeles, all (150+) sites of the City of Los Angeles Department of Water and Power electrical stations, and people at work at more than 225 sites in Los Angeles. For the designer, what was a part of the otherwise ignorable street furniture of a city now might become thematic and meaningful. So corner shopping malls, often populated by at least one storefront church, are ubiquitous, meaningful, and vital parts of an urban design.

We have developed a system of using large numbers of smartphones, ones that have good video, good communication (3G, WiFi), and GPS, into an urban tomography system (Figure 18.1). One can send out a smartphone-equipped crew to a complex urban situation (a market, an event, a busy street), and in short order get a many-faceted video document about that place, and its rituals and experiences (in effect, crowdsourcing). The big problem, indicated above, is how to manage and employ such a rich corpus of media.

We have also developed a system for accurate aural documentation of urban places in surround sound, using a quite portable surround-sound microphone and recording system, creating Dolby Digital 5.1 records that may be played on any a home-theater system that can play DVD movies.

And, as indicated above, we have re-photographed many of Marville's *c.* 1870 photographs of Paris (an enormous systematic corpus, unique for any city until the twenty-first century), studying how the urban built environment persists and is revised (Figure 18.2). Moreover, we have employed Google Street View to get the best armchair re-view of those same images (http://www.usc.edu/sppd/parismarville will link to the map).

One of the crucial lessons we have learned from these various documentation efforts is that *one has to go to the site, one has to see it and experience it in actual time and place, in person.* Only then can one have a sense of what is important, a sense of what one does not want to leave out, a sense of what one has not even imagined beforehand. You need to be there. You are then

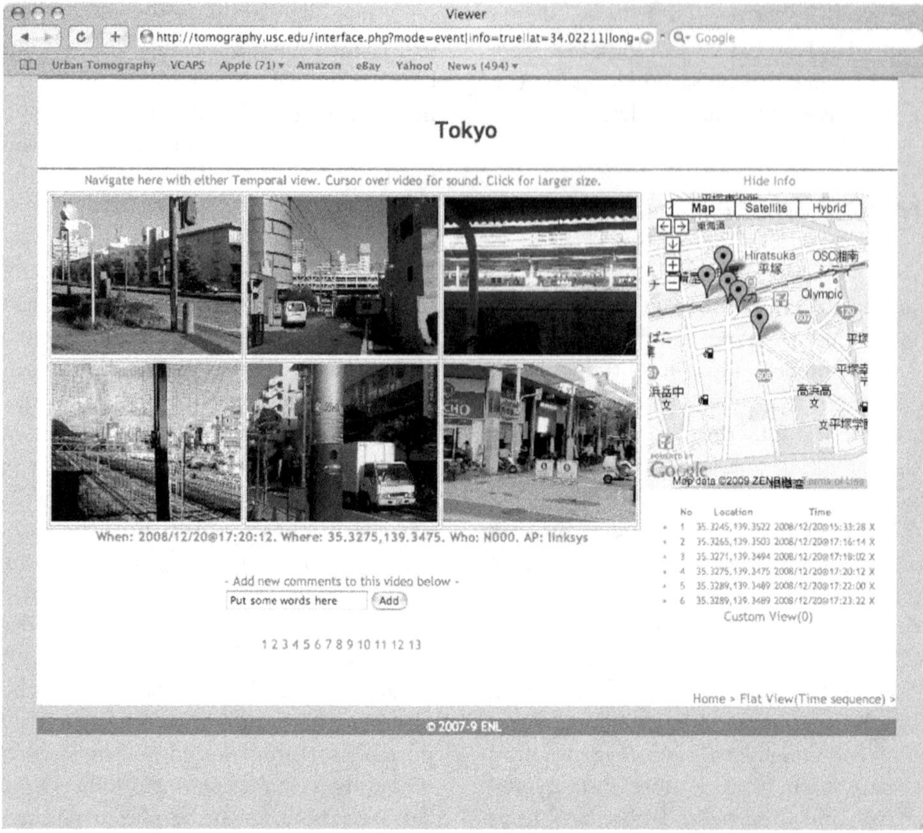

Figure 18.1 Multiple views of Hiratsuka, a suburb of Tokyo. Source: Kazuma Kazeyama and www.maps.google.com.
Note: Multiple views using the Urban Tomography smartphone system.

more reliable in imagining that place or that situation in the past or how it might be altered for the future. Even accurate aural recordings are not the same as being there, for visual and other sensory information affects what we hear and appreciate.

The urban sensorium

Actual places are not only seen, but are experienced as dynamically-changing aural, haptic, olfactory, and gustatory sites. We now can do a good job of documenting and reproducing the aural, but not the other three (except verbally, as in novels and poetry).

Surround-sound recordings and presentation can be highly accurate and verisimilitudinous – if that is the aim of the recording engineer, although for most cinema and music purposes, accuracy is not a primary concern. In cinema, ambience is created rather than accurately presented. Background and competing sounds are managed and overlaid. Yet, as indicated earlier, it is now possible to make surround-sound recordings that are true to the original experience, however "too real" that recording might sound to someone who is listening to her

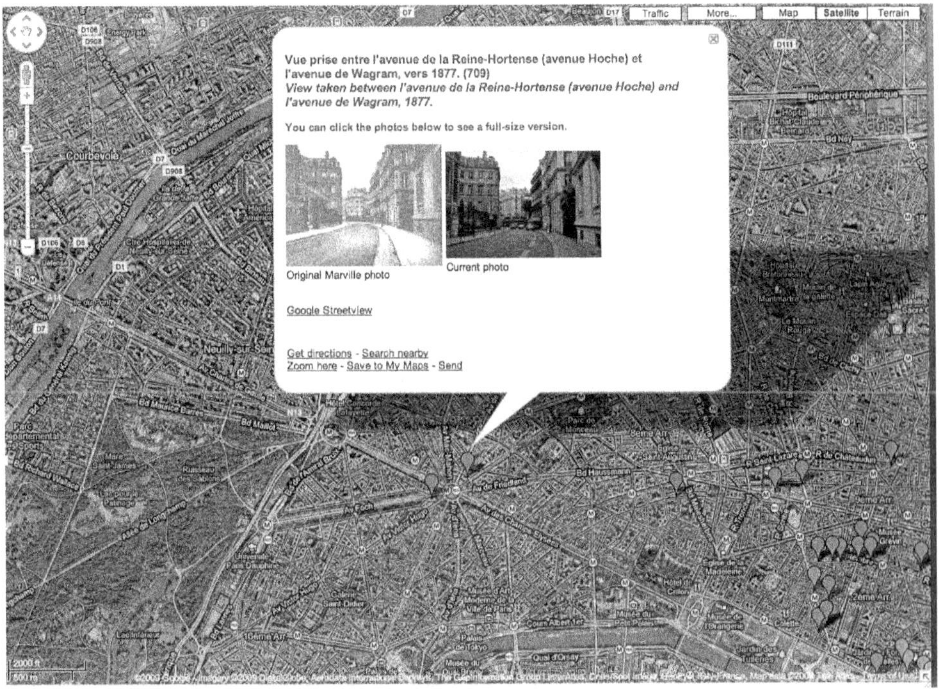

Figure 18.2 Paris Marville Google Map. Source: Kazuma Kazejama and www.maps.google.com.
Note: Notice the link to Street View.

ambience with some care. In actual experience, we tend to filter out lots of "noise" and competing sounds. But a recording won't be accurate unless it includes all those apparently-interfering seemingly-less-important sounds.

Similarly, we actually walk through a site or experience a place dynamically. A built model or a computer simulation is convincing when we can walk through it, or fly over it. It is now possible to use a computer animation program such as Google *SnatchUp* or *Maya* to make a simple model of a street, even decorating the buildings with correct facades, and then walk through or fly through the setting. In fact, in many urban places, you can start with Google *Maps* to get an outline of the buildings, Street View to get the facades, and then use simple cubes in *Maya* to model the street. While *Maya* itself has a steep learning curve when it is used for animation, it is comparatively easy to use it to make a flythrough or walkthrough movie of a setting. SketchUp is even easier.

Storytelling and experience

Actual experience of life is conveyed by telling a story. A picture is only worth a thousand words when those words are expressed in an involving way. The collection of media representations of an urban design setting becomes useful when it is built into a story. So, hundreds of images of storefront churches might become a story of ethnic religiosity in a modern age. And given the contentious nature of urban development, a corpus of media will be built into many stories, stories that are intentionally selective in their use of the corpus of images and movies and sounds. So, when we do documentation employing media tools, we are in effect creating a variety of potential stories. And even if we are biased in the stories we have in mind, the media documents allow for counter-stories largely because pictures, movies and sound tracks include much more than we might anticipate or control (even when they are artificially produced, as in Hollywood motion pictures).

Again, the key words are *storytelling* and *experience*. Urban tomography allows for the richness of documentation that enables telling more complex stories that encompass wider ranges of experience. Media documents here are not meant to be singular, aura-filled works of art. Rather they are presentations of the world we make to ourselves, presentations that allow us to imagine how it might be different, in many different ways.

Notes

1 Tufte (1990) speaks of "small multiples" in much the same sense.
2 And we may review those slices to develop a richer sense of the whole. But, no matter what, not everything will be documented. We will miss, in effect, anticipating future questions or at least potential areas of interest.
3 See, for example, Zola's *La Curée* (The Kill) 2008.
4 Pitt (2008) provides wonderful details.
5 *QuickTime*, in particular, now allows for multiple audio tracks keyed to different speakers, and inexpensive computer audio systems allow readily for at least seven or eight such tracks.
6 By the way, all of these approaches are operative in artificial intelligence research and arts.

References

Bruner, J.S., Goodknow, J.J., and Austin, G.A. (1956) *A Study Of Thinking*, New York: Wiley.

d'Alembert, J.L.R. (1995) *Preliminary Discourse to the Encyclopedia of Diderot*; trans. R.N. Schwab, Chicago: University of Chicago Press.

Daley, E. (2003) "Multimedia Literacy," *Educause*, (March-April): 33–40.

Diderot, D. (1993) *A Pictorial Encyclopedia of Trades and Industry*, New York: Dover.

Kostof, S. (1991) *The City Shaped*, New York: Thames and Hudson.

Pitt, L. (2008) *Paris, Un voyage dans le temps*, Paris: Parigramme.

Ruscha, E. (1966) *Every Building on the Sunset Strip*, Los Angeles: Edward Ruscha.

Sokolowski, R. (2000) *Introduction to Phenomenology*, Cambridge: Cambridge University Press.

Tufte, E. (1990) *Envisioning Information*, Cheshire, CT: Graphics Press.

Whyte, W.H. (1988) *City: Rediscovering the Center*, New York: Doubleday.

Zola, É. (2008) *The Kill (La Curée)*; trans. Brian Nelson, New York: Oxford University Press.

Further reading

Daley, E. (2003) "Multimedia Literacy," *Educause*, (March–April): 33–40. A concise description of multimedia literacy.

Hales (2006) *Silver Cities: Photographing American Urbanization, 1839–1939*, Albuquerque: University of New Mexico Press. A fine work on the history of how photography and urbanism worked together.

Krieger, M.H. (2011) *Urban Tomographies*. University of Pennsylvania Press. Also a series of articles with collaborators, which appeared in the *Journal of Planning Education and Research*: 24(2004): 213–215; 27(2007): 228–239.

Sampson and Roudenbush (1999) "Systematic Social Observation of Public Spaces: A New Look at Disorder in Urban Neighborhoods," *American Journal of Sociology*, 105: 603–651. Shows just what systematic photographic surveys can do for social scientific investigations.

Sokolowski, R. (2000) *Introduction to Phenomenology*, Cambridge: Cambridge University Press. A nice introduction to phenomenology,

Tufte, E. (1990) *Envisioning Information*, Cheshire, CT: Graphics Press. On information graphics.

Whyte, W.H. (1988) *City: Rediscovering the Center*, New York: Doubleday. A pioneering work in urban documentation.

19

Visualizing change

Simulation as a decision making tool

Peter Bosselmann

Urban designers reason with change. Thus an anthology on urban design includes reflections on how professionals use representations to make a selection from the richness and complexity of cities to simulate proposed changes. Like in many other professions, they use simulations to show the eventual effects of alternative conditions and courses of action. Throughout history and across disciplines, simulations have been used to forecast conditions that might become reality; that is, if present assumptions about the future continue to hold true. The applications of simulations are broad and have grown in engineering, design and planning as well as navigational training, medicine and education. Fundamentally, two types of simulations are possible: existing and future urban conditions can be explained as concepts or as experiences (McKechnie 1977). Conceptual simulations convey abstract forms of information. Perceptual simulations convey an experience (see Figure 19.1).

Leonardo da Vinci's map of Imola from 1502 was an early, probably the first, accurate example of a conceptual representation, a map made to aid Cesare Borgia in the conquest and subsequent repair of the town's fortifications (Pinto 1976). At the time of the Italian Renaissance, no one could have seen a town like Imola as a true orthographic projection. For Borgia, the map was an image of a town in the palm of his hand that held great strategic promise. On the opposite end of the concept-percept continuum, Filippo Brunelleschi's 1415 perspective of the Baptistery San Giovanni in Florence is believed to be the first perceptual representation that accurately simulates an experience of the three dimensional world depicted on a two dimensional surface. Brunelleschi had painted the perspective on a wooden board, drilled a hole into the board at the center of the perspective and asked viewers to step up to the doorway of the cathedral Santa Maria del Fiori, where he had painted the view. When asking the viewer to look through the hole in the board with the painted side pointing away from the viewer, the viewer – with one eye closed – would see the baptistery in reality. Brunelleschi would raise a mirror to intersect the line of sight and the viewer would see the reflection of the painting in the mirror, just as a modern day viewer would observe a scene in the viewfinder of a mirror-reflex camera. Upon lowering the mirror, the viewer would again see the baptistery in reality (White 1976).

The two historic examples demonstrate simulation's close ties to technology, but

Figure 19.1 Conceptional Representation of City Form, Venice Biennale. Source: Peter Bosselmann.
Note: From the 2006 Venice Biennale, comparing the densities of the world's largest cities.

also to the need to verify the information that is depicted. For Brunelleschi it was the invention of large format silver-plated glass that made possible his representation of human experience. For Da Vinci, the transfer of the magnetic compass to the western world from China made possible what we would call today "geo-referencing" Imola's location on the surface of the earth, and the placement of all elements of the town onto a spatial grid, a polar grid in his case.

Reality, existing or future, cannot be represented in its entirety. Its representation can only involve selective aspects. What is selected from reality, and what is left out, can significantly influence the outcome of simulations. Since simulations remain abstractions of reality, does the simulated world behave in the same manner in the real world as it appears to behave in simulations? The answer to this question is important for urban designers who use simulations to explore the implications of policy and decision-making on the form of cities.

If response equivalence between simulated and real world experiences cannot be guaranteed, simulations would have no credibility, could be misleading and should not be used in decision making processes. For both forms of simulation, conceptual and perceptual, validations remain a necessity. Advances in technology have not changed the need for veridicality, unless we are interested in simulating deceptions, to persuade, to advertise or to create fiction. Widely used to manipulate audiences' attitudes, simulations can influence audiences to adopt a favorable view – the view of the simulators or their clients; simulations can heighten human experience above and beyond the experience of the everyday world. Designers, planners, their clients and politicians are not immune to the use of simulation as a tool to deceive and to manipulate. Arguably the latter is simulation's chief purpose.

Discussing simulations offers a rich array of subjects. A chapter in the context of this book has to narrow its scope. Thus the chapter concentrates on simulation as a modeling activity. We are interested in simulations that allow urban designers and others to gain knowledge about the elements of urban structures, how elements of the structure perform, and how compatibly they fit within existing physical, social and economic conditions. We focus chiefly on perceptual simulations, those that convey future experiences, because this genre has developed rapidly through computer modeling and digital image processing, visualization, and animation applications. Given the increased accessibility and the frequently persuasive application of perceptual simulations, readers, who might be skeptical about the intent of perceptual simulation, would need answers to the question whether it is possible to produce simulations that have the same documentary quality as the abstract plan, section and elevation diagram, which are customarily used by the design profession.

The abstract diagrams designers use to generate form are not well understood outside the profession. Even inside the profession, the diagrams and their underlying concepts, such as allowable floor area ratios, land-coverage, density, as in units or people per acre, rate of absorption, and mix of uses are all terms in need of interpretation that first need to be modeled in order to be understood. Simulations turn such abstract concepts and transform them towards the realm of the concrete; images that a person can look at, imagine what it might be like to be next to, move through, or look out from. Through simulations, urban form and associated conditions become more understandable. Because models allow for greater clarity, models and simulations are useful for explaining urban conditions to those who may not otherwise understand the implications of decision-making, such as politicians, community representatives, and the news media – in other words, the public at large (Appleyard 1977). Simulations alone cannot claim to deliver judgment about good performance, fit or compatibility, for the evaluators will make such judgments. But simulations make possible an open, public discussion among evaluators about the nature of change, its perceived degree of faithfulness to a recognized tradition or a conscious break with tradition – toward new beginnings.

Modeling future experiences

All references cited thus far, including those that reflect on historic developments, originated in the 1970s. The decade saw a curiosity about the origins of graphic conventions. This trend happened at a time when computational techniques started to emerge that made possible representation of the sensory world in digital form. First computerized sound, and shortly thereafter digital imagery, became commonplace (Mitchell 1992). These technological advances coincided with a growing awareness of what we call today urban ecology and sustainability of cities and landscapes. The two trends seem unrelated, but can quickly be brought together when coupled with the related concern for improved methods for visualizing change and with measuring the impacts of change on the environment.

However, independent of technological advances, by the 1970s, the design profession was also forced to realize a lack of rigor (Jencks 1977 and 1988) in representing proposed changes in cities, especially when the proposed changes were controversial. Today's reader can only cringe to learn that plans to route a four lane highway through Washington Square in Manhattan were not only very real in the minds of people like Robert Moses, but were also proposed without consultation of those whose lives would be affected. Or, alternatively, today's reader can feel much gratitude towards activists like Jane Jacobs who clashed with Moses over the prospect of pulling down 14 blocks in the heart of Greenwich Village. This type of controversy, described by Anthony Flint (2009), and played out in many cities during the post World War II decades, gave the impetus to set up laboratories dedicated to improved public communication of large scale engineering, design and planning projects. Indeed, the then newly enacted environmental impact reporting procedures mandated full disclosure of large-scale design and engineering projects, thus forcing professionals to reflect on the veracity of their predictions and the graphic methods used to illustrate the outcome of their designs (Smardon *et al.* 1986, Bosselmann 1993).

In this context, matters of human perceptions and cognition, the knowledge domain of psychology, became relevant to the design professions (Craik and Feimer 1988). Among the skills psychologists had to offer was their ability to measure human responses to the environment. Members of the public place greater scrutiny on simulations, as do planning commissioners, entitlement lawyers, and community activists when professional media are used for presenting evidence in public discourse and decision making.

The central question was: "Will people reliably react to simulated scenes in much the same manner as they would react to an experience of the real world?" Answers to this question involved a large-scale validation project sponsored by the National Science Foundation and carried out at the University of California's Environmental Simulation Laboratory starting in 1972 (Bosselmann and Craik 1987). Residents and nonresidents were randomly selected to tour a suburban environment complete with shopping centers and office parks, followed by the screening of a virtual drive through the same area. In addition to a battery of questionnaire surveys, some subjects saw the virtual tour and not the real world tour, and some saw both in the sequence described or vice versa. The experiment concluded that simulations can be surrogates of a real world experience. This meant that ideally the simulations should not be presented in static form, but as dynamic animations, produced in a manner that comes close to human experience, moving through space and time.

The experiment also acknowledged that subjects who were unfamiliar with the setting reported close to equivalent experiences after the real world tour, after watching a movie made of the same tour and after watching a tour of a virtual, simulated world, or vice versa. But for subjects familiar with the simulated world, the equivalence of the two experiences was not as strong. For them the real world setting had social meaning that could not readily be simulated. Thus the validation experiment touched upon findings about *sense of place.* Much at the same time, theories about *place* had emerged, first in geography (Tuan 1977), and somewhat later in the field of psychology (Sime 1986) claiming that places in cities, neighborhoods and landscapes take on meaning, subject to social dynamics, familiarity, memory, attachment and dependencies.

The validation project also confirmed a number of earlier theories, first J.J. Gibson's ecological theory of perception that explains the evolution of human vision (Gibson 1979). He was the first to remind us that perception is a dynamic process, which operates under constantly changing conditions and frequently in motion over time. Therefore decisions made based upon direct observation of the real world will differ from those made after viewing visual media that represent conditions frozen in time and in a static state such as plans and single images.

A second theory of perception confirmed by the validation project was Egon Brunswick's probabilistic theory (Brunswick 1956): The observer builds up a repertoire of probabilities that provides likely conclusions by combining trustworthy clues to give an educated guess about the true nature of a situation or place. The probabilistic theory likens the process of perception to an optical lens with the environment on one side and the observer on the other. The observer becomes active in recombining the visual clues, in focusing and testing the validity of what is seen through the lens. The validity of the observed information is strengthened when the observer has access to accuracy tests that verify what is observed through independent means.

The politics of simulations

Admittedly, for the everyday user of simulations, perceptual theories of simulations would be of limited use, if it were not for the fact that simulations are produced in a highly politicized milieu. Change in cities will always be associated with controversy. Especially when large projects are considered, proponents and opponents compete for public attention, appeal to decision makers and will treat information about change selectively, emphasizing its benefits or detriments. Only narratives and imagery are made public that portray a proposal at its very best, or worst, depending on who is preparing the case. For an outsider, the credibility gap appears obvious and the difference between the real and the imagined can at times be comical, but for the actors involved the matter is deadly serious, because much can be at stake. Therefore, anybody interested in reducing the credibility gap for the benefit of a more open debate would call for simulations that are more veridical. Simulations should be representative of the changes that a new project will impose on existing and future conditions. If possible, simulations should consider cumulative changes, without exaggerating or diminishing the impacts of change. And the 3D models used to produce simulations should be open to accuracy tests.

Realistically, such work could not be expected from proponents or from opponents, but could only be performed by individuals outside the controversy. Therefore, early simulation laboratories emerged at research universities. The laboratory at Berkeley was not the first of its kind, that credit goes to the University of Lund in Sweden (Acking and Küller 1973). The early simulation laboratories, which also included facilities at universities in the Netherlands and in Germany, were started by professionals in academia with the goal to improve citizen participation in planning and design.

The participatory theme was still dominant in the establishment of a more recent group of laboratories first in Tokyo one at Waseda University, another at Keio University. Modeled after the Berkeley laboratory, the objective for these two laboratories was to develop participatory techniques that would allow citizens' groups to understand the consequences of Japan's unified urban planning law on the development dynamics in their community. As a result of simulation studies, citizens' groups have successfully argued for exemption from the unified urban planning law and in favor of what in Japan is called "Detailed District Planning Law," a set of rules more closely tied to existing social and economic activities (Satoh 2007). An earlier laboratory was started at the New School in New York City in the late 1980s to simulate large-scale development projects in New York City (Kwartler and Longo 2008).

The most recent laboratory was established in Milan, Italy with the purpose of examining a scale of urban development projects that is still relatively new in the largely horizontal European Cities. In cities like Milan, Copenhagen, Rotterdam, St. Petersburg and elsewhere, the existing urban block structure is challenged by the insertion of large, frequently very high buildings that defy integration into the existing city fabric because they are frequently gated or enclosed by walls to create a controlled environment for administrative or commercial activities, like the *Garibaldi Republica* projects in Milan or the proposed *Gazprom* tower in St. Petersburg (Bosselmann 2008).

The early laboratories used physical models to test building and urban design proposals. In the mid-1980s, when computer-modeling applications became available, a technology transfer took place. Most three-dimensional urban modeling is done with Geographic Information Science (GIS) or Computer Aided Design (CAD) applications, but physical models are still used for detailed discussions and presentations to the public (Bosselmann 2007).

Computer-based modeling is not only available at selected university locations, but with only minor capital investments, it has become widely accessible. A form of consultancy has emerged that furnishes developers and their architects with their own simulation studies. In San Francisco, as in most major cities, a sizeable industry of lawyers, designers and technical support staff has grown around this activity. Their prime occupation is in assisting developers to successfully navigate through the project approval process. Equipped with their own digital models, developers routinely try to persuade decision-makers to increase entitlements. Not that developers and their technical support staff would openly lie; they simply distort the truth by presenting information selectively, showing the proposed project from only the most opportune angle, or leaving out important aspects of its context. Also, a developer has no interest in showing the effects of cumulative change. That is, to show what would happen if neighboring properties receive similar increases in entitlement. Such images and narratives could be a major distraction from a single developer's proposal and might easily lead to a negative decision for the developer.

The current era of planning deregulation should not be interpreted as having brought to an end all balanced public communication about change in cities. Simulations will always play a role in urban transformations. Change in cities is the process of becoming different. In an age of rapid urbanization in some parts of the world, shrinking or dispersed cities in others, few urban research topics could be more important than to explore the meaning of change.

Simulating magnitude, rate and the nature of change

Three dimensions of change are important and they can be simulated: magnitude, rate and nature of change. The last is the most interesting and I will save it for the conclusion. The first two appear obvious and can be measured with relative ease.

When talking about the magnitude of change in urban design we refer either to the diminishing, yet more frequently, increasing size of urban form. Simulation here is used to show the implications of decision-making. Setting allowable building heights, quantities such as density, allowable floor area to land ratios, and setbacks from public rights of way, are subject to public approval processes frequently expressed as part of detailed area plans. All can be simulated, both as abstractions and at a level that comes close to human experience.

For more than half a century, the vantage point from Treasure Island (see Figure 19.2 and Figure 19.3) has been used to test proposed changes to allowable building heights in San Francisco and the compliance of such changes with general plan policy. The policy calls for a downtown skyline configuration that resembles the shape of a hill, a constructed hill, and compatible with the hills of San Francisco's natural topography. The policy was widely discussed and voted on prior to the adoption of the General Plan's urban form element (City of San Francisco 1974). Under this general policy, the actual building height dimensions matter less; more important is the contribution each set of proposed buildings will make to the shape of the "downtown hill."

Looking at the second to last frame (Figure 19.2e), the tower on the extreme left side of the frame is clearly not in compliance with the general plan policy. Constructed in 2006, the tower near the Bay Bridge sets a negative precedent, a violation of an

(a) The skyline in 2005

(b) Simulated view of buildings possible under the 2004 Transbay Terminal Area Plan

(c) Simulated view of approved buildings on a site next to the Transbay Planning Area

Figure 19.2 Magnitude of Change – Rate of Change, San Francisco abandons its downtown hill configuration. Source: Peter Bosselmann.

important policy. The proposed tall tower in the center of the last frame, not approved as of 2009, also illustrates a magnitude of change that stands out. The sequence of frames illustrates that the magnitude of proposed changes in downtown San Francisco has challenged city government to re-examine an established policy. Simulations are useful in this context. As pointed out earlier, simulations cannot, in and by

(d) Buildings possible under the 2005 Rincon Hill Area Plan

(e) A rendered version of all potential buildings

(f) The skyline in 2009

Figure 19.2 Continued.

themselves, provide judgments, but they open up the discussion to those who would otherwise not contribute on the implication of the decisions that need to be made.

The rate of change describes the dynamics of change over time. The same sequence of images in Figure 19.2 can be used to illustrate the pace, or rate of change. In downtown San Francisco, the change

Figure 19.3 San Francisco skyline with proposed Transit Tower. Source: Peter Bosselmann.

simulated was expected to happen gradually, but cumulatively. However, the images brought to mind questions a fundamental assumption: Will current trends continue? Will San Francisco's economy continue to absorb as much floor space as shown? Will the pace of development slow down or stop altogether? As time passed, the parties involved were forced to re-examine their assumptions about the demand for space and the financing necessary to build it. Without simulations, few would have asked such questions when these projects were discussed. Even after viewing the simulations not enough people did ask; but many people have asked since.

Animating the change to the San Francisco skyline over time is a useful tool in understanding the pace of change. Some viewers may side with the proponents of change, watching with civic pride; others side with opponents, startled over the city's proposed rate of development. The animation provides clarity about the collective history and future of San Francisco and what might be at stake when thinking about the essential structure of their city.

A city can be understood as a product of history, as traces of the past are inescapably ingrained in the dynamics of urban form. Simulations can lead to a discovery of a city's essential structures. Such structures include elements of city form that have mutated through time, but constantly adapted to change, thus remaining viable in their contribution to the fabric of a city. Many examples of urban patterns come to mind, where simulation can be used to explain morphological processes to inform design principles that support the essential structure of a city. The San Francisco skyline with its building contributing to a constructed hill is a result of such a morphological process. The block structure of the city is another; perimeter blocks with buildings facing streets that sometimes defy topography. In Manhattan, it is the ridgeline that travels mainly above Sixth and

Seventh Avenue down the center of the island forming peaks and saddles. In Tokyo, the upper plain and the waterlogged lower plane form the essential structure of the city. Milan's essential structure is made of blocks, open in the center and located along tree-lined *parterres.* The center of London is in the process of re-inventing its essential structure, where Saint Paul's dome is no longer used as the central reference point. Shanghai appears to be further along in having an array of peculiar shapes for its essential urban structure. The words used here are simple sketches, intended to evoke images. The actual structures carry more meaning, because every city has such spatial structures that define urban form; in some cities they are stronger than in others. Simulation can be a tool that communicates the shared appreciation of city form, a tool that can measure how change will contribute or alter form in terms of magnitude and pace.

Finally, discussions about the *nature of change* are tied less to quantities like magnitude or rate of change, but more to values. What good is change in cities, if it is not for greater livability, vitality and a greater *sense of place* for most? *Nature of change* reflects on how change influences the human experience of cities, including their sense of beauty. Generally, good experiences come to mind, but the bad cannot be ignored. Simulating the nature of change is best done in a communal setting with models and images on display that trigger a discussion among participants who share the use of a district, a neighborhood or street; again Shigeru Satoh's work at the Waseda Simulation Laboratory in Tokyo is relevant here: The large department stores of Tokyo's Ginza district no longer attracts the same number of customers as in the past, and the owners of the stores have contemplated reusing their land to build hotel towers. Japan's unified planning law would even encourage such transformations by compensating the owners of adjacent smaller parcels that would need to be absorbed into the new developments. The property owners of the smaller parcels organized together with the smaller merchants and restaurants owners and with the help of simulations have successfully developed a detailed Ginza based planning code that allows for transformations, but does not jeopardize the existence of the supporting commercial activities and their property configuration (Satoh 2007).

To design places that bring about attachment, dependency, and identity clearly goes beyond the setting of dimensions. This is where simulations can play an important role. A group of people can gather around a model and discuss the future of a neighborhood or district. Designers participating in such a discussion might be tempted to think, first of all and quite literally, about the dimensions of a place. This is understandable because designers create spatial geometries and they define proximities and place objects in space. It is only natural for designers to believe that decisions about the correct spatial dimensions influence how people act in space, both functionally as well as emotionally. But residents and people with a vested interest will predictably add additional meaning to such a discussion. Tony Hiss describes a bond that exists between a person and a particular setting (Hiss 1990). This means that an individual has made an emotional investment in a place. Clearly, such a bond is associated with a person's life cycle. Also, place attachment, dependency, and identity depend not only on one particular experience, but also on an ongoing relationship with a physical setting that in most cases is shared with other people. In this context gender, race, and income are important, as are exposure, familiarity, choice, and cultural norms. That said, the disciplines of psychology and geography have produced little empirical research that suggests what physical characteristics are likely to contribute to *sense of place*. Simulations, on the other hand, will focus the discussion away from the general to the specifics of places. Simulations are produced with the fundamental premise that it is possible to take parts of a city into a laboratory in order to experiment with its elements. Such experiments make an important contribution to the political discourse about change to city form.

In this essay I have argued that it is possible for those who produce simulation to act both as agents for change and stay committed to the *sense of place* because urban designers have a special skill to communicate abstract concepts so others can imagine what life in the contemporary city could be like.

References

Acking, C.A. and Küller, R. (1973) "Presentation and judgment of planned environments and the hypothesis of arousal," in W.F.E. Preiser (ed.) *Environmental Design Research Vol. 1.*, Stroudsburg, PA: Dowden and Hutchinson: 72–83.

Appleyard, D. (1977) "Understanding Media: Issues, theory and a research agenda." In Altman, I. and Wohlwill, J.F. (Eds.) *Human Behavior and Environment*, Vol. 2 New York: New Plenum.

Bosselmann, P. (1993) "Dynamic Simulations of Urban Environments." In Marans, R.W. and Stokols, D. (Eds.) *Environmental Simulations, Research and Policy Issues*, New York: Plenum Press.

—— (2007) "The Nature of Change," *Teritorio* 43.

—— (2008) *Urban Transformation – Understanding City Design and Form*, Washington, DC: Island Press.

Bosselmann, P. and Craik, K. (1987) "Perceptual Simulations of Environments." In Bechtel, R.B., Marans, R.W., and Michelson, W. (Eds.) *Methods in Environmental Behavior Research*, New York: van Nostrand Reinhold.

Brunswick, E. (1956) *Perception and the Representative Design of Psychological Experiments*, Berkeley, CA: University of California Press.

City of San Francisco, Department of Planning, (1971) "The Urban Design Plan", adopted as an element of the City's General Plan on 26 August 1971. San Francisco, CA.

Craik, K. and Feimer, N.R. (1988) "Environmental Assessment." In Stokols, D. and Altmann, I. (Eds.) *Handbook of Environmental Psychology*, New York: John Wiley & Sons Inc.

Flint, A. (2009) *Wrestling with Moses*, New York: Random House.

Gibson, J. J. (1979) *An Ecological Approach to Visual Perception*. Boston, MA: Houghton Mifflin.

Hiss, T. (1990) *The Experience of Place*. New York: Alfred A. Knopf.

Jencks, C. (1977) *The Language of Postmodern Architecture*. New York: Rizzoli.

—— (1988) *Architecture Today*. New York: H.N. Abrams.

Kwartler, M. and Longo, G. (2008) *Visioning and Visualization: People, Pixels and Plans*, Cambridge, MA: Lincoln Institute of Land Policy.

McKechnie, G.E. (1977) "Simulation Techniques in Environmental Psychology." In Stokols, D. (Ed.). *Perspectives on Environment and Behavior: Theory, Research, and Applications*, New York: Plenum Press, 169–189.

Mitchell, W. (1992) *The Reconfigured Eye, Visual Truth in the Post Photographic era*, Cambridge, MA: MIT Press.

Pinto, J. (1976) "Origins and Development of the Ichnographic City Plan," *Journal of the Society of Architecture Historians*, 35(1): 35–40.

Satoh, S. (2007) "Creating Community Through Machidukuri with The Help of Visual Simulation," *Teritorio*, 43: 24–26.

Sime, J.D. (1986) "Creating Places or Designing Spaces?" *Journal of Environmental Psychology*, 6: 49–63.

Smardon, R., Palmer, J. and Fellmann, J. (Eds.) (1986) *Foundation for Visual Project Analysis*, New York: John Wiley and Sons Inc.

Tuan, Y.F. (1977) *Space and Place: The Perspective of Experience*, Minneapolis, MN: University of Minnesota Press.

White, J. (1976) *The Birth and Rebirth of Pictorial Space*, Boston, MA: Boston Book and Art Shop.

Further reading

Bosselman, P. (1997) *Representation of Places, Reality and Realism in City Design*, Berkeley, CA: University of California Press. On Simulation.

Bishop, I. and Lange, E. (Eds.) (2005) *Visualization in Landscape and Environmental Planning*, New York: Taylor and Francis. On overviews of simulation and visualization.

Cullen, G. (1959) *Townscape*, London: Architectural Press. One technique that has proven powerful and is still used in dynamic simulations and visualizations is *serial vision*, first documented by Gordon Cullen.

Kemp, M. (1990) *The Science of Art: Optical Themes in Western Art from Brunelleschi to Seurat*, New Haven, CT: Yale University Press. On historic references to the emerging graphic conventions during the Italian renaissance.

20

City design in the age of digital ubiquity

Eran Ben-Joseph

> The most profound technologies are those that disappear. They weave themselves into the fabric of everyday life until they are indistinguishable from it.
>
> (Weiser 1991: 94)

As the digital revolution advances, the fundamental nature of urban design will also change. Digital technology first made its impact on the workplace in the 1980s with the advent of the personal computer, which vastly increased productivity and facilitated production of goods and services. In the 1990s the spread of the Internet changed the nature of work, consumption, communication and entertainment. Now, with the diffusion of wireless communications and pervasive computing, digital technology is moving into the fluid realm between the home, the workplace and the public social sphere where urban life occurs.

By 2008 more than half of the world's population owned a mobile phone, (4.1 billion), up from one billion in 2002, with developing nations one of the fastest growing markets. Internet use has doubled since 2002; now almost a quarter of the world's population is connected to the web compared to 11 percent seven years ago. Hundreds of cities are in the process of providing ubiquitous public wireless access for their citizens (ICT Development Index International 2009). This infrastructure of data and digital models describing our world can be metaphorically described as an urban nervous system. In the future, flows of real-time data will enable us to be sensed as a living system and to respond to changing conditions, yielding tremendous efficiencies and a higher quality of life.

Digital information is both transforming urban life and creating new possibilities to understand and support city planning and design. Sophisticated models allow the simulation and computation of the most varied city and landscape parameters, from traffic conditions to wind velocity and air quality. Designers and planners now have a range of powerful tools, whether developed for general use or specifically for planning purposes, to help with designing and visualizing the implications of decisions, and also communicating the logic behind decisions to others.

Understanding and analyzing urban functions in real-time has already impacted and benefited urban management and design. In the area of transportation planning and traffic management, digital technology is used to track the movement of vehicles in real time which leads to changing signage and lane markings to maximize efficient use of the road network, and to implement congestion pricing as was implemented in central London. Cities such as San Francisco are experimenting with the use of wireless sensor networking (WSN) to streamline parking congestion by assigning empty spaces to incoming vehicles thus eliminating searches (Economist 2008). Similar benefits across a range of urban functions, such as smart electric grids and infrastructure systems, can increase efficiency and reduce the cost of physical construction for upgrading these urban systems.

The existence of ever-present communications and information technology also provides many possibilities for the public to organize themselves both as a market and political force to effect change. This can be seen in communities where digital storytelling and web-networks have become a means to reassert local needs (West Philadelphia Landscape Project 2009). In urban planning, the exigency of including public opinions as well as technical analyses in a time of rapid urban change is challenging our ideals of deliberative planning process. New media and digital interfaces may provide an answer to this dilemma. Through sophisticated visual models of environmental, transportation, and other proposed features, the design impacts of alternatives are made accessible to the layperson in real time science. This technology provides an alternative to the typical disjointed and often removed planning and design mode by allowing a wide constituency to participate instantaneously in the planning process. The need remains, however, for the designer to use his creativity to anticipate forms of urban space shaped by a participatory process in a pluralistic setting.

Urban design and digital interfaces

Both the physical and social implications of the mediated city provide opportunities to invent and deploy new ubiquitous digital tools. These are characterized by new forms of interactions and delivery systems that seamlessly interact with one another in a multiplicity of ways. Some of the most intriguing technological developments are occurring in the realms of Human-Computer Interactions (HCI), Augmented Reality (AR) and bottom-up, Internet delivery models such as Semantic Web, and Web 2.0.[1]

Three important factors in the practice of urban design will greatly benefit from these improved models and information delivery systems: *collaboration*, *cognition*, and *creativity*.

Collaboration

It is has been argued that traditional methods of public participation in urban planning often fail to achieve their goals (Innes and Booher 2004: 419–436). A preferred alternative is a direct, face-to-face form of multi-party problem solving dubbed "collaborative participation." This approach is distinguished from traditional public participation in that it allows different individuals and interest groups to interact directly with designers and decision-makers in round table or charrette type formats. Collaborative participation often uses neutral facilitation and/or creative and informal techniques such as role-playing, open-ended conversation, or manipulative Participatory 3-D Modeling (P3DM) to facilitate consensus through dialogue and physical design oriented actions (Roberts 1997: 124–132; Kellam 2008).

A key component of these improved models are information delivery systems that facilitate communication, harness collective knowledge, and build capacity for the end-user to engage in collaboration and data manipulation (O'Reilly 2005). Often referred to as Web 2.0, these advances allow user-generated "mash ups" which combine data from different sources to provide a unique service of interest to specific communities. The fast and powerful capacity to handle complex situations, has also resulted in a wide adoption of Web based Geographic Information Systems (WGIS), and agent-based simulation environments (UrbanSim, NetLogo, AnyLogic) by spatial planners and designers. Adding to these WGIS tools groupware, wiki-style environments, further enhances the possibilities for collaboration. Such models allow experts to create easily accessible frameworks that the general public can adopt to populate with their own content. These in return increase spatial data access and dissemination, and allow online exploration and geovisualization, thus opening up new possibilities for participation by soliciting opinions and incorporating local knowledge often in real-time.

Google Earth

Google's array of software tools is an example of these digital interfaces. Designed to create, store, and communicate rich spatial media online, these tools have evolved into one of the most powerful user-generated, intuitive online spatial devices. SketchUp, one of Google Earth's instinctive 3-D modeling interfaces, allows even novice spatial thinkers to create 3-D models. Its online user-generated repository, 3-D Warehouse, contains thousands of useful models from specific buildings to street furniture such as bike racks, benches, curbs and trees. Many specific models of buildings or urban design projects are geo-tagged, enabling them to be viewed in situ using the Google Earth interface. As technology commentator Tim O'Reilly puts it, "It becomes clear that Google Earth is not just a data visualization platform. It's a framework on which hundreds of different data layers can be anchored" (O'Reilly 2006).

It is clear that such interfaces are fast becoming a substitute for desktops and local servers housing data sets that are accessible only to other parties who share the same software. The Google StreetView project hints at the future potential of these tools, offering immersive panoramic accounts of street scenes around the world. One could only anticipate that in the near future users will be able to place a 3-D/designed project into StreetView to better understand the contextual implications of a proposal.

With time we can expect functions found in each separate tool to gradually merge and play a greater role in urban design and planning. We expect the spatial collaborative process for professionals to be facilitated by these new tools for imagining, producing, displaying, and reacting to spatial configurations. Collaboration and real-time public input will converge into a seamless process where the gap between the professional and the user disappears.

WikiCity

Unlike Google Earth and other bottom-up generated mapping, the WikiCity concept may offer a new paradigm of collaboration. With the spread of real-time location-sensitive sensor-based data sources, tracking and instant input can generate data to determine space usage patterns and social behavior. An example of such a concept is the tracking of mobile phone use and its interpretation into live mapping (SENSEable City 2009). In Mobile Landscape: Graz, cell phone use in the city of Graz was tracked and mapped. Beyond anonymous tracking, the experiment also allowed willing users to trace their own movement through the city. The traces of each registered user were then drawn showing the speed and pathways of movement through space and time. While still in its experimental stage, the act of mapping activities while generating live input has become a new way for interpreting and responding to the city's events and actions. On the one hand, the ability to track and map individuals provides an analytical mechanism to further understand the urban condition in real-time. On the other hand, it provides feedback, making the user an active participant rather than just a passive and observed entity.

It is the growing deployment of individual wireless devices and sensors into an integrated "mash," working together as a system, that makes the "real-time city" useful for design and planning. While currently such data are typically centralized and processed, in the future such intelligence constructed from a bottom up process is likely. Would it be possible then to envision more dynamic and adaptive planning practices, which may include, for example, a feedback loop of remotely sensed data reflecting urban "experiences" for decision-making processes? (Ratti and Berry 2007) (see Figure 20.1).

Such new features may challenge current design and planning practices, as Anthony Townsend (2000: 87) noted:

> Massive decentralization of control and coordination of urban activities threatens the very foundations of city planning – a profession based upon the notion that technicians operating from a centralized agency can make the best decisions on resource allocation and management and act upon these decisions on a citywide basis.

Professional interoperability

The ability of diverse organizations and professionals to work together on complex projects is the essence of urban design. Managing technical know-how, from engineering and design to financial and political factors, impacts performance and results. Digital tools such as Computer Aided Drafting (CAD) have been used to design and manage building projects for

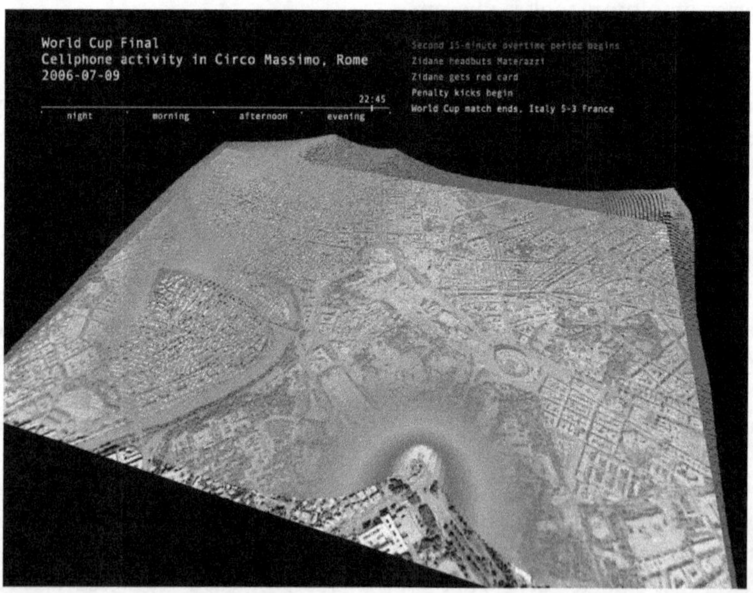

Figure 20.1 Real Time Rome showing people's location during the World Cup final soccer match between Italy and France on July 9, 2006. Source: ©SENSEable City Lab – used by permission.
Note: Real time Rome is an example of real-time mapping showing movements of users based on mobile phone use.

many years. With added 3-D capabilities these tools have also been very useful in producing accurate renderings and fly-through simulations. Recent developments in such software have also brought about capabilities for simulating the construction process, enhancing visualization and coordinating across disciplines. Building Information Modeling (BIM) digital tools allow for a virtual information model (in place of a traditional construction document package of drawings and specifications) to be handed from the primary designer to consultants (surveyors, engineers, cost estimators and others) and finally to the clients and contractors. At each stage discipline-specific knowledge and tracking of changes can be added, reducing loss of information and creating a coordinated effort in managing the project. Unlike typical management and drafting tools, BIM allows for complete 3-D modeling and simulation of the construction process. BIM has helped designers such as Frank Gehry to convert innovative and complex design ideas into constructed projects at various scales (Gehrytechnologies 2009).

The ability of these tools to integrate across platforms and disciplines, and to link visual elements with data, will enhance design comprehension and decisions. One could see the use of these tools to demonstrate a project's life cycle and the ability to extract information about how it performs. An example of this could be a facility management department turning to its virtual model to find leaking infrastructure rather than trying to locate it physically. The model could provide information on the type and size of pipes, or even the part number needed, all before actually retrieving it.

The connection between such visual models and urban databases could also offer an interesting merger with the legal controls of urban design and planning. In the future, structured text documents such as specifications, zoning or design codes could be virtually tagged to a visual (and also virtual) model. One could then retrieve this information as various elements of the city are viewed, allowing links to standards or codes and an understanding of the design's legal framework.

Cognition

The abundance of both real-time and recorded information at our fingertips offers tremendous opportunities for the initial understanding of place. Street Views remotely controlled webcams, photosharing, Google *Maps* and Google Earth allow for better incorporation of local knowledge as well as for immediate comprehension of the locality. Yahoo's TagMaps and Google *Maps* services, which integrate Flickr or Picassa photos in a geospatial framework, provide a richer experience of visual evidence in a study area, as well as facilitating user contributions and documentation of places or issues of concern.

Remote sensing

The ability to remotely sense and see actual site conditions brings immediate and tangible information to the designer. Webcams are a good example. A vast network of webcams provides a constant feed of images that can be monitored, controlled remotely and accessed 24 hours a day. Many of these provide rich, uninterrupted data on the changing physical and environmental conditions, as well as on social and cultural use. Sites such as EarthCam essentially work as a clearinghouse for worldwide maps of web cameras that show a range of global locations – from surf conditions in Hawaii to crowd conditions at the Plaza Mayor in Madrid, Spain (Earthcam 2009). While the use of webcams for security or monitoring traffic conditions is widely exploited, recent trends indicate a strong potential for companies in design and construction of the built environment. Construction webcams can remotely monitor and archive progress over the Internet, thereby eliminating the need for multiple site visits. Observed activity made available by a live webcam trained on public spaces can give new meaning to the social study of spaces, as originally used by William H. Whyte in New York City in 1970s (Whyte 1980). For example, the Demonstrate project set a web camera over a UC Berkeley Plaza for six weeks. The camera, which was accessible to anyone on the Internet, allowed online participants to participate in discussion prompted by the observed activities in the plaza (Demonstrate 2009).

Beyond images and tagged information, other environmental conditions are also gaining a larger platform on the web. One example is the site Pachube, a web service that enables one to connect, tag and share real time sensory data from objects, devices, buildings and environments from around the world. By sharing real-time environmental data via the internet one can both capture input data as well as facilitate information sharing. For example, live streaming data on air quality, temperature, or light levels can be accessed for wide ranging locations, including not only established government and educational monitoring stations, but also from amateur collaborators who may track conditions in their immediate environment (Pachube 2009).

Virtual and tangible

While remote sensing informs and increases our understanding of existing conditions, one of the great challenges that urban planners and designers face is integrating and communicating spatial concepts and design ideas into these digital tools. Generally, the interface through which design ideas are presented and manipulated has seen little development. Few platforms exist that allow immediate, real-time, and seamless changes in response to public or professional input. Often, several different modes of representation must be utilized within a project to convey different types of information and aspects of the design. It is this separation between various representative forms that increases the cognitive load on both the designer and the audience, who must draw relationships between dislocated pieces of information. Ideally, these digital tools would communicate proposed changes and make their impact easily understood. These systems could be used not only as tools for design professionals but also as an interactive application to enrich communication and learning within the design process. The integration of such envisioning tools into the decision-making process will provide professionals the ability to make better judgments while incorporating various stakeholders' expectations.

Two examples may provide some clues as to the future of such digital immersion tools: Virtual and Mirrored Cities and Tangible Infoscapes. Virtual Cities combines 3-D digital models, with aerial photographs and street level video to create an urban model that can then be used for interactive fly, drive and walk-through demonstrations. It can be a realistic model of an existing, historic or imaginary place. The Urban Simulation Team at the University of California, Los Angeles is building a real-time virtual model of the entire Los Angeles basin. The model can scale from satellite to street level views accurately enough to allow the signs in shop windows and the graffiti on the walls to be legible. Beyond an accurate depiction of the city, Virtual LA is also used to model new designs and place them into community context. In the case of new transit lines and stations, the team used the model to create visualizations of right-of-way alternatives and to model new transit-related commercial development to be shown to the client and the community.

Virtual simulation opens the possibility for displaying and querying real-time data. For example, in Los Angeles, the team has been in discussions with the City about the feasibility of using the model in conjunction with Global Positional System (GPS) transponders to accurately locate and remotely manage emergency response vehicles in real-time. Complex modeling and possible scenarios can be tested and seen with much greater accuracy (Urban Simulation Team 2009). In Virtual London, a 3-D digital model of central London, with geophysical spatially tagged attributes has helped to forecast flooding as a way to understand issues of climate change. The model has also been tagged with air pollution, land use, and retail data in surface form, and is used as an interactive geographic information system viewable as a 3-D database (Batty and Hudson-Smith 2005) (see Figure 20.2).

With the integration of new web interfaces, virtual cities (worlds) have been mirrored on the internet. These mirrored cities (worlds) bring visualization to the point where users can freely experiment, interact and voice opinions within these fictional environments. In the case of Virtual London, as well as in other limited experiments, virtual cities and virtual designs have been integrated into mirrored worlds such as Second Life. While in some cases the mirroring allows any Second Life user to explore the city, interact with others and leave comments, in other instances it has been used to solicit direct design input. The Boston Redevelopment Authority created a virtual representation of a park in Second Life to solicit direct feedback from residents. According to those who were involved, the design process was aided by the use of Second Life because it decreased the divide between designers and the public. For instance, at one meeting, a group of teenagers requested a basketball court, but when a full-scale court was sketched in by the designer, the group realized that there was no way it could fit. This process also helped with the placement of public art and water features as well as the location of parking,

Figure 20.2 University College London 3-D virtual model. Source: © Centre for Advanced Spatial Analysis University College London – used by permission.
Note: Model simulating a 10-meter rise in sea level in the city. Combined with Second Life, it allows users to interact with the model as avatars.

which was always mentioned as a thorny problem without solutions during previous non-Second Life public meetings (Freeman 2006, Knack 2009).

While urban simulations, such as those described previously, have progressed at an impressive rate over the last decade, they are still confined to two-dimensional (2D) flat interfaces such as screens. As such, they leave much to be desired from the perspective of both the end user and observer. Thus, they lack the immediate, tangible interaction that one gets with touching graspable, physical objects. One area of research that investigates the integration of the "real world" and computational media is Computer-Augmented Environments or AR. The most common AR approach is a visual overlay of digital information onto real-world imagery with see-through, head-mounted or hand-held display devices or video projections. Several researchers have created AR-based urban design and planning support systems. The Human Interface Technology Laboratory New Zealand (HIT Lab NZ) has developed virtual and augmented reality interfaces as well as collaborative interactive applications. One of their experiments, BenchWorks, is an Augmented Reality Aided Urban Design platform. It combines optical, magnetic and real physical models. Users wearing Head Mounted Displays (HMD) are able to virtually insert and manipulate objects within the physical model (Seichter *et al.* 2007).

Similarly, the University College London's VR Centre for the Built Environment has created an Augmented Round Table for Architecture and Urban Planning. Using see-through augmented reality glasses, the table generates virtual models of a design scheme being discussed. Twin cameras fitted to the glasses and computer vision techniques provide head tracking for the users, as well as tracking real-world placeholder objects on the table that allow the users to interact with the virtual model. The specialized eyeglasses and the real world placeholders mean that all members of the round table have equal access to the design user interface. Through simulations and visualizations of various performances, consequences of the proposed design changes can be viewed and evaluated as the design is being manipulated (VR Centre For The Built Environment 2004).

At the Massachusetts Institute of Technology's Media Laboratory, virtual visualization and digital information has been extended to everyday physical objects and environments. The Tangible User Interface's (TUIs) distinct approach is in its focus on graspable physical objects for input rather than on enhanced visual devices. Combining these devices with urban planning discourse has produced new tools essential for the understanding of place making. The Illuminating Clay, for example, allows designers to manipulate 3-D models of landforms and objects upon which visual data is projected as the shape is formed. As the clay surface changes its shape by the touch of the users, data such as topography, slope, aspect, cut and fill or travel time, is calculated and projected on the surface. A perspective window screen also allows users to explore the clay model from a person's height. The result is a powerful simulation tool that provides access to a full efficacy of computational resources in a manner that is comfortable and intuitive (see Figure 20.3).

While complex manipulative virtual reality interfaces are still in their infancy, integrating real-time information into everyday objects is fast gaining popularity. Finitude by Mobilizy, for example is a mobile travel guide for cell phones based on location-based Wikipedia and other web content. Worldwide points of interest may be searched by GPS or by address and displayed in a list view, map view or cam view on one's cell phone screen as it is pointed at an object (see Figure 20.4).

Another application that explores the combination of wearable computing and everyday objects is the Sixth Sense.

Figure 20.3 Illuminating Clay combines physical models and digital information. Source: © Eran Ben-Joseph.
Note: Illuminating clay is an example of Tangible User Interfaces (TUIs), which allow for design and simulations that are comfortable, intuitive and collaborative.

Figure 20.4 Wikitude by Mobilizy displays information on a mobile device as one points it at an object. Source: © Mobilizy – used by permission.
Note: This is one example of augmented reality integrated with mobile devices. In this case it is used as a mobile travel guide for cell phones based on location-based Wikipedia and Qype content. World-wide points of interest can be searched by GPS or by address and displayed as one points it at an object.

Developed as a prototype by the Media Lab at MIT, it allows the user to project information from a cell phone onto any surface – walls, the body of another person or one's own hand. With hand gestures one can manipulate and interact with the data, all without the need to look at the cell phone screen. Information thus becomes more useful to people in real time, with minimal effort, and in a way that does not require any behavior changes (Mistry 2009).

The connections between the virtual and the actual are promising both for the designer and the layperson. The intersection between digital and tangible offers a seamless way to create new associations between input and output, as well as to palpably present complex analytical concepts. The promise of these systems will eventually be tested by the quality of resulting designs. This will be especially interesting with regard to the divergence between reality and its depiction in virtual worlds. Not everything that happens in Second Life can be replicated in the real world. In fact, as a virtual landscape, Second Life is designed to function in ways that are not possible in real life. The laws of physics do not apply as avatars can climb up tall objects with ease, and even teleport and fly. Nor do real-world regulations apply to Second Life, such as zoning or other municipal bylaws that would restrict a particular design. While digital tools like Second Life may not be the best in restricting design input to realistic ideas allowed by existing regulations, they do open the door for new and imaginative ideas by a wide range of users.

Creativity

Creativity is considered an essential component of human intelligence and an intrinsic part of design. But can new digital tools, such as those described previously, aid in the creative process, or even become artificially creative? There are a number of projects in Computational Creativity that attempt to recreate creativity in computers. AARON for example is a Cybernetic computer which creates original paintings (Kurzweil CyberArt 2009). At the Austrian Research Institute for Artificial Intelligence researchers are teaching a computer to play the piano like a human, not only copying but also finding patterns in performances and composing new unique scores (Austrian Research Institute for Artificial Intelligence 2009).

Artificial Intelligence and Computational Creativity imply a reduction of creative events to a sequence of rules and conventions that can then be interpreted in a new way. This is where the new realm of "shared intelligence," manifested in mashed webs and Web 2.0, can provide a boost to creativity. The omnipresent data created and shared on the web is introducing a form of collective intelligence which is bound to shape the way we approach design and bring new resourcefulness.

The formation of a collective constellation of computers, networks, services and users that organize themselves without a conscious directive, is often referred to as Cloud Computing (Vaquero *et al.* 2009). Beyond the technical ability of these systems to be dynamically scalable and offer network resources not owned by a single entity, they provide a way to create shared logic. The idea of using the individual knowledge of users and capturing it so that new patterns in the data emerge may seem counterproductive; however, as Hudson-Smith *et al.* explain: "This slightly surprising notion is based on the fact that although a large number of individual estimates may be incorrect, their average can be closer to the mark than any individual estimate" (Hudson-Smith *et al.* 2008).

Data sets created by users not only add new forms of intelligence but they can also produce by-products not previously obtainable. While individuals are interacting and synthesizing, the available data sets can generate such products. Creative Computing and Machine Learning may produce it involuntarily. An example of this process can be seen in the work of image/object recognition. Recent developments in this field have yielded interesting results in describing the contents of an image. Using tags is a leading approach to organizing photo collections. The user navigates the collection by typing a query tag and reviewing the retrieved images. An alternative method is to have users label features within images by tracing over each feature to create a geometrical (shaped) knowledge base of objects. For example, users can trace around a lamppost and identify it as well as adding to it other information such as its style and height. These collaborative labeling includes internet sites such as ESPgame, LabelMe, and Mechanical Turk (GWAP 2009; LabelME 2009; Mechanical Turk 2009). With hundreds of thousands of images' features being labeled and geometrically defined, automated systems become capable of learning and defining associations to untagged images. In the case of LabelMe such experimentation has led to computerized ability to depict scenes with absolute 3-D information.

Similarly, with facial recognition software, object recognition will allow users to access a vast amount of image data and organize it (or have the computer organize it) through numerous associations. For example, images can be arranged and retrieved according to their association with a particular land use such as a store, mid-rise housing, etc. Or they can be stratified and recalled according to a specific feature or color, for example a lamp post, a pine tree or a wooden bench, providing a new vocabulary and typology to be used as inspiration for design and discovery.

Creativity and innovation can also be enhanced through the tools by which one interacts with information and data. At present most access to digital information occurs primarily through Graphical User Interfaces (GUIs). The omnipresent interface of computing as we know it, was probably devised in 1981 with the Xerox Star workstation. It established the "desktop metaphor" which simulated an interaction between a working page on a bitmapped screen with a pointing device (mouse), windows and icons. However, traditional GUIs may not be the best format for design work as they still represent a separation between the digital and the tactile – a crucial element in the creative process. It is widely agreed that graspable interfaces or tangible user interfaces (TUIs), such as the Illuminating Clay, enable new and richer forms by extending the design space onto physical objects. The integration of digital information and physical forms by TUIs makes digital information directly manipulable with our hands. Such physical interactions are very common for designers who work with models and physical 3-D representations. The unique ability of TUIs to seamlessly integrate digital information, ideas and input is a key in changing one's perception and increasing cognitive information. Tools such as Illuminating Clay or G-Speak do not only yield a creative and enjoyable design process, but also provide the ability to test ideas and observe the resulting impacts in real time, allowing both the designer and the public to be better informed and involved (Oblong 2009).

Limitations and considerations

Digital technology is transforming our cities and the way we interact with them. The future will bring further significant and enticing changes. A closer intersection will be seen between the virtual and the physical, information will be imbedded into forms and environments, and bottom-up interconnected systems will form new kinds of intelligence.

Yet these promises also bring uncertainties and raise challenges. What will be the effects of economic and cultural differences on the use and spread of such tools? How can we reduce the digital divide and assure equal access to information and communication technology, and the equal acquisition of related skills? How will the increase of surveillance capabilities and the potential loss of privacy influence public willingness to engage with new digital technologies? What will be the role of professionals and decision makers in the face of growing self-organized public user interfaces? Will digital tools and ICT erase disciplinary boundaries, or increase them? How can accuracy, truth and legitimacy of data and information be guaranteed? Will the sheer free-flowing volume of information and data create an information overload that will overwhelm planning processes, thus preventing decision-making?

Such questions and challenges should play an important role in the development of new digital tools. These questions and challenges should also be an integral part of ongoing research as to how behavior and policy making will be shaped by technology. Still, there is no doubt that this new dynamic is transforming and will transform the planning and design process. Decision-making will become more adept at measuring and predicting outcomes, recognizing unintended consequences, and fine-tuning development strategies. Urban planning and design can be transformed from a process that is superfluous and static to a process that is dynamic, decentralized, participatory, and self-correcting.

Note

1 Human–Computer Interaction (HCI) is the study of interaction between people (users) and computers. It is often regarded as the intersection of computer science, behavioral sciences, and design. Augmented reality (AR) is a field of computer research that deals with the combination of real-world and computer-generated data (also referred to as virtual reality). The Semantic Web provides a common framework that allows data to be shared and reused across application, enterprise, and community boundaries. It is a collaborative effort led by W3C. Web 2.0 is the second generation of the World Wide Web, especially the movement away from static web pages to dynamic and shareable content and social networking.

References

Austrian Research Institute for Artificial Intelligence. (2009) Online. Available HTTP: <http://www.ofai.at/research/impml/index.html> (accessed 14 May 2009).

Batty, M. and Hudson-Smith, A. (2005) "Imagining the Recursive City: Explorations in Urban Simulacra," *Working Paper Series*, London: University College London Centre for Advanced Spatial Analysis: Working Paper 98. Online. Available HTTP: <http://www.casa.ucl.ac.uk/publications/workingPaperDetail.asp?ID = 98> (accessed 5 November 2009).

Demonstrate. (2009) Online. Available HTTP: <http://demonstrate.berkeley.edu/> (accessed 14 May 2009).

Earthcam, (2009) Online. HTTP: <http://www.earthcam.com/> (accessed 14 May 2009).

Economist (2008) "'Spot Prices' in Technology Monitor" *The Economist*. Sept 17, Online. HTTP: <http://www.economist.com/displaystory.cfm?story_id = 12236749> (accessed 10 June 2009).

Freeman, G. (2006) "Jackson Heights: Playing Games in the Park," *The New York Times*, June 18. Online. HTTP: <http://www.nytimes.com/2006/06/18/nyregion/thecity/18wire.html?_r = 3> (accessed 14 May 2009).

Gehrytechnologies, (2009) Online. HTTP: <http://www.gehrytechnologies.com/> (accessed 14 May 2009).

GWAP, Online. HTTP: <http://www.gwap.com/gwap/about/> (accessed 14 May 2009).

Hudson-Smith, A., Batty, M., Crooks, A., and Milton, R. (2008) "Mapping for the Masses: Accessing Web 2.0 through Crowdsourcing," *Working Paper Series*, London: University College London Centre for Advanced Spatial Analysis: Working Paper 143. Online. Available HTTP: <http://www.casa.ucl.ac.uk/publications/workingpapers.asp> (accessed 5 November 2009).

ICT Development Index International (2009) *Measuring the Information Society*, Geneva, Switzerland: The Telecommunication Union.

Innes, J. and Booher, D. (2004) "Reframing Public Participation: Strategies for the 21st Century," *Planning Theory and Practice*, 5(4): 419–436.

Kellam, M. (2008) "Residents get to give input about 'Little NoHo'," *LA Daily News*, 11 October 2008. Online. Available HTTP: <http://www.nowpublic.com/culture/little-noho-model-draws-some-attention-oblate-spheroid> (accessed 10 May 2009).

Knack, R. (2009) "The Next Level: Second Life put to the test," *Planning*, March. Online. Available HTTP: <http://planning.org/planning /2009/mar/nextlevel.htm> (accessed 10 March 2009).

Kurzweil CyberArt Technologies, (2009) Online. Available HTTP: <http://www.kurzweilcyberart.com/index.html> (accessed 14 May 2009).

LabelME, Online. Available HTTP: <http://labelme.csail.mit.edu/> (accessed 14 May 2009).

Mechanical Turk, Online HTTP: <https://www.mturk.com/mturk/welcome> (accessed 14 May 2009).

Mistry, (2009) Online. HTTP: <http://www.pranavmistry.com/projects/sixthsense/index.htm#ABOUT> (accessed 14 May 2009).

Oblong, (2009) Online. HTTP: <http://oblong.com/> (accessed 14 May 2009).

O'Reilly, T. (2005) "What Is Web 2.0?, Design Patterns and Business Models for the Next Generation of Software," *O'Reilly News*, Online. HTTP: <http://www.oreillynet.com/pub/a/oreilly/tim/news/2005/09/30/what-is-web-20.html> (accessed 10 May 2009).

—— (2006) "Google Earth, SketchUp, and Second Life," *O'Reilly Radar*, Online. HTTP: <http://radar.oreilly.com/archives/2006/06/google-earth-and-sketchup.html> (accessed 10 May 2009).

Pachube, Online. HTTP: <http://www.pachube.com> (accessed 14 May 2009).

Ratti, C. and Berry, D. (2007) "Sense of the City: Wireless and the Emergence of Real-Time Urban System" in V. Châtelet (ed.) *Interactive Cities*, Orléans: Editions HYX. pp

Roberts, N. (1997) "Public Deliberation: An Alternative Approach to Crafting Policy and Setting Direction," *Public Administration Review*, 57(2): 124–132.

Seichter, H., Dong, A., Vande-Moere, A. and Gero S. J. (2007) "Augmented Reality and Tangible User Interfaces in Collaborative Urban Design," *CAAD futures,* Springer: Sydney, Australia (7) 3–16.

SENSEable City Lab at MIT, (2009) Online. Available HTTP: <http://senseable.mit.edu/> (accessed 14 May 2009).

Townsend, A. (2000) "Life in the Real-Time City: Mobile Telephones and Urban Metabolism," *Journal of Urban Technology*, 7 (2): 85–104.

Urban Simulation Team, (2009) Online. Available HTTP: <http://www.ust.ucla.edu/ustweb/about_us.html> (accessed 14 May 2009).

Vaquero, L. M., Rodero-Merino, L., Caceres, J., Lindner, M. (2009) "A break in the clouds: towards a cloud definition," *ACM SIGCOMM Computer Communication Review archive*, 39 (1).

VR Centre for The Built Environment (2004) *Augmented Round Table for Architecture and Urban Planning*, Online. Available HTTP: <http://www.vr.ucl.ac.uk/projects/arthur/> (accessed 10 March 2009).

Weiser, M. (1991) "The computer for the 21st Century," *Scientific American*, 265 (3): 94–104.

West Philadelphia Landscape Project, (2009) Online. Available HTTP: <http://www.wplp.net/> (accessed 14 May 2009).

Whyte, W.H. (1980) *Social Life of Small Urban Spaces*, Washington D.C.: Conservation Foundation.

Further reading

Fisher, P. and Unwin, D. (2002) *Virtual Reality in Geography,* London: Taylor & Francis. Covers all the major uses and methods of virtual reality used by geographers and planners.

Harkin, J. (2009) *Cyburbia: The Dangerous Idea that's Changing How We Live and Who We Are,*

London: Little, Brown. Historical overview of the digital age's origin and promises.

Kwartler, M. and Longo, G. (2008) *Visioning and Visualization: People, Pixels, and Plans,* Cambridge, MA: Lincoln Institute of Land Policy. (Principles, techniques, and cases based on their professional experiences in developing public involvement processes that are used to apply information technology to planning and design.)

McCullough, M. (2004) *Digital Ground: Architecture, Pervasive Computing, and Environmental Knowing*, Cambridge, MA: MIT Press.

Whyte, J. (2002) *Virtual Reality and the Built Environment,* Oxford: Architectural Press. Guide to the practical uses of virtual design, construction, and management. Providing an overview of industrial applications for virtual reality.

Part 5

Process

Introduction

Not much literature has focused on the process of urban design and its relationship to the final design outcome. This is partly because an influential school of thought has always considered the process of design as a "black box" phenomenon – that often leads to a "eureka" moment that takes place in the designer's head, a result of her intuition and creativity. Under this view, there is nothing much that external circumstances or factors can do to better the process of design so that it achieves a better form (Osborn 1963; Broadbent 1966). The opposing view comes from rationalists, who view the process of design as a "glass box" – a process that is completely explicable and can be affected and made better, if only it follows a series of logical steps (Jones 1992).

Assuming that the black box/glass box metaphors represent polar opposites, and the process of design falls somewhere in between the two poles, are there guidelines, norms, or frameworks, which can help designers achieve a better urban form? How can we increase the safety net against mediocre urban design and enhance the likelihood of good design? How do we increase the possibility that the process of design will lead to a good city form? What is the right balance between prescription (expressed in standards, design codes, and guidelines) that many planners seem to favor, and unrestrained creativity often cherished and sought by architects? These are some of the questions that authors in this section seek to address.

Some would argue that a process that is more likely to lead to many different creative ideas and novel concepts is more likely to achieve a better outcome. Alternatives are valued in the rationalistic model of design, guided by the belief that the likelihood of finding a better solution increases with the number of alternatives generated (Banerjee and Loukaitou-Sideris 1990). Are then processes such as design competitions and design charrettes, which lead to a more exhaustive sets of alternative solutions or involve brainstorming by multiple individuals, more likely to produce a better design outcome? Some of the chapters in this section address these questions.

Some would also argue that a good process is one that leads to a good "fit" between user needs and urban form. But unlike most architectural problems that have a clear user/client, urban design problems deal with public, private, and group interests often in conflict with each other. Additionally, urban design frequently leads to redistribution or regulation of territorial power, control, and rights of different social groups. Some would then grant that a good process has to be a democratic one, which allows the expression and mediation of these different interests, and even leads to considerations of fairness, equity, and distributive justice in a specific solution. How can urban design processes incorporate public involvement and deliberation? What are the challenges and opportunities for public participation in urban design? This section responds to these questions.

The first two essays in this section focus on a compilation of tools that are intended to inform and guide the process of design so that it reaches a specified set of ends. William Baer delineates the different hues of design standards, from those which only aspire to "satisfice" and hope to avoid worst case scenarios in the design praxis, to those that wish to encourage creativity and innovation. He explains the trade-offs involved in choosing each type of standard. Matthew Carmona draws from experiences in the United Kingdom to explore the relationship between the application of design guidance (and in particular design code) and design quality. He argues that while the design code is only one tool in the armory of designers, codes can contribute to better designs if certain fundamental factors are included in the coding process.

The next two essays look at processes that contribute to a more exhaustive search for solutions to urban design problems. Ute Lehrer examines urban design competitions, using the competition of Potsdamer Platz in Berlin as a case study. She finds that in addition to offering a wider range of ideas and a fairer process than commissioned projects, competitions have the added benefit of raising public awareness about the politics of large scale projects. They also bring media attention, which helps marketability. In the end, however, the process of competitions "can be only as good as their program, their jury, the selection of the architectural firm, and the local conditions that tie all these components together." Doug Kelbaugh describes the process of design charrettes, which he finds as contributing to good design because of its democratic intent, participatory mode, intensive brainstorming, and avoidance of "unduly political pressure" that often takes place in commissioned urban design projects. He argues that as a design process, charrettes consistently produce more imaginative solutions than conventional design consultations.

For Jeffrey Hou, author of the last essay in this section, a virtuous urban design process has to be democratic. He gives an account of the rise of the importance and the different forms of citizen participation in design, but also the challenges that it faces. He argues that new technological tools are helping to open up the urban design discourse to larger audiences transforming the conventional structures of the participatory model to more inclusive practices of "citizen design."

References

Banerjee, T. and Loukaitou-Sideris, A. (1990). "Competitions as a Design Method: An Inquiry." *Journal of Architectural and Planning Research*, 7(2): 114–31.

Broadbent, G.H. (1966). "Creativity," in Gregory, S. (Ed.) *The Design Method*. London: Butterworths.

Jones, C.J. (1992, 2nd edition). *Design methods*. New York: John Wiley and Sons, Inc.

Osborn, A.F. (1963). *Applied imagination*. New York: Scribener's Sons.

21

Customs, norms, rules, regulations, and standards in design practice

William C. Baer

One facet of design praxis not well publicized yet increasingly important is the specific manner of expressing norms and standards governing professional practice so that they best accomplish their purpose. Building codes, zoning ordinances, and historic preservation overlays are examples. Their internal structure can be designed (Baer 1997). Rules and regulations of this sort have an internal organization or grammar and *meta-rules* for its organization. These meta-rules apply to the standards and regulations that govern different professions, so they are a common ground for the city-building professions of architecture, civil engineering, landscape architecture, and planning.

Adoption of professional standards was part of an international movement at the end of World War I to incorporate a more "scientific" basis underlying planning and housing proposals, based on efficiency and an increasing industrialization and standardization, as well as on design (Lebas *et al.* 1991: 250–251). The initial purpose was to protect the public's health, safety and welfare by prohibiting unsafe construction practices through use of building codes, and incompatible land uses through use of zoning ordinances. In the main these regulations were couched in the negative, thus prohibiting, for example, unsafe construction practices, or harmful industrial uses in a residential area (Skitowski and Ohm 2006).

With those concerns under control, however, today society seems less satisfied with regulations generated from the often-cramped admonition of the negative, which mostly fail to account for design considerations. It seems that society recently has become more venturesome in regulatory undertakings and more willing to conduct new experiments in approaches. In turn, design professionals are now seeking to impose their own views through such mechanisms as well. New urbanism, form-based zoning ordinances, and design guidance are examples (see chapters by Talen and Carmona in this volume, and also Parolek *et al.* 2008). But how should these new regulations be formulated? What should design professionals have in mind regarding specific approach and wording when they become involved in devising and modifying them?

Terminology and purpose

There are a number of words that mean approximately the same thing – devices to guide human behavior – and it is not clear which is the broadest term. Here we will use "rule" as the most generic; use "regulations" to mean government-issued rules, and use "standards" to mean a profession's internally devised rules. (Government regulations often incorporate some professional standards making them public standards, too.) Fundamentally, rules are merely human devices to translate between desired ends and possible means (Simon 1969). Standards and regulations tend to be about preferred means, but now and then also about preferred ends. Sometimes it is not clear what they are about. This latter is surprising, but the candid admission by the International Conference of Building Officials that sometimes they did not know the *intent* of some parts of their voluminous building code is a case in point (ICBO 1995: vi.). Perhaps that unfortunate ignorance is more common than we suppose in the development of a variety of standards and regulations.

A framework to view professional practice

Professional mind-set surrounding the standard

First, there is the basic mind-set behind any rule. Out of all possible behaviors in a situation, or of all possible results from an activity, the rule at minimum attempts to eliminate the unhelpful ones, and especially the harmful ones. Since professionals render a service that requires them to exercise judgment, and whose purpose is to benefit clients, professional standards at their most basic are exemplars of cautious behavior, designed to avoid great risk. By risk we do not mean "known variation," but rather uncertainty, ignorance, incomplete knowledge, and ambiguity (Shapira 1995). In this sense professionals assume that the public may accept professional decisions that tend only to "satisfice" on the up-side, that is accept decisions that may be only "good enough" at their best but not incurring excessive costs to achieve, while virtually requiring professional standards to devote disproportionate attention to "worst case scenarios" so as to prevent them (Shapira 1995). This emphasis is in part a function of the degree to which the practice affects public health, safety, and welfare. Medical doctors as a rule follow this caution more than designers, whose impact on public health, safety, and welfare is more indirect.

Usually (and some of these terms will be expanded upon later on) *prescriptive* standards practice a mini-max (minimize maximum regret). These are cautious, tried, and true approaches that have been shown by years of experience to be safe and not to include many unintended and unwanted side effects. Because in general people are risk averse they tend to prefer a small amount of an almost sure thing from a professional standard than only a small chance at a very large gain. More recently, a *performance* approach to standards has attempted a maxi-min strategy (maximize minimum reward) to increase the gain from following the differently oriented standard at presumably only slightly higher risk. This latter approach seeks to encourage creativity and innovation in meeting a requirement rather than merely invoking the duller "tried and true" (not-permitted-to-experiment) approach. Nevertheless, the initial enthusiasm for performance has been dampened by the subsequent realization that the performance desired is often presented in such a limited and circumscribed way that one can fulfill it without being aware of unintended and unwanted side effects not mentioned in the performance rule.

Casting the basic thrust of the standard: negative vs. positive

Standards are typically couched in the negative (proscriptive) or the positive (prescriptive). In effect they say, "Do not do the following: (that is, a list of actions not to undertake, or, alternatively things not to produce) but everything else is all right." Alternatively, standards can be couched in the positive: a list of actions or things to produce, one of which must be undertaken but nothing else is permitted. Generally, couching rules in the negative requires less knowledge of a variety of causes and effects. Historically, negative rules have been based on the lower expenditure of time, effort, and money. It is often a strategy of "good enough." We know that "X" works, so why spend additional efforts seeing what else might work?

Unfortunately, the world is not this simple or clear-cut as negative or positive, black or white. Sometimes situations, our knowledge about them, and the accompanying regulations are grey, merely cautionary rather than prohibitory, or merely advisory or recommendatory rather than mandatory, requiring independent judgment by both the user of the standards and the recipients of its outcome.

Characteristics of standards

Apprehending the world

Given these two general settings for rules, their effort to be safe and sure, and their negative or positive cast, there are three dimensions to them. The first pertains to the world as sensed and appreciated in the design sense, and to the world as acted upon. Products (a stock) are at one end of the dimension – that is the world as sensed – with process or procedures (a flow) at the other. In other words, the dynamics of a situation gives rise to the statics of the results of those dynamics, and they in turn produce further dynamics and then another result. The substantive attributes of what a profession works on determines how it apprehends and then represents the world to achieve its best effects. Sentential (word sentences) are one common means; maps, diagrams, and pictures are another. Professions resort to one or the other to suit the particular circumstance (Larkin and Simon 1987). As Simon (1969) explained:

> These two modes [process and product] of apprehending structures are the warp and weft of our experience. Pictures, blueprints, most diagrams, and chemical structural formulas are state descriptions. Recipes, differential equations, and equations for chemical reactions are process descriptions. The former characterize the world as sensed; they provide the criteria for identifying objects, often by modeling the objects themselves. The latter characterize the world as acted upon; they provide the means for producing or generating objects having the desired characteristics. (Simon 1969: 111)

Problem solving, Simon goes on to note, "requires continual translation between the state and process descriptions of the same complex reality." This *translation* works roughly as follows: a problem is posed by giving a state (or outcome) description of the solution. The task next is to discover a sequence of processes that will produce or render the goal state (or the desired outcome) from the initial state (or the status quo), or, alternatively, to discover the process that will lead to one or several intermediate states, from which another process takes us to the final desired state. A constitution is a process description of steps to be taken for how to achieve good government; a blueprint is a state depiction of component parts that when assembled and built deliver the desired building.

Choosing where to modify an aspect of the world by the standard

The second dimension to rules has inputs at one end, outputs at the other. The focus here is on means versus ends, or in the terminology to be followed here, on prescriptions versus performance. Prescriptions concentrate on what should go into an effort without describing the result; performance describes what the result should be without describing how it should be achieved. The latter is preferred to the extent it allows the designer creativity in devising the means to achieve the goal, but things are not that simple. Experimentation with the performance approach revealed that there were problems (not encountered when using prescription) of: first, means-ends chains (achieving a sub-goal as a means to achieve an intermediate goal, as a means to achieve the primary goal) (March and Simon 1958, Hattis 1972); second, negative side-effects where the goal is achieved but so are other unwanted achievements (i.e. meeting the performance requirement of "self-extinguishability" nevertheless allowed plastic manufacturers to introduce products which gave off toxic fumes) (Baer and Banerjee 1977); third, problems of interface between non-specified parts, where the choice of meeting one goal does not mesh with the choice made in meeting another (Hutcheon 1972), hence lack of the supposed freedom to innovate (i.e. presumably unfettered performance is in fact constrained by the above considerations) (Baer and Banerjee 1977: 206); fourth, expensive testing of the results, or not being able to know the true effects from a "solution" until passage of many years (Wright 1983: 101, 104); and fifth, "the supremacy of second rate materials" when workers are unfamiliar with the first-rate, i.e. they do not know how to effectively work with or implement the new, innovative solution (Van Court 1972: 946).

How specific should we be?

The third dimension relates to the specificity with which a profession can express the end-points of the other two dimensions – *criteria* versus *standards* (the regrettable use of the word "standard" for both a particular aspect and the larger realm is virtually unavoidable given usage in the English language, so we will use italics to mean the sub-aspect of the larger framework for standards). *Criteria* are relative while *standards* are absolute. Both need elaboration because they are often not distinguished in normal discourse, but they can have precise differences that are important to distinguish for rule making.

Ideally, all professional standards should be so clear that degrees of the attributes should be capable of being divided up into objective units of measure, and then some dividing line established along those degrees to distinguish good and bad situations, or professionally acceptable and unacceptable ones. In other words, a normative evaluation is superimposed upon the objective measures so that all those in a profession can agree on what is what. Figure 21.1 illustrates this point.

In Figure 21.2 we illustrate two types of *standards*, those employing criteria, and those using "standards." The top part of Figure 21.2 shows criteria in the form of values to be maximized or minimized. The precise amounts are vague but stress that more is desirable, or that less is desirable. In the bottom part of Figure 21.2, we see a situation where the good and the bad are clear-cut, and professionals can resort to *standards*. The type illustrated

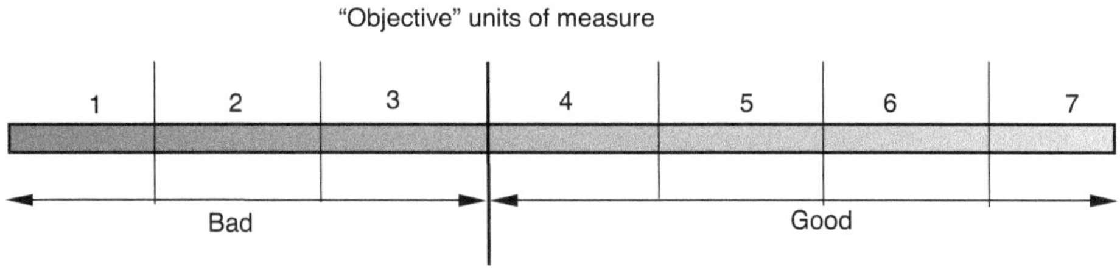

Figure 21.1 Normative evaluation superimposed upon "objective" measures.

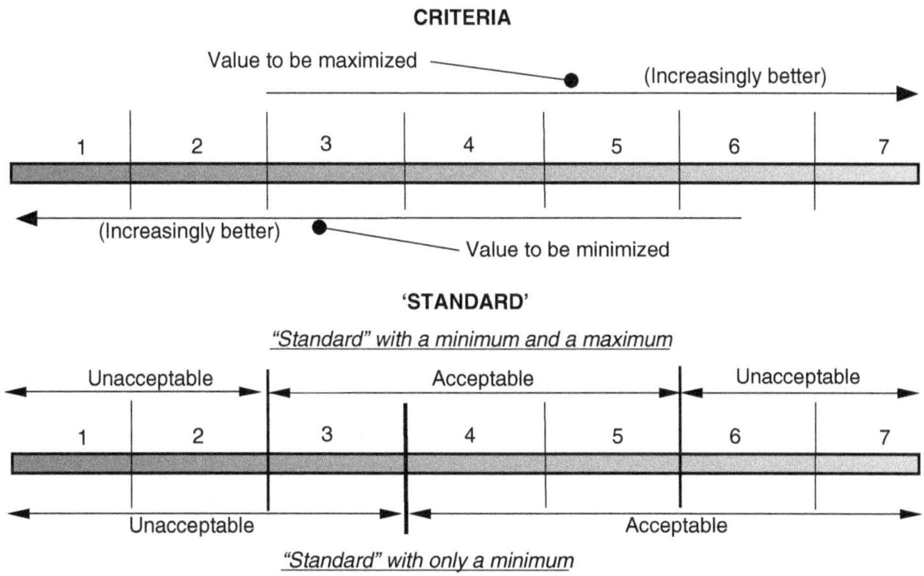

Figure 21.2 Differences between criteria and standards.

here uses either one cut-off point (Acceptable/Unacceptable) or two (Unacceptable/Acceptable/Unacceptable). Following Boyce (1970) we can say that *standards* are the most precise normative judgment involving a measurement. In short, standards are *decision rules* about the acceptability of a thing or condition in the face of uncertainty (Baer and Banerjee 1977; Baer 1986).

While we have shown seemingly "hard" numbers for our demarcations of the attribute in Figure 21.1, we must be cautious about using numbers along a scale for an attribute and what they can say. A scale can consist of nominal classifications, say Elizabethan style, Victorian, Modern, and Post Modern but other than being different we cannot say in the abstract much about the scale and the desirability of being one place along it vis-à-vis the other places. Alternatively, the objective measures might be ordinal – they can be ranked, but how much difference there is between the first and the second, might be unknown, or three times more than between the second and third, but we have no way of establishing those degrees of difference. A more precise number scale, but still leaving something to be desired in terms of drawing conclusions from it and making judgments,is an interval scale. A certain distance along it means the same thing on different parts of the scale, but "0" does not mean an absence of the thing measured. Fahrenheit and Celsius temperature scales are examples. Finally, the attribute might be measured in terms of a ratio scale, where the same distance along the scale at different points means the same thing, "0" means the absence of the thing being measured, and we can say that a "2" of something along the scale is twice (or half) a "1," and that a "3" (twice) is (or half) a "6." Planners have been guilty of confusing these kinds of scales in the past in their pursuit of making objective judgments about various social conditions (Hodge 1963). Often, however, we do not possess an exact understanding of the phenomena that we are working with (see Figure 21.2).

Rather than standards, professionals must accept *criteria*, where we don't exactly know where we are, or where we should be along the dimension of the attribute, but we can agree either that more of the attribute is probably better than less or vice versa, but we cannot be more precise than that. A criterion creates a continuum, or penumbra within which a decision must take place. It establishes a direction or vector ("as much space as possible") along which one should aspire. In Boyce's (1970) terms, a criterion is a value to be minimized or maximized. Independent professional judgment plays a strong role here, rather than a book of standards that even a non-professional might look up and understand. Prescription and performance "standards" and criteria can be *combined* as follows:

A. *Performance criteria*: Point to some goal vector, a direction of aspiration, rather than to some precise target or threshold for achievement. Allowing parking to be at grade or below ground and covered or uncovered, is an example. In effect some kind of parking is required but its exact characteristics are not prescribed.

B. *Performance standards*: Indicate levels for satisfactory and unsatisfactory achievement, and describe the desired performance of the system in clearly demarcated terms without specifying how to achieve the goal. An instance is *requiring a parking space for every 1,000 sq. ft. of development*. It shows how the parking space is measured and performs vis-à-vis the use space.

C. *Prescriptive criteria*: Emphasize the means to accomplish an end that may or may not be well-identified, but the details of the means are open to interpretation, e.g. *parking entrances to subterranean garages should be located as close as possible to the side or rear of each lot. "As close as possible,"* is how the entrance is to be located, but that phrase is open to considerable interpretation as far as the actual performance goes.

D. *Prescriptive standards*: These rules are the most precise of all. They avoid dilemmas of means-ends chains that performance standards engender, for they concentrate exclusively on particular means, completely and precisely specified. An example is requiring that *where an alley is present, services and trash containers shall be located on the alley.* We know how to provide for services and trash containers, but is convenient access the goal? Or is reducing possible noise and disturbance from trash pickup the goal? Either of which might be achieved by other means.

Figure 21.3 diagrams the three dimensions of professional standards taken together. The figure helps make clear the eight arenas with their different characteristics that exist in a profession's efforts to distinguish the acceptable from the unacceptable practice. It also illustrates the variety of configurations that these professional standards can come in. At the upper left corner, for instance, the sub-cell depicts prescriptive process criteria. The semblance

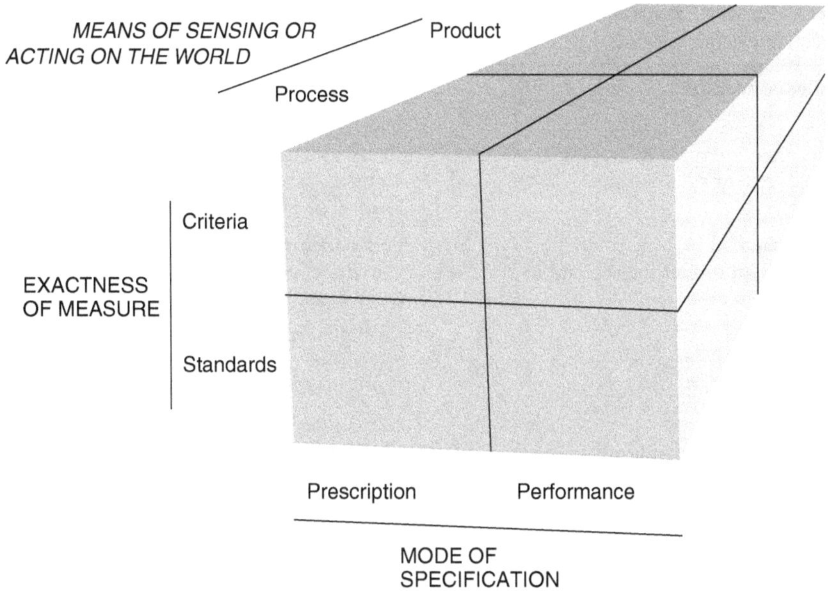

Figure 21.3 Eight permutations of rule forms.

of the processes is both well known and exacting, yet ultimately somewhat indeterminate in the particulars. An example is specifying the particulars of design competitions. In practice, standards and regulations are not pure examples of one or the other. *Different forms of the above eight possibilities may be strung together (concatenated) in a paragraph, sometimes even in a single sentence with several clauses.*

Table 21.1 provides examples for each of the eight cells to bring this rather abstract "theory" down to earth, to impart a flavor for what is meant by the framework.

Strategies in using different combinations of rule components

Professional standards and official regulations are comprised of components for dealing with risk. Their combinatorial design represents attitudinal stances toward risk-taking and risk aversion in light of societal objectives. These combinations also reflect views on cost-effectiveness and opportunity costs from alternative ways of achieving those goals. Accordingly, there are implicit strategies that underlie the choice of one or several of these combinations.

Choosing between process and product

The issue is contextual, depending largely on technical knowledge of the phenomena and current knowledge of cause and effect. (1) The choice is probably due less to risk than to relevance. (2) The distinctions between process and product are more likely ones of degree or emphasis than absolutes. Product and process are likely to be intertwined, linked in chains (associated with a means-end chain) (see Simon 1969). (3) Professionalism is largely concerned with teaching skills and expertise over use of artifacts (e.g. formulas, blueprints) that capture, mimic, or amplify process and product.

Table 21.1 Types of rules, standards, and regulations related to urban development and the nature of their construction.

Embodies a product	**Examples**
Performance criteria (outputs often interpretive because of nominal or ordinal scale of measurement)	Vision statements Certain sections of building codes Contract for professional services Form-based codes (secondary orientation)
Performance standard (little leeway for output interpretation)	Performance zoning Certain sections of building codes Form-based codes (secondary orientation)
Prescription criteria (inputs often interpretive because of nominal or ordinal scale of measurement)	Most general plans Certain sections of building codes Form-based codes (primary orientation)
Prescription standard (little leeway for input interpretation)	Certain sections of building codes Most zoning ordinances Form-based codes (primary orientation) Maps, diagrams, blueprints
Embodies a process	**Examples**
Performance criteria (outputs often interpretive because of nominal or ordinal scale of measurement)	Building specifications Contract for professional services
Performance standard or protocol (little leeway for output interpretation)	Contract for professional services with tightly demarcated time frame
Prescription criteria (inputs often interpretive because of nominal or ordinal scale of measurement)	Action plan Form-based code transect
Prescription standard or protocol (little leeway for input interpretation)	Action plan Form-based code transect

Choosing between prescription vs. performance

Here there are two considerations: (1) the uncertainty of those who formulate the standards and regulations about cause-effect relationships; and (2) the uncertainty about the level of knowledge and skills possessed by people who apply the standards and regulations, and the willingness of those who must conform to them. The prescriptive mode implicitly assumes that satisfactory (not necessarily optimal) solutions are specified for known problems. These solutions also avoid unwanted side-effects. Change can only come upon re-writing the code. There are few incentives to search for improvements.

By contrast, the performance specification, being goal-oriented, allows innovation in the means. It too is risk averse, but provides incentives for finding better solutions. The performance approach cannot easily control for unintended side-effects.

There must be extensive (and often expensive) testing. Moreover, because solutions may be innovative and not fully tested, officials assume greater risk in approving it – and being wrong – than insisting upon the prescribed solution, which, even if wrong, absolves blame (Baer 1986).

Choosing between criteria vs. standards

The use of criteria has to do with knowledge of cause and effect, and thresholds of occurrence. Criteria suggest that there are no known critical thresholds or target ranges to be achieved – only that more (or less) of the value achieved (avoided) is better. Advantages are their ability to sustain relevance during a period of change, and to encourage effort (e.g. no "satisficing," where minimums, often for cost reasons, become maximums in practice). Disadvantages are the uncertainty of the degree to which the criteria should be met. Transaction costs may be high as criteria invite disagreement over acceptability of a given action. However, in the political realm, the ambiguity of criteria may be an advantage, allowing agreement on the regulation in the first place because of its vagueness (Hetzel *et al.* 1993).

Standards by contrast, suggest greater technical understanding of cause-effect relationships and the relatively greater importance of the precise degree of behavior or result being specified. Critical thresholds are established that must be achieved (or avoided). But their very precision may make difficult any political agreement over their specification.

Reflections on past practice

What has been the general history of urban design practice in light of these classifications? Emily Talen's (2009) wide ranging account of codes in urban history reveals that they are largely prescriptive standards oriented to a final *product*. Apparently, this approach has appeared to be the most logical and appropriate to people down through history, so it is perhaps no surprise that the new form-based codes revert to the same set. Perhaps because Christopher Alexander's alternative approach (1977, 1979 and 1987) is such a departure it has caused so much attention (Mehaffy 2008: 62). In intent, Alexander's approach consists of largely prescriptive criteria, but which are oriented toward a *process*. Moreover, he explicitly adds a step – describing the process's purpose or goal, which in his case is to create a product, a city that is of a "whole" (Alexander 1987). Note that wholeness is not a goal standard; it is a goal criterion, left undefined. But his addition shows what is missing in traditional urban codes. What is their purpose? What larger societal goal will be attained if the codes are followed, if the individual aspects of the prescriptive standards for a product are met or achieved? Presumably, the answer is something along the lines of "increasing people's health, safety, and welfare or well-being" from perceiving and sensing a well-formed neighborhood and city, but this end or purpose is apparently thought so basic that it is rarely stated. Is that omission helpful?

Conclusions

Systematizing the way to think about rule formulation helps provide clear professional norms, regulations, and standards for design. It also helps in communicating across professions. The proliferation of technical knowledge and professional "know-how" about our urban environment has led to specializations in urban professions unheard of at the outset of the twentieth century when the world was still largely rural and agrarian. These professions must communicate with one another over an urban project and its design. Professional standards, their framework and grammar, offer a helpful means to understand the emphases of the city-building professions as they collaborate on aspects of urban design.

References

Alexander, C. (1977) *A Pattern Language: Towns, Buildings, Construction*, New York: Oxford University Press.

—— (1979) *A Timeless Way of Building*, New York and Oxford: Oxford University Press.

—— (1987) *A New Theory of Urban Design*, New York and Oxford: Oxford University Press.

Baer, W.C. (1986) "Expertise and professional standards," *Work and Occupations*, 13: 532–552.

—— (1997) "Toward the design of regulations for the built environment," *Environment and Planning B: Planning and Design*, 24: 37–57.

Baer, W.C. and Banerjee, T.K. (1977) "Environmental research, environmental design, and the 'applicability gap'" in P. Suedfeld, J.A. Russell, L.M. Ward, F. Szigeti and G. Davis (Eds.) *The Behavioral Basis of Design*, Book 2: Session Summaries and Papers. Proceedings of the Seventh International Conference of the Environmental Design Research Association, Vancouver, British Columbia, Canada, Stroudsburg, PA: Dowden, Hutchinson & Ross, 203–210.

Boyce, D.E. (1970) "Toward a framework for defining and applying urban indicators in plan-making," *Urban Affairs Quarterly*, 6(2): 145–171.

Hattis, D.B. (1972) "The relationship of the performance concept to the planning process – developing performance requirements for community mental health centers" in B.E. Foster (ed.) *Performance Concept in Buildings (2 vols.)*, Washington, DC: National Bureau of Standards Special Publication 361.

Hetzel, O.J., Libonati, M.E. and Williams, R.F. (1993) *Legislative Law and Process*, Charlottesville, VA: The Michie Company.

Hodge, G. (1963) "Use and misuse of measurement in planning," *Journal of the American Institute of Planners*, 29(2): 112–121.

Hutcheon, N.B. (1972) "Report of the rapporteur," in B.E. Foster. (ed.) *Performance Concept in Buildings (2 vols.)*, Washington, DC: National Bureau of Standards Special Publication 361, February.

International Conference of Building Officials (ICBO) (1995) *Handbook to the Uniform Building Code: An Illustrative Commentary*, Whittier, CA: International Conference of Building Officials.

Larkin, J. and Simon H.A. (1987) "Why a diagram is (sometimes) worth 10,000 words," *Cognitive Science*, 11: 65–100.

Lebas, E., Magri, S. and Topalov, C. (1991) "Reconstruction and popular housing after the First World War: a comparative study of France, Great Britain, Italy and the United States," *Planning Perspectives*, 6: 249–267.

March, J.G. and Simon, H.A. (1958) *Organizations*, New York: Wiley.

Mehaffy, M. W. (2008) "Generative methods in urban design: a progress assessment," *Journal of Urbanism*, 1(1) (March) 57–75.

Parolek, D.G., Parolek, K. and Crawford, P.C. (2008) *Form-Based Codes: A Guide for Planners, Urban Designers, Municipalities, and Developers*, Hoboken, NJ: John Wiley & Sons.

Shapira, Z. (1995) *Risk Taking: A Managerial Perspective*, New York: Russell Sage Foundation.

Simon, H.A. (1969) "The architecture of complexity" in H.A. Simon (ed.) *The Sciences of the Artificial*, Cambridge, MA: MIT Press, 84–118.

Skitowski, R.J. and. Ohm, B.W (2006) "Form-based land development regulations," *The Urban Lawyer*, 38 (1): 163–172.

Talen, E. (2009) "Design by the rules: the historical underpinnings of form-based codes," *Journal of the American Planning Association*, 75, 2 (Spring) 144–160.

Van Court, D.P. (1972) "Discussion" in B.E. Foster (ed.) *Performance Concept in Buildings (2 vols.)*., Washington, DC: National Bureau of Standards Special Publication 361, February.

Wright, J.H.G. (1983) *Building Control by Legislation: The UK Experience*, Chichester: John Wiley & Sons.

Further reading

Baer, W.C. (1997) "Toward the design of regulations for the built environment," *Environment and Planning B: Planning and Design*, 24: 37–57. Covers a wide scope of regulating the built environment and approaches to them by different fields.

Hodge, G. (1963) "Use and misuse of measurement in planning," *Journal of the American Institute of Planners*, 29, 2, 112–121. Discusses some elementals of measurement overlooked by planners and designers in developing rating scales; includes planning examples.

Simon, H.A. (ed.) (1969) *The Sciences of the Artificial*, Cambridge: MIT Press. A seminal work on human constructs.

Talen, E. (2009) "Design by the rules: the historical underpinnings of form-based codes," *Journal of the American Planning Association*, 75, 2 (Spring) 144–160. A wide-ranging look at how earlier civilizations besides the early United States have approached the basic problem.

22

Decoding design guidance

Matthew Carmona

This chapter focuses on the use of design guidance as a tool in the design/development process. It begins with a short introduction to design guidance as a generic type, revealing its variety and distinguishing characteristics. Based on empirical research in England derived from a national pilot program, the chapter goes on to examine one particular form of guidance – the design code. The aim here is to explore in greater depth the relationship between design guidance and the broader design and development process. The discussion recognizes that the nature and limitations of all forms of design guidance need to be fully understood before they are applied in practice.

What is design guidance?

At its most basic, design guidance can be defined as a generic term for a range of tools that set out design parameters with the intention of better directing the design of development. Different countries have different traditions and use different forms of guidance to greater or lesser degrees. Design guidance of various descriptions is very popular in continental Europe, for example the German Bebauungsplans which represent sophisticated site-specific tools for guiding the urban structure of developments, whilst in France typo-morphological guidance is commonly used to understand and respond to the character of larger historic areas. In Australia, Victoria's Rescode provides a state-level design guide for residential developments, while in the US, the New Urbanists' Transect offers a generic form of design guidance offering prescriptive design solutions for all types of development across the continuum from city core to countryside.

In the UK, if one asked "What is design guidance?" the detailed and unwieldy residential design guides produced by local authorities up and down the country since the 1970s would come to mind; the Essex Design Guide being the most famous (see http://www.the-edi.co.uk/?section = publications_EDG). These forms of guidance were, and still are, produced by the public sector to guide the design of (predominantly) housing developments across entire counties. Yet design guidance does not have to take this form. It does not have to be produced by the public sector; it can relate to all types of development, and rather than generic guidance for all areas within an administrative jurisdiction, it can be customized to guide development for specific areas or sites.

Reflecting this diversity, there has been a proliferation of types of design guidance including: local design guides, design strategies, design frameworks, design briefs, development standards, spatial master plans, design codes, design protocols, and design charters. These terms are often confusing, poorly defined and over-lapping, and despite attempts to classify them in relation to one another (e.g. Carmona 1996), their sheer variety only helps to illustrate the ambiguity of design guidance as a design/ development tool, and the confusion that can too easily result from its use.

In this chapter, no attempt will be made to discuss each of these types of design guidance. Instead, by way of example, discussion will focus on one particular form of design guidance – the design code. Accordingly, it is first necessary to briefly put some flesh on the bones of the definition proposed at the start of this chapter, by discussing the nature and diversity of design guidance generically as a tool, starting with what design guidance is not.

Design guidance is not a legally defined and binding ordinance or policy, because these tools suggest an element of enforceability that the term "guidance" cannot possess. Instead, guidance suggests advice rather than compulsion. Second, it cannot be a "blue-print," because "guidance" equally suggests a sense of direction for, but not an end solution to, a design problem. Finally, guidance cannot simply be analysis such as site or character appraisals, as analysis in isolation does not suggest a design direction at all, only information that might be useful in establishing one. As such, it is not always immediately apparent how design guidance fits into the range of tools available to those in the development process.

Kevin Lynch's (1976: 41–55) four modes of action for public authorities – diagnosis, policy, design, and regulation – for example, make no reference to guidance. In fact, aspects of design guidance will often have a role in each of Lynch's modes, and the boundaries between guidance and at least the first three modes will not be clear. Some forms of policy may contain guidance; some design guidance will contain site or character appraisal information, and seemingly fixed design schemes may be open to interpretation as successive phases of a development come forward.

The characteristics of design guidance

Despite the ambiguity and the surfeit of labels for different forms of design guidance, it is possible to classify design guidance through a number of its characteristics:

Subject matter – classifying by subject matter is the most obvious and straightforward, in other words by land use, location (suburban, urban, rural), or development issue (e.g. infill sites, shop fronts, building additions, etc.). Some forms of design guidance may deal with more than one of these.

Context type – a related issue is the context to which guidance pertains, and in particular its relative sensitivity, for example whether concerned with extensive new-build sites, in-fill development in established urban areas, or change within a historic setting.

Scale of application – a further related issue concerns the scale of application; whether dealing with strategic design concerns such as infrastructure provision, urban design issues (space networks, public realm, mix of uses, etc.), or questions of architecture and detailed landscape design.

Governance level – in the UK, design guidance is produced at all levels from central government and its various agencies, to regional and sub-regional authorities, to local authorities. This can produce complex regimes of policy and guidance that are sometimes conflicting and repetitious.

Generic vs. specific – a related question is application, whether guidance relates to specific and well-identified sites, or is generic, relating to large areas (e.g. a whole local authority) and undefined sites. Generally, the smaller the scale of application, and the lower the governance level, the greater the degree of specificity.

Level of detail – different forms of design guidance vary considerably in terms of their level of detail, from broad aspirational principles of "good" design, to very detailed guidance on particular aspects of a design problem. The level of detail will even vary within a particular guide, from subject to subject.

Level of prescription – to some degree the level of specification will depend on the level of importance attached to a particular design concern, which may also be reflected in the way guidance is expressed. Although design guidance should remain advisory, some aspects may be expressed with a greater or lesser degree of conviction than others: "developers should normally..." as opposed to "developers might consider..."

Ownership – whether instigated and owned by a public or private organization offers a further means to classify guidance. Typically design guidance is associated with the desire of public sector agencies to improve (in the public interest) the design of private sector development. But design guidance is also produced by the private sector both to guide an enterprise's own developments and to shape the inputs of different corporate partners into a common project; for example where different home builders are working on neighboring phases of a larger development. As in the public sector, the contents and style will vary from case to case.

Process or product – a critical distinction will reflect the relative emphasis in guidance on the design, development and regulatory processes as opposed to the desired products or outcomes. Design guidance typically incorporates both sets of concerns, although some will focus solely on one or the other.

Medium of representation – a final classification might reflect the medium through which guidance is represented, be that traditional printed form, or through more interactive electronic and web-based means. This will not necessarily change the content of guidance, but will determine its style and most likely how and by whom it is used.

The above distinctions are demonstrated in Table 22.1 for three very different (but historically influential) examples of design guidance in the UK. Unfortunately, knowing that a great variety of design guidance exists is of little value unless users understand, first, why different forms of guidance are used and, second, their problems and potentials. The first of these questions appears simple; all forms of design guidance exist for one purpose, to inform the process of design so that it is more likely to achieve a specified set of design ends. Thus guidance can be deemed successful if these outcomes are better than would have been achieved without it.

The goals envisaged for design guidance, however, may vary, depending on the ambitions of its instigators and the nature of the development context; whether the intention is to establish minimum desirable thresholds for quality or to raise the bar and strive for superior design. The former – a safety net approach – may be the limited ambition of a generic design guide or a guide in an area beset by poor quality development. The latter – a springboard to excellence – should be the case for site-specific guidance or for guidance in an area where stakeholders are already committed to achieving better quality (see Table 22.1). Although not mutually exclusive, these aspirations would depend on the nature of likely users, the extent to which they are receptive to the content

Table 22.1 Design guidance compared.

	Canary Wharf Design Guidelines (1987)	*Hulme Guide to Development (1994)*	*Essex Design Guide (2005)*
Generic 'type'	Design code	Design strategy/code	Local design guide
Subject matter	Commercial office and public realm	Residential development and public realm	Residential and mixed use areas
Context type	New build brownfield	Clearance and regeneration	Infill and new build greenfield
Scale of application	Architecture and landscape	Urban design	Urban design, architecture, landscape
Governance level	n/a (enterprise zone)	Local	Sub-regional
Generic vs. specific	Specific	Specific	Generic
Level of detail	Highly detailed	Broad principles	Comprehensive coverage
Level of prescription	Highly prescriptive	Advisory	Advisory
Ownership	Private	Public, quango	Public, local government
Process or product	Product	Product	Process and product
Medium of presentation	Traditional	Traditional	Traditional
	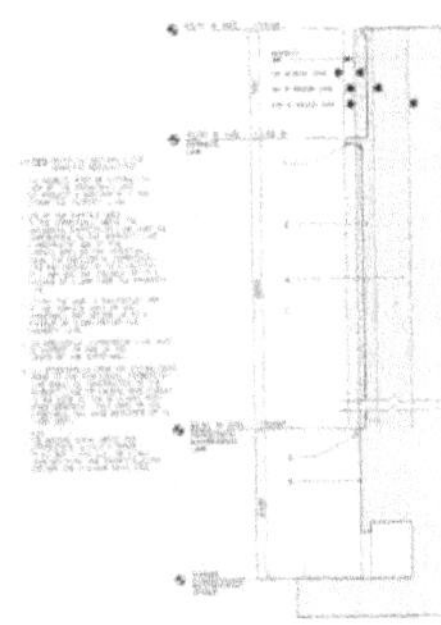		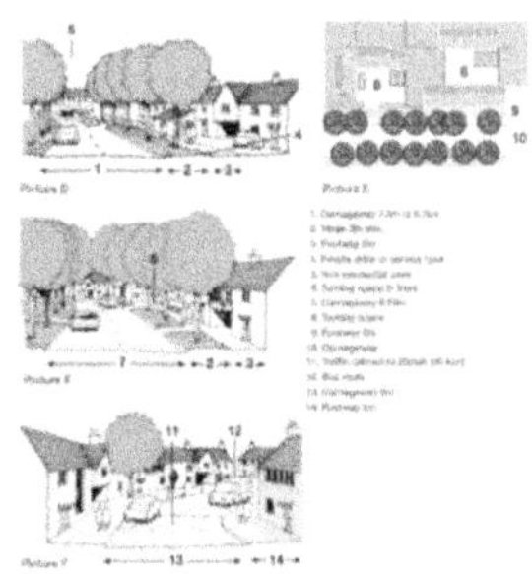
Goals	Higher quality	Threshold quality	Threshold quality

of guidance, and on the balance of power between stakeholders (particularly between public and private sectors) within the development process (Bentley 1999: 28–43).

All this implies that the nature of the development process and how design guidance is used within it needs to be fully understood. This is best discussed through focusing on a particular type of design guidance: the use of design codes in England. By this means it will be possible to clarify the problems and potentials of at least this one form of design guidance, used in one context, and also to extrapolate some wider lessons of relevance elsewhere.

Design codes in England – a national pilot program

No one sets out to create poorly laid out, characterless places, yet throughout the world much of what is built today continues to display these characteristics. In England, for example, recent analysis of new-build housing schemes across the country has revealed consistent failures to deliver even basic design aspirations, such as distinguishing between public and private realms; letting public space and buildings, rather than highways, dictate layout; and taking advantage of the positive characteristics of sites (CABE 2004,

2005, 2007). Driven by concerns over quality, coupled with a national need to deliver more housing, in 2004 the national government launched an extensive pilot program aimed at assessing the potential of design coding to deliver better quality development. This national pilot program involved the detailed monitoring and evaluation of nineteen development projects over a two-year period (Carmona and Dann 2006) and revealed a range of potential benefits of design codes, including:

- Better designed development, with less opposition locally and a more level playing field for developers
- Enhanced economic value derived from the positive sense of place that better quality design can deliver
- Less uncertainty with the planning process and a resulting positive climate for business investment
- Streamlined regulatory processes, saving time and money for developers and local authorities alike
- A more coordinated development process, built on consensus instead of conflict.

On the face of it such benefits might seem puzzling when many of the generic development standards used to guide the design of the sorts of sub-standard schemes referred to above could be described as coding – of sorts. Regulations for building control, highway design standards, density, and open space standards used by many local planning authorities fall into this category. Most of these, however, are limited in their scope and technical in their aspirations and are not generated out of a physical vision or understanding of a particular place. Instead, these types of generic development standards are about achieving minimum thresholds across the board and apply to whole administrative areas. They are what Ben-Joseph (2005) has described as the hidden codes of the city. Research has suggested that the slavish adherence to such guidance is a direct cause of much bland and unattractive development (Carmona 2001).

Site-specific design codes, by contrast, are a distinct form of detailed design guidance that stipulates the three-dimensional components of a particular development and how these relate to one another without establishing the overall outcome. The aim is to provide clarity over what constitutes acceptable design quality for a particular site or area. Used in this way, and unlike generic development standards, design codes can provide a positive statement about the qualities of a particular place (see Figures 22.1 and 22.2).

Why choose codes?

In England today, national planning policy requires that "Planning authorities should plan positively for the achievement of high quality and inclusive design for all development" (ODPM 2005: para 34). In the residential sector, the increasing imperative to deliver better quality design has led to a decline in the traditional way of doing business. That typically saw developers ignoring local policy and guidance, submitting sub-standard planning applications, then using their often considerable resources to battle their way through the permissive national planning appeals process in order to obtain planning permission (Carmona 2001).

Today, instead, most large-scale residential or mixed use development proposals are preceded by the preparation of detailed design guidance in order to create the confidence that design quality will be forthcoming. Such guidance may be of several types, for example a detailed master plan, or a loose development framework followed by more detailed

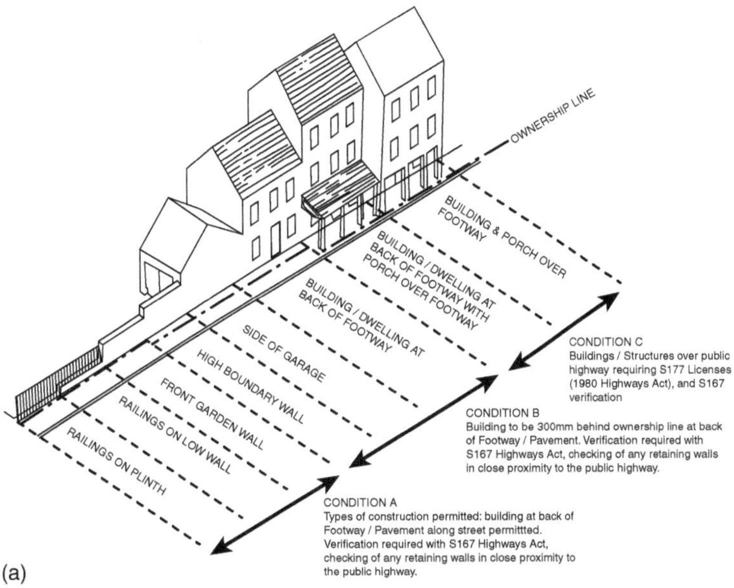

(a)

(b)

Figure 22.1 Example case study – Swindon.

Note: The Swindon Southern Development Area project is a large-scale urban extension on a site of 309 hectares in the western corridor of the town of Swindon. 4,500 homes were proposed plus a mixed-use street, schools, employment and park and ride facilities in a master plan that had outline planning permission. The developer, through a collaboration agreement with the council (in its role as majority landowner) led the preparation of the design code with its consultant code designers. The vision for the code was set out as part of the master planning process whilst the code was intended to put the master plan into effect. The master plan proposed a contemporary reinterpretation of a traditional Wiltshire settlement with traffic subjugated to pedestrian movement, a human scale, and a continuous street network. The code elaborated the vision, by defining appropriate references for built form character, for example a materials palette. It placed particular stress on typical street sections and plans, on sustainability, the design of the public realm, and a traditional approach to the architecture. The code followed intensive discussions between the code designer and the planning and highways authorities to agree on coding principles. A planning condition to the detailed application requires that the code be approved by the council before construction begins. The code will also be a part of land sales agreements.

(a)

(b)

Figure 22.2 Example case study – Newhall.

Note: Newhall is the first phase of a proposed urban extension of 2,800 homes and supporting amenities, with outline planning permission for 440 dwellings on a site of 17.4 hectares. The development is promoted by the landowner, New Hall Projects Ltd, a firm with a vision for a contemporary extension to Harlow. The scheme follows the preparation of a planning and design brief which was agreed with the council. An outline application was then submitted for the first phase. A planning condition to this required the approval of a detailed master plan, a requirement that is being met through the preparation of design codes. The codes are being prepared for each parcel of land that is marketed, and these form part of the brief to potential developers. The code designer assesses developer submissions and takes an active role in making sure that the master plan vision is achieved. In later phases a joint venture arrangement has been established between the landowner and parcel developers in order to retain more control over the final outcomes.

development briefs for each phase of development. Although different, each form of guidance will share many of the same costs and benefits of design coding. The final choice of which form of design guidance to use, is best left to local preference, but findings from the national pilot program showed that design codes can be distinguished from other forms of detailed design guidance because of their particular ability to:

- Establish high quality design aspirations in a manner that allows their consistent application across successive phases of large sites

- Provide a robust form of design guidance that, because of its relative prescription, is difficult to challenge at appeal
- Test, develop, and deliver the site-specific vision (usually contained in a master plan) by designing and fixing the "must-have" design parameters of a scheme
- Create a level playing field for development interests, based on their willingness and ability to deliver high quality design.

Of these, perhaps the key strength of design codes is their ability to coordinate design across the successive development phases of large sites in order to deliver a coherent design vision. As such, they are most valuable when sites are either: large (or multiple smaller adjacent sites) that will be developed in phases over a long period of time, in multiple ownership, or likely to be developed by multiple development and design teams.

Where do codes fit within the development process?

If design codes are the guidance of choice, the next question is how should they operate? Production of a new development involves many disparate processes and design codes may play a role in each:

- *Design processes* – design codes are tools to set the detailed urban design parameters of projects across the different scales of design intervention, from street and block sizes and layouts to landscape and architectural concerns, towards a coordinated vision of place.
- *Development processes* – because of the detailed up-front work required for their preparation, the design phase of codes offers an opportunity for stakeholders to explore and negotiate different design options and their associated costs.
- *Planning processes* – the preparation of design codes provides an opportunity for planning authorities to engage directly in the design process, rather than reactively responding to already completed development proposals. They also offer a ready means against which to evaluate and monitor detailed planning applications.
- *Adoption processes* – design codes have a role in the legal adoption by the state of highways, open space, drainage and other infrastructure. They enable these processes to be coordinated with design, development and planning matters at an early stage, thereby avoiding possible conflicts later in the development process.

Through the national pilot program it was possible to identify a common set of phases involved in successful implementation of design codes. Although the process is essentially linear (Figure 22.3), it is often necessary to return to and refine earlier decisions in the light of later information. In summary it incorporates:

1. Initiating the code – defining an agreed process and establishing leadership arrangements.
2. Coordinating inputs into the coding process – the skills, financial resources, and the roles and relationships of various actors who will in turn design and implement the code.
3. Appraising the local context for coding – including existing policy and guidance or consents already covering the site, the character of the site, and any existing physical vision such as a masterplan.

4 Designing the code – devising, structuring, writing and illustrating the content and expression of the code.
5 Formalizing the code – giving the code institutional status by adopting for planning, highways or other purposes, or by other means such as tying it to a land sales agreement.
6 Implementing the code – using compliance with the code as the basis for selecting design and development teams for individual land parcels, to inform the site design process, and also for assessment and regulation of the resulting proposals.
7 Managing code compliance – via monitoring and enforcement processes to evaluate performance of the code in order to refine it, and through use of the code for project aftercare.

The creation and use of a design code also draws from and feeds into the broader development process (see Figure 22.3). In reality every development process is different and its various phases do not always follow a neat sequence. Nevertheless, it is valuable to consider the coding and development processes together in order to understand how the code can be informed by the wider processes of development. Importantly, code preparation will draw information from other development stages (e.g. master planning and community engagement), and likewise, once prepared, the code will feed into and inform later development stages (e.g. parcel design or detailed approvals).

The stakeholders, roles and motivations

The central role of the design code within the development process means that it brings together a wide range of individuals and organizations with a stake in the development outcomes. These can be divided into two groups: the "coding team," which comprises the full range of professional stakeholders involved in producing and using the code, and "wider interests," such as the local community (Table 22.2). The coding team can be broken down into four sets of interests: land, design, development and public interests. The national pilot program suggested that understanding the intersecting roles and primary motivations of these groups is the key to forging a successful coding process. Individually they will vary (see Table 22.2), but collectively they will encompass:

- The delivery of high quality design;
- Optimizing investment returns – a necessary pre-condition;
- Creating a predictable and efficient development process – to facilitate the necessary investment;
- Delivering planned development capacities – e.g. through determining densities, use mixes, etc;
- Achieving key technical design parameters – whilst avoiding their over-dominance in design outcomes;
- Establishing consensus over the development.

Arguably, therefore, to succeed, design codes will need to address these collective motivations. But not every scheme that is subject to a design code will follow the same process, and the roles of key stakeholders will vary correspondingly. For example, whether public (see Figure 22.1) or private (see Figure 22.2) sector stakeholders lead, the process may determine who takes which role within the coding team. Certain roles can also be combined in single stakeholders, for instance: local authorities with appropriate skills in-house may take on the role of code designer; landowners may act as the master-developer; or the master-developer may subsume the role of parcel developer.

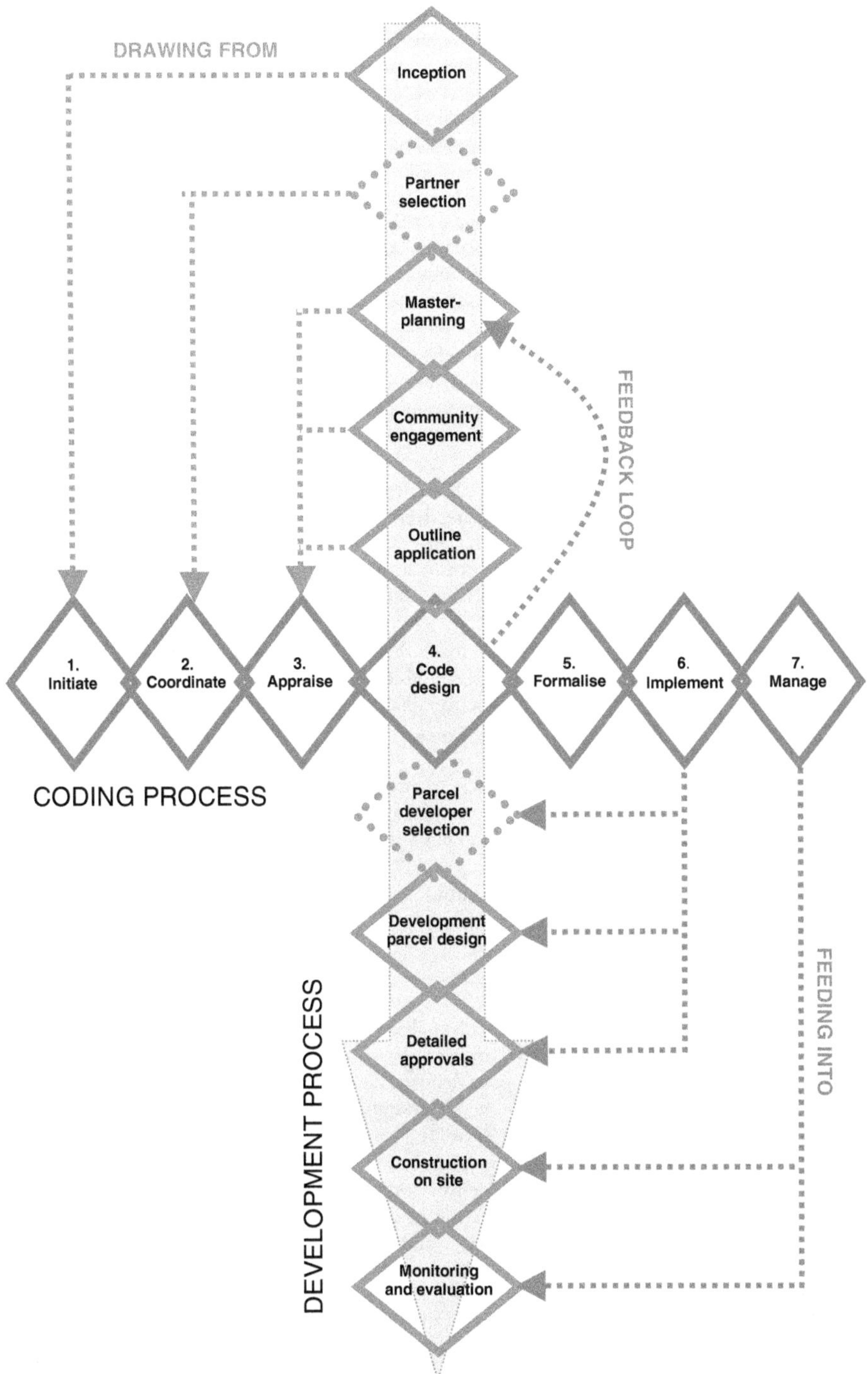

Figure 22.3 Coding and the development process.

Table 22.2 The roles and motivations of key stakeholders within a typical coding process.

Groups	*Interests*	*Stakeholders*	*Prime motivations*	*Key potential stakeholder roles include*
Coding team	**Land interests**	Landowner	To get the land developed and make a profit	Establishing aspirations from the start for design quality, using freehold rights throughout to guarantee delivery against the design code
		Master-developer	To maximise site potential and thereby long-term profit	Initiating the site-based vision and code design process through appointment of designers, and subsequently assessing parcel development proposals against the code
		Funding agency	To deliver a return on public investment	Using landownership and funding powers to deliver the requisite skills, resources and know-how for a high quality coding process, and effective assessment and enforcement
	Design interests	Masterplanner/ framework designer	Within client objectives to deliver a coordinating design vision	Preparing the masterplan or development framework as a strong vision for the long-term development of a site(s), reflecting any existing policy and guidance, local consensus on the vision and the client's brief
		Code designer	To make the design vision deliverable	Coordinating different interests as a basis to prepare the design code as a means to implement the essential principles contained in the masterplan/vision
	Development interests	Parcel developers	To maximise site potential and thereby profit	Developing proposals and achieving consents to deliver on site a development parcel within the masterplan/ vision
		Registered social landlords (RSLs)	To house social tenants	If involved, developing proposals and achieving consents for the delivery on site of a development parcel – or part thereof – within the masterplan/vision
		Parcel designers	Within client objectives to deliver a viable design solution	Creatively interpreting the code and masterplan to develop high quality designs for individual land parcels and their constituent buildings, spaces and areas
	Public interests	Planning authority	To protect and deliver complex economic, social and environmental public interest objectives	Establishing aspirations from the start for a high quality development, initiating or playing a role in initiating the masterplan/vision and design code, and administering the development control and any enforcement processes on the basis of the code
		Highways authority/agency	To deliver a safe and efficient movement network	Playing a role in design code production, revising and updating existing highways standards as necessary, and assessing and adopting the infrastructure that results from development

Table 22.2 (*Continued*)

Groups	*Interests*	*Stakeholders*	*Prime motivations*	*Key potential stakeholder roles include*
		Environment Agency	To protect local environmental resources	Approving discharge from drainage facilities (i.e. SUDS), and advice on incorporation in the design code
		Building control	To satisfy technical building regulations	Approving parcel proposals against the national building regulations, and advice on incorporation and adaptation for the design code
Wider interests	**Private interests**	Utilities providers (including water)	To establish an efficient and profitable utilities network	Adopting service infrastructure, and advice on incorporation of requirements in the design code
		Local councillors	To satisfy statutory obligations whilst protecting local voter interests	Establishing design aspirations in advance of development interest, approving masterplan/vision and design code and delegating authority to officers to manage the delivery
	Community interests	Existing community	To protect and enhance local amenities (and often property values)	Engaging in the masterplanning / vision making process through serious and significant involvement
		Future occupiers	To meet future community needs	Involvement through normal planning processes and engagement in long-term management and maintenance processes on the basis of the design code

Seven fundamentals of coding

The national pilot program revealed seven further fundamental factors for the success of coding projects which begin and end with a commitment to design quality.

Urban design first

The achievement of good urban design should be the primary objective of all involved in the preparation and use of design codes. Increasingly, a compatible range of urban design principles are being advocated in practice manuals (e.g. Llewelyn-Davies 2007). These look beyond narrow debates about architectural aesthetics, and also reject purely technical design solutions. The goal of sustainability in particular needs to inform almost every aspect of code production, from considerations of density and mixed-use to the use of particular building materials or the choice of species in landscape design. It also implies a concern for social and economic sustainability, where good quality urban design has an important role to play in promoting social inclusion and economic revitalization across spatial scales.

Setting quality thresholds

Design codes should establish the essential unifying elements of "place," encouraging and enabling interpretation around that theme. First, they can set clear thresholds below which quality should not fall by providing both the parameters for design and the criteria against which formal assessments of the quality of proposals can be made. These criteria need to be expressed with a clarity and comprehensiveness that will allow proposals to be assessed in an objective manner. Second, codes can inspire those who design with them to strive for better design than they otherwise would do. Just as the constraints and opportunities of the site or the clients' brief provide a focus around which designers will creatively develop proposals (RFAC 1994: 69), so should the content of design codes, providing the freedom to innovate within the clearly established and unifying parameters of place.

Investing up front

The preparation of design codes involves a significant up-front commitment of time and resources by all parties. In the UK today, code or no code, such an up-front investment is to be expected for the major development proposals where design codes are typically used. The national pilot program suggested that design coded schemes enhance sales values and increase land values which more than compensate for the additional resources required during the design process. For the public sector, many potential "sticking-points" will be resolved during the coding process that would otherwise require negotiations during the processing of the planning application. Codes simply redistribute the time and resources required from both the public and private sectors – effectively front-loading them – rather than significantly adding to them.

Rules for delivery that build upon a spatial vision

Design codes are effective tools to help interpret, articulate and deliver the design vision expressed elsewhere, typically in a master plan or development framework (Table 22.3). As such, codes need to be built upon the firm foundation of a robust vision that has been tested for its technical and financial feasibility. Usually the vision will be prepared for a particular site, but sometimes it may apply to a wider area containing a number of development sites. Design codes themselves vary considerably along a continuum from those that significantly develop the core urban design principles of a design vision that otherwise remains largely conceptual, to those that only express (in a technical sense) the detailed design principles that are already established in the vision. Codes are equally valid at all positions along the continuum, whilst the level of detail and prescription across codes, or from coded element to coded element will be a matter for local decision.

A collaborative environment and a partnership of interests

A strong commitment to collaboration between partners and within organizations is a pre-requisite for successful and efficient coding. Designs of very different character and quality can still be produced using the same design code, emphasizing the critical importance of other factors as well, namely the quality and commitment to achieving excellence of all members of the coding team, and the resources at their disposal to secure this. Critical to the success of such a partnership is a core three-way relationship between the key public sector, land and the design interests. If a strong three-way relationship can be forged early on, then a commitment to the design code can be developed and maintained across these stakeholders, thus obviating any negative external pressures later in the process.

The importance of clear and effective leadership

Clear leadership is critical to effective coding, for keeping up the momentum and

Table 22.3 Design codes, building on the site-based vision.

Scales of action	*Masterplan*	*Design code*
Settlement pattern	Major infrastructure	Major roads, bridges, public transport network, design principles for combined heat and power systems
	Structure planting	Continuity, species, relation to topography
	Water management	Drainage, recycling, reed beds, water features
	Road and cycle network	Road types, hierarchies, dimensions, capacities and characters, cycle network continuity
	Open space network	Standards, open space typology and features, connectivity
	Character areas	Centres and sub-centres, walkable catchments, parcel size and sub-divisions
Urban form	Connections	Edge treatments, boundaries
	Street network	Urban grain, grid types, connectivity
	Block pattern	Block form, privacy distances, interiors
	Building lines	Frontage continuity, set backs
	Plot form	Plot size, width, adaptability
	Building location	Orientation, position on plot, overlooking and overshadowing, natural surveillance
	Density contours	Plot ratios, dwelling per hectare, intensification nodes
	Views and vistas	Relation to topography, corridors, backgrounds
Urban space	Open space	Standards, types, forms, layout, access, landscape, planting, management
	Public space	Patterns, types, enclosure ratios, forms, layout, connection, uses, management
	Carriageways	Road tracking, junctions, road specifications, traffic calming, services routing, servicing
	Cycle and footpaths	Path specifications, cycle track specifications, paving, kerbs, gutters, road markings, other details
	Public/private space	Principles for courtyards, mews, cul-de-sacs, covered streets, arcades, colonnades,
	Private gardens	Standards, back gardens, front gardens, roof gardens, landscaping
	Play spaces	Standards, types, equipment, management
	Parking	Standards, car parks, parking courts, on-street types and treatments, overlooking, lighting, landscaping
Local character	Building forms	Bulk, massing, heights, storey heights, forms building envelopes, plan depths, adaptability
	Building types	Detached, semi-detached, terraced / town house, flats, fronts and backs
	Building frontage	Active frontage, entrance frequency, architectural styles, features, proportions, rhythms, expression, window/wall ratios, materials, colours, balconies, porches, signage, shop-front design
	Mix of uses	Distribution, proportions, mixing – vertical, horizontal
	Townscape features	Eave lines, rooflines, chimneys, corner treatments, landmark/background treatments, focal points, advertising
	Heritage assets	Integration, preservation, management
	Street trees	Species, numbers, placements
	Soft landscape	Standards, planting species, biodiversity, lawns and verges, planting beds and areas, planters
	Public realm	Street furniture, bollards, boundary treatments/materials, public art, fountains, paving materials, colours, utilities equipment, street lighting, amenity lighting, bus shelters, CCTV, public toilets, cycle storage and parking

Table 22.3 (*Continued*)

Scales of action	*Masterplan*	*Design code*
Technical factors		Environmental standards and energy efficiency Access standards and disabled parking Refuse storage and recycling Tenure mixing, affordable housing Management and maintenance issues

Note: It will not always be necessary to include all these elements in a particular masterplan or design code.

making decisions. More often than not, successful examples of coding are characterized by one party or another being strongly motivated to achieve quality and acting in effect as a design champion. This leadership can come from landowners, master-developers, local authority officers, funding agencies or code designers, or a combination thereof. Political leadership is also required. Involving key local political decision-makers early within the coding process can help to gain political support, lead to a smoother planning process, and will give local politicians the necessary confidence to delegate decision-making authority to their professional advisors on the basis of the agreed design codes.

No substitute for skills – a multi-disciplinary approach

Design codes require the exercise of advanced design skills throughout the process of their preparation and use. Unlike other processes of development, coding distributes the creative input across three phases of design. The quality of the development is dependent upon the quality of the area or site-based spatial vision, the quality of the code itself, and the quality of the parcel or scheme design. This compares favorably with other design intensive approaches such as development based on a detailed master plan where the design endeavor is split between two phases of design (master plan and parcel design). In the UK it has marked a major advance on what has been the dominant model for large-scale residential development, where the basic design parameters are established to gain the outline planning permission after which a specialist layout designer prepares detailed schemes for each parcel of land based on standard housing units and technical, generic, development standards.

To code or not to code?

Throughout the national pilot program, arguments for and against the use of design codes raged in the British professional press: that they would stifle design creativity; be excessively bureaucratic and restrictive; and only deliver traditional design solutions (see Carmona 2010). Just like any other form of detailed design guidance, if design codes are poorly designed, or inappropriately used, then they may be part of the problem, and not the solution. However, international experience, for example in Germany and the Netherlands (see Carmona and Dann 2006: 232–234), and now evidence from the UK, suggests that these misconceptions have little basis in fact.

Used correctly, codes have a particular role to play in helping to deliver design quality for types of development – particularly large-scale predominantly residential developments – where it has typically been lacking in the recent past. The seven fundamentals discussed above relate directly to design coding, but also, in their essence, to other forms of design guidance. Design codes are not alone as tools with a role to play in enhancing design quality, and are certainly not appropriate for all forms of development. However, where they are, the evidence now suggests that they can make a real contribution to raising the bar and delivering a better quality built environment.

References

Ben-Joseph, E. (2005). *The Code of the City: Standards and the Hidden Language of Place Making*, Cambridge MA: MIT Press.

Bentley, I. (1999). *Urban Transformations: Power, People and Urban Design*, London: Routledge

Carmona, M. (1996). "Controlling Urban Design – Part 1: A Possible Renaissance," *Journal of Urban Design*, 1(1): 47–73.

—— (2001). *Housing Design Quality, Through Policy, Guidance and Review*, London: E.&F.N. Spon.

—— (2009). "Design Coding and the Creative, Market, and Regulatory Tyrannies of Practice," *Urban Studies*, 46(12): 2643–2667.

—— (2010). Coding for Creativity and Value, *Urban Studies*.

Carmona, M. and Dann, J. (2006). *Preparing Design Codes: A Practice Manual*, London: Department for Communities and Local Government.

Carmona, M., Blum, R., Hammond, L., Stevens, Q., Dann, J., Karski, A., Pittock, C., Rowlands, S., Stille, K. (2006a). *Design Coding in Practice, An Evaluation*, London: Department for Communities and Local Government

Carmona, M., Marshall, S. and Stevens, Q. (2006b). "Design Codes: Their Use and Potential," *Progress in Planning*, 65(4): 209–289.

Commission for Architecture and the Built Environment (CABE) (2004). *Housing Audit, Assessing the Quality of New Homes, London, the South East and the East of England*, London: CABE.

—— (2005). *Housing Audit. Assessing the Design Quality of New Homes in the North East, North West and Yorkshire and Humber*, London: CABE.

—— (2007). *Housing Audit. Assessing the Design Quality of New Homes in the East Midlands, West Midlands and South West*, London: CABE.

Llewelyn-Davies (2007). *Urban Design Compendium 1*, London: English Partnerships & Housing Corporation.

Lynch, K. (1976). *Managing the Sense of a Region*, Cambridge, MA: MIT Press.

Office of the Deputy Prime Minister (ODPM) (2005). *Planning Policy Statement 1: Delivering Sustainable Development*, London: ODPM.

Royal Fine Art Commission (RFAC) (1994). *What Makes a Good Building? An Inquiry by the Royal Fine Art Commission*, London: RFAC.

Further reading

Ben-Joseph, E. (2005). *The Code of the City: Standards and the Hidden Language of Place Making*, Cambridge MA: MIT Press. Examination of the relationship between standards and place making. It includes a historical overview of the evolution of codes and standards and an analysis of their impact on urban form.

Bentley, I. (1999). *Urban Transformations: Power, People and Urban Design*, London: Routledge. An inquiry of how people use and transform urban environments, including specifics for making better cities.

Carmona, M. (2001). *Housing Design Quality, Through Policy, Guidance and Review*, London, E.&F.N. Spon. Examination of how the public sector can utilize policy tools to achieve higher quality residential developments.

Carmona, M. and Dann, J. (2006). *Preparing Design Codes: A Practice Manual*, London, Department for Communities and Local Government. (http://webarchive.nationalarchives.gov.uk/+/http://www.communities.gov.uk/publications/citiesandregions/preparingdesigncodes) (accessed 2 September 2010). A practice guide providing direction for the use of design codes.

23

Urban design competitions

Ute Lehrer

> The procedure can be only as good as its sponsors, participants, jurors, experts – women and men equally. It is an illusion to wanting to guarantee highest quality *through competitions.*
>
> (Becker 1992: 249; italics in original)

> Although competitions attract publicity, (...) publicity does not necessarily translate into a successful building.
>
> (Nasar 1999: 3)

Competitions have marked the history of planning, urban design and architecture in many ways. The earliest records we have are from 448 BC for a memorial at the Acropolis, but we find indications for their continuous use throughout history (Spreiregen 1979: 299). The current system has its roots in the nineteenth-century French Beaux Arts tradition that was appropriated throughout the world – in some places more intensively (e.g. Switzerland) than in others (e.g. the United States of America). It has been argued that a well functioning competition system leads to better urban design (Alexander and Witzling 1990) and that "those countries whose architectural output is of the highest quality and is broadly applied have a great number of competitions" (Nasar 1999:6).

Today, competitions are often used to find not only the best possible design, but also to draw attention to a specific development. While investors typically prefer direct hiring over competitions because they have a greater control over the decision-making process and its final outcome, they also recognize the benefit of competitions with regard to attracting public attention. The physical articulation of a building or the spatial layout of an area is only one of the outcomes of a competition process. Indeed, using the urban design competition as a vehicle to draw public attention to a project – and turning the process into a spectacle – appears to become sometimes as important as the outcome. This spectacularization of the building process relies on the production of physical images and on their intrinsic and perceived meaning (Lehrer 2006). In addition, urban design competitions are increasingly appropriated as a tool for the recomposition of the actors who are involved in the economic, social, cultural, and political production of space, as well as to achieve a symbolic transformation of a place in cities that try to achieve global status. Both the recomposition of actors as well as the spectacularization of the building process have become key components of urban design competitions worldwide.

This chapter starts with a brief discussion of the main purpose of urban design competitions, their structure and different formats and routines, as well as their advantages and disadvantages. These arguments are further illustrated in a brief presentation of the two-stage competition for the redevelopment of historical war-damaged Potsdamer Platz in Berlin. The chapter concludes with an analysis of the competition at Potsdamer Platz and some thoughts about how competitions can be improved.

The institution of urban design competitions

Urban design competitions as part of the production process of built environments, are generators of both ideas for and images of urban environments, and are an efficient instrument to find the best proposal for a stated design question. They are different from architectural competitions in so far as they deal directly with public and private interests and contribute to a "redistribution or regulation of territorial power, control and rights of different social groups" (Banerjee and Loukaitou-Sideris 1990: 125). Therefore successful architects not only need to possess good design skills but also have to have a fundamental understanding of social, economic and political relations. Urban design competitions are particularly well suited in complex situations and have the potential to not only provide legitimacy for urban development processes but also to advance city-building processes by offering a variety of proposals and innovative ideas to a jury, who then can elaborate about the merits of each proposal. Thus, they are considered to be more democratic than direct hiring.

Competitions in general are an advanced design method because they generate a range of alternatives and perspectives to a stated program. While competitions come with a certain price, they are also described as cost-effective and can become "a publicity and fund-raising vehicle" for projects (Seidel 1990: 173). In addition, they have the capacity to educate the general public about the importance of good urban design and therefore foster an environment of high standards.

The process of competition itself is also an image producer: first, by announcing the terms of the competition (program); second, by opening or limiting the list of participants invited to compete (form of competition); third, by selecting and promoting the composition of the jury (jury selection process), fourth, by selecting the winning team and scheme; and fifth, by presenting the winner to the public and press (public presentation of outcome).

The production process of built environments undergoes a number of different stages – from the idea, to the design and possible revisions, to the construction, to the finished product. Some stages may follow certain, fairly standardized routines, while others are often dependent on the idiosyncrasies of a particular project and process. Regulations are set in place in order to control the process and the outcome. In the case of urban design competitions it is important that the decision-making process is fair and can be reviewed by all interested parties. Therefore, competitions expose and make public the traditionally confidential relationship between the client and the architect. Because of the public nature of a competition and its potential for direct image production, the design program has to be made publicly accessible, and the winning entries have to be shown in an exhibition. The media, of course, plays a crucial role in providing a platform for open debate.

Types of design competitions

Design competitions vary in their procedures and impact over the final shape of the built environment, and can be divided into two fundamentally different approaches: *ideas competition* (or also referred to as concept competition) and *project competition* (or also referred to as implementation competition). The choice or suitability of the type of competition depends on the kind of tasks to be achieved. Both types of competition can take place independently or complementarily, but if both approaches are chosen, the ideas competition always takes place before the project competition. A third type, on-site charrette competitions, are known for their high degree of effective community participation in the design process. Charrettes are not discussed here, but are presented in Chapter 24 of this volume by Doug Kelbaugh.

Ideas competitions are used to stimulate discussions and encourage innovation. They represent a practical tool, particularly in first, large-scale projects where the building volume will be much greater than the immediate need of space of the investors (for example, at Potsdamer Platz, Berlin; see Lehrer 2003); second, where the future site of a complex is not clearly defined (for example, the competition for the Wexner Center in Columbus, Ohio; see Nasar 1999); or third, where the land is not in the possession of the sponsor of the competition (for example *Hauptstadt Berlin,* 1957–58; see Geisert 1990). An ideas competition is a design exercise exploring a range of possibilities that are conceptual in nature. In order to implement these ideas, a further competition or direct hiring of an architectural firm is needed. Ideally, entries for ideas competitions not only analyze the spatial relationship of the new development to the rest of the urban landscape, they also shed light on the economic and social impacts of such developments.

The project competition has two main purposes: (a) to find a good design scheme (the program), and (b) to facilitate the implementation of the proposal in the most efficient and successful way. In contrast to the ideas competition, this type of competition is about making concrete and buildable plans. In fact, this procedure is very similar to the direct hiring of an architect, with the difference that direct hiring generates a single proposal, while a competition yields a number of different approaches and alternatives to choose from.

The chosen format of a competition dictates the eligibility of participants. Most of the literature distinguishes competitions into open, limited, and invited (Nasar 1999: 22). These different formats vary greatly in their advantages and disadvantages.

Open design competitions have a significant history in the European culture of competitions because they usually produce the most diverse solutions to a stated problem (Becker 1992). However, they are less common in North America and other parts of the world because it is often assumed, wrongly, that open design competitions would be inefficient, time consuming, and too expensive (Strong 1976). Open competitions are seen as advantageous because of the direct selection process of a proposal by a jury, and for the innovative ideas that often result from them. In addition, because they are open, they not only receive a large variety of entries but quite often also generate innovative proposals. Because the entries are anonymous, young architects also have a chance to be successful and win a competition, but even more importantly, an open competition provides them with a platform for developing their skills. Because there is no limitation on who can participate, sometimes open design competitions are faced with serious logistical problems due to the sheer amount of entries. For example, the ideas competition for the German parliament at *Spreebogen* in 1993 received 835 entries from 44 countries (Straub 1993); the World Trade Center Memorial Competition in 2003 had a total of 5,201 entries from 63 nations. Both competitions had such a huge participation rate that it was difficult to find enough space for exhibiting all the entries for jury review. Some open competitions with international participation, are now treated as milestones in architectural history, such as the 1922 competition for the Chicago Tribune Tower which had 263 entries from 23 countries (Solomonson 2003) or the competition for the Palace of the Soviets in Moscow with 160 entries from 24 countries. In both cases, it was not the winning design that made history but rather the submissions that showed innovative design schemes that became classics in the education of future generations of urban designers. Milestones for urban design were, among others, the competitions for the Ringstrasse Vienna in 1858, for Grossberlin in 1910, and for Brasilia in 1956.

Limited competitions are more common where certain criteria need to be fulfilled, for example geography (e.g. only licensed architects within a certain region can participate), nationality (e.g. only US citizens could submit entries for the US Embassy in Berlin), age group (e.g. only architects under 40 are allowed to participate in "Europan," a Europe-wide open competition (see Strong 1976: 115–116), or stage within the profession (e.g. only architectural students can submit entries). The benefit of a limited competition is that it reduces the number of entries while remaining open to all architectural firms or groups who fulfill certain criteria. Such limitation can nevertheless be a drawback, particularly when the entries all represent rather similar or even parochial approaches. Various countries in Europe prefer to allow participation to only domestic architectural firms. However, over the past few years, the European Union has exercised increased pressure to open up competitions to all architectural firms who reside in Europe.

Invited competitions are considered a compromise between the format of an open competition and the direct hiring of an architectural firm. The outcome of invited competitions is relatively predictable in terms of design approaches since one knows more or less the various approaches that the invited architectural firms will take. This stands in clear contrast to open competitions where there is a wide range of design suggestions (see Table 23.1). Invited competitions provide the fewest possible alternatives with the greatest possible guarantee of architectural attention. But there is a certain elitism encouraged by invited competitions, particularly in the case of projects for large corporations, where the same architectural firms are repeatedly invited. One could go so far as to argue that invited competitions are exclusionary practices that monopolize the ability to be hired for representative jobs to a handful of architects worldwide; such as Daniel Libeskind, Zaha Hadid, Herzog and De Meuron, Richard Rogers, Norman Foster, Renzo Piano, Arata Isozaki, Rem Koolhaas, and Frank Gehry. Indeed, some professional associations have

Table 23.1 Selection process in relation to certainty of outcome.

Importance of selection process	*Uncertainty of outcome*		
	low	*medium*	*high*
low			Open competition
medium		Limited competition	
high	Invited competition		

raised their voices against invited competitions because of their restrictive limitations and privileging of a selected few star architects and their firms.

The importance of the program

Whatever the process is, be it ideas competition and/or project competition or direct hiring, programming is the most important part of a successful design and building process, and the quality of the entries stands in direct relation to the quality of the program. The program translates the sponsor's expectations into requirements to guide the participants of the competition. A well-defined program is particularly important for complex and unusual design problems and for cases where rapid transformations have changed the physical, social, political and economic landscape. The program not only guides the design, it may also set limits to the potential responses. By convention, the program defines the terms of eligibility and selection of the entries and, therefore, cannot be changed during the competition period, since direct communication between client and architects or designers is not allowed.

The risk of receiving unsuitable proposals due to a misdefinition of the project can be minimized by including in "the formal programming process also [...] the opportunity for input and interaction with a wide variety of interested parties and potential users" (Alexander and Witzling 1990: 97). One way to do this is by what Nasar calls "pre-jury evaluation," a method that scientifically studies "popular opinions about design entries prior to the jury deliberations" (Nasar 1999: 3). Yet, sometimes, design competition entries purposefully violate some aspects of the program. Projects that violate the program usually cannot receive a prize; they, however, can be recognized officially for their superior design approach. Violations against the program will most likely take place in the first round of a two-stage competition. Architects who find flaws in the program or who see a need to go beyond the directives of a specific program can influence the design program for the second stage of such a competition. Lucio Costa's schematic design for Brasilia, the new capital of Brazil, which ignored the programmatic details of submission, is a good case in point (Lang 2005).

Competitions vs. direct hiring

No matter their format, competitions are generally preferred over direct hiring in Europe because of their many positive benefits. As already mentioned, competitions provide a format to find optimal solutions for a stated problem; they are open and democratic; they give access to smaller and less known offices to participate in the process and offer young architects a platform to exercise their talents. This stands somewhat in contrast to the situation in North America, where the professional associations often argue against competitions because the amount of work that goes into a competition is rarely sufficiently compensated (Loukaitou-Sideris and Banerjee 1998). Therefore, they see competitions as an exploitative system, where the sponsor pays relatively little and ends up with a whole range of ideas.

Competitions differ from direct hiring of an architectural firm in many ways. First, and in contrast to direct hiring procedures, which usually focus on the credentials of the firm, a competition provides the sponsor with a wide range of ideas and proposals; this is particularly true when the chosen format is an open competition. While a range of ideas does not necessarily guarantee ideal and innovative solutions, competitions have allowed young and upcoming architects and planners to have a chance to compete with established professionals.

Significant contributions to the evolution of the profession have often come from the hands of young architects. The outcome of the open competition for the Vietnam memorial in Washington, D.C. in 1981 is a classic example where the winner among 1,421 entries was the architectural student Maya Lin, who had the most convincing proposal.

Second, unlike direct hiring, where communication between the client and the architectural firm is key, there is little to no exchange between the two parties during a competition. In fact, it is even disallowed. Therefore, a well-defined program is of utmost importance to the success of the competition.

Third, the winning design is a result of a selection process that includes a number of voices, and hence is not reduced to the opinion of the client only. The composition of the jury and the influence that certain individuals can have in the decision-making process affects the selection of the winning proposal.

Fourth, the type of competition may define further contract negotiations. A first prize in an ideas competition rarely comes with a promise for a building contract. Thus, in the case of Potsdamer Platz, Hilmer and Sattler won the first prize in the first round of the ideas competition, but did not get any commission. In contrast, a first prize in a project competition usually means that, if anything gets built at all, it has to be according to the design of the first-prize winner.

Fifth, the most significant difference between ideas and concept competitions in contrast to direct hiring is that the process with all its different steps – from announcing the competition to selecting a proposal – becomes just as important as the final product. One has to acknowledge that there exists also a long-standing tradition of corporations hiring renowned architects as a guarantee for not only receiving (hopefully) well-designed spaces but also public attention right from the beginning of the design and construction process.

Despite some shortcomings, competitions are nevertheless an interesting tool in the production process of built environments because of two additional aspects. First, different actors, such as investors, developers, professional organizations, politicians, and citizen interest groups, have an influence on the outcome by expressing their opinion during the decision-making process of a competition. At first sight, decision-making processes in competitions seem to be only about the selection of a design, but they are embedded into routines and conventions of specific places. With the increased potential of media exposure of competitions, actors and their decisions are greatly exposed to the public gaze. With that comes also a higher level of public scrutiny and accountability. This is very different from the practice of direct hiring of an architect where the decision-making process involves fewer people and very few opportunities for public engagement. With some caution, I therefore want to suggest that competitions can be viewed as a more democratic process than the direct hiring because they generally involve state representatives, investor(s), and professionals of urban design and architecture. At least at a superficial level this process allows space for different perspectives and opinions. Particularly in the case of large-scale projects, which have a strong impact not only on surrounding neighborhoods but also on the overall urban fabric, planning and design processes ought to be as open for participation as possible. Second, competitions incorporate almost all aspects of the production of space, including the marketability of the building process. Hence, one of the positive side effects of a competition is that the structure of the competition itself (particularly for invited competitions) works as a valuable strategy for public relations. This is a lesson that investors eventually begin to understand. Instead of, or in addition to, putting money into an extensive advertising strategy, the sponsor can use the urban design competition as a marketing strategy to generate exposure and public acceptance. Depending on the nature of the project (size, amount of money involved, complexity), the reputation of participating design firms, and the specific circumstances (scandals, dramas, public opposition, etc.), competitions usually find relatively large press coverage not only in the professional journals and magazines but also in the regular daily newspapers. This wide dissemination of images of the built environment creates free advertisement for the sponsor of the competition and the developer of the project, as well as a broader platform for public debate, often inspired more by the image and less by the substance.

The redevelopment of Potsdamer Platz, Berlin, demonstrates in great detail these two additional characteristics of competitions as well as the role of decision-making processes under the veil of democracy and as a marketing strategy. In the following section, I will present the case of Potsdamer Platz, where investors who knew how to produce cars and airplanes or provide entertainment and other services, found themselves in the situation of using architectural and urban design competitions for the first time, and quite successfully applied this instrument to gain more media attention.

Competitions at Potsdamer Platz

How does the concrete example of Potsdamer Platz fit into the previously discussed abstract forms of competitions? What are some lessons that can be learned from this case study in regards to competitions and image production? The section that follows explains by example that powerful investors are also confronted with local routines and conditions and can be kept in check by local actors. In other words, architectural and urban design competitions for large-, and even small-scale projects force investors into a regulatory regime where they do not have absolute power. The competitions at Potsdamer Platz, embedded in their specific circumstances, demonstrate both the politics that are involved in place-making mechanisms and the power of different actors at specific times. At the same time, they make evident that competitions play a major role in the image production of the built environment. Additionally, Potsdamer Platz illustrates the advantages and disadvantages of competitions as regulatory mechanisms in finding appropriate solutions. "Solutions" is used here in a double meaning. While competitions are a means to find the most pleasing physical shape for a specific program, quite often competitions also serve to re-evaluate a specific program. This is more often the case in ideas competitions than in project competitions, and can lead to a situation in which urban designers are dissatisfied with the programming, and where they decide either not to participate in a competition altogether or to violate purposefully the program of the competition in their design proposal.

After 1989, Berlin's Potsdamer Platz, which was divided by the wall for almost thirty years, became a prime focus for multinational investors wishing to create and locate in a new business center in the geographical middle of the reunified Berlin (Figure 23.1). Sony from Japan, Daimler Benz and the Swedish-Swiss Corporation Asea Brown Bovery bought large parts of the land and wanted to turn the former wasteland into a profitable real estate project as fast as possible. Since the city of Berlin was the previous land owner, it could put conditions into the contract when selling the land to the three investors, demanding that they conduct an urban design competition. The competition at Potsdamer Platz demonstrates the interplay between these more global actors with the local actors and conditions.

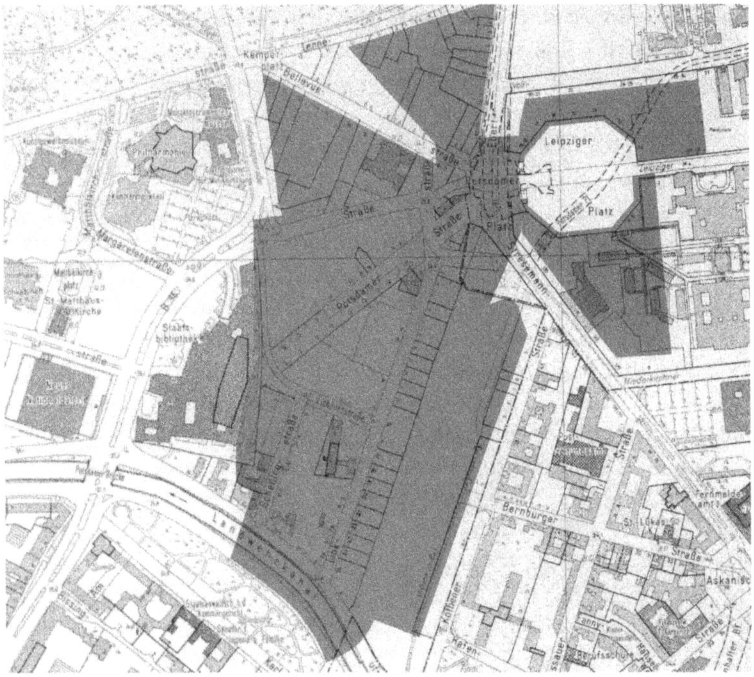

Figure 23.1 Potsdamer Platz and Leipziger Platz site.

Very soon after the fall of the Wall, a group of architects, planners, and politicians established themselves as a loose group that was able to influence planning processes and define the aesthetics of the New Berlin. The urban design competition was central to the consolidation of this group and their influence in planning and design processes in the rest of the city (Lehrer 2003).

The redevelopment of Potsdamer Platz was articulated in a two-stage competition, first an idea competition, and then a project competition, on the basis of two factors. First, because of its size and location the redevelopment would have a major impact on the city overall and on the surrounding neighborhoods specifically. A competition is the standard procedure in German planning practice in this kind of context. Second, holding a competition represented a political compromise among different governmental factions, as the land deal was a heavy political issue in Berlin's "Red-Green" coalition in the summer of 1990. As a middle ground of the divergent positions, it was written as a clause into the contract when the State of Berlin sold the land to Daimler-Benz, that there had to be a two-stage competition.

Not surprisingly, the investors preferred a direct hiring of an architectural firm to a competition, as they did not quite understand why a competition was a useful tool in finding an appropriate design scheme for the area. Their understanding was that as the owners of the land, they should be able to do more or less whatever they pleased. Another group that was unhappy about the chosen format were the *Architect Guilds of Berlin* who contested the format of a limited competition and who argued that only an open competition could provide a fair and democratic process, particularly because of the complexity of the situation, the size and geographical location of the project, and its symbolic importance.

The two-stage limited competition took place between 1991 and 1993 and brought a lot of media attention to Berlin. The first stage was an ideas competition, sponsored by the Senate of Berlin in collaboration with the three adjacent districts of the area. The main goal of the competition was to integrate the area into the polycentric fabric of Berlin and to reduce the mono-functionality of the adjacent cultural forum by creating new spaces for offices, shopping, retail, entertainment, and spaces for public as well as social functions. The program also asked for a clear spatial definition of Potsdamer Platz and its links to Leipziger Platz, as well as traffic connections. While the investors' main interest was a high floor-area ratio of 5.0, the sponsors' intention was to respect the historical development of Berlin, to connect Potsdamer Platz with the rest of the city, and to build a symbolic place reflecting Berlin's new role in the global economy.

At first, the competition was supposed to be open but due to political shifts it was limited to sixteen invited architectural offices. The majority of these selected firms were sympathetic to the dominant architectural approach, pushed by Berlin's head planner and known as "critical reconstruction." Only less than a handful of the invited architectural firms were known for progressive or radical architecture, and it was very clear from the beginning that the outcomes would stay more or less within a preferred framework. This was to the dislike of Rem Koolhaas, one of the jury members, who attacked the decision in a newspaper article, in spite of an unwritten code of conduct in competitions not to question the collective decision in public. In addition, the developers presented a counter-proposal that had been clandestinely designed by Richard Rogers and which was not part of the official ideas competition. It was a strong demonstration of the investors not trusting the previously agreed and politically sanctioned way of planning processes at Potsdamer Platz. Both the counter-proposal and Koolhaas' public outcry were highly unusual and undermined the established rules of how to proceed.

The second stage was held as a project competition with the goal to develop a master plan for each of the three sites, which would then be followed by proposals for individual buildings by commissioned architects. The sponsors of the competitions were the Senate of Berlin, the district Tiergarten as well as the individual investors: Daimler-Benz, Sony, and Asea Brown Boveri with Terreno (A+T). This time the investors were allowed to sit on the jury with voting power, of which they took advantage by trying to influence the outcome of the competition. By then, the investors had become aware of the ability of the competition to produce images that could reflect beneficially on them. Daimler-Benz, who had just finished a new and rather mediocre headquarters building in Stuttgart where they had hired an architect without investing any time into a competition, realized that they could take advantage of the competition format and turned their critique of Berlin's city-building processes into a pretty sophisticated media campaign by transforming their construction site into a permanent spectacle (Lehrer 2006).

The final result of this two-stage competition shows the influence and the position of the various actors over the course of the two years duration of the competition process. While the first stage of the competition was strongly shaped by the official position of Berlin's Building Director, Hans Stimmann, and his preference for "critical reconstruction," the second stage showed a much stronger influence of the individual sponsors. The investors at the Asea Brown Boveri with Terreno A+T site were not very much engaged in defining the outcome of the competition. Their interest was mainly speculative in nature and not so much geared toward receiving corporate-style architecture at this location. The way that the buildings responded to the streetscape, the kind of materials used, and the architectural forms all took

Figure 23.2 A+T buildings at Potsdamer Platz, Berlin. Source: Ute Lehrer.

up elements of what was referred to by Berlin's building culture as "European." (Figure 23.2)

On the other extreme is the project for Sony, which clearly speaks the language of American corporate culture (Figures 23.3 and 23.4). Sony was not ready to make any concessions in terms of architectural approach and obviously was powerful enough in the selection of the invited architectural firms for the competition. With the exception of two offices, all invitees were from abroad and most of them were known to favor trophy building. The selected design of Helmut Jahn can be called an "American" approach of designing urban space: a superblock comprised of an assemblage of buildings, creating an indoor plaza which is covered by a spectacular glass ceiling, and a skyscraper as the focal point toward the surroundings. This building complex is identified with the American city because of the semi-controlled, consumer-oriented spaces it creates with an atrium in the middle as

Figure 23.3 Sony Headquarters at Potsdamer Platz, Berlin. Source: Ute Lehrer.

Figure 23.4 Part of the Sony Complex, Potsdamer Platz, Berlin. Source: Ute Lehrer.

well as through the applied architectural language of skyscraper, mega-complex and the choice of a glass-and-steel façade.

The third case, the site of Daimler-Benz, demonstrates elements from both design principles. The overall structure of the site with its distinct parcels of land and the height limitations for most of the buildings can be called "European." "American" style, however, is expressed by the three skyscrapers at the ends of the site as well as the superblock that is created between a number of buildings by covering the street with a glass ceiling, and turning the outside space into indoor space in the form of a shopping mall. Daimler-Benz as an investor was in a more delicate situation than the other two corporations, because of its nature and what it represented. As a multinational corporation that has its roots in Germany, Daimler could not act as the outsider as Sony did, threatening to pull out of the deal if they were not allowed to put forward their architectural preference, or as

Figure 23.5 Debis at Potsdamer Platz, Berlin. Source: Ute Lehrer.

the speculative and almost faceless investor, as the A+T group appeared during the competition process. Daimler-Benz had to engage and respect to a certain degree the local discourse about architecture. However, since Daimler-Benz wanted to have a "corporate business card," a representative site for their subsidiary *debis,* they also could not leave the design solely up to the decision of the competition jury. Therefore, they put forward some of the architectural firms they wanted to invite for their project competition. What Daimler-Benz ended up with is a collage city, which borrows elements from both the European and North American city (Figure 23.5).

Conclusion

An analysis of the two-stage competition at Potsdamer Platz reveals that the instrument of urban design competition can become more than just a means for finding a suitable design for a place – it can also work as a catalyst in the negotiation process of different actors. A positive reading would then be that this institutional setting forces unequal partners to learn from each other's expertise and points of view, and therefore might help to increase the overall quality of city building. In addition, the media attention that design competitions receive helps to raise the awareness of the general public about the aesthetics and politics around city-building processes and, therefore, helps to foster a public discourse, which may be of assistance to the overall outcome for a project of the size, nature, and complexity of Potsdamer Platz.

Throughout the competition, the three multinational investors – Daimler-Benz, Sony and ABB – were portrayed as the actors genuinely interested in transforming the wasteland at Potsdamer Platz into usable, and therefore profitable, urban space. The rhetorical message of these multinational corporations towards the community at large was that by building at Potsdamer Platz and by creating jobs, they were helping Berlin achieve a significant role within the global economy. In their opinion, the symbolic meaning of redeveloping Potsdamer Platz was unquestionably a sign of progress. Design competitions can become a means for shifting public attention from substantive elements of the project – such as the nature of the business conducted there, the kinds of jobs created, the effects on local traffic flows, etc. – to issues about image and aesthetics. Hence, the investors benefit by focusing the media's attention away from potentially controversial issues.

The design competition at Potsdamer Platz defined not only the spatial articulation of the site but also the role of investors, planners, city officials, architects, and interest groups. The tensions that arose in finding the most appropriate design language for this site should be seen in the larger context of Berlin's search for its new identity. The conflict over the design was not limited to architectural style; the whole debate struggled between insiders and outsiders, local and foreign/global actors, as well as about favoritism, nepotism and different levels of power. Therefore, the dispute over which architectural language was the proper one for the site at Potsdamer Platz was in fact not only about the architectural style per se, but also about the identity shift that Berlin was undergoing. It is only by taking this into consideration that we can understand the reason for the heated discussion about the European versus American city among the group of experts and politicians. As the design outcome for the sites of Daimler-Benz, Sony as well as A+T indicates, there is a clear relationship between the engagement of a sponsor of a competition and the winning proposal. But what it also shows is the relevance of the image that is carried with a specific design approach. Potsdamer Platz played a central role in this battle over style that was central in Berlin's way of developing its new image after the fall of the Wall.

The case of Potsdamer Platz can be used as an example of the benefits, challenges and loopholes of urban design competitions. While one would hope that design competitions are the fairest process for finding the best solution for a stated program, competitions have also limitations. As the quote at the chapter's introduction indicates, competitions can be only as good as their program, their jury, the selection of the architectural firm, and the local conditions that tie all these components together. The case of the World Trade Center in New York, where Libeskind's winning design was subordinated under SOM after pressures by the developers, demonstrates the fragility of a fair competition system. Nevertheless, we want to conclude that urban design competitions usually actively contribute to a better urban design outcome because they start with the understanding that there is not just one but a number of approaches with different merits and drawbacks. It is via the selection process by the jury, the discussion in the media, and the response by the general public that solutions can be found to urban design questions that are responsive to the complexity of cities and their societal needs.

References

Alexander, E. and L. Witzling (1990). "Planning and Urban Design Competitions: Introduction and Overview," *Journal of Architectural and Planning Research*, 7(2): 91–104.

Banerjee, T. and A. Loukaitou-Sideris (1990). "Competition as a Design Method: An Inquiry," *Journal of Architectural and Planning Research*, 7(2): 114–131.

Becker, H. (1992). *Geschichte der Architektur- und Städtebauwettbewerbe*. Stuttgart, Berlin, Köln: Verlag W. Kohlhammer, Deutscher Gemeindeverlag.

Geisert, H. (ed.) (1990). *Hauptstadt Berlin – Internationaler städtebaulicher Ideenwettbewerb 1957/1958*, Berlin: Argon Verlag.

Lang, J. (2005). *Urban Design: A Typology of Procedures and Products*, Architectural Press.

Lehrer, U. (2003). *Image Production and Globalization: City-Building Processes at Potsdamer Platz*, Chicago: UMI.

—— (2006). "Willing the Global City: Berlin's Cultural Strategies of Interurban Competition After 1989" In: N. Brenner and R. Keil (eds.), *The Global City Reader*, Routledge, 332–338.

Loukaitou-Sideris, A. and T. Banerjee. (1998). *Urban Design Downtown: Poetics and Politics of Form*. Berkeley, CA: University of California Press.

Nasar, J.L. (1999). *Design by Competition: Making Design Competition Work*. Cambridge: Cambridge University Press.

Seidel, A. (1990). "Design Competitions Receive Mixed Reviews," *Journal of Architectural and Planning Research*, 7(2): 172–180.

Solomonson, K. (2003). *The Chicago Tribune Tower Competition: Skyscraper Design and Cultural Change in the 1920s*. Chicago: Chicago University Press.

Spreiregen, P. (1979). *Design Competitions*. New York: McGraw-Hill.

Straub, D. (1993). *Spreebogen Berlin: Internationaler städtebaulicher Wettbewerb Spreebogen – Ergebnisse der Überarbeitung*. Berlin: Senatsverwaltung für Stadtentwicklung und Umweltschutz.

Strong, J. (1976). *Participating in Architectural Competitions: A Guide for Competitors, Promoters, and Assessors*. London: The Architectural Press Ltd.

Further reading

Nasar, J.L. (1999). *Design by Competition: Making Design Competition Work*. Cambridge: Cambridge University Press. Arguing in favor of the involvement of citizens in the decision making process, the author credits the instrument of competition for finding a good urban design, but also cautions against the effect of publicity that a competition generates.

Spreiregen, P. (1979). *Design Competitions*. New York: McGraw-Hill. While this book is already over three decades old, it is still the standard text for urban design competitions. The book addresses the origin, various formats and outcomes of competitions.

24

The design charrette

Douglas S. Kelbaugh

In the long history of humankind ... those who learned to collaborate and improvise most effectively have prevailed.

(Charles Darwin)

The term "charrette" has been revived in both the academy and the profession from the tradition and lore of the École des Beaux-Arts. It originally referred to a wagon that was dispatched by the professor to pick up student drawings. To be "en charrette" was to work until the very last moment, even to the point of jumping on the wagon to finish a drawing. Although the practice and term fell out of favor with the rise of Modernism, in the last few decades the word has been recoined to mean an intense design workshop. Unlike the traditional charrette, which was a solo architectural effort in an authoritarian regime, the contemporary version is usually a team effort urban in scale, participatory in mode, and democratic in intent.

There are varying definitions of charrettes and a range of types, with different purposes and methodologies. The most succinct description is *an illustrated brainstorm*. They have been described as the best way to get the most creative proposals for addressing the most challenging problems from the most accomplished designers in the most compressed period. Two basic types have emerged: the competitive charrette, in which *multiple* schemes are developed for the same site by different teams, and the collaborative charrette, in which a *single* scheme is developed by teams that work on different aspects (land use, transportation, etc.) or subareas of the same site or that work on separate sites. Academic charrettes, which are run by architecture and planning schools and are open-ended in purpose, tend to be the first type, while professional practice generally favors the second type.

This chapter focuses on the competitive charrette and is based on the two dozen academic examples at the University of Washington (UW) and the University of Michigan (UM) that I have organized and led in as many years. Starting almost by accident in 1985 while I was Architecture Chair at UW, the workshop quickly evolved into a four- or five-day, intensive design workshop that brought together three or four teams to generate and present different visions for a particular site. Several thousand students, faculty, guest professionals, and consultants participated in the workshops and a total of some 10,000 people attended the public presentations at the end of these events. A score of booklets was published and distributed, as well as a couple of books written.

When I moved to the University of Michigan in 1998 to become the Dean of what soon was to be renamed the Taubman College of Architecture and Urban Planning, I transplanted the annual charrette. It soon took root in Detroit and in the college and continued during the subsequent decade. Eight of the ten charrettes organized at UM focused on large, vacant or under-utilized sites in the central city, especially ones that needed or would benefit from redevelopment in the immediate or near future (Figure 24.1). In a large city with a small planning and development department, the UM charrette became what some citizens described as the most important annual event in the public discourse on the future of the city. The UW workshops were generally held on campus, but the UM student and faculty participants decamped 45 miles from the campus in Ann Arbor to downtown Detroit.

Types of charrettes

The charrettes typically dealt with an urban design issue, project, or site of civic importance. Several ariants emerged: some sought to test and illustrate new public policies or design ideas on real sites. Others responded to requests for help from community/civic organizations or government agencies; while a third type explored a particularly glaring problem or promising opportunity offered by a specific site. Many charrettes were hybrids, for example testing a new idea (e.g. TOD) on an empty or underutilized site.

They consistently advanced creative solutions on real sites for real clients and users, as opposed to being a theoretical or academic exercise for the sake of the students' education (although their pedagogic benefits were manifold). The time horizon of the proposals was ten, twenty or more years, depending on the site and the team. The level of feasibility, i.e. the practicality of realizing the design proposals, also varied from project to project and from team to team. Some design proposals were unrealistically ambitious or visionary, but most tended to seek the responsible middle ground and find the sweet spot between an inspiring vision and a workable proposal.

Why this balance and moderation prevailed in these charrettes is difficult to say, especially during an era of rather extravagant, hyperbolic design and rarefied design theory in architecture schools. It might be explained by the fact that the audience

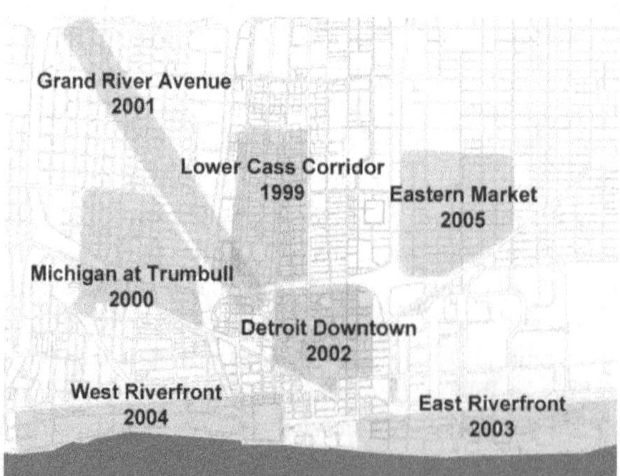

Figure 24.1 Sites of University of Michigan charrettes. Source: Doug Kelbaugh.

and clientele consisted of local leaders and the general public rather than fellow academicians – the laity rather than the choir, as it were. Additionally, team leaders tended to be more mainstream than avant-garde, but even when the latter were invited, they tended to tone down their designs and rhetoric. Another factor was the inclusion of urban planners and urban planning students, who tended to be more pragmatic and process-oriented (although often more idealistic about social and environmental causes), as were the landscape architecture practitioners and students.

Participants

Participants were divided into teams, which were led by one or two visiting professionals (architect, urban designer, landscape architect, or urban planner, some of whom were both professionals and academicians),[1] one or two local design professionals, and a design faculty member or two. This group co-led a team of 10 to 15 graduate students from architecture, urban design, urban planning, or landscape architecture programs in the host university and sometimes from other local universities. Occasionally, business, law, and public policy students would join the teams. A handful of local high school students were often invited as understudy members of the teams. Most teams operated like temporary offices with the professionals and faculty members acting as design partners and the students as the design and production teams, although the roles were fluid and modes varied with the composition of the team and its leadership. Depending on the site and program, sometimes stakeholders and local citizen volunteers became active team members, but typically they acted as consultants or observers (due to the extended duration and technical skills needed for a charrette).

The event

The workshop typically consisted of the following stages:

Getting oriented/fact-finding

Prior to the event, a packet of information about the charrette process and the project and site was sent to the participants. The on-site process began with a morning-long bus and/or walking tour of the site and environs, guided and narrated by local residents and professionals. After lunch, there was an afternoon of briefings by community leaders, land owners, government officials, and business leaders, as well as financial and technical consultants. These speakers were an indispensable part of the program and carefully chosen based on the problems and opportunities suggested by or latent in the project or site. There were as many as a dozen five to fifteen-minute presentations, and sometimes a longer keynote talk. Urban historians, commercial experts, real estate developers, and public artists would sometimes participate in various consulting and speaking capacities. In some cases, residents of the area would speak and stay on as working team members.

Getting started/brainstorming

Following the briefings, the teams would work independently and intensely for the following three days in an atmosphere of friendly and open competition. First they would discuss and distill what they found to be the most important and salient information provided during the briefings or from any relevant data or literature made available to the teams or gleaned from the web. In some cases, students would prepare in advance by doing preliminary research and analyses in their studio courses before the charrette. Each team collectively brainstormed ideas based on what they perceived to be the needs and opportunities of the site, as well as the advice and information offered by experts, residents, stakeholders and consultants (Figure 24.2). The mix and interaction of design professionals, faculty, and students was vibrant, creative, and productive. The high combustion chemistry between them produced ideas and designs of varying merit, but there were always compelling and imaginative schemes that conventional, linear consulting would rarely if ever generate.

The teams usually engaged in no-holds-barred discussions while they considered and tested ideas from any and all of their members. Good team leaders would work to get everyone involved in the discussions. Initially, no idea or question would be too radical, too pointed, too extraneous or too obvious. On the one hand, there can be a dampening effect when all the team members are from a single discipline, because competitive colleagues are often too embarrassed to ask basic questions. On the other hand, different disciplines have different vocabularies that must be bridged. Studies have shown that complex problems are often better solved by diverse teams of people with different levels and areas of expertise than by teams of experts from a single field, because there is less fear of appearing unintelligent, and the brainstorming is more freewheeling. In any case, many design and planning concepts would exfoliate in this exploratory and fertile stage.

These charrettes typically did *not* have any written program or problem statement. Such a defining document was usually beyond the scope and capacity of the organizers, and was thought to be too prescriptive and limiting or lending a false sense of precision to the work. Occasionally research was done in advance and a skeletal program with numerical maxima and minima was issued. More often general recommendations and mandates were given, such as "prioritize retail" or "maximize the number of housing units." The intent was to have each team tease out what *they* determined to be "the highest and best use" of the site, as well as how to

Figure 24.2 A Detroit charrette visit from Senator Levin. Source: Doug Kelbaugh.

phase development. This laissez-faire approach to program and phasing worked very well for most of the charrettes I have participated in. In fact, it may be a reason for their success.

Distilling the options

As mutually acceptable ideas were generated, team leaders would often sort themselves and the students into sub-teams for additional research and for the development of options, which would be periodically presented to other members of the team in pin-ups. About halfway through the process, usually toward the end of the second day, options would be winnowed down and an overall strategy would emerge by consensus. If no clear consensus emerged in time, team leaders adopted a strategy based on one of the contending ideas or one they personally found the most promising. Some team leaders pushed their own strategy from the outset, but the overwhelming majority of teams incorporated ideas generated by students. In a few cases, there was a required pin-up at the end of the second day for all the teams to compare and coordinate their proposals, a practice needed in collaborative charrettes in which teams are working on *different but abutting* areas. However in competitive charrettes, these interim reviews were both too time consuming and thought to dumb down the gene pool of ideas.

Designing and producing

Toward the end of the second day, if progress was satisfactory, the process would change quickly and dramatically from expansive brainstorming to a disciplined focus on the production of drawings, images, and text (and occasionally a video or a physical model). The modality would diametrically shift from that of a wide and inclusive funnel for collecting ideas to a circumscribed team effort that executed the design and illustrated it with a myriad of images. In a sense, the funnel would be turned upside down about half way through the workshop. The second part of the charrette was usually a feverish team effort (Figure 24.3). It was a race, sometimes exhilarating and sometimes panicky, to develop and represent the creative explosion of ideas from the first half. However, important ideas sometimes cropped up later in the process, making the scramble to the deadline all the more intense.

Presenting to the public

The workshop would culminate with a public event that included a posted exhibit of the drawings and occasional models, a reception, and a fifteen to twenty minute presentation by each team – all at a prominent venue within or near the study area (Figure 24.4). The general public, stakeholders, business and institutional leaders, government officials, and the media would be notified by printed and email invitations, as well as word of mouth. The crowds ranged from 200–400 people at the downtown charrettes to 100–150 at the suburban ones. The media coverage would typically include local TV stations (both live and subsequently on talk shows) and newspapers. Shortly after the event, CDs containing the presentations (originally color slides, later digital) from both the initial briefings and the team presentations would be distributed to key parties and the media. At the end of the semester, a report, a full-color printed booklet (30–100 pages) detailing the design proposals would be published and hundreds of complimentary copies distributed to the participants, sponsors and larger audience. More than just a chronicle and archive of the event, these publications sometimes helped catalyze the adoption and implementation of proposed concepts and designs (more so in

Figure 24.3 Four teams at a Detroit charrette in full swing. Source: Doug Kelbaugh.

Figure 24.4 Several team leaders prepare for the public presentation. Source: Doug Kelbaugh.
Note: Michigan Senator Carl Levin is briefed by Lance Brown of CUNY, while Roy Strickland, Director of Taubman College's Urban Design program, and the author look on.

Seattle, which was in better economic condition than in Detroit).

Site selection

Site selection is extremely important, as the charrettes were usually more driven by site constraints and opportunities – as noted earlier, what the site "wanted to be" – than by a particular program or project. The site was typically chosen in consultation with government, civic and community leaders, although the availability of funds or sponsors sometimes influenced the selection process. The criteria would vary from time to time, but some were not negotiable: it should present a timely problem of significant enough size and scope to warrant the mobilization of many participants and resources, and the site and topic should be consequential in social, environmental, and planning terms. If a charrette also answered an urgent need or seized a glaring opportunity, so much the better. And, if it was likely to influence or trigger actual development, better yet.

The UW charrettes were sited throughout the Seattle region, although two were in Italy and there was a joint one with UM and the School of Architecture in Ahmedabad, India. The majority of the sites were relatively open and/or marginal areas, preferably under-utilized and ripe for development or redevelopment. They usually ranged from fifty to five hundred acres – large and open enough to exercise the full range of the design talent and experience assembled, and small enough to be handled in four or five days. The UM sites were typically several hundred acres and concentrated in central Detroit, plus two suburban outliers on much larger sites. Charrettes were typically more appropriate for large, open sites that lend themselves to bold concepts and broad-brush schemes. One charrette that looked at an immense site of several thousand acres for an "aerotropolis" around the Detroit International Airport pushed the practical limits of a four-day workshop. On the other hand, sites of less than fifty acres were also suitable if they challenged and kept the teams busy.

Sites set in the midst of mature neighborhoods or districts were generally avoided to keep demolition of buildings and displacement of people and businesses to a minimum. Typically, more populated and built-out sites, which often have nuanced social issues and smaller scale opportunities, were more effectively approached as semester-long design studio problems or as research projects. The slower pace of studios and research enabled them to be more patient, more inclusive of community involvement, and more sensitive to the microsurgery needed. In any case, gentrification is a chronic and morally troubling problem, with trade-offs that seem to be unavoidable – a structural issue in a market economy with little if any housing and business subsidies.

Sponsors and funders

The sponsors and supporters of these academic charrettes tended to be public agencies, organizations, foundations, or institutions, rather than the private sector. Generally, sponsors were also funders, although there were some honorary sponsors. Because UW and UM are public universities with service missions, the sponsors or clients generally belonged to the public or nonprofit sectors. A private philanthropist or developer have occasionally sponsored or helped underwrite a charrette, as long as it was understood that academic freedom could not be compromised. Often the number of sponsors and contributors would grow to a half dozen or more. In the UM case, a longstanding partnership was established with Detroit Edison, the regional gas and electric utility, as well as with a private donor who started a small charrette endowment. Some local foundations also contributed grants from time to time.

The level and type of donation varied with the donor, with some giving cash grants and others donating or discounting in-kind services (e.g. food, transportation), equipment (e.g. computers, phones, cameras) or space (e.g. workshop space, venue for public presentations, hotel rooms). Although the college and university were funders of last resort, they always contributed staff time, equipment, media relations, vehicles, etc., as well as the long hours (late into the night, sometimes all night) of faculty, staff, and especially students.

With academic charrettes, outside funding was almost always needed, because university funds were perennially insufficient or non-existent for such outreach.[2] The budgets for the UW events in Seattle were typically $10,000 to $15,000. The UM charrettes in Detroit were larger and typically had a budget of $50,000 or more, plus in-kind contributions from the regional gas and electricity company that annually provided space, equipment, and food. One of the major cost differences was that in the Detroit events, all the students and sometimes local professionals, who numbered 60 or more, were provided hotel lodging and meals for four or five days and nights. Also, the demand and the cost for computing and printing equipment steadily increased over the years, as production switched from handicraft to machine. Funds were annually raised from the office of the university president and/or provost, local donors, corporations, foundations, and agencies to cover the entire cost of the event and follow-up publications. Cities and public agencies also contributed funds, although the City of Detroit, which was financially strapped, was never asked for any financial or in-kind contributions. In the early years, willing sponsors and funders were harder to find than in later years, when interested communities and agencies sometimes asked and even competed to be selected and local professionals volunteered to work for little or no honorarium.

Although costly to mount, it can be argued that the market value of charrettes is considerably greater than the expenditure. Indeed, to conduct a similar event entirely with paid professionals and staff would cost several hundred thousand dollars. The free labor of the students and reduced fees of the professionals, many of whom would have normally commanded much higher daily rates, made the charrette a relative bargain for the sponsors and funders, as well as a gift to the city. And the community groups or developers would probably not have been able to attract some of the illustrious designers and planners that the university-affiliated workshops were able to recruit.

Non-academic charrettes

There is also a parallel tradition of non-academic, professional design workshops/charrettes, which are sponsored by private developers or by organizations such as the AIA (American Institute of Architects), ULI (Urban Land Institute) and NEA (National Endowment for the Arts). The AIA has been sponsoring R/UDAT (Regional/Urban Design Assistance Teams) workshops for forty years. These three or four-day grassroots events have helped more than 140 communities address issues of urban growth and land use, inner-city neighborhoods, downtowns, environmental degradation, waterfront development, and commercial revitalization. More than 500 professionals representing over forty disciplines have donated more than $3.5 million of their time. A more recent AIA variant, the SDAT (Sustainable Design Assessment Team) brings together multidisciplinary teams of professionals from across the country to help communities seeking to improve their sustainability. The ULI Advisory Services provides advice to sponsors on land use and real estate development issues. Established in 1947, this fee-based program has convened over 500 three-to-five-day panels of volunteer ULI members in the US and other countries. The Mayor's Institute on City Design is a partnership program of the National Endowment for the Arts, the American Architectural Foundation, and the United States Conference of Mayors. Since 1986, the Institute has offered interdisciplinary design workshops that have helped 800 mayors better understand and lead urban design in their cities.

Charrettes sponsored by developers are typically led by architecture and urban planning firms. DPZ (Duany Plater-Zyberk) has a long track record of starting their projects with a charrette.[3] These collaborative charrettes are usually a week or longer in duration and result in a single proposal, which is developed by a single, multidisciplinary team of firm employees, consultants, and local players. They include community break-out groups and an interim public review, as well as a final presentation to the community. In other cases, local or state governments may sponsor workshops, such as the Envision Utah process that Fregonese-Calthorpe Associates led in 1998, and a flurry of community charrettes along the Gulf Coast that followed Hurricane Katrina. The regional scale exercises may occur over many months with multiple workshops and public presentations (some broadcasted on community TV). All these community events build on a long architectural tradition of public and private workshops extending back four decades to ones led by architects like Bill Caudill of CRS (Caudill, Rowlett, Scott) and Charles Moore.

Benefits, contributions, pitfalls

These compressed, adrenaline-driven brainstorms have always been highly creative. They consistently generate more imaginative ideas and proposals than conventional, linear design consulting would likely produce. The chemistry of multi-generational collaboration *within* teams and febrile competition *between* teams in a competitive charrette engenders remarkable levels of invention and production and, seemingly without fail, produces unpredictable, imaginative and compelling proposals. However, their fertile results should be seen as more illustrative than definitive, and are only *one*, early step in the longer planning and development process.

The Seattle and Detroit charrettes ran the gamut of sites and programs, envisioning development where there was a hole in the urban fabric, where there were poorly utilized and under-populated areas, or where empty land offered entirely new and exciting opportunities. And they have had positive, intended and unintended consequences. For example, the 1988 charrette on a greenfield site along an existing rail line south of Seattle resulted in *The Pedestrian Pocket Book,* a national best seller in urban design and architecture. The small book helped jumpstart TOD (Transit Oriented Development), which has since become a well-known and influential strategy for planning and development in general and for New Urbanism in particular (see chapter by Stefanos Polyzoides). As an indirect result of its ten charrettes, Taubman College opened in downtown Detroit a community design center, which provides *pro bono* or low-cost community design services, offers architecture classes to high school students, and hosts community meetings and smaller design workshops.

Because they were primarily or completely underwritten by third party sponsors and essentially gifts to the public, teams are not beholden to or unduly influenced by political pressure. Visiting professionals and students, on top of bringing fresh eyes to a problem, are neutral and unencumbered by local political knowledge or allegiances. This design freedom and autonomy is conducive to a healthy and open-minded visioning process. When commissioned or sponsored by an agency, organization or developer who already has a plan in mind or in hand, the process can be compromised. This is not to say that a general vision or previous studies could not be shared in advance with the participants. Indeed previous analytical or design work on the site was always made available, if not actually presented to the teams in the briefing session. Suggestions and certain provisos and mandates were often appropriate, as well as briefings about the socio-political and economic-financial terrain and constraints.

These charrettes produced considerable local buzz and publicity. They were usually followed-up with presentations to community groups and stakeholders, and the results were often published in the local print media and aired on the local electronic media, including TV interviews and talk show appearances. This attention helped precipitate the commissioning of further studies and/or built projects. Although the charrette proposals were rarely if ever literally or fully implemented, they did concretely influence subsequent planning and development.

They consistently generated *visions* for the public and provided palpable imagery and new ideas for public discussion, dissemination and digestion, as well as adoption by the community. They have rejuvenated and elevated public consciousness in positive, proactive, and provocative ways that seem to be widely understood, respected, and appreciated. Of course, there was always the danger of raising expectations too high within the community and the public at large. Accordingly, caution and discretion were used in the presentation, publication, and dissemination of the results.

There were, to be sure, other external and internal problems and challenges: Some teams proposed unrealistic, extravagant and unfeasible schemes; some students found the format too fast, too difficult or too disorganized and unevenly paced; other students felt their ideas were under-appreciated or overlooked altogether (on the other hand, many students volunteered and were eager to participate, even as a curricular overload); some faculty were annoyed that some students had to miss their classes on campus or were bothered that considerable staff time and resources were deployed to plan and mount the events; sometimes team co-leaders did not "gel" well, which was counterproductive as well as frustrating for team members; and a few charrettes produced more heat than light. Ironically, the UM charrettes were often appreciated and valued more outside than inside the school. Indeed, as noted earlier, some groups and organizations requested, even competed, to have them in their communities, and the university was often happy to be associated with and take credit for the community service.

Indeed, the charrette can be a highly effective technique to enlarge the range and type of ideas for a project or site – ideas that can later be modified, tempered, amalgamated, implemented, or discarded. It is also a collaborative, democratic, transparent, fun and engaging way to help stakeholders – community residents, property owners, municipal officials, government agencies, institutions, and developers – to develop a sense of shared ownership and common vision essential to moving projects forward in a democratic society. The neutrality and respect of a university allow it to put forth and test bold, new ideas with the public in a non-threatening way. In short, charrettes have been successful in jumpstarting new development; consolidating diverse projects; gathering data and citizen input;

expanding public consciousness and imagination; and promoting ideas and visions.

Academic value

At the University of Michigan, despite their challenges and shortcomings, the charrettes were of significant academic value. Uniquely among curricular initiatives they embodied in a single event, the university's tri-partite mission of *teaching, research,* and *service.* They offered a rich opportunity to *teach* students invaluable lessons in design and planning, as well as in working closely with top local and visiting practitioners and academicians. They also provided students a chance to interact with their own instructors on a more protracted basis. Indeed, there was healthy socializing among faculty, professionals, and fellow students around shared meals and festivities. Many of the professionals claimed to learn from the experience and, despite the modest honoraria, some asked to be invited back.

The group dynamics were for some students a baptism by fire into the challenges, pleasures and benefits of interdisciplinary teamwork, to a degree that many of the students, especially the architecture students, too rarely experience in their design studios. As such, charrettes should arguably become more commonplace and central in architecture and planning programs. They seem especially important in urban design programs. If not required of architecture and urban planning students, the experience should be available on an elective basis.

In addition to their pedagogic benefits, charrettes are also a form of *research,* in that they explore and test prevailing and new methodologies, as well as proposed solutions to particular problems and opportunities. They are also clearly and emphatically a form of *service*, with thousands of hours of student, faculty and staff sweat equity offered pro bono to the community. They provide a transparent public forum and visible event with which the university can effectively partner with the community to envision and discuss its future, while advancing the local and national dialogue on the city.

On top of honoring the triple university mission of teaching, research, and service, charrettes have also been profoundly interdisciplinary, typically with faculty, students, and professionals from up to a half dozen disciplines. They have nurtured and cross-fertilized academic life and the educational experience by bringing together a diverse mix of people and ideas to address common issues.

For all these reasons, they are an excellent investment of institutional, economic and human resources. They also make compelling sense as an integral part of design and planning education. They can be a stunning, synergistic, and lasting contribution by universities to their communities. Rarely do so many factors fall on the positive side of the ledger for the university, the design professions, and the community. And, if Darwin was correct, charrettes nurture and teach collaborative skills and habits essential to our surviving and thriving.

Notes

1 The visiting team leaders included such academic/professional leaders as, in no particular order: Alex Krieger, Anne Whiston Spirn, Jonathan Barnett, Rich Haag, Laurie Olin, Anne Vernez-Moudon, Elizabeth Plater-Zyberk, Peter Calthorpe, Harrison Fraker, Michael Dennis, Linda Jewel, Andres Duany, Michael Pyatok, David Sellers, Don Prowler, Walter Hood, Dan Solomon, Joseph Esherick, Lee Copeland, Ellen Dunham-Jones, Ken Greenberg, Gary Hack, Henning Larsen, Mary-Ann Ray, Michael Speaks, Stefanos Polyzoides, Elizabeth Moule, Max Bond, Michael Sorkin and many other distinguished designers and planners.

2 An exception were several charrettes organized by Patrick Condon at the University of British Columbia, which were underwritten by an endowment.

3 Bill Lennertz, a former member of the DPZ firm, has founded the National Charrette Institute, which can be visited at www.charrretteinstitute.org.

Further reading

Condon, P. (2008) *Design Charrettes for Sustainable Communities*, Washington, DC: Island Press. Theory and details of design charrettes based on University of British Columbia experience.

Farr, D. (2008) *Sustainable Urbanism, Urban Design with Nature*, Hoboken, NJ: Wiley Eco-design of neighborhoods and cities, with case studies.

Kelbaugh, D. (1997) *Common Place, Toward Neighborhood and Regional Design*, Seattle: University of Washington Press. Theory, design and policy based on a decade of design charrettes in Seattle.

Lennertz, W. (2008) *The Charrette Handbook: The Essential Guide for Accelerated, Collaborative Community Planning*, Chicago: APA Planners Press. A comprehensive, step-by-step guide and reference for non-academic charrettes.

25

Citizen design

Participation and beyond

Jeffrey Hou

Urban design has historically been the domain of design and planning professionals. The scope and complexity of projects and policies that consider form, functions, and finances frequently require specialized knowledge. Reinforced by the legacy of institutional and professional elitism in design and planning, the practice often shuns the participation and knowledge of ordinary citizens. The complex mechanisms that drive many large-scale projects also veil them from easy scrutiny. As a result, although urban design in essence addresses the making of the public realm, it was not until recent decades that its practice has become more open to the involvement of the public.

In the decades since the 1950s, increasing criticisms toward the impacts of large development on communities, environment, and historic character of the city have subjected urban design projects to closer examinations and often heated debates. In recent years, incentives and subsidies for private development along with the gentrification of downtown and subsequent displacement of vulnerable populations have brought further public attention and contestation among different interest groups. In some cases, participation has helped produce improved design or policies that address multiple needs and interests. In other instances, poorly conceived or managed public processes have resulted in dissent and protests, legal challenges, project delay, and increased cost to municipalities and developers.

How has urban design practice addressed public involvement and deliberation? What are the challenges and opportunities for public participation in urban design? What are the current trends and directions in both theory and practice? In this chapter, using the experience of North America and more specifically the United States as a case study, I examine the growing complexity and challenges of public participation in the increasingly diverse and pluralized urban settings. In reviewing the recent discourses and practices, I also comment on the trends of democratic urban design, which has evolved from an institutionalized participatory model to a more inclusive practice of "citizen design."

Rise of citizens in urban design

In recent decades, various forms of participatory design and planning have emerged in cities around the world. In the United States, although citizen participation has been a cornerstone of the nation's democratic traditions (Day 1997, Hester 1999), public participation in contemporary urban design practice did not emerge until the 1950s, as demolition of inner-city neighborhoods drew growing citizen opposition. In the name of slum clearance and urban renewal, once functioning neighborhoods were bulldozed to make way for construction of new housing and freeways funded with federal dollars. These large-scale projects provided some of the first instances of contemporary urban design experiments, but with the heavy cost of displaced communities and businesses and the destruction of the spatial and social fabric of cities. The citizens' revolt against these projects helped push through the establishment of legal requirements for citizen participation in the US. The 1954 Urban Renewal Act came with legislatively mandated participation. The Economic Opportunity Act of 1964 established the Community Action Program and introduced the requirement of "maximum feasible participation." Through the Model Cities program of 1966, citizen participation became a requirement in the planning and implementation of federally funded urban programs. In 1969, citizen participation also became required for all phases of the planning process under the Federal-Aid Highway Act (Weiner 1999).

As participation became an institutionalized requirement, the professions of architects, landscape architects, and planners responsible for the design of these projects also went through a moral and ethical reawakening. In planning, Davidoff (1965) proposed the approach of advocacy planning to replace planning as a technocratic exercise. He argued that planners should serve as advocates for disadvantaged groups, and that plural plans should be presented to the public (Davidoff 1965). In architecture, the community design movement has sought to address equity and justice issues in urban communities since the 1960s, through a growing network of community design centers (Curry 2004). In landscape architecture, community participation is seen as an essential ingredient of making successful urban open space (Francis 2003). The reawakening in professional discourse was accompanied by a shift in scholarly research toward human perceptions and experiences. In *The Death and Life of Great American Cities* (1961), Jane Jacobs reminded planners and designers that a vibrant city life results from diverse human activities. Other notable examples, such as Kevin Lynch's *The Image of the City* (1960) and William H. Whyte's *Social Life of Small Urban Spaces* (1980) also focused on the experiences of users in the city.

The discourse of citizen participation and the focus on human experiences of place has, to a considerable extent, re-oriented the elitist tendency of urban design practice in the US. It also has forced a rethinking of the fundamental values of design in favor of human scale, social interactions, and cultural practices, as well as democratic process and justice in the making of urban environments.

Established practices

Today, a few decades after the turbulent period of urban renewal, public participation has become a common practice in urban design. Throughout North America, citizens are engaged in multiple levels of decision-making concerning land uses, scale and types of development, and form and bulk of buildings. Over the years, a variety of participatory techniques have been introduced to engage the broader public. In addition, methods of consensus building, conflict resolution, and organizational participation have augmented the participatory process. The following highlights a set of common practices in the US that are also in use in other parts of the world.

Legal mandates and processes

Citizens in the US today have the ability to influence a variety of urban design choices – zoning ordinances, transportation decisions, preservation of landmarks and districts, etc. – through propositions and initiatives, as well as electing civic leaders who share their preferences. Unsatisfied with the making of specific decisions, citizens can also file lawsuits against the decision-making bodies. These legal challenges can result in delay and significant changes to the scope, scale, and other aspects of a project. Using these established channels of participatory democracy, citizens can shape policies and project decisions that influence urban design outcomes. In Seattle, for instance, citizens passed an initiative in 1971 to preserve the historic Pike Place Market against large-scale commercial redevelopment. The initiative created a historic district, mandating the preservation of the character and activities of the market and the surrounding area.

Public review and comment

As part of the institutionalized participatory process, reviewing and commenting on specific projects and proposed policies remain a common opportunity for citizen input. In most North American cities, relevant agencies notify residents and the public of proposed land use changes and allow for comments during a specified period. Public hearings or meetings are often required for projects to receive approval from communities, neighborhood councils, and/or municipal authorities. Special commissions, oversight committees, task forces, and review boards are commonplace in evaluating specific aspects of projects or policies. The positions are elected or appointed by elected officials. When given proper authority, these processes can significantly alter the scale and character of a project. On the other hand, without legal authority or transparency, advisory reviews can serve only as token gestures of participation.

Visioning exercises

In neighborhoods and cities across North America, visioning activities that invite participation of residents, the business community, and other stakeholders are now a widely used approach to generate ideas and bring public consensus for future choices. Facilitated typically by trained professionals these events allow community members to generate visions and suggestions that can influence the character of specific projects or areas. Done effectively, the process can mobilize public support and in turn political will for implementation. In Youngstown, Ohio, a rustbelt city facing economic and urban decline, urban designers organized a series of gradually expanding workshops, culminating in a large public meeting that engaged civic leaders and the public to identify strengths, weaknesses, opportunities, and threats facing the city. The process not only helped clarify the diverse perspectives and challenges, but also deepened the involvement of participants in the collective vision (Faga 2006).

Project design

In many public or community-driven projects today, it is often possible for citizens and stakeholders to participate directly in the design process. In these cases, citizens and stakeholders are invited to express their preferences in the early phase of a project. Later, they might respond to, and sometimes vote on proposed design alternatives. Various techniques are now available and used to solicit opinions and engage citizens. They include charrettes (see chapter by Kelbaugh), focus groups, game simulation, group interaction, public forum, and workshops (Sanoff 2000). Others include walking tours, interviews and questionnaires, visual preference surveys, and a growing variety of digital simulation tools (see chapter by Ben-Joseph). In these projects, participation can help generate a preferred design and serves to generate community support.

Advocacy and organizing

In most metropolitan areas in North America today, a growing number of non-profit advocacy organizations have influenced urban design decisions on issues ranging from transportation and business revitalization to historic preservation, ecological conservation, environmental justice, and climate change. By articulating the importance of specific issues as well as impacts of proposals and by sometimes participating in the political process, these organizations are able to influence public opinions and shape decision-making. In the San Francisco Bay Area, for example, groups such as Greenbelt Alliance, Trust for Public Land, and Urban Ecology have been active participants in shaping land use and urban design decisions in different municipalities in the region. In New York City, Project for Public Space has developed a database of projects around the world, and advocates for specific principles of design and development of urban public space (see chapter by Kathy Madden).

Direct actions

Aside from these formal and institutional mechanisms, individual citizens and community groups may also undertake direct actions to influence the design of neighborhoods and urban spaces. Community gardening, for example, has been a powerful tool for local residents and communities to transform vacant urban sites into productive lands and gathering places. They provide food and sometimes income to communities and individuals who have limited employment opportunities (Hou *et al.* 2009). Community art projects have also been an effective means for neighborhood improvement and community building. In North Philadelphia, the Village for Arts and Humanity has engaged local youth and children in transforming the blighted neighborhood through educational programs and outdoor art installations that occupied vacant lots once used for illicit activities.

Challenges facing participatory urban design

Despite its growing presence, participatory urban design continues to face a wide variety of challenges in North America. While most practitioners, theorists, and the public generally support the moral and intrinsic value of participation, few are satisfied with the actual processes and outcomes. Ironically, as opportunities to participate are more widely available today, citizens do not necessarily choose to participate. The barriers range from the limitations of institutionalized participation to the challenges of engaging an increasingly heterogeneous urban populace.

Bureaucratization

Legal mandates of participation represent an important breakthrough in favor of the rights of citizens. However, the institutionalization of participation has also negative consequences. For example, Arnstein (1969) characterizes aspects of institutionalized participation as tokenism. Francis (1999) observes that contrary to its original moral purpose, participation is often structured to satisfy mandated requirements and is not intended to fully engage the community. Innes and Booher (2004: 419) argue that legally required methods of public participation in government decision-making in the United States do not achieve genuine participation in planning, "the methods often pit citizens against each other, as they feel compelled to speak of the issues in polarizing terms to get their points across." Hou and Rios (2003) also argue that the increasingly bureaucratic and standardized participatory practice fails to reflect the vitality and complexity of urban places in North America.

Self-interests

Rather than pursuing collective goods, public participation in the US has increasingly become a tool for citizen groups to protect and advance their self-interests. Hester (1987) argues that while community participation has become more mainstream in professional practice, it is more productive in defending exclusionary groups than promoting the public good. In contrast to the Civil Rights era, he observes that citizens today are more motivated by shortsighted self-interests, coupled with splintering of public goods (Hester 1999). Unlike the more focused struggle for accountability in the 1950s, there is now a broader array of interests in today's urban design projects and processes, ranging from developers and businesses to residents and environmental groups. As participation becomes institutionalized and more narrowly focused, the procedures more often serve the louder voices rather than engaging the different parties in a more inclusive and substantive dialogue.

Multiculturalism

As today's cities and neighborhoods become more socially and culturally diverse as results of immigration and changing social values, engaging multicultural groups in urban design presents another profound challenge. The established practice of participation, focusing on a narrow set of procedures, is ill prepared to deal with the multicultural complexity of contemporary cities. On the one hand, the formalized rules and procedures tend to encourage certain groups while marginalizing others (Hillier 1998; Tauxe 1995). On the other hand, professionals have to confront with diverse communication styles, cultural nuances and conceptions of issues (Briggs 1998; Umemoto 2001).

Privatization

Similar to the challenge of diversity, participatory design mechanisms have lagged behind the increasingly complex public/private processes that produce many of today's urban design projects. With the financial constraints of many municipalities today, the number of privately funded (yet publicly subsidized) urban amenities have been on the rise. The projects include sports stadiums, large urban parks, and a variety of privately owned public spaces. Although design guidelines do exist in many municipalities, these projects often escape complete public scrutiny. Eager to get these projects built, municipalities seem to prioritize private investment over public accountability and transparency. The results have been encroachment of private and corporate interests into the public realm, in forms of branding and control of public spaces, "fortressed downtowns" (Davis 1992), "liminal space" (Zukin 1991), "polarized downtowns" (Loukaitou-Sideris and Banerjee 1998), and "city as a theme park" (Sorkin 1992). As private investments return to urban cores in North America, urban design through the improvement of open space, streetscapes and other amenities, has often facilitated the process of gentrification. However, as the process is driven mostly by private capital, there is little that individual citizens can do to challenge it.

Design aesthetic

The question of whether public participation produces "good urban design" has been a perennial debate among professionals and theorists. Hester (1999) argues that adversarial planning and litigations have often created uninspiring places, perpetuating the status quo. In a study that evaluates the quality of participatory design in the Boston Southwest Corridor, Crewe (2001) finds designers frequently complaining about the "hodge podge" of citizen interests to be threatening to the overall functional coherence and image of the project. However, she also finds that the professionals' views change when different criteria for "good design" are applied (Crewe 2001). In a professional culture that favors and rewards signature design, participatory design continues to face the challenge of acceptance by the professional community.

Mistrust

The persistence of mistrust among citizens, professionals, developers, and government authorities presents yet another challenge for democratic urban design. Given past frustrations, citizens are often suspicious of the effectiveness and intention of participatory process in terms of its actual impact on decision-making. On the other hand, developers and government authorities are concerned with the uncertainty of public processes. Meanwhile, design professionals are caught between the interests of their clients and the ethics of serving the broader public good. As a result, the prevalent tendency has been to fall back on established, bureaucratic processes of participation to satisfy the legal requirement. Faga (2006: 207) observes professionals' desire to control the public process by making sure that "the public has only a limited amount of choices, otherwise they won't be able to decide." Such tendency exacerbates the ongoing mistrust among the different players in urban design.

Directions for the future

Despite these limitations and challenges, participation in one form or another continues to play an important role in shaping the landscapes of urban America. While often criticized for not meeting its intended purposes, participation is also seen as a necessary and unavoidable part of urban design practice today (Faga 2006). Kaliski (2005) argues that, far from being *ad hoc*, new layers of mandated public input are actually creating a better urban form. In many parts of the world, in the face of top-down decision-making, citizen involvement in urban design is a goal that many activists and professionals are striving for. Given the continued significance of citizen participation, what are the current trends in the field?

New tools and technology

Commenting on the urban design process in Los Angeles, Kaliski (2005) observes a rise of "citizen experts," armed with access to information, as well as planning and design knowledge. These citizen experts play an increasingly active role in determining the evolution and design of the city (Kaliski 2005). While the precise role of citizen experts is still to be defined, there is indeed a wider array of tools available for citizens to take part in understanding and evaluating urban design issues and proposals. The Internet and search engines have made information instantaneously available at the fingertips of citizens. Free online resources are enabling citizens to visualize and manipulate outcomes of urban design. More complex programs allow simulated walkthroughs and produce suitability maps and visualization of different planning scenarios. In Portland, Oregon, citizens armed with new skills have presented alternative guidelines for infill development (Snyder 2006). In New York City, communication technologies made it possible to organize a large-scale workshop with thousands of participants to envision the rebuilding of Lower Manhattan following 9/11 (Faga 2006). These new tools and technologies have the potential to make citizens become informed and engaged, while pushing the temporal, spatial, and social boundaries of participation.

Expanding participation

In recent years, the discourse and practice of participatory design have continued to expand. To overcome the parochially focused practice of participation, Hester (1999) argues for a more holistic and inclusive view. He suggests a visionary synthesis that links participation across geographic, cultural, and class barriers, and thus achieves "local checks" with "regional balances" (Hester 1999: 24). Similarly, in place of the narrowly focused participation, Hou and Rios (2003) present a community-driven model that recognizes and engages a broader set of actors and processes in collaborative planning and design. At a different level, Hou and Kinoshita (2007) argue for the importance of informal processes in participatory urban design. They find that activities such as neighborhood events (Figure 25.1), tours, and even meals and personal conversations can overcome limitations of institutionalized participation by animating interactions, building trust, and creating new meanings and social relationships among the diverse actors (Hou and Kinoshita 2007).

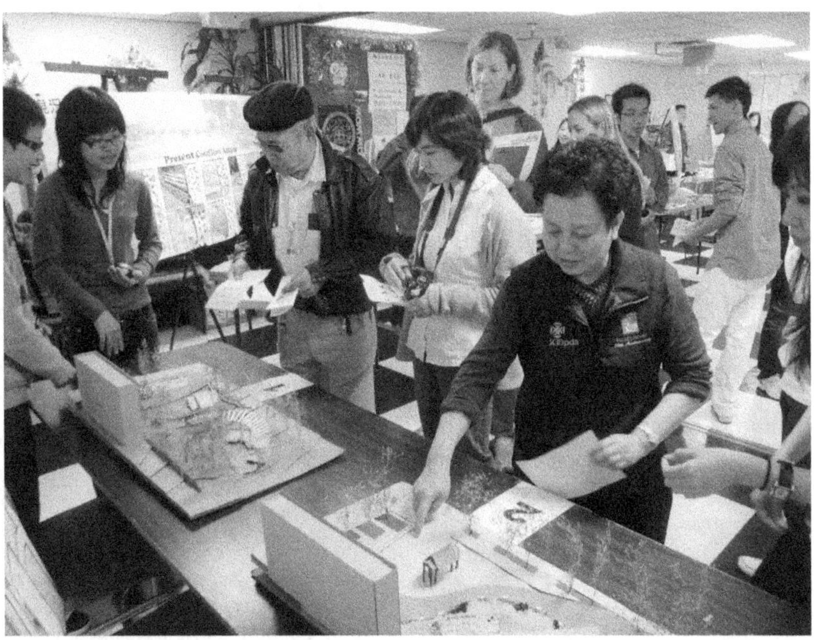

Figure 25.1 Open house in Seattle's Chinatown International District. Source: Jeffrey Hou.
Note: Open houses provided opportunities for conversation between neighborhood residents and designers concerning proposed design.

Citizenship and empowerment

Increasingly, participation in urban design is seen not as only creating and improving places, but also building communities and forging new identities and responsibilities. In this vein, Rios (2008) proposes a "polity approach" in urban design as a vehicle to encourage, facilitate, and organize groups toward collective action. Rather than working within the "confines of aesthetic, technical and/or economic consideration," the polity approach changes social and institutional relations in and through the production of public space (Rios 2008: 213–214). Instead of treating participants as clients or consumers, the approach recognizes them as full citizens (Rios 2008). The polity approach is particularly important in the context of engaging multicultural and marginalized groups. It addresses the critique by Juarez and Brown (2008) that most of the mainstream work in involving marginalized groups produces extraction of information rather than empowerment.

Asset-based approach

Having emerged from the field of community development, asset-based approach has important implications for future directions of participatory urban design, particularly in the context of distressed neighborhoods. Instead of dwelling on the community's needs, deficiencies and problems, asset-based approaches focus on existing capacities in the community to effect change (Kretzmann and McKnight 1993). One example of asset-based approach is the Neighborhood Matching Fund program in Seattle. Since 1989, the program has supported thousands of community-initiated improvement projects, from building new parks and playgrounds and restoring streams and wetland to developing neighborhood plans (Diers 2004). In addition to cash, the program allows volunteer hours to be counted as a match. This enables many disadvantaged communities to utilize their available human resources and build relationships and capacities that enable them to improve their neighborhoods and take on further actions (Figure 25.2).

Collaboration

As urban design involves an increasing number of actors and organizations, the civic model of urban design is shifting toward not only a participatory one but also a collaborative one with designers and stakeholders working as more equal partners. Collaboration is important, as urban design increasingly has to address multiple values, competing interests, social and economic conflicts, cultural differences, and institutional complexities. Innes and Booher (2004) argue that the current legally

Figure 25.2 Belltown P-Patch, citizen-initiated project in Seattle. Source: Jeffrey Hou.
Note: Supported by the Neighborhood Matching Fund, the Belltown P-Patch is one of many citizen-initiated projects in Seattle that enhance the livability and appearance of the city.

required public participation is insufficient and inadequate in dealing with the new level of complexity in the public arena. They identify research showing that collaborative participation can solve complex and contentious problems and create an improved climate for future actions (Innes and Booher 2004).

Public space activism

Demonstrating the ability of individuals and small groups to affect changes, a growing number of cases have emerged in North America and around the world in which collective efforts by citizens and professionals contribute to the remaking of urban spaces (see Hou 2010). Starting often outside the regulatory domain, some of these efforts have begun to be recognized as legitimate ways of improving the urban environment. In Portland, Oregon, some street intersections were repainted and transformed by citizens into community gathering places. Recognizing their role in traffic calming, neighborhood improvement, and community building, the City has since legalized such efforts. In Los Angeles, Latina/o immigrants have adapted and transformed vacant storefronts, streetscapes, and private yards into active places for businesses and social interactions. These new urban vernaculars constitute what Crawford (1999) calls "everyday urbanism," a practice that challenges the formal and mainstream practice of urban design.

Non-profit practice

In recent years, a new crop of alternative organizations has emerged that facilitate the involvement of citizens as well as volunteering professionals in urban design. Organizations like Public Architecture in San Francisco and Architecture for Humanity (through its Open Architecture Network) are helping to facilitate involvement of design and planning professionals in assisting local communities through pro bono and other types of services. Similar to the more traditional neighborhood design centers and university-based service-learning programs, the non-profit practice can make design more accessible to the underserved communities, and has the potential to better engage local citizens in design and planning processes. They also transform the role of professionals from providers of technical service into active citizens themselves.

From participatory design to citizen design

Today, the most powerful and distinct expression of citizen involvement in urban design occurs not in the legally required public meetings, but in streets, ballot boxes, reclaimed sites, and community gardens. It occurs through networks of individuals and groups facilitated by emails, instant messages, blogs, and social networking sites, with the help of a growing array of design and visualization tools. As a result, urban design today is no longer an exclusive domain of professional architects, landscape architects, and planners, but is instead a publicly negotiated process that involves a high number of individuals, interest groups, and public agencies. Rather than passive recipients of information, citizens are playing a more active role in urban design through individual and collective actions.

While Arnstein's notion of "citizen control" (1969) once seemed like a far-fetched ideal, citizen power is now being realized through a variety of means and expressions. From coast to coast in the US, citizen-initiated efforts are transforming the contemporary urban landscapes. In Oakland, California, a community organization initiated the first transit-oriented development in the region by transforming a transit parking lot into a mixed-use, office and retailed development that provided jobs and public spaces for the local neighborhood (Figure 25.3). In Seattle, citizens twice voted to impose taxes to support the acquisition, development, and programming of green and open spaces in the city. In addition to predetermined sites, portions of the funds allow citizen groups to initiate their own projects. In New York City, the innovative High Line began with a grassroots campaign that later won the support of City and State authorities. In Atlanta, the plan to redevelop BeltLine, a 22-mile long historic rail segment that surrounds the core urban area, grew out of a graduate architecture thesis (Gravel 2008). While participation in the traditional sense remains a critical ingredient in these efforts, the making of these initiatives has gone far beyond participation in a narrow sense.

This expanding role and capacity of citizens in urban design today suggests the emergence of a more inclusive practice of "citizen design" as distinct from the conventional model of participatory design. The practice of citizen design moves beyond participation as a legal, and procedural requirement. It sees urban design not as an exclusive realm of professional practice but as a field in which citizens can exercise their full rights and responsibilities, as well as their new skills and knowledge. For institutions and the professionals, the rise of citizens means that they have to engage citizens and community stakeholders as equal partners in the design and planning process. It requires understanding the more complex and fluid processes of engagement, navigating the social and cultural nuances, building alliances and partnerships, and expanding the repertoire of participation. For the municipal authorities in

Figure 25.3 Fruitvale Village, Oakland, California. Source: Jeffrey Hou.
Note: Connecting the BART station to the center of the nearby neighborhood, the Fruitvale Village was initiated by the Unity Council, a non-profit community development and service organization in Oakland, California.

particular, it involves creating institutional mechanisms that support not only citizen involvement but also citizen initiatives. For practitioners, educators, public agencies, as well as the public, it requires a re-envisioning of urban design not simply as a technical and professional exercise but a public and democratic practice in its fullest sense.

References

Arnstein, S.R. (1969) "A Ladder of Citizen Participation," *Journal of the American Institute of Planners*, 35(4): 216–224.

Briggs, X. (1998) "Doing Democracy Up-Close: Culture, Power, and Communication in Community Building," *Journal of Planning Education and Research*, 18: 1–13.

Crawford, M. (1999) "Introduction" in J. Chase, M. Crawford, and J. Kaliski (eds) *Everyday Urbanism*, New York: Monacelli Press: 8–15.

Crewe, K. (2001) "The Quality of Participatory Design: the Effects of Citizen Input on the Design of the Boston Southwest Corridor," *Journal of the American Planning Association* 67(4): 437–455.

Curry, R. (2004) "Community Design Centers" in B. Bell (ed.) *Good Deeds, Good Design: Community Service through Architecture*, New York: Princeton Architectural Press: 61–70.

Davidoff, D. (1965) "Advocacy and Pluralism in Planning," *Journal of the American Institute of Planners*, 31(4): 331–338.

Davis, M. (1992) "Fortress Los Angeles: the Militarization of Urban Space," in M. Sorkin (ed.) *Variations on a Theme Park: the New American City and the End of Public Space*, New York: Hill and Wang: 154–180.

Day, D. (1997) "Citizen Participation in the Planning Process: An Essentially Contested Concept?" *Journal of Planning Literature*, 11(3): 421–434.

Diers, J. (2004) *Neighbor Power: Building Community the Seattle Way*, Seattle: University of Washington Press.

Faga, B. (2006) *Design Public Consensus: the Civic Theater of Community Participation for Architects, Landscape Architects, Planners, and Urban Designers*, New York: John Wiley & Sons, Inc.

Francis, M. (1999) "Proactive Practice: Visionary Thought and Participatory Action in Environmental Design," *Places*, 12(2): 60–68.

—— (2003) *Urban Open Space: Designing for User Needs*, Washington, DC: Island Press.

Gravel, R. (2008) "Designing Infrastructure/ Designing Cities" in B. Bell and K. Wakeford (eds.) *Expanding Architecture: Design as Activism*, New York: Metropolis Books: 140–145.

Hillier, J. (1998) "Beyond Confused Noise: Ideas Toward Communicative Procedural Justice," *Journal of Planning Education and Research*, 18: 14–24.

Hou, J. (Ed.) (2010) *Insurgent Public Space: Guerrilla Urbanism and the Remaking of Contemporary Cities*, London: Routledge.

Hou, J. and Rios, M. (2003) "Community-driven Placemaking: the Social Practice of Participatory Design in the Making of Union Point Park," *Journal of Architectural Education*, 57(1): 19–27.

Hou, J. and Kinoshita, I. (2007) "Bridging Community Differences through Informal Processes: Reexamining Participatory Planning in Seattle and Matsudo," *Journal of Planning Education and Research*, 26(3): 301–313.

Hou, J., Johnson, J. and Lawson, L. (2009) *Greening Cities, Growing Communities: Learning from Seattle's Urban Community Gardens*, Seattle: University of Washington Press.

Hester, R.T. (1987) "Participatory Design and Environmental Justice: Pas De Deux or Time to Change Partners?" *Journal of Architectural and Planning Research*, 4(4): 289–299.

—— (1999) "A Refrain with a View," *Places*, 12(2): 12–25.

Innes, J.E. and Booher, D.E. (2004) "Reframing Public Participation: Strategies for the 21st Century," *Planning Theory & Practice*, 5(4): 419–436.

Jacobs, J. (1961) *The Death and Life of Great American Cities*, New York: Random House.

Juarez, J.A. and K.D. Brown. (2008) "Extracting or Empowering? A Critique of Participatory Methods for Marginalized Populations," *Landscape Journal*, 27(2): 190–204.

Kaliski, J. (2005) "Democracy Takes Command: the New Community Planning and the Challenge to Urban Design," *Harvard Design Magazine*, Spring/Summer: 20–26.

Kretzmann, J.P. and McKnight, J.L. (1993) *Building Communities from the Inside Out: A Path*

Toward Finding and Mobilizing a Community's Assets, Evanston IL: Institute for Policy Research.

Loukaitou-Sideris, A. and Banerjee, T. (1998) *Urban Design Downtown: Poetics and Politics of Form*, Berkeley, CA: University of California Press.

Lynch, K. (1960) *The Image of the City*, Cambridge, MA: MIT Press.

Rios, M. (2008) "Envisioning Citizenship: Toward a Polity Approach in Urban Design," *Journal of Urban Design*, 13(2): 213–229.

Sanoff, H. (2000) *Community Participation Methods in Design and Planning*, New York: John Wiley & Sons, Inc.

Snyder, K. (2006) "Putting Democracy Front and Center: Technology for Citizen," *Planning*, July: 24–29.

Sorkin, M. (ed.) (1992) *Variations on a Theme Park: the New American City and the End of Public Space*, New York: Hill and Wang.

Tauxe, C.S. (1995) "Marginalizing Public Participation in Local Planning: An Ethnographic Account," *Journal of the American Planning Association*, 61(4): 471–482.

Umemoto, K. (2001) "Walking in Another's Shoes: Epistemological Challenges in Participatory Planning," *Journal of Planning Education and Research*, 21: 17–31.

Weiner, E. (1999) *Urban Transportation Planning in the United States: A Historical Overview*, Westport, CT: Praeger.

Whyte, W.H. (1980) *The Social Life of Small Urban Spaces*, Washington, DC: Conservation Foundation.

Zukin, S. (1991) *Landscapes of Power: From Detroit to Disney World*, Berkeley, CA: University of California Press.

Further reading

Bell, B. and Wakeford, K. (eds.) (2008) *Expanding Architecture: Design as Activism*, New York: Metropolis Books. Innovative practices by design activists, community design centers, non-profit organizations, and educators to engage communities and the public in addressing a wide variety of issues ranging from reutilization of urban lands to affordable housing.

Bloom, B. and Bromberg, A. (2004) *Belltown Paradise/Making Their Own Plans*, Chicago: WhiteWalls, Inc. Features several case studies of quirky yet effective citizen-driven projects in Europe and North America.

Faga, B. (2006) *Design Public Consensus: the Civic Theater of Community Participation for Architects, Landscape Architects, Planners, and Urban Designers*, New York: John Wiley & Sons, Inc. Features case studies of public processes in several high-profile urban design projects, and explores the complexity in building consensus in public decision-making concerning urban design.

Hester, R.T. (2006) *Design For Ecological Democracy*, Cambridge: MIT Press. Focuses on the connections between democratic design practice and the ecological integrity of cities and regions.

Sanoff, H. (2000) *Community Participation Methods in Design and Planning*, New York: John Wiley & Sons, Inc. A comprehensive survey of both the history of citizen participation in design and planning and specific methods and techniques.

Part 6

Components

Introduction

In this section we will consider the following questions: How is urban design contributing to forming and transforming the different components of the city? How has the design of downtown and suburban areas, residential, commercial, and cultural districts, and public spaces changed over the years? What are the emerging trends in the design of these different urban milieus, and how do they address physical, social, or economic goals?

Collectively, the chapters in this section paint a clear picture of urban design's aspirations in shaping the form of the city. Design has traditionally been used as an apparatus for the achievement of physical and aesthetic goals: to create attractive new urban areas or revitalize formerly obsolete city districts. These aesthetic improvements are often linked to economic aspirations. In an era of globalization and intense competition among cities, urban design has also been thought to contribute to a city's economic prowess by boosting its identity and uniqueness and building its reputation as a livable and attractive place. A downtown entertainment or retail district, a cultural complex, or a new park may serve as attractors for residents and tourists and their dollars. Urban design is supposed to contribute to the enhancement of the quality of life of city residents by increasing the linkages and connections within and between areas, making streets more walkable, and increasing the functionality, comfort, and aesthetics of public spaces. More recently, urban design is also expected to support the lofty goals of urban sustainability and health by promoting greener and energy-efficient neighborhoods that provide opportunities for active living.

The chapters that follow also reveal some of the tensions encountered in the design of the different components of the city. For example, as Ajay Garde makes clear, neighborhood design may have been guided by the desire to enhance feelings of neighborliness and communal interaction in school settings, parks, and other public spaces, but this has often led to exclusivity encouraged by design. Similarly, as John Archer argues, the design of suburbs has been confronted by the tension of "mediating the politics of public realm versus private property." This tension is even demonstrated in the design and functions of public spaces, which according to Mark Francis, have often promoted more "parochial" or "community-private" needs than more inclusive settings. And while the rhetoric of cultural institutions may refer to the democratic goals of integrating culture to everyday life, Carl Grodach argues that their urban design is often guided by desires for distinctiveness and recognition, and urban forms that stand distinct and separate from the rest of the city. As Ann Forsyth mentions, another tension rises from debates that tout the merits of tradition versus newness, expressed in the selection of vernacular or modern design styles. More recently, the tension between desires for globalization and calls for expressing local identities and needs have been reflected in the built environment, notably in the design of consumption spaces. As Klaus Kunzmann explains, such spaces often adopt similar themes or design clichés in order to blend better in the global landscape, even though differing consumption needs and desires for city identity may be calling for more local and contextual design solutions. Elizabeth Macdonald refers to the tension that may arise from efforts to break away from well established design standards and norms in order to incorporate additional and alternative uses in previously mono-functional spaces. Lastly, Gary Hack discusses how the tradition and vision of urban design has always been geared towards the creation of permanent and stable landscapes, which may be in contradiction to the realities of ever-evolving and changing cities.

Drawing mostly from the experiences of US cities, the first essay in this section reviews the different models of downtown urban design, and explains how they were the products of different socio-economic circumstances. Urban design and civic architecture have always played a major role in downtown imagery. In that sense, downtowns are products of purposeful design actions that have followed particular goals. In the later part of the twentieth century, municipalities have used urban design as a medium to revitalize their central business districts and attract mostly upscale residents, employees, and visitors. This chapter discusses the major themes that capture the logic of downtown design as well as the challenges that it is facing.

Since the rise of modern suburbs three centuries ago, the discourse on their design has drawn from eighteenth-century English aesthetic ideals of the Picturesque and nineteenth-century American ideals of pastoralism, but also individualism and republican virtue. John Archer discusses how such notions continue to influence suburban design, but also traces the later consequences on their built form of automobile dependency, building standardization, and mass production. These have contributed to ordinary and often criticized settings, which nevertheless have kept attracting the preferences of consumers. While New Urbanists propose to redesign suburbia using prescriptive sets of codes and design guidelines, the author argues for a "recasting" of suburban design to allow for more flexibility and diversity in the design of suburban dwellings based on the revealed preferences of residents.

New towns and large-scale planned communities may represent opportunities for urban designers to have a "clean slate" on which to express a comprehensive ideal of a "good city form." Urban designers have chosen to do so in different contexts and different ways. Ann Forsyth explores the variations of new towns in different countries and the various traditions that have influenced their design. A variety of physical, social, economic, and – more recently – ecological goals have often accompanied the design of new towns. Promotion of a town's self-sufficiency, jobs-housing balance, social integration and interaction among residents, regional economic development, but also the achievement of a low-carbon footprint and high energy efficiency, have been some of the lofty goals of new town design, which have been met with various degrees of success.

Neighborhoods constitute key components of the urban landscape. In contemporary times, neighborhood design has been influenced both by social values as well as the economic imperatives of the market. Ajay Garde traces these influences giving a critical analysis of both earlier patterns of neighborhood design as well as more contemporary trends. He argues that the conceptualization of recent models of neighborhood design (such as planned unit developments, New Urbanist neighborhoods, and the LEED-ND rating system for sustainable neighborhoods) are quite reminiscent of Clarence Perry's Neighborhood Unit paradigm, which was conceived as a design response to the problems of urban development.

Spaces of consumption – for shopping, entertainment and cultural activities – are indispensable components of cities. Urban design has played an important role in creating landscapes of consumption to attract consumers. Drawing mostly from the Western European city, Klaus Kunzmann details the type and variety of its consumption spaces. He explains that the design of such spaces is often guided not only by profit-making imperatives of the market but also by municipal aspirations for revitalizing obsolete central cities or strengthening the image and identity of a district or city. This leads to the employment of design strategies for the creation of spaces considered as appealing to consumers. These include theming and the creation of "set pieces" borrowed from different places or times or the development of flagship buildings by "starchitects."

Similarly, renowned architects are often commissioned by cities to design iconic cultural complexes to boost their image and global identity. This trend has intensified following the "Bilbao effect," where the building of a Guggenheim Museum by Frank Gehry brought international notoriety and tourist revenues to a formerly sleepy Basque town. Carl Grodach gives an overview of the recent trends in the design and planning of cultural complexes. He traces how changes in cultural institutions run parallel to shifts in urban economies which emphasize consumption, services, and knowledge-based industries, and the emergence of urban development strategies that wish to capitalize on these changes. Using four case studies, he details how urban design is employed as a tool that mediates between public and commercial culture in the context of contemporary cultural institutions.

Streets are a ubiquitous component of the city and represent an important part of the public realm. The design of streets has long been determined by traffic engineering standards, which privilege the efficient circulation of vehicles, resulting in single-use spaces, devoid of pedestrian activity. Elizabeth Macdonald explores the new opportunities and emerging trends in street design that aspire to create streets which are pedestrian- and biker-friendly, accommodate different transportation modes, have a "green" infrastructure, and provide the flexibility of accommodating different uses, depending on the time of the day.

The design of public spaces represents an important focus for urban design. Still, as Mark Francis argues, the record of urban design in creating successful and democratic public spaces is rather mixed. While well-intended, mixed-use spaces incorporate public spaces, too often these end up in accommodating only a homogeneous public. Public spaces in such complexes are available only to tenants and are not inclusive of the general public. In contrast to mixed-use spaces, Francis juxtaposes the concept of "mixed-life places," which he defines as "public spaces that are at once diverse, democratic, inclusive and memorable." He articulates a typology as well as the necessary ingredients for the design of such spaces.

This section concludes with a chapter by Gary Hack, who turns our attention to a component of the city that is rather neglected by urban designers: the urban flux. This includes the temporary artifacts that decorate urban environments: the billboards and other advertisements, digital signs, seasonal decorations and lighting, temporary art, construction scaffolding, and the like. Such elements have the potential to create more spontaneous, dynamic, and unpredictable experiences which may be more exciting than those stimulated by orderly, overly designed, and tightly regulated places. While careful to articulate the legal and ethical boundaries of flux, Hack prompts urban designers to consider the opportunities of using flux elements in cities.

26

Downtown urban design

Anastasia Loukaitou-Sideris
and Tridib Banerjee

From ancient times and in every part of the world cities have derived their essential identities from their centers. Serving civic, economic, and cultural functions, city centers invariably and symbolically compose the most prominent urban district as the control and command post of a city's economy, the house of its governance, the hub of its cultural institutions, and the core of its identity. Over the last century the center of the city has experienced major changes involving the remolding of its design and form, character, and social meaning. Some of these changes had to do with the transformed nature of the economy, with the way people lived and the built environment was produced (Sudjic 1992). Drawing mostly from the experience of American downtowns, this chapter will give an overview of the changes that have resulted in different models of central city design. While we will briefly discuss the origin of the urban form and the development of downtowns of earlier eras, we will mostly focus on contemporary times and seek to critically review and anticipate the implications of socio-cultural and economic trends in the urban design of downtowns.

From town square to central business district

The pre-capitalist city had a semantic unity; it was organized around a center within which the specific social practices of politics, religion, and commerce were exercised (Gottdiener 1986). In colonial cities, this center typically comprised of the square, plaza, common green, or place d'armes surrounded by major public buildings (church, town hall, meeting house, customs house) as well as the shops and dwellings of prominent citizens. The indigenous population of course were not included in this scheme and consigned to peripheral land of lower importance (Lynch 1981). The square at the core served as the "generative space" of the entire colonial settlement, which was laid out from the center outward (Suissman 1993). The central square thus served as the focal space, morphologically derived from the grid, as was customary in cities built according to the Laws of the Indies (Heckscher 1977).

As towns progressively became centers of a mercantile economy, an identifiable central business district (CBD) emerged, shaped by demands for shipping, storage, distribution, and administration. The CBD became the physical expression of a progressive and profound restructuring that by the middle of the nineteenth century would convert the mercantile colonial town into an industrial capitalist center (Soja 1989). The first American CBD emerged in New York around Wall Street, and because of its location at the southern tip of Manhattan's elongated shape, became known as "downtown" (Ford 1994), with the terms "midtown" and "uptown" used for sections to the north. Soon similar business districts developed in Boston, Philadelphia, and Chicago. By the mid-nineteenth century little had remained of the colonial square's original concept or its functions. New financial institutions (banks, insurance, and trust companies) and merchant show rooms required more space than those that the four sides of the square offered, and found it along the linear corridors of Main Streets – the commercial spines of downtown districts. The term "downtown" which grew out of the geography of Manhattan became a generic expression in the American cities, while CBD or the "center" remained the common expression for cities elsewhere in the Western world.

The early American CBDs were dense and compact. The concentration of activities in a limited space – the effect of agglomeration economy – resulted in a mixed pattern of land uses. Warehouses stood next to offices, and department stores abutted train depots. This was essentially a walking downtown since its area rarely exceeded eight blocks square. Dense rows of low and medium-rise (up to 10 stories) buildings formed continuous facades along the sidewalks. At the edges of downtown and in close proximity, one could find both slums for poor immigrants and fashionable streets for wealthy citizens.

Designing the hub of the city

This rather *ad hoc* and disorderly form of early CBDs would soon be inadequate and inefficient for the emerging role of downtown as the core of an expanding city. Indeed, in the second half of the nineteenth century, new transportation technologies enabled downtown to develop as the hub of a radial urban form (Loukaitou-Sideris and Banerjee 1998). Street car systems converging on downtown emphasized its primacy as the all-too-important focus of the industrial metropolis, reinforcing its monocentric form. New construction technologies that enabled skyscraper development were responsible for the newfound verticality of downtown skyline. The morphology of downtown also changed during that era, consolidating smaller lots into larger blocks, closing alleys, widening streets, and later developing superblocks by consolidating multiple street blocks.

Grand office buildings, hotels, theaters, libraries, and railway terminals served thousands of office workers, shoppers, and visitors each day. This prominent role of downtown and the accumulation of wealth it represented invoked a grand civic design, inspired by the City Beautiful Movement at the turn of the twentieth century. The City Beautiful vision advanced by Burnham's plan for Chicago further glorified the center. A vision of a grand public realm that emphasized the centrality and primacy of the CBD included a monumental civic design: the convergence of grand avenues on city landmarks and the construction of imposing public buildings and spaces. This was consistent with the emerging culture of downtown. The public came to downtown to have a night out, to go to shows, or visit the cinema. The decorum of visiting downtown required formal attires (Forsher 2003). For the first twenty years of the twentieth century the leadership of major cities – Washington DC, Cleveland, and Chicago – initiated and partially carried out large scale design proposals for respective downtowns. These plans strived for a formalistic reconstruction of city centers, inspired by the neoclassical, Beaux Arts tradition. In contradistinction to the perceived disorder and chaos of the emerging industrial city, City Beautiful plans emphasized aesthetics and visual order, using Baroque design principles of balance, axial order, and hierarchical arrangements of space (Foglesong 1986; Bennett 1990).

But City Beautiful plans did not, in the minds of many, solve the problems that were plaguing CBDs: congested streets, incompatible land uses, and lack of light and air, exacerbated by an increasing stock of high-rise buildings. These concerns resulted in an abrupt switch from a design paradigm that emphasized aesthetics to one that called for rationality and efficiency in the spatial organization of urban form. The new model, identified by some as City Functional (Hall 1989), did not utilize architectonic means but relied instead on new regulatory mechanisms – zoning of land uses, restriction of building heights, establishment of building envelopes and setbacks – to shape the urban form that emphasized downtown's primacy (Loukaitou-Sideris and Banerjee 1998).

The vision for the center of the American metropolis that the City Functional espoused echoed early modernist influences from Europe. This was the time (early 1920s) that Le Corbusier was advancing his vision of *La Ville Contemporaine* – with magnificent towers, sweeping open space, and multi-lane highways converging at its center (see also chapters by Birch and Fishman in this volume). The City Functional model of urban design represented a transition to the modern era, initiating paradigms that were followed and accentuated after World War II. The division of downtown into functional cells defined by zoning was a rational solution meant to safeguard and enhance the value of prime real estate. This was the dawn of what Christine Boyer (1983) would refer to as the "rational city" (see also Banerjee 2009).

Renewing downtown

Downtown real estate, however, would drastically lose its value during the Great Depression of the 1930s as the construction activities came to a standstill. The North American city center began to lose progressively its predominant role in daily life, as expanding automobile use and a multibillion-dollar program of national highways allowed more people to live far away from downtown after World War II. Businesses followed residents, as department stores, movie theaters, and offices found a more profitable market in the suburbs.

The understanding that downtown was a declining center and an area in crisis struck city councils and planners in the 1950s and 1960s. A series of master plans were prepared that called for the renewal of blighted downtowns. By the end of the 1950s, some 700 downtown plans had emerged (Frieden and Sagalyn 1989), committed to converting the ailing core into a modern and efficient business center. The new vision of the modernist city required drastic changes in the urban form. Urban renewal efforts involved extensive demolition of older structures considered blighted, the street grid was obliterated by new and wider thoroughfares, the block system was altered, and landmark buildings of the past (city hall, court house, railroad terminal) were dwarfed by corporate highrise towers and megastructures. Distances between activities increased, and curb cuts to underground parking garages made sidewalks unfriendly to pedestrians. There were not many pedestrians left on the streets anyway, as white-collar professionals were pulled into underground plazas and overhead skywalks.

According to the modernist vision, land use and transportation had to be integrated, and the downtown had to be designed as a unified whole (Loukaitou-Sideris and Banerjee 1998). Urban design plans of the 1960s promoted the construction of inner-loop freeways encircling the CBD and parking facilities at the downtown periphery, while modernist complexes of office buildings, shopping centers, and civic structures surrounding open-air plazas were designed at the downtown core. Megastructures such as Renaissance Center in Detroit, Prudential Center in Boston, Embarcadero Center in San Francisco, Charles Center in Baltimore, California Plaza in Los Angeles, and Gateway Center in Pittsburgh cut off existing streets to create superblocks of buildings and open spaces that were dramatically different in scale from downtown spaces of earlier eras. Contact with the surrounding city fabric was minimal; transportation linkages (parking, subway stations) were often under the building; retail activity was arranged around private courts in enclosed and sometimes sunken plazas. This was in sharp contrast to street-oriented retailing that characterized earlier downtowns.

Downtown woes

The city center that emerged from the urban renewal schemes of the 1950s, 1960s, and 1970s was dressed in corporate attire but the glittering skyscrapers of downtown could not mask some lingering problems. For one, years of suburbanization and decentralization had effectively challenged the primacy of the city center. In the early years of the twentieth century, US downtowns were competing with each other to build the tallest building, the most luxurious department store, or the grandest hotel. Fifty years later however, downtowns found that their competition was coming from outlying suburban centers. The suburban shopping mall had become an important alternative to downtown's Main Street and department stores. The residential core of the center that could have provided a critical mass of downtown users and shoppers had been obliterated by urban renewal.

The renewal of US downtowns was fueled by federal policies but was primarily carried out by corporate investment and real estate interest. But the increased reliance on private initiative and funding resulted in an uneven development and polarization of many city centers. Corporate investment concentrated on building a new downtown on cleared land, creating exclusive settings accessible to certain segments of the public. In many US centers, this has resulted in a polarized downtown: an old and derelict part populated by the homeless, the poor, and the immigrants, and a new and glamorous part housing the corporate world. The division between the new and old, wealthy and poor, native and ethnic created a segregated urbanism of two downtowns existing in close proximity, one at the CBD or downtown core, the other at the margin or downtown frame often occupying the abandoned but historic spaces of an earlier era (Loukaitou-Sideris and Banerjee 1998).

The reliance on private investment shifted the design initiative to the private sector. Cities became increasingly dependent on the private sector for providing downtown amenities, relying on incentive zoning formulas and other zoning regulations, as well as entitlement processes to negotiate the outcome of design (Kayden 2000). When we wrote *Urban Design Downtown* in 1998, we found that development in most US downtowns was characterized by a market-driven urbanism, carried on by private sector initiative and public sector approval. In contrast to the modernist urban design plans of the previous era that vied for comprehensiveness, development in downtowns at the end of the twentieth century was incremental and episodic, relying on specific catalytic projects – often "signature buildings" by "starchitects." The private nature of urban design and development led to a fragmented and disjointed urban form, what we called a *collage downtown*, which did not meet urbanistic goals of attention to context, livability, coherence, and linking of districts through pedestrian connections.

Downtowns in the late twentieth and early twenty-first century

By the end of the twentieth century, the US downtown still continued to be the major employment center in its region, albeit a weak one, in a polycentric and fragmented metropolis (Fogleson 1986), and still struggling with the inexorable centripetal tendencies of metropolitan decentralization. Local government functions and related services remained the major tenants of downtown office buildings. The original corporate headquarters of the industrial economy continued to leave downtown and locate in peripheral areas. Holding the center together became a major challenge as cities continued to subsidize downtown commercial development in the form of land write down, FAR bonuses, vacated street rights of way, and the building of parking structures, pedestrian malls, light rail stations, and other such improvements. Transit malls like those of Denver's and Portland's or Minneapolis' celebrated Nicolett Mall (Figure 26.1), artfully integrated vehicular and pedestrian traffic to create a pedestrian and shopper-friendly downtown core. In addition, some cities undertook extensive pedestrian improvements to develop skyways and bridges to protect the shoppers and workers from the harsh elements (as in the case of Seattle or Minneapolis) or just from traffic (as in Los Angeles), but in the process robbing the street life at the ground level.

Some cities also developed festival marketplaces that were often built around a theme to attract a consuming public back to the city center. Following the successful example of Quincy Market in Boston that used the shell of an earlier market to house a series of restaurants and retail shops, some cities started revamping and utilizing existing assets or recycling infrastructure of the industrial age (old markets, warehouses,

Figure 26.1 Nicollet Transit Mall, Minneapolis. Source: Tridib Banerjee.

industrial waterfronts, etc.) to bring more vibrancy into downtown areas. Thus, San Francisco remodeled an old chocolate factory to create Ghirardelli Square, Baltimore revamped its waterfront with the building of Harborplace, and so did New York with its South Street Seaport and the historic Fulton fish market. More recently New York has reenergized the High Line converting an old railway line into an elevated park. There is no doubt that such projects have helped attract more visitors – natives and tourists – to parts of downtown, but some of them often operate merely as "tourist bubbles" (Judd 1999) effectively blocking out more problematic downtown areas and accentuating the polarization of downtown. Similarly, while new office towers came to be occupied by new clients – mainly financial and professional services – sometimes serving as headquarters of banks, insurance companies, or financial institutions thus bringing new life to downtown, the "frame," i.e. the peripheral areas of the CBD languished.

In cities like Detroit the decline of the automobile industry and the large scale abandonment led to a wholesale decanting of jobs and population. With no future economic revitalization in sight many shrinking cities of the Rustbelt sought to draw national sport franchises by building sports stadiums and entertainment complexes or convention centers in the immediate periphery of downtown, areas otherwise dominated by parking lots, abandoned industries or warehouses. The idea has been tried out in Sunbelt cities as well, with San Francisco and Los Angeles erecting new sports and entertainment complexes, hoping to boost their downtown economy and stimulate additional development.

The faith that many cities have placed in the ability of cultural industries to act as tools of economic development and revitalize previously decaying central areas (see Florida 2005) has led the municipal governments to pursue flagship cultural projects (see also chapter by Grodach). The Disney Hall in Los Angeles and Seattle's new Art Museum are prime examples of this strategy that uses spectacular architecture and high-powered artistic endeavors to attract suburbanites and their dollars for an evening (or more) back to the city. The Moscone Center in San Francisco and the proposed redevelopment of Lincoln Center in New York are larger scale urban design projects motivated by the same objective. Design of such facilities often uses spectacular architecture, neon lights, and digital billboards to generate the necessary "buzz" deemed appropriate for the center of the city (Currid 2007). The transformation and rebirth of Times Square from a seedy and run-down place to the media-center of the world is a prime example of such strategy. Planners and urban designers hope that such cultural and/or media-centered projects can also act as catalysts attracting more development in downtown, including residential projects.

Indeed residential projects in the "frame" or "core" of downtown are becoming quite commonplace in most American cities. Even some of the older districts of downtowns, languishing previously, are seeing a rebirth like the Center City in Philadelphia that has undergone a major revival as the older townhouses have gentrified, and prosperous and young professionals have found the urbanism of the old city appealing. Loft living, which was commonplace among artists, painters, and sculptors has become an attractive housing option for many professionals in an emerging "back to the city" mood. Older industrial and warehouse districts are going through such residential revival in areas like the South of the Loop in Chicago, the warehouse district fronting the Los Angeles River in downtown Los Angeles (Figure 26.2), or the redevelopment of Denny Regrade which metamorphosed from parking lots

Figure 26.2 Lofts in downtown Los Angeles. Source: Tridib Banerjee.

and low-rise commercial structures to a thriving in-town residential district near downtown Seattle. At times, these are complemented by the provision of open space amenities that can make downtown residential life all the more pleasurable, such as the Millennium Park in Chicago or the proposed Cornfield Park in Los Angeles.

From the late 1980s through the 1990s infrastructure investments, or rather reinvestments – thanks to the ISTEA money created by the US Congress to fix America's aging infrastructure – further helped to transform older downtowns into more attractive locations for new businesses and residents. These investments helped to bring not only light rail and commuter train lines to downtown, but also to create new public spaces by tearing down older and now obsolete freeway infrastructure. Thus, in San Francisco, the two-level Embarcadero Freeway was taken down to create a magnificent pedestrian promenade along San Francisco's waterfront stretching from the Fisherman's Wharf to the Bay Bridge. The four-block stretch of Octavia Street in San Francisco's Hayes Valley is a successful example of reclaiming freeway space and transforming it into a residential corridor with ample public space and separate pedestrian and bicycle rights of way (see Bosselmann 2009 and also the chapter by Macdonald in this volume). The Big Dig in Boston, truly a megaproject in its scope, involved taking down the raised Central Artery and putting all through traffic underground in downtown Boston, and thus creating some seven acres of linear open space.

The logic of downtown urban design

The logic of the American downtown in modern times derives from the essential premise of land economics – that is, the location that has the highest accessibility and centrality would command the highest land rent or price, because the demand is highest for the central location as it maximizes exposure, customers, and other business opportunities. The expression "the 100 percent corner" commonly used in the real estate business refers to this central location, and is actually measured by the highest volume of pedestrian traffic count. The rent declines as one moves away from the center and as the overall accessibility of the location diminishes. This process results in a rent gradient, based on competitive bidding for land balanced against the transportation cost which increases as one moves further out from the center. The bid-rent function, as it is commonly known in the economic literature (Alonso 1964), leads to a three-dimensional conical form of curved surface, where businesses and industries would locate in the center, and housing and other related services will locate further away from the center. Higher density housing will be closer to the center occupying less land but gaining transportation cost advantage, while the lower density housing consuming more land will be located further away from the center where land is cheaper, but transportation cost is higher.

Historically this logic has explained the mono-centricity, as expounded further by the concentric ring theory proposed by sociologists Park and Burgess or the sectoral model of urban growth as amended by Hoyt. Chicago has been the early inspiration for these models, supported by other cities of the Midwest and the Atlantic Seaboard more generally. The generic urban form models described the monocentric cities as "radial" or "radio-centric" (Lynch 1990), where the distribution of population and activities are dominated by a single center. Subsequently the monocentric model of urban form and growth was countered by alternatives which emphasized the emergence of multiple centers, albeit secondary to the central business district, in many cases in direct competition with the main center, and thus vitiating its primacy. In the sociological literature, the work of Harris and Ullman (1945) made this argument, inspired by the emergent polycentric form of the Los Angeles landscape. This was also supported by urban economists' explanation of how the bid-rent curve could result in a multi-centric pattern, where other centers would emerge as the travel distance and travel cost increase with the expansion of the urban space of a growing metropolitan area. Even the typology proposed by Lynch (1981) included such form descriptors as "grid city," "galaxy," and "polynucleated net." Nevertheless, even this polycentric city model has been recently put into question by urban economists (see Gordon and Richardson 1996).

The corresponding challenges for urban designers have also followed a paradigm shift. Urban design initially had as its context a monocentric city where people came to work, shop, and attend to various civic and cultural affairs, and which used to be the seat of local government and the locus of corporate wealth and influence, as evident in the concentration of corporate headquarters. At this time, the task of urban design was to regulate the height and bulk of buildings and emphasize the civic design and the public realm, promoting the images of corporate and political power structures and downtown real estate interests. The public coffer had surpluses to pay for cultural and civic buildings. The Army Plaza of Brooklyn designed by Frederick Law Olmstead and Calvert Vaux, and built in the latter part of the nineteenth century is a case in point. So is the Civic Center Plaza in San Francisco, and the grand axis of Denver's Civic Center Park anchored by the State Capitol on the one side and the City County building on the other, with the historic library located on one side of the park. The civic design of that era also included major museums, concert halls, libraries, and the like. Corporate capital paid for office towers, department stores, theaters and entertainment centers, and railroad stations or port facilities. The train stations of Chicago, New York, Kansas City, Los Angeles, and Washington, D.C. are all examples of civic design of that era that continue to be used today as major venues of public life, albeit under different circumstances.

The scope of urban design came to be redefined as the primacy of the center began to decline in most major metropolitan areas, with the decentralization of jobs and economic activities, and the fragmentation of the metropolitan political space (Fogleson 1986). Multiple centers appeared in the metropolitan landscape challenging the primacy of the original downtown by offering alternative location options for businesses and industries. Freeway development exacerbated the centrifugal tendencies, accelerating the exodus of the middle class from the central city to the suburbs. Large suburban shopping malls essentially obviated the original shopping draw of downtowns, as did the drive-in theaters and suburban cinemas for downtown entertainment attractions. As industries moved out the role of downtown as the primary employment center became tarnished as well. The downtown began to accumulate empty shells of industrial structures, warehouses, along with department stores, cinemas, and office complexes of the earlier era. The east side of downtown Los Angeles remains a classic example of this narrative of decline and obsolescence. The flight of jobs and middle-class citizens from the central city left behind pockets of poverty and the attendant pathology of crime, drugs, and homelessness.

As this scenario unfolded in city after city throughout the country, the structural decline and decay came to be identified with urban blight, the only remedy to which was seen as urban renewal. Fond of organic analogs, many planners and policy makers saw pockets of poverty, dilapidation, and abandonment as "cancer" that can only be treated by surgical excision or large-scale clearance and rebuilding. Urban design became an important mechanism for the urban renewal program, as the designers offered the visions for rebuilding downtown in an attempt to recover its old glory. The modernist legacy of Haussman and Corbusier inspired clearance of the existing fabric of the city, and replacing it with large developments in consolidated super-blocks that protected the pedestrians from the cars and traffic (see also the chapter by Fishman in this volume). The Bunker Hill development is a classic example of the urban design of renewal. Baltimore's Charles Center, Boston's West End and the Government Center, and San Francisco's Embarcadero Center are also examples of this modernist urban design associated with urban renewal. The Federal housing program enlisted urban designers to build large public housing projects to house the inner city poor in the fringes of downtown. Later, they would prove to be a major failure in social engineering, and would be demolished unceremoniously – St. Louis' Pruitt Igoe project being a case in point.

The future of downtown design

As Sites (2003) has effectively argued drawing from New York's redevelopment experiences in the South Side, much of the contemporary redevelopment of urban spaces, especially in the downtowns of global cities, reflects a form of "primitive globalization." He argues that cities that are trying to compete in a global economy are building new downtowns and rebuilding old ones to attract global capital and businesses in a competitive international marketplace. In this model of primitive globalization the public sector plays a critical role in building the physical infrastructure and spaces that could draw global capital. The same theme is echoed by a comparative study by Savitch and Kantor (2002) where they define the process of antecedent urban transformation as a case of "global sweep" (i.e. the interests of capital) with a "local broom" (i.e. demolition and redevelopment of the older building stocks – and not inconsequently, the resident population). Some of the examples of downtown re-design we have cited earlier are cases in point. The urban design in a globalizing era has so far been one of mimicking some universal global style often set by the signature architecture of international "starchitects." In the absence of more authentic expressions of urban design, this trend is likely to continue. Two factors may help to stem the tide in the mature Western economies, and at least one of them may have similar influences in the emerging economies of the developing world as well.

The first factor has to do with the aging baby-boomers (see Myers 2007) in America, and in fact in most parts of the developed world. It is generally expected that a vast proportion of this age group may choose to leave the low density suburbs for more compact central city neighborhoods which are likely to offer more activity and consumption choices for an aging population. This may require urban designers to think of designing downtowns as "elder-friendly communities" (Alley *et al.* 2007) with appropriate amenities and facilities. More generally, it seems, the young, single, and professional class is choosing to live in downtowns or adjacent neighborhoods, and we are beginning to see a significant rise in new residential construction or adaptive reuse of older industrial and office buildings, as we have previously indicated.

The second factor that may further augment the trend of living in downtown is the emerging "peak-oil" or "post-oil" scenarios for urban transformation. Diminution of energy resources, according to this scenario, will compel urban residents to drive less, live in compact cities, and thus consume less energy. Indeed climate change and global warming initiatives at the national, state, and local level might force these trends even sooner through legislative and fiscal policies. Already planners and urban designers are beginning to respond to these imperatives. The Bay Terminal area redevelopment planning currently underway in San Francisco, is already considering urban design measures to increase pedestrian amenities, walkability and bikeability of the next phase of downtown addition, with integrated energy district planning and other such innovations. There are ample indications that we are at the threshold of a new era of imaginative urban design based strictly on sustainability goals rather than continuing with glamorous corporate complexes of the current era of primitive globalization.

References

Alley, D., Liebig, P., Pynoos, J., Banerjee, T., and Choi, I.H. (2007). "Creating elder friendly communities: Preparations for an aging society" *Journal of Gerontological Social Work*, Housing Special Issue, 49(1/2): 1–18.

Alonso, W. (1964). *Location and Land Use; Toward a General Theory of Land Rent*, Cambridge, MA: Harvard University Press.

Banerjee, T. (2009). "US planning expeditions to post-colonial India: From ideology to innovation in technical assistance." *Journal of the American Planning Association*, 75(2): 193–208.

Bennett, L. (1990). *Fragments of Cities: The New American Downtowns and Neighborhoods*. Columbus: Ohio State University Press.

Bosselmann, P. (2009). *Urban Transformation: Understanding City Design and Form*. Washington, DC: Island Press.

Boyer, C.M. (1983). *Dreaming the Rational City: The Myth of American City Planning*. Cambridge: MIT Press.

Currid, E. (2007). *The Warhol Economy: How Fashion, Art, and Music Drive New York City.* Princeton, NJ: Princeton University Press.

Florida, R. (2005). *Cities and the Creative Class.* New York: Routledge.

Foglesong, R. (1986). *Planning the Capitalist City: The Colonia Era to the 1920s.* Princeton, NJ: Princeton University Press.

Ford, L. (1994). *Cities and Buildings: Skyscrapers, Skid Rows and Suburbs.* Baltimore and London: The Johns Hopkins University Press.

Forsher, J. (2003). *The Community of Cinema: How Spectacle and Cinema Transformed American Downtown.* Westport, CT: Praeger.

Frieden, B. and Sagalyn, L.B. (1989). *Downtown Inc.: How America Rebuilds Cities.* Cambridge, MA: MIT Press.

Gordon, P. and Richardson, H. (1996). "Beyond polycentricity: the dispersed metropolis, Los Angeles, 1970–80." *Journal of the American Planning Association*, 62(3): 289–295.

Gottdiener, M. (1986). "Recapturing the center: A semiotic analysis of the shopping mall." In M. Gottdiener and A. Lagopoulos *The City and the Sign: An Introduction to Urban Semiotics.* New York: Columbia University Press.

Hall, P. (1989). "The turbulent eighth decade: Challenges to US city planning." *Journal of the American Planning Association*, 55 (3): 275–282.

Harris, C.D. and Ullman, E.L. (1945). "The nature of cities." *Annals of the American Academy of Political and Social Sciences*, 242: 7–17.

Heckscher, A. (1977). *Open Spaces: The Life of American Cities.* New York: Harper and Row.

Judd, D. (1999). "Constructing the tourist bubble." In D. Judd and S. Fainstein (Eds.) *The Tourist City.* New Haven, CT: Yale University Press.

Kayden, J. (2000). *Privately Owned Public Space: The New York Experience.* New York: New York City Department of City Planning and the Municipal Art Society of New York.

Loukaitou-Sideris, A. and Banerjee, T. (1998). *Urban Design Downtown: Poetics and Politics of Form.* Berkeley, Los Angeles, London: University of California Press.

Lynch, K. (1981). *A Theory of Good City Form.* Cambridge, MA: MIT Press.

—— (1990). "The pattern of the metropolis." In T. Banerjee and M. Southworth (Eds.) *City Sense and City Design: Writings and Projects of Kevin Lynch.* Cambridge, MA: MIT Press. 47–64.

Myers, D. (2007). *Immigrants and Boomers: Forging a New Social Contract for the Future of America.* New York: Russell Sage Foundation.

Savitch, H.V. and Kantor, P. (2002). *Cities in the International Marketplace: The Political Economy of Urban Development in North America and Western Europe.* Princeton, NJ: Princeton University Press.

Sites, W. (2003). *Remaking New York: Primitive Globalization and the Politics of Urban Community.* Minneapolis, MN: University of Minnesota Press.

Soja, E. (1989). *Postmodern Geographies: The Reassertion of Space in Critical Social Theory.* New York: Verso.

Sudjic, D. (1992). *The 100 Mile City.* London: A Deutsch.

Suissman, D. (1993). "Plaza mexicana." *Places*, 8(3): 4–19.

Further reading

Fogelson, R. (2003). *Downtown: Its Rise and Fall, 1880–1950.* New Haven, CT and London: Yale University Press. A detailed account of the history of the American downtown until the middle of the twentieth century and an analysis of the causes that led to its decline at that time.

Ford, L. (2003). *America's New Downtowns: Revitalization or Reinvention?* Baltimore, MD: The Johns Hopkins University Press. A review and evaluation of the urban form of sixteen American downtowns.

Frieden, B. and Sagalyn, L.B. (1989). *Downtown Inc.: How America Rebuilds Cities.* Cambridge, MA: MIT Press. Details the negotiation and deal-making between public and private sectors that characterized the rebirth of downtown after the 1970s.

Isenberg, A. (2005). *Downtown America: A History of the Place and the People who Made It.* Chicago: Chicago University Press. A social and cultural history of US downtowns in the twentieth century.

Loukaitou-Sideris, A and Banerjee, T. (1998). *Urban Design Downtown: Poetics and Politics of Form.* Berkeley and Los Angeles: University of California Press. A critical appraisal of downtown urban design and an account of how US downtown projects are conceived, scripted, developed, and used.

27

Suburbs

Rus in urbe, the picturesque, and selfhood

John Archer

The genesis of suburbia as a modern planning type is traceable to the Thames Valley west of London at the outset of the eighteenth century: here one finds clusters of compact, detached bourgeois dwellings set at a distance from the pressure and congestion of city life, designed to embrace the landscape and provide for domesticity and nonproductive leisure activity. These early suburbs were built in a piecemeal, accretive fashion over time; with the rise of a bourgeois economy, such enclaves soon grew and multiplied, to accommodate the growing entrepreneurial-professional middle class. By the late eighteenth century, developers were preparing formal plans for suburban subdivisions, and by the mid-nineteenth century the suburb had become an acknowledged, distinct planning paradigm. During this process, the understanding of what that paradigm entailed, both socially and aesthetically, steadily coalesced around a set of expectations and conventions that has informed the theory and practice of suburban design ever since.

From the very beginning, a foundational concept to the design of suburbia has been the melding of country and city. Commonly expressed in the eighteenth and nineteenth centuries by use of the Latin phrase *rus in urbe*, the intention and expectation is to combine the beauty and authenticity of nature with the sophistication and amenity of the city, minus the disadvantages of either environment, in one location. Among the foremost techniques for realizing this end has been the picturesque, the English eighteenth-century landscape aesthetic that valorized a contemplative, aestheticized, nonproductive engagement with nature. The siting of the dwelling with relation to the landscape, the views of the landscape to be enjoyed from inside the dwelling, and the opportunity for passive or mildly active engagement with the landscape (such as strolling) all are central components of this aesthetic that have remained integral to suburban design ever since, especially in Britain and America. Persistent features such as individual front lawns and back yards, bay and picture windows, and curving residential streets without sidewalks, all are progeny of this aesthetic.

Nineteenth-century American architects and planners not only capitalized on the picturesque, but also designed suburbs to exemplify political ideals of the new nation, including republican virtue and pastoral repose: suburbs were the environment best suited to fostering virtuous family life, separation from the moral corruption and pollution of the city, and immersion in a bountiful, restorative landscape.

No less important is the proliferation of the single-family house, the history of which as a building type is intimately connected with the rise of suburbia (Archer 2005). As the instrument of choice by which the upwardly mobile entrepreneurial-professional bourgeoisie has commonly demarcated selfhood, family, and neighborhood, ownership of a single-family house in a suburb has continued for three centuries to confirm arrival at, and membership in, a new, elevated social status. In these varying respects – aesthetic, moral, political, and social – the practice of suburban design from the beginning of the eighteenth century established enduring expectations and perceptions of what is at stake in suburbia, and of how it should continue to be designed.

City Beautiful and Garden City

Starting in the late nineteenth century two major campaigns to reform suburban design – under the aegis of the City Beautiful and Garden City movements – sought, on the one hand, to compensate for the worsening social deterioration of the industrial-capitalist city, while also affording opportunity for civic and social betterment through aesthetic refinements. Through techniques such as neoclassical building facades, formal geometric street plans, and pastoral park landscapes, these movements sought to inculcate respect for civic order and a taste for genteel leisure in the urban population. As Richard E. Foglesong has shown, early twentieth-century City Beautiful planners eagerly advocated such aesthetic strategies as instruments of benevolent social control, such that "control by design experts" ultimately could overcome deficiencies of the market system in fashioning urban space. More generally, as Foglesong notes, turn-of-the-century City Beautiful aesthetics sought to transform the entire city plan into a physical apparatus for advancing social and political goals – for example, at a time of large-scale immigration, inculcating in the urban citizenry a respect for country, American culture, and capitalism (Foglesong 1986: 134, 125). For planner Charles Mulford Robinson, design bore an efficacy that bordered on the eugenic: "civic art properly stands for more than beauty in the city. It represents a moral, intellectual, and administrative progress as surely as it does the purely physical." And "when it comes to the homes of the workers," the aesthetic caliber of the built environment "may be supposed to influence the battle, to help the forward or retrograde movement of the race." Aesthetic control thus became a crucial factor in the larger enterprise of civic reform. Robinson envisaged the public appreciating "the value of an authoritative aesthetic control," and presumably acceding, gratefully, to its imposition (Robinson 1909: 17, 229, 21).

Suburbs were among the targets of this "civic" agenda, particularly with respect to the "poor taste of untrained individualism" (Robinson 1909: 21) that could be found in the often mismatched scales and styles of suburban housing produced by small-scale builders and developers. The remedy was to cede control to professionals trained in aesthetics, who could coordinate building designs and standards. Thomas Adams, the first manager of Letchworth Garden City and later director of the Regional Plan of New York, mistrusting small-scale builders' capacity to conform to an appropriate aesthetic, recommended large-scale professional intervention: "until more architects are employed to design the smaller buildings, which constitute the greater part of cities, the standard of civic architecture will be low" (Adams 1934: 323).

A central goal of the City Beautiful civic agenda was to strengthen the sense of *community*. One root of the problem lay in the longstanding, uncomfortable tension in American culture between shared community interests and the private rights of individuals and their property. At issue (then as now) was the question of whose aesthetics would prevail: the homeowner's, in pursuit of self-expression and self-realization? Or an overarching community aesthetic that would subordinate the expression (and some rights) of the individual to the articulation of a common or collective whole? In the early decades of the twentieth century, buoyed by the spirit of municipal reform, many came to the conclusion that to subordinate private interest to community benefit was the best way to serve the interests of all. Robinson set the tone of the discussion in 1903: "The exterior of your home ... is not private property." Although he recognized the need to balance "civic art" with the "rights of privacy," Robinson nevertheless argued that in some situations the whole could be aesthetically more efficacious than the sum of its parts: "the individual residents ... are to be encouraged ... to co-operate, that there may be a harmonious result and that each effect may be heightened by its neighbours" (Robinson 1909: 230, 239–240, 234–235). Thomas Adams came to an even more pointed conclusion: "to improve public taste, ... [i]ndividualism must be controlled." Presaging the rise of private homeowners' associations later in the twentieth century, he proposed that individualism be harnessed not by state or municipal authority, but instead privately, "through cooperative action" in the form of "associations of individuals" (Adams 1934: 116–117).

Contemporary with reformist City Beautiful planning, the Garden City movement also sought to harness individualism in the service of community. As detailed in his manifesto *Garden Cities of To-morrow* (1902), Ebenezer Howard's strategy for eradicating the evils of the industrial city involved planting lower-density "garden cities" of 30,000 inhabitants on 1,000 acres, each surrounded by another 5,000 acres of agricultural land (Figure 27.1). These lower-density, mixed-use, walkable settlements not only would unite the best characteristics of "Town" and "Country" (echoing the classic suburban principle of *rus in urbe*), they also would resolve the competing interests of individual and collective in a community form that would embody what he termed "social individualism" (Howard 1902).

Key to realizing these social and ideological goals was a set of specific design strategies: surrounding the community with a broad expanse of greenbelt, providing low-density housing made up of small scale units (detached, semidetached, and short terraces), connecting buildings to landscape with picturesque plantings, employing vernacular architectural styles, grouping buildings in organically shaped, closely focused clusters, and laying out streets in non-orthogonal patterns, all to replicate a likeness of rural community that would foster neighborliness, mutual support, and cooperation.

The influence of Garden City principles has been profound: starting with the work of Richard Barry Parker and Raymond Unwin in English Garden Cities, followed by the writings and designs of John Nolen in the United States, their progeny include the postwar English New Town movement, as well as pioneering twentieth-century American suburbs such as Radburn, New Jersey, and the 1930s Greenbelt towns, 1960s suburban new towns such as Reston, Virginia, Columbia, Maryland, and Irvine, California, and, as detailed below, many aspects of New Urbanist and related practices since the 1990s.

In the early twentieth century, suburbs optimistically appeared to hold the promise of effecting more balanced relations with nature and more cooperative relations among people. More recently, as detailed in the next section, suburbia progressively became the problem, as the apparatus of the American Dream appeared

Figure 27.1 Letchworth Garden City, England. Source: John Archer.

to host ever more serious social, aesthetic, and environmental concerns. Nevertheless many of the primary objectives identified by suburban critics and designers in the early twentieth century have continued to remain central to the present: merging country and city, balancing community and self interest, mediating the politics of public realm versus private property, and enlisting aesthetics and design as key instruments for effecting civic, moral, and ecological improvement.

Confronting mass production

The rise of standardized products and processes in homebuilding during the 1920s, and mass production and prefabrication techniques in the 1930s and 1940s, effected widespread changes in the design and appearance of suburbia, precipitating a critical discourse of disdain and pathology that still informs the discourse on suburbia today. In the eyes of City Beautiful and Garden City designers, suburbia, if configured according to the proper aesthetic paradigms, could serve as an engine of social progress and cultural rapprochement. But the economics of land subdivision into relatively small single-family parcels for the expanding postwar middle class market, and of cost reduction by producing houses in ever greater numbers, frustrated such expectations. Aesthetics as a "civic art" gave way to the pragmatics of social mobility and the allure of new lifestyles. As mortgages with low down payments and long repayment periods became the standard way to purchase a home, houses became exchangeable commodities. Still an assertion of individual status, selfhood, and distinction, but also interchangeable and transferable consumer products, suburban houses in their generic similarity and narrow stylistic range were now simply inadequate to realize the sort of aesthetic improvement that earlier reformers had desired.

Critics did not mince words. As early as 1950 Robert Moses penned a bombastic essay titled "Build and Be Damned," referring to "monotonous new communities" and "clusters of little pastel houses," in which he decried the "horrors being perpetrated by uncontrolled boom building" (Moses 1950). Frequently suburbia's detractors borrowed ammunition from the mid-century wave of Marxist and Frankfurt School critiques of mass culture. In the same way that many critics pointed to media such as television, film, popular fiction, popular music, and advertising as evidence of the demise of American culture (Rosenberg and White 1957), others pointed to suburbia – a mass product for a mass audience – as inverse proof of what City Beautiful and Garden City designers had proposed: that the absence of proper aesthetics bespoke, and even engendered, social and cultural dysfunction. Erich Fromm charged in his bestseller *The Sane Society* (1955) that characteristics of tract housing such as open floor plans, picture windows, and close quarters contributed to a lack of privacy that, in turn, thwarted "the opportunity to fashion a truly individual self." Consequently life in suburbia had become a matter of "giving up oneself, becoming part and parcel of the herd, and liking it" and, worse, the social and political dimensions of "virtue" had been lost, in favor of a private compulsion "to be adjusted and to be like the rest" (Fromm 1955: 153, 159, 158). Popular tirades like John Keats's *The Crack in the Picture Window* (1956) and John McPartland's tale of alcoholism, infidelity, and abuse in *No Down Payment* (1957) made into a film the same year, offered salacious portraits of suburbia as a terrain of frustration, fear, loathing, and alienation. Lewis Mumford's 1961 indictment of suburban uniformity, conformity, and prefabrication in *The City* reverberated in Malvina Reynolds' 1962 song "Little Boxes," followed in 1964 by Peter Blake's polemic *God's Own Junkyard* (Mumford 1961: 485; Blake 1964: 8, 20), all of them nevertheless missing the point that Americans were voluntarily moving to suburbia at an accelerating pace, and liking it.

Some critics were less persuaded than others. William H. Whyte's analysis of Park Forest, published in 1953 in *Fortune*, was equivocal. Noting that those who had built this tract development had intentionally incorporated some variation in the façades of individual houses, he found that residents largely rejected the opportunity for further differentiation: as he put it, "in some areas residents have apparently agreed to unify the block with a common design and color scheme for garages and such. In such blocks an otherwise minor variation [thus] becomes blatant deviance." This last word, "deviance," illustrates a crucial point: for while later critics would seize on such comments as proof of pervasive pressures to conform (or else be branded as "deviant"). Whyte was after something else, namely the residents' commitment to community. For although residents exhibited a wide range of interests and tastes, he also noted that the prevailing ethic (and aesthetic) was not so much a matter of keeping up with the Joneses, as an inclination "to keep *down* with them." Those who did seek some degree of distinction risked breaking the bonds of community that Whyte found otherwise so persistent (Whyte 1953b: 188; Whyte 1953a: 86). Still, the author of *The Organization Man* was never entirely sanguine about suburban conformity, and his effort at even-handedness was a distinct exception to the ensuing tide of condemnation (for another exception see Herbert J. Gans's favorable assessment of Levittown, New Jersey [Gans 1967]).

Critics notwithstanding, the growth and expansion of suburbia has continued to accelerate. As detractors have noted, design has become less concerned with aesthetics per se than with "following the market" and "building what sells." Ordinarily this entails a standardized design typology (for example, ranch house or bungalow, with interior features such as a "great room" or cathedral ceiling) that facilitates popular lifestyle activities, and stylistic conventions (colonial, Tudor, Victorian, etc.) that signal degrees of taste and status – in sum, a highly commodified design vocabulary. Still, such are the terms in which suburbia remains the principal apparatus through which Americans constitute themselves as communities, neighborhoods, families, and individuals, as public citizens and as private selves, and pursue the American Dream. Critics continue to deplore the social and environmental consequences, well exemplified by James Howard Kunstler's acerbic *Geography of Nowhere* (1993). Dolores Hayden indicts sprawl – and thus much of suburbia – as "socially destructive. It intensifies the disadvantages of class, race, gender, and age by adding spatial separation. Sprawl is politically unfair as well as environmentally unsustainable and fiscally shortsighted" (Hayden 2004: 11). Such is the contested arena in which parties with very different interests and very different objectives – planners, developers, builders, designers, marketers, buyers, residents, critics, and the public at large – all converge to debate, revise, and implement the course of suburban design. The following sections delineate principal threads of recent debates.

Establishing control

In observing the reluctance of Park Forest residents to differentiate their houses significantly from one another, William H. Whyte in 1953 put his finger on the very balancing act that developers of many communities have since confronted: to what degree, and how, can design restrictions be imposed to advance a common good, such as defining and sustaining community? Controls may be applied to appearance, use, change, and growth, at scales ranging from the individual house and yard to regions spanning multiple counties. In addition, access may be restricted by walls and gates.

Restricting appearance, growth, and access – ordinarily through master planning and gated entrances – are not new approaches. Both techniques have roots in the nineteenth century. Llewellyn Park, New Jersey (1856) is an early example. Master planning has been a key element in government-sponsored suburbs since the 1920s and 1930s (e.g. Radburn, the Greenbelt towns), and it became instrumental to suburban developments of much greater scale in the 1960s (e.g. Rancho Bernardo, California [1961], 6107 acres, or Irvine [1960] over 50,000 acres). In the southwestern United States, typically three-quarters or more of all new developments are now master planned communities. There and in other portions of the United States, gated communities account for a third to as much as 80 percent of new housing (Low 2003: 15).

Master-planned and gated communities often are designed, and marketed, as oriented toward certain social class and lifestyle aspirations (Blakely and Snyder 1997), securing privacy and safety (Low 2003), and especially maintaining property values. But they also bring limitations, making it difficult or impossible to introduce changes in taste or lifestyle, expand the family or extend the household beyond the nuclear family (e.g. through additions to a house), or develop community bonds beyond the gated confines. In addition, the popularity of master-planned and gated communities has resulted in some more profound shifts in the American social landscape – consequences not seen, or foreseen, in the nineteenth or early twentieth centuries. First, resources and advantages become alienated from the public realm, especially insofar as amenities provided for master-planned and gated communities, such as community and recreation centers, become sequestered as part of what Lyn Lofland terms the "parochial realm" (Lofland 1998). Second, in decisions regulating matters such as aesthetic criteria or personal activities and pursuits, the determination hinges as much, or more, on property values rather than civic improvement or personal opportunity (McKenzie 1994). And third, the landscape increasingly becomes segregated into monocultures – not unlike suburbs of the eighteenth and early nineteenth centuries – defined by age, economic class, aesthetic tastes, and lifestyle pursuits, contributing little to personal growth or societal betterment. Rather, as one astute critic has written, suburbia promises "the potential to inhabit and create your own private fiction, individually crafted, physically enhanced, and relentlessly adhered to" (Dewey 1997: 263).

Recreating history, anchoring community

Historically the growth of suburbia has been a combination of two distinct geographic processes – outward expansion from built-up centers, gradually encompassing existing small towns and villages; and erecting new tracts and communities that, when they do furnish their own civic and commercial spaces, have tended to forgo town centers for malls, office parks, and other single-purpose zones. By the 1960s a significant countertrend took hold in new development, as large master-planned communities were partitioned into individual "villages," each having separate access to the arterial road system, and its own village center with shops, community facilities, and sometimes offices. To accommodate the growing popularity of automobile transportation, planners sought to maximize vehicular circulation while maintaining small-town scale by building road systems at multiple scales (arterial, feeder, and local), serving both to separate residential villages from, and connect them with, town and commercial centers. Already seen in Ebenezer Howard's diagrammatic plan for multiple garden cities, such hierarchy was fundamental to the design of the 1930s Greenbelt towns, as well as 1960s and 1970s new towns such as Reston, Columbia, Irvine, and The Woodlands, Texas.

While the 1960s new-town model has informed much of suburban design ever since, critics have continued to regret its dependence on the automobile, lack of neighborhood identity, and absence of "main street" environment. Drawing on these concerns as well as growing concerns over the declining state of civic engagement in America, Andres Duany and Elizabeth Plater-Zyberk proposed in "The Second Coming of the American Small Town" (1992: 48) that "Americans need to be reacquainted with their small-town heritage and to be persuaded of the importance of protecting the human habitat every bit as rigorously as the natural habitat." They recommended a wholesale recasting of planning practices and standards to echo patterns that were successful in historic American cities such as Alexandria and Charleston, early twentieth-century suburbs such as Coral Gables, and the work of planners such as Raymond Unwin and John Nolen. Among their recommendations are mixed-use development, mixed densities, street networks that accommodate cars but do not privilege them, walkability, attractive and reliable transit, and design codes that implement such goals (Duany *et al.* 2000).

A key element of many reformist agendas for suburbia is a sharply diminished role for the automobile, in favor of pedestrian oriented "neotraditional towns." In "The Traditional American Town" (1993) Peter Calthorpe offers a somewhat more sanguine take on the automobile, acknowledging that "[w]e cannot simply return to a time [when] people walked, the shopkeepers lived upstairs, and neighbors were all on [a] first-name basis." Indeed, "the auto, suburbia's form giver, will not retreat" (Calthorpe 1993: 21). Calthorpe has nevertheless been a leader in articulating the importance of public transportation as an integral element of the circulation network. The practice of orienting suburbs around public transportation lines dates well back into the nineteenth century, but since the early 1990s planners have taken new interest in "transit oriented developments" (TODs) that afford greater building and population density, thus enhancing activity at the pedestrian scale and ultimately making transit integral to the economic and social vitality of a community (Calthorpe 1993:43; Cervero *et al.* 2004; chapter by Polyzoides in this volume).

With respect to the built fabric of the community, Calthorpe proposes that "certain design principles" are "both timeless and contemporary." These include "walkable streets [that lead] to close and useful destinations," lined by sidewalks, trees, porches, balconies, and entries, rather than garages. Along with restoring such amenities to common spaces, Calthorpe also argues for restoration of the "traditional Commons," a place that "defines the meeting ground of a neighborhood and its local identity," and which "should be brought back to the center of our communities" (Calthorpe 1993: 21, 23).

Drawing on such concerns for traditional, small-town life and neighborhood civic engagement, Duany, Plater-Zyberk, Calthorpe, and others founded the Congress for the New Urbanism 1993 as a vehicle for shaping public policy and reforming planning practice, focusing on the rehabilitation of existing cities and suburbs as well as changing the conventions of new construction. Anchored in a Charter of twenty-seven principles published in 1996, New Urbanist practice focuses on compact, pedestrian-oriented, mixed-use planning, while simultaneously enlisting design (often anchored in specific codes and regulations) as a critical means to enhance community and civic participation.

The design codes and principles for any given New Urbanist community are nominally based on a vocabulary anchored in local or regional traditional historical practice – a set of principles that, having proved successful in a past era, are proffered as paradigmatic for the present as well, such that when codified and replicated in the present they can restore past measures of neighborliness and civic engagement (Langdon 1996). Critics, on the other hand, charge that importing such "timeless" principles from prior eras amounts to conservative nostalgia, and that strict administration of design codes affords only a static, stultifying, artificial aesthetic (Crawford 2005: 19, 22). Often "history" or "heritage" is identified as essential to affording a new settlement a sense of place or community. Yet efforts to manufacture such histories, while widespread, are often overzealous. The master-planned community of The Villages, west of Orlando, for example, attempts to establish its "hometown" identity via faux-historical markers, maps, and texts throughout its town center that narrate an entirely fictitious history (Bartling 2008). More common, and more pragmatic, have been widespread and generally successful "placemaking" efforts to retrofit existing suburbs with the town centers that they never had, or outfitting new TOD hubs with commercial centers analogous to those at transit stops in streetcar suburbs of past eras (Bohl 2002).

Clustering, sprawl, and greening

In advancing controls on design, growth, and the automobile, in efforts to maintain tradition and enhance community, suburban design since the 1960s has remained an essentially conservative practice, anchored in long-standing paradigms that date to the nineteenth and early twentieth centuries. As it became apparent in the early 1960s that unchecked development posed significant threats to the American landscape, concern for land conservation became part of the planning discussion. An early publication by the American Society of Planning Officials thus proposed that instead of subdividing a large tract of land entirely into large private lots, the process of "cluster subdivision" would allow a number of houses to be grouped closely together, reserving the remaining land as swaths of undeveloped terrain that could afford aesthetic, recreational, and ecological benefits (Rosenthal 1960). Conservationist approaches in this vein have informed a number of master-planned communities since, such as Irvine, (1960), Prairie Crossing, Grayslake, Illinois (1995; Figure 27.2), and Jackson Meadow, Marine-on-St. Croix, Minnesota (1998). Comparable efforts on a broader, regional, scale, such as Portland, Oregon's Urban Growth Boundary, extend the Garden City greenbelt principle across multiple counties, demarcating developable land from terrain to be preserved as rural. Portland's boundary, dating from the early 1970s, encompasses approximately a quarter million acres. Inside the boundary, policy favors smaller lot sizes, higher building densities, and extensive transit-oriented development. This has remained the most prominent and effective growth boundary in the country, although higher land and housing prices within, and a consequent leapfrogging of development beyond the peripheral protected areas, have partly undermined the boundary's conservationist purpose.

On balance, conservation efforts have been all but overwhelmed by the accelerating growth of suburban sprawl, widely criticized for contributing to environmental devastation, exacerbating social injustice, and aesthetic blight. Nevertheless sprawl is not without its defenders (Bruegmann 2005); and Joel Kotkin's vision of the "new suburbanism" takes critics and reformers

Figure 27.2 Prairie Crossing, Grayslake, Illinois. Source: John Archer.

to task for being over-ambitious, arguing that single-family houses, yards, and automobile dependency are here to stay. Still, his remedies vary little from New Urbanist convention: he advocates affordability, density, "clustering of services," and "village-building," much of which focuses on new or restored town centers (Kotkin 2005: 21–31).

Ecological issues, ranging from the loss of wetlands and productive farmland to traffic congestion and excessive carbon production, have become increasingly influential in shaping the course of suburban design. Efforts to address these concerns, commonly under headings such as "sustainable design," "smart growth," or "green building" (Van der Ryn and Calthorpe 1986; Beatley 2004; Girling and Kellett 2005; Low *et al.* 2005; Bullard 2007; Friedman 2007; Gause 2007) confront the challenging paradox that growth, by definition, is at odds with the planet's ecological balance. Nevertheless there are successful strategies that at least ameliorate the imbalance, including new building and (preferably) retrofitting at greater densities, mixed land and building use, reducing the length and frequency of automobile trips, reducing household energy and water use, minimizing waste, building from resources that are renewable and require less carbon output, and conserving land.

Recasting design

The urgency behind such strategies – stemming from the increasing cost and scarcity of resources, and progressive degradation of the environment – demands substantial changes to the way suburbia is planned, designed, and lived. Key to those changes will be rethinking the manner in which the most common elements of suburbia – dwellings – are conceived, designed, and used. The anticipated shift toward more mixed use developments and greater building densities necessitates innovative, pragmatic approaches to facilitating residents' personal and social aspirations as well as sustaining their practical, everyday lives, and doing so in a continually flexible manner. Instead of imposing a prescriptive, unified aesthetic regime – a common strategy in the present culture of master planning, codes, and regulations – dwelling design may be approached more broadly in terms of praxis, that is, as the activities, pursuits, and goals through which suburbanites conduct their everyday lives. The patterns and objects of everyday life offer a basis for articulating a different kind of design aesthetic, one that instead of favoring static monocultural tracts based on past paradigms, explicitly affords opportunity to incorporate (rather than resist) such factors as difference, growth, and change – all of which will continue to proliferate, rapidly, on scales ranging from the personal to the global, and which suburbia needs to accommodate.

Two approaches to design are especially germane in this context. The first, arising from scholarship on everyday life (de Certeau 1984; Lefebvre 1991–2005), explores the complex range of purposes that dwellings serve in people's everyday living. House, yard, neighborhood, leisure, automobiles, and the vast range of consumer goods are integral parts of an intricate social, economic, and aesthetic nexus in which people are continually building their lives, through a constant stream of activities and practices that articulate identity, difference, connection, and community, and advance their goals and pursuits (Miller 1987; Rojas 2003; Crawford 2005). As residents' interests, expectations, and the realities of daily life grow and change over time, they are progressively less well served by a narrowly defined, even fictitious vision of community that is manufactured out of design codes or nostalgic paradigms. They are better served by a vision that recognizes complex ways in which community is constructed and modulated, often at the micro level, in everyday activities (e.g. ways in which the yard and garage are used) and the objects used to carry out those activities (e.g. product choices) – the sort of complex everyday idiosyncrasies that are so effectively portrayed in photographic work by Bill Owens (1973) or Gregory Crewdson (2008), among others. Consistent with that vision is community design that not only supports mixed uses, but also incorporates mixed constituencies, mixed goals and visions, and mixed social practices, lived and produced through the terms of mixed aesthetics. This is notably not an appeal for aesthetic anarchy; rather, it is an appeal to identify the complex ways in which suburbanites already live and thrive, to recognize the quotidian tastes and everyday practices through which those lives are articulated, to understand how to nourish those tastes and practices, and to provide for healthy change over time. Suburbia already successfully sustains a wide range of such tastes and practices. The challenge is to help them flourish better, in a way that enhances the ongoing vitality of the whole.

The second, related approach to recasting design is an appeal to incorporate flexibility, building the capacity for progressive and deliberate change into both the fabric of what is built as well as the rules according to which things get built. Municipal zoning and community master plans have long been effective at limiting or preventing change, both short- and long-term. Yet as Avi Friedman has noted, because zoning regulations (and, by extension, many master plans) incorporate few mechanisms for negative feedback (and thus for progressive change), they not only institutionalize mistakes but also neglect the dynamic possibilities of local cultures (Friedman 2002b: 17–19). The consequence is that residents whose interests, circumstances, or aspirations change often find that their best option for pursuing those changes is simply to move away – thus undoing any social capital they may have produced so far, while simultaneously reinforcing the perception of suburbia as little more than a commodity wasteland. The remedy is to devise open mechanisms of governance that better facilitate change. Friedman also advocates "build[ing] greater flexibility into the design process" itself, as a proactive means to address unpredictable, yet inevitable, change in economic relations, environmental concerns, family size and structure, and the relation between work and home (Friedman 2002b: 173–175; for steps in this direction see Friedman 2002a, Schmitz *et al.* 2003: 12) – to which one might add ever-accelerating changes in technology, media, leisure, and taste. Renée Chow, approaching the question of dwelling design from a complementary perspective, argues that the traditional understanding of the house as a self-enclosed shell or volume is equally myopic: it only consigns us to replicate the "fragmented, unintelligible, and largely underused landscapes that constitute a significant proportion of today's residential environment." She proposes to recast the dwelling and surrounding landscape in terms of intersecting threads of fabric, with warp and weft together producing a weave that is durable but also flexible and alterable over time (Chow 2002: 34–35; see also Girling and Kellett 2005).

These and other designers and critics make the case that suburban design not only needs more deliberate flexibility, but also must be proactive in more thoroughly engaging, and facilitating, the everyday practices, energies, and aspirations of those who choose to live there. Challenges such as resource shortages and climate change are global in scale; through design that is flexible, and that focuses on everyday praxis, suburbia can respond effectively at the local level and personal scale.

Acknowledgment

I am grateful to Holley Wlodarczyk for invaluable research assistance on this essay.

References

Adams, T. (1934) *The Design of Residential Areas*, Cambridge, MA.: Harvard University Press.

Archer, J. (2005) *Architecture and Suburbia*, Minneapolis, MN: University of Minnesota Press.

Bartling, H. (2008) "A Master-Planned Community as Heterotopia," in M. Dehane and L. De Cauter (eds.) *Heterotopia and the City*, London: Routledge.

Beatley, T. (2004) *Native to Nowhere*, Washington, D.C.: Island Press.

Blake, P. (1964) *God's Own Junkyard*, New York: Holt, Rinehart and Winston.

Blakely, E. and Snyder M.G. (1997) *Fortress America*, Washington, D.C: Brookings Institution.

Bohl, C.C. (2002) *Place Making*, Washington, D.C.: Urban Land Institute.

Bruegmann, R. (2005) *Sprawl: A Compact History*, Chicago: University of Chicago Press.

Bullard, R.D. (ed.) (2007) *Growing Smarter*, Cambridge, MA: MIT Press.

Calthorpe, P. (1993) *The Next American Metropolis*, New York: Princeton Architectural Press.

Cervero, R. Murphy, S., Ferrell, C., Tsai, Y.-H., Arrington, G.B., Boroski, J., Smith-Heimer, J., Golem, R., Peninger, P., Nakajima, E., Chui, E., Dunphy, R., Myers, M., McKay, S., and Witenstein, N. (2004) *Transit-Oriented Development in the United States*, Washington, D.C.: Transportation Research Board.

Chow, R.Y. (2002) *Suburban Space: The Fabric of Dwelling*, Berkeley, CA: University of California Press.

Crawford, M. (2005) "Everyday Urbanism" in R. Mehrota (ed.) *Everyday Urbanism*, Ann Arbor, MI: University of Michigan.

Crewdson, G. (2008) *Beneath the Roses*, New York: Abrams.

de Certeau, M. (1984) *The Practice of Everyday Life*, trans. S. Rendall, Berkeley: University of California Press.

Dewey, F. (1997) "Cyburbanism as a Way of Life" in N. Ellin (ed.) *Architecture of Fear*, New York: Princeton Architectural Press.

Duany, A. and Plater-Zyberk, E. (1992) "The Second Coming of the American Small Town," *Wilson Quarterly*, 16: 19–50.

Duany, A., Plater-Zyberk, E., and Speck, J. (2000) *Suburban Nation: The Rise of Sprawl and the Decline of the American Dream*, New York: North Point Press.

Foglesong, R.E. (1986) *Planning the Capitalist City*, Princeton, NJ: Princeton University Press.

Friedman, A. (2002a) *The Adaptable House*, New York: McGraw-Hill.

—— (2002b) *Planning the New Suburbia: Flexibility by Design*, Vancouver: UBC Press.

—— (2007) *Sustainable Residential Development*, New York: McGraw-Hill.

Fromm, E. (1955) *The Sane Society*, New York: Holt, Rinehart and Winston.

Gans, H.J. (1967) *The Levittowners*, New York: Pantheon.

Gause, J.A. (ed.) (2007) *Developing Sustainable Planned Communities*, Washington D.C.: Urban Land Institute.

Girling, C. and Kellett, R. (2005) *Skinny Streets and Green Neighborhoods*, Washington, D.C.: Island Press.

Hayden, D. (2004) *A Field Guide to Sprawl*, New York: Norton.

Howard, E. (1902) *Garden Cities of To-morrow*, London: Sonnenschein.

Keats, J. (1956) *The Crack in the Picture Window*, Boston: Houghton Mifflin.

Kotkin, J. (2005) *The New Suburbanism*, Costa Mesa, CA: The Planning Center.

Kunstler, J.H. (1993) *The Geography of Nowhere*, New York: Simon & Schuster.

Langdon, P. (1996) "The New, Neighborly Architecture," *American Enterprise*, 7: 41–46.

Lefebvre, H. (1991–2005) *Critique of Everyday Life*, trans. J. Moore, London: Verso.

Lofland, L.H. (1998) *The Public Realm*, Hawthorne, New York: Aldine de Gruyter.

Low, N., Gleeson, B., Green, R., and Radovic, D. (2005) *The Green City: Sustainable Homes, Sustainable Suburbs*, New York: Routledge.

Low, S. (2003) *Behind the Gates*, New York: Routledge.

McKenzie, E. (1994) *Privatopia*, New Haven, CT: Yale University Press.

McPartland, J. (1957) *No Down Payment*, New York: Simon and Schuster.

Miller, D. (1987) *Material Culture and Mass Consumption*, Oxford: Basil Blackwell.

Moses, R. (1950) "Build and Be Damned," *Atlantic*, 186: 40–42.

Mumford, L. (1961) *The City in History*, New York: Harcourt, Brace & World.

Owens, B. (1973) *Suburbia*, San Francisco: Straight Arrow Books.

Reynolds, M. (1962) "Little Boxes," copyright by Schroder Music Co. (ASCAP), renewed 1990.

Robinson, C.M. (1909) *Modern Civic Art* (3rd edn), New York: G.P. Putnam's Sons.

Rojas, J. (2003) "The Enacted Environment" in C. Wilson and P. Groth (eds.) *Everyday America*, Berkeley: University of California Press.

Rosenberg, B. and White, D.M. (1957) *Mass Culture*, New York: Free Press of Glencoe.

Rosenthal, J.K. (1960) *Cluster Subdivisions*, Chicago: American Society of Planning Officials.

Schmitz, A., Engebretson, P., Merrill, F.L., Peck, S.E., Santos, R.L., Shewfelt, K., Stein, D., Torti, J., and Utter, M.A. (2003) *The New Shape of Suburbia*, Washington, D.C.: Urban Land Institute.

Van der Ryn, S. and Calthorpe, P. (1986) *Sustainable Communities*, San Francisco: Sierra Club Books.

Whyte, W.H. (1953a) "The Outgoing Life," *Fortune*, 48 (July): 84–89, 156–160.

—— (1953b) "How the New Suburbia Socializes," *Fortune* 48 (August): 120–122, 186–190.

Further reading

Blauvelt, A. (ed.) (2008) *Worlds Away: New Suburban Landscapes*, Minneapolis, MN: Walker Art Center. Catalog of an exhibition on suburban art, culture, and design.

Chase, J.L., Crawford, M., and Kaliski, J. (eds.) (2008) *Everyday Urbanism*, expanded ed., New York: Monacelli Press, 2008. Essays exploring design in the context of everyday life and neighborhood interests.

Hayden, D. (2003) *Building Suburbia*, New York: Pantheon. A history of American suburbia, 1820–2000.

Teaford, J.C. (2008) *The American Suburb: The Basics*, New York: Routledge. Discusses suburban demographics, policy concerns, and planning.

28

Planned communities and new towns

Ann Forsyth

Urban designers often dream of creating new environments from scratch, environments that can demonstrate the potential for urban design and physical planning to solve important problems related to urban life. Planned residential and mixed use neighborhoods and larger scale new towns seem to offer that opportunity. While comparatively rare in practice, and frequently not achieving the aspirations of their planners and designers, proposals for new towns litter the intellectual history of urban design. This chapter explores the diversity of new towns in terms of how new towns are defined, the traditions they draw on, the issues they engage with, and the variations among new towns in different countries. In the United States, typically private developers have built the new towns but elsewhere governments have had a major role. New towns have achieved high visibility in different parts of the world in different periods – Britain in the 1940s, Sweden in the 1950s, the United States in the 1960s, and China, in recent years. While overall new towns have housed relatively few people, they have been important locations for innovation in design over the past decades. Contemporary new towns promise to continue that tradition with experiments in circulation planning, ecological design, social organization, and aesthetics.

Definitions

There are many examples of planned cities in history, for example the Spanish colonial cities in Latin America or administrative cities in China. However, the twentieth century saw an expansion of new town developments created as post-colonial capital cities, to provide worker housing, as part of planned decentralization processes, and generally as designed alternatives to manage metropolitan development.

The terms planned communities and new towns have no consistent definition although typically they include dimensions such as size, comprehensiveness, and level of planning. They range from designed neighborhoods of a few hundred people to large scale, multi-use, self-contained developments in the hundreds of thousands. However, while there are no definitive criteria, many people distinguish between communities on the basis of size, scope of planning, and location. For example, in the garden city tradition within planned community practice, a garden suburb is a neighborhood of a few hundred to a few thousand houses; a garden city typically has around 30,000 people and adheres to the tenets laid out by Ebenezer Howard (below), and a new town may be even larger and needs to be self contained in some ways. Such garden cities and new towns are large enough to have a mix of activities and have something of the self sufficiency of a free-standing town or small city in terms of a full range of residential and employment opportunities, a mix of ages and incomes, and access to social and cultural resources (Forsyth 2003). However some may consider planned residential suburbs of substantial size to be new towns in the sense of being comprehensively designed and developed.

An alternative way of distinguishing among the different examples of large scale developments is between ordinary or incremental suburbs; master planned suburbs, and neighborhoods that have a strong urban design character but may be as small as a few thousand units; packaged suburbs that are in the tens of thousands, that have received some urban design attention, but may be largely residential at the outset (e.g. a Levittown); and new towns or new communities that are both larger and demonstrate substantial design quality and a mixture of uses.

Location also is a means of differentiating among such developments. One 1970s typology of new towns, or in US terminology "new communities," distinguished between: *freestanding or self-contained* new towns in isolated locations; *satellite* new towns with a comprehensive mix of activities but in the orbit of a metropolitan area; *new-towns-in-town* or large and comprehensive redevelopment areas towards the centers of large urban areas; and *add on new towns or growth centers* adding a large development to an existing small town (Griffin 1974).

Planned communities and new towns also reflect different underlying motivations on the part of their developers, designers, and planners. A major reason for the past century has been to redirect growth caused by migration to large cities, providing alternatives both to existing core cities and suburban sprawl (Osborne and Whittick 1977). This has made satellite new towns an important urban type. However, in this same time period many have wanted to demonstrate the potential for comprehensive planning and design to create healthier and more functional environments, spur regional economic growth, test new building and planning approaches and technologies, consolidate a claim to territory (e.g. in Israel), or provide a sense of national or regional identity as in the case of planned capital cities (Forsyth 2003; Osborne and Whittick 1977).

Traditions

In design terms, new towns draw on a number of urban planning and design traditions. Those in the garden tradition are perhaps best known through the work of Ebenezer Howard, whose garden city idea was immensely influential in the early twentieth century, reflecting a strong element of social concern including local participation, social welfare, and social mix. The garden city idea attracted many adherents among professionals and civic groups and became the basis of an international movement for better planning (Hall and Ward 1998). However, the modernist movement in urban design also proposed new towns, in this case more influenced by the power of new technologies such as the automobile and prefabricated building (Gropius 1945). Each form of new town promoted more orderly development, higher quality design, and local access to services. As new towns were built around the world, their planners also engaged with innovative design ideas. For example Radburn-style planning was much used, and refined, in planned community and new town contexts. In addition, due to their powerful proponents and symbolic importance, planned capital cities provided important practice settings for key design thinkers (e.g. Doxiadis in Islamabad and Le Corbusier in Chandigarh) (Vale 1992).

It is undeniable, however, that most histories of new towns center around the garden city tradition. In 1898, Ebenezer Howard's *Tomorrow: A Peaceful Path to Real Reform*, later known as *Garden Cities of Tomorrow*, proposed the new synthesis of a utopian vision in the form of what he called a garden city. These relatively small cities of 30,000 people residing on 1,000 acres would be surrounded by a farming population of 2,000 on farm land. Cities would be self-contained but linked by railways and canals to a network of other such cities in a constellation of cities he called the social city. A slightly larger central city would be at the core of the social city. This would famously allow residents of the social city to experience the best of the town and country (Howard 1902; Forsyth 2003).

The garden city idea evolved in following decades reflecting changing urban circumstances, differing national contexts, and early experiences building such cities. However, Howard's work is still important in demonstrating a number of the key dimensions or concerns that have been the focus of subsequent developments in new towns and even the less formulaic planned communities. A major concern of the garden city movement was the problem of increasing city size and how to balance the opportunities of the city with the benefits of rural life that at the time of Howard were quite recent memories. The garden city solution was not to change existing cities but to build new, more human scaled environments, at some distance from the existing and congested urban areas. In these smaller urban areas people could live in closer contact with nature, using new forms of cooperatively provided services, and with an overall improved quality of life (Howard 1902).

Optimal city size was a key area of discussion, and generally over the following decades the size considered to be necessary for self sufficiency increased from the tens of thousands to the hundreds of thousands. While the search for an optimal city or metropolitan size has largely been abandoned, those advocating self contained areas typically, though not always, propose population ranges over 100,000. Many planned communities, however, are far smaller. Smaller developments, albeit not self-contained, are easier to build and market, and can still demonstrate innovations in physical design and neighborhood planning.

A second key issue dealt with by Howard and still central in later periods has been how to combine the benefits of urban and rural lifestyles in a way that enhances quality of life. One of the enduring attractions of suburban development is that it has the potential to do just this – placing residents close to the jobs, recreational, educational, retail, and cultural opportunities of central cities but also in leafy surroundings evoking something of rural or natural aesthetic. The garden city proposed a particularly compact and self-contained version of this that would minimize some of the problems that suburban areas could suffer from – for example in a garden city with a balance of housing opportunities, jobs, and population groups, there would be little need for long commutes.

A related issue was the problem of paying for urban services, and here Howard proposed corporate ownership by special garden city companies with land leased to residents (even those owning their housing). Rents would pay for the initial land costs but also provide money for services; in the longer term, once the initial loan was paid off, this would provide a substantial income stream for services such as pensions and sickness benefits. This kind of cooperative ownership was seldom achieved. Nevertheless, early garden cities did have a variety of approaches to tenure. More contemporary planned communities typically have resident or homeowner associations, or public authorities, that collectively manage common areas and varying amounts of collective services. These organizations are seldom as innovative as Howard's proposal, with many in the United States receiving substantial criticism for their onerous regulations (Howard 1902; McKenzie 1994; Forsyth 2003).

The obvious inequalities of London were the key to inspiring Ebenezer Howard to propose the early garden city proposal, and the earliest large scale garden cities in the post-Howard era were all in the London area (Howard 1902; Osborne and Whittick 1977; Hall and Ward 1998). However, this rich tradition drew on earlier nineteenth-century experiments designing company towns, colonial settlements, and leafy upper middle class suburbs along with Christian socialism and generally utopian thought (Howard 1902, 2003; Ward 2002). Garden city associations sprung up around the world, and by the 1920s, garden cities and smaller neighborhood-scale garden suburbs had been developed in different countries. They ranged from the philanthropically sponsored garden suburb of Forest Hills Gardens in suburban New York (1910-), the early public housing neighborhood of Daceyville Garden Suburb in suburban Sydney, Australia (1912-), the privately developed middle-class enclave of Den-en-Chōfu in suburban Tokyo (1918-), to cooperative and worker initiated developments of Floréal and Logis in Brussels, Belgium (1921-) (see Hall and Ward 1998; Forsyth 2003; Ward 1992).

From the 1920s such garden city philosophies and practices found key promoters in the United States in the form of the Regional Planning Association of America (RPAA). This group met for about 10 years in the 1920s and 1930s and involved important urban intellectuals such as Lewis Mumford, Clarence Stein, and Henry Wright. Important for this tradition was their neighborhood prototype, Radburn, New Jersey. Along with the contemporary but distinct idea of the neighborhood unit, proposed by Clarence Perry, a resident of Forest Hills Gardens (see above), these two North American ideas influenced new towns and planned neighborhoods around the world (Perry 1939; Stein 1957).

The key common theme in Radburn planning and the neighborhood unit idea was to respond to the growing dominance of automobile transportation by creating enclaves that tamed the car within their boundaries but used the car to link to the surrounding metropolis. The Radburn idea was more physical, creating superblocks where houses fronted onto interior green open space and pedestrian networks with cars relegated to the rear of the houses in culs-de-sac functioning as service courts. The neighborhood unit idea sprung from more sociological understandings of contemporary life. Perry was interested in creating environments centered around community facilities – hence the now classic idea of designing neighborhoods of 6,000 to 10,000 people centered around an elementary school. He was also influenced by sociological literature that proposed that in order to foster public participation and social interaction people needed relatively homogenous enclaves (Silver 1985; Biddulph 2000). These were not models for complete new towns and some of their core ideas were dropped when they were replicated (people in the US tended to disregard the back to front arrangement of houses in pure Radburn planning; those in Britain disregarded the proposed socially mixed, rather than socially homogenous, neighborhood units) (Biddulph 2000; Thorns 1976, 72). However, it was attractive to those building new towns or redeveloping old ones to have a defensible way of breaking down the scale of the city into smaller units that seemed to have benefits for phasing construction, providing services, and developing a sense of place or community identity.

While many studies found that a sense of community was difficult to develop, promoters still hoped for such outcomes (Alexander 1972; Biddulph 2000; Gans 1968; Keller 1968; Thorns 1976; Forsyth 2003).

While the garden city tradition is important, it is not the only source for new towns. Modernist developments – such as Brazil's capital city Brasilia, some of Stockholm's transit-oriented suburban new towns of the 1950s (if these commuter suburbs can be considered new towns), or Cumbernauld in Scotland – featured high rise housing blocks and higher overall densities than were typical in early garden suburbs (Holston 1989; Godschalk 1967). Other new towns, particularly those by private developers, represented a more basic sense of nostalgia for small town life – with Columbia, Maryland being a key example. Many of the larger traditional neighborhood (new urbanist) developments are falling into this category (Forsyth 2005). Further, many new towns combine several traditions – for example Radburn planning and modernist architecture.

Urban design issues

In terms of urban design, planned communities and new towns typically have a range of aims related to local *physical design and aesthetics* (architecture, landscape, the pattern of blocks and streets), *social concerns* (population mix, community facilities, designs believed to foster interaction), *economic activity* (shopping, local employment, regional economic development), *transportation* (internal circulation, self-containment and accessibility, the balance of different transportation modes), and *nature* (planting, access to the outdoors, local agriculture, energy, and other topics related to the ecological footprint of developments) (Forsyth and Crewe 2009). There is also the major issue of building a large scale development from scratch. Such concerns have developed over time – concerning nature for example, local agriculture and access to the outdoors were more important in early garden cities than the idea of the ecological footprint that had yet to be properly conceptualized. However, the palette of planning responses has not changed much over the past century – typically designers divide new towns into neighborhoods, create relative self-containment at several scales, balance development and open space, and deal with the car by either banishing or embracing it. This is perhaps not surprising as basically all the contemporary transportation technologies had been invented by the end of the nineteenth century, and the same can be largely said of the other dimensions. Much discussion has occurred, however, about which issues are most important and which design strategies most effective, with some proving to be ineffective (e.g. the idea that merely putting people close to each other would cause them to become friends), and some arguably counterproductive (Biddulph 2000). Out of many potential urban design topics, the discussion below highlights a handful of these key issues.

Physical design and aesthetics

One topic of enduring interest in urban design in the past century has been the issue of the modern versus traditional aesthetics. Should buildings, landscapes, and urban designs reflect a contemporary response to new technologies, social concerns, and artistic practices or are some kinds of environments and scales of design more appropriately treated using more traditional approaches and patterns. In architecture, newness has been valued by tastemakers for most of the century; in urban design the situation has been more balanced with some approaches such as Radburn planning envisaging a new kind of environment separating the car from the pedestrian; and others such as the European urban villages movement taking a more traditional approach to both movement patterns and architecture; in this case valuing highly connected street patterns and traditional styles and materials. Some planned communities and new towns have tended toward the traditional – exemplified by the arts and crafts aesthetic of Letchworth, the first garden city. However, there are very many counter examples – from Soviet-era suburban satellites to modernist Swedish new towns. Some more contemporary designs are seen as less than successful – for example the grandiose scale of the core of Brasilia. However, many innovative designs have aged well – for example the transit, pedestrian, and cyclist oriented design of Almere (planned in the 1970s) outside Amsterdam in the Netherlands.

Many planned communities have substantial *social* aims, and a key one has been promoting social interaction through urban design. The form of social interaction hoped for has varied – from intense friendships to neighborly nodding and sharing community facilities. Many planned communities also want to promote social mix – the mixing of dwellings of people who differ on important social characteristics such as income, race, and age. The specific mix has varied however – many European new towns primarily emphasized worker housing with a relatively small number of professionals; many US planned communities have aimed at the middle and upper middle class with only small amounts of lower cost housing. Significant research in the 1950s and 1960s failed to show substantial or intense interaction among people who did not share something in common, such as education, although there are other benefits from having diverse people share a location such as better service provisions for those with low incomes (Keller 1968; Thorns 1976; Biddulph 2000). Indeed a phenomenon called the "new town blues" provides a counter example in terms of planned communities fostering social interaction. This term refers to the loneliness and disillusionment of recent movers who go to planned communities hoping to make social connections and solve personal problems (Godschalk 1967). However, positive social outcomes are still touted as important aims of many planned communities, for example the new urbanist emphasis on neighbors seeing each other on front porches.

At a larger scale, many planned communities make contributions to regional *economic development* as in the cases of numerous company towns and regional growth centers such as Ciudad Guayana in Venezuela (Osborne and Whittick 1977; Peattie 1987). New towns typically aspire to providing jobs for residents – in fact for many, a planned residential community that does not provide such jobs is not a new town but a master planned community or packaged suburb. This is perhaps why the new urbanist movement of the last decades fits uneasily in this tradition. Many such developments are too small or have too few jobs to be a new town but are rather new neighborhoods or new commuter suburbs; others are planned as extensions of existing areas rather than whole new towns.

At a still smaller scale, planned neighborhoods often tout the ability for people to walk to shops. New towns have aimed to go further, typically proposing to have enough jobs for their residents. This has been a challenge, made more complex by the difficulty of ensuring that people in the new town actually work there given other opportunities. That has raised a number of transportation issues, particularly the balance between pedestrians and cyclists, transit, and cars. Cumbernauld in Scotland, for example, planned in the 1950s, completely separated pedestrians and cars, placed shops in a town center surrounded by higher density residential areas

Figure 28.1 Cumbernauld, Scotland. Source: Ann Forsyth.
Note: Cumbernauld was designed to take population from Glasgow. It featured complete separation of vehicular traffic and pedestrians which for some time reduced accident rates.

and jobs on the outer edge (Figure 28.1). For a while it had one of the lowest accident rates in Britain (Forsyth and Crewe forthcoming). Many new towns in Europe and Asia are transit-oriented making driving difficult (Figure 28.2).

In the latter part of the twentieth century the idea of design with *nature* has become increasingly important. It is also a subtly different idea about nature compared with earlier periods transforming from a concern with rural landscapes and resources to a new interest in ecology. In terms of more ecologically planned new towns of the more contemporary period, there are two distinct patterns of such planned communities. The first, based on what has been termed the compact-city idea, proposes a high-density, energy and land efficient city. Singapore's new towns would be an example. The second proposes a lower-density approach with on-site water treatment, native plantings, and close connection between people and plants; such developments are often smaller (Crewe and Forsyth forthcoming). While the two approaches can be combined, this is typically expensive – for example onside water filtration in a high density environment – so this represents an area of tension within the new town planning field.

Finally, there is the issue of planning a new town from scratch. This raises a number of important issues, and for a period in the last century the new town development process was the subject of some attention (Bailey 1973; Golany 1976). There are several key issues. One of the most important is if the town should be conceived of as a whole or as a set of parts added together. As the approach typically combines these two, planners need to decide how large and self-contained each part is, how it is linked to other parts, the importance of larger centers, and how the new town should link to other areas. In terms of building a new town, the classic problem is that a great deal of infrastructure

Figure 28.2 Tsukuba Science City on the outskirts of Tokyo. Source: Ann Forsyth.
Note: The image shows its multimodal transportation system; however the train took several decades to arrive.

needs to be in place before people arrive but until enough people arrive, there is not a revenue stream to pay for it. This is one reason that the privately developed new towns of the United States have not had large proportions of low-income housing – they had to raise revenue early selling to the middle class (or richer). Something similar happens with jobs. Company towns have jobs in place early but often lack a variety of options; other new towns need to attract industries. Without significant government intervention, and even with it, this can take years. These have design implications because the new town needs to be designed to function while under construction as well as when fully built – a difficult challenge.

Conclusions

Apart from a brief federal interest in new towns in the 1970s, in the United States new towns have tended to be built by the private sector, at the neighborhood scale (Burby and Weiss 1976). Europe and Asia have been key locations for government-sponsored developments, although there are also important examples elsewhere, particularly in the Middle East and North Africa. Some of these programs are focused on promoting higher quality design, others have regional planning aims. While many have drawn on the garden city tradition it has been modified for different people and places – for example over the course of the British New Towns Program of the 1950s through the 1970s, successive waves of towns tended to be larger in size. The French, starting their program later, proposed that the new towns around Paris have populations in the hundreds of thousands each (Hall and Ward 1998, 97; Forsyth 2003). The Soviet Union had a large program of planned new cities and satellite towns, though how many were self contained new towns is difficult to assess (Osborn and Whittick 1977; US/USSR 1981). New community developments at the neighborhood scale, however, remain popular and prolific.

Currently, one of the largest programs for constructing planned communities is in China. Large new town style developments designed and developed by teams from the United States and Europe have received much attention. Those with exotic themes have also had a high profile – for example the nine European themed new towns around Shanghai, in fact themed neighborhoods, typically in larger developments. However, these are a small part of a much larger program of new town building with many of the developments touting ecological credentials. It will be important to assess these developments over time.

Earlier in the twentieth century, garden city and new town advocates imagined such developments might be a significant component of urban development. This has not been the case. However, they have provided important models for comprehensive physical planning and urban design and continue to capture the imagination of urban designers interested in building concepts from the ground up.

References

Alexander, C. (1972, orig. 1966). "The City is Not a Tree." In Bell, G. and Tyrwhitt, J. (Eds.) *Human Identity and the Urban Environment.* Harmondsworth: Pelican.

Bailey, J. (Ed.) (1973). *New Towns in America: The Design and Development Process.* New York. John Wiley and Sons.

Biddulph, M. (2000). "Villages Don't Make a City." *Journal of Urban Design* 5(1): 65–82.

Burby, R. and Weiss, S. (1976). *New Communities US.* Lexington, MA: Lexington Books.

Crewe, K. and Forsyth, A. (Forthcoming). "Compactness and Connection in Environmental Design: Insights from Ecoburbs and Ecocities for Design with Nature." *Environment and Planning B.*

Forsyth, A. (2003). "New Towns." In Christensen, K. and Levinson, D. (Eds.) *Encyclopedia of Community.* Thousand Oaks, CA: Sage.

—— (2005). *Reforming Suburbia.* Berkeley, CA: University of California Press.

Forsyth, A. and Crewe, K. (2009). "A Typology of Comprehensive Designed Communities since the Second World War." *Landscape Journal* 27(2): 56–78.

—— (Forthcoming). "New Visions for Suburbia: Reassessing Aesthetics and Place-making in Modernism, Imageability, and New Urbanism." *Journal of Urban Design.*

Gans, H. (1968). *People and Plans.* New York: Basic Books.

Godschalk, D. (1967). "Comparative New Community Design." *American Institute of Planners Journal* 33(6): 371–387.

Golany, G. (1976). *New-Town Planning: Principles and Practice.* New York: John Wiley and Sons.

Griffin, N. (1974). *Irvine: The Genesis of a New Community.* Washington, DC: Urban Land Institute.

Gropius, W. (1945). *Rebuilding our Communities.* Chicago: P. Theobald.

Hall, P. and Ward, C. (1998). *Sociable Cities: The Legacy of Ebenezer Howard.* Chichester: John Wiley.

Holston, J. (1989). *The Modernist City: An Anthropological Critique of Brasilia.* Chicago: University of Chicago Press.

Howard, E. (1902, orig. 1898). *Garden Cities of Tomorrow.* London: Sonnenschein.

—— (2003). *Tomorrow: A Peaceful Path to Real Reform.* Original edition with commentary by Hall, P., Hardy, D., and Ward, C. London: Routledge.

Keller, S. (1968). *The Urban Neighborhood: A Sociological Perspective.* New York: Random House.

McKenzie, E. (1994). *Privatopia: Homeowner Associations and the Rise of Residential Private Government.* New Haven, CT: Yale University Press.

Osborn, F. and Whittick, A. (1977). *The New Towns: The Answer to Megalopolis.* Third edition. London: Leonard Hill Books.

Peattie, L. (1987). *Planning: Rethinking Ciudad Guayana.* Ann Arbor, MI: University of Michigan Press.

Perry, C. (1939). *Housing for the Machine Age.* New York: Russell Sage Foundation.

Silver, C. (1985). "Neighborhood Planning in Historical Perspective." *Journal of the American Planning Association* 51(2): 161–174.

Stein, C. (1957). *Toward New Towns for America.* New York: Reinhold.

Thorns, D. (1976). *The Quest for Community: Social Aspects of Residential Growth.* New York: John Wiley.

US/USSR New Towns Working Group (1981). *Planning New Towns: National Reports of the US and USSR.* Washington, DC: US HUD, Office of International Affairs.

Vale, L. (1992). *Architecture, Power, and National Identity.* New Haven, CT: Yale University Press.

Ward, S. (ed). (1992). *The Garden City: Past, Present and Future.* London: E. &F.N. Spon.

Ward, S. (2002). *Planning the Twentieth-Century City.* Chichester: John Wiley.

Further reading

Alonso, W. (1977, orig. (1970)). "The Mirage of New Towns." In Allen, I. (Ed.) *New Towns and the Suburban Dream.* Port Washington, NY: Kennikat Press. A classic critique of the idea of new towns.

Eichler, E. and Kaplan, M. (1967). *The Community Builders.* Berkeley, CA: University of California Press. An early study of developers of large-scale planned communities.

Fishman, R. (1977). *Urban Utopias in the Twentieth Century.* Cambridge, MA: MIT Press. Examines the work of Howard as well as Wright and Le Corbusier.

Phillips, D. and Yeh, E. (Eds.) (1987). *New Towns in East and South-east Asia.* New York: Oxford University Press. An important collection on Asian new towns.

Smookler, H. (1976). *Economic Integration in New Communities.* Cambridge, MA: Ballinger. Part of the University of North Carolina studies on planned communities focusing on economic issues.

29

Neighborhood spaces

Design innovations and social themes

Ajay Garde

Neighborhood spaces constitute a key component of urban form. Reformist ideas, social values, and market innovations have influenced the design of neighborhood spaces to evolve over time. The diffusion of these innovations in the design of neighborhood spaces has, in turn, influenced the form of the city. What constitutes a neighborhood, however, depends on how one conceptualizes it, and thus remains a focus of academic discourse. This chapter begins with a brief discussion of alternative and competing constructs of a neighborhood. Does a neighborhood comprise a discrete physical territory or is it a physically more amorphous concept with a distinct social and economic notion associated with the idea of a community? Are there certain distinguishing characteristics that differentiate neighborhood spaces from other urban areas? The text following this discussion reviews the innovations and current practices in the design of neighborhood spaces. This narrative includes influential ideas and variants of the original concepts and the evolution of ideas in the design of neighborhood spaces. The chapter concludes with a brief discussion of some of the noteworthy differences across the important innovations in neighborhood design.

Defining the neighborhood

The concept of a neighborhood as a distinctive space specific to our residential life experience is important to the field of urban design. For one, it is a useful construct for subdividing larger urban areas, such as a city or a region, into smaller areas that are more conducive to planning, designing, and managing the complex urban environment. The idea of using neighborhoods as a way of organizing and managing places goes back to the cities of antiquity. Gordon (1946) stated that the kinship network and the family system in ancient China comprised the neighborhoods. The plans of the ancient Greek cities of Miletus and Thurium reveal that the principles of planning a city involved the creation of socially segregated neighborhoods (Mumford 1961). Today the concept of the neighborhood continues to influence the planning and design of new development projects in urban and suburban areas, and the idea of neighborhood as a spatial unit remains a common feature in municipal level planning.

Several academics have examined the concept of neighborhood and have proposed different definitions. While the common characteristics of most definitions of neighborhood include territory and/or inhabitants, there is no consensus on a definition or agreement on what makes a neighborhood. According to Mumford (1954), neighborhoods are created whenever a group of people shares a place. Park (1915) observed that proximity and neighborly contact among the inhabitants of an area may lead to the social and political control of a geographical area and that this territorially controlled geographical area amounts to a neighborhood. According to this view, the subareas of cities are different from one another physically, socially, economically, or politically, and thus acquire neighborhood identity. More specifically, the identity of a neighborhood is a function of its noticeable physical components such as its distinctive style of architecture or discernible geographical boundaries such as a freeway, shoreline, or the edge of a park that demarcate an area.

Sometimes, the identity of a neighborhood reflects its distinctive characteristics, including the ethnicity, religion, or sexual orientation of its inhabitants. Chinatown, in addition to the Little Italy, Little India, Little Japan, Little Manila, and Little Saigon neighborhoods, are examples of ethnic neighborhoods that have formed in several large cities around the world. As another example, Hillcrest, a neighborhood in San Diego, California, derives its identity from its gay and lesbian community. Additionally, income characteristics of residents are sometimes used to identify "working class" or "affluent" neighborhoods. Moreover, places with a name known to their inhabitants, such as the hundreds of named neighborhoods in the city of Chicago, are seen as neighborhoods even though the boundaries and names of such neighborhoods are subject to change over time for a variety of reasons, including immigration and gentrification. While we commonly associate neighborhoods with residential areas, sometimes they receive their designation for the purposes of funding and governance. For instance, the 1999 Charter Reform of Los Angeles led to the formation of 89 certified neighborhood councils to make government more responsive to local needs (City of Los Angeles Department of Neighborhood Empowerment n.d.).

Kotler (1969) proposed a political definition of neighborhood and suggested that, when a group of residents in an area come together for shared political goals, the boundaries of the geographical area from which the residents come together can be identified as a neighborhood. Thus, when people come together to protest a proposed freeway going through a residential area, this area can be identified as a neighborhood. However, Kotler's definition of neighborhood is not very helpful for planners and designers because the boundaries of a neighborhood are subject to change, depending on the nature of activity and the participation of inhabitants. Suttles (1972) argued that the notions of neighborhood are intrinsic to what he calls "social construction of communities." He suggested that planned residential areas aspire to cultivate this social construction of communities through clearly defined boundaries, well-cultivated and publicized identity of place that often includes a name, and homogeneity or a certain "cultural unity" of residents that involves screening of the residents by the realtor. He observed that residents in some areas may feel more "invested" in part because they share some common interests and "look out for each other." Thus, these residential areas may be seen as "neighborhoods." This is not always the case, however. In residential areas where people do not share common interests, they may "limit their liabilities" by minimizing their "investment" so that they can "pull out" if physical conditions deteriorate or if the property values decline.

Neighborhoods also can be identified from the users' perception of an area, based on their familiarity with and frequency of use of the area. According to Keller (1968), small villages can be viewed as neighborhoods because of their familiarity. The frequency of use of certain common areas such as a community center, places where people shop or play, or the area encompassing their social networks also can signify a sense of neighborhood. Indeed, it is possible to ask residents to draw a map of the areas with which they are most familiar, or places that they most frequently visit, and then use these collective maps to identify the potential boundaries of a neighborhood. Alternatively, residents could be asked to draw the boundaries of the neighborhood on a map.

Residential areas that involve certain contracts or rules or that entail shared responsibilities can be identified as neighborhoods. Housing developments that allow shared ownership of common facilities, along with the individual ownership of the private dwelling, are generally known as common interest developments (CIDs). Usually, CIDs are self-governed by the community or homeowners associations that manage and maintain the shared facilities such as streets, parks, open spaces, clubhouses, recreation centers, and the like. Every homeowner in a CID is a member of the homeowners association, which is governed by a set of bylaws and rules known as the Covenants, Conditions and Restrictions (CC&Rs). These CIDs are not characterized by specific size or type of housing and may include a variety of housing types, from multifamily to single-family homes. The number of housing units in a CID can vary from a few homes to large complexes of thousands of units. While all CIDs involve some shared responsibilities, they are not necessarily identified as neighborhoods. Whether a particular CID can be identified as a "neighborhood" depends on how we define the concept of neighborhood. Additionally, CIDs with large number of units can be subdivided into several neighborhoods.

The concept of CID began in the 1900s with the intent to include within residential areas certain common amenities such as swimming pools, tennis courts, and community centers, which could be used only by the residents. Later, increasing demand for higher security, ease of maintenance, and exclusivity contributed to the proliferation of CIDs. In addition, limited revenues available to local governments for infrastructure development contributed to the promotion and development of CIDs. The most common type of CID is the planned unit development (PUD), a concept introduced in the 1960s. A PUD is a comprehensively planned development that provides more flexibility and better control in overall site planning and in organizing the buildings and open spaces, as compared to neighborhoods that are not comprehensively planned. Some CIDs, such as the retirement communities or age-restricted neighborhoods, are planned and designed to cater to the lifestyle and daily needs of those groups.

From a neo-liberal perspective, a neighborhood can be identified as a territory that is shaped by a certain assignment of property rights over public and private goods. According to Webster (2003), a neighborhood can be seen as an economic entity in which shared public and private attributes within a certain territory are governed by formal and informal contracts. Given this understanding, the concept of a neighborhood may suggest equilibrium and efficiency in the production and consumption of public and private goods. This concept of the neighborhood as an economic entity is useful in understanding the reassignment of property rights from the general public to the residents within a CID. Thus, the "rights" to the public amenities in a city, such as parks, playgrounds, public open spaces, and streets that were normally accessible to all, are reassigned within a CID as exclusive facilities available only by contract. These contracts and rules govern the use of common areas and amenities and influence the behavior of inhabitants. Lofland (1998) identifies these spaces as "parochial" insofar as they are available only by contract to a certain community.

The terms "neighborhood" and "community" often have been used interchangeably, in part because the search for a good neighborhood is simultaneously a search for a good community. Additionally, the term "sense of community" is frequently used in planning documents, in part due to the popular idea that physical planning can create a sense of community. However, both concepts are complex and involve multiple meanings that go beyond the specific social or physical dimensions that represent them. Although the concept of neighborhood generally comprises a certain territory and proximity of inhabitants, even when the boundaries of this territory are not clearly defined, the idea of community is not necessarily limited to proximity and it is possible to have a community without propinquity, as Webber (1964) has argued. Thus, it is possible to engender a sense of community in neighborhoods through the empowerment of inhabitants. Indeed, a sense of community can emerge among the residents of an area as a result of their involvement in local activities. However, most researchers agree that physical design, by itself, cannot necessarily produce a sense of community. Meanwhile, scholars interested in the phenomenon of "place attachment" have studied the physical characteristics of urban environments and, in some cases, have found certain features of the physical environment contributing to a sense of community (Talen 2000). Even so, according to Sennett (1977), developers, designers, and planners try to create a sense of community through physical design, in response to the erosion of civility and public life in urban and suburban areas. He argues that, in reality, they are producing mainly gated communities, CIDs, and enclaves of different kinds that further contribute to the formation of "clubs" that intensify social exclusion and segregation.

Designing the neighborhood

Neighborhood unit

Several approaches have emerged in the physical planning and design of neighborhood spaces. The neighborhood unit concept is considered an important innovation, one that was originally conceived as an organization of a residential environment that would serve the quotidian needs of its residents. The neighborhood unit was a precursor to current practices such as the planned unit developments and CIDs that make up a significant proportion of residential arrangements in today's suburbs. Formulated in the 1920s by Clarence Perry, the neighborhood unit concept demonstrated how a model layout plan and a set of normative principles can be used for the planning and design of good residential environments. Perry (1929) identified four essential components of a neighborhood unit: a centrally located elementary school, small parks and playgrounds, small stores and shops, and a residential environment in which all public facilities should be within safe pedestrian access. In addition, Perry (1939) specified six physical attributes, including the size, boundaries, open spaces, institutional sites (e.g. a school), local shops, and internal street system that, in an appropriate relationship, organize these four elements. He proposed that the size of the neighborhood unit should be determined by the population needed to support one elementary school and thus the population density of the proposed neighborhood should determine the area needed to develop it. He further suggested that the school should be located in the center so that the children need not walk more than half a mile, which in turn led him to conclude that a neighborhood unit should have an area of about 160 acres, because the desirable population density was 10 units/acre. In the formulation of the neighborhood unit concept, considerable emphasis was placed on separation of land uses as well as segregation of pedestrian and vehicular traffic so that children could avoid crossing the major streets. The traffic arterials therefore could not cut through the neighborhood unit and were instead used to define its boundaries. Local neighborhood parks and recreation spaces, for which Perry specified a minimum of 10 percent of the total land area, also served to maintain an inwardly-oriented, self-sustained planning unit (Dahir 1947).

The neighborhood unit concept was a planning response to the problems associated with the rapid transformation of urban life and society during the late nineteenth and early twentieth centuries. At that time, rapid urban growth in cities resulted in overcrowding, communicative disease, as well as crime and social pathology. Confronted with such malaise and malady, reformers in the late nineteenth century sought to identify ways to address these problems by seeking order and a development pattern for urban areas. Perry's inspiration came from reformers such as Jane Addams, Robert Park, and

Figure 29.1 Neighborhood unit diagram. Source: Committee on Regional Plan of New York and Its Environs. *Regional Survey of New York and Its Environs* (Vol. 7). New York (1929) – used by permission.
Note: Conceptual diagram of the neighborhood unit as originally conceived for the New York Regional Plan.

John Dewey, who argued that some degree of stability and continuity is necessary for the development of individuals in a rapidly urbanizing and transforming society and emphasized the role of place and community in the life of urban residents. Perry was responding to the needs of the family life when he developed the plan for a neighborhood unit, with the elementary school as its central focus (Figure 29.1), and argued that the elementary school could serve as a community center (Dahir 1947). The layout of the neighborhood unit attempted to encourage village-like interactions among the inhabitants and sustain primary associations of individuals and the needs of an average family life in a rapidly urbanizing and transforming society. Perry further suggested that the neighborhood unit should serve as a building block to organize the larger city. Uniformity in size and type of dwelling units, encouraging social homogeneity of each community, also ensured market success.

The neighborhood unit concept became quite popular and received endorsement from many quarters. The Committee on the Hygiene of Housing of the American Public Health Association (APHA) adopted this concept for setting "healthful and hygienic" standards for residential environments (American Public Health Association 1948). In the late 1940s, planning and design ideas of the neighborhood unit were adopted, modified, and institutionalized by various public agencies and professional organizations, including the American Society of Planning Officials (Banerjee and Baer 1984). While designers and planners promoted the neighborhood unit concept for the purpose of creating and sustaining a sense of community, private developers and lending institutions saw it as a means for protection of property values, and public agencies promoted its purpose of protecting public health, safety, and welfare. New developments in several other countries also adopted these ideas.

Perry's proposal incorporated some of the "family life values" promoted by contemporary reformers. Although the imperatives of such social concerns were not always grounded in empirical findings, the paradigm nevertheless received widespread institutional acceptance. Local planning manuals and zoning ordinances throughout the United States adopted the neighborhood unit principles. Inadequacies of the paradigm were determined later, including the belief that the formulation of the model was flawed (Banerjee and Baer 1984). In a similar vein, Willmott (1962) criticized the assertion that residential subunits or "neighborhoods" in the British new town of Stevenage, England, contributed to "neighborliness." Critics have argued that social homogeneity was an underlying theme in the formulation of the neighborhood unit, and the implementation of the concept ultimately protected lenders' interests more than promoting the possibilities of socially heterogeneous neighborhoods (Isaacs 1948).

The neighborhood unit concept nevertheless was an influential idea and a precursor of current practices in neighborhood design. Different combinations of the physical design attributes of the neighborhood unit continue to influence the planning and design of residential environments, even though the central location of the school and the size of residential development are no longer considered essential. Gated communities seemingly have emerged as a variant of the neighborhood unit idea and have gained popularity in residential developments in the United States as well as in several other countries, including Argentina, Brazil, Canada, and Mexico. Gated communities are residential neighborhoods with boundary walls and restricted entry, guarded by private security. In some residential areas that have a high crime rate, residents are permitted to close the street and provide gates at the access, primarily for security reasons.

A significant proportion of gated communities, however, are private developments.

According to Blakely and Snyder (1997), private gated communities suggest separation, distinction, exclusion, and protection. The physical manifestations of these characteristics comprise isolated location, exclusive high-income housing, privatized amenities, surveillance cameras, on-site security systems, and fortress-like boundary walls. In many of these gated communities, private agencies provide their own security, street maintenance, parks, recreation facilities, garbage collection, and other services. The community amenities that are within the gated areas, such as parks, swimming pools, and tennis courts are not accessible to outsiders. Usually, these gated communities are designed as inward-oriented neighborhoods with considerable uniformity in the size and type of housing, intended to achieve social homogeneity, which is one of the underlying themes of the neighborhood unit concept. Additionally, the physical characteristics of gated communities that comprise clearly defined boundaries, community recreation areas, and internal street system restricting through traffic are variants of the neighborhood unit concept. In short, a gated community is a physical manifestation of the neighborhood unit concept, with gates and boundaries but without the school.

New Urbanism

New Urbanism is a recent reform movement in urban design, which grew out of an intellectual rejection of suburban sprawl and the ubiquitous rise of the non-place "edge-city" phenomena (Congress for the New Urbanism 2000; Duany and Plater-Zyberk 1991; Garreau 1991; Katz 1994). New Urbanist principles emphasizing physical design as a tool for improving the quality of life of urban and suburban areas gained considerable popularity in the 1990s. Advocates of New Urbanism claim that their neighborhood development strategies can minimize deterioration of environmental quality, support place-based economy, and promote social equity, as compared to conventional suburban developments that produce sprawl and aggravate environmental, social, and economic problems. New Urbanists expect to minimize environmental deterioration by reducing land consumption, reducing the number and length of automobile trips, and conserving energy. They hope to support place-based economy by addressing locally defined needs, such as the jobs-housing balance, in mixed-use and mixed-income developments. They also anticipate that they can promote social equity by providing low-income people with equitable access to housing in New Urbanist projects, which in turn will provide them with equitable access to better environmental, economic, and social resources.

New Urbanist projects strive to integrate a mix of land uses, a compact urban form, an interconnected network of streets and blocks organized around a neighborhood center, a variety of housing types and densities, and a pedestrian-oriented design with an emphasis on providing civic spaces and amenities within walking distance (Figure 29.2). The expected benefits of New Urbanist projects include efficient use of land due to the higher densities and smaller lots as well as preservation of the ecological quality of neighborhoods, districts, and regions (Calthorpe 1993). Proponents have argued that New Urbanist projects are economically more efficient to build, as compared to conventional suburban subdivisions (New Urban News 2001). They also claim that New Urbanist projects improve social life and enhance a sense of community among residents. While such claims of advocates of New Urbanism are undergoing empirical scrutiny, many such projects classified as New Urbanist have received various

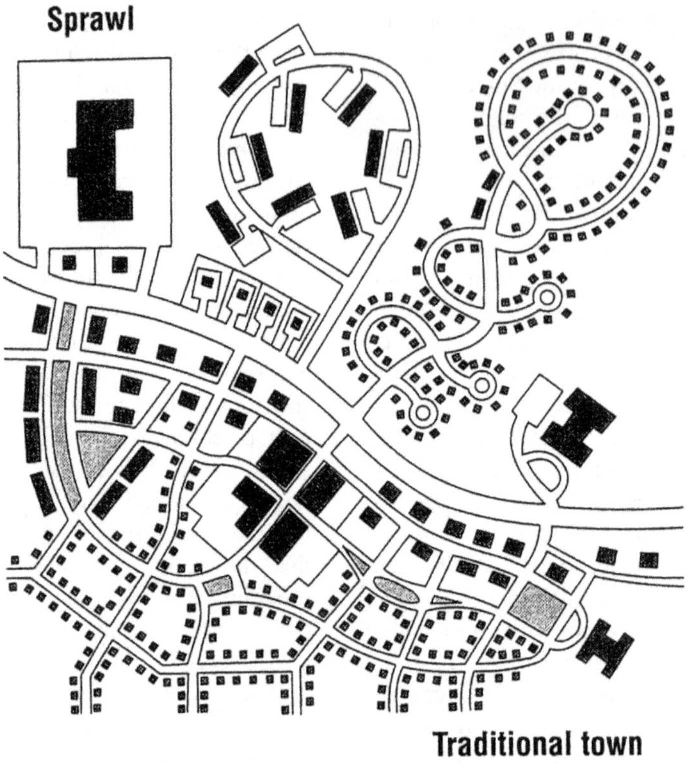

Figure 29.2 Conventional and neo-traditional suburban development. Source: © Duany Plater-Zyberk & Co – used by permission.
Note: Conceptual plan comparing the conventional suburban development pattern as "sprawl" and the New Urbanist development pattern as "traditional town."

forms of institutional endorsements and are being built in the United States. Many US municipalities have adopted New Urbanist principles in new suburban developments, urban redevelopment and in-fill projects, and urban transit-oriented developments. The design concepts of New Urbanism have influenced projects in several other countries, including Poundbury, near Dorchester, England; Kemer, near Istanbul, Turkey; the Gorbals, in Glasgow, Scotland; and Puri Jaya, in Tangerang, Indonesia (New Urban News 2001).

Arguably, the planning and design of New Urbanist developments draw considerably from the idea of the neighborhood unit without explicit acknowledgement (see Duany and Plater-Zyberk 2003). While the Charter of the New Urbanism (Congress for the New Urbanism 2000) also addresses the design of communities at the block level and at the regional level, it is the neighborhood level principles that draw significantly from the neighborhood unit. In fact, most of the ideas of the neighborhood unit concept, with the exception of the theme of social homogeneity, are retained, and additional principles are included in the Charter. For instance, the idea of a pedestrian-friendly development organized around community facilities that serves the daily needs of its residents can be traced to the neighborhood unit concept. New Urbanist principles also suggest that neighborhoods should be designed in such a way that children could walk or bike to school. The New Urbanist principles go beyond the neighborhood unit concept by proposing diversity in the size and type of housing and by encouraging socially heterogeneous neighborhoods.

The design concepts and planning ideas of New Urbanism are beginning to influence public policy. Policy initiatives such as "smart growth" have incorporated several principles of New Urbanism (see chapter by Aseem Inam). Although in most suburban areas, zoning ordinances and subdivision regulations permit only low-density, single-family subdivisions, these regulations are being revised in many regions that are experiencing substantial population growth. It is expected that such revised regulations will support New Urbanist-type higher-density, mixed-use, and mixed-income developments. The New Urbanist design movement has spawned several derivative concepts associated with specific design principles that are discussed by Ivonne Audirac in a separate chapter in this book.

Sustainable neighborhoods

In recent years, amid concerns about global warming and the large carbon footprint associated with sprawl, designers, developers, and planners have been trying to improve the sustainability of individual buildings as well as neighborhood-scale projects. Several approaches, including a number of Leadership in Energy and Environmental Design (LEED) rating systems, have been developed to evaluate projects using a set of criteria that, according to proponents, contribute to their livability and sustainability. The LEED for Neighborhood Development (LEED-ND) rating system is one such approach that has gained considerable popularity for promoting sustainability, through physical planning and design, in neighborhood development projects.

Three organizations, the US Green Building Council (USGBC), the Congress for the New Urbanism (CNU), and the Natural Resources Defense Council (NRDC) have collaborated to develop the LEED-ND rating system. Thus, the LEED-ND rating system builds on the New Urbanist principles. If a project meets the prerequisites and earns the credit points needed to reach specified thresholds, it receives a "sustainable" certification. The LEED-ND is a voluntary and market-driven approach to encourage neighborhoods to "reduce land consumption, reduce automobile dependence, promote pedestrian activity, improve air quality, decrease polluted stormwater run-off, and build more livable, sustainable communities for people of all income levels" (USGBC 2007: 1).

The rating system comprises specific categories that include prerequisites as well as optional criteria for which credit points could be earned. In addition, the rating system provides an opportunity to earn points for features that are not part of the rating criteria. For example, the smart location and linkage category emphasizes locating the project on a site that is within or near already developed communities with an existing infrastructure, which provides opportunities to use public transportation and walking. The neighborhood pattern and design category focuses on physical planning and design concepts, such as compact development, walkable streets, and diversity of uses, and urban infill projects. The green construction and technology category focuses on construction technology and management techniques that improve energy- and water-efficiency of the project and thus result in the reduction of its carbon footprint. In addition, the rating system provides credit for design innovation and for addressing criteria that are important to the region. Neighborhood development projects are required to register with the USGBC and submit the necessary documentation for verification and certification. A project must meet each prerequisite and earn the minimum total credit points for achieving LEED-ND certified, silver, gold, or platinum certification. Several neighborhood development projects have been certified using the rating system (USGBC n.d.).

Many researchers have examined the issue of sustainable development and have proposed different methods to promote sustainable neighborhoods. While there is a general agreement about a broad definition of sustainable development, there is no consensus on planning and design features that make a neighborhood sustainable. However, advocates argue that the LEED-ND rating system will have a considerable influence on the diffusion of certain planning and design concepts into neighborhood development projects that will make them more livable, healthy, and sustainable (Smart Growth Network 2006; US Environmental Protection Agency n.d.; USGBC 2007).

Indeed, the physical design of a neighborhood can contribute positively to improving the energy- and water-efficiency of the project by using techniques such as solar orientation, on-site energy generation using solar energy, heat island reduction, wetland and water body conservation, and the like. While the physical design of a neighborhood does not determine social behavior, certain characteristics of the built environment can contribute positively by encouraging or discouraging walking, biking, or riding mass transit and thus influencing certain behaviors, which in turn may contribute positively to public health. The Active Living and Healthy Cities and Communities initiatives, which emphasize pedestrian-oriented neighborhood design with trails, bike lanes, and wider sidewalks connecting parks and open spaces, are important approaches in this context and are discussed by Marlon Boarnet and Lois Takahashi in a separate chapter in this book.

Conclusions

This chapter described different conceptualizations of neighborhood spaces. While the common characteristics of most definitions of neighborhood include territory and/or inhabitants, there is no consensus on what makes a neighborhood. This is because it is difficult to assess characteristics such as geographic boundaries as well as the fact that physical, social, political, or economic components that are commonly used to identify a neighborhood do not always overlap. Indeed, the concept of neighborhood has contributed significantly to the debates on physical and social determinism. However, the idea of designing the neighborhood as a unit has endured over time, in part because residents' quality of life and homeowners' property values are affected by design, and because the neighborhood unit can be used as a building block for the development of cities.

The conceptualizations of the neighborhood unit, New Urbanism, and the LEED-ND rating system were discussed next. New Urbanism and the LEED-ND rating system are reminiscent of the neighborhood unit concept that was conceived as a response to the problems of urban development of the past. Some of the recent versions of neighborhood space design have put considerable emphasis on spatial insulation, social exclusion, and economic distinction, which are noticeable in the design of gated communities and in some CIDs.

While these conceptualizations of neighborhood spaces share some common attributes, there are some important differences that are noteworthy. The neighborhood unit idea was conceived as an inward-oriented geographic unit that emphasizes homogeneity of residents. New Urbanist projects are expected to include a variety of housing to encourage and accommodate a diverse group of residents. The LEED-ND rating system is expected to go beyond New Urbanism and promote projects that improve energy- and water-efficiency of the neighborhood and result in the reduction of the neighborhood's carbon footprint. Whether the LEED-ND rating system influences the design of the neighborhoods to make them more sustainable and livable for people of all income levels remains to be seen. Meanwhile, the planning and design of neighborhoods will continue to evolve and influence the form of cities.

References

American Public Health Association (1948). *Planning the Neighborhood: Standards for Healthful Housing*, Chicago, IL: Public Administration Service.

Banerjee, T. and Baer, W. C. (1984). *Beyond the Neighborhood Unit: Residential Environments and Public Policy*, New York: Plenum.

Blakely, E. and Snyder, M. G. (1997). *Fortress America: Gated Communities in the United States*, Washington, DC: Brookings Institution Press.

Calthorpe, P. (1993). *The Next American Metropolis: Ecology, Community and the American Dream*, New York: Princeton Architectural Press.

City of Los Angeles Department of Neighborhood Empowerment (no date) *Neighborhood Council Directory*, Available HTTP: <http://done.lacity.org/ncdatabase/nc%5Fdatabase%5Fpublic/> (accessed 7 July 2009).

Congress for the New Urbanism (2000). *Charter of the New Urbanism*, M. Leccese and K. McCormick (eds.), New York: McGraw-Hill.

Dahir, J. (1947). *The Neighborhood Unit Plan, its Spread and Acceptance: A Selected Bibliography With Interpretive Comments*, New York: Russell Sage Foundation.

Duany, A. and Plater-Zyberk, E. (1991). *Towns and Town-making Principles*, New York: Rizzoli.

—— (2003). "Lexicon of the new urbanism," in Watson, D., Plattus, A., and Shibley, R. (Eds.), *Time-saver Standards for Urban Design*, New York: McGraw-Hill.

Garreau, J. (1991). *Edge City: Life on the New Frontier*, New York: Doubleday.

Gordon, N. J. (1946)."China and the neighborhood unit," *The American City*, 61: 112–113.

Isaacs, R. (1948). "The 'neighborhood unit' is an instrument of segregation," *Journal of Housing*, 5: 215–219.

Katz, P. (1994). *The New Urbanism: Toward an Architecture of Community*, New York: McGraw-Hill.

Keller, S. (1968). *The Urban Neighborhood: A Sociological Perspective*, New York: Random House.

Kotler, M. (1969). *Neighborhood Government*, Indianapolis, IN: The Bobbs-Merrill Company.

Lofland, L. (1998). *The Public Realm: Exploring the City's Quintessential Social Territory*, New York: Walter de Gruyter, Inc.

Mumford, L. (1954). "The neighborhood and the neighborhood unit," *Town Planning Review*, 24: 256–270.

—— (1961). *The City in History: Its Origins, Its Transformations, and Its Prospects*, New York: Harcourt, Brace and World.

New Urban News (2001). *Comprehensive Report and Best Practices Guide*, Available from New Urban Publications Inc., P.O. Box 6515, Ithaca, NY 14851.

Park, R. E. (1915). "The city: suggestions for the investigation of human behavior in the city environment," *The American Journal of Sociology*, 20(5): 577–612.

Perry, C. A. (1929). "The neighborhood unit" (Monograph I). In *Neighborhood and Community Planning, of the Regional Survey of New York and Its Environs* (Vol. 7), New York: Committee on Regional Plan of New York and its Environs.

—— (1939). *Housing for the Machine Age*, New York: Russell Sage Foundation.

Sennett, R. (1977). *The Fall of Public Man*. New York: Knopf.

Smart Growth Network. (2006). *This is Smart Growth*. Available HTTP: <http://www.smartgrowthonlineaudio.org/pdf/TISG_2006_8–5x11.pdf>(accessed 15 January 2007).

Suttles, G. D. (1972). *The Social Construction of Communities*, Chicago and London: The University of Chicago Press.

Talen, E. (2000). "The Problem with Community in Planning," *Journal of Planning Literature*, 15(2): 171–183.

US Environmental Protection Agency. (n.d.). *Environmental benefits of smart growth*. Available HTTP: <http://www.epa.gov/smartgrowth/topics/eb.htm> (accessed 28 February 2008).

United States Green Building Council (USGBC) (2007). *LEED for Neighborhood Development Rating System, Pilot Version*. Available HTTP: <https://www.usgbc.org/ShowFile.aspx?DocumentID = 2845/> (accessed 25 September 2007).

—— (n.d.) *LEED for Neighborhood Development registered Pilot Project List updated 6/24/09* Available HTTP: <http://www.usgbc.org/ShowFile.aspx?DocumentID = 3546> (accessed 13 July 2009)

Webber, M.M. (1964). "The urban place and the non-place urban realm" in Webber, M.M., Dyckman, J.W., Foley, D.L., W. L.C. Wheaton, C.B. Wurster (Eds.) *Explorations into Urban Structure* (pp. 79–153), Philadelphia, PA: University of Pennsylvania Press.

Webster, C. (2003). "The Nature of the Neighbourhood." *Urban Studies*, 40(13): 2591–2612.

Willmott, P. (1962). "Housing Density and Town Design in a New Town," *Town Planning Review*, 33: 114–127.

Further reading

Blakely, E. and Snyder, M. G. (1997). *Fortress America: Gated Communities in the United States*, Washington, DC: Brookings Institution Press. A critical assessment of the proliferation of gated communities and the increasing social and spatial segregation in American society.

Congress for the New Urbanism (2000). *Charter of the New Urbanism*, Leccese, M. and McCormick, K. (Eds.), New York: McGraw-Hill. A detailed discussion of each of the New Urbanist principles for designing the region, the neighborhood, and the block, with illustrative examples.

Keller, S. (1968). *The Urban Neighborhood: A Sociological Perspective*, New York: Random House. A summary of the literature on "neighbors and neighboring" and a critique of physical planners' attempts to design the "neighborhoods" using the neighborhood unit concept.

Perry, C. A. (1929). "The Neighborhood Unit" (Monograph I). In *Neighborhood and Community Planning, of the Regional Survey of New York and Its Environs* (Vol. 7), New York: Committee on Regional Plan of New York and its Environs. A proposal for designing neighborhoods using certain physical design principles to establish an order and a development pattern for the urban areas.

Suttles, G. D. (1972). *The Social Construction of Communities*, Chicago and London: The University of Chicago Press. An analysis of urban residential areas, in terms of social control and territoriality, which contribute to "community differentiation" and characterize the neighborhoods.

United States Green Building Council (2007). *LEED for Neighborhood Development Rating System, Pilot Version*. Available HTTP: <https://www.usgbc.org/ShowFile.aspx?Document ID = 2845/> (accessed 25 September 2007). A comprehensive approach for evaluating neighborhood-scale developments with operationally defined criteria to be used for the LEED-ND certification.

30

Spaces of consumption

Klaus R. Kunzmann

The city offers multiple settings and occasions for consumption. Consumption has always been a key dimension of urban development and design (Simmel 1971; Sombart 1967; Braudel 1986; Brewer and Trentman 2006). Investment in spaces of consumption has considerable impact on the form and appearance of cityscapes. Consumption includes not only products, such as food, apparel or jewelry, but also entertainment and leisure. In the post-industrial city, where consumer preferences and lifestyles are scripted and stage-managed by the media, urban design plays an increasingly critical role in attracting consumers to the city. Consumers today may be young or old, highly educated or illiterate, black or white, healthy or chronically sick, affluent or deprived. In the cosmopolitan city they search for the latest fashion, for culturally defined or fair-trade products, for local or imported goods, for opportunities and locations to consume and enjoy as individuals and in groups. All this is reflected in the spatial organization of the built environment.

Whether regenerating city centers, revitalizing city quarters and brownfields or developing suburban communities, architects and planners as well as developers and investors have certain images and functions in mind, when deciding on locations and projects and in specifying their functions and style. At the beginning of the twenty-first century, the outcome of such a diversity of criteria is visible on the urban form of cities, whether they are located in Asia, North America, Europe or Australia. With global communication and logistic flows, cities gradually converge in style and appearance, following architectural or design gurus, the media, and marketing rationales. Benefiting from advanced building technologies and new materials, city builders promote their international models and projects for the consumer city. Local traditions are preserved only when there is a whiff of profit.

It is the consumer oriented central city with its shopping and entertainment districts, cultural quarters, flagship projects, and public parks which brands the image of cities globally. International tourists are attracted to cities that have successfully conserved their particular urban heritage and added new iconic buildings to existing cultural facilities (Stadbaukultur NRW 2003). Generally, mayors, city managers and developers are more interested in leaving their footprints on a city by commissioning signature projects to internationally renowned architects. The outcome is often a series of urban theme parks, which display traditional and modern architecture to frame sequences of public spaces allowing tourists to stroll *(flânerie)* and consume locally produced products and renowned international brands. In Europe, it is the traditional, more or less authentic, European city which is reinvented by carefully controlling its built environment. Elsewhere, creative copies of such cityscapes and iconic images attract visitors as well as investors, event managers and journalists. The more attractive these central cities are, the more educated middle-class households in the suburbs show an interest to flee the suburban blues, thus contributing to growing re-urbanization flows in the first decade of the new millennium.

This chapter on consumption and urban design will focus on two themes, on the spatial configuration and rationale of consumption in the city region, and on the urban design implications and dimensions of consumption. The chapter, however, will not deal with the economic, social and cultural dimensions of consumption. This has been done in a number of recent publications (see Corrigan 1997; Slater 1997; Warde 1997; Clarke 2003; Clarke *et al.* 2003; Jayne 2006; Sassatelli 2007).

In recent years consumption even got an environmental dimension, when biologically produced food became a fashion of middle-class urbanites, who are concerned with the protection of natural resources. Once such concerns become more popular in the post-industrial society, even new spatial configurations of consumption in the city region may evolve, promoting regional production, new economic circuits, and urban agriculture. However, this dimension of consumption, as well as the whole phenomenon of gentrification and its linkages to consumption are not treated equally in this chapter, as it would require a much more in-depth discussion about its underlying reasons, the difficulties to guide and balance urban development processes, and the balance between the negative social and the positive economic and design implications. Gentrification has been extensively covered in other publications (see Smith 1996; Smith and Williams 1986; Madanipour in this volume).

This chapter evolves around five threads:

1. Similar to infrastructure, consumption locations in a city structure and the urban landscape are key components of city building.
2. Consumption in the postmodern city is a complex amalgam of shopping, leisure and entertainment of economic, cultural, and social dimensions. The consumption profile of a city shapes identity building and contributes to city branding.
3. Despite the accelerated globalization of trend-setting media and the often similar branding and marketing strategies, consumption cultures and patterns are still different in North America, Asia and Europe, though such differences tend to slowly decrease.
4. Urban design for consumption spaces requires a comprehensive understanding of the city as a vivid cultural, social and economic entity.
5. Demographic changes and altering consumption cultures in Europe favor re-urbanization processes and tend to support the renaissance of the compact city.

Types and variety of consumption spaces

The location and appearance of consumption spaces in a city region vary considerably, ranging from high-end luxury shopping boulevards and galleries to low-end strip malls in the suburbs, from "airport cities" to gentrified or rundown commercial corridors, and from corner stores to street vendors. Following the logic of the market, the advice of marketing gurus, and the consumption behavior of citizens, consumption is always concentrated along boulevards and shopping streets, around public places, at railway stations or bus stops, at leisure grounds, and tourist spots. Consumption spaces are located where people have to wait, where they change modes of transport, where they enjoy vistas, where scarce or surplus time represents an essential element of city life. In market economies, urban quarters without consumption spaces either accommodate various forms of public services or have purely low-density residential functions.

Consumption spaces represent a hierarchical system. The city core is the highest rank in the complex hierarchy of urban consumption spaces, followed gradually by lower-order centers, such as the neighborhood shopping center or the corner shop. The logic of this system is based on the mobility of consumers, who have daily, weekly, and monthly demands for products, but also for window-shopping, flânerie, and entertainment. Thereby individual time budgets play a role as does accessibility. The hierarchy of consumption is defined by the size and importance of urban places. This hierarchy of consumption locations is still predominant in Europe, where 2000 years of urban history have created cultural townscapes, reflecting traditions of architecture, crafts, trade, and social life. Many of these urban places were and still are centers of consumption for citizens, and the more prominent the city is as a cultural asset and tourist spot, the more the city center, with its streets, boulevards or narrow lanes, galleries and plazas serves as the consumption heart of the city. The more such city centers accommodate residential and diverse cultural functions, the more attractive they become for the newly "invented" creative urban class (Florida 2006).

The dominant types of consumption spaces in a city are briefly sketched as follows:

Shopping streets, boulevards and plazas: shopping streets and public plazas are the most common consumption spaces in cities. They are ubiquitous in the metropolitan area, and rents for shops and spaces for services follow a hidden agenda of purchasing power and related ranking. Until today, the main street and central piazza of renaissance cities in Europe, as well as the boulevards, such as the ones that Baron Haussmann deliberately cut into the historic city fabric of Paris, remain the principal loci of consumption. Other such examples include the Königsallee in Düsseldorf, the Kurfürstendamm in Berlin, the Bahnhofstrasse in Zürich or the Ramblas in Barcelona – all of which have cult status among global travelers. This is also true for the grand arcades and galleries such as those in Brussels, Milan, or Paris – places designed in the late nineteenth century to attract *flâneurs* even when temperature or rain makes leaving the home uncomfortable. Such boulevards and galleries are settings for the affluent urban consumer and a favorite place for street musicians, street vendors, but rarely beggars, who are not welcome at such upscale spaces. There the global brands display their fashionable products while law and finance firms usually occupy the upper floors. Renowned consumption spaces in the city core top the ranking list of urban shopping streets, while hierarchical categories of other such streets characterize the quality of the respective urban district, which they serve. Thereby the shopping street leading to the central railway station is a particular type of consumption space, offering affordable consumer goods for arriving and departing travelers.

Central city malls and arcades: a recent response to the dreary suburban shopping mall is the central city shopping arcade. In a way, the air-conditioned arcade is reinventing the famous central city department store, which was an urban icon of the late nineteenth and early twentieth century, and which itself was a reinterpretation of the oriental bazaar. The shopping arcade is a kind of late twentieth – early twenty-first century reincarnation of famous department stores such as Galleries Lafayette in Paris or the KADEWE in Berlin. Such arcades are huge shopping malls built into the existing urban fabric. To a certain extent, they are a response to the complaints about the negative economic and social effects of the suburban shopping malls, imported from the suburbs of the US. From street level, the new central city mall looks like a huge regular shop (Figure 30.1). Once entering the air-conditioned complex, three or more shopping levels accommodate a broad selection of shops, services, cafés and restaurants. In contrast to the pedestrian-oriented public shopping streets, these are private consumption spaces, controlled by private security guards. In their contracts the individual shop owners are usually required by the mall management to redecorate their shops after a certain time (typically five years) to comply with the changing times and consumer values. Not complying with changing consumer desires is fatal for individual shops in such arcades. These new urban arcades have contributed much to the renaissance of city centers in many European cities. The shopping arcade "Das Schloss" in Berlin is one such example.

Markets: traditional consumption spaces in cities all over the world are markets. In European, Middle Eastern and Asian cities, markets are spaces, where trading and shopping are combined with meeting and watching people. Markets, in addition to their trading and shopping functions have an essential social function. In the modern post-agrarian and post industrial city, markets have an additional entertaining or even emotional function. They are locations where citizens and tourists stroll around and may buy things which catch the eye. In recent times, consumption trends towards healthy organic foods have revived farmer markets at prominent city locations, where specialized farmers and

Figure 30.1 Das Schloss: Shopping Arcade in Berlin. Source: Wang Fang – used by permission.

traders are offering regional food for more affluent and health conscious middle class households.

Tourist circuits: as a rule, the central city shopping precincts are also the circuits of city tourists, particularly if they include attractive public and cultural spaces, such as museums, historic buildings, and renowned architectural landmarks, though some of the goods offered here for tourist consumption differ from those consumed by the locals. Ideally, tourist circuits combine spaces for cultural education, entertainment, and consumption. The more such areas represent the expected cultural cliché of a city, in terms of visual appearance and local products including local food, the more economically successful they are, and the more they add to the city's international profile. The Mozartgasse in Salzburg, Gamla Stan in Stockholm, Oxford Street in London or the Ramblas in Barcelona are pertinent examples.

Entertainment districts: a particular category of consumer spaces in the city are entertainment districts. They range from the traditional red-light districts behind the railway station (famous examples are to be found in Paris and Amsterdam) to entertainment quarters, which are the night target of the young jet set in a city region, where individuals seek distraction and social contacts after work. Not quite attractive during daytime, such quarters flourish after sunset. Decorated by colorful lights, and featuring clubs, theaters, galleries, bars, fashion and design outlets, they become the stage for those who want to see and wish to be seen.

Ethnic consumer islands: cosmopolis, the life space of the post-modern multicultural society, has brought about a diversity of fragmented cultural and consumer spaces to the city region. The Jewish quarters were the first, though forced, ethnic quarters in European cities. Centuries later, Chinatowns evolved in cities around the world, where Chinese goods and memories are sold. Prominent examples of such enclaves are in San Francisco, Singapore and London. Here Chinese traders and migrants opened restaurants and shops in a few blocks at the edge of a city, where property values were low. More recently, these little Chinatowns have become targets of international tourists searching for cheap authentic restaurants. While ethnic quarters (Little Italy, Koreatown, Germantown or Spanish pueblo) are quite common in North American cities, they are a more recent phenomenon in European cities. With growing migration flows, ethnic quarters continue to evolve all over city regions in Europe where migrant communities seek access to food and other consumer goods which originate from their home countries. The Turkish quarter in Berlin or the Vietnamese district in Paris are such examples.

Street vending: street vending is the lowest order in the hierarchy of consumptions spaces. In cities of the Third World it is a common feature of city life. There, vendors are selling products they produce themselves, bring in from rural areas or sell on behalf of middlemen and trading companies specializing in low end products, such as textiles, household wares, watches or gimmicks. As a rule, the use of public space along busy streets or popular plazas is clearly regulated. In European cities, street vending is visible in multiple forms. It takes place in tourist circuits, offering souvenirs, real or fake antiques and cheap food, on market days in small country towns, displaying textiles and second hand books, and on special occasions, such as Christmas and local festivals, offering food and products related to the particular event.

This system of inner city shopping streets, plazas, markets and tourist circuits is paralleled by a system of shopping malls and specialized consumer stores in the sprawling suburban hinterland of a core city. There are three types of out-of town consumption spaces, suburban (greenfield) shopping centers, out-of-town hypermarkets and airport cities.

Suburban shopping malls: suburban shopping malls can be found in city regions, where infrastructure-led urbanization processes have resulted in accelerated suburbanization and hyper-urbanization. In Europe, Stockholm saw such developments in the early 1960s and Paris in the late 1960s. In both cities, they were deliberate measures to plan for balanced spatial, polycentric development. In some cities in the US and China, which lack traditional town centers, these suburban centers serve also as suburban community centers in addition to their consumption functions. These centers offer huge parking spaces, food courts and facilities for children. In most cases, their design quality is very low. (A remarkable exception is the shopping center designed by the British architect David Chipperfield in the outskirts of Bern, Switzerland.) Their functional basic architecture is decorated with glass, tiles, or stucco. Such centers do not need to comply with design regulations of neighboring quarters since they are typically distanced from the rest of the city fabric. The developer's ambitions do not go beyond sound engineering and short-term profitability during the first ten years of the building's life cycle. Usually, after seven to twelve years, these centers have come into age, causing the owners to review the original concept and re-model them to maintain their profitability or even demolish them.

Hypermarkets: scattered around the urban fringe or located along highly accessible transport corridors linking the city core with the suburban belt, "big box" retail outlets offer construction materials (e.g. Home Depot in the US or Hornbach in Germany), furniture (IKEA in Sweden and many European and American cities, or Porta in Germany), carpets, automobiles, or last year's fashions in factory outlets.

Aerovilles: in recent times, airports have become a different type of city center of the global society and international tourist community. The functional design of airport terminals is gradually being modeled after successful shopping centers, with a diversity of shops, cafés, and restaurants. The airport in Singapore features even a spa landscape and a large swimming pool (Sauter-Servaes and Rammler 2002). Such settings help passengers to kill time between flights or if their departure is delayed, seduce travelers to buying luxury goods, exploit the holiday mood of tourists, give business travelers a chance to do some last minute shopping for family or friends, or allow airport employees to shop for everyday goods. Munich, Vienna, Zurich or Dubai airports are pertinent examples. The design of the airport "boulevards" follows the latest fashion in shop design, giving the shoppers the impression that they are urban flâneurs in Paris, London, Milan or New York. As a rule, after a decade, with changing design fashions, the shopping arcades at these airport cities (or aerovilles) renew themselves. The life cycle of the airport shopping centers is even shorter than that of shopping centers in the city.

The previous selection of consumption locations and spaces in the city region is drawn mainly from a European perspective. The categories could be complemented by other types of consumption spaces reflecting cultural differences and traditions or different local socio-economic environments. Consumption types in Asia, Latin America or Africa may vary.

One additional consumption space is becoming increasingly important in the electronic age: *Virtual shopping.* Young people with limited budgets for consumption are developing new approaches to balance their consumption dreams with the reality of their bank accounts. They explore the shopping and evening event arenas of their cities to get a feeling of what is "in" and what they absolutely "must" have to be a part of the milieus to which they wish to belong. Equipped with such real world information, they do their shopping on the internet, to compare prices and save money. Time budgets play a big role in such consumer attitudes. Virtual shopping, however, adds an additional burden to real-world consumption spaces. Thus more and more, city centers and shopping precincts become nicely decorated information arenas and fairs, where fashionable goods are displayed to attract the eye, where coffee shops and open-air restaurants of spaces for communication and relaxation, while the act of buying takes place elsewhere, on the internet or in the cheaper out-of-town Big-Box and suburban shopping mall. This is one reason why the average turnover of shops in city centers is very high, because the margins between success and failure are indeed very small. In the end, the only such consumption spaces to survive are those that benefit from long owner-occupied traditions, or from being an outlet of a brand chain at an attractive consumption location in the city region.

Design dimensions of consumption

Buildings and locations for consumption follow the rules of the market. Consumption spaces in a city are in the hands of developers, investors and banks, as well as city governments, regulating the use of private and public space. As a rule, city governments and local administrations do not have much influence on the design of consumption spaces, once they have given public approval for city development concepts or accepted an investment project in principle. In conservation areas only, they can and do exert strong design control in order to protect the urban heritage and the native spirit of a city. The regulation of consumption locations and spaces is a key determinant in city building processes. While consumption habits and patterns are slowly converging across cultural traditions and boundaries, regulatory frameworks for consumption spaces still differ from country to country, with manifold implications on location, design and accessibility rules.

Public debates regarding the location and types of spaces of consumption in the city region mainly center around two discourses:

1. There is the discourse about the role of consumption for revitalizing the central city. This discourse is often linked to re-urbanization phenomena and to strategies of urban containment.
2. There is the debate on identity and image and on the role of city branding and the theming of consumption spaces, which some times includes a debate over international versus local styles in urban design.

Obviously these two discourse arenas in urban development processes are highly interrelated and have complex mutual implications. Though with different emphasis, the discourses vary depending on the socio-cultural traditions, economic conditions, political and administrative milieus, and academic discourses present in a particular region or country.

City renaissance, urban heritage, and consumption

While North America and Asia are theming their consumption spaces along European models to raise their attractiveness,

European cities are trying to preserve their cultural heritage, knowing that it is their most precious territorial capital that they have in times of globalization. Conflicts continuously arise in European cities, when developers in central cities are confronted with rigid heritage regulations. All over Europe, projects have been realized in central cities, which accommodate consumption functions in historic buildings or precincts. In most cases, however, only the decorative historical façade is kept to pretend a conservation ethos. Backyards or courtyards are covered with glass, and parking structures are deeply buried under the basements of existing historical buildings. Such examples can be found in many European cities, such as in the historical heart of Lille and other French cities, in Potsdam, Germany or in Glasgow, UK. A particularly controversial project has been the former residence of the Duke of Brunswick in Brunswick, Germany, which was demolished during World War II. An influential developer specializing in shopping centers (ECE) has restored and maintained the edifice and converted it into the front building of a huge shopping mall, notwithstanding considerable opposition from the media and architectural critics (Figure 30.2). Another prominent example of converting a historical building into a multi-function complex, including a shopping mall, is the Fiat factory in Lingotto, Turin in Italy, a historic building of high architectural value.

Such flagship developments go hand in hand with trends of re-urbanization and urban renaissance, with the rediscovery of the traditional city centers as consumption and entertainment spaces. This is paralleled by gentrification processes caused by changing demographic structures, new consumption values and new generations

Figure 30.2 Façade of war-demolished Schloss in Braunschweig. Source: ECE, Schloss-Arkaden, Braunschweig, Germany – used by permission.

of more cosmopolitan urbanites, and the notorious "creative class," coined by Richard Florida (Florida 2006). Young educated couples without children prefer to live in the central city, where they can combine housing with work, shopping, and entertainment. Simultaneously, local economies are changing from large industrial mass production to creative industries in small and medium enterprises, knowledge industries, and consumer services. Such enterprises prefer to locate in the built-up fabric of inner cities, where trendsetting creative milieus and specialized clusters evolve. The complex mix of urban activities (housing, working, shopping and entertainment) for citizens as well as for urban tourists supports such trends of urban renaissance. Such consumption-driven re-urbanization trends can be observed in many cities around the world, from Pittsburgh and Boston to Stockholm and Copenhagen. It is true for cities in North America, as for cities in Australia, Japan and Europe, though the reasons and conditions may vary from region to region.

It is urban competition and international benchmarking which cause cities to profile their cityscapes, to continuously update their appearance for the international real estate market, as well as for city marketing brochures, trendsetting media, and tourist brochures. The visual quality of a city, primarily its city centre, is essential for attracting talents, investors, developers and tourists. Promoting architectural icons designed by mainstream trendsetting architects is one approach to attract outside interest and strengthen local identity. This approach has been followed by the public sector, which commissions international flagship design for cultural consumption icons, such as museums and opera houses, as well as for fairs and airports, which represent the entry points to a city for business travelers and tourists. In contrast, the private sector typically has much less ambition when it comes to the design of consumption spaces in the city. However, there are exceptions in Europe, such as the Westside shopping center by Daniel Libeskind in Bern, or the eye-catching Selfridges department store in Birmingham, UK designed by Future Systems Architects.

In recent years, complex shopping malls have mushroomed at strategic locations in central cities, benefiting from the cityscape and ambience of the central business districts. Prominent examples can be found in Los Angeles (*The Grove*), San Diego (*Horton Plaza*), in Beijing (*The Place*) (Figure 30.3), and all over the European continent, with Zürich (*Sihl City*), Bern (*Westside*), Berlin (*Potsdamer Arcaden*), and Oberhausen (*Centro*) as particularly spectacular examples.

In all these consumption flagships, shopping is combined with food and entertainment. These new urban centers offer relaxation from daily stress in a secure environment for different (mostly well-to-do) social and age segments. The architectural scenery of such places reflects the Zeitgeist of urban design and the dreams of developers, designers, and marketing specialists. Their architectural styles range from modernist architecture to neo-traditional consumer landscapes, where functional requirements are hidden behind stylish decoration and impressive stage settings.

Experience, however, shows that the space in between such buildings is often neglected, even though it may be as important for attracting consumers as the building itself. Public piazzas and sidewalks in central cities (and not only there) are an essential element of successful consumption spaces in a city. This is why pedestrianized shopping streets appeared in European cities – the Ljinbanan in the war-demolished central city of Rotterdam claims to be the European prototype. The pedestrianized, car-free space in between the buildings raised the attractiveness of the shopping precinct considerably.

Figure 30.3 The Place: Chaoyang district, Beijing. Source: Klaus R. Kunzmann.

Figure 30.4 Qianmen Street, Beijing. Source: Wang Fang – used by permission.

The public space used as an urban stage was essential for the accomplishment of Kärntnerstrasse in Vienna or the unexpected success of Third Street Promenade in Santa Monica, California. The newly opened shopping street in the city of Beijing has been designed and developed following such models (Figure 30.4).

However, in Europe opinion about pedestrianized streets has changed in recent years. Shopping streets are preferred again, where streetcars or tramways (but not cars!) run through the previously pedestrian space to demonstrate urbanity.

Today, it is the mix of traditional architecture shaping the European city over centuries with new additions and adaptations attracting the eye and the camera, along with the steady flow of strolling and consuming citizens and tourists, which makes the urban spirit of a city. Thereby consumption is a key attraction factor in this urban mix; it keeps a city alive, vibrant, and globally attractive.

International versus local styles in design for consumption

When deciding on regulations for new urban projects and seeking opinions and support from local citizens, the controversy between international and local style dominates the debate. Politicians and citizens usually prefer the latter, while architects and their intellectual and academic environment opt for international style, expecting "signature" architectural solutions, beyond functional steel structures and curtain walls. It is a never-ending controversy, wherever mediocre and profit-oriented design appears in a city.

Two concerns, however, are usually dominating such controversies: the concerns for identity and image. In times of growing globalization and internationalization of life styles and products, identity has become a key concern. Maintaining the identity of a place is seen as a move to guard against the negative implications of globalization. When it comes to consumption, it is the local or regional product, together with the built and natural environment, which is seen as the asset assuring such identities.

Image is the other concern. Urban competition has made the external image of a city a key asset for urban success. To a great extent, a city owes its image to its iconic buildings and quality of public spaces at its center. Tourists (and tourist guides) judge a city accordingly (Judd and Fainstain 1999). International media play an influential role in the making of a city's image. Cosmopolitan journals covering fashion, architecture, design, food, traveling and real estate, communicate urban images to their readerships. Thereby urban image and consumption are greatly interlinked. *Shopping in ...* is a favorite feature of such journals and other international media. While applauding the local style, they contribute to the internationalization and disneyfication of consumption spaces.

In order to attract consumers, theming has become a key concept in design for consumption (Gottdiener 1997). The Disney Corporation has been one of the first to apply theming of sites and buildings as an essential principle. Disneylands around the world have been built by architects from the US headquarters of the Disney Corporation, drawing on design concepts that target the imagination of children and their parents and grandparents. The outcome is design clichés, which show little respect for local context. Themed hotels are added to the iconic Main Street, to demonstrate cosmopolitanism. Thus, Disneylands in Los Angeles, Orlando, Florida, Paris, Tokyo, Hong Kong, and soon Shanghai look quite similar and are not worth mentioning from an architectural perspective. Theming is also the axiom of the casino moguls in Las Vegas,

Figure 30.5 Venusfort, Tokyo. Source: Klaus R. Kunzmann.

who are pioneers in globalizing consumption spaces (Gottdiener *et al.* 1999). Their projects are targeting successful gamblers to spend their fortunes, or unsuccessful ones to console their losses by buying luxury goods or eating in designer restaurants. Simulacra of architectural highlights, such as the Venetian in Las Vegas, recently even replicated in Macao, attract consumers to huge shopping arcades, and are capturing gamblers, shoppers and curious *flâneurs* in their physically built virtual worlds (Kunzmann 2002, 2004). Strolling around a sterilized copy ofVenice, without smelling waste water and not being cheated by tourist guides, street vendors, and burglars, seems to be quite a successful recipe to raise consumption and profits. Both Disneyland and Las Vegas have inspired investors around the world, particularly in Asia. The Grove, a successful central city shopping center in Los Angeles for example, is designed by using bits and pieces of imagined European architecture. Similar projects can be found in Chinese and Japanese cities. The Italian style shopping mall Venusfort in Japan is one example (Figure 30.5).

One more aspect is essential when designing consumption spaces in central cities. Streets and open spaces around consumption places may host events, seasonal fiestas, demonstrations, performances, and public viewing. Only with such urban ephemera the quality of spaces becomes apparent. Such uses change with the seasons and with the quality of the events. The more careful these spaces are designed with regards to the needs of their users, the more the surrounding consumption spaces benefit from their proximity to them.

Conclusion

Regardless of the implications of globalization and the convergence of consumption values, consumption patterns in regional environments still differ considerably. Consequently urban development of consumption spaces may differ from country to country, along with associated local, regional, or national regulations governing the expansion of consumption spaces, their zoning, shopping times and accessibility. International institutions such as the European Commission and neoliberal promoters of free trade and single global markets aim to converge such traditions and issue regulatory frameworks to support the gradual adaptation of regional regulations to international standards, pretending they are impediments to economic development and wealth.

Consumption in the city is clearly more then shopping for daily or weekly needs.

> Cities are molded by consumption and consumption has been molded by cities. ... Consumption stands at the intersection of different spheres of everyday life - between the public and the private, the political and the personal, the individual and the social"; consumption is a means and motor of social change; an active ingredient in the construction of space and place; and central to our identities. Consumption has multiple political, economic, social and cultural roles, and that is in the morphology of cities that its expression is most explicit. (Jayne 2006, 214)

It is the consumption dimension that this chapter wanted to contribute to this book on urban design.

References

Braudel, F. (1986) *Die Dynamik des Kapitalismus*, Stuttgart: Klett Cotta.

Brewer, J. and Trentmann, F. (2006) *Consuming Cultures, Global Perspectives. Historical Trajectories, Transnational Exchanges*. Oxford: Berg.

Clarke, D.B. (2003) *The Consumer Society and the Postmodern City*. London: Routledge.

Clarke, D.B., Diel, M. and Hosiaux, K. (Eds.) (2003) *The Consumptiom Reader*. London: Routledge.

Corrigan, P. (1997) *The Sociology of Consumption*. London: Sage.

Florida, R. (2006) *Cities and the Creative Class*. London. Routledge.

Gottdiener, M. (1997) *The Theming of America: Dreams, Visions and Commercial Spaces*. Boulder, CO: Westview Press.

Gottdiener, M., Collins, C.C.D., and Dickens, D.R. (1999), *Las Vegas, The Social Production of an All-American City*. Oxford: Blackwell.

Jayne, M. (2006) *Cities and Consumption*. London: Routledge.

Judd, D.R. and Fainstain, S.S. (Eds.) (1999) *The Tourist City*. New Haven, CT: Yale University Press.

Kunzmann, K.R. (2002) "The Future of the European City: Qingdao, Celebration or Las Vegas?" in: Henning, T. (Ed.). *The Copenhagen Lectures: Future Cities*. Fonden Realnia: Copenhagen; 91–108.

—— (2004) "Venice, Venice, and Venice, Three Realities of the European City" in Koll-Schretzenmayr, M., Keiner, M. and Nussbaumer, M. (Eds.) (2004) *The Real and Virtual Worlds of Spatial Planning*. Berlin: Springer,

Sassatelli, R. (2007) *Consumer Culture, History, Theory and Politics*, London: Sage.

Sauter-Servaes, T. and Rammler, S. (2002) *Delaytainment an Flughäfen*. Berlin: WZB Working Papers.

Simmel, G. (1971) *On Individuality and Social Form Chicago*. Chicago University Press.

Slater, D. (1997) *Consumer Culture and Modernity*. Cambridge, UK: Polity Press.

Smith, N. (1996) *The New Urban Frontier. Gentrification and the Revanchist City*, London: Routledge.

Smith, N. and Williams, P. (1986) *Gentrification of the City*. Boston: Unwin Hyman

Sombart, W. (1967) *Luxury and Capitalism*. Ann Arbor, MI: University of Michigan.

Stadtbaukultur, N.R.W. (2003) *Shopping Center Stadt: Urbane Strategien für ine nachhaltige Entwiclkung*. Gelsenkirchen: Europäisches Haus der Stadtkultur.

Warde, A. (1997) *Consumption, Food and Taste*. London: Sage.

Further reading

Clarke, D.B. (2003) *The Consumer Society and the Postmodern City*, London: Routledge. Working through the controversial ideas of the consumer society's most influential theorists, Jean Baudrillard and Zygmunt Bauman, this book assesses the ways in which consumerism is reshaping the nature and meaning of the city, examining the nature of consumption and its increasing centrality to postmodern society.

Gottdiener, M. (1997) *The Theming of America: Dreams, Visions and Commercial Spaces*, Boulder, CO: Westview Press. An investigation into why the built environment is increasingly cluttered with shopping malls, theme parks, and fast food franchises. It contains a brief history of consumerism and marketing in the United States and embarks on a tour of the contemporary landscape of themed commercial environments.

Jayne, M. (2005) *Cities and Consumption,* London: Routledge. Providing a critical review of the ways in which urban development has been conceptualized, this book critiques urban regeneration initiatives, examines ordinary and spectacular consumption, and describes the relationship between consumption and development of the modern and postmodern city.

Sassatelli, R. (2007) *Consumer Culture, History, Theory and Politics.* London: Sage. An historically grounded and theoretically informed discussion of contemporary consumer culture as well as a critical understanding of its diversity, reach, and ambivalence. It is concerned with the practices, discourses, and institutions which have brought about and characterize consumer culture.

Sombart, W. (1907/1967, English translation) *Luxury and Capitalism*, Ann Arbor, MI: University of Michigan. A classic text about the role of luxury products in shaping cities, arguing that luxury drives the economy and generates urbanization, and at the same time, serves as one of the inherent limits behind the haunting economic crisis cycles of capitalist societies.

31

Cultural institutions

The role of urban design

Carl Grodach

Governments and private patrons have long sponsored cultural institutions – public or nonprofit organizations engaged in artistic, intellectual, and educational activities such as museums of art, culture, history, or science and performing arts organizations – to symbolize their wealth, civic commitment, and devotion to the fine arts. Today, however, as cities have become more focused on economies of culture, consumption, and entertainment, they are increasingly likely to commission a museum or performing arts center to demonstrate their global city status and spark economic revitalization. Simultaneously, cultural institutions have evolved from classically-inspired, inwardly-focused temples of art and history into more commercially viable, publicly accessible, and broadly appealing destinations. Cultural institutions have sought to reach out to new audiences both by attempting to represent the specific interests and experiences of diverse groups and by defining new, shared experiences that revolve around consumption and spectacle, all while retaining their cultural authority and role as tastemakers. Although the iconic architecture of some cultural institutions has tended to dominate the media spotlight (think of the Guggenheim Bilbao), urban design plays an important role in negotiating these competing objectives and in promoting the use of cultural institutions as an economic development and urban revitalization tool. Such tasks represent an important facet of urban design work, and are increasingly relevant as public life takes place less in traditional forms of public space – squares, plazas, promenades – and more in semi-private spaces of consumption and entertainment. As such, cultural institutions offer a rich site in which to study how urban design is used to adapt to changes emanating from the local and global levels and the associated effects on the use and meaning of urban space.

To this end, the chapter provides an overview of the defining and emerging trends in the design and planning of contemporary cultural institutions. The first section sets the stage through a brief review of the primary roles, characteristics, and urban design of early public museums and arts theaters. The following section examines how changes taking place inside and outside cultural institutions beginning in the 1960s altered their design and mandate. Finally, the discussion turns to the emergence of cultural institutions as sources of urban revitalization. Drawing on four representative examples, this section considers the role of urban design in each project. The conclusion summarizes the challenges urban designers face as they attempt to navigate the terrain between public and commercial culture in the context of contemporary cultural institutions.

Precedents

Public cultural institutions originated in Europe during the eighteenth century. A key example is the Louvre, which was formed during the French Revolution in a former palace to publicly display the royal art collection and create a defining symbol of national heritage and identity (Duncan 1995). Decades later, the first permanent public museums appeared in the US such as the Metropolitan Museum of Art and the Boston Museum of Fine Arts and, by the early twentieth century, virtually every major city in the US had constructed their own temple of culture often modeled on the Louvre, down to its classically inspired architecture (Steffensen-Bruce 1998). This cultural building boom was triggered by rapid urbanization, which produced a concentration of wealthy patrons and visitors and engendered a perceived need to educate the masses. Although created by affluent individuals rather than the state, like their European counterparts, these museums and performance theaters ultimately provided a social space dedicated to cultivating and demonstrating an appreciation of the fine arts that, in turn, reinforced class distinctions (Bennett 1995; Bourdieu 2007; DiMaggio 1982). Still, museums in particular functioned not simply as exclusive bastions of high culture, but also served as a space for the upper classes to inculcate in the country's growing working classes and immigrant populations an appreciation of the arts, social etiquette, and civic virtue (Bennett 1995; Steffensen-Bruce 1998). Like Fredrick Law Olmstead's vision for Central Park, many elites and social reformers viewed museums as a space that could counter the negative effects associated with industrialization and urbanization and improve the cultural life of US cities, though one largely defined by the bourgeoisie.

Architecture and urban design supported these roles. Intended as a refuge from the coarse realities of daily life in the industrial city, many early public cultural institutions were located in parks (Steffensen-Bruce 1998), and numerous European cities confined the fine arts to dedicated compounds such as the Museumminsel (Museum Island) in Berlin. Across the US, City Beautiful planners and architects in cities relied on fine arts institutions to reinforce their design objectives. Typically arranged with other public buildings at the terminus of a grand boulevard and surrounded by fountains, plazas and sculpture, cultural institutions served as monumental anchors of their grand plans dedicated to urban order and beauty (Wilson 1989). As in other public buildings of the time, architects employed a mix of neoclassical and Beaux-Arts styles to recall ancient, yet highly evolved civilizations and Enlightenment philosophies as well as to signal the aura of the art inside (Steffensen-Bruce 1998; Yanni 1999). These districts physically isolated the fine arts and culture from more lowbrow recreational activities and the surrounding city where most people lived and worked.

The segregation of high culture culminated in the 1950s and 1960s with projects like Lincoln Center in New York and subsequent imitators such as the Kennedy Center in Washington, DC and the Los Angeles County Performing Arts Center. Their urban design layout is one of fragmentation rather than linkage: the isolated cultural compound typically contains a collection of fine arts institutions that face an interior plaza turning their backs to the surrounding streets. Often, the site is elevated above street level with underground parking so that visitors can directly enter and exit the complex without coming into contact with the surrounding urban environment. This introverted design was as much in keeping with traditional notions of public cultural institutions as it was a response to the declining conditions of many central cities in the US. At the same time, Lincoln Center marked the beginning of a shift toward thinking about cultural institutions as a source of urban attraction and renewal. Robert Moses expressly conceived of the cultural complex as a catalyst for raising property values and attracting higher-end uses to the Upper West Side of Manhattan and, for the first time, demonstrated the potential of culture as a tool for generating wealth by altering the character of a place. Still, the isolated cultural enclave remains a fixture in many cities today and some new museums such as the Getty Center in Los Angeles, with its hilltop campus looking down on the city below, continue to follow this model.

Contemporary trends

As cities began to recognize cultural institutions as a source of urban renewal, their mandate, function, and physical form began to change. Although in some regards they remain houses for high culture for affluent audiences, since the 1960s cultural institutions have worked to alter their traditional mandate. For one, rather than solely focusing on arts and history from a classical European vantage point, museums and art centers have slowly responded to criticisms of elitism and unequal representation. Many have attempted to reflect the cultural expressions of different groups in their programs, engage a wider range of communities in dialog, and, at times, provide a staging ground to debate pressing local and global issues (Karp *et al.* 2007; Karp *et al.* 1992; Loukaitou-Sideris and Grodach 2004; Simpson 1996).

Public cultural institutions have sought to become more inclusive not only due to political demands, but also due to financial pressure. Keenly aware that their traditional patron base is ageing, they have sought ways to cultivate new patrons and appeal to an increasingly diverse society. A performing arts center today will often present jazz ensembles, modern dance, world music, and Broadway musicals alongside the symphony, opera, and ballet. In conjunction, cultural institutions have initiated more revenue-producing activities. Cafés, high-end restaurants, stores, and merchandising have all become standard components as have blockbuster exhibitions on subjects ranging from Vincent Van Gough to King Tut to Star Wars. Many offer an eclectic combination of activities from all-night art parties to family programming. At the same time, however, many continue to market themselves as a source of prestige and cultural capital to retain upscale patrons and attract corporate sponsorship (Wu 2002). Cultural institutions have become permeated with acknowledgments to the contributions of their corporate benefactors. In some instances, even the exhibitions have become commercial vehicles. Witness a recent show on the work of Takashi Murakami created by MOCA in Los Angeles, and now traveling the world, that features a functioning Louis Vuitton store in the center of the exhibition. Given the emphasis on corporate display and marketing, if cultural institutions are in fact public spaces, they have more in common with shopping malls or branded entertainment destinations than with New York's Central Park.

Perhaps the most widely noted change in contemporary cultural institutions has been the emphasis on iconic buildings designed by star architects (Evans 2003; Hamnett and Shoval 2003). In resolute contrast to the old palaces and neoclassical designs, architects have responded to the ambitions of contemporary cultural institutions with landmark structures that are intended to attract media attention and visitors from around the world. Frank Lloyd Wright's Guggenheim in New York and the Sydney Opera House are archetypes of today's cultural spectacles like the Guggenheim Bilbao and the Tate Modern in London. Moreover, in order to compete, cultural institutions established in earlier eras such as the Louvre and 1960s-era projects like Lincoln Center, discussed below, have undergone or are planning costly renovations and expansions to better integrate their cultural compounds with the surrounding city and make them more inviting to visitors.

While some view these changes as evidence of the increasing commodification of social life, others argue that they have enabled cultural institutions to serve as public spaces where many different people can hang out, socialize, and take part in a variety of activities (Gurian 2006). As Malcolm Miles (2005: 891) points out, the "Tate [Modern] has moved the cultural centre of London ... less by converting the city's diverse publics to modern art than by becoming a new social space, a place to meet, eat, buy books and be seen." Urban design plays a central role in this shift toward more open and convivial consumption destinations. Today's cultural institutions contain public spaces, grand entrances, public gardens, and amphitheaters not only to accommodate larger audiences, but also to extend the entertainment experience beyond the museum walls and attract a more diverse audience. Colorful banners advertising exhibitions and playful public art – a giant Jeff Koons puppy or a Claes Oldenberg clothespin – adorn large plazas filled with programmed activities such as live music or dance lessons. Indeed, all of these buildings and additions, at least outwardly, denote a shift away from the deference to elite culture and national heritage and toward a new openness rooted in consumer culture. Indeed, as the Murakami/Louis Vuitton exhibition and the turn to high-profile architecture suggest, cultural institutions have not necessarily turned away from their role in defining and reproducing symbolic capital as they have reframed this role around high style, design, and consumption. In turn, as they alter their prestige image, they have become reframed as a tool to reinvent the city rather than provide a fortress against it. I turn to this development in the remainder of the chapter.

Cultural institutions, urban revitalization, and public space

As cultural institutions have reorganized their mandate and function they have become more attractive to cities not so much to improve local access to the arts or educational opportunities as to provide a source of urban revitalization and branding (Hamnett and Shoval 2003; Strom 2002). Since the 1970s, as governments around the world have become more focused on consumption and entertainment as a route to reviving their dilapidated central city spaces, they have commissioned hundreds of major cultural projects with the intention of creating a cosmopolitan city image and attracting private sector investment. As a result, many cities have eschewed the development of isolated cultural compounds in favor of iconic flagship cultural buildings that are intended to anchor larger mixed-use redevelopment districts. Indeed, cities increasingly consider cultural institutions as ideal catalytic projects – buildings that incrementally generate and influence the character of new development (Attoe and Logan 1989). However, as some of the examples below demonstrate, with the emphasis on eye-catching architecture, project planners often fail to consider the wider urban design context within which the building is located. This oversight has not only generated concern for the role of urban design in facilitating catalytic development (Sternberg 2002), but has led some to question the concept of catalytic planning and to call attention to the wider socio-economic factors surrounding cultural facility planning (Grodach 2008).

Much of the urban design literature that addresses culture-led redevelopment draws on established theory like Kevin Lynch's (1960) concept of legibility and Jane Jacobs' (1961) conditions for city diversity (mixed use spaces, diverse building types, small blocks, and density). McCarthy (2006), van Aalst and Boogaarts (2002), and Wansborough and Mageean (2000), for example, emphasize the need for producing a distinct place identity, actively used public spaces, and locating cultural institutions among diverse land uses and buildings. Some propose standards for the size, height, and density of cultural districts and the buildings and streets within them (Montgomery 2003). Other work focuses specifically on the cultural facility. Sternberg (2002), for instance, proposes guidelines for "cultural catalysts," emphasizing the importance of a location among existing commercial spaces, a pleasant pedestrian environment that is linked to these commercial sites, and good transportation access. However, while attention to urban design as an economic development strategy has assisted some cultural projects to remake struggling places into lively destinations, as the following examples show, the costs associated with creating a space oriented toward tourism, consumption, and gentrification are often not part of the planning process. Further, despite the problems associated with the traditional cultural compounds and campus settings, due to the interest in creating a defined context for tourism and investment, many cultural institutions continue to be designed in isolation, disengaged from the city.

Centre Pompidou

One of the earliest attempts to develop a cultural catalyst is the Centre Pompidou, which opened in 1977 in response to the French government's mandate to "democratize" culture. In this regard, the cultural center is significant for merging the purported goal of enhanced public participation in the arts and culture with the goal of urban revitalization. To this end, rather than emulating the architectural styles of past cultural palaces, the architects Richard Rogers and Renzo Piano designed a "flexible container" that could accommodate multiple opportunities to experience art, education, and entertainment and in which "there is no obvious hierarchy which separates art and learning from more everyday activities" (Richard Rogers Partnership 2008). The primary role of urban design, therefore, was not to create a walled-off cultural campus, but to bring new life to the adjoining Beaubourg and Les Halles neighborhoods. This is accomplished through a vast public plaza, which attracts people to chat, people-watch, and listen to street performers (Figure 31.1). In addition, Pompidou is located near two metro stops, surrounded by cafés, restaurants, and shops, and within relatively easy walking distance of other Parisian icons like the Louvre and Notre Dame. In these ways, Pompidou meets Jacob's (1961) conditions for city diversity and Sternberg's (2002) qualifications for a successful cultural catalyst.

By most accounts, the cultural center's appealing public space, spectacular architecture, and mix of activities has been a huge draw, attracting an astounding five million visitors annually (Centre Pompidou 2006) as well as sparking major

Figure 31.1 Centre Pompidou, Paris. Source: Studio Piano and Rogers, Architects – used by permission.

private investment in the area. However, critics charge that the success of this *grand projet* has merely catalyzed gentrification and that the project was only possible through the extensive capital investment of federal and Parisian authorities. Further, rather than democratizing the restricted realm of high culture, studies have found that most visitors are highly educated, young professionals or foreign visitors who prefer to frequent the art center's plaza and tour the building rather than spend time in the galleries and research center (Heinich 1988). In effect, while Pompidou has produced a lively urban space and catalyzed commercial development, it has not necessarily built a truly public realm, nor has it expanded the public's appreciation of fine art. Rather, urban design and culture have been more successful at remaking urban space for the enjoyment of tourists and middle class visitors and not necessarily for revitalization that incorporates existing residents or the public at large.

Guggenheim Bilbao

Whereas the Pompidou was charged with remaking an urban district, Spanish authorities in Bilbao envisioned a museum that would put their city on the global cultural map. Opening in 1997, the Guggenheim Bilbao, like the Pompidou, emphasized innovative architecture by a star architect, and was largely financed by public sources expressly for urban revitalization. In this regard, it too has been a huge success attracting nearly one million visitors annually (Plaza 2006) and generating a purported $1.36 billion since its opening (Guggenheim Museum Bilbao 2005). Frank Gehry's design is widely regarded as garnering the attention of the popular media and luring cultural tourists from around the world even more so than the world-renowned Guggenheim collection on display. However, while the Guggenheim Bilbao has effectively functioned as a beacon to private investment and global tourism, this has come at the price of gentrification and displacement in other parts of the city (Vicario and Martinez Monje 2003). Further, critics charge that, with its imported art and architecture, the museum is detached from its locale, doing little to connect with residents and local artists or reflect Basque art, history, and culture (Ceballos 2004; Guasch 2005).

While many point to the museum building and programming as evidence of these circumstances, urban design plays an important supporting role in defining this global cultural space. The Guggenheim anchors one end of the nearly 350,000 square meter (85 acre) Abandoibarra redevelopment area, a former shipyard that fronts the Nervión River in the city center. The quasi-public redevelopment authority, Bilbao Ría 2000, conceived and implemented the development of this mixed-use cultural and business district based on a master plan by César Pelli, Diana Balmori, and Eugenio Aguinaga to "enhance the competitiveness of the city" and "attract international investments" (Bilbao Ría 2000; Rodriguez *et al.* 2001: 173). The museum itself is surrounded by park space and pedestrian paths that stretch along the river and provide excellent views of the green hills that surround the city. These features and three tram stops are intended to link the Guggenheim with other high-profile buildings designed by well-known architects in the project area including a Sheraton Hotel (by Ricardo Legorreta), Zubiarte shopping center (by Robert Stern), the Deusto Library (by Raphael Moneo), and the Euskalduna Conference and Concert Hall located at the other end of the redevelopment area.

While the project design creates a distinct, coherent identity and pedestrian environment within the Abandoibarra, it also serves to physically isolate the museum within a portion of the larger redevelopment area as well as distinguish the project itself from the surrounding city. The Guggenheim is hemmed in on two sides by the river and a raised street, while a large open space separates the museum from the rest of the project (Figure 31.2). Although the museum borders the city streets outside the project boundaries, the orientation of the project contrasts starkly the existing street grid and, therefore, forms more of a contrast between the redevelopment area and the surrounding urban fabric than integrating the new and the old. The resulting detachment from the city is reinforced by the fact that, other than the museum's own restaurant and shop and an immediately adjacent tapas bar, there are currently no commercial or residential spaces within the project let alone within eyesight of the main entrance. Although residential buildings and an office tower (to be designed by Pelli) are planned for the area, the primary commercial spaces in the Abandoibarra are and will be contained within the Zubiarte shopping mall. Further, it is likely that the completion of these features will only reinforce the identity of the project as a self-contained and staged public space aimed at global tourists and the business class. While the Guggenheim Bilbao is the most iconic emblem of the "new Bilbao," the overall project design reinforces this image and emphasizes the distinction from rather then the engagement with the rest of the city (Bilbao Ría 2000).

The Harley-Davidson Museum

Although not as well-known as the preceding examples, the Harley-Davidson Museum (HDM) is emblematic of many contemporary flagship cultural institutions aspiring to catalyze urban revitalization. HDM forms a large-scale complex (130,000 square feet) that includes galleries, archives, office, restaurant, café, retail, and special event space and is located on a 20-acre former industrial site along a river just outside downtown Milwaukee.

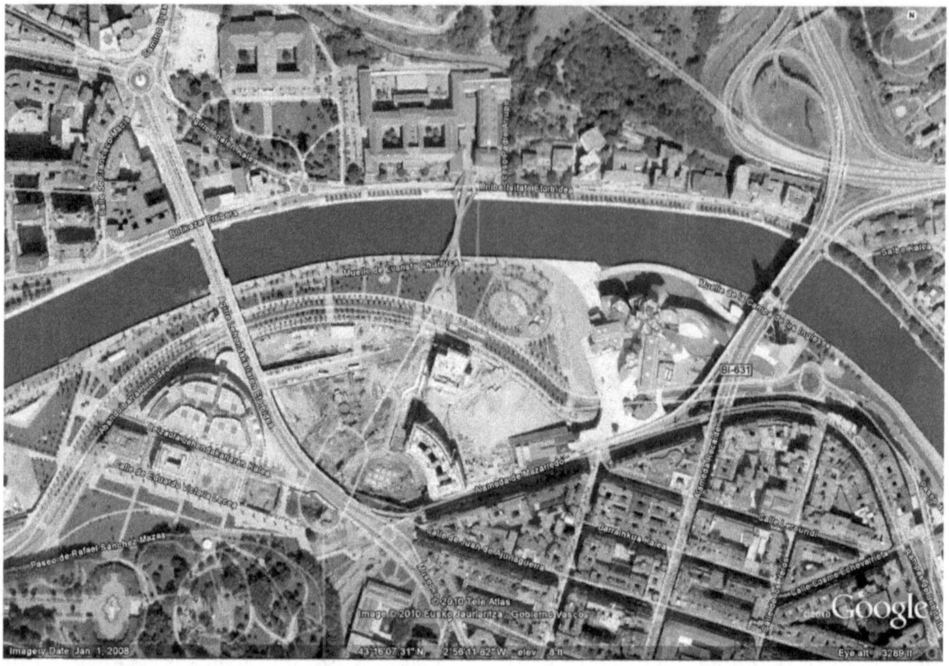

Figure 31.2 Abandoibarra redevelopment area, Bilbao. Source: www.maps.google.com
Note: Guggenheim Bilbao at left midground.

The museum, which opened in 2008 in the city of the company's founding, is described by the mayor as "a catalyst for the entire city" (Goldberg 2006). For the company, the museum is intended to be the physical manifestation of its well-known brand and a pilgrimage site for the legions of devotees to the Harley motorcycle. Just as the exhibitions and architecture tell the story of Harley-Davidson and its role in spawning motorcycle culture, urban design is an integral dimension in reproducing the Harley "brand community" – a group of people that affiliate with and bond through their shared interest, rituals, and traditions rooted in a product, which in turn serves to build customer loyalty (Fourier *et al.* 2000). As such, the museum's role as a source of product and city branding merges with its function as a social meeting ground.

What is interesting about this museum is that it attempts to accomplish these goals less through architectural spectacle than through urban design. A key inspiration for the firm responsible for the design of the museum and its grounds, Pentagram Design Studio, was the bike rallies that take place in cities and small towns around the world where often thousands of bikers meet, inspect each other's motorcycles, and socialize in a concentrated area. "These rallies are such an essential part of the Harley-Davidson experience that we felt it was essential to create a place that captured their spirit, but where those who are new to Harley-Davidson would feel welcome" (Pentagram Design Studio 2008). To this end, the designers sought to create an "urban experience" and an area that "felt like a neighborhood within the city" (Pentagram Design Studio 2008). Rather than designing one large facility that opens onto a large plaza, they divided the museum into three smaller buildings arranged along a set of "streets" that continue the existing street grid through the site and form a central public space between the museum buildings, which they refer to as a "hot spot" (Figure 31.3). The streets are intended to connect the museum to the surrounding city and provide a context for social interaction. In these ways, urban design assists an image-conscious corporation in tying consumers to their product and forms a social space for both Harley community members and the general public.

However, extending the social interaction and the associated purchasing power of museum visitors beyond the site will likely be a challenge given its isolation. A major issue is that the extension of the street grid through the site is largely figurative because the museum is surrounded by water on three sides, and only one street actually crosses the river to downtown and the gentrifying Historic Walker's Point area nearby. Not only is street access poor, but the area is bounded by major freeways, which hinder the already weak pedestrian connectivity. While project designers have worked to integrate the museum into the city fabric, site restrictions place the HDM in a rather self-contained milieu, albeit one that speaks the new language of mixed-use, public space and sociability rather than the seclusion of a cultural compound. For the city of Milwaukee and Harley-Davidson, this may matter little if the museum carves out a new niche destination and catalyzes wider development as the Guggenheim did for Bilbao. The case of the Harley-Davidson museum underscores a problem inherent to most catalytic projects – that contemporary urban design, no matter how publicly-minded, is crafted within a distinct and bounded space rather than within a more comprehensive plan and so winds up serving largely as a means of reinforcing distinction for its sponsor, in this case, a global corporation.

Lincoln Center redevelopment

Lincoln Center established the concept of the post World War II cultural citadel intended to lure suburbanites to the city center. Nearly 50 years later, the 16-acre cultural complex is engaged in a $1.2 billion overhaul intended to address its inward-focused design and respond to "changes in the needs and interests of the public" (Lincoln Center for the Performing Arts 2008). As the architects of the redevelopment plan, Diller Scofidio + Renfro, state "the challenge is to interpret

Figure 31.3 Harley-Davidson Museum site plan. Source: Craig Wilson, Kite Aerial Photography – used by permission.
Note: The image can be accessed at http://mavspace.uta.edu:443/grodach/Urban Design Images.

the genetic code of this 'Monumental Modernism' into a language for younger, more diverse audiences following several generations of cultural and political change. We would like to turn the campus inside-out by extending the intensity within the performance halls into the mute public spaces between those halls and the surrounding streets" (Diller Scofidio + Renfro 2004).

A central focus is on West Sixty-fifth Street, which bisects the campus and is lined by seven of the resident organizations (Figure 31.4). Here the architects will reduce the street width, enlarge sidewalks, redesign the existing blank street wall of cultural facilities, and incorporate a new retail store. The recently completed first phase entailed the transformation of Alice Tully Hall into an iconic gathering space. The central focus of the renovation and 100,000 square foot expansion is the enlarged lobby and bar area framed by a three-story glass entrance. Hanging above the lobby and visible from the street are two cantilevered spaces: a Julliard dance studio and the "donor's gallery," a space reserved for major patrons. Outside, the "grandstand," a raised, wedge-shaped seating area allows passersby to sit and view activity in the lobby, donor's gallery, and dance studio (Figure 31.4). Facing the theater across Sixty-fifth Street will be a sloping grassy public space, which forms the roof of a new signature restaurant. The redevelopment plan also calls for opening Lincoln Center's classically-ordered central spaces by expanding the primary entrance and main plaza on Columbus Avenue, placing a grove of trees off the main plaza, and creating a new promenade by dropping the existing service street below the plaza. Free Wi-Fi access and various information displays, including a scrolling LED text embedded into the stairways at the project entrances, will be located throughout the complex. Finally, a new visitor center is planned across from the main campus, which the Center describes as a "hub of civic and cultural activity" and "a portal to the artistic, cultural and daily life of the West Side" (Lincoln Center 2008).

With its contradictory blend of populism, elitism and consumerism and an emphasis on merging high design and technology with a formal public space, the Lincoln Center redevelopment may well be the ideal embodiment of the changing image of cultural institutions as urban entertainment destinations and social spaces. The classical public space is reinforced through the promenade yet this is filled with visual stimuli and information suggesting a frenetic urbanism and, simultaneously, is accompanied by smaller public spaces meant to encourage more casual, intimate use. There is a new high-end restaurant but it is capped off by a public lawn and the redesign of Tully Hall creates a lively set of spaces that cleverly turns people-watching into a spectator sport. However, this arrangement perhaps unintentionally recalls historical contradictions inherent to cultural institutions. The Tully Hall "grandstand," which sets up a view of theater-goers mingling in the lobby, Julliard performers, and the donors perched above in their private space is like peering into a lavish shop window, though it is high design and consumption rather than high culture that is on display and just out of reach. Perhaps most significant, Lincoln Center's walls are only brought down now, 50 years after it was established to gentrify Manhattan's Upper West Side. By speaking only to general concepts of publicness and not to this fact, the new design may contribute to further erasing the area's history and risks producing metaphors of openness and transparency without necessarily diversifying Lincoln Center's audience or recognizing the undemocratic foundation on which it is built.

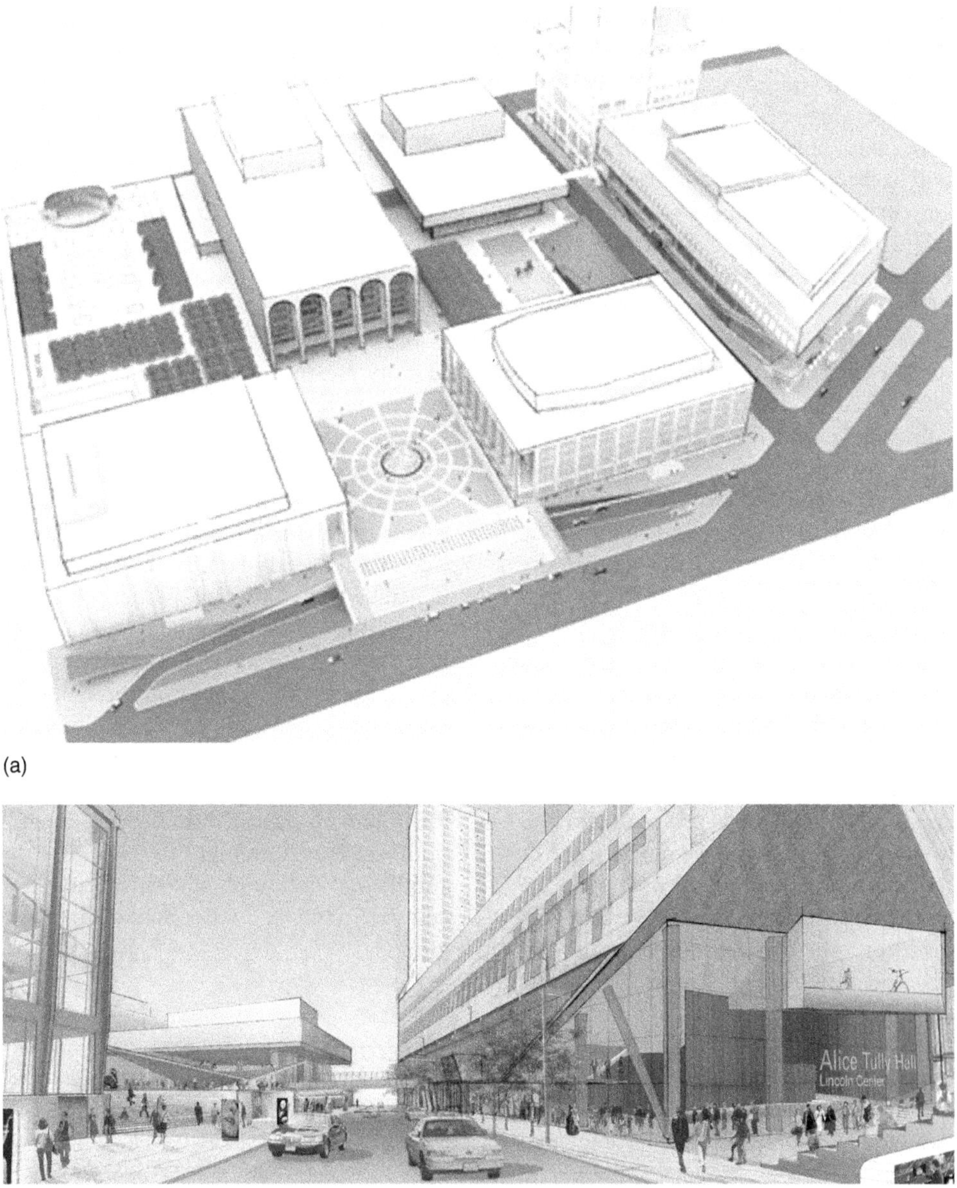

(a)

(b)

Figure 31.4 Lincoln Center Redevelopment, 65th Street Panorama. Source: Diller Scofido + Renfro in collaboration with FXFOWLE Architects – used by permission.

Conclusion

While some have proclaimed that the global recession portends "the end of the Bilbao decade," if history is any indication, the drive to produce iconic cultural destinations will not become a thing of the past (Campbell 2009). Rather, because it is unlikely that new sources of public funding will enable cultural institutions to reduce their dependence on the revenue they acquire as tourist attractions, they will likely retain and possibly even expand the emphasis on large, flashy architecture and consumer experiences. For their part, cities will likely continue to seek imageable projects in the competition for tax revenues and global capital. As both the Harley-Davidson museum and the redevelopment of Lincoln Center suggest, the design of future cultural institutions may change to some extent, but this does not mean the end of cultural institutions as we have come to know them. In fact, public cultural institutions have long adapted to changing times by adding new layers to their mandate rather than totally shifting directions. Today most pursue multiple, often contradictory functions: they seek to represent diverse histories and cultures and cultivate new audiences while serving an elite patron base; they provide an entertainment and consumer experience while educating visitors; and they attract economic activity while focusing on "art for arts sake."The financial pressures and newfound populism that have engendered this reorganization have in turn repositioned cultural institutions as an attractive mechanism for urban redevelopment. Although cultural institutions today seek to portray an image of openness and approachability, they continue to draw their power from their distinction from everyday life.

Urban design today, as historically, has largely supported this arrangement. Whereas in the past design assisted in reinforcing the separation of high culture from daily life, more recently it has become a tool to carve out a space that is distinct from the locale and geared toward the attraction of global capital, tourism, and consumption as the Guggenheim Bilbao aptly illustrates. Although, as Gurian (2006) asserts, cultural institutions do indeed provide a space for social interaction and mixing, it is rarely the sort of integrated, central public space that reflects the diverse city. Even when they purportedly pursue and achieve this feat, the potential for gentrification is great, as has occurred around the Centre Pompidou. Although the designers of the Harley-Davidson Museum have possibly created a new model for cultural institutions in the "post-Bilbao" era, the impetus is to provide an attractive, urban context for corporate rather than civic aspirations and a public whose shared experiences and identity are rooted in consumption and a corporate brand. Similarly, as the Lincoln Center redevelopment shows, designing more integrated and accessible spaces can entail blurring claims to democratize culture with objectives related to consumption and image. In fact, each of these cases illustrates the difficulty of designing for competing objectives while undoing the historical segregation of culture within the city. Particularly because cultural institutions have become emblematic public spaces, urban designers must experiment with new ways of creating opportunities for public engagement and social interaction as they attempt to achieve the conflicting ambitions of tourist attraction, cultural temple, and public meeting grounds.

References

Attoe, W. and Logan, D. (1989). *American Urban Architecture: Catalysts in the Design of Cities,* Berkeley, CA: University of California Press.

Bennett, T. (1995). *The Birth of the Museum,* London: Routledge.

Bilbao Ría (2000). <http://www.bilbaoria2000.org/ria2000/ing/zonas/zonas.aspx?idZona = 1>. (accessed 27 December 2008).

Bourdieu, P. (2007). *Distinction: A Social Critique of the Judgment of Taste.* Cambridge, MA: Harvard University Press.

Campbell, R. (2009). "Marking the End of 'The Bilbao Decade': Times Dictate a Shift Away from Vanity Projects," *Boston Globe* 11 January. <http://www.boston.com/ae/theater_arts/

articles/2009/01/11/marking_the_end_of_the_bilbao_decade/?page = full> (accessed 17 January 2009).

Ceballos, S. (2004). "The Role of the Guggenheim Museum in the Development of Urban Entrepreneurial Practices in Bilbao," *International Journal of Iberian Studies*, 16(3): 177–186.

Centre Pompidou. (2006). *Centre Pompidou Activity Report.*

Diller Scofidio + Renfro, Fox and Fowle Architects, Street of the Arts, Lincoln Center, New York (2004) <http://www.arcspace.com/architects/DillerScofidio/lincoln/index.htm> (accessed 11 May 2009).

DiMaggio, P. (1982). "Cultural Entrepreneurship in Nineteenth-century Boston: The Creation of an Organizational Base for High Culture in America," *Media Culture and Society*, 4: 33–50.

Duncan, C. (1995). *Civilizing Rituals: Inside Public Art Museums*, London: Routledge.

Evans, G. (2003). "Hard Branding the Cultural City-From Prado to Prada," *International Journal of Urban and Regional Research*, (27)2: 417–430.

Fournier, S., McAlexander, J., Schouten, J. and Sensiper, S. (2000). *Building Brand Community on the Harley-Davidson Posse Ride*, Cambridge, MA: Harvard Business School.

Goldberg, R. (2006). "Harley Museum to Include Three Buildings, 'Urban' Look," *Business Journal of Milwaukee* <http://milwaukee.bizjournals.com/milwaukee/stories/2006/02/20/daily42.html.> (accessed 27 December 2008).

Grodach, C. (2008). "Museums as Urban Catalysts: The Role of Urban Design in Flagship Cultural Development," *Journal of Urban Design*, 13(2): 195–212.

Guasch, A. M. (2005). "Global Museums versus Local Artists: Paradoxes of Identity between Local and Global Understanding." In Guasch, A.M. and Zulaika, J. (Eds.), *Learning from the Bilbao Guggenheim*, Reno, NV: Center for Basque Studies, University of Nevada.

Guggenheim Museum Bilbao (2005). *Guggenheim Museum Bilbao Annual Report.* <http://www.guggenheim.org/annual_report/2005/GMB.pdf> (accessed 31 July 2007).

Gurian, E. H. (2006). *Civilizing the Museum: The Collected Writings of Elaine Heumann Gurian*, London: Routledge.

Hamnett, C. and Shoval, N. (2003). "Museums as Flagships of Urban Development." In Hoffman, L., Fainstein, S. and Judd, D. (Eds.), *Cities and Visitors: Regulating People, Markets, and City Space*, Malden, MA and Oxford: Blackwell Publishing.

Heinich, N. (1988). "Pompidou Center and its Public: The Limits of a Utopian Site." In Lumley, R. (Ed.), *The Museum Time Machine: Putting Cultures on Display*, London: Routledge.

Jacobs, J. (1961). *The Death and Life of Great American Cities*, New York: Vintage.

Karp, I., Kreamer, C. and Lavine, S. (Eds.) (1992). *Museums and Communities: The Politics of Public Culture*, Washington, DC: Smithsonian Institution Press.

Karp, I., Kratz, C. and Szwaja, L. (Eds.) (2007). *Museum Frictions: Public Cultures/Global Transformations*, Durham, NC: Duke University Press.

Lincoln Center for the Performing Arts (2008). *Lincoln Center Unveils Dynamic Design For Vibrant Community Space* By Tod Williams Billie Tsien Architects <http://www.lincolncenter.org/press–release/PR_Visitor CenterDesign_MicroPark_7-16-08_FINAL.pdf> (accessed 11 May 2009).

Lincoln Center for the Performing Arts (2008). <http://www.lincolncenter.org/load_screen.asp? screen = transforming> (accessed 27 December 2008).

Loukaitou-Sideris, A. and Grodach, C. (2004). "Displaying and Celebrating the 'Other': A Study of the Mission, Scope, and Roles of Ethnic Museums in Los Angeles, *The Public Historian* 26(4): 49–71.

Lynch, K. (1960). *The Image of the City*. Cambridge, MA: MIT Press.

McCarthy, J. (2006). "Regeneration of Cultural Quarters: Public Art for Place Image or Place Identity?" *Journal of Urban Design*, 11(2): 243–262.

Miles, M. (2005). "Interruptions: Testing the Rhetoric of Culturally-led Urban Development," *Urban Studies*, 42(5/6): 889–911.

Montgomery, J. (2003). "Cultural Quarters as Mechanisms for Urban Regeneration. Part 1: Conceptualizing Cultural Quarters," *Planning Practice and Research*, 18(4): 293–306.

Pentagram Design Studio (2008). <http://blog.pentagram.com/2008/07/new_work-harleydavidson-museum-2.php#more> (accessed 22 December 2008).

Plaza, B. (2006). "The Return on Investment of the Guggenheim Museum Bilbao," *International Journal of Urban and Regional Research*, 30(2): 452–467.

Rodriguez, A., Martinez, E. and Guenaga, G. (2001). "Uneven Redevelopment: New Urban Policies and Sociospatial Fragmentation in Metropolitan Bilbao," *European Urban and Regional Studies*, 8(2): 161–178.

Richard Rogers Partnership (2008). <http://www.richardrogers.co.uk/render.aspx?siteID = 1&nav IDs = 1,4,22,818> (accessed 22 December 2008).

Simpson, M. (1996). *Making Representations: Museums in the Post-Colonial Era,* London: Routledge.

Steffensen-Bruce, I. (1998). *Marble Palaces, Temples of Art: Art Museums, Architecture, and American Culture 1890–1930*, London: Associated University Press.

Sternberg, E. (2002). "What Makes Buildings Catalytic? How Cultural Facilities Can Be Designed to Spur Surrounding Development," *Journal of Architectural and Planning Research* 19(1): 30–43.

Strom, E. (2002). "From Pork to Porcelain: Cultural Institutions and Downtown Development," *Urban Affairs Review*, 38(1): 3–21.

Van Aalst, I. and Boogaarts, I. (2002). "From Museum to Mass Entertainment: The Evolution of the Role of Museums in Cities, *European Urban and Regional Studies*, 9(3): 195–209.

Vicario, L. and Martinez Monje, P. (2003). "Another 'Guggenheim Effect'? The Generation of a Potentially Gentrifiable Neighborhood in Bilbao," *Urban Studies*, 40(12): 2383–2400.

Wansborough, M. and Mageean, A. (2000). "The Role of Urban Design in Cultural Regeneration," *Journal of Urban Design*, 5(2): 181–197.

Wilson, W. H. (1989). *The City Beautiful Movement*, Baltimore, MD: Johns Hopkins University Press.

Wu, C-T. (2002). *Privatizing Culture: Corporate Art Intervention since the 1980s*, London and New York: Verso.

Yanni, C. (1999). *Nature's Museums*, London: Athlone Press.

Further reading

Bennett, T. (1995). *The Birth of the Museum*, London: Routledge. Thorough study of the origins, evolution, and functions of museums.

Evans, G. (2003). "Hard Branding the Cultural City-From Prado to Prada," *International Journal of Urban and Regional Research*, (27): 2: 417–430. Discussion of urban branding and marketing through flagship cultural projects and architecture.

Grodach, C. (2008). "Museums as Urban Catalysts: The Role of Urban Design in Flagship Cultural Development," *Journal of Urban Design* 13(2): 195–212. Analysis of the role of urban design in culture-led redevelopment projects.

Karp, I., Kratz, C. and Szwaja, L. (Eds.) (2007). *Museum Frictions: Public Cultures/Global Transformations,* Durham, NC: Duke University Press. Third volume of work on politics of museum display and community interaction.

Steffensen-Bruce, I. (1998). *Marble Palaces, Temples of Art: Art Museums, Architecture, and American Culture 1890–1930*, London: Associated University Press. Comprehensive history of early museum architecture and urban planning.

Sternberg, E. (2002). "What Makes Buildings Catalytic? How Cultural Facilities Can Be Designed to Spur Surrounding Development," *Journal of Architectural and Planning Research* 19(1): 30–43. Offers guidelines for cultural facilities to catalyze development.

Strom, E. (2002). "From Pork to Porcelain: Cultural Institutions and Downtown Development," *Urban Affairs Review*, 38(1): 3–21. Explanation of the rise of cultural institutions as tools of urban revitalization.

32

Streets and the public realm

Emerging designs

Elizabeth Macdonald

It is a pretty good guess that when people think of their city's public realm, they mostly think of parks and public buildings. And yet, in almost any North American city, the amount of land occupied by public parks, public buildings, and public squares does not equal the amount of land occupied by public rights-of-way. Streets typically represent 25 to 35 percent of all developed land in American cities. Overwhelmingly, that space is owned and controlled by the city, and more to the point, the public. The public can and does determine how the immense amount of space in public rights-of-way shall be used, and how it is designed. Of course, "the public" is not a homogenous entity. Who can and does participate in the public decision-making regarding street design has during the last century come to be very limited. All too often the missions, goals, plans, professional norms and street design standards promulgated by bureaucratic institutions, such as state transportation agencies and city planning or public works departments, trump the wishes of local community members, especially those in poorer neighborhoods who do not have the power to fight them. Nonetheless, the public has the ultimate say over publicly owned lands, and through political processes can influence what decisions are made regarding street design.

In part, the high percentage of urban land is in streets because of the network nature of public ways, and hence their ubiquity. But it is also because North American streets tend to be quite wide by world standards; 60 feet, 80 feet, 100 feet, and even 125 feet are common widths for residential and commercial streets alike, with some streets up to 200 feet wide or more. These street widths stem from the nineteenth-century desire to ensure light and air to dwellings and workplaces of the industrial city, and were further reinforced by the twentieth-century need to accommodate more and more automobiles.

For many older urban areas, city founders were largely responsible for street layouts and patterns, street sizes and, in some cases, even detailed street designs. This was certainly the case in Savannah, Georgia where in 1733 James Oglethorpe laid out one of the most memorable street and block patterns in a US city. Frequently, surveyors were the street layout professionals. Later, however, it was transportation engineers who determined street designs. Today, city governments have opportunities to redesign the cross-sectional configurations of urban streets, though seldom to change their overall patterns in major ways. For new urban areas, local governments generally leave the planning of street patterns and individual street designs to land developers, who most often use bureaucratically generated street patterns and design standards to determine the shape and form of the future public realm of streets.

What follows in this chapter is an exploration of the multiple dimensions of streets and street design relative to the public realm. The exploration begins with a discussion of the limited purposes currently assigned to streets, specifically the focus on accommodating motorized vehicles. This is followed by discussion of the opportunities inherent in streets and innovative approaches to their design. Next, the discussion shifts to the major challenges faced by designers and communities interested in implementing innovative street designs. In conclusion, the issue of values is raised as a central concern.

The status quo: a limited purpose public realm

An abundance of space notwithstanding, the public realm of street rights-of-way is, for the most part, designed and used primarily for very limited purposes. Think about a typical American urban street: two narrow concrete walkways edging a relatively wide asphalt roadway. The sidewalks may or may not be lined with trees. Where they exist, trees are usually widely spaced, not very big, and confined to small patches of open ground. Sidewalk edges are very often cluttered with signs and signals to direct drivers, and contain street lights directed toward the vehicle paths. The roadway itself is channelized into vehicle travel and parking lanes by painted pavement markings. Lately, on a few streets, there are also painted lanes for bicycle movement.

American streets have been engineered mostly for movement. But we know that movement is only a part of what the public realm and public rights-of-way are about. As has been said elsewhere, streets. "... moderate the form and structure and comfort of urban communities. Their sizes and arrangements afford or deny light and shade ... they may have the effect of focusing attention and activities on one or many centers" (Jacobs 1993: 3–4). Streets allow people to be outside. They are places of social and commercial encounter and exchange. They are places where you meet people, where kids create games and play. They are political spaces. They are places for light and air. And they can or should be well-functioning ecological spaces. And yet, most American streets have not been designed to address, let alone embrace, non-movement functions.

In fact, even the movement functions of streets are by and large narrowly conceptualized. Generally speaking, since the advent of automobiles and their widespread private ownership by the 1930s, and with the corresponding ascendance of traffic engineers as the designers and keepers of public streets (taking over a role held by landscape architects during the late 1800s and early 1900s), there has been an accumulation of street design standards and professional norms that privilege the easy movement of motorized vehicles. They recommend geometric configurations of roadways that supposedly increase auto and truck safety, such as wide travel lanes, but in reality do not necessarily achieve it, as will be discussed later.

Urban streets in the US have to a great extent become single-purpose spaces, and this is the result of an engineering construct called the Functional Classification of Streets. This system separates streets into different types according to the motorized vehicle movement and property access functions they are supposed to perform (ITE 1992; AASHTO 2004). Basically, the method assigns specific movement and access functions to each of five street types: freeways, expressways, arterials, collector streets, and local streets. Within each type, the functions are inversely correlated; that is, the higher the movement function, the lower the access function. Traffic engineers have developed street design standards for each functional classification based on criteria deemed appropriate for each. The criteria include access control, design speed, design volume, level of service, and highway capacity. The choice of criteria and the way in which they are applied to the classifications reflect an emphasis on vehicle flow, an orientation toward eliminating potential movement conflicts, and a basic philosophy of not questioning the preeminent needs of vehicular traffic. For instance, the Institute of Transortation Engineers (ITE) *Traffic Engineering Handbook* recommends that some limitations on access be applied to all street classifications because of concerns over "the quality and safety of traffic flow." In addition, the criteria for design speed, traffic volume, level of service, and capacity, all recommend erring on the side of excess for all classifications – essentially over designing streets in terms of lane widths and capacity in order to ease vehicle movement and accommodate possible future traffic demands. This has resulted in public rights-of-way which have an unbalanced distribution of space and are made up of rather alarming expanses of roadway asphalt.

Consider a typical 100-foot wide arterial street carrying four lanes of vehicle traffic, two in each direction, a dedicated left turn lane, and two lanes of parking. It is very typical for the travel lanes and the left turn lane to be 12 feet wide, and the parking lanes to be 10 feet wide. These dimensions result in a total roadway width of 80 feet, leaving just 20 feet to be divided between the two sidewalks. So, 80 percent of the right-of-way is devoted to cars and trucks and only 20 percent remains for other uses.

Opportunities for streets as public space

"Times," though, as the song goes "they are a changing." New generations of city dwellers, urban designers, transportation planners, environmentalists, bicyclists, public transit advocates, and mostly lay people have been leading a charge for creating public streets that really are the public realm – multi-functional places that serve many community needs and desires. Why should drivers be encouraged to go fast on a local shopping street when a slower vehicle movement pace would be more comfortable for pedestrians and might make local commercial offerings more visible and accessible? Why shouldn't there be trees in the public realm to give shade in summer and help clean the air and slow water run-off? Why shouldn't there be unpaved ground and permeable paving surfaces so that rain can percolate into the local groundwater and nourish local plants, rather than being whisked away to storm water sewers, especially given growing concerns over groundwater depletion in so many urban areas? Why not favor transit vehicles over autos, given how much less space they require to transport equal numbers of people and their lesser environmental impacts.

And why shouldn't walkers and bicyclists of all ages be truly thought about and catered to, not only for equity reasons but also because of public health needs? Large numbers of Americans, especially children, suffer from obesity, and research tells us that lack of daily physical activity is a major reason why. The problem is so severe that epidemiologists warn that we face a diabetes epidemic in the coming decades (Frumkin *et al.* 2004). The linear nature of streets makes them ideal places for walking, jogging, and biking, and their ubiquity means that they could be readily available to everyone. However, researchers find that many people would only use such paths if they are attractive, comfortable, and safe (Saleens and Handy 2008; Lee and Moudon 2008). Fortunately, this is not so hard to achieve. In terms of physical elements that can be provided in the public realm, it means trees (for shade, aesthetics, and perhaps protection from vehicles), clearly marked paths that motorized vehicles cannot intrude upon, well-designed crosswalks at intersections, and traffic calming.

And what about places to sit and socialize with neighbors and visitors and for kids to play? Particularly in already developed neighborhoods, it is unlikely that new public open spaces will be created because of the expense and difficulty of acquiring land. For many neighborhoods, the already existing public space in streets may represent the best viable option for new parks. Many street rights-of-ways are so wide that parks can be built within them and still leave room for vehicles. Indeed, the list of possibilities and opportunities is long, and certainly, there is much headway to be made. The remainder of this chapter explores possibilities for innovative street design and highlights some of the challenges that exist today.

Innovative street designs

While the public realm of streets often is not likely to be made wider, the good news is that the opportunities are many and exciting for innovative redesign of streets in existing or newly urbanizing areas. Urban designers around the world are taking up the challenges. Consider here only a few examples.

Traffic calming

Traffic calming is perhaps the best way to make street space available and comfortable for non-motorized movement and for social activities of all kinds. While roadway speed bumps have proved to be a highly effective traffic-calming device, and are widely used, other more innovative approaches not only slow traffic but also increase the pedestrian realm.

Sidewalk widening at street corners and mid-block, and the accompanying roadway narrowing that help calm traffic, are being implemented in many different ways. The people of San Francisco use the Duboce Triangle area as their model when pressing for such improvements. Done in the late 1960s and early 1970s, the much too wide streets were realigned with perpendicular rather than parallel parking stalls. Good-sized corner and mid-block sidewalk widenings of a plaza scale were added with sitting spaces and new trees. Residents took it from there, adding more planting and seating. A few trees, planted at strategic locations in the center of the roadway of several streets, let everyone know that this is not a hurry-up area.

In some neighborhoods of many European cities – the Mazarin District in Aix-en-Provence, inner-city areas of Amsterdam, throughout Bruges, to name but a few – the crosswalks at intersections or mid-blocks (as in downtown Dubai) are raised to the height of the sidewalks, serving as a speed-bump and at the same time, announcing to all that the intersection is a pedestrian realm, and indicating to motorists and bicyclists alike that they need to slow down for their own comfort as well as for the safety of pedestrians. The crosswalks that cross major traffic streets typically are not raised, letting drivers and pedestrians alike know that vehicle movement is important on those streets.

Some cities, particularly many in the Netherlands, have instituted shared space residential street designs that are designated as wholly pedestrian realms into which slow-moving vehicles are permitted. Called *Woonerfs* (a Dutch term) these street designs slow vehicle travel by alerting drivers, through signage and the lack of curbed roadways, that pedestrians (especially kids) will be present and have rights to use the whole right-of-way. Design elements, such as strategically placed trees, planted areas, or parking spaces, are typically used to create meandering paths for vehicle travel, to ensure that vehicle movement will be slow.

Increasingly, one finds an absence of curbs that delineates who should travel where within rights-of-way in commercial areas with particularly high pedestrian volumes but where for some reason vehicles cannot be altogether excluded. Rather, vehicles of all types and people are allowed to mix. Vehicles do not move fast because drivers must watch out for other legitimate users of the pedestrian realm. Examples can be seen throughout Granville Island, in Vancouver, and on any number of old streets in European cities, such as the very old Via dei Giubbonari, in Rome.

The Rue des Petits Carreaux (Figure 32.3), in Paris, is a recently built example of this interesting street type. A shopping street with residences above, located near the Rue du Rivoli, it is a 35-foot wide basically pedestrian street, but one on which some autos and motorcycles are permitted. The ends of the street are controlled by movable bollards, common on many European streets that allow only for vehicles owned by local shopkeepers or residents. The original central roadway, where most people walk, is 15 feet wide, bounded by modestly raised sidewalks. Cars are few and move slowly, in one direction. At one intersection, with Rue Leopold Bellam, there are three sidewalk cafes at the corners, and so the walking space is not the whole width of the right-of-way. One hardly notices the few, slow-moving cars engulfed by the crowds of walkers.

Multiway boulevards

Multiway boulevards are streets that can handle large amounts of relatively fast-moving through traffic, where it is necessary to do so, while also graciously accommodating a good balance of other uses (Jacobs *et al.* 2002). These streets have a relatively wide central roadway for the through traffic, flanked by narrow one-way side access roadways for slow-moving local traffic and parking. The roadways are separated by raised medians (or malls) containing one or more rows of closely spaced trees and often pedestrian walkways and transit stops. If well-designed, this configuration creates extended traffic-calmed pedestrian realms that extend to the outer edges of the side medians.

A number of communities are currently contemplating implementing multiway boulevards, both to replace unwanted urban freeways and to restructure suburban strip arterials. San Francisco recently built a multiway boulevard for the first purpose. Until the Loma Prieta earthquake of 1989, San Franciscans, particularly residents of the Hayes Valley neighborhood, had to endure the Central Freeway, a massive double-decked affair. Picture two wide decks of freeway built above a local street, Octavia Street, blocking views of City Hall along the way, with at least four sets of on- and off-ramps to connect to the ground. Octavia Street, dank and dark under the freeway, was a place for prostitution and drug-trafficking. But given a second chance by damage done by the earthquake, and led by local neighborhood activists, San Franciscans voted to tear the freeway structure down and replace it with a multiway boulevard (Figure 32.1). The right-of-way left by the freeway was enough to carry three lanes of through traffic in each direction, two local access lanes, one on either side of the through lanes and separated from them by a walking strip with plantings and two tree-lined sidewalks, and still allowing some space for new housing along one of the access roads. In total, there are five rows of new trees. Cyclists use the slow-moving side access lanes. Best of all, the last block on Octavia Boulevard is capped by a new, small park, named Patricia's Green in honor of a local activist, bordered by narrow access ways. The new park, including a children's play area, has

Figure 32.1 Octavia Boulevard, San Francisco. Source: Judith Stilgenbauer – used by permission.

become the center of the rejuvenated neighborhood, with new and busy eating establishments that help define its edges.

Multi-modal streets

Multi-modal streets are designed to provide a better balance of right-of-way space to movement by non-motorized modes, primarily walking and biking, and transit. A recently coined term for a mode balanced street is "complete street." Recently, with federal level encouragement, many American states have adopted complete streets policies. New York City is at the forefront of implementing them through its recent development of a citywide complete streets plan.

Multi-modal street designs are becoming common in Northern Europe – Copenhagen and Amsterdam (Figure 32.2) come particularly to mind – and they are also showing up in other venerable European locales. Paris is perhaps the most elegant of the world's cities. Most Parisian streets – from the café-lined grand boulevards like Boulevard Saint-Germain to the memorable tree-lined Avenues like Avenue Montaigne, to the countless narrow medieval streets throughout the historic heart of the city are full of life, and, thankfully still endure (Figure 32.3).

But do they? In fact, Parisian urban designers and planners have been tinkering mightily with their streets to encourage more bicycling. And much of it is high quality tinkering. Here we look at two streets.

Boulevard Magenta (Figure 32.4) has a right-of-way approximately 95 feet wide. Once a four-lane street, plus parking lanes and typically wide sidewalks with one line of trees on each, it has been reconfigured with fourteen-foot sidewalks that maintain the old trees and lights for pedestrians in line with the trees, a five-foot dedicated bike path along each walk, beyond the trees, a parking lane with new trees and light poles, and a ten-foot wide bus and taxi lane that is separated by a curb from two traffic lanes in the center. Thus, the

Figure 32.2 Street in Amsterdam. Source: Elizabeth Macdonald.

Figure 32.3 Rue des Petits Carreaux, Paris. Source: Elizabeth Macdonald.

Figure 32.4 Boulevard Magenta, Paris. Source: Elizabeth Macdonald.

street has now become a multi-modal street with less dedicated space for autos and much more consideration for movement by public transit and bicycles.

Or consider Avenue Jean Jaurès, a major shopping street in a local residential area, with a right-of-way of approximately 100 feet. Once four traffic lanes plus two parking lanes, it now has two traffic lanes, one parking lane that alternates from one side to the other, block by block, two bike lanes, 15-foot wide planting areas, and wide sidewalks. The planted areas, beyond shrubbery, have two lines of trees, while the other side has one line. All in all, it is a very imaginative design that seems to have attended to a variety of interests: cyclists, merchants, environmentalists, pedestrians, and still leaving room for the automobile.

Transit priority streets

Transit priority streets are designed to privilege the movement of transit vehicles over other traffic, and to be pedestrian-friendly because walking is associated with transit use. Typically, these streets have special lanes dedicated just for transit. Sometimes taxis are allowed to use these lanes as well.

One can look at Curitiba, Brazil's so-called "structurals" for fine examples of transit priority streets. (Others have adapted the Curitiba model, with modifications; Bogotá is an example.) The structurals, made from adapting already existing streets for the most part, have bi-directional central lanes dedicated to buses and taxis. A raised island separates the transit realm from a parking and local auto travel lane, and then there are the sidewalks. The parking spaces are on the left side of the auto travel lane so that transit platforms can replace them at stops, leaving only one lane to cross for transit riders. Innovative raised glass "boarding tubes" on the platforms facilitate quick passenger boarding, allowing the bus transit system to operate at subway volumes.

Land uses (houses and offices) for the depth of one block along the structurals are very dense, purposely so to encourage transit use. The next streets, parallel to the structural, are one-way for autos and for more buses. Land uses beyond these one-way streets are legislated to very low densities. The whole integrated transit-land use-automobile idea is extremely simple, recalling Soria y Mata's early lineal city concept.

In Amsterdam, the new IJ Burg district, being built on filled "islands" and expected to have 45,000 residents and 12,000 jobs, is structured around a linear transit priority street that is the only through street. It has central dedicated bi-directional streetcar tracks, flanked on each side by a narrow one-way vehicle travel lane, and then a wide walk for pedestrians and cyclists.

Flexible streets

Flexible streets are streets that are purposefully designed to accommodate different uses at different times of the day, week, or year, depending on the needs and desires of the local community. Castro Street, in Mountain View, California, is a fine example of a flexible street. Designed in the 1980s by the San Francisco urban design firm of Freedman, Tung and Bottomley, and built in the 1990s, it was and is the city's main street. It is a street of subtle changes in level and strategically placed design elements within a strong, clear, overall design concept (Jacobs 1993: 168–169).

Sidewalks, parking, and the roadway are on three different levels. Sidewalks, set at the highest level, effectively separate pedestrians from automobile traffic and permit a good view to the other side of the street. A parking apron, with trees planted along its outer edge, toward the roadway, provides a flexible transition zone that is clearly in the domain and at the pace of pedestrians. The slight elevation change and the line formed by the rolled curb parking apron clearly separate it from the street. Equally significant to the availability of various levels for different paces of movement, is the availability of the parking realm for outdoor restaurants or cafés, exhibition spaces, and special community events. Add low walls with seating in many places between the sidewalks and parking/event areas and you have the makings of a fine, flexible street.

Green streets

We are facing huge environmental crises in the world today, with such potentially disastrous consequences for all forms of life, let alone human life, that it seems almost crazy not to use publicly owned resources, such as streets, to the fullest extent possible to address environmental needs. Because streets are distributed throughout the city, any ecological benefits that they are designed to have can also be widely distributed. Consider just four possible ecological roles of public streets: protecting and maintaining ground water supplies, counteracting the urban heat island effect, combating global warming, and providing wildlife habitat and corridors.

Typical asphalt and concrete street surfaces create rainwater run-offs that contribute concentrated pollutants to storm water drainage sewers, pollutants that are commonly expelled untreated into urban water bodies. Minimizing non-porous surfacing on public right-of-ways and providing green swale retention basins, where plants work to filter out harmful toxics, would maximize on-site rainwater infiltration and so help recharge groundwater reservoirs. Porous paving materials direct water to where it ought to go.

Minimizing heat-absorbing surfaces on streets and planting an abundance of large shade trees help reduce the urban heat island effects experienced in cities. Urban heat islands are domes of warm air that hover over urbanized areas. They are created because the hard surfaces prevalent in cities – concrete, asphalt, and stone – absorb the sun's rays causing both surface and ambient temperatures to rise. Cities are known to become as much as 20°F warmer than surrounding rural areas, which is particularly problematic in warm climate areas but can have negative affects everywhere. Large street trees help shade hard street surfaces from the sun's rays, thereby decreasing heat build-up, and they contribute direct cooling effects through evapotranspiration. Especially in summer, this cooling can be very important for pedestrians (Streiling and Matzarakis 2003).

Planting street trees that add up to real urban forests is a strategy to combat global warming because trees absorb carbon dioxide from the atmosphere as they grow, storing it in roots, trunks, branches, and leaves. (Of course, reducing use of carbon dioxide producing energy sources is the most important strategy.) Furthermore, the web of greenery in street tree forests creates habitat for birds, insects, and animals, and provides corridors that animals can use with some safety to move between outlying open spaces and city parks.

Professionals know how to design "green" streets to serve these ecological roles: trees, planted swales, and earth (Portland Metro 2002a, 2002b). What is needed is the community desire and political will, spurred on relentlessly by responsible professionals, to disinvest cars of some of the space that has been given to them and make "green" use of the re-claimed space.

The spatial challenge

Given all the possibilities for urban streets, the challenge lies in finding space for all or most of the legitimate users without making the public rights-of-way excessively wide. Why are wide rights-of-way a problem? First, streets give scale to a city. In older cities, built when transport was by foot or horse-drawn carriage, blocks were not too big and intersections were many. A comparison of old versus new cities and suburbs shows the trend that started in the early twentieth century, toward larger blocks, wider streets, fewer public rights-of-way, and fewer intersections. When city blocks get bigger, the ability for people to get around gets worse. Limited street routes contribute to vehicle congestion and to a lack of walkability. Why have big blocks come about? Because of a desire, on the part of redevelopment agencies and property developers, to allow for larger scale developments, bulkier buildings, and more control. The result is centralized power and ownership and decrease of small scale ownership and hence participation in urban life.

Second, when streets are wider than necessary they create waste of space and energy. Consider for a moment the impacts over one square mile of 60-foot versus 50-foot wide rights-of-way. Assuming blocks 400 feet long and 200 feet wide, there is a difference of about 15 fewer blocks with the larger right-of-way. Depending on parcel sizes that are used, the difference in the number of single family lots varies from 300 fewer parcels with 40-foot wide lots, to 480 fewer parcels with 25-foot wide lots. With 80-foot rights-of-way, there are 41 fewer blocks, and 820 to 1,312 fewer parcels.

These are conservative examples. It takes little imagination to think of the consequences wide streets have in terms of less and more costly housing, more expensive public infrastructure, less accessible community services, more required travel times between places, more fuel usage, more carbon dioxide pollution, less likelihood of local shops within walking distance because of lower density, and on and on. Even over a moderately extended urban area, the cumulative waste can be huge.

The "standards" challenge

The main challenges to implementing innovative street designs that enlarge the public realm possibilities of urban streets lie in the entrenched engineering design standards and norms that so heavily influence the form of urban streets. While it may be relatively easy to modify design standards for local residential streets, if residents demand that it be done, changing the design standards for heavier trafficked streets has, and will likely continue to be, extremely challenging. Four design norms that make urban arterial streets problematic from a public space perspective are the following.

High speed limits and wide travel lanes

Many arterial streets have posted speed limits of 30 to 40 miles per hour, meaning that their design speed is even higher. And yet, research shows that traffic accidents involving pedestrians grow exponentially more severe as speeds increase (Hamilton-Baillie 2004). Indeed most pedestrian fatalities occur on non-local roads and speed is the main culprit (Litman 1999). The human body is built to withstand impacts of roughly 20 miles per hour (mph), a person's maximum running speed. Pedestrians hit by cars going 20 mph or less generally survive (a 95 percent chance). Those hit by vehicles traveling at higher speeds are likely to be killed (an 80 percent chance).

Even when arterial street speed limits are posted at 25 mph, drivers often drive faster because the street design encourages it. Automobiles are typically six to seven feet wide. Trucks and buses may be eight to nine feet wide. And yet, design standards for arterial streets call for providing 12-foot wide lanes where possible (the same width as recommended for freeway lanes!) to create a more forgiving and supposedly safer driving environment. But, research shows that wider lanes encourage faster driving speeds, resulting in a less safe pedestrian environment (Fitzpatrick *et al.* 2000).

Intersection designs that privilege ease of car and truck movement

It has become normal for the geometric design of intersections along arterial streets to be based on the turning movement requirements of the largest trucks. This means wide radius curb corners, which result in a longer intersection crossing distance for pedestrians, hence increasing their traffic exposure risk. Other intersection designs meant to ease auto movement, such as dedicated right turn lanes, may result in sidewalk narrowings at corners and the loss of bicycle lanes near intersections.

Lack of safe pedestrian crossings

Many arterial streets have been purposely designed to have long blocks, in order to minimize the traffic interruptions caused by intersections. Other arterial streets have been designed so that only some intersections are controlled by traffic signals, often just the major cross streets. Along with the high traffic volumes, the long blocks and unsignalized intersections make street crossings difficult for pedestrians. While marked crosswalks would seem to help the situation, by themselves they do not necessarily improve crossing comfort and safety, especially on multi-lane streets, because drivers all too often are not vigilant of them. Some research has shown that compared to unmarked crosswalks, marked crosswalks at mid-block locations or unsignalized intersections on multi-lane streets are associated with more pedestrian fatalities, due to the combination of an increased false sense of safety on the part of pedestrians and actual low crosswalk compliance rates on the part of drivers. So, it has been considered safer not to mark such crosswalks. However, recent research shows that multiple traffic-calming features, such as in-roadway flashing lights in combination with marked crosswalks, greatly increase driver compliance (Fitzpatrick *et al.* 2006; Godfrey and Mazzella 2000; Huang and Cynecki 2001).

Restrictions on street trees

There are so many restrictions against planting trees along sidewalks that it is a wonder we have any trees in cities at all. Engineering geometric design policy manuals, such as those of the American Association of State Highway and Transportation Officials (AASHTO), recommend designing street intersections with clear sight triangles in order to improve a driver's ability to see potential conflicts with other vehicles before entering an intersection. These triangles extend hundreds of feet beyond the intersection. Within the clear sight triangles, the recommended design solution is to eliminate any object above sidewalk level that would intrude into the sight triangle and interfere with a driver's vision, *where practical* (AASHTO 2004). Traffic and highway engineering textbook examples describing the clear sight triangle concept generally show diagrammatic plan views of intersections with sidewalk trees indicated as the objects to be eliminated from the sight triangle. In the diagrams, trees are represented as solid circles, which implies they are solid cylinders going all the way to the ground (Garber and Hoel 1997). This representation is of course unrealistic because street trees are typically trimmed to be high branching. Although the intent of the clear sight triangle idea is to eliminate physical elements from a driver's cone of vision, which operates in a three-dimensional world, the triangle is conceptualized in two-dimensional terms rather than three-dimensional terms. In reality, the part of a street tree that would intrude on a driver's central cone of vision is the trunk, a relatively thin vertical element.

In practice, the engineering policy recommendations regarding intersection clear site triangles, and the embedded assumptions that street trees must be eliminated from them, has resulted in many cities adopting street design standards that include large set-back restrictions on sidewalk trees at intersections. These often apply regardless of how a given intersection is controlled, while similar restrictive regulations are not put in place for other objects that commonly occur on sidewalks near intersections, such as newspaper racks, traffic signal poles, streetlights or parking meters. Furthermore, urban street design ordinances generally do not require holding on-street parking spaces back a large distance from an intersection, so in practice parking spaces often come right up to the stop limit line or backside of the crosswalk. Local ordinances typically restrict placing trees near parking meters, roadway signs, or driveway intersections.

However, recent research using advanced computer modeling techniques suggests that street trees – if properly selected, adequately spaced, and pruned for high branching – do not create a strong visibility problem for drivers entering an intersection (Macdonald 2006). The research suggests that deciduous street trees planted close to intersections and spaced as little as 25 feet apart, pruned so that horizontal limbs and leafing start well off the ground, do not constitute a visibility safety hazard on urban streets. Rather, on-street parked cars, particularly large ones, create substantially more problems with visibility.

A question of values

In the end, taking fuller advantage of the public space and ecological opportunities presented by urban streets will require a shift in public values. In this age of peak oil, the question is not how to continue designing streets to accommodate cars as usual, but rather how to re-design them for a new, more sustainable, more pedestrian-oriented age.

References

American Association of State Highway and Transportation Officials (AASHTO) (2004). *A Policy on Geometric Design of Highways and Streets*, Washington, DC: American Association of State Highway and Transportation Officials.

Fitzpatrick, K., Carlson, P., Brewer, M., and Wooldridge, M. (2000). "Design Factors That Affect Driver Speed on Suburban Streets" *Transportation Research Record* 1751:18–25.

Fitzpatrick, K., Turner, S., Brewer, M., Carloson, P., Ullman, B., Trout, N., Park, E.S., and Whitacre, J. (2006). *Improving Pedestrian Safety at Unsignalized Crossings.* Washington, DC: NCHRP.

Frumkin, H., Frank, L., and Jackson, R. (2004). *Urban Sprawl and Public Health: Designing, Planning, and Building For Healthy Communities*, Washington, DC: Island Press.

Garber, N.J. and Hoel, L.A. (1997). *Traffic and Highway Engineering*, 2nd Edition, Boston: PWS Publishing Company.

Godfrey, D. and Mazzella, T. (2000). *Success in Redesigning Main Streets for Pedestrians*, City of Kirkland, Washington.

Hamilton-Baillie, B. (2004). "Urban Design: Why Don't We Do It in the Road? Modifying Traffic Behavior through Legible Urban Design" *Journal of Urban Technology*, 11(1): 43–62.

Huang, H.F. and Cynecki, M.J. (2001). *The Effects of Traffic Calming Measures on Pedestrian and Motorist Behavior*, edited by T.-F.H. R. Center. McLean, VA: US Department of Transportation.

Institute of Transportation Engineers (ITE) (1992). *Traffic Engineering Handbook*, Washington, DC: ITE.

Jacobs, A. (1993). *Great Streets*, Boston, MA: MIT Press.

Jacobs, A., Macdonald, E., Rofé, Y. (2002). *The Boulevard Book: History, Evolution, Design*, Boston, MA: MIT Press.

Lee, C. and Moudon, A.V. (2008). "Neighborhood Design and Physical Activity" *Building Research & Information*, 36(5): 395–411.

Litman, T. (1999). *Traffic Calming Benefits, Cost and Equity Impacts*, Victoria Transport Policy Institute, December 7.

Macdonald, E. (2006). *Street Trees and Intersection Safety* (Working Paper 2006–11), Berkeley, CA: University of California at Berkeley Institute of Urban and Regional Development.

Portland Metro (2002a). *Green Streets: Innovative Solutions for Stormwater and Stream Crossings*, Portland, OR: Metro.

—— (2002b). *Trees for Green Streets: An Illustrated Guide to Selecting Street Trees that Reduce Stormwater Runoff from Streets and Improve Water Quality*, Portland, OR: Metro.

Saelens, B.E. and Handy, S.L. (2008). "Built Environment Correlates of Walking: A Review" *Medicine & Science in Sports & Exercise*, 40 (7S):S550-S566.

Streiling, S. and Matzarakis, A. (2003). "Influence of Single and Small Clusters of Trees on the Bioclimate of a City: a Case Study" *Journal of Arboriculture*, 29(6): 309–316.

Further reading

Beatley, T. (2000). *Green Urbanism: Learning from European Cities*, Washington, DC: Island Press. The multiple dimensions of green urbanism, including the transportation dimensions.

Engwicht, D. (1999). *Street Reclaiming: Creating Livable Streets and Vibrant Communities*, Gabriola Island, BC: New Society, 1999. Innovative street design approaches and practical strategies.

Gehl, J. and Gemzoe, L. (2000). *New City Spaces*, Copenhagen: Danish Architectural Press. Case study examples of innovative street redesigns around the world.

Gehl, J. and Gemzoe, L. (1999). *Public Spaces – Public Life*, Copenhagen: Danish Architectural Press. The pedestrianization of central Copenhagen.

Moudon, A.V. (Ed.) (1991). *Public Streets for Public Use*, New York: Columbia University Press. An extended treatment of issues related to streets and the public realm.

Southworth, M. and Ben-Joseph, E. (1997). *Streets and the Shaping of Towns and Cities*, New York: McGraw-Hill. A history of street standards and their influence on the form of the urban public realm.

Sucher, D. (2003). *City Comforts: How to Build an Urban Village*, Seattle. WA: City Comforts Inc. Inexpensive strategies of traffic calming on streets and creating livable neighborhoods.

33

Mixed-life places

Mark Francis

A few weeks after two teenagers shot and killed twelve of their classmates and a teacher at Columbine High School in April 1999, residents of Littleton, Colorado gathered for a memorial to mourn their loss. Lacking a central public space in their community, the only place they could find large enough was the parking lot at the local shopping mall. Too many communities today are like Littleton, lacking public space that supports local life and community-wide events. The shopping mall has become the de facto town center in many small or suburban communities. In larger cities, public spaces are more specialized in both design and management resulting in a lack of place that brings together the diverse publics now common in American cities. There are important exceptions, such as historic large parks and some large downtown plazas, but community-wide public places are too few or unsuccessful.

Historically, public spaces have represented important assets for cities. Public space, as defined by the Danish urban designer Jan Gehl (1987), is the space and life between buildings. Every time you leave a building, you are in an open space either designed or natural. Gehl does not distinguish between private and public ownership of spaces but regards spaces outside or between buildings as part of the public realm. While a somewhat simplistic definition especially with so many specialized or privatized spaces today, it captures well the essence of pubic open space. The types of public spaces that fit Gehl's definition include parks, plazas, streets, sidewalks, waterfronts, urban gardens, and natural areas.

A more complete definition of public space, however, is that of an open, publicly accessible place where people go for group or individual activities (Lynch 1990; Carr *et al.* 1992). Formed naturally over time or deliberately by design and policy, public spaces are the physical settings where public life takes place (Figure 33.1). There are hopeful signs that public life in cities is increasing today with more demand for nearby public spaces that provide opportunities for gathering, watching or just hanging-out.

Public spaces have been expanding in numbers in traditional types such as parks, plazas and streets and more innovative varieties such as community gardens, greenways, and waterfront areas (Ryan 2006; Wooley 2003; Czerniak and Hargreaves 2007). In recent decades, there has also been a rapid growth in research and literature on public spaces including empirical studies of their use (Cooper-Marcus and Francis 1998; Shaftoe 2008), cultural role (Low *et al.* 2005; Harvey and Fieldhouse 2005) and meaning (Carr *et al.* 1992; Hajer and Reijndorp 2001; Carmona, *et al.* 2003). In addition, analysis and urban

Figure 33.1 Outdoor café in Oslo reflects the unidimensional users of too many public spaces today. Source: Mark Francis.

design methods to make good public spaces have also improved (Loukaitou-Sideris and Banerjee 1998; Wooley 2003; Zeisel 2006; Ward-Thompson and Travlou 2007).

Well designed and successful public spaces facilitate human use, provide basic rights of access and publicness, and are meaningful and memorable for urban residents. However, not all public spaces contain these successful ingredients, as William H. Whyte (1980) and many other researchers have found. Moreover, some contemporary developments result in public spaces that are exclusionary and not accessible to different groups. In this chapter, I draw from my research and design work of over thirty years to rethink how we can best create good urban places and suggest alternatives to current practice.

Here I critique one of the most sacred principles in new urbanist development today – the concept of mixed-use. I argue that mixed-use, while well intended as a way to create a positive public realm, has too often become harmful, working against many of the essential democratic place qualities needed in city design. It often results in public spaces that are parochial (Lofland 1989) and open only to homogeneous and narrow groups of urban residents and visitors. My purpose here is to raise some concerns and limitations of mixed-use projects and to suggest an alternative perspective on the theory and practice of making good public spaces. As a more useful construct, I propose the concept of what I call "mixed-life." In this chapter, I explore the ingredients and qualities of public spaces that are at once diverse, democratic, inclusive and memorable. The chapter points out the qualities inherent in mixed-life places and highlights examples of urban design and planning that support such thinking.

Public space in urban design

A central purpose of urban design is the creation of public space. Drawing from historic precedents of cities where squares, streets, and open spaces form a strong public realm, designers and planners have worked to incorporate public open space as a central feature in new urban development. Yet the results of urban design practice in creating successful public space are mixed. For example, much public space today has become privatized limiting access or use to specialized groups of users (Banerjee 2001). Additionally, poorly conceived and designed spaces have resulted in conflicts between different user groups that in turn have led to increased control and surveillance of spaces.

Considerable differences still exist in how to conceptualize and include public space in urban design projects. While mixed-use developments do not always promise to provide successful public spaces, new urbanist thinking is largely based on creating a dynamic public realm. Mixed-use projects attempt to create positive public spaces through a combination of form based codes, design guidelines, and a mix of specific land uses. Yet the efforts of new urbanists to create a diverse and vibrant public realm have been limited (Day 2003; Brill 2001).

Social theorists and some designers have argued for a more inclusive approach to designing for public life. Social scientist Michael Walzer (1986), for example, distinguishes between two distinct types of urban space. One is "single-minded" space designed for one particular usage or activity. The other he characterizes as "open-minded spaces," designed for "a variety of uses including unforeseen and unforeseeable uses, used by citizens who do different things and are prepared to tolerate, even to take an interest in, things they do not do." Walzer goes on to suggest that open-minded space is a "breeding ground for mutual respect, political solidarity, and civil discourse" (p. 472).

Landscape architect Walter Hood (1997) suggests that good public spaces evolve over time and should provide for spontaneous activity. He states "social injustices are created when certain uses are ignored or not provided for ... sometimes causing conflicts when unprogrammed uses occur" (p. 8). He encourages designers to create a physical framework to allow for common daily activity and practices to emerge. His design projects provide for more spontaneous and improvisational activity. Architectural theorist Charles Jencks has characterized improvised design as "enformality." He states that this "is more than a style and approach to design ... a basic attitude towards the world, of living with uncertainty, celebrating flux and capturing the possibilities latent within the banal" (Jencks 1993, p. 59). Urban design practice according to these views should be willing to entertain new ideas and approaches and be tolerant to a range of known activities while also being receptive to unknown uses and users.

The limits of mixed-use

> Mixed-use development reverses the planning trends of the past half-century, which were largely designed to segregate uses. Mixed-use developments provide an interesting place to live, and promote a sense of community. They provide a degree of safety with 24 hour "eyes on the street." They also provide places for people to shop, work and recreate in close proximity to housing – through some means other than the automobile. (National Governors' Association 1999)

As an antidote to the delineation of the city as a collection of single-use territories, the concept of mixed-use has emerged as a way to guide new urban development. In response to the complaints that single-use patterns often lack richness and density, mixed-use is frequently intended to build community and promote walking, biking and community life. Many urban designers have adopted this "new urbanist" view, and some city officials have followed by incorporating mixed-use into requirements and codes for city development.

A typical mixed-use project today consists of ground floor retail with housing or office space above and an expanded sidewalk, a small plaza or larger public space (Schwanke 2003). Some communities are attempting to develop entire districts or neighborhoods as mixed-use areas. Much of the conceptual thinking behind mixed-use is to increase density in towns and cities to promote walking and biking. While viewed by some developers as financially risky, mixed-use nevertheless has become widely accepted.

I argue that while often well intended, mixed-use has led more to separation and segregation of people and activities than a diverse mix of life. Too often mixed-use projects become havens for young and often childless professionals and exclude new residents, families, and people of different cultural backgrounds (Low *et al.* 2005). Public open spaces in mixed-use developments are too often privatized and under the surveillance and tight control of managers and property owners (Banerjee 2001; Kohn 2004). They largely fail to provide for design and management, and often preempt a full range of uses and users (Francis 1989). They result in privatized spaces that present an idealized and artificial form of public life (Lofland 1989; Brill 2001). While some researchers and designers have tried to develop other more inclusive terms such as "multi-use," "lifestyle centers," or "third places" (Oldenberg 1999), there is a need for a fresh concept to guide the making of urban places.

The problem with mixed-use development as commonly implemented today is that they are mostly real estate projects focused on profit rather than creating community or social diversity. Rarely do mixed-use projects create finely grained public places, a result of misguided intention and poor programming and design. Too often built projects include expensive housing above commercial establishments such as Starbucks and other national chains. They frequently serve as a veil or mask for other agendas such as land speculation and gentrification. The developments are class- or ethnicity-specific, exclusive of important populations such as children, the elderly or newer immigrants.

An example of the limits of mixed-use is the Pearl District in Portland, often heralded as the model of large-scale urban development (Figure 33.2). Despite its good intentions, this area demonstrates the difficulty of translating social and environmental goals into effective policies and codes. While economically successful, most of the housing built in the Pearl District consists of studio or one-bedroom units targeted for young professionals with no children. As a result, the City of Portland had to close its only school in the Pearl District due to the lack of children in the neighborhood. This has forced the city to revise its planning guidelines to promote more family-friendly development in this and other neighborhoods.

There are efforts underway to rethink the mixed-use concept (Herzog 2006) and explore urban design alternatives (Ryan 2006). Ellen Dunham-Jones and June Williamson (2008) for example, offer suggestions on how to retrofit existing shopping malls and suburban shopping centers to better integrate them with the rest of the city fabric. They argue that good design integrates surrounding neighborhood streets to promote local walkability, provide greater access to transit, and reduce parking ratios. One example of this new way of thinking is Northgate Mall in Seattle, one of the original enclosed shopping malls designed by planner Victor Gruen.

Figure 33.2 The Pearl District in Portland, Oregon. Source: Mark Francis.
Note: While widely heralded by planners, it has largely failed to attract a diversity of residents and users.

Built on top of an undergrounded Thornton Creek, the recently remodeled mall has added 100,000 square feet of commercial space, two plazas and other uses such as family-oriented restaurants and events. Planners are trying to make the mall better connected to surrounding areas to promote easier access on foot and by bicycles. In addition, part of Thornton Creek is now open to the sky and reveals a series of rain gardens on the mall's site. While redevelopment efforts such as this are encouraging, especially in the integration of social, economic and environmental concerns, there is a need for a new and enlarged social and cultural construct to guide the making of urban places.

Mixed-life, not mixed-use

> This ubiquitous principle (mixed-use) is the need for cities for a most intricate and finely grained diversity of uses that give each other constant mutual support, both economically and socially. (Jacobs 1961, p. 14)

As an alternative to mixed-use, I suggest a more useful and inclusive concept to guide future urban design and development – mixed-life spaces. I define mixed-life spaces as those settings, designed or natural, that support a diversity of people, experiences, and meanings. They are public spaces where a variety of people feel safe and comfortable "hanging out" and are the hallmark of good and healthy urbanism. The precedents for this concept can be found in the work of Jane Jacobs (1961), Kevin Lynch (1984), William Whyte (2001), Randy Hester (2006), Michael Brill (2001), Ray Oldenberg (1989), and Richard Sennett (2008) among others.

Jane Jacobs, considered by some as the first and most influential advocate for mixed-use, offers a broader view of the role of mixed-use urban places. In her seminal book *Death and Life of Great American Cities* (1961), Jacobs mentions mixed-use numerous times, and devotes an entire chapter to it. She states:

> The district, and indeed as many of its internal parts as possible, must serve more than one primary function; preferably more than two. These must insure the presence of people who go outdoors on different schedules and are in the place for different purposes, but who are able to use many facilities in common. (Jacobs 1961, p. 152)

She goes on to suggest how designers and planners can best use this concept in practice:

> I think that unsuccessful city areas are areas which lack this kind of intricate mutual support, and that the science of city planning and the art of city design, in real life for real cities, must become the science and art of catalyzing and nourishing these close-grained working relationships ... (Jacobs 1961, p. 14)

What Jane Jacobs describes as "diversity" approximates more the concept of mixed-life than contemporary mixed-use developments, which have largely failed to meet her standards of urbanism. Mixed-life has the promise to create this fine-grained relationship that many urban designers seek for our cities.

Mixed-life is not seen as an all-encompassing concept or a call for a universal mix of lifestyles, income classes, ethnic groups, and stages in lifecycle. It recognizes that conflicts do exist between the demands and preferences of different groups and their cultural practices in public open spaces (Low *et al.* 2005). Understanding user and cultural difference can lead to more diverse public space and more inclusive public life. Too often design attempts to "design out" rather than embrace these conflicts. Good design and responsive management can negotiate and mediate these differences.

What enables mixed-life places?

What qualities are needed to create mixed-life places? While all may not be present in one place, they include accommodation of use in ways that offer comfort, relaxation, active engagement, passive engagement, discovery, and fun (Carr *et al.* 1992; Francis 2003). They often provide for public access, freedom of action, choice, user claim and control, and symbolic or in some cases real community ownership. They also create memorable experiences such as meaning derived from group and individual connections. They connect rather than divide, provide for appropriation and spontaneity of activity, and are adaptable. They are not static places but part of an ongoing effort to shape a positive and healthy public realm.

Mixed-life places are centers. Randy Hester (2006: 23–24) suggests,

> Centers are good places where people gather to undertake different activities. Good centers are intense concentrations of different uses – commercial, civic, residential, transportation, religious and educational – that attract people from different income levels, gender groups and life-cycle stages. The activities feed off each other, each inviting more users by proximity to others, and seemingly incompatible uses invite diverse publics to the same place. Successful centers do not exclude by subtle

design symbols of discrimination. Instead they exclude inclusivity.

Nature is a powerful enabler of mixed-life. Mixed-life places include some form of constructed nature such as plantings, native vegetation or natural systems such as water features or sustainable storm water management. Natural elements are one of the most effective ways to make the built environment healthy and engaging for city residents. Research has clearly shown the psychological and physiological benefits of access to and contact with nature as part of everyday life (Kaplan *et al.* 1998; Moughtin *et al.* 2009).

Mixed-life places are also not discriminating or judgmental. They can also provide opportunities for active engagement such as recreation, exercise, skateboarding, gardening or food production. They also frequently have a compassionate constituency of people involved over a long period of time in the design, planning, management or use of the place and willing to fight for its preservation or improvement.

Not all places can or even should be mixed-life places but every community should have at least one. For example, in designing Central Park and Davis Farmers' Market in Davis, California, I found that the goal of mixed-life helped me formulate a design program and master plan that included diverse elements such as a community garden, farmers' market, children's plaza, teen center and an area where homeless people could gather without making other users feel uncomfortable. Central Park (Figure 33.3) brings nature and culture together in what Randy Hester has called a kind of "civic meadow" (Hester 2006). Hester observes that "centeredness" is part of Central Park's design success. "Activities are arranged around the open lawn that measures 120 feet by 300 feet. Activities are placed in the trees that surround the lawn. This creates a unity among disparate activities by calming the noise of the boisterous and constraining the impulse to over schedule activities ... The center enables" (p. 24). The result is a place where the entire community comes together for everyday activities, bi-weekly markets or citywide demonstrations and festivals.

Mixed-life places are what I have come to think of as "democratic space" (Francis 1991). Good public spaces are those where people feel comfortable to hang out without the need to buy things or use laptops or cell phones. They do not just mix uses

Table 33.1 Some principles of mixed-life places.

- Are naturally diverse
- Are accessible
- Are used for different purposes
- Feel safe and comfortable
- Blend uses and activities
- Connect rather than divide
- Invite active engagement and appropriation
- Invite people of all ages
- Attract the entire community
- Are places to hang-out in
- Are joyful places
- Are memorable
- Are "democratic space"
- Are resilient and can change over time

Figure 33.3 Central Park, Davis, California. Source: Mark Francis.

but blend them into a seamless ecology of activities and experiences. They are what could be called "free space." They are what Jan Gehl (1987) calls "go to" rather than "go through" places.

The users of mixed-life places are not the usual occupants found in mixed-use projects. They can include new and long-term residents and recent immigrant groups. They also include women and the elderly. Children and teenagers are essential to good public places. Important stakeholders also include space managers, government officials, local business owners, but also people passing through and enjoying the space and residents who live near by.

A typology of mixed-life places

How do you know a mixed-life place when you see it? When I began to look for examples, I first thought they would be hard to find. But there are encouraging signs with many built or planned projects especially by landscape architects and some urban designers. Indicators include: people using adjacent streets, more people walking or on bikes than in cars, nearby or visible public transit, and children and teenagers without their parents but in the presence of others.

They include new types of urban projects emerging in response to failures of mixed-use such as community gardens, farmers' markets, flea markets, and work live housing. I have found more landscape examples than buildings with landscape architects taking the lead in making mixed-life spaces. There are positive signs of inventive designs and unique projects being developed.

Today we see the emergence of a wider and encouraging array of urban projects beyond those that have resulted from the typical mixed-use formula. They tend to be more diverse in architectural and landscape architectural character than typical mixed-use projects, and include a wider variety of activities. Examples include community gardens, farmers' markets, and flea markets, which attract a diverse segment of the public (Hou *et al.* 2009; Francis *et al.* 1984). Used clothes stores, ethnic food, art and music festivals, sales of locally

Table 33.2 A typology of mixed-life places and some movements that support them (with selected examples).

- **Urban Parks and Squares** (Pioneer Courthouse Square, Portland, OR; Takachi Ecology Park, Obihiro City, Japan; Davis Central Park, Davis, CA)
- **Large Parks** (Central Park, New York City; Orange County Great Park, Irvine, CA; Crissy Field in San Francisco, CA; and the proposed Fresh Kills Park on Staten Island, NY)
- **Farmer's Markets** (Davis Farmer's Market, Davis, CA; Greenmarkets in New York City; Ferry Building Market, San Francisco, CA)
- **Flea Markets and Street Vending** (found in or around many communities)
- **Food/Art Festivals** (Santa Barbara Cabrillo Art Fair, CA)
- **Street Fairs** (Bay to Breakers Race, San Francisco, CA)
- **Demonstrations** (such as on the Mall and in Lafayette Park in Washington, DC; Market Street, San Francisco, CA)
- **Urban Gardens** (Boston Public Garden, Boston, MA)
- **Community Gardens** (Clinton Community Garden, New York City; Karl Linn Community Garden, Berkeley, CA)
- **Community Greening Projects** (Guerilla gardening in London, UK and Los Angeles, CA; Sacramento Tree Foundation)
- **Urban Waterfronts** (Hudson River Park, New York City; Cheonggyecheon River Restoration Project, Seoul, Korea; Malmo Waterfront Park, Sweden)
- **Greenways** (Highline Park Project, New York City; Davis Greenway, Davis, CA; Brooklyn-Queens Greenway, NYC)
- **Skate Parks** (Santa Barbara Skate Park at Stearn's Wharf, Santa Barbara CA; Burnside Skate Park in Portland; and the Westblaak Skatepark in Rotterdam, The Netherlands)
- **Natural Areas/Open Landscapes** (cliffs, forests, meadows, wetlands, etc.)
- **Natural Beaches** (East Beach in Santa Barbara, CA; Copacabana, Rio de Janeiro, Brazil; Jones Beach, Long Island, NY)
- **Urban Beaches** (Harbor Bath, Copenhagen, Denmark; Spree Bridge Bathing Ship, Berlin, Germany; Seine Urban Beach, Paris, France)
- **Parking Parks/Gardens** (ReBar and Trust for Public Land)
- **Ecological Neighborhoods** (Village Homes, Davis, CA; Hammarby Sjöstad, Stockholm, Sweden)
- **Democratic Streets** (Octavia Blvd, San Francisco, CA)
- **Green Infrastructure Projects** (Portland Rain Gardens, OR; Bioswales; Complete Streets)
- **Green Movements that Support Mixed-Life Places** (Sustainable Cities; Healthy Cities; Active Living; Green Economy; Environmental Justice; and Urban Agriculture)

or regionally produced goods, also can attract and support mixed-life.

Other recent examples of mixed–life projects include large parks (such as Crissy Field in San Francisco and the proposed Fresh Kills Park on Staten Island), greenways (such as the Highline Park Project in Manhattan), and farmer's markets (such as the Greenmarkets in New York City and the Davis Farmer's Market). They also include urban gardens such as community gardens organized by community greening groups in most cities, and more ad hoc efforts such as guerilla gardening taking over vacant lots and street corners in London and Los Angeles (Francis *et al.* 1984). River and waterfront development such as the Hudson River Parkway along the west side of Manhattan, the 3.5 mile Cheonggyecheon River Restoration Project in Seoul Korea, and the Malmo Waterfront Park in Sweden are all examples of mixed-life places constructed in the last decade or so. Their design elements are a diversity of programmed areas, a predominance of natural features, design for diverse user groups, and opportunities for active engagement with the environment.

Natural or not designed places such as wild or natural areas remaining in cities such as woods, cliffs and forests are also examples. Beaches such as East Beach in Santa Barbara, California and Copacabana in Rio de Janeiro, Brazil bring diverse users and activities to cities. Efforts to construct new urban beaches and swimming areas on rivers and harbors such as the Harbor Bath in Copenhagen, Denmark and Spree Bridge Bathing Ship in Berlin, Germany are further examples.

Urban squares such as Pioneer Square Courthouse in Portland, Oregon and the redesigned Union Square in San Francisco are further examples. There are also a growing number of urban parks such as Central Park and Davis Farmer's Market in Davis, California and Takachi Ecology Park in Obihiro City, Japan that provide central gathering spaces and create new town centers. Even more specialized spaces such as skate parks are creating more diverse life in cities such as the Skateboard Park at Stearn's Wharf on the Santa Barbara, California waterfront, Burnside Skate Park in Portland, Oregon and the Westblaak Skate Park in Rotterdam, the Netherlands.

There are signs that mixed-life can also be a part of new or redesigned neighborhoods. Examples of this include Village Homes, an ecological community in Davis, California designed in the late 1970s, and Hammarby Sjöstad, a large-scale ecological development in Stockholm, Sweden. Elements that make them mixed-life areas are sustainable design elements such as jointly owned common areas, accessible open channel drainage and wetland areas, native and edible landscape features, and opportunities for residents to participate in managing the built landscape.

Streets and left over spaces such as street corners and vacant lots are also being transformed in some cities into mixed-use places. Examples include streets such as the work of ReBar and the Trust for Public Land to claim parking spaces as open space. A number of efforts directed at traffic calming and to put streets on "traffic diets" are being used to make streets more pedestrian and bike friendly. These movements include "Streets as Places" that create activity areas for people to gather on streets, "Complete Streets" that make a more equitable balance between pedestrians and bicyclists with automobiles, and "Shared Space" that reduce the separation of cars from pedestrians (see also chapter by Macdonald in this volume). New forms of transit and movement are also playing a role including shared cars and free use of bicycles now found in some European cities.

Advances in sustainable design including stormwater management, urban forestry, and green roofs are creating exciting opportunities for making mixed-life places. In addition, emerging movements such as sustainable cities, healthy cities, green economy, environmental justice, and urban agriculture are also encouraging more ecological, inclusive, and just public places.

Creating mixed-life places: the role of form and urban design

Urban design can play a critical role in making diverse public spaces. While some of the examples cited are the result of citizen design or even non-planning, clearly design and planning are essential in creating good public places. If mixed-life places can be designed, then there is also a role for city planning policies and codes to encourage these types of projects. Clear design guidelines and supportive public policies are critical and can contribute to lasting success.

The idea of public space matters in urban design practice. Some have suggested concepts such as vibrancy (Gehl 1987) or conviviality (Shaftoe 2008) to explain the expanding role of public spaces in cities. I would go even further, offering the concept of energetic public spaces – places that are designed and managed to be deliberately open, inclusive and diverse. Through design, management and public intent, they are intentionally created as mixed-life places. They are not just open to different publics and possibilities of unknown activities but also vital and outgoing in their social intent (Figure 33.4). While this may seem overly nostalgic or even deterministic, I believe that public space design and management needs to be more proactive and intentional in their efforts to create true mixed-life places (Francis 1999). This reflects an understanding that public space is a social necessity, a right of publicness, and an opportunity for people to have meaningful life enhancing experiences (Carr *et al.* 1992; Parkinson 2009).

Furthermore, users must be able to exert some control over places they use (Francis 1989). Good public spaces are frequently the result of community participation

Figure 33.4 Central Park, Davis, California, was purposefully designed as a mixed-life place.
Source: Mark Francis.

where local residents take ownership and stewardship of the projects (Hester 2006). While city and regional governments can encourage the development of these projects with adequate funding and supportive policies, their success relies ultimately on a strong local constituency for their implementation and ongoing management.

Mixed-use alone does not insure mixed-life. It can contribute but only in concert with other factors such as deliberative concern for social justice and fine-grained design. Inventive and compelling form is also needed. Many of the examples cited above had visionary designers and decision makers willing to take risks and try new things.

Mixed-life places provide an important area for future research and design practice. There are many opportunities for developing new case studies (Francis 2001) and asking basic questions about urbanism in all its forms. Useful methods of analysis already exist including observation, interviews, post occupancy evaluation, GIS mapping, and studies of programming and participation (Zeisel 2006). There is also the need for continued longitudinal research to monitor the success of public spaces. There are many new graduate theses and dissertation topics to explore here.

We also need more built examples especially at the building, community or neighborhood scale. We need a better understanding of the role of programming, participation and management in making mixed-life places. A wider typology of spaces is also needed. Proactive practice that engages children, youth, immigrants, and new residents in making projects is essential.

Mixed-life offers a promising yet challenging realm for future urban design practice. As cities continue to increase in diversity and complexity, designers and planners need to invent new forms of urban places and test their effectiveness in creating exciting and memorable urban life.

Acknowledgments

I would like to thank Stan Jones, Randy Hester, Lyn Lofland, Ray Lorenzo, Marcia McNally, Dean MacCannell, Ib Omland, Michael Rios, Leanne Rivlin, LeRoy Tonning, and the editors of this book who provided useful comments on the development of the ideas presented in this chapter.

References

Banerjee, T. (2001). "The Future of Public Space: Beyond Invented Streets and Reinvented Places." *Journal of the American Planning Association*, Vol. 67, No, 1, pp. 9–24.

Brill, M. (2001). "Mistaking Community Life for Public Life." *Places*, 14, 2: 48–55.

Carmona, M., Heath, T., Oc, T., and Tiesdell, S. (2003) *Public Places-Urban Spaces.* Oxford: Architectural Press.

Carr, S., Francis, M., Rivlin, L. and Stone, A. (1992). *Public Space.* New York: Cambridge University Press.

Cooper-Marcus, C. and Francis, C. (1998) *People Places: Guidelines For Urban Open Space.* New York: John Wiley and Sons.

Czerniak, J. and Hargreaves, G. (2007). *Large Parks.* New York: Architectural Press.

Day, K. (2003). "New Urbanism and the Challenges of Designing for Diversity," *Journal of Planning Education and Research*, Vol. 23, No. 1, pp. 83–95.

Dunham-Jones, E. and J. Williamson (2008). *Retrofitting Suburbia: Urban Design Solutions for Redesigning Suburbs.* New York: Wiley.

Francis, M. (1989). "Control as a Dimension of Public Space Quality." In Altman, I. and Zube, E. (Eds.) *Public Places and Spaces. Human Behavior and Environment*, Volume 10. New York: Plenum.

—— (1991). "The Making of Democratic Streets." In Vernez Moudon, A. (Ed.). *Public Streets For Public Use.* New York: Columbia University Press, pp. 23–39.

—— (1999). "Proactive Practice: Visionary Thought and Participatory Action in Environmental Design." *Places*, 12(1): 60–68.

—— (2001). "A Case Study Method for Landscape Architecture." *Landscape Journal*, 19(2): 15–29.

Francis, M., L. Cashdan and Paxson, L. (1984). *Community Open Spaces*. Washington, DC: Island Press.

Gehl, J. (1987). *Life Between Buildings.* New York: Van Nostrand Reinhold.

Hajer, M. and Reijndorp, A. (2001). *In Search of the New Public Domain*. Rotterdam: Nai Publishers. 2001.

Harvey, S. and Fieldhouse, K. (2005). *The Cultured Landscape*. London: Routledge. 2005.

Herzog, L.A. (2006). *Return to the Center: Culture, Public Space, and City Building in a Global Era.* Austin, TX: University of Texas Press.

Hester, R. (2006). *Design for Ecological Democracy*. Cambridge: MIT Press.

Hood, W. (1997). *Urban Diaries.* Washington, DC: Spacemaker Press.

Hou, J., Johnson, J. and Lawson, L. (2009). *Greening Cities, Growing Communities: Learning from Seattle's Urban Community Gardens*. Seattle, WA: University of Washington Press.

Jacobs, J. (1961). *The Death and Life of Great American Cities*. New York: Vintage Books.

Jencks, C. (1993). *Heteropolis: Los Angeles, the Riots and the Strange Beauty of Heteroarchitecture.* London: Academy Editions.

Kaplan, R., Kaplan, S., and Ryan, R. (1998). *With People in Mind: Design and Management of Everyday Nature.* Washington, DC: Island Press.

Kohn, M. (2004). *Brave New Neighborhoods: The Privatization of Public Space.* New York: Routledge.

Lofland, L.H. (1989). "The Morality of Urban Public Life: The Emergence and Continuation of a Debate." *Places*, 6(1): 18–23.

Low, S., Taplin, D. and Scheld, S. (2005). *Rethinking Urban Parks: Public Space and Cultural Diversity.* Austin: University of Texas Press.

Lynch, K. (1984). *Good City Form.* Cambridge: MIT Press.

—— (1990). "The Openness of Open Space." In Banerjee, T. and Southworth, M. (Eds.), *City Sense and City Design: Writings and Projects of Kevin Lynch*. Cambridge: MIT Press, pp. 396–412

Moughtin, J.C., Laurea, P.S., and McMahon, K. (2009). *Urban Design and the Therapeutic Environment.* London: Architectural Press.

National Governors' Association. (1999) "Principles of Smart Growth". Washington, DC: NGA.

Oldenberg, R. (1999). *The Great Good Place.* Cambridge: Da Capo Press.

Parkinson, J. (2009). "Symbolic Representation in Public Space: Capital Cities, Presence and Memory." *Representation*, 45(1): 1–14.

Ryan, Z. (2006). *The Good Life: New Public Spaces for Recreation*. New York: Van Alen Institute.

Schwanke, D. (2003). *Mixed-Use Development Handbook.* Washington, DC: Urban Land Institute.

Sennett, R. (2008). *The Uses of Disorder: Personal Identity and City Life*. New Haven, CT: Yale University Press.

Shaftoe, H. (2008). *Convivial Urban Spaces: Creating Effective Public Places*. London: Earthscan.

Walzer, M. (1986). "Pleasures and Costs of Urbanity." *Dissent*, 33: 470–475.

Whyte, W. (1980). *The Social Life of Small Urban Spaces*. New York: Project for Public Spaces.

Wooley, H. (2003). *Urban Open Spaces*. London: Taylor & Francis.

Further reading

Carmona, M., Heath, T., Oc, T., and Tiesdell, S. (2003). *Public Places, Urban Spaces.* London: Architectural Press. A comprehensive discussion of urban dimensions that support good public space.

Carr, S., Francis, M., Rivlin, L., and Stone, A. (1992). *Public Space.* New York: Cambridge University Press. An overview of public space and public life including needs, rights and meanings.

Cooper-Marcus, C. and Francis, C. (Eds.). (1997). *People Places: Design Guidelines for Urban Open Space.* New York: Wiley. A classic overview of creating new public space based on user needs.

Czerniak, J. and Hargreaves, G. (2007). *Large Parks.* New York: Princeton Architectural Press. A review of historic and contemporary parks typically as large as New York City's Central Park.

Francis, M. (2003). *Urban Open Space.* Washington, DC: Island Press. Provides an overview of the principles needed to design open spaces for user needs.

Gehl, J. (1987). *The Life between Buildings.* New York: Van Nostrand Reinhold. The classic study of public space use and design drawn from observations of central Copenhagen.

Loukaitou-Sideris, A. and Banerjee, T. (1998). *Urban Design Downtown: Poetics and Politics of Form.* Berkeley, CA: University of California Press. A useful guide to how urban design shapes public space.

Madanipour, A. (2003). *Public and Private Spaces of the City.* London: Routledge. A useful examination of the role of public space in city development.

Ward-Thompson, C. (2007). *Open Space: People Space.* New York: Taylor & Francis. A review of contemporary research and design practice on public open spaces.

Whyte, W. (2001). *The Social Life of Small Urban Spaces.* New York: Project for Public Spaces. A classic study by a sociologist on the use of public open spaces including ingredients needed to create successful places.

Zeisel, J. (2006). *Inquiry by Design.* New York: Norton Press. A useful guide to environmental psychology methods useful in evaluating public spaces.

34

Urban flux

Gary Hack

Most urban design is devoted to assuring stability, coherence and predictability of urban settings. Design guidelines and zoning rules limit changes to the streetscape, and enforce a degree of conventionality on what gets built. At the same time, many of the most interesting places in cities are just the opposite: disordered, unpredictable, changing at a rapid pace, and open to individual initiatives that constantly reshape them.

Ask a teenager in Tokyo where she likes to go to meet others, and Shibuya will be the first place that comes to mind, not the ordered but dull commercial streets and boulevards of the city. In Philadelphia, it will be Manayunk or South Street, or in Los Angeles, Melrose Avenue. Ever changing Times Square is the top tourist destination in New York, and the city's emerging hot neighborhoods include the Lower East Side and Chelsea, places where the disorder of the new contrasts sharply with the past. Clearly something is missing in the way we think about urban design if we can not account for the places residents – and many designers – find most interesting.

Permanence versus flux

The current mindset of urban design has been shaped by an *architectural* vision that seeks to design urban environments in all their detail. The underlying idea is that a stable framework for urban life will offer a semblance of continuity in the face of the ever changing occupants and activities of the urban milieu. To make this approach instrumental designers have devised a variety of techniques – design guidelines, form based development regulations, signage controls, pattern books, design review processes and panels, among others.

The notion of promoting environmental stability conveniently intersects with most designers' natural desire to leave a lasting imprint on the city, but it would be a mistake to dismiss this simply as professional hegemony. In most places, there is a broad constituency that supports tight controls over the built fabric of cities, rooted in the desire to protect property values and ward off new arrivals who may have opposing views about how to live (Costonis 1989). The popularity of historic districts, which "freeze" environments in a particular era, and conservation districts that aim to preserve a specific character in the built fabric of areas, are evidence of this subtext. Planned unit developments, where form, materials and colors are all subject to design review, have proved more saleable than their uncontrolled counterparts. Property owners and developers understand that tightly controlled environments carry less risk over time than places subject to a thousand whims.

The approach of thinking of urban design as architecture writ large long predates the emergence of professionals who call themselves urban designers. The projects of Sixtus V in Rome have had a powerful influence on urban form, as Edmund Bacon's (1968) *Design of Cities* attests. The aspiration of creating a grand baroque ensemble – wide streets and avenues, with vistas across the city, lined by controlled facades, terminating in monuments or squares or public buildings – dominated urban design for more than two centuries. The re-planning of Paris, Berlin and many other European cities in the nineteenth century provided powerful models for the new world, captured in Werner Hegemann's and Elbert Peets' (1922) monumental book on urban design.

The architectural vision of urban design arrived on American shores via Daniel Burnham's plans for the Worlds Columbian Exposition in 1893 (Larson 2003). The first full blown example of an architectural image applied to an entire city was Burnham's subsequent Plan for Chicago, a collaboration with Edward Bennett (Burnham and Bennett 1909). The Diaspora of the École des Beaux Arts, notably Paul Cret (Grossman 1996) and Jacques Gréber (1920), and the peripatetic practices of Burnham and Bennett spread the ideas of the City Beautiful to countless cities, promoting civic centers, cultural districts, university campuses, and city plans organized around a fully designed public realm. Few grand plans were actually completed, but where they were, as in Philadelphia's Benjamin Franklin Parkway (Brownlee 1989) or San Francisco's civic center, they became the centerpieces of their cities. More importantly, the early design-centered plans became the basis for zoning in many communities, institutionalizing the desired control over the built form of places.

The most recent manifestation of this line of thought is the New Urbanism movement (Duany *et al.* 2003), a nostalgic throwback to a simpler era when settlements were "towns" which could be designed with one hand. Sketches of urban ensembles by Leon Krier (2009), whose ideas are direct descendents of nineteenth-century European urbanism, are almost purely architectural, rarely including trees or showing spaces occupied by humans, and never incorporating a sign or advertisement to signal that the buildings have uses within them. Krier's Poundbury, a new English neighborhood built with such a mindset, has proved to be just as dull as the sketches (Figures 34.1 and 34.2). New Urbanist codes specify not only the form of buildings and places, but the types of fences, porches and materials, the slope of roofs and their gables, and other details (Parolek *et al.* 2009). These rules are generally accompanied by the appointment of a town architect, or creation of a design review panel that may occasionally allow the easing of standards – as in permitting several architects at Seaside to build modernist homes for themselves. But such exceptions are seen as rare departures from the consistent mannerist fabric of the community.

This approach is not without its critics in the architectural world. *Learning from Las Vegas*, by Robert Venturi, Denise Scott Brown and Steven Izenour (1972), was one of the first explorations of the idea that environments are communication devices, with ever changing messages. In their subsequent books and projects, they have argued that the new potentials of media open the possibilities of environments where signs, symbols and popular associations merge with the built fabric (Venturi and Scott Brown 2004). There has been considerable debate over the issue of authenticity in virtual environments such as those produced by Disney, or in the themed casinos of Las Vegas (Sorkin 1992). Films, television, and video games are virtual test beds of ideas about environments, with quick turnarounds about

Figure 34.1 Leon Krier's sketch for Dorset Village of Poundbury. Source: Leon Krier – used by permission.
Note: It exemplifies the architectural preference for stability of the environment.

Figure 34.2 Neighborhood in Poundbury, England. Source: Duchy of Cornwall, photo: Ian Skeller – used by permission.
Note: A completed section of Poundbury shows just how dull the environment can be if no room is left for residents' or tenants' additions.

human responses. Philosophers including Jean Baudrillard have argued "the real no longer exists" (Baudrillard 1994), and a body of thought has emerged centered on virtual reality, simulacra and simulation.

Rem Koolhaas and other architects have argued that design regulations constitute a straitjacket to creativity, and are undemocratic in the way that they stifle free expression. Designers like Frank Gehry have sought to push and deform envelopes to register a greater range of meanings. Daniel Libeskind and others have sought to produce narrative buildings that provide an accessible context for the activities they house.

Even before the architects, artists explored new visions rooted in flux and change in urban life. Beginning in the 1950s, the neo-dada Fluxus movement created art that captured the experience of the moment, using all the available dimensions of artistic experience – sight, sound, events and spectacles, and the absence of each (Friedman *et al.* 2002). Fluxus, spearheaded by John Cage, George Brecht and George Maciunas, among others, realized art in the form of scores for performances and any objects left behind were valued only as traces of memory of what had just transpired (Higgins 2002). Other Fluxus artists appropriated handbills, postage stamps, utilitarian clothing and other everyday objects as artistic media that could be reproduced and distributed as propaganda for change. Performance art has blossomed. Graffiti art has become respectable, with individual artists recognized for their signature tags. Billboards, particularly those sitting idle, are highly sought after as a venue for propaganda art, following Jenny Holzer's lead (Holzer and Smith 1998). The notions of accidental, unpredictable and astounding have become mainstream in the arts.

What all these movements have in common is an emphasis on constantly changing experience, with environments and the artifacts they contain as the props for the performance. The intervention of artists may be fleeting or may leave a permanent record, and is judged by whether it captures the attention of the viewer, stimulates thought and conveys meaning.

At the same time, new technologies have emerged which dematerialize previously stable aspects of the environment. It is possible to construct entirely pixilated facades of buildings that are changeable in an instant; these can be programmed for artistic or commercial purposes. Vegetated facades of buildings can be planted to change seasonally and blur the line between structure and landscape. Transit vehicles can become gigantic mobile billboards, and pixilated billboards are towed through city streets, avoiding any restrictions on signage. City streets are transformed by weekend festivals and events, and become a stage for city life (Schuster 2009). In an era of short takes, the constant search for sensory stimulation and dominance of the visual channel, flux presents a powerful tool for shaping impressions of places.

The possibilities of flux

Walk the commercial streets of most Asian cities that have not lived with the tight yoke of regulation and it is often difficult to detect the buildings that lie behind the maze of signs, canopies, street furnishings, sales racks, vending carts and other intermediaries between the pedestrian and merchant, not to mention sounds, smells and human contacts. The narrow streets of Hong Kong's Tsim Sha Tsui district epitomize this (Figure 34.3): they are high expression zones, with merchants shouting for attention through every available channel. The street changes its character from hour to hour as new merchandise is moved out into the street, and from day to night, as signs are illuminated, music increases in volume, and the number of hawkers multiplies.

Figure 34.3 Street in Hong Kong. Source: Gary Hack.

Even in tightly regulated Tokyo, the hot spots are areas with the greatest accumulation of flux: Shibuya and Shinjuku (Figure 34.4) are among the top destinations, dominated by electronic billboards and signs, but also Akihabara, the electronics center with streets crowded by neon green signs for new gadgets, the Ginza and especially its alleys with changeable decorations and displays arching across the facades, and thousands of minor streets devoted to local shopping or entertainment where temporary displays and seasonal decorations dominate.

In Manila, Hong Kong, and Bangkok, and many other cities, elaborate seasonal Christmas decorations transform every commercial district, beginning at least two months before the holiday and extending a full month after. It may seem odd to encounter gigantic snowmen on a balmy January day along Sukhumvit Road, but this is about participating in a worldwide commercial phenomenon that is only incidentally religious. Not to be outdone, many American homes and neighborhoods compete for "best decorated" prizes each holiday season.

Such additions to the environment resonate directly with observers and have a temporal dimension that is difficult to achieve through architecture alone. Perhaps the most memorable city images are the temporary memorials expressing collective grief over the loss of lives or valued places. The posters of loved ones on walls and barricades surrounding the ruins of the World Trade Center were an immediate outpouring of the city's sentiments over the loss of friends and neighbors. Who will forget the memorial to Princess Diana that was created spontaneously outside

Figure 34.4 Shinjuku, Tokyo. Source: Gary Hack.

Kensington Palace, or the memories and thousands of votive candles left in Madrid's Atocha Station honoring those lost in the March 2004 bombing?

As we walk through most cities, we often fail to notice another form of flux – the temporary elements added to the environment as buildings are constructed, restored or renovated, or streets are dug up for new transit lines. Buildings are covered with shrouds to mask their unfinished appearance; construction canopies protect pedestrians or adjacent buildings; temporary barriers protect zones set aside for construction vehicles, and a variety of temporary lighting, directional signage and other artifacts may be installed. Because these elements are "temporary" they usually escape the tight scrutiny that more permanent elements receive. Gigantic advertising banners often grace the facades or canopies during the renovation process. And many of these additions are far from temporary, lasting for years on complex construction projects.

One common strategy is to render the underlying building in a new light through its construction shroud. The multi-year renovation of the Doges Palace in Venice "peeled back" the façade to show the paintings inside (Figure 34.5); the classical face of the Madeleine Church in Paris was rendered in pointillist dots (Figure 34.6), that merged optically when viewed from Place de la Concorde. Michael Graves' creative scaffolding for the multi-year restoration of the Washington Monument emphasized its modular character, rather than obelisk form (Figure 34.7); when it was removed, the structure seemed altogether too smooth and unitary.

Of course temporary elements are commonly handled in the most banal way – Jersey barriers, plywood or chain-link construction fences, rental scaffolding that makes for difficult passageways. But there

Figure 34.5 Doge's Palace, Saint Mark's Square in Venice. Source: Gary Hack.
Note: A construction shroud that "peeled back" the façade to expose the paintings within provided a new vision for St Mark's Square, Venice, during the multi-year renovation of the Doge's Palace.

Figure 34.6 Madeleine Church, Paris. Source: Gary Hack.
Note: For almost a decade, the Madeleine church in Paris was covered by a pointillist shroud that assured that its image was covered but not lost.

Figure 34.7 Temporary scaffolding on the Washington Monument. Source: Gary Hack.

is plenty of scope for creativity in the design of the temporary environment. Christo has made an artistic career of wrapping structures of all types, adding to their mystery while highlighting their existence. And the bright red InfoBox that graced Potsdamer Platz in Berlin during its lengthy reconstruction period – providing an unparalleled viewing platform – clearly merited a place as a permanent fixture.

A variety of new technologies are animating previously permanent elements of the city in powerful ways. LED arrays on the facades of buildings, such as the CIRA Center in Philadelphia, allow the façade to be programmed to celebrate each day – Gigantic "P"s for Phillies victories, a pink ribbon for the Race for the Cure, a waving flag for the Fourth of July, and other symbols for Christmas and events throughout the year. In Beijing, the Green Pix Zero Energy Media Wall takes this idea to the extreme (Figure 34.8), covering the entire face of a large office building with an array of pixels dense enough to transmit moving images. Avoiding the criticism that such displays are wasteful, solar receptors are embedded on the façade and generate enough energy to run the displays.

Interior public spaces with giant high density digital arrays have become the meeting points in several cities. The grand public space at the Comcast Centre in Philadelphia has over 200 hours of creative programming, ranging from rifs on the building and its occupants to travelogues of city places; the gigantic array can also download real time images from around the globe and even become invisible by simulating the wood panels that line the space beyond. Visitors and residents of Chicago flock to the Crown Fountain at Millennium Park, where artist Jaume Plensa has embedded rotating LED displays of dozens of faces of city residents on two giant stellas – images that disgorge water from the lips of the images on an unpredictable schedule. The power of personification is thoroughly engaging.

We often ask artists to encapsulate their special visions of places in situ in the form of public art. At its best, public art can symbolize ideas and sentiments that go beyond buildings and landscapes – humor, irony, a lost past, a wider geography – or add a new layer of experience. Imprinting the glass facades of the Minneapolis Public Library with woodland forest images totally changes the context of the building, as does the shrink wrap images of musicians that cover the façade of that city's Orchestra Hall.

It is also possible to create collective public art that engages hundreds, even thousands of residents of a place in its creation. Incorporating signs, symbols, or objects created by school children or area residents is a way of creating an immediate attachment to a place. Walls for communication

Figure 34.8 The GreenPix Zero Energy Wall in Beijing. Source: ©Simone Giostra & Partners/Arup/Ruogu – used by permission.
Note: The GreenPix Zero Energy Wall in Beijing employs LED lighting powered by solar cells integrated into the wall. It is capable of being programmed as an artistic element in the city.

may seem anachronistic in a world of bits, but as the Democracy Wall in Beijing and the painting of the Berlin Wall demonstrated, communicating in public is quite different than vicarious contact. It can be a threat to the social order, or a sign of hope for change. It is validated by words and expressions of those who encounter it. It is not accidental that important political events often take place in public spaces.

Flux as a force for civic improvement

The most obvious locale for urban flux is at the pedestrian crossroads of cities, where communication and celebration will reach the largest number of people. Perhaps the most prescient example is the re-emergence of Times Square in New York, through a deliberate program of re-energizing the area with signs, lights and images (Sagalyn 2001).

By the 1970s Times Square had become a cesspool of vice, crime and sex oriented retailing – a place most New Yorkers and tourists avoided. Several attempts were made to "clean up" the area, by proposing the construction of sedate new office buildings, a trade mart and other "respectable" uses, eliminating several of the ill reputed theaters and the chaotic array of marquees, signs and other flux that had remained from the years that it was the Great White Way. These plans failed, fortunately, and planners dug deep in the collective memory of Forty-second Street and reimagined a new Times Square filled with activities again, but based on brands and flagship media outlets that had instant recognition. The plan for Forty-second Street Now (Stern and Kalman 1993), proposed that the lower levels of all structures in Times Square be covered with signage, the gaudier the better. To carry out the plan the city created new zoning for the area that for the first time *required* that signage be a part of every building plan. Super signs were mandated on all large buildings, which need to incorporate a minimum of 50 sq ft of signage for each linear foot of street frontage of their zoning lot, and slightly less onerous requirements apply to buildings on small lots (New York City Planning Commission).

Over less than a decade, Times Square re-emerged as *the* place to visit in New York City. While several new structures were constructed, benefiting from the financial incentive of incorporating outdoor advertising space, dozens of existing buildings also added new eye catching displays, and captured the emerging spirit of place (Figure 34.9). ABC opened its morning show broadcast set to the street, and magnified its programs on a gigantic lighted animated screen visible to all in the Square. NASDAQ signed up for space in a new structure at 4 Times Square which came with a full block super sign, reputed to be the world's largest, programmed to show an ever changing animated array of financial news, advertising and announcements. Not to be outdone, Reuters created programmable signage running the full 40 storey height of its new structure directly across the square.

The race to outdo others and attract the eyeballs of Times Square visitors continues to this day. The signage allowed has been profitable enough to convince several owners of sites to forego building tall buildings, creating instead street level retail topped by elaborate staging structures that support signage – Western stage sets on steroids (Figure 34.10). Other elements of flux have been added to the square including a

Figure 34.9 Times Square, New York. Source: Gary Hack.

Figure 34.10 Times Square, New York. Source: Gary Hack.
Note: Television studios on the street, visible to pedestrians, with programming and zipper news leads beamed constantly to visitors in Times Square.

TKTS booth, designed as a viewing platform for experiencing the full scope of the dazzling displays. Times Square has re-emerged as Media Central, with an international audience.

Flux can also help transform neglected or forgotten neighborhoods of cities. Philadelphia's Mural Arts Program, led by Jane Golden, has created over 2,500 murals over the past 25 years in transitional areas of the city, providing an identity and "face" for neighborhoods often thought of (by outsiders) as threatening or abandoned (Golden *et al.* 2002) (Figure 34.11). Their canvasses include blank party walls, underpasses, retaining walls, the surface of parking lots, gigantic oil tanks and other artifacts visible to all. Neighborhood residents, especially youth, are involved in conceptualizing the mural and working with an artist on its execution. They frequently depict local residents or traditions, celebrating those who have risen above the neighborhood's reputation – Dr. J in his North Philadelphia neighborhood (Figure 34.12), Mario Lanza and Frank Rizzo in South Philadelphia, but also high school kids, grandmothers, community leaders, and local heroes. Sometimes the murals speak to shared sensibilities, as in the Dominican neighborhoods of North Philadelphia where murals recall the verdant landscapes of the home country of many of the residents. In other cases they celebrate the area's diversity, or heroic efforts to rid the neighborhood of crime, or larger traditions such as the legacies of the civil rights movement. Mural programs can transform the perception of neighborhoods in far less time than it takes to change its physical fabric. When permanent buildings are added to the neighborhood, local residents often press for the murals to remain as visible symbols of the area's history.

The Badlands of North Philadelphia have spawned another collective art project that is transforming a badly deteriorated neighborhood house by house, lot by lot. (Wener *et al.* 2001) The Village of Arts and Humanities began in 1986 with a charismatic leader, Lilly Yeh, an artist with a vision. Partnering with several gang leaders in the neighborhood, the immediate objectives were to reclaim an abandoned house to conduct after school arts programs,

Figure 34.11 Murals in a North Philadelphia neighborhood. Source: Gary Hack.
Note: Murals in this North Philadelphia neighborhood remind recent immigrants of Dominica, and the possibilities of restoring the area into a verdant home.

Figure 34.12 Mural in a North Philadelphia neighborhood. Source: Gary Hack.
Note: Dr. J, who was a hero to his Philadelphia fans, personifies the potentials of young people growing up along these deteriorated streets.

and clean up the litter strewn lots that surrounded it. As art teachers, students and neighborhood residents were enlisted in the cause, they elevated their sights to creating a park and gathering space where events could be held and the arts displayed (Figure 34.13). The wonderfully exuberant park incorporated waste materials, using mosaics in undulating walls that Gaudi would be proud of, and importing an imagery of Dagon figures and other African themes. It also spoke to the hopes, aspirations and heritage of the neighborhood residents. As the art center and park became a magnet for community life, new projects were begun on adjacent blocks, and the village now encompasses about 50 blocks, 12 art parks, over 20 acres of vacant land under management, a variety of afterschool programs, an arts residency program and constant events attracting new people to the neighborhood. Grassroots efforts are improving many US neighborhoods, but what distinguishes the village is its emphasis on leading with the arts and physical changes that immediately transform the image of the area.

As the Philadelphia examples show, urban flux can communicate in powerful ways that change perceptions, and help build attachment between residents and their environments (Figure 34.14). The arts are not the only ways to accomplish this. Ethnic groups adopt neighborhoods and change them through subtle but instantly recognizable ways that conform to their norms. Latino neighborhoods in LA and Cuban areas of Miami are immediately identifiable by the colors, awnings, signage and displays, smells and sounds that quickly dominate commercial areas, as well as by the ways that public spaces are appropriated and used. The Charlotte Street neighborhood in the South Bronx, an area that was "renewed" with single family housing, has been adopted mainly by Puerto Rican residents whose signature security grilles and lavish bougainvilleas now lend an unmistakable character to the area. These changes, largely below the radar screen of urban designers, zoning and design regulations, offer immediate and low cost ways of signaling territories and appropriating spaces.

Some ethical and legal issues with flux

Urban flux is not without controversy. Reacting to the fact that "temporary" signage attached to scaffolds of buildings under renovation seems to persist well after construction is complete, a group in

Figure 34.13 Village of Arts and Humanities, Philadelphia's Badlands. Source: Gary Hack.

Figure 34.14 Melrose Avenue in Los Angeles. Source: Gary Hack.
Note: Graffiti artists have been coopted in re-imaging Melrose Avenue in Los Angeles.

Greenwich Village proposed a ban on temporary signage. An environmental activist group in Philadelphia, SCRUB (Society Created to Reduce Urban Blight), has tackled flux on a number of fronts: fighting building wraps on the grounds that they are signage in another form; seeking to prevent the display of large winning entries of beer-drinker "art competitions" sponsored by brewers (whose name is prominently displayed); and trying to stop the installation of street furniture (news stands, benches, transit shelters) covered by advertising, which they regard as "clutter."

Mural programs face other issues, including the question of artists' rights, ownership of the work, and long term maintenance responsibilities. There is often controversy over the subjects of murals, and first amendment rights are frequently debated. A recent controversy before the historical commission in Philadelphia involved neighbors' objections to a mural they found offensive – they were not objecting to the presence of a mural, but the specific content. There is not unanimity over the view that the shared environment ought to be the medium for either self-expression or statements about group identity. The issue of commercial speech as it relates to art remains difficult to disentangle.

Regarding flux as public art, whether or not created by recognized artists, carries with it certain obligations. Many states and national governments protect the right of artists to have a say over the long term disposition of their works. The US Visual Artists Rights Act (17 USC Section 106A) prohibits without the consent of the artist " ... any intentional distortion, mutilation or other modification of [an artist's work] which would be prejudicial to his or her honor or reputation." This includes destruction of the work, although the act contains significant exceptions that allow the relocation of art when buildings are destroyed, and specifically includes "work for hire." Ironically, the designers of the buildings which incorporate public art do not enjoy the same protection of their creative works.

The difficulty of regulating flux stems from how slippery it is to define, and how arbitrary it is to set limits on what is temporary. Is a fully pixilated façade to be treated as a building element or a sign? Is it a commercial billboard or work of art – how does one predict the uses it will be put to over time? Are there limits to how long a construction wall or building shroud may be maintained without being subject to review of its contributions to the cityscape?

The larger issue is whether it is possible, or desirable, to regulate urban flux. One point of view asserts that regulations should provide maximum latitude for personal, commercial and artistic expression, that the city's central purpose is to stimulate communication. Forcing people to live and work behind the facades of anonymous buildings, it is argued, is the definition of alienation. Observing the multi tasking of urban residents plugged into their iPhones and iPods, while reading newspapers, riding on transit and navigating crowds, it is obvious that the tolerance (even desire) for sensory input has increased exponentially.

The opposite view is that urban society exists because there are commonly accepted standards for public behavior, speech and sensory experiences, and if these are absent, the result is anarchy, or at a minimum, cacophony, with the wealthiest (read commercial) and loudest voices dominating the scene. As evidence, they point to the threat felt by New York City residents in the 1980s having to use graffiti covered subways and walk streets assaulted by vendors and hawkers. Sensory overload can be uncomfortable, if not threatening. Stanley Milgram's studies on sensory overload suggest that city residents adopt a variety of strategies for coping, including filtering out the environment and avoiding places of high overload (Milgram 1970). "Cities are more than occupied billboards," writes Joseph Rykwert. "Advertising physically separates us from the lived experience of urban fabric, however ugly or beautiful" (Rykwert 2009).

Clearly there is plenty of middle ground between these extremes. We may wish more stability in our residential areas where personal investments in homes and neighboring relationships are affected by the immediate environment. In commercial and entertainment areas of cities, we may be more tolerant of rapid change – a large function of shopping is, after all, entertainment, and stimulation need not stop at the door of the shop, theater or performance venue. But even in the areas we wish to control most, residents desire some avenue for personal expression. Shaping the landscape that borders on streets has largely fallen outside design review, but in urban areas other instruments of change may be equally important.

Implications for urban design

The recognition that urban flux has an important social function in cities can help suggest new approaches to how they are designed and regulated. Fine cities evolve, with multiple layers of meaning. Too much design attention has been devoted to creating the most durable elements of city fabric and too little to the layers of immediate experience, added or programmed later by the occupants of places. Several principles might help urban designers adjust their sights:

- *Every environment should have zones of expression.* Designers ought to be less prescriptive or deterministic, and leave more decisions to the ultimate occupants of places. Urban design should recognize that places are designed continuously over their life; they are never in stasis and always in flux. Triaging the impermanence of environment is essential, distinguishing between the long lasting elements, those that are tied to the life of the occupants, others that change with the seasons, and still others that change weekly, daily or hourly. Designers need to explore multiple scenarios for variable cycles of change, anticipating and accommodating the unpredictable.

- *Distinguish high flux from low flux areas.* It is not useful to adopt and apply a similar strategy for regulating the development of the city for all areas. Low flux areas need to be more tightly controlled (while allowing clear zones of expression); other areas need to be left largely open to the push and pull of owners and occupants. Sign controls, now typically applied across an entire municipality, need to be differentiated by area, and may be totally inappropriate in some. In some instances, flux may be required (as in signage at Times Square), while in other areas restrained.
- *Every city needs a communications Centrum.* Not every city can support a Times Square, but every city needs a place to go where residents can be connected in public to each other and the larger world. Electronic media make this possible in central public spaces through the installation of electronic billboards. If these are created privately (on buildings or advertising billboards), cities should insist upon a portion of the time being reserved for community messages and as an artistic outlet – much as they have in granting television concessions. More modest elements are also possible, like the Charlottesville Community Chalkboard, outside city hall in this Virginia community, which allows individuals to scrawl messages (they are removed after a week) and organized groups to communicate their preoccupations to the politic. The European tradition of a *kunsthalle* – a place for communication and display of the artistic expressions of a community – needs to be reinvented for public spaces.
- *Flux can accelerate change in transitional areas* As I have noted, flux such as murals and public space improvements can shift the perceptions of an area more quickly and at less cost than wholesale renewal of an area. In some instances it can accelerate private renewal; in others it can set the stage for coordinated redevelopment efforts.
- *Designing flux is an important creative act.* There are dozens of neglected venues for creative design of the city, including creating a palette for a local commercial district, designing staging for a construction project, co-opting artists to render a construction wall as an element of public art, to name just a few. Young artists and designers have discovered installation art, and the ever changing city offers an endless set of sites for their efforts. We need to think of cities as being designed continuously, rather than with one bold plan.

References

Bacon, E.N. (1968). *The Design of Cities.* New York: Viking.

Baudrillard, J. (1994). *Simulacra and Simulation.* Ann Arbor, MI: The University of Michigan Press.

Brownlee, D. (1989). *Building the City Beautiful: The Benjamin Franklin Parkway and the Philadelphia Museum of Art.* Philadelphia: University of Pennsylvania Press.

Burnham, D.M. and Bennett, E. (1909). *Plan for Chicago.* Chicago: City Club.

Costonis, J. (1989). *Icons and Aliens: Law, Aesthetics and Environmental Change.* Urbana, IL: University of Illinois Press.

Duany, A., Plater-Zyberk, E. and Aluminan, R. (2003). *Civic Art: Elements of Town Planning.* New York: Rizzoli.

Friedman, K., Smith, O., and Sawchyn, L. (2002). *The Fluxus Performance Workbook.* Performance Research e-Publication.

Golden, J., Rice, R., Kinney, M., Graham, D., and Ramsdale, J. (2002). *Philadelphia Murals and the Stories They Tell.* Philadelphia: Temple University Press.

Gréber, J. (1920). *L'Architcture aux Etats Unis.* Paris, Payot.

Grossman, E.G. (1996). *The Civic Architecture of Paul Cret.* Cambridge: Cambridge University Press.

Hegemann, W., Peets, E. and Platus, A. (1998; republished from the 1922 edition by Hegemann and Peets) *American Vitruvius: An Architect's Handbook of Civic Art.* New York: Princeton Architectural Press.

Higgins, H. (2002). *Fluxus Experience.* Berkeley, CA: University of California Press.

Holzer, J. and Smith, E.A.T. (1998). *Jenny Holzer (Contemporary Artists).* New York: Phaidon Press.

Krier, L. (2009). *The Architecture of Community.* Washington, DC: Island Press.

Larson, E. (2003). *The Devil in the White City: Murder, Magic and Madness at the Fair that Changed America.* New York: Crown Publishers.

Milgram, S. (1970). "The Experience of Living in Cities," *Science*, 167: 1461–1468.

New York City Planning Commission, *New York City Zoning Ordinance, Special Times Square regulations*, Section 81–732 (a). 3

Parolek, D.G., Parolek, K. and Crawford, P.C. (2009). *Form Based Codes: A Guide for Planners, Urban Designers, Municipalities, and Developers.* New York: Wiley.

Rykwert, J. (2009). "Cities Are More Than Occupied Billboards." *The Architects Journals.* (6 February 2009). <http://www.architectsjournal.co.uk> (accessed 18 August 2010).

Sagalyn, L.B. (2001). *Times Square Roulette: Remaking the City Icon.* Cambridge, MA: MIT Press.

Schuster, J. M. (2009). "Planning for Creative Places." In Hack, G., Birch, E., Sedway, P., and Silver, M. (Eds.), *Local Planning: Contemporary Principles and Practice.* Washington: ICMA.

Sorkin, M. (Ed.) (1992). *Variations on a Theme Park: The New American City and the End of Public Space.* New York: Hill and Wang.

Stern, R. A.M. and Kalman, T. (1993). *42nd Street Now: A Plan for the Interim Development of 42nd Street.* 42nd St Development Project.

Venturi, R. and Scott Brown, D. (2004). *Architecture as Signs and Systems: For a Mannerist Time.* Cambridge, MA: Harvard University Press.

Venturi, R., Scott Brown, D., and Izenour, S. (1972). *Learning from Las Vegas.* Cambridge, MA: MIT Press.

Wener, R., Axelrod, E., Farbstein, J. and Shibley, R. (2001). *Placemaking for Change: The 2001 Rudy Bruner Award for Urban Excellence.* Rudy Bruner Foundation.

Further reading

Costonis, J. (1989). *Icons and Aliens: Law, Aesthetics and Environmental Change.* Chicago: University of Illinois Press. An excellent discussion of aesthetic policymaking, the underlying social and psychological values, and the use of the legal system to enforce aesthetic standards.

Golden, J., Rice, R., and Yant Kinney, M. (2002). *Philadelphia Murals and the Stories They Tell.* Philadelphia: Temple University Press. A richly illustrated account of the Philadelphia Mural Arts program and its process of engaging residents of low income areas in transformative murals.

Sagalyn, L.B. (2003). *Times Square Roulette: Remaking the City Icon.* Cambridge, MA: MIT Press. A detailed account of the remarkable transformation of Times Square, including the role of urban flux in creating and reshaping its image.

Venturi, R. and Scott Brown, D. (2004). *Architecture as Signs and Systems.* Cambridge, MA: Harvard University Press. Thought provoking essays, and examples of work, by the earliest advocates of incorporating large scale graphics and displays in architecture.

Part 7

Debates

Introduction

As we noted in our introduction, and as apparent in the chapters included in Part 2; covering the roots, influences, and pedagogic traditions of urban design, the practice of urban design has a strong normative bent. Some of this derives from the aesthetic, romantic, and visionary traditions of architecture and landscape architecture. But we must also acknowledge the more recent influences of the humanities and critical social sciences, which have focused on the human causes and consequences of the built environment. They have further strengthened the normative stance of urban design by complementing intuition with reason. When Kevin Lynch (1981) spoke of normative theory as one of three types of planning and design theories (the other two being about decision-making and the urban phenomenon), he was simply formalizing this tradition.

Yet, there is also a positivist side of urban design that derives its imperatives from the scientific, rational, and technical traditions in architecture, civil engineering, and landscape architecture, significantly augmented in recent decades by the analytical and empirical traditions of social sciences.

These two traditions in urban design offer a healthy and continuing dialectic in the field, but also define the scope and essential matrix of many of the continuing debates and unresolved tensions in setting the current and future directions of urban design practice. The chapters in this section represent some of these controversies, although not strictly as a pure dialectic but more as a derivative of the pure form. As we will see in the following chapters, some of the debates are within the normative tradition, while others are entirely within the positivist tradition, but with normative overtones. Take for example, the sprawl versus compact city debate that the Ewing, Bartholomew, and Nelson chapter revisits and updates. While the protagonists of sprawl see it as a rational outcome, their case for it is supremely normative steeped in the neo-liberal market tradition. The case for compact cities, while intrinsically grounded in the normative tradition of intuition and reason, attempts to construct its arguments from empirical observations and analysis. The chapter by Southworth and Ruggeri on the theme of place and identity in the emerging global city, on the other hand, focuses on the unresolved issues about place and non-place, or how place identities are formed or constructed.

This debate remains mainly within the normative tradition, but only occasionally encountering some positivist claims as in Mel Webber's (1976) construct of non-place urban realm, which is considered quite prescient in the context of contemporary information and communication technology revolution. Let us consider briefly the terrain of the outstanding debates as captured by these chapters.

Perhaps the most significant debate to date on the future of contemporary urban form involves the proposed model of compact city as an alternative to the ubiquitous urban sprawl. Reid Ewing, Keith Bartholomew, and Arthur C. Nelson begin their chapter by revisiting an earlier debate triggered by a provocative piece written by Gordon and Richardson (1997) making a case for urban sprawl as a market outcome representing aggregate consumer preferences. In the same journal issue, Ewing (1997) had responded to their article with a critique of the Los Angeles-style sprawl, and why such urban outcome is untenable as a social choice. The authors disagreed on just about everything. The utilitarian logic and the libertarian ethos of the Peter Gordon and Harry Richardson (1997) article stood in stark contrast to Ewing's social wellbeing and distributive justice-based arguments against urban sprawl. Although couched in positivist arguments, both remained inherently normative, but neither tried to resolve the debate against the criterion of Pareto superiority, the hallmark of public decision-making in the Western liberal democracies. The debate, in the Gordon and Richardson terms, is really a variant of the market versus planning debate, which they argued forcefully a few years earlier in the same journal with a rebuttal from one of their colleagues (see Richardson and Gordon 1993; Banerjee 1993).

A major element of the debate involved the concept and measurement of urban sprawl, with Ewing arguing that different interpretations can be obtained with different measures of sprawl. In this current chapter Ewing, Bartholomew, and Nelson revisit the debate with new and more precise measurements of urban sprawl and new empirical findings linking issues of public health, energy consumption, and carbon footprint. They also report a growing level of acceptance of the compact city urbanism among the American public. They argue that given the results available today, the case for compactness is getting increasingly stronger, and thus, one could argue, as a Pareto superior outcome, although that case has not been made analytically yet. Nevertheless, the detractors of the compact city do not yet seem ready to concede the point, as evidenced in a recent book by historian Robert Bruegman (2005) which questions the efficacy and user appeal of the compact city.

The chapter by Ali Madanipur addresses a different type of tension that has been around for a while in both developed and developing worlds, but becoming more critical in the growing multicultural contexts of cities in a globalizing world. While the global economy has no doubt done much good in many corners of the world, it has also brought about income polarization and marginality in social and ethnic terms in many Western cities. As Madanipour argues, the multicultural urbanism of these cities is being defined by two parallel and related phenomena – gentrification on the one hand, and social exclusion on the other. Gentrification often leads to the displacement of poorer people from their inner city and older neighborhoods which are often the target of redevelopment. The emerging spaces of habitation or commerce of the new global upper class typically exclude the poorer social and ethnic classes by creating physical barriers. Thus it becomes an urban design issue since the design interventions leading to these urban transformations contribute to gentrification and social exclusion.

Other authors have documented these effects in different settings. Sites (2003), for example, defines the process as a form of "primitive globalization," as seen in the clearance and gentrification of the Lower East Side in New York. Similarly, in a comparative study of ten different cities of Europe and North America, Savitch and Kantor (2002) characterize the process as an outcome of the combination of "global sweep" and "local broom." Madanipour sees this tension as a challenge that urban design should address, as a form of public intervention in modulating the social exclusion and segregation outcomes of the market processes.

If social exclusion and segregation are outcomes of globalization, another effect of the same process brought to the fore is the theme of the Michael Southworth and Deni Ruggeri chapter, and this has to do with notions of place, non-place, and place identity. The authors begin the chapter drawing from Manuel Castells' (1996) work who suggested the rise of the information society is creating "spaces of flow" which are increasingly supplanting the more traditional "spaces of places". The Southworth and Ruggeri chapter reviews the various conceptions of place identity, including the notion of the non-place urban realm, the difference between image and identity, and how place identity is actually experienced. They address the tension between the high design and the vernacular, and the question of identity and authenticity. They argue that we should consider place identity as a multi-faceted gradient and provide examples of such identity gradient in practice. They reject the place/non-place dichotomy and suggest a nuanced approach in urban design seeking to create place identity in the future metropolis. The gradient approach, they believe, is appropriate in negotiating the range of options possible between the existing places and future opportunities.

If the chapter by Ewing, Bartholomew, and Nelson has made a case for the compact city, the chapter by Ivonne Audirac addresses the tension that continues between the alternative and competing models of the compact city, or what she calls the tension between the old and new urbanism. Clearly the dominant compact city model is the one defined by the popular New Urbanism movement, whose roots and development she reviews in some detail. But as she points out, even within this movement, exist at least two major variants, with some intellectual tension between their foci and directions. The Traditional-Neighborhood Development (or TND) derived its inspiration from Leon Krier's influential work and became popular in the east coast cities, while Transit Oriented Development (or TOD) became more popular in the west coast. Interestingly while the TND idea took root in the older and already denser east coast urban areas, its self-contained urban modules depended mainly on the automobile as the mode of access and linkage. TODs in contrast, while nurtured in the automobile dependent urbanization of the west, emphasized transit linkage and ridership as the basis of their success in achieving compact urbanism.

The debates between the new and old urbanism, according to Audirac, revolve around several issues. One such issue is the deterministic and essentialist position taken by their protagonists. Another issue has to do with the lack of appreciation, if not the denial, of the metropolitan landscape and the regional economy and landscape. Audirac points out that the existing urbanism of sprawl or what she calls "old urbanism," which new urbanists are trying to fix or replace, is actually a result of earlier models of urbanism – ranging from the CIAM movement, neighborhood unit formula, and various CIAM principles. Ultimately, she argues that the debate may come to a close if the new urbanism can adapt itself to the changing realities of urban transformations.

In her chapter, Emily Talen focuses on another dimension of the current tension between old and new urbanism, namely the traditional zoning and other implementation measures versus the new form-based codes or FBCs, which are increasingly seen as the New Urbanist options to obtaining desirable urban outcomes. She traces the historiography of zoning in the US cities and what the form based zoning purports to achieve. She explains the idea of Transect, a popular planning tool for New Urbanists, and also a mechanism through which they expect to scale up to the larger landscape at the regional level.

Clearly, much of these movements began with explicit rejection of the contemporary modes of land use control that have traditionally defined the urban form of American cities. The protagonists expect that the form-based zoning or the transect principle will eventually dictate the future growth and development of cities and maximize such welfare goals of community, distributive justice, and sustainability. Meanwhile, some of these debates will continue no doubt, but we expect them to be better informed by the research and workshop experiences involving the public.

References

Banerjee, T. (1993). "Market planning, market planners, and planned markets." *Journal of the American Planning Association* 59(3): 353–360.

Bruegman, R. (2005). *Sprawl: A Compact History*. Chicago: Chicago University Press.

Castells, M. (1996). *The Rise of the Network Society*. Oxford: Blackwell Publishers.

Ewing, R. (1997). "Is Los Angeles–Style Sprawl Desirable?" *Journal of the American Planning Association*, 63: 107–126.

Gordon, P. and Richardson, H. (1997). "Are Compact Cities a Desirable Planning Goal?" *Journal of the American Planning Association*, 63: 95–106.

Lynch, K. (1981). *A Theory of Good City Form*. Cambridge, MA: MIT Press.

Richardson, H. and Gordon, P. (1993). "Market planning, oxymoron or common sense?" *Journal of the American Planning Association*. 59(3): 347–352.

Savitch, H. V. and Kantor, P. (2002). *Cities in the International Marketplace: The Political Economy of Urban Development in North America and Western Europe*. Princeton, NJ: Princeton University Press.

Sites, W. (2003). *Remaking New York: Primitive Globalization and the Politics of Urban Community*. Minneapolis: University of Minnesota Press.

Webber, M. (1976). "The Urban Place and the Non-Place Urban Realm." In Weber, M. et al. (Eds.) *Explorations in the Urban Structure*. Philadelphia: University of Pennsylvania Press.

35

Compactness vs. sprawl

Reid Ewing, Keith Bartholomew, and Arthur C. Nelson

In 1997, the *Journal of the American Planning Association* published a pair of point-counterpoint articles now listed by the American Planning Association as "classics" in the urban planning literature. In the first article, "Are Compact Cities Desirable?" Peter Gordon and Harry Richardson argued in favor of urban sprawl as a benign response to consumer preferences. In the counterpoint article, "Is Los Angeles-Style Sprawl Desirable?" the lead author of this chapter argued for compact cities as an alternative to sprawl. Gordon and Richardson, and Ewing disagreed about nearly everything: the characteristics, causes, and costs of sprawl, and the cures for any costs associated with sprawl.

This debate articulated the main disagreement in academic research about the nature and degree of interaction between land use and transportation. In the intervening years, scholars have developed measures for sprawl and quantified its impacts for the first time. The smart growth, new urbanist, and transit-oriented development movements have come of age in reaction to sprawl. Lifestyle changes and demographic shifts have begun to favor compact development, infill development, and walkable communities. Climate change, and its relationship to urban development patterns, has emerged as the main planning challenge of the twenty-first century. Concerns about rising obesity rates have led to a new partnership of public health professionals and planners under the rubric of active living. Many states and cities in the US have undertaken smart growth initiatives (see chapter by Inam in this volume). Light rail transit has been built in many American cities, while major highway expansion projects have been put on hold.

Today, the case for sprawl seems dramatically weaker, and the case for compact communities dramatically stronger, than they did 12 years ago. It is time to reprise the debate based on new research and writing on these and related topics. This chapter will summarize the literature since 1997 in each of the four areas contested by these two articles.

Characteristics of sprawl

Both articles used the term "compact" to describe one end of the development continuum, and "sprawl" to define the other end. Depending on the context, Gordon and Richardson (1997) equated compact development to high density or monocentric development, arguing that a city like Los Angeles is in fact compact by virtue of its high average density (Figure 35.1). This is a most unfortunate characterization of compactness. High density is still not the preferred living arrangement for most Americans; and monocentric development is an anachronism, as central business districts have become just one of many centers in large metropolitan areas. Density is only one dimension of sprawl, and that endless, uniform density is itself a hallmark of sprawl.

Gordon and Richardon (1997) sometimes equated sprawl to "low density," and other times to "dispersed," "decentralized," "polycentric," or "suburban" development. In contrast, Ewing (1997) defined sprawl explicitly as one of three forms: first, leapfrog or scattered development, second, commercial strip development, or third, large expanses of low-density or single-use development. This definition comports with popular notions of sprawl. But even this definition has its limitations, and was expanded to include any development pattern characterized by poor accessibility and lack of functional open space. In Ewing's view, compact development was anything that didn't fit this definition, meaning a development pattern with contiguity, strong centers, mixed land uses, medium to high densities, good accessibility, and permanent open spaces.

Measuring sprawl

Since 1997, the broader definition of sprawl has been operationalized in quantitative measures developed by ourselves and others. The first attempts to measure the extent of urban sprawl were crude. Several researchers created measures of urban sprawl that focused on density (Fulton *et al.* 2001; Malpezzi and Guo 2001; Nasser and Overberg 2001; Lopez and Hynes 2003; Burchfield *et al.* 2005). Density has the big advantage of being easy to measure with available data. Judged in terms of average

Figure 35.1 Endless Los Angeles. Source: Reid Ewing.

population density, Los Angeles looks compact. Another notable feature of these studies was the wildly different sprawl ratings given to different metropolitan areas by different analysts. With the exception of Atlanta, which always ranked as very sprawling, the different variables used to measure sprawl led to very different results. In one study, Portland was ranked as most compact and Los Angeles was way down the list. In another, their rankings were essentially reversed (Glaeser *et al.* 2001; Nasser and Overberg 2001).

Meanwhile, some scholars started developing more complete measures of urban sprawl. Galster *et al.* (2001) disaggregated land use patterns into eight dimensions: density, continuity, concentration, clustering, centrality, nuclearity, mixed use, and proximity. Sprawl was defined as a pattern of land use that has low levels in one or more of these dimensions. Each dimension was operationally defined, and six of the eight were quantified for 13 urbanized areas. New York and Philadelphia ranked as the least sprawling of the 13, and Atlanta and Miami as the most sprawling.

Since then, Galster and his colleagues have extended their sprawl measures to more than 50 metropolitan areas confirming the multidimensional nature of sprawl. In one study, metropolitan areas were ranked in 14 dimensions, some related to population, others to employment, and still others to both (Cutsinger *et al.* 2005). The 14 dimensions, which were reduced to seven factors through principal components analysis, tended to cancel out each other. Metropolitan areas ranking near the top on one factor were likely to rank near the bottom on another. Los Angeles, for example, ranked second on both "mixed use" and "housing centrality," but forty-eighth on "proximity" and forty-ninth on "nuclearity." Given so many overlapping variables, this type of analysis can get confusing.

Ewing, Pendall, and Chen (2002) also developed sprawl indices that like Galster's are multidimensional, but are more focused and demonstrate wider degrees of variability between metropolitan areas. They defined sprawl as any environment with first, a population widely dispersed in low-density residential development; second, a rigid separation of homes, shops, and workplaces; third, a lack of major employment and population concentrations downtown and in suburban town centers and other activity centers; and fourth a network of roads marked by very large block sizes and poor access from one place to another. These indices were used to measure sprawl for 83 of the nation's largest metropolitan areas. All sprawl indices were standardized, with mean values of 100 and standard deviations of 25. The indices were constructed so that the more compact a metropolitan area was, the larger its index value. More sprawling metropolitan areas had smaller index values. Thus, in the year 2000, the relatively compact Portland, Oregon, metropolitan area had an index value of 126, while the slightly smaller Raleigh-Durham metropolitan area had an index value of 54 (Figure 35.2). Los Angeles ended up near the middle of the pack, with an index of 102. Satellite photographs show the relatively compact Portland, Oregon, metropolitan area at the top, and the sprawling Raleigh-Durham metropolitan area at the bottom. Photographs are at the same scale. The patterns of development are dramatically different even at this scale.

Causes of sprawl

Conceiving sprawl differently, the two earlier articles cited different reasons for its proliferation. To Gordon and Richardson (1997), sprawl was a reflection of market forces. By their reasoning, consumers and businesses prefer outlying locations where land is inexpensive and congestion moderate. Modern telecommunications make

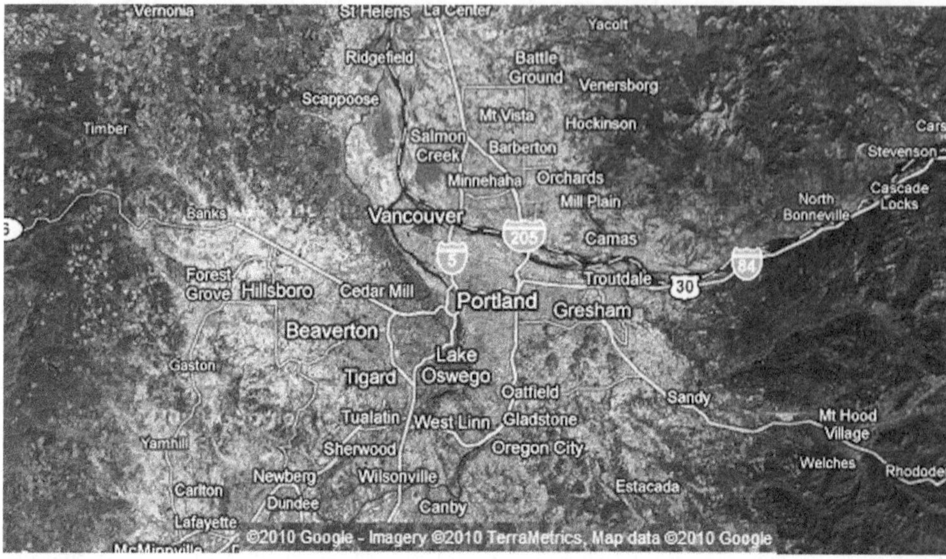

(a)

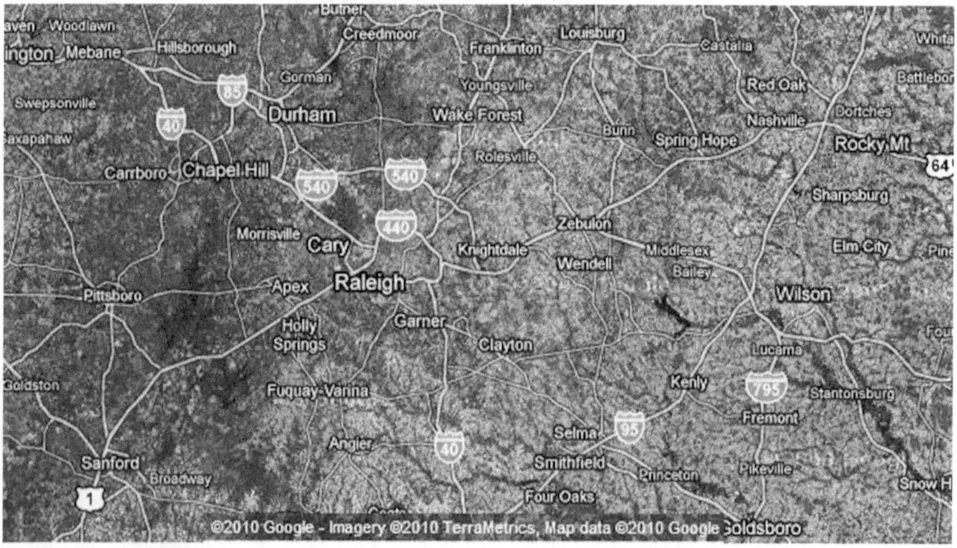

(b)

Figure 35.2 Satellite Images of Portland (a) and Raleigh (b). Source: www.maps.google.com

clustering of businesses unnecessary. The low cost of automobile travel allows people to live far from their places of work and shopping. The resulting decentralized settlement patterns are economically efficient, and the only sources of market failure – that might render settlement patterns inefficient – are subsidies for the automobile (encouraging long-distance driving) and local land use regulations (discouraging higher densities and mixed uses).

In contrast, we view land markets as fraught with imperfections that induce sprawl. Perfectly functioning markets require many buyers and sellers, good information about prices and quality, homogeneous products in each market, no external costs or benefits, and so forth. Land markets meet none of these requirements. The rate of land appreciation is uncertain, causing land speculation and (where speculators guess wrong or land becomes legally encumbered) scattered development. Owner-occupied housing is subsidized through the tax code, a public policy that particularly benefits suburban residents who are primarily homeowners. Outlying development is subsidized through utility rate structures that are independent of distance from central facilities. The land market is rife with externalities, and government regulation may introduce additional market distortions by holding down densities and segregating land uses.

Consumer preferences

Given the choice between low-density suburban living and high-density urban living, most Americans will choose the former. However, compact alternatives to sprawl come in many forms, and these forms collectively have more than "boutique appeal" (Gordon and Richardson's [1997] dismissive term). Studies show that with a more complete set of housing choices compact development can hold its own in the marketplace.

Perhaps the best national assessment of the demand for compact development is the National Survey on Communities, conducted for Smart Growth America and the National Association of Realtors (Belden *et al.* 2004). In this survey, respondents were given a choice between communities labeled "A" and "B." Community A was described as having single-family homes on large lots, no sidewalks, shopping and schools located a few miles away, commutes to work of 45 minutes or more, and no public transportation. In contrast, community B was described as having a mix of single-family and other housing, sidewalks, shopping and schools within walking distance, commutes of less than 45 minutes, and nearby public transportation.

Overall, 55 percent of Americans expressed a preference for community B, the smart growth community. This community appealed to 61 percent of those who were thinking of buying a house within the next three years. Commuting time had a significant influence on respondents' preferences. About a third of the respondents said they would choose the smart growth design if commutes were comparable, while another quarter preferred such a design if it also meant being closer to work.

Bolstering these results, a national consumer survey by the global public relations company Porter Novelli found that 59 percent of US adults now "support the development" of compact communities (defined in detail in the survey itself). Half would now be interested in living in a compact community (Handy *et al.* 2008). Levels of support were high among all groups except rural residents. More impressive than the absolute levels of support was the increase in support between survey years 2003 and 2005 by a statistically significant 15 percent. The authors attributed the increase to the media's attention to sprawl and its costs.

Shifts in the real estate market are evident already. Downtown and in-town housing tops the list of hot markets each year in the Urban Land Institute's *Emerging Trends in Real Estate* (ULI and PricewaterhouseCoopers 2005, 2006, 2007). In 2003, for the first time in the country's history, the sales price per square foot for attached housing – that is, condominiums and townhouses – was higher than that of detached housing. Because the demand is greater than the current supply, the price-per-square foot values of houses in mixed-use neighborhoods show price premiums ranging from 40 to 100 percent, compared to houses in nearby single-use subdivisions (Leinberger 2008).

In light of changing residential preferences and demographics, Nelson (2006) projects that in 2025, the demand for attached and small-lot housing will exceed the current supply by 35 million units (71 percent), while the demand for large-lot housing will fall short of the current supply (Figure 35.3).

Public subsidies

Consumer preferences help explain suburbanization and decentralization of activities within metropolitan areas, but they cannot explain the extent of dispersal, the absence of mixed land uses, and the loss of valuable natural areas. We must look to market failures to explain these phenomena.

The Office of Technology Assessment (1995) lists all manner of subsidies that result in urban sprawl. The biggest are subsidies for the highway system. If motorists had to cover the full social costs of automobile use – including vehicle emissions, free parking, uncompensated accident costs, military presence in the Middle East, and other external costs – they would likely opt for residential, work, shopping, and other locations that require a fraction of their current travel. This is what happens in Europe, where gasoline prices are about three times higher than in the US.

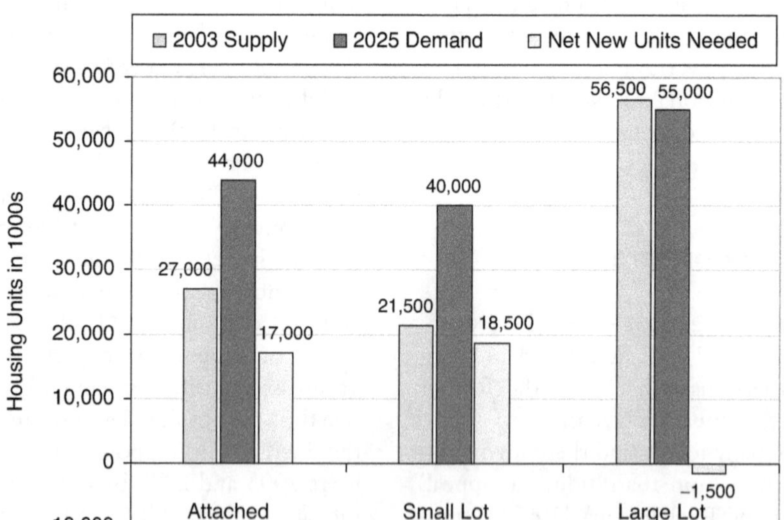

Figure 35.3 2003 Housing supply vs. 2025 housing demand in the US. Source: A.C. Nelson (2006).

Since 1997, at least 14 studies have estimated the true social cost of automobile use. Arguably the most careful study was conducted by Mark Delucchi for the Federal Highway Administration (Delucchi 1998). Nationwide totals from that study have been converted to a per vehicle mile basis by Todd Litman (2009) of the Victoria Transport Policy Institute (Table 35.1). With a subsidy of just 30 cents per vehicle mile, the average auto user would have to pay an additional gas tax of $6 per gallon (20 mpg × 30 cents/ gallon) to internalize the external costs of automobile use.

We recently got a glimpse of what would happen if users were forced to pay the full cost of their automobile use out-of-pocket. With the 2007 spike in gasoline prices, the prices of suburban and exurban homes dropped considerably, but centrally located properties enjoyed price appreciation (Cortright 2008).

Land use regulation

Does government regulation of land introduce market distortions? The literature since 1997 seems to suggest so, but there are also caveats. In his book *Zoned Out: Regulation, Markets, and Choices in Transportation and Metropolitan Land Use*, Jonathan Levine (2006) argues that extensive use of restrictive low-density zoning constrains the exercise of a free market in real estate development and that denser urban development would result from a more open market. In other words, one of the causes of sprawl is our current system of locally controlled zoning regulations that essentially mandate the construction of sprawl. This is a contention loudly voiced by new urbanist planners as well. Levine supports this contention in a number of ways, some more compelling than others. First, he cites studies showing that municipal zoning lowers densities below market levels and creates more exclusivity in suburban communities (p. 52). Much of this evidence, however, is out-of-date. Literally hundreds of jurisdictions around the nation have amended their zoning ordinances in the intervening years to allow dense, mixed-use, transit-oriented development at least in certain districts. Somewhat more convincing is the evidence from a survey of land developers, who perceive land use regulation as the principal barrier to "alternative development" forms, particularly dense development (pp. 127–132). Nearly half of the developers proposing such alternatives report that they have been turned down by planning authorities. The caveat here is that land developers

Table 35.1 Estimates of motor vehicle costs.

Cost item	*Example*	*Per vehicle mile*
Personal nonmonetary costs of using motor vehicles	Motorist personal travel time and accident pain and suffering	17.4–25.5¢
Private-sector motor-vehicle goods and services	Vehicle expenses, paid travel time	40.2–45.3¢
Bundled private sector costs	Parking subsidized by businesses	2.7–9.4¢
Public infrastructure and services	Public roads, parking subsidized by local governments	5.3–8.8¢
Monetary externalities	External accident damages, congestion	3.4–6.2¢
Nonmonetary externalities	Environmental damages, crash pain	10.4–25.2¢
	Total	$0.79–1.20

Source: Litman (2009).

may be rather biased when it comes to dealings with local government. A third block of evidence comes from household preference surveys in Atlanta and Boston (pp. 149–162). Levine finds that a significant number of households in Atlanta live in neighborhoods that are less walkable than they would ideally choose. There are two caveats here. One is the close correspondence between residential preferences and residential character in Boston, which also has land use regulation. Apparently, regulation has not produced such a mismatch in the Boston metropolitan area. The other caveat is the inability to discern the cause of the mismatch in Atlanta. All we know is that walkable neighborhoods are undersupplied in Atlanta. This could as well be due to, for example, the failure of the development community to keep pace with changing residential preferences.

Costs of sprawl

Urban planners are ultimately less interested in development patterns, per se, than in the costs and benefits of one pattern versus another. That is to say, there are no inherently good or bad patterns, only good or bad outcomes. The loaded term "sprawl" has come to be applied to certain development patterns because of their documented negative outcomes. The following review focuses on costs of sprawl that have been extensively researched since the 1997 point-counterpoint articles.

Vehicle miles traveled

In 1997, Gordon and Richardson (1997) claimed that "... the link between high-density development and reduced VMT (vehicle miles of travel), and hence reduced energy consumption, is by no means clear." This statement was simply incorrect. By 1997, a number of studies had linked density and other dimensions of compact development to lower VMT. The statement is even less defensible today. The potential to moderate travel demand through changes in the built environment is now the subject of more than 200 empirical studies. Indeed, it has become the most heavily researched subject in urban planning. There are at least 10 surveys of this literature (Handy 1996; Badoe and Miller 2000; Cao *et al.* 2009; Crane 2000; Ewing and Cervero 2001; Heath *et al.* 2006; McMillan 2005; Saelens *et al.* 2003; Stead and Marshall 2001; Saelens and Handy 2008).

In travel research, urban development patterns have come to be characterized by "D" variables. The original "three Ds," coined by Cervero and Kockelman (1997), are density, diversity, and design. The Ds have multiplied since Cervero and Kockelman's original article, with the addition of destination accessibility and distance to transit (Ewing and Cervero 2001).

For 14 carefully controlled travel studies, Ewing and Cervero (2001) synthesized the literature by extracting elasticities of VMT and vehicle trips (VT) with respect to the first four Ds – density, diversity, design, and destination accessibility. From the elasticities in Table 35.2, we would expect a doubling of neighborhood density to result in approximately a 5 percent reduction in both VT and VMT, all else being equal. Note that the elasticity of VMT with respect to destination accessibility is larger than the other three combined, suggesting that areas of high accessibility – such as center cities – may produce substantially lower VMT than dense mixed-use developments in the exurbs. The effects of the four Ds captured in this table are additive: Doubling all four Ds would be expected to reduce VMT by about one-third.

A caveat associated with this research is the possibility of self-selection. In theory, the documented association between the

Table 35.2 Typical elasticities of travel with respect to four D variables.

	Vehicle trips (VT)	*Vehicle miles traveled (VMT)*
Local density	–.05	–.05
Local diversity (mix)	–.03	–.05
Local design	–.05	–.03
Regional accessibility	.00	–.20

Source: Ewing and Cervero (2001).

built environment and travel choices could as well be due to people who are predisposed to walk or use transit selecting to live in places where these choices exist, rather than the built environment actually influencing people's travel choices. Does residential choice come first, and travel choice or some other outcome follow (environmental determinism)? Or do people's propensities for travel determine their choice of residential environment (self selection)? Between environment and attitude, which drives behavior?

At least 38 studies using nine different research approaches have attempted to test and control for residential self selection (Cao *et al.* 2009). Nearly all of them found "resounding" evidence of statistically significant associations between the built environment and travel behavior, independent of self-selection influences: "Virtually every quantitative study reviewed here, after controlling for self-selection through the various ways discussed earlier, identified a statistically significant influence of one or more (built environmental) measures on the (travel behavior) variable of interest" (Cao *et al.* 2009, 389).

Oil dependence and climate change

Gordon and Richardson (1997) pointed to the "global energy glut," the weakness of the OPEC cartel, and the low real price of gasoline as evidence that energy impacts of sprawl were not worth worrying about. They argued that advances in vehicle emission control technology would solve our air quality problems. In contrast to this argument, Ewing (1997, 114) countered that:

> While the best case envisioned by [Gordon and Richardson] has the real price of gasoline holding steady, it is the worst case that worries others. ... The fact that the most recent large-scale war fought was in the Persian Gulf is itself a testament to the risk of relying on the political stability of this region for a commodity [oil] so essential to economic activity. ... Being unregulated, carbon dioxide emissions represent a bigger threat to national welfare than do regulated emissions. There is now a near-consensus within the scientific community that carbon dioxide build-up in the atmosphere is causing global climate change, and that the long-term effects could be catastrophic.

Twelve years after the exchange, there seems to be little doubt that the "worst case" is upon us. Our dependence on foreign oil has never been greater. Gasoline recently peaked at an all-time high of more than \$4 per gallon, CO_2 concentration in the atmosphere is the highest it has been in the past 20 million years, and the fingerprints of climate change are everywhere (Emanuel 2005; Westerling *et al.* 2006; Madsen and Figdor 2007).

Worldwide demand for oil continues to grow as Asia and the rest of the world follow the auto-centric ways of the US. The reference projection from the Energy Information Agency (EIA) estimates that US petroleum use will increase 8 percent between 2005 and 2030, and China's will increase by a staggering 78 percent, bringing it to within 70 percent of the US consumption level (EIA 2008b). Worldwide demand in 2030 is expected to be approximately 112.5 million barrels of oil per day. Petroleum production levels, however, are not expected to keep pace with this growing demand. EIA projects that world production of conventional oil in 2030 will be only 102.9 million barrels per day – about 10 million barrels short of demand (EIA 2008a).

Peak oil production ("peak oil") occurred at a national level in countries around the world, with peaks in the lower 48 United States, Alaska, and Mexico occurring in 1971, 1989, 2004, respectively (Zittel and Schindler 2007). Researchers now are engaged in estimating when the phenomenon will occur at a global scale. The US Government Accountability Office (2007) puts the date sometime before 2040. Regardless of the peaking date, a constant factor running through the analyses is that oil prices will increase significantly; the only real debate is how fast (Haubrich and Meyer 2007).

The literature establishing connections between energy consumption and urban form, already well-established in 1997, has continued to expand (Alberti 1999; Andrews 2008; Bento *et al.* 2003; Burchell *et al.* 1998; Cooper *et al.* 2001; Ewing and Rong 2008; Kenworthy and Laube 1999; Saunders *et al.* 2008; US. Environmental Protection Agency 2003). The results of these more recent studies further confirm that compact development patterns are substantially more energy-efficient than low-density sprawl. With the arrival of peak oil, compact development practices are likely to become more important for national energy policy.

Climate change, closely tied to fossil fuel consumption, has emerged since 1997 as the leading environmental issue facing the planet. Greenhouse gas concentrations have risen from pre-industrial levels of approximately 280 parts per million (ppm) CO_2 equivalent (CO_2e) to 430 ppm CO_2e (Stern 2007). Rather than slowing down, the growth of atmospheric CO_2 seems to be speeding up as a result of the expanding global economy, the increasing carbon intensity of the global economy, and a decline in the efficiency of CO_2 sinks on land and oceans (Canadell *et al.* 2007). The result is climate change. "Warming of the climate system is unequivocal, as is now evident from observations of increases in global average air and ocean temperatures, widespread melting of snow and ice, and rising global mean sea level" (IPCC 2007: 5).

Among the benefits of compact development, perhaps the most important are greater energy security and reduced carbon footprint. Compact development can reduce fuel consumption and CO_2 emissions by 20 to 40 percent as compared to sprawl (Ewing *et al.* 2008).

Physical activity, obesity, and public health

The point-counterpoint articles did not address the impacts of sprawl on physical activity, obesity, or public health generally. This was 1997, only a year after the Surgeon General's report on *Physical Activity and Health* and several years before the Robert Wood Johnson Foundation launched its active living initiatives in response to the US obesity epidemic and related diseases. Historically, urban planners had focused strictly on travel, while physical activity researchers had focused on leisure-time activity. Only since about 2000 have the two fields converged.

The literature on the built environment and physical activity has expanded exponentially as funding has become available for active living research. The tremendous volume of research has generated a review of the many literature reviews (Gebel *et al.* 2007). Results of this research clearly show an association between the built environment and physical activity levels.

In 2003, sprawl was related for the first time to physical activity, obesity, and chronic diseases (Ewing *et al.* 2003). After controlling for age, education, other sociodemographic and behavioral covariates, adults living in sprawling countries walked less, weighed more, were more likely to be obese and to suffer from high blood pressure than those living in compact counties.

Seventeen of 20 subsequent studies have established statistically significant links between some aspect of the built environment and obesity (Papas *et al.* 2007), and correlations between sprawl and obesity have been affirmed (Committee on Physical Activity, Health, Transportation, and Land Use 2005; Frank *et al.* 2004; Kelly-Schwartz *et al.* 2004; Lopez 2004; Sturm and Cohen 2004; Cho *et al.* 2006; Doyle *et al.* 2006; Ewing *et al.* 2006; Rundle *et al.* 2007; Joshu *et al.* 2008).

The possibility of self-selection has been raised in this literature as well. Two studies have garnered media attention by contending that residential self-selection, not environmental determinism, accounts for the relationship between sprawl and obesity (Plantigna and Bernell 2007; Eid *et al.* 2008). Both conclude that people with higher body mass indices choose to live in sprawling neighborhoods. One can understand indifference to the built environment on the part of overweight individuals, but actual preference for places that preclude physical activity, when these people don't plan to engage in it anyway, defies logic.

Other costs

Recent research also shows that compact development outperforms sprawl in the following areas:

- income growth (Nelson and Foster 1999; Nelson and Peterman 2000);
- central city economic health (Nelson *et al.* 2004a, Dawkins and Nelson 2003);
- protection of farmland (Nelson and Sanchez 2005);
- racial integration (Nelson *et al.* 2004b; Nelson *et al.* 2005); and
- residential neighborhood quality (Nelson *et al.* 2007)

Cures for sprawl

The only policy intervention endorsed by Gordon and Richardson (1997) was the imposition of congestion charges and emission fees as shadow prices for external costs of automobile use, specifically for delay and air pollution imposed on others. But while congestion pricing and emission fees have been touted by economists for decades, politicians have not exactly rushed to meter their constituents' travel.

The first federal congestion pricing demonstration program, from 1973 through 1978, produced no demonstrations. The well-funded federal Congestion Pricing Pilot Program, first authorized in 1991 and reauthorized in 2005 with a bigger budget, has produced very little other than express toll lanes for solo drivers. A change of name, to the Value Pricing Pilot Program, has failed to overcome resistance to charges for something that has always been free, road use.

A recent case in point is the failed effort of New York City to adopt an area-wide congestion pricing program. The program would have charged $8 for automobiles to enter a specific zone in a southern portion of Manhattan. Despite the power of Mayor Michael Bloomberg, as well as years of planning, the New York City Congestion Pricing Plan died a political death at the hands of the State Assembly. Area-wide congestion pricing is a good idea whose time, apparently, has not come.

Our answer to sprawl is proactive planning of the type found almost everywhere except in the United States (but beginning to appear here out of necessity). What Gordon and Richardson (1997) refer to as "command-and-control" policies is really just planning. The common practice of local governments in the US, to wait for property owners to come forward with rezoning requests, is not planning but reacting, and hardly command-and-control.

Proactive plans should be supplemented by policies that reward good development and discourage bad. In the first wave of growth management nationally, the concern was how much growth would be allowed. In the second wave, the focus shifted to where and when growth would be permitted, and who would pay for it. The third wave is upon us, shifting the emphasis to what kind of growth is allowed or encouraged. Oregon's Transportation Planning Rule, New Jersey's State Plan, California's Climate Change – Smart Growth Act (SB 375) are examples of initiatives to upgrade the quality of development, wherever and whenever it should occur.

Lest this answer to sprawl appear hopelessly European, Ewing (1997) cited an example from the United States. It wasn't Portland, OR, Arlington County, VA, or San Diego, CA, leaders in this type of planning. They could be dismissed as

Figure 35.4 Public–private partnership at Baldwin Park, California. Source: Baldwin Park Development Corporation – used by permission.

anomalies. It wasn't Charlotte, NC, Denver, CO, or Salt Lake City, UT. They hadn't embraced transit and transit-oriented development at the time of the point-counterpoint. Rather it was Orlando, FL, home of Mickey Mouse and Shamu the killer whale. The point was that if Orlando could do it, anyone could.

Orlando government had entered into a partnership with the owners of multiple tracts southeast of the city. Through a cost-sharing arrangement, the partnership had prepared a master plan and development standards for the 12,000-acre site. The plan, development standards, and financial and administrative incentives are being used to encourage compact, mixed-use development where sprawl would otherwise almost surely occur.

An even better example of a public–private partnership is the new town in-town of Baldwin Park, built on the site of the former Naval Training Center in Orlando (Ewing 2007). The city chose a master developer with a proven track record in reusing a military base. The two have been partnering ever since. In Baldwin Park, a public-private partnership has taken the place of Gordon and Richardson's unfettered market (Figure 35.4). Here, a handshake has replaced the invisible hand so revered by these two economists.

As the nation grows to 400 million by 2040, *how* it grows will affect the quality of life and economic well-being of Americans. The evidence is mounting that compact development, done right, can confer more benefits with fewer costs than urban sprawl. Concerns about energy, health, climate change, along with the shifting demographics have resolved the debate of compactness vs. sprawl.

References

Alberti, M. (1999). "Urban Patterns and Environmental Performance: What Do We Know?" *Journal of Planning Education and Research*, 19: 151–163.

Andrews, C.J. (2008). "Greenhouse Gas Emissions along the Rural-Urban Gradient." *Journal of Environmental Planning and Management*, 51(6): 847–870.

Badoe, D.A. and Miller, E.J. (2000). "Transportation-Land-Use Interaction: Empirical Findings in North America, and Their Implications for Modeling." *Transportation Research Part D*, 5: 235–263.

Belden, Russonello, and Stewart (2004). *National Survey on Communities*. Washington, D.C.: National Association of Realtors and Smart Growth America. www.brspoll.com/Reports/Smart%20Growth.pdf. (accessed 18 August 2010).

Bento, A.M., Cropper, M.L, Mobarak, A.M., and Vinha, K. (2003). *The Impact of Urban Spatial Structure on Travel Demand in the United States*. Washington, DC: World Bank.

Burchell, R.W., Shad, N., Listokin, D., Phillips, H., Downs, A., Seskin, S., Davis, J.S., Moore, T., Helton, D. and Gall, M. (1998). *The Costs of Sprawl – Revisited*. Washington, DC: Transportation Research Board.

Burchfield, M., Overman, H.G., Puga, D. and Turner, M.(2005). "Causes of Sprawl: A Portrait from Space." *Quarterly Journal of Economics*, 121(2): 587–633.

Canadell, J.G., Le Quéré, C., Raupach, M., Field, C.B., Buitenhuisc, E.T., Ciais, P., Conway, T.J. Gillett, N.P., Houghton, R.A., and Marland, G. (2007). "Contributions to Accelerating Atmospheric CO_2 Growth from Economic Activity, Carbon Intensity, and Efficiency of Natural Sinks," *Proceedings of the National Academies of Science, Early Edition*. www.pnas.org/cgi/reprint/0702737104v1 (accessed 18 August 2010).

Cao, X., Mokhtarian, P.L., and Handy, S.L. (2009). "Examining the Impacts of Residential Self-Selection on Travel Behaviour: A Focus on Empirical Findings." *Transport Reviews*, 29(3): 359–395.

Cervero, R. and Kockelman, K. (1997). "Travel Demand and the 3Ds: Density, Diversity, and Design," *Transportation Research D*. 2, 199–219.

Cho, S., Chen, Z., Yen, S.T., and Eastwood, D.B. (2006). "The Effects of Urban Sprawl on Body Mass Index: Where People Live Does Matter,"

The *52nd Annual ACCI Conference*, Baltimore, Maryland, March 15–18.

Committee on Physical Activity, Health, Transportation, and Land Use (2005). *Does the Built Environment Influence Physical Activity? Examining the Evidence – Special Report 282*. Washington, DC: National Academies of Science.

Cooper, J., Ryley, T., and Smyth, A. (2001). "Energy Trade-offs and Market Responses in Transport and Residential Land-use Patterns: Promoting Sustainable Development Policy." *Urban Studies*, 38(9): 1573–1588.

Cortright, J. (2008). *Driven to the Brink. How the Gas Price Spike Popped the Housing Bubble and Devalued the Suburbs*. CEOs for Cities, Portland, OR. http://brokerinsider.com/pdf/driventothe brinkfinal.pdf (accessed 18 August 2010).

Crane, R. (2000). "The Influence of Urban Form on Travel: An Interpretive Review." *Journal of Planning Literature*, 15(1): 3–23.

Cutsinger, J., Galster, G., Wolman, H., Hanson, R., and Towns, D. (2005). "Verifying the Multi-Dimensional Nature of Metropolitan Land Use: Advancing the Understanding and Measurement of Sprawl." *Journal of Urban Affairs*, 27(3): 235–259.

Dawkins, C.J. and A.C. Nelson (2003). "Statewide Growth Management Policy and Central City Revitalization." *Journal of the American Planning Association*, 69(4): 381–396.

Delucchi, M. (1998) *The Annualized Social Cost of Motor-Vehicle Use in the United States, Based on 1990–1991 Data: Summary of Theory, Methods, Data, and Results*, Institute of Transportation Studies, University of California, Davis.

Doyle, S., Kelly-Schwartz, A., Schlossberg, M., and Stockard, J. (2006). "Active Community Environments and Health: The Relationship of Walkable and Safe Communities to Individual Health." *Journal of the American Planning Association*, 72(1): 19–31.

Eid, J., Overman, H.G., Puga, D., and Turner, M.A. (2008). "Fat City: Questioning the Relationship Between Urban Sprawl and Obesity." *Journal of Urban Economics*, 63: 385–404.

Energy Information Agency (EIA). (2008a). *International Energy Outlook 2008*. Washington, DC: US Department of Energy.

—— (2008b). *World Liquids Consumption by Region, Reference Case, 1990–2030*. Washington, DC: US Department of Energy.

Emanuel, K. (2005). "Increasing Destructiveness of Tropical Cyclones over the Past 30 Years." *Nature*, 436: 686–688.

Ewing, R. (1997). "Is Los Angeles–Style Sprawl Desirable?" *Journal of the American Planning Association*, 63: 107–126.

—— (2007). "Finding Happiness in Public-Private Partnerships: The Case for Case Studies." *Planning*, January, 53.

Ewing, R. and Cervero, R. (2001). "Travel and the Built Environment: A Synthesis." *Transportation Research Record*, 1780: 87–114.

Ewing, R. and Cervero, R. (2010). "Travel and the Built Environment: A Meta-Analysis," *Journal of the American Planning Association*. 76(3): 265–294.

Ewing, R. and Rong, F. (2008). "The Impact of Urban Form on US Residential Energy Use." *Housing Policy Debate*, 19(1): 1–30.

Ewing, R., Pendall, R., and Chen, D. (2002). *Measuring Sprawl and Its Impact*. Washington D.C: Smart Growth America.

Ewing, R., Brownson, R.C., and Berrigan, D. (2006). "Relationship between Urban Sprawl and Weight of United States Youth." *American Journal of Preventive Medicine*, 31(6): 464–474.

Ewing, R., Schmid, T., Killingsworth, R., Zlot, A., and Raubenbush, S. (2003). "Relationship between urban sprawl and physical activity, obesity, and morbidity." *American Journal of Health Promotion*, 18(1): 47–57.

Ewing, R., Bartholomew, K., Winkelman, S., Walters, J., and Chen, D. (2008). *Growing Cooler: The Evidence on Urban Development and Climate Change*, Washington, DC: Urban Land Institute.

Frank, L.D., Andersen, M.A., and Schmid, T.L. (2004). "Obesity Relationships with Community Design, Physical Activity and Time Spent in Cars." *American Journal of Preventive Medicine*, 27(2): 87–96.

Fulton, W., Pendall, R. Nguyen, M. and Harrison, A. (2001). *Who Sprawls Most? How Growth Patterns Differ across the US*. Washington, DC: Brookings Institution.

Galster, G., Hanson, R., Ratcliffe, M., Wolman, H., Coleman, S., and Freihage, J. (2001). "Wrestling Sprawl to the Ground: Defining and Measuring an Elusive Concept." *Housing Policy Debate*, 12(4): 681–717.

Gebel, K., Bauman, A.E., and Petticrew, M. (2007). "The Physical Environment and Physical Activity: A Critical Appraisal of Review

Articles." *American Journal of Preventive Medicine*, 32(5): 361–369.

Glaeser, E., Kahn, M., and Chu, C. (2001). *Job Sprawl: Employment Location in US. Metropolitan Areas*, Washington, DC: icy, Brookings Institution.

Gordon, P. and Richardson, H. (1997). "Are Compact Cities a Desirable Planning Goal?" *Journal of the American Planning Association*, 63: 95–106.

Handy, S.L. (1996). "Understanding the Link between Urban Form and Non-work Travel Behavior." *Journal of Planning Education and Research*, 15(3): 183–198.

Handy, S., Sallis, J.F., Weber, D., Maibach, E. and Hollander, M. (2008). "Is Support for Traditionally Designed Communities Growing? Evidence from Two National Surveys," *Journal of the American Planning Association*, 74(2): 209–221.

Haubrich, J.G. and Meyer, B. (2007). *Peak Oil*. Cleveland, OH: Federal Reserve Bank of Cleveland.

Heath, G.W., Brownson, R.C., Kruger, J., Miles, R., Powell, K.E., Ramsey, L.T., and the Task Force on Community Preventive Services. (2006). "The Effectiveness of Urban Design and Land Use and Transport Policies and Practices to Increase Physical Activity: A Systematic Review." *Journal of Physical Activity and Health*, 3: 55–76.

Intergovernmental Panel on Climate Change (IPCC). (2007). *Climate Change 2007: The Physical Science Basis, Summary for Policymakers*. Working Group I contribution of the Intergovernmental Panel on Climate Change: Fourth Assessment Report. www.ipcc.ch/.

Joshu, C.E, Boehmer, T.K., Ewing, R., and Brownson, R.C. (2008). "An Examination of Personal, Neighborhood and Urbanization Correlates of Obesity in the United States." *Journal of Epidemiology and Community Health*, 62: 202–208.

Kelly-Schwartz, A.C., Stockard, J., Doyle, S., and Schlossberg, M. (2004). "Is Sprawl Unhealthy: A Multilevel Analysis of the Relationship of Metropolitan Sprawl to the Health of Individuals." *Journal of Planning Education and Research*, 24, 184–196.

Kenworthy, J. and Laube, F. (1999). "A Global Review of Energy Use in Urban Transport Systems and its Implications for Urban Transport and Land-Use Policy." *Transportation Quarterly*, 53(4), 23–48.

Leinberger, C.B. (2008). *The Option of Urbanism: Investing in a New American Dream*, Washington, DC: Island Press.

Levine, J. (2006). *Zoned Out: Regulation, Markets, and Choices in Transportation and Metropolitan Land Use*, Washington, DC: Resources for the Future.

Litman, T. (2009). *Transportation Cost and Benefit Analysis II – Literature Review*, Victoria Transport Policy Institute, Victoria, British Columbia. http://www.vtpi.org/tca/(accessed 18 August 2010).

Lopez, R. (2004). "Urban Sprawl and Risk for Being Overweight or Obese." *American Journal of Public Health*, 94(9): 1574–1579.

Lopez, R. and Hynes, H.P. (2003). "Sprawl in The 1990s: Measurement, Distribution, and Trends." *Urban Affairs Review*, 38(3): 325–355.

Madsen, T. and Figdor, E. (2007). *When It Rains, It Pours: Global Warming and the Rising Frequency of Extreme Precipitation in the United States*. Boston, MA: Environment America Research & Policy Center. http://www.environmentamerica.org/uploads/oy/ws/oywshWAwZy-EXPsabQKd4A/When-It-Rains-It-Pours–US–WEB.pdf. (accessed 18 August 2010).

Malpezzi, S. and Guo, W. (2001). *Measuring "Sprawl": Alternative Measures of Urban Form in US. Metropolitan Areas*, Madison, WI: Center for Urban Land Economics Research, University of Wisconsin.

McMillan, T.E. (2005). "Urban Form and a Child's Trip to School: The Current Literature and a Framework for Future Research." *Journal of Planning Literature*, 19(4): 440–456.

Nasser, H.E. and Overberg, P. (2001). "What You Don't Know about Sprawl: Controlling Development a Big Concern, but Analysis Has Unexpected Findings," *US Today*, February 22, 2001.

Nelson, A.C. (2006). "Leadership in a New Era." *Journal of the American Planning Association*, 72(4): 393–407.

Nelson, A.C. and Foster, K. (1999). "Metropolitan Governance Structure and Economic Performance." *Journal of Urban Affairs*, 21(3): 309–324.

Nelson, A.C. and Peterman, D.R. (2000). "Does Growth Management Matter?" *Journal of Planning Education and Research*, 19(3): 277–286.

Nelson, A.C. and Sanchez, T.W. (2005). "The Effectiveness of Urban Containment Regimes in Reducing Exurban Sprawl." *DISP 160*: 42–47.

Nelson, A.C., Sanchez, T.W. and Dawkins, C.J. (2004a). "Urban Containment and Residential Segregation: A Preliminary Investigation." *Urban Studies*, 41(2): 423–440.

Nelson, A.C., Burby, R.J., Feser, E., Dawkins, C.J., Quercia, R., and Malizia, E. (2004b). "Urban Containment and Central City Revitalization." *Journal of the American Planning Association*, 70(4): 411–425.

Nelson, A.C., Dawkins, C.J. and Sanchez, T.W. (2005). "The Effect of Urban Containment and Mandatory Housing Elements on Racial Segregation in US Metropolitan Areas," *Journal of Urban Affairs*, 26(3): 339–350.

Nelson, A.C., Dawkins, C.J. and Sanchez, T.W. (2007). *Urban Containment and Society*. Hampshire: Ashgate.

Office of Technology Assessment (1995). *The Technological Reshaping of Metropolitan America*. Washington, DC: Congress of the United States, 193–218.

Papas, M.A., Alberg, A.J., Ewing, R., Helzlsouer, K.J., Gary, T.L., and Klassen, A.C. (2007). "The Built Environment and Obesity: A Review of the Evidence." *Epidemiologic Reviews*, 29(1): 129–143

Plantinga, A.J. and Bernell, S. (2007). "The Association between Urban Sprawl and Obesity: Is it a Two-Way Street?" *Journal of Regional Science*, 45(3), 473–492.

Rundle, A., Roux, A.V.D., Freeman, L.M., Miller, D., Neckerman, K.M., and Weiss, C.C. (2007). "The Urban Built Environment and Obesity in New York City: A Multilevel Analysis." *American Journal of Health Promotion*, 21(4): 326–334.

Saelens, B.E. and Handy, S. (2008). "Built Environment Correlates of Walking: A Review." *Medicine & Science in Sports & Exercise*, 40(S), S550-S567.

Saelens, B.E., Sallis, J.F., and Frank, L.D. (2003). "Environmental Correlates of Walking and Cycling: Findings from the Transportation, Urban Design, and Planning Literatures." *Annals of Behavioral Medicine*, 25(2): 80–91.

Saunders, M.J., Kuhnimhof, T., Chlond, B., da Silva, A.N. (2008). "Incorporating Transport Energy into Urban Planning." *Transportation Research Part A*, 42(6): 874–882.

Stead, D. and Marshall, S. (2001). "The Relationships between Urban Form and Travel Patterns. An International Review and Evaluation." *European Journal of Transport and Infrastructure Research*. 1(2): 113–141.

Stern, N. (2007). *The Economics of Climate Change: The Stern Review*. Cambridge: Cambridge University Press. www.hm-treasury.gov.uk/independent_reviews/stern_review_economics_climate_change/stern_review_report.cfm. (accessed 18 August 2010).

Sturm, R. and Cohen, D. (2004). "Suburban Sprawl and Physical and Mental Health." *Public Health*, 118(7): 488–496.

ULI–the Urban Land Institute and PricewaterhouseCoopers LLP (2005). *Emerging Trends in Real Estate 2005*. Washington, DC.

—— (2006). *Emerging Trends in Real Estate 2006*. Washington, DC.

—— (2007). *Emerging Trends in Real Estate 2007*. Washington, DC.

US Environmental Protection Agency (2003). *Characteristics and Performance of Regional Transportation Systems*. Washington, DC.

US Government Accountability Office (GAO). (2007). *Crude Oil: Uncertainty about Future Oil Supply Makes it Important to Develop a Strategy for Addressing a Peak and Decline in Oil Production*. Washington, DC: Author. Retrieved March 21, 2008, from http://frwebgate.access.gpo.gov/cgi-bin/getdoc.cgi?dbname = gao&docid = f:d07283.pdf.

Westerling, A.L., Hidalgo, H.G., Cayan, D.R., and Swetnam, T.W. (2006). "Warming and Earlier Spring Increase Western US Forest Wildfire Activity." *Science*, 313: 940–943. http://www.sciencemag.org/cgi/rapidpdf/1128834.pdf. (accessed 18 August 2010).

Zittel, W., & Schindler, J. (2007). *Crude Oil: The Supply Outlook*. Ottobrunn, Germany: Energy Watch Group. Retrieved June 30, 2008, from http://www.energywatchgroup.org/fileadmin/global/pdf/EWG_Oilreport_10-2007.pdf.

Further reading

Ewing, R., Bartholomew, K., Winkelman, S., Walters, J., and Chen, D. (2008). *Growing Cooler: The Evidence on Climate Change and Urban Development*. Washington DC: Urban Land Institute. Aimed at the climate change dimensions of urban development and transportation, the book also covers the more general subject of land use–transportation interactions and other allied topics, including market demand for Smart Growth, self-selection bias, and government policy responses.

Ewing, R. (2009). *Pedestrian- and Transit-Oriented Design: A Primer for Smart Growth.* Washington DC: Urban Land Institute and American Planning Association. Comprehensive summaries of the land use–transportation literature with a particular focus on urban design related dimensions; it also contains original research into the quantitative and qualitative impacts of urban development forms.

Levine, J. (2006). *Zoned Out: Regulation, Markets, and Choices in Transportation and Metropolitan Land Use.* Washington DC: Resources for the Future. Provides a thought-provoking critique and redefinition of the role that regulation plays in real estate markets, particularly as it pertains to land use-transportation policy.

Newman, P. and Kenworthy, J. (1999). *Sustainability and Cities: Overcoming Automobile Dependence.* Washington, DC: Island Press; and Boarnet, M. and Crane, R. (2001). *Travel by Design: The Influence of Urban Form on Travel.* Oxford: Oxford University Press. A pair of readings presenting significantly different perspectives on land use–transportation influences.

36

Living together or apart

Social mixing, social exclusion, and gentrification

Ali Madanipour

In recent decades, cities have undergone dramatic social, economic and cultural changes, triggering new tensions and debates about the production and use of urban space (Atkinson and Bridge 2005). As urban populations grow, diversify, and fragment, what should be the response of urban design? Should city designers accommodate and facilitate the growing social and economic differentiation through segregation and gentrification, or should they be expected to encourage social mixing and integration? Are social exclusion, gentrification, and displacement inevitable outcomes of development and revitalization of cities? This chapter starts with an outline of the changing urban conditions, followed by an examination of the spatial implications of social diversity and polarization; whether urban design should facilitate different social groups to live together or live apart and the intended or unintended consequences of these strategies, which include segregation, exclusion, social mixing, gentrification and displacement.

Changing societies, transforming cities

During the past decade, the populations of the largest US cities have all grown (US Census Bureau 2009). In 2007–2008, out of 75 US cities with populations larger than 200,000, only 11 lost population. In some US cities, such as Washington, DC and Atlanta, central cities have grown faster than the suburbs, appearing to reverse a longstanding trend (Frey 2009). For the first time since the 1930s, New York in the past two decades has once again led the city population growth in the United States (National Atlas 2009). In Europe, while some post-industrial and post-socialist cities continue to shrink, there is a process of re-urbanization, partially reversing the counter-urbanization process which characterized the 1970s (Buzar *et al.* 2007; Champion 2001). In the UK, London and smaller cities in the south and east have grown rapidly in the last three decades. Meanwhile, after long periods of decline, large UK cities in the north and west have also displayed a recent recovery in population trends, with modest population increases in their central areas (Parkinson 2006; Pointer 2005; Urban Task Force 1999). These demographic changes run parallel to economic, political and social change and their associated spatial transformation.

The last three decades could be characterized as the ascendancy of market-based solutions to economic and social problems.

The global economic crisis of 2008 initially appeared as an end to an era that started a generation ago, proposing a model of development that would shrink the state and stimulate the economy through deregulation. In a new international division of labor, manufacturing industries were relocated to regions with low production cost, leading to a decline of these industries in the first-wave industrial countries, where the services sector filled the gap in employment. Meanwhile the new information and communication technologies indicated the possibility of a new rationale for western economies, proposing what is sometimes described as knowledge-based economy. This new model emphasized knowledge and innovation as its primary assets, information and communication technologies as its tools, high-skilled workers as its elite workforce, financial industries as its engine, and free market as its operational mechanism. Cities are envisaged as firms engaged in international competition with other cities to attract resources.

Socially, cities have experienced structural changes at different levels (Buzar *et al.* 2007; Champion 2001; Glasmeier 2005). Globalization has fuelled international migration, helping urban populations to grow and diversify. More people live longer, and smaller households live in more diverse arrangements. Lifestyle changes have diversified ways of living, moving away from the more standardized routines of the past. Consumption has been a driving force in the economy of cities, leading to the creation of spaces and activities that attract new visitors and residents. In these cities, socio-economic and cultural diversity has created a mosaic of difference, which is simultaneously celebrated and feared. In the industrial economy, workers organized, unionized, and obtained welfare provisions and support mechanisms, thus reducing the gap between the highest and lowest paid workers (Esping-Andersen 1999). However, the nature of work has changed in the new service economy and under the conditions of globalization, where support mechanisms have weakened, leading to a pervasive sense of fragmentation and insecurity. The effects of economic liberalization and the reduction in the size and roles of the state have been painful for the low-income, low-skilled groups. Deindustrialization had also serious negative effects on the life chances of high skilled industrial workers, whose skills may no longer be required. The result is an increase in social inequality across the advanced industrial countries in the past two decades (OECD 2008). As globalization and marketization have intensified, the gap between rich and poor and the number of people below the poverty line have both increased.

The combined effects of the growth and diversity of urban populations, economic restructuring and political liberalization, new technologies, and increased social inequality have created new urban conditions, with their own spatial manifestations. As social inequality has risen, the possibility of different social groups living together is contested by the unleashing centrifugal and dividing forces. While there is some evidence of reduced segregation in the US cities with small black populations, the underlying structure of racial segregation has hardly changed (Glasmeier 2005; Smelser *et al.* 2001). Meanwhile, economic restructuring and transformation of welfare provision have led to a rise in social and spatial segregation and exclusion in European cities (Madanipour *et al.* 2003). Public policy and urban design are challenged to respond to social segregation and gentrification, which are the results of struggle over space. During the past two decades, New York has once again led the urban population growth in the United States, triggering heated debates about the character of the city's social and spatial change, as exemplified by the gentrification of Times Square

Figure 36.1 Times Square, New York. Source: Ali Madanipour.

shown in Figure 36.1 (e.g. see Smith 1998; Makagon 2004; Miller 2007; Berman 2009).

The capitalist industrial society had produced a distinctive urban space (Lefebvre 1991). It had gathered large numbers of workers in growing cities, concentrated in dense working class neighborhoods near industrial workplaces, living according to rigid temporal and spatial routines. It had produced central areas as the places of command for this economy (Loukaitou-Sideris and Banerjee 1998). As they grew in size, and new transport and construction technologies emerged, cities adapted accordingly. The adjustment was of two kinds. One kind, which started to emerge from the latter part of the nineteenth century but intensified from the middle of the twentieth onwards, involved suburbanization of the middle class with a distinct preference for single family homes (Fishman 1987). The associated garden city movement (Howard 1960), and its subsequent forms such as new towns and more recently new urbanism, advocated tidying up suburbanization. The other kind of adjustment involved was through large-scale redevelopment of cities, as the modernists had advocated (Le Corbusier 1971). The form of most contemporary cities reflects both trends: suburbanization and the renewal of the center, both favoring the growing urban middle class. Public policy and subsidy helped the development of homes and highways in the suburbs as well as restructuring of the city's center with major motorways and housing renewal programs. However, when the industrial city declined, the emerging service city began to produce its own space, where two primary trends of segregation and gentrification mirror the two main forms of urban development in the industrial city, i.e. suburbanization and urban renewal.

The fear of the city is rooted in the nineteenth-century metropolis, which generated a sense of pride but also alienation and fear (Briggs 1968). Rich and poor lived in parallel worlds in different enclaves, and poor working and living conditions kept the city always on the edge of revolution (Engels 1993). Large cities were formed of strangers, uprooted from their rural communities, and living in an anonymous world, in which traditional forms of social cohesion and control could not work anymore. To cope with the information and emotional overload, the city people withdrew into a psychological bubble made of an indifferent attitude (Simmel 1950). This was now a society, rather than a community (Tönnies 1957), and heterogeneity was one of its fundamental features (Wirth 1964). These trends re-emerged in the late twentieth century after a new wave of laissez faire has unveiled and magnified some of the deeply embedded features of the modern city.

As the gap between the rich and poor has widened, crime rates have risen in historic terms,[1] and globalization has further diversified the urban population; the sense of insecurity and the fear of "the other" have been magnified. In this context, two opposing visions of the city emerge. On the one hand, cities are perceived and portrayed as dangerous places that should be avoided, resulting in citadel building and fortress mentality, withdrawal from public spaces, and the proliferation of gated neighborhoods (Blakely and Snyder 1999). On the other hand, the city's newly reasserted economic and social significance demands a more positive attitude towards urban living, more tolerance towards difference and living with others. Two sentiments are discernible at the opposing ends of a spectrum: welcoming strangers or keeping them away. The spatial implications of these sentiments are either spatial integration or segregation in urban space, or in other words, living together or apart.

Living apart? Social segregation and exclusion

Keeping strangers away through spatial segregation is a longstanding trend in city design and development, which is now being used in new forms. Physically, it has been translated into creating spatial enclaves with the help of spatial barriers between social groups, using distance and walls to keep them apart. The standard spatial barrier in the modern city has been distance. The development of transport technologies has allowed people to move away from the city, stretching and fragmenting the city into separate districts and neighborhoods, creating a distance between social classes. The land and property markets, with the help of land use zoning and public policy have further consolidated and institutionalized this social geography (Cullingworth and Cave 2003; Keating and Krumholz 1999). Distance, however, appears unable to control the fear of the city, and a much older device is being employed: the wall. In ancient cities, walled enclaves of citadels protected the elite from the rest of the city, which itself was walled off from the threats of the outside world. With improved security and the sense of social equality, walls had come down in modern times. But now that deep fears and insecurities are back in the globalizing city, walls have become commonplace, such as in the urban regeneration project in Dublin, Ireland, depicted in Figure 36.2, which has created a new higher-income neighborhood next to a lower-income area. The mistrust between the two neighborhoods is so high that they are separated by a wall, dubbed as the Berlin Wall by those on the outside.

The notion of defensible space coined by Oscar Newman is a popular argument for social control through urban design. Initially it grew out of a critique of public housing schemes for low-income households and a praise of private control over space. "All Defensible Space programs," Oscar Newman wrote (1996: 9), "have a common purpose: They restructure the physical layout of communities to allow residents to control the areas around their homes." Walls and gates, supported by

Figure 36.2 Urban generation project in Dublin, Ireland. Source: Ali Madanipour.

security guards and cameras when affordable, represent the ultimate form of control. For the better-off, these devices offer a sense of security, a status symbol, and a mechanism to enhance their property value. The worse-off, however, worry about such practices, which could exaggerate their vulnerability, as gating becomes de facto "a device for either locking them in or locking them out" (Newman 1996:61).

At the other end of the spectrum from the new citadels are the ghettos, which are spatial manifestations of social exclusion. Social exclusion is a complex and multidimensional process, in which an individual may suffer from lack of access to resources, to decision making, or to cultural exchange. When these forms of systematic disadvantage converge in space, acute forms of social exclusion occur, as common for many deprived neighborhoods (Madanipour 2007). The ghetto is a discrete component of an urban settlement, typically occupied by a minority group. Trapped in a highly deprived area, either by choice or by force of the social and economic conditions, the ghetto residents are subject to various forces – political decisions, cultural preferences, technocratic know-how, and market forces – that dictate their life chances and choices, either from the outside compelling a certain behavior, or from the inside, asserting individual choice and identity (Spear 1967). As social inequality and cultural diversity intensify, the danger of social exclusion increases for larger numbers of people, threatening to fragment the urban society into a collection of unrelated pieces (Vergara 1999).

Many urban design texts and projects advocate the partitioning of the city into distinctive neighborhoods, a trend rooted in a long tradition of seeing neighborhoods as a constituent part of the city life and a unit of city building (Park *et al.* 1984; Mumford 1954; Lynch 1979). In their critique of liberal political processes, communitarians have favored neighborhoods as a basis for social integration. With the rise of environmental concerns, the idea of city neighborhoods has been embraced by different shades of opinion. Breaking the city into manageable and distinctive districts appeared good for the market, and building cities by neighborhood looked like a dream for large scale developers. Furthermore, it reflected, and reproduced, the city's fragmented social structure (Madanipour 2001).

Should planners and designers take the centrifugal forces of economic stratification and cultural differentiation for granted, or should they try to confront these forces? The outcome of accepting and consolidating differentiation in urban space, as imperatives of market forces and preferences of some social groups, is that it could generate parallel worlds, whereby different ethnic groups and social classes may live in their own enclaves. While these enclaves may be empowering for some, as a context in which to feel safe, they may also exacerbate social exclusion and inequality, whereby some groups are trapped within a limited range of places and activities (Blakely and Snyder 1999; Vergara 1999). The basic element of any response to these challenges should include equality and freedom of choice for individuals, particularly for those in weaker social groups. Are people living in enclaves out of choice or in response to some force? Where would they choose to live if they were free from discrimination, poverty and fear?

Living together? Social mixing, gentrification and displacement

An alternative approach to the challenges of diversity has promoted social mixing and welcoming strangers. In response to social segregation and exclusion, a school of thought emphasizes the important role that strangers play in an objective analysis of society (Simmel 1950; Schutz 1970), and advocates designing cities through the social and cultural engagement of strangers and their spatial integration (Jacobs 1961; Sennett 1993; Hillier and Hanson 1984). Rather than subdividing the city into distinctive citadels, these authors promote accessibility, overlap, and spatial openness (Madanipour 1996) in the hope that such measures could obviate festering social conflicts in urban space escalating into a permanent strife. Urban design, it is argued, can help facilitate peaceful co-existence, particularly in accessible public spaces of the city (Madanipour 1999).

Many urban planners and designers favor the promotion of mixed communities, using the related urban design concepts that create public spaces, mixed land uses, mixed-tenure housing, and mixed neighborhoods. Protagonists argue that these measures can confront social fragmentation and exclusion by creating vibrant and integrative urban areas. As an alternative to the segregated and stratified urban spaces that emerged through suburban sprawl, mixed income areas could prevent radicalization and conflict by promoting contact and engagement between cultural and ethnic groups. Rather than segregating different social groups into identifiable enclaves, leading to a potential Balkanization of the city, mixing people through urban planning and design can open up a pathway to social integration. Social mixing is promoted as a response to the decline and ghettoization of public housing areas, paving the way for social inclusion and social sustainability. Integrative places, however, are difficult to create and maintain, and in the face of the centrifugal trends of social polarization and inequality, social mixing may lead to gentrification, becoming more complicated than it seems at first.

Gentrification is a shift in the control and use of space from lower-income to higher-income social groups. It tends to occur at the neighborhood level, changing the conditions and character of an area through alteration in land use and population. Gentrification can be conceptualized as a consequence of competition and conflict over land as a finite resource, within the context of structural economic transformation and social change. Gentrification is a result of several factors: demand for prime urban space from high-income groups, profit-seeking by land and property markets, and public sector economic regeneration policies ostensibly for fighting social segregation and physical corrosion.

Gentrification is partly driven by the logic of production of space, in which the market looks for opportunities for investment and profit (Lefebvre 1991; Smith 1996). Observers have written about the return of the capital to the city, and the conflict between exchange value and use value (Boyer 1990; Logan and Molotch 1987). Those who invest and deal in land and property are interested in its monetary value and return on their investment, and not necessarily in who uses it and how. Through the anonymity of money and the complexity of financial institutions and markets, investors may not even know where their investment has been made, or may see property like any other asset, which can be sold and exchanged in the marketplace. In contrast, residents and users of the place may not have a financial stake and are only interested in it as their home, playground, etc. In market economies these values may coincide in many instances. But when new development or a transformation of an existing area occurs, or land and property is exchanged in the market, this tension arises.

Gentrification is also partly driven by the logic of demand for space (Ley 1996). As new demographic patterns, lifestyles, and work patterns have emerged, and new economic activities such as financial services, media and cultural industries are based in cities, the workers in these sectors wish to live nearer their workplaces. In response, city center affluent housing has grown and gentrification of working class areas has accelerated (Hamnett 2003).

Gentrification is also partly driven by public policy (Cameron 2003; Punter 2009). Governments are actively involved in the regeneration of cities, realizing the economic and social significance of cities, reclaiming legitimacy after the earlier deterioration of urban environment through disinvestment, and enabling investment in land and property markets, from which large tax revenues could be made. As public-private partnerships have spread as the model of urban development, government agencies have a direct role in gentrification. It is seen as an inevitable, or even desirable, outcome of urban regeneration, aiming to eradicate the signs of poverty and decline, which would include the presence of the poor and their associated services and spaces. The word gentrification is no longer taken to have a negative meaning; instead it has been embraced as a positive outcome of major regeneration projects. For example, Rappongi Hills, an upmarket shopping and entertainment complex in central Tokyo, Japan, has been promoted as a gentrification project (Figure 36.3).

Gentrification is sometimes the by-product and sometimes the driving force of urban change. For three decades, redevelopment of post-industrial cities has involved attempts to eradicate the traces of the industrial era, as exemplified by the transformation of Covent Garden and the docklands in London, creating new spaces for new activities to replace the old, with much debate and controversy. The shift tends to be painful and controversial, especially when the area is densely populated, and claims and counter-claims of different social groups over land are resolved in favor of the more powerful voices (Slater 2006; Lees and Ley 2008). In the countries where the issues of race have been entwined with cities, the result has been even more controversial. The wave of gentrification and displacement, which has been intensified through schemes such as HOPE VI in the United States, has created a widespread backlash. It has brought together different strands of resistance,

Figure 36.3 Rappongi Hills complex in Tokyo. Source: Ali Madanipour.

exemplified in a network of grassroots organizations called The Right to the City, which aims at forming a coordinated response to gentrification and displacement, and changing the terms of the urban policy debate towards paying more explicit attention to racial justice and economic democracy (Right to the City 2009).

Is gentrification an inevitable sign of progress and improvement for cities? If cities are to be given a new lease of life, they need investment. If high income groups move into an area, it has been argued, they bring with them part of the necessary investment, associated with improvement in urban amenities and services. However, mixing may occur only physically, but not necessarily generating social and cultural links between different groups. Resentment and alienation, rather than integration, may be the outcome. Investment in the urban environment is often followed by a rise in rent levels, making it difficult for some people and activities to survive without public subsidies. As a place goes upmarket through regeneration and renewal, it becomes unaffordable for some of its inhabitants, who may be displaced against their wish. Even publicly funded programs of urban improvement have historically, and in their recent and current reincarnations, caused displacement for poor urban communities. In developing new suburbs, much has been said about promoting mixed communities, without sufficient signs of success in encouraging different income groups to want, or be able, to live together. Social segregation may be exclusionary, but social mixing is also more complex than it seems, with potential detrimental impacts on weaker social groups.

Should urban design be intertwined with gentrification or segregation? Since the 1980s, urban design has been involved in shaping regeneration projects and guiding investment in the physical transformation of cities going through economic restructuring. As the process has been led by, or favored, the market, it has been unable or unwilling to deliver public goods and services. Urban designers have been among the campaigners for public space as an important public good; at the same time making public space has become an essential ingredient in the gentrification process (Madanipour 2010). The question for urban designers and planners is whether a city be partitioned along the lines of diversity, or be envisaged as a complex space with overlapping layers. Is social mixing possible under the conditions of heightened social inequality, and can such mixing lead to social tolerance or to inexorable displacement of some people and consequent resentment? As private investment has become the main form of financing city building, should urban space reflect the demands of investors and their buyers, or should it also take into account the needs and aspirations of those with weaker economic, political or cultural positions?

Conclusion

The general context of the city is changing, which includes the transformation of the urban economy from the prevalence of industries to services, changing relations between public and private sectors, the changing nature of work and technology, increased social diversity, and intensified social inequality. As the urban space is reconfigured to adjust to and facilitate further changes, a key debate has been about the spatial organization of the city: should the increasingly fragmented population be encouraged to live together or apart? Should the role of urban design be accepting social fragmentation as a given and trying to design appropriate spaces, or should urban design try, in its limited ways, to overcome fragmentation and segregation? Urban designers have come up with different answers to these questions, but owing to contextual conditions, the consequences of their decisions are often more complicated than they predicted: hopes for creating safe and familiar neighborhoods could lead to further fragmentation and tribalization of the city, and hopes for obtaining social mixing and peaceful coexistence could lead to displacement and gentrification. The damaging consequences of social inequality are well-known, and while urban design cannot change the fragmented social context, it needs to be aware of the potential consequences of the solutions it embraces. A key test for any urban design solution would be social equality and individual choice; whether the weaker social groups are forced to accept changes in their conditions or they too have access to a degree of choice about their urban environment.

Note

1 Although the past decade has witnessed the fall of the US crime rates from their peak in the 1990s, the rates are still very high in historic terms. For example, the rate of violent crime in 1960 was 160.9 per 100,000 population, as compared to 466.9 in 2007. It has fallen from the peak of 757.5 in 1992, but it is still high in historic terms, and enough to engender a sense of insecurity. See the Bureau of Justice Statistics at the US Department of Justice http://bjsdata.ojp.usdoj.gov/dataonline/Search/Crime/State/RunCrimeStatebyState.cfm. (accessed 18 August 2010). In the UK, a similar trend is observable, which shows a historic rise, but a declining crime rate since the mid-1990s. See the UK's Home Office report http://www.homeoffice.gov.uk/rds/pdfs08/hosb0708.pdf. (accessed 18 August 2010).

References

Atkinson, R. and Bridge, G. (Eds.) (2005). *Gentrification in a Global Context: The New Urban Colonialism*, London: Routledge.

Berman, M. (2009). *On the Town: One Hundred Years of Spectacle in Times Square*, London: Verso.

Blakely, E.J. and Snyder, M.G. (1999). *Fortress America: Gated Communities in the United States*, Washington, DC: Brookings Institution Press.

Boyer, M. C. (1990). "The Return of Aesthetics to City Planning," In Crow, D. (Ed.) *Philosophical Streets: New Approaches to Urbanism*. Washington, DC: Maisonneuve Press. 93–112.

Briggs, A. (1968). *Victorian Cities*, Harmondsworth: Penguin.

Buzar, S., Ogden, P., Hall, R., Haase, A., Kabisch, S., and Steinfuhrer, A. (2007). "Splintering Urban Populations: Emergent Landscapes of Reurbanization in Four European Cities," *Urban Studies*, 44(4): 651–677.

Cameron, S., (2003). "Gentrification, Housing Re-differentiation and Urban Regeneration: 'Going for Growth' in Newcastle upon Tyne," *Urban Studies*, (40)12: 2367–2382.

Champion, A.G. (2001). "A Changing Demographic Regime and Evolving Polycentric Urban Regions: Consequences for the Size, Composition and Distribution of City Populations," *Urban Studies*, 38(4): 657–677.

Cullingworth, B. and Caves, R. (2003). *Planning in the US: Policies, Issues and Processes*, Second Edition, London: Routledge.

Engels, F. (1993). *The Condition of the Working Class in England*, Oxford: Oxford University Press.

Esping-Andersen, G. (1999). *Social Foundations of Postindustrial Economies*, Oxford: Oxford University Press.

Fishman, R. (1987). *Bourgeois Utopias: The Rise and Fall of Suburbia*, New York: Basic Books.

Frey, W. (2009). *Big City Populations Survive the Housing Crunch*, Washington, DC: The Brooking Institution, http://www.brookings.edu/opinions/2009/0701_housing_frey.aspx, (accessed 27 July 2009).

Glasmeier, A. (2005). *An Atlas of Poverty in America: One Nation, Pulling Apart, 1960–2003*, University Park, PA: The Pennsylvania State University.

Hamnett, C., (2003). *Unequal City: London in the Global Arena*, London: Routledge.

Hillier, B. and Hanson, J. (1984). *The Social Logic of Space*, Cambridge: Cambridge University Press.

Howard, E. (1960). *Garden Cities of To-morrow*, London: Faber & Faber.

Jacobs, J. (1961). *The Death and Life of Great American Cities*, New York: Vintage Books.

Keating, D. and Krumholz, N. (Eds.) (1999). *Rebuilding Urban Neighborhoods: Achievements, Opportunities and Limits*, Thousand Oaks, CA: Sage.

Le Corbusier, (1971). *The City of To-morrow, and its Planning*, London: The Architectural Press.

Lees, L. and Ley, D. (2008). "Introduction to Special Issue on Gentrification and Public Policy," *Urban Studies*, 45 (12): 2379–2384.

Lefebvre, H. (1991). *The Production of Space*, Oxford: Blackwell.

Ley. D. (1996). *The New Middle Class and the Remaking of the Central City*, Oxford: Oxford University Press.

Logan, J. and Molotch, H. (1987). *Urban Fortunes: The Political Economy of Place*, Berkeley, CA: University of California Press.

Loukaitou-Sideris, A. and Banerjee, T. (1998). *Urban Design Downtown: Poetics and Politics of Form*, Berkeley, CA: University of California Press.

Lynch, K. (1979). *The Image of the City*, Cambridge, MA: MIT Press.

Madanipour, A. (1996). *Design of Urban Space: An Inquiry into a Socio-Spatial Process*, Chichester: John Wiley.

—— (1999). "Why are the Design and Development of Public Spaces Significant for Cities?" *Environment & Planning B: Planning and Design*, 26, 879–891.

—— (2001). "How Relevant is 'Planning by Neighborhoods' Today?," *Town Planning Review*, Vol.72, No.2, pp.171–191.

—— (2007). "Social Exclusion and Space." In LeGates, R. and Stout, F. (Eds.), *The City Reader*, Fourth Edition, London: Routledge, 158–165.

—— (2010). *Whose Public Space? International Case Studies in Urban Design and Development*, London: Routledge.

Madanipour, A., Cars, G. and Allen, J. (Eds.) (2003). *Social Exclusion in European Cities*, London: Routledge.

Makagon, D. (2004). *Where the Ball Drops: Days and Nights in Times Square*, Minneapolis, MN: University of Minnesota Press.

Miller, K. (2007). *Designs on the Public: The Private Lives of New York's Public Spaces*, Minneapolis, MN: University of Minnesota Press.

Mumford, L. (1954). "The Neighbourhood and the Neighbourhood Unit," *Town Planning Review*, 24: 256–270.

National Atlas (2009). *Population Change and Distribution: 1990 to 2000*, http://nationalatlas.gov/articles/people/a_popchange.html, (accessed 27 July 2009).

Newman, O. (1996). *Creating Defensible Space*, Washington, DC: US Department of Housing and Urban Development, Office of Policy Development and Research.

OECD, (2008). *Growing Unequal? Income Distribution and Poverty in OECD Countries*, Paris: OECD.

Park, R., Burgess, E. and McKenzie, R. (1984). *The City*, Chicago: University of Chicago Press.

Parkinson, M. (2006). *State of English Cities*, Volume 1, London: Office of the Deputy Prime Minister.

Pointer, G. (2005). *The UK's Major Urban Areas*, http://www.statistics.gov.uk/downloads/theme_compendia/fom2005/03_FOPM_UrbanAreas.pdf, (accessed 27 July 2009).

Punter, J. (Ed.) (2009). *Urban Design and the British Urban Renaissance*, London: Routledge.

Right to the City (2009). *The Right to the City*, (http://www.righttothecity.org/, (accessed 1 March 2009).

Schutz, A. (1970). *On Phenomenology and Social Relations, Selected Writings*, Chicago: The University of Chicago Press.

Sennett, R. (1993). *The Conscience of the Eye: The Design and Social Life of Cities*, London: Faber & Faber.

Simmel, G. (1950). *The Sociology of Georg Simmel*, New York: The Free Press.

Slater, T. (2006). "The Eviction of Critical Perspectives from Gentrification Research," *International Journal of Urban and Regional Research*, 30(4): 737–757.

Smelser, N., Wilson, W.J. and Mitchell, F. (Eds.) (2001). *America Becoming: Racial Trends and their Consequences*, Washington, DC: National Academy Press.

Smith, N. (1996). *The New Urban Frontier: Gentrification and the Revanchist City*, London: Routledge.

—— (1998). "Giuliani Time: The Revanchist 1990s," *Social Text*, 57. 16(4): 1–20.

Spear, A. (1967). *Black Chicago: The Making of a Negro Ghetto, 1890–1920*, Chicago: University of Chicago Press.

Tönnies, F. (1957). *Community and Society*, New York: Harper and Row.

Urban Task Force, (1999). *Towards an Urban Renaissance*, London: Spon.

US Census Bureau, (2009). *New Orleans was Nation's Fastest-Growing City in 2008*, Press Release, 1 July 2009, http://www.census.gov/Press-Release/www/releases/archives/population/013960.html, (accessed 27 July 2009).

Vergara, C.J. (1999). *The New American Ghetto*, New Brunswick, NJ: Rutgers University Press.

Wirth, L. (1964). *On Cities and Social Life: Selected Papers*, Chicago: University of Chicago Press.

Further reading

Blakely, E.J. and Snyder, M.G. (1999). *Fortress America: Gated Communities in the United States*, Washington, DC: Brookings Institution Press. A critique of fortress urbanism as the extreme form of socio-spatial segregation.

Carr, J. and Kutty, N. (Eds.) (2008). *Segregation: The Rising Cost for America*, New York: Routledge. Collection of articles providing a critique of segregation on the grounds of social justice and economic wellbeing.

Freeman, L. (2006). *There Goes the 'Hood: Gentrification from the Ground Up*, Philadelphia: Temple University Press. Examination of gentrification from the perspective of residents of two black inner city neighborhoods in New York.

Lees, L. and Ley, D. (2008). Introduction to special issue on gentrification and public policy, *Urban Studies*, Vol.45, No.12, pp. 2379–2384. This is a special issue of the journal *Urban Studies* devoted to the subject of gentrification.

Smith, N. (1996). *The New Urban Frontier, Gentrification and the Revanchist City*, London: Routledge. Influential book articulating a radical critique of gentrification.

Urban Task Force (1999). *Towards an Urban Renaissance*, London: Spon. A major document produced by a committee led by the architect Richard Rogers, setting out the vision of design-led renaissance for British cities with social and land use mix.

37

Beyond placelessness

Place identity and the global city

Michael Southworth and Deni Ruggeri

With the onslaught of global communication and economic forces, place identity and place attachment – two key concepts in urban design – have taken on new importance. Cities are torn between the necessity to be a part of the world network and the need to preserve their uniqueness and cultural roots. While new symbols of progress have erased traditional environments, innovative definitions of community have emerged. In the pre-industrial city, physical proximity and neighborhood form were crucial in defining who we were, both as individuals and as community members. The automobile and information technology have altered these dynamics. As metropolises expand, massive developments consume open land with little concern for locale, nature, and tradition. In this context, how people relate to the places in which they live and what kind of bonds they develop with them have become vital questions for urban planners and designers.

The rise of the information society has led to a redefinition of identity and the factors that affect it. Manuel Castells describes a new techno-economic paradigm in which location no longer matters for industries and services: "social meaning evaporates from places, and therefore from society, and becomes diluted and diffused in the reconstructed logic of a space of flows ..." (Castells 1989: 348–349). Instant global communications leads societies to adopt the same ideas. As a reaction to an anonymous, mass-produced landscape, place identity may be more important than ever to provide a sense of stability, meaning, and settings for face-to-face interaction. He considers the trend to be very destructive, and sees the need to preserve or reestablish local identity. "At the cultural level, local societies, territorially defined, must preserve their identities, and build upon their historical roots, regardless of their economic and functional dependence upon the space of flows" (Castells 1989: 350).

This chapter discusses various conceptions of place identity, as well as its prospects in the global metropolis, and proposes a new definition of it as a multifaceted gradient. Does place identity really matter in a global society? What type of place identity emerges out of the contemporary global landscape? What is the role of planners and designers in shaping place identity today?

Interest in place identity has a long history. Phenomenologist Martin Heidegger feared the impacts of technology on "authenticity." For him, "dwelling" was the essence of living, building, and planning, but was threatened by modernity. His language is difficult, but worth the challenge: "The nature of building is letting dwell. Building accomplished its nature in the raising of locations by the joining of their spaces. *Only if we are capable of dwelling, only then can we build.* ... Dwelling ... is *the basic character of Being in keeping with which mortals exist*" (Heidegger 1971: 160). In his 1951 lectures "Bauen Wohnen Denken" (Building Dwelling Thinking) and "Das Ding" (The Thing) Heidegger spoke of the "thingness of things" and the loss of "nearness." As transportation and communication technology overcame distance, a uniformity of near and far resulted, with everything "lumped together into uniform distancelessness" (Heidegger 1971: 166). The process destroys what he called "nearness" or the distinctness of local places. The issues that Heidegger raised are major challenges in urban experience today.

Identity and imageability

Urban designers and architects tend to think of place identity mainly in terms of its physical form. Kevin Lynch's pioneering work on place identity advanced theory and methods for studying perception of the physical environment and provided a normative framework for creating memorable places. In *Image of the City* Lynch proposed that identity, along with structure, helps create "imageability," the city's "high probability of evoking a strong image in a given observer" (Lynch 1960: 9). Lynch defined "identity" as "the extent to which a person can recognize or recall a place as being distinct from other places – as having a vivid, or unique, or at least a particular, character of its own" (Lynch 1981: 131). Identity is not merely a quality of the physical place, but is also a function of the person. However hard a designer might work to create a distinctive and memorable place, whether or not it has strong identity is dependent upon the observer – her culture, purpose, and mood.

In *Townscape* Gordon Cullen explored a related but more descriptive and less analytical approach to place identity. Through a perceptive set of observations and vivid sketches, he demonstrated how chaotic urban environments might be corrected to establish a coherent sequence of "vistas" providing a framework for users' perception of urban scenes. Through artful use of texture, color, street furniture, light, view, and built form he showed how engaging and cohesive urban places might be created (Cullen 1961).

Many designers still seem to believe that identity resides exclusively in the physical environment. New Urbanists have attempted to codify place identity by establishing rigid design codes that assure human-scaled streets and buildings. However, the codes are typically generic with formulaic conceptions of regional vernacular and little diversity of expression. Because they do not evolve from local culture, they compromise vernacular design traditions. Moreover, they have been largely insensitive to topography, vegetation, and hydrology, fundamental features in the establishment of healthy place identity.

Place identity and individual experience

Identity in urban design is not just about eye-catching memorable form. Although the sensory characteristics of a place may get our attention, identity is more than skin deep. Charles Moore's Piazza d'Italia in New Orleans had striking form, yet meant little to the neighborhood it was in. What a place means to people is a deeper level of identity. Meaning or significance may result from personal experiences with a place: the market where we shop every

Saturday, or the neighborhood where we grew up. Aside from Boston's obvious monuments such as the Hancock Tower or Richardson's Trinity Church, many other landmarks may be largely invisible, or at least without meaning to newcomers. But for a lifetime resident of Beacon Hill, nearly every building and space may be packed with memories of people and events, present and past. Social activity, even if temporary, can enhance place identity, as in Seattle's Pike Place Market (Figure 37.1) and the New Orleans Mardi Gras, which are vivid in people's minds because of their street life. Place significance may also result from historic or political events. But places with strong public identity need not have strong visual identity. The birthplace of the Free Speech Movement on Sproul Plaza at UC Berkeley is nearly invisible, marked by a small bronze circle in the pavement that designates a column of air where one is free to think or speak anything (Figure 37.2). Similarly, "People's Park" is another "sacred place" for many people in Berkeley, the site of a bloody conflict between students and police in 1969. It symbolizes a major turning point in protests against the Vietnam War, yet it is a visually nondescript urban park. While strong visual form is not essential for identity, it can provide a framework for attaching meanings. Place identity has greatest power when visual form, individual and social meaning come together. According to Lynch (1960: 119), "(S)ense of place in itself enhances every human activity that occurs there, and encourages the deposit of a memory trace."

Yi Fu Tuan writes of places as multinucleated constellations of experiences (Tuan 1977: 183). Other researchers have described place-identity as the understanding of who we are in relationship

Figure 37.1 Pike Place Market, Seattle. Source: Michael Southworth.
Note: Its social life is an important ingredient of its identity.

Figure 37.2 Free Speech Movement site at UC Berkeley. Source: Michael Southworth.
Note: The Free Speech Movement birthplace has strong political identity but is nearly invisible.

with the places in which we live (Opotov and Clayton 2003). Environmental psychologist Harold Proshansky has attempted to explain place identity and the role of personal experiences in shaping it. He takes as a point of departure Erik Erikson's (1959) "reflected appraisal" mechanism – the process by which a child establishes his identity by distinguishing himself from other people – and extends this appraisal to include places, objects and even abstract ideas about places. Thus, place identity is the result of a constant, and often subconscious negotiation between individuals and the potpourri of experiences, objects, and even idealized places they encounter during their lives (Proshansky *et al.* 1983). Proshansky's work finally recognized the importance that places – real or fictitious – play in shaping our identity as individuals. His research also began to suggest that since places are shared by multiple individuals, place identity is naturally linked to social identity.

Identity and "non-places"

Shortly after Lynch launched his work on the city image, Melvin Webber startled planners with the possibilities of a non-place society. He developed a communications theory of urban form based on transportation and communications technology, as well as institutional changes, which were making possible sprawling metropolises

(Webber 1963). Rather than being exclusively place-based, urban form could also be thought of as social relationships, economic patterns, or transportation nodes and networks:

> "... it might encourage us to see urbanity – the essence of urbaneness – not as buildings, not as land use patterns, not as large, dense, and heterogeneous population aggregations, but as a quality and as a diversity of life that is distinct from and in some measure independent of these other characteristics. Urbanity is more profitably conceived as a property of the amount and the variety of one's participation in the cultural life of a world of creative specialists, of the amount and the variety of the information received" (Webber 1964: 88).

For Webber, urban form needed to go far beyond its spatial qualities to include flows of information, money, people and goods, as well as activity patterns. The classic spatially defined city with clear edges and open land on the periphery was an outmoded concept that constrained communications and cultural development. Thus, he hypothesized the "nonplace community," interest-based communities not tied to place: "It is the accessibility rather than the propinquity aspect of "place" that is the necessary condition [for community]" (Webber 1964: 109). He concluded that planners must "free themselves from the obsession with placeness and unless they can come to view the urban communities as spatially extensive, processual systems in which urbanites interact with other urbanites wherever they may be. For it is interaction, not place, that is the essence of the city and of city life" (Webber 1964: 147). Webber anticipated globalization by several decades, although he did not specifically discuss the global city or telebusiness in his writings, which predate the Internet and the dramatic emergence of global commerce.

Insiders versus outsiders

Donald Appleyard, a contemporary of Webber on the faculty at UC Berkeley, had diametrically opposed views on the significance of place. Steeped in Lynchian imageability theory, he advanced research on image development and the meanings of place through many studies with human subjects. For him, places – spatial locations in all of their social and sensory diversity – were the framework for life and could not be replaced by nonplace communications. His research revealed that people see and value places differently, depending upon factors such as their experience, culture, and class (Appleyard 1976 and 1979). Anticipating recent research from cultural geography and anthropology on identity and identity politics (Castells 1997, Anderson *et al.* 2003), Appleyard suggested that access to shared meanings is political, and that designers have a key role in determining which meanings get divulged and who the insiders and outsiders in a place may be. In his final unfinished manuscript, "Identity, Power, and Place," he delved into the many meanings and scales of "place" – from home and neighborhood, to tourist city. He argued that we cannot understand the social symbols of places we do not know. As familiarity decreases we rely on "stereotypical cues and categories ... insubstantial myths and impressions that are picked up from the news media, novels or academic studies, we depend more and more on surface appearance to learn what is going on ... [Even] places in our own cities where other subcultures live are distant and foreign to our understanding" (Appleyard 1979: 151).

By suggesting that identity means more than just legibility and that it is essentially socially constructed, Appleyard showed the limitations of physical imageability as a proxy for place identity. If we want to truly tackle the issue, we must wear multiple hats and become acquainted with the research methods that best allow us to tap into these meanings, and to understand how they are generated, shared and perpetuated. In a highly mobile and diverse culture, place identities will be multiple and constantly shifting. The role of urban designers is to uncover the meanings that are embedded in the places we work on, rather than limit ourselves to defining the official image.

Environmental psychologists Stokols and Schumaker (1981) coined the term "social imageability" to indicate the shared meanings and discourses that lead to place attachment. Researching the effects on identity of displacement and detachment from familiar places, sociologist Melinda Milligan has identified what she calls "locational socialization," through which one's active involvement with a place generates shared meanings (Milligan 1998 and 2003: 383). These meanings are layered onto a place, and it is in the very act of embedding these meanings that place identity and place attachment emerge and are shared with others.

High design versus the vernacular

The shaping of place identity has become a major focus in architectural and urban design. Does the work of "Starchitect" designers like Frank Gehry or Richard Meier result in a stronger sense of identity? Or does it simply contribute to situating us within the realm of anonymous "global" cities? Corporations, and even cities, seek to establish international brand identity to gain recognition and advantage in the global marketplace. Such branding is never based on authentic place and culture, but consciously avoids local references so as to attract attention in global media space. Like corporations, art institutions now see the museum building as a logo that can capture worldwide attention and strengthen its stature and finances. Gehry's Guggenheim Museum in Bilbao, Spain not only brought acclaim to the museum, but also transformed an unknown city into a tourist destination. International airports are also designed in global placelessness; regardless of location, one experiences the same environment, the same food, the same shops, and the same procedures. There can be advantages: finding the familiar in strange places can be reassuring. Las Vegas is certainly the placeless iconic city par excellence. Ignoring its own ecology and traditions, it has stolen the "authentic" identities of a dozen other cities. Its economy depends upon consumption of the placeless spectacle.

In contrast to "starchitect" designers, landscape architect Randy Hester has done much to uncover and make meaningful the idiosyncratic, yet authentic, qualities of places, and in particular the places of everyday life. Rather than imposing normative design models on towns, his design process engages community members in discovering the places they know and love, and then uses them as the foundation for future development. "Sacred places" do not need to be imageable to have a strong identity. In fact, they may be humble laundromats, cafés, post offices, boardwalks, a park bench, or even one's workplace environment (Hester 2006). Hester's realization raises the issue of the conflict that exists between high design and everyday, vernacular landscapes (Clay 1994; Groth 1997).

Identity and authenticity

Identity can also be interpreted as authenticity, as the quality of a place being unique, distinctive, and rooted in the locale. Geographer Edward Relph describes authentic places as being generated unselfconsciously and without theoretical pretense by individuals working alone or in small community groups over long spans of time. "The end result is places which fit their context and are in accord with the intentions of those who created them, yet have a distinct and profound identity that results from the total involvement of a unique group of place-makers with a particular setting" (Relph 1976: 68). Ancient Italian hill towns and preindustrial English villages epitomize these qualities.

What does "authentic" mean in today's postindustrial, global city? Is it even possible? Authenticity in place making is a slippery concept. A building or landscape that one makes oneself, using local materials, following local traditions would be considered "authentic" by most standards. This is the way most towns and cities were produced before industrialization. But industrialization brought the possibility of new materials and processes, as well as new ideas about form and style imported from distant places and cultures that might be seen as "inauthentic." Things get complicated when over time, the inauthentic is absorbed into local traditions and thus acquires a degree of authenticity. For example, the Ranch house, the product of the Modernist search for a "living machine" that could be standardized and prefabricated, has now become part of the American vernacular (Hess 2005).

In his book *Place and Placelessness*, Relph bemoans the loss of "placeness" in the postindustrial city. For Relph, like Heidegger, the tenets of Modernism, along with transportation and communications technology, have nearly eliminated the possibility of "authentic" places. Post-industrial landscapes have been shaped by processes, materials, and regulations that have nothing to do with specific places, and in fact may be found nationally or even globally. For example, the landscapes of tourism such as invented pioneer villages, seaside resorts, or Disney World may have a strong identity, but not authenticity. For Relph these are "other-directed places" that have been made to attract outsiders and have been subject to "Disneyfication," "museumisation," or "futurisation" (Relph 1976: 92–105). In the late post-industrial city, the inauthentic has often become desired as a symbol of achievement. The upscale suburbs of Beijing and Shanghai emulate California gated suburbs such as Irvine with faux Victorian and Mediterranean mansions and three-car garages. In Bangalore and Beijing, Western style shopping malls and housing projects have replaced the traditional street bazaar and *hutong*. Is the search for place identity a lost cause in the global city?

Place identity as a multi-faceted gradient

Relph and others see the world in dualities; places either have identity or they do not, they are either authentic or inauthentic. The dichotomy of place versus placelessness does not capture the complex and multi-faceted contemporary city, which presents many degrees and shades of "placeness," whether urban, suburban, rural or natural, old or new. Traces of placeness can be seen everywhere, and designers need to become more sophisticated at dealing with this gradient of placeness.

Much academic discussion exists on the new geographies of the global city, from the iconic global spaces *à la* Chicago's Millennium Park that respond to the needs for international visibility and city imaging (Vale and Bass-Warner 2001), to the ethnic neighborhoods of the inner city, to the polluted post-industrial sites being regenerated into everyday landscapes or precious ecological habitats, to the low-density spaces of suburbia. These landscapes make up the fabric of the contemporary city and contribute to its complex identity, yet not all of them receive the attention of urban designers. In particular, the residential neighborhoods at the urban edge seem to have been dismissed by urban designers as "placeless" and inauthentic. This can only partially be explained by designers' predisposition for dense, traditional urban forms. Another explanation may be methodological, as lower density, segregated environments may be too difficult to visualize using traditional design methods, such as figure ground diagrams, sections, elevations, and axonometrics.

We suggest that place identity should be thought of as a gradient that includes several dimensions and should be as complex as the processes at play in every neighborhood. It should account for aesthetic appeal and imageability, but be expanded to include social considerations, the discourses and meanings that are shared by community members. It should consider responsiveness to context, but also that context has been expanded by the automobile and by the "information society" to include geographies once impossible to connect. It should value authenticity of forms, but also realize that preindustrial authenticity may not be possible. The identity gradient should discern between native and non-native, between designs that respect pre-existing natural conditions and those that erase them. It should also include considerations of process, valuing community involvement over the interests of a few, and vernacular and self-built landscapes over those designed by star-architects and their wealthy clients.

The identity gradient in practice

Irvine

Irvine, California illustrates how such an identity gradient might play out in the lived experiences of residents and their narratives about place identity. Irvine is a new town of 200,000 people developed in the late 1960s on what was once the most productive agricultural landscape in Southern California. Its identity as a new town drew on Kevin Lynch's imageability theory and on the resulting normative framework of nodes, edges, paths, districts and landmarks, implemented through the work of many well known architects, urban designers and landscape architects, including Hideo Sasaki, Peter Walker, and Ian McHarg. The Irvine Company, the original planner and sole developer for the new town, employed urban design as an instrument to promote a new place identity and position Irvine as an alternative to the Los Angeles "sprawl." The planners hoped that Irvine's landscape would help shape a sense of belonging in the new residents and create a distinctive place where residents would live in close contact with nature (Ruggeri 2009).

With its profusion of Mediterranean references in both the mass-produced housing and urban design features, Irvine epitomizes Edward Relph's "inauthenticity" and is disliked by many planners and architects, who consider it the quintessence of placelessness (Soja 1992). To some, Irvine appears as a more refined or simply better maintained version of traditional suburban sprawl. Wide arterial streets, a form of mostly single family homes, auto-oriented shopping areas, schools and neighborhood parks make up the physical form of Irvine like many other suburbs. However, it is in the unique combination of these elements that Irvine distinguishes itself. Physical elements were choreographed in a master plan grounded in the theory of imageability, with large setbacks along roads acting as strong edges, and existing windbreaks acting as pedestrian paths. This landscape appears in residents' descriptions of their favorite places. A survey revealed that four out of ten would regret leaving Irvine's landscape and open spaces behind, an indicator of strong place identity and attachment (Pretty *et al.* 2003). However, the landscape

Figure 37.3 Fourth of July celebration in Woodbridge, Irvine. Source: Deni Ruggeri.
Note: Social life is as important to Irvine's identity as visual form.

is not the only thing residents are attached to. Neighboring activities, social interaction and community events, sports events and school activities are also cited as elements that would be difficult to give up (Figure 37.3) (Ruggeri 2009).

The resonance of urban design elements in the narratives of Irvine residents explains only a small part of its place identity. The shared values of its residents are almost as important in shaping Irvine's identity as its visual form. These include the pragmatic embracing of cleanliness, orderliness and maintenance and the emphasis placed on a well-funded and organized school system. Education is a fundamental value for half of Irvine residents, and people are willing to live in a somewhat regimented master planned environment to ensure a brighter future for their children. Also important in defining Irvine's place identity gradient is the shared belief that this landscape breeds success and embodies the traditional American image of a self-made citizen whose wealth is the result of dedication to family, education, and hard work. It is out of the combination of all of these qualities – aesthetic, functional, social, and value-driven – that Irvine's identity emerges and positions itself against other suburban communities (Ruggeri 2009).

One of the lessons one can learn in Irvine is that while the traditional schemes urban designers have used to find and evaluate place identity may be helpful, they are not fully adequate. Irvine's landscape does not fit Relph's place/non-place model. Although it may have been designed mainly to promote imageability, its landscape also sets the stage for a lifestyle of satisfaction, convenience, and a shared search for success, both financial and educational. Similarly, his discussion on authenticity seems to be irrelevant to the people of Irvine. Their identity has little to do with the physical appearance of its landscape and the fact that they live in a Mediterranean, California-Ranch or

Craftsman-style home. It is inside the apparently "inauthentic" architecture of Irvine's homes, in the spaces between, in the public parks and playgrounds, along the pedestrian paseos connecting each neighborhood to the schools and other village facilities that Irvine's identity gradient finds its true expression, and that is where urban designers should look for it.

Millennium Park

Seen through the lens of the identity gradient, places that designers value may reveal unexpected negative dimensions and a much less clear identity. Millennium Park in Chicago illustrates the point. The park, recently designed by landscape architect Kathryn Gustafson, is undoubtedly a popular success. Since its opening in 2004 it has been visited by millions of tourists and hailed by many in the design community as one of the most successful urban parks of our times. It features a collection of eye-catching spaces, including a stage for concerts and public events, a botanical garden with fields of native grasses mimicking the Midwestern prairies, a plaza acting as a pedestal for a giant reflecting sculpture, and a playground with water spurting out of digital totems showing ever-changing images of people of all ethnicities.

From a user perspective, the park is experienced as a sequence of imageable fragments rather than as a memorable whole. Despite being well used, the park is far from democratic, being scrutinized by security cameras and police patrols. From the standpoint of its sustainability, the botanical garden includes ornamental grasses and native plants, while the rest of the park features large expanses of turf and formal plantings. Frank Gehry's stage does attract the attention of visitors, but its use is regulated by an official schedule of events. Every square foot of the park is programmed, leaving little to chance or spontaneity. Finally, everything about the park is big, making visitors feel dwarfed by the scale of many of its elements. Only in the botanical garden, the least costly of its elements does one have the opportunity to touch and feel. The park seems to have been designed as a corporate logo, a spectacle for the global citizens and corporations, who contributed half of its 475 million price tag. Is Millennium Park authentic? Insofar as it is an expression of our global world, it is. There is something truly "Chicagoan" about it, since the city has a tradition of high design and innovation. But is it the expression of the city's complex identity as a melting pot? Clearly, the answer is "no."

Kenzo Tange's new town of Librino

The new town of Librino in Sicily also illustrates the complex facets of place identity. Designed in 1971 by Kenzo Tange, the new town was a satellite city planned to handle Catania's future growth. Tange was chosen for his international reputation, which would have ensured immediate visibility and identity for the largest public housing development in Italy. His Modernist plan embodied the belief in a metabolic architecture, "a dynamic environment that could live and grow by discarding its outdated parts and regenerating newer, more viable elements" and buildings "that could cope with the problems of our rapidly changing society, and at the same time maintain stabilized human lives" (Dahiden 1972). The plan, Tange's interpretation of Le Corbusier's "towers in the park," fit perfectly the emerging identity of Catania as "Southern Italy's Milan," and its role as an economic engine for the region.

Kenzo Tange's plan for Librino consisted of a polycentric city of ten self-contained neighborhoods connected by a loop of high-speed arterial roads. The designer gave careful consideration to topography, hydrology, and climate, preserving valleys as corridors, both for automobiles and storm water, while residential buildings were placed on high ground to give everyone the best possible view and solar exposure. Librino's open space framework featured green fingers and piazzas in the inner portions of each superblock linked by elevated pedestrian walkways. A central green corridor of native vegetation became the main framework for the new town.

The implementation of Tange's ideas proved to be a challenge, as his plan failed to address the many pre-existing settlements. The architect joined forces with a local engineer, who was left in charge of its implementation. As a result of this process, many of the old farmhouses were preserved and attempts were made to integrate them into the neighborhoods.

Despite all efforts, the plan was not successful. In contrast to an initial vision that stressed the central open space as a structuring element, the city and public housing authorities concentrated their funds on the construction of housing and portions of the road system, leaving the open space unimproved. Once built, the towers were squatted by homeless families, and the inner blocks that police could not access attracted criminals. The new town is now home for 80,000 people, many of whom have built their own homes either illegally or in variance to the original plan. The Modernist architecture that was supposed to make Librino an international architectural Mecca failed to achieve many of its goals (Figure 37.4).

Imageability and sense of orientation in Librino are limited, facilitated only by the presence of natural features such as views of the volcano Etna and the sea. Attempts to create an imageable environment through plantings of specimen trees and shrubs have had limited success. Children love the schools and public facilities of

Figure 37.4 Librino, Italy. Source: Deni Ruggeri.
Note: The ancient olive groves play a greater role in Librino's identity than the modernist towers.

Librino, but have a hard time getting oriented and must rely on automobile transportation to get to and from school (Figure 37.5). While a few community groups exist, the lack of a physical center for the new town, a deliberate planning choice by Tange, makes it difficult for people to identify themselves as citizens of Librino. Even after 30 years, there is still no name to identify a Librino resident, and even if there were such a name, more than half of the residents would refuse to use it (CEDOC 2008).

Although the landscape may have little imageability, on any given day many people can be seen on the streets, unimproved green areas and the few sports facilities that do exist. Surprisingly, many in Librino have developed an attachment to the place they call home, particularly among the younger generation (CEDOC 2008). Half of the adults and three-quarters of adolescents declared that they would not choose to live anywhere else. This paradox may be explained by social imageability and the social interactions that take place in its unimproved landscape. It raises the issue of whether an imageable, well-maintained environment is at all necessary. Answering this question would require more research, but no one is really interested in finding out the truth. Both the development company and the city of Catania are still focused on completing the original plan and improving the public image and outsiders' perceptions of Librino, rather than that of its residents. The insider versus outsider debate raised by Donald Appleyard thirty years ago still remains unresolved in Sicily's only new town.

Urban design and place identity in the future metropolis

In a highly mobile society in which social networks span the globe, place identity

Figure 37.5 Child's drawing of Librino. Source: Deni Ruggeri.
Note: The child's drawing reflects the importance of everyday places in Librino's place identity.

may be more important than ever before. It can provide a sense of security and stability in a transient, rapidly evolving society. A meaningful spatial framework can also enhance social communications and coherence, fostering social and community development.

If we accept the idea that identity is a gradient, rather than a simple place/non-place dichotomy, what does this mean for designers? Such a definition complicates the work of urban design, but we believe that by looking at identity as a multi-faceted gradient we can improve our work. It helps us to be contextual in our evaluation of what "fits" in a neighborhood. It brings us closer to the perceptions of residents, who see identity not as black or white, but as a gestalt. True, physical imageability is important to the establishment and preservation of identity, but one must not forget that social imageability is just as critical in defining what we identify with.

A new and more nuanced definition of place identity is needed to bridge the gap between those who see place identity in binary terms, and those who believe it exists everywhere. It can be found by looking at a range of places, from the historic downtowns of our cities to the everyday landscapes of suburbia, using a variety of methods, including physical form analysis, observations, interviews and other sociological methods. This new definition should consider the need for memorable and imageable environments, expressions of shared social values, new forms of non-place communities, and the multiple mechanisms by which meanings are embedded and communicated in the landscape.

As anonymous mass-produced environments and global iconic forms are rapidly replacing vernacular form, the natural setting of cities and regions remains a fundamental source for local and regional identity. The natural landscape and ecological systems can provide a framework for authenticity and sense of place and must be integrated into the built city. Landscape architect Michael Hough emphasizes that both nature and culture must be considered in design and planning. To achieve regional identity he stresses the importance of knowing the place well and fitting designs to the people. It is also important to connect with history and to provide environmental education. We must practice sustainable design and do as little as possible, avoiding big projects that are more likely to compromise regional identity (Hough 1990). Ecological design can establish deeper levels of regional and local identity that will persist over time.

Seeing identity as a gradient requires a fundamental shift in the way urban design is taught and practiced. As Donald Appleyard emphasized, urban designers must also become sophisticated in reading the landscape not only for visual clues, but also for its social messages and meanings. They must travel to sites, talk with residents, and listen to their stories. Good identity-forming urban design comes from direct experience of both the landscape and its people. Urban design curricula must include social science methods, in addition to more traditional design methods. Identity cannot be captured simply through snapshots, but requires a continuous engagement with the place and its life. It should come from the bottom up, rather than top down, from the users, rather than the experts, from incremental projects, rather than large-scale urban renewal programs.

One final concern has to do with the resilience of place identity over time. Regardless of what the ingredients of this identity gradient may be, it is important that urban designers plan for its natural evolution. In Irvine, Librino, and many other designed landscapes, the original identity is often maintained through rigid regulations preventing personal expression in favor of the preservation of the initial concept. As a result, design integrity is seen as a sign of a healthy identity, while changes and adaptations are interpreted as signs of an unhealthy, degraded place identity. Maintaining a gradient of identity is a much more complex endeavor, as it cannot rely purely on the maintenance of an original form. It must include considerations of social, economic and cultural processes needed to successfully manage the evolution of the cities and neighborhoods we design, allowing them to change and adapt to future conditions, while maintaining their essence. Only this can insure that the place identity resulting from our designs will be stronger, more imageable, and ultimately more sustainable.

References

Anderson, K., Domosh, M., Pile, S. and Thrift, N., (Eds.) (2003). *Handbook of Cultural Geography,* Thousand Oaks, CA: Sage.

Appleyard, D. (1976). *Planning a Pluralist City: Conflicting Realities in Ciudad Guayana,* Cambridge, MA: MIT Press.

—— (1979). "The Environment as a Social Symbol: Within a Theory of Environmental Action and Perception," *Journal of the American Planning Association,* 45(2): 143–153.

Castells, M. (1989). *The Informational City: Information Technology, Economic Restructuring, and the Urban-Regional Process,* Oxford: Blackwell.

—— (1997). *The Power of Identity,* Oxford: Blackwell.

CEDOC (2008). *Il Quartiere di Librino: Percezione ed Atteggiamenti sulle Problematiche Infrastrutturali e dei Servizi,* Catania, Italy: Università di Catania.

Clay, G. (1994). *Real Places: An Unconventional Guide to America's Generic Landscape.* Chicago: University of Chicago Press.

Cullen, G. (1961). *Townscape,* London: Architectural Press.

Dahiden, J. (1972). *Urban Structures for the Future,* New York: Praeger.

Erikson, E. (1959). *Identity and the Life Cycle,* New York: W.W. Norton Company.

Groth, P. (1997). "Frameworks for Cultural Landscape Study," in P. Groth and T. Bressi (Eds.) *Understanding Ordinary Landscapes,* New Haven, CT: Yale University Press: 1–24.

Heidegger, M. (1971). *Poetry, Language, Thought,* Albert Hofstadter (trans.), New York: Harper & Row.

Hess, A. (2005). *The Ranch House,* New York: Harry N. Abrams.

Hester, R. (2008). *Design for Ecological Democracy,* Cambridge, MA: MIT Press.

Hough, M. (1990). *Out of Place: Restoring Identity to the Regional Landscape,* New Haven, CT: Yale University Press.

Lynch, K. (1960). *Image of the City,* Cambridge, MA: MIT Press.

—— (1981). *A Theory of Good City Form,* Cambridge MA: MIT Press.

Milligan, M. (1998). "Interactional Past and Potential: The Social Construction of Place Attachment," *Symbolic Interaction,* 21, 1: 1–33.

—— (2003). "Displacement and Identity Discontinuity: The Role of Nostalgia in Establishing New Identity Categories," *Symbolic Interaction,* 26, 3: 381–403.

Opotov, S. and Clayton, S. (Eds). (2003) *Identity and the Natural Environment,* Cambridge, MA: MIT Press.

Pretty, G.H., Chipuer, H.M., and Bramston, P. (2003). "Sense of Place Amongst Adolescents and Adults in two Rural Australian Towns," *Journal of Environmental Psychology,* 23: 273–287.

Proshansky, H.M., Fabian, A.K. and Kaminoff, R. (1983). "Place-Identity: Physical World Socialization of the Self," *Journal of Environmental Psychology,* 3: 57–83.

Relph, E. (1976). *Place and Placelessness,* London: Pion Limited.

Ruggeri, D. (2009). *Place-Identity, Attachment and Community Attitudes in Suburbia: The Irvine Ranch Case,* Dissertation, Berkeley, CA: University of California.

Soja, E.M. (1992). "Inside Exopolis: Scenes from Orange County," in M. Sorkin (ed.) *Variations on a Theme Park,* New York: Hill and Wang, 94–122.

Stokols, D. and Shumaker, S.A. (1981). "People in Places: A Transactional View of Setting," in J. Harvey (ed.) *Cognition, Social Behavior, and the Environment,* Hillsdale, NJ: Lawrence Erlbaum: 441–488.

Tuan, Y.F. (1977). *Space and Place. The Perspective of Experience,* Minneapolis, MN: University of Minnesota Press.

Vale, L. and Bass-Warner, S. Jr. (Eds.) (2001). *Imaging the City: Continuing Struggles and New*

Directions, Cambridge, MA: Center for Urban Policy Research.

Webber, M. (1963). "Order in Diversity: Community without Propinquity," in Wingo, L. (Ed.) *Cities and Space: The Future Use of Urban Land,* Baltimore, MD: Johns Hopkins Press.

—— (1964). "The Urban Place and the Nonplace Urban Realm," in Webber, M., Dyckman, J., Foley, D., Guttenberg, A., Wheaton, W., Bauer Wurster, C. (Eds.) *Explorations into Urban Structure,* Philadelphia PA: University of Pennsylvania Press: 79–153.

Further reading

Appleyard, D. (1979). "The Environment as a Social Symbol: Within a Theory of Environmental Action and Perception," *Journal of the American Planning Association* 45(2): 143–153. A pioneering treatise on the connections between the physical environment and its social meaning, this article develops a communications model of environmental action and perception.

Hough, M. (1990). *Out of Place: Restoring Identity to the Regional Landscape,* New Haven, CT: Yale University Press. A provocative book arguing that we should give more emphasis on restoring the diversity inherent in the ecological systems and human communities.

Opotov, S. and Clayton, S. (Eds.) (2003). *Identity and the Natural Environment,* Cambridge, MA: MIT Press. An examination of the ways in which our self identity influences our relationship with nature, and vice versa.

Relph, E. (1976). *Place and Placelessness,* London: Pion Limited. A pioneering analysis of the attributes that generate a sense of place or, its opposite, placelessness.

38

Old vs. new urbanism

Ivonne Audirac

Among the various urban design discourses that claim an antidote to the social, environmental, economic, and aesthetic malaise of American conventional urbanism, New Urbanism stands tall and confident against the "city functional" promoted by the Congrès Internationaux d'Architecture Moderne (CIAM). New Urbanism has emerged at the confluence of several postmodern movements critical of modern rationalism, the ecological ravages of industrialism, urban social inequality and segregation, and automobile-dependent cities. The Charter of the New Urbanism (CNU) has become the overarching design episteme for a new regionalism, good city form, sense of community, ecological sustainability, public health, social justice, civility, and democratic participatory planning (CNU 1999; Duany *et al.* 2000; Calthorpe 2001; Calthorpe and Fulton 2001; Farr 2007).

In reviewing New Urbanism's roots, trends, and debates as alternative to the "Old Urbanism," this chapter argues that the allure of the "new" in New Urbanism stems from the movement's ability to repackage 1960s critique of automobile-centric urbanization[1] with the 1991 Rio-Summit's sustainable development ideals of ecological, economic, and social sustainability. United by a shared vision of the power of traditional neighborhood development (TND) and transit oriented development (TOD) to wean urban lifestyles from automobile dependence and to infuse neighborhoods with the public life and community amenities absent in conventional suburbia, TND and TOD make pedestrian mobility and metrics the core and essence of a new language of place making. Despite their common goals, TND and TOD solutions differ in their regional design approaches to interlacing with the Old Urbanism's regional mobility network: The rural-urban *transect* (Duany and Plater-Zyberk) and the transit-based *urban network* (Calthorpe) respectively offer ideal models of metropolitan gradients and circulation corridors. In practice, both approaches must deal with the logic and metrics of a regionalism of swift flows where, since early on in the twentieth century, motorized mobility has been the overarching goal, and transportation engineering bureaucracies, rather than architect-planners, have been the de-facto urban designers. In addressing this challenge, the two approaches resort to different design tactics and "paradigm" politics engendering at the same time a host of controversies and debates at the core of CNU, some of them reminiscent of CIAM.

Roots and influences

New Urbanism's principles and urban design manifesto coalesced in 1993 in the Charter of the Congress for the New Urbanism (CNU), which convened a network of likeminded architects, urban designers, planners, developers, lawyers, public officials, and citizen activists engaged in righting the wrongs of CIAM, Euclidian zoning and post-war urbanization. New Urbanism is not a monolithic top-down organization; there are several strands of New Urbanism. However, its roots can be traced to two distinct 1980s urban design perspectives: Traditional-Neighborhood Development (TND), which emerged in the US east coast bearing Léon Krier's influential imprint,[2] and Transit-Oriented Development (TOD) born in the west coast from an array of ecological critiques of modernist design and town planning, many of them extending Jane Jacobs and her contemporaries' influential diatribe against modern, auto-centric city building. Both perspectives denounce urban sprawl as manifested in segregated land uses that make driving indispensable and in the proliferation of single-family-home subdivisions devoid of a public realm, sense of place, and a sense of community. Both prescribe diverse, pedestrian-friendly neighborhoods with mixed-land use destinations accessible within a five or ten minute walk. But, although by the end of the 1990s the two approaches were viewed as fungible community design innovations, their design solutions and normative strategies differ in the way they organize the center and in the way they articulate the neighborhood with the rest of the urban fabric, that is, their corresponding regionalisms.

In its well known prototypical formulation, TND converges on an identifiable town center laid atop a dense grid of boulevards, narrow streets, alleyways, and plazas, while TOD focuses on a transit station to which all major streets converge. TND and TOD are similarly compact with enhanced public realms of pedestrian-friendly frontages, tree-lined streets, and front porches, while parking and garages are tucked in alleyways behind houses and buildings. However, whereas TOD, organized around transit stations, depends on high quality transit service for effective substitution of car trips for transit trips – a cornerstone of the environmental sustainability claims of the movement – TND relies on existing arterials and freeways for metropolitan travel and public-realm aesthetics rather than green considerations. This fundamental distinction stems from variations in the critique of modern urbanism and from the ideal models from which their designers drew inspiration.

Traditional neighborhood development

Andrés Duany and Elizabeth Plater-Zyberk's Town of Seaside, in the Florida Panhandle (Figure 38.1), is the American mise-en-œuvre of Léon Krier's critique of modern urbanization and the destruction of traditional European city patterns. Krier (1984) denounced the industrial capitalist city, large corporations, planning and land use zoning and the automobile as destroyers of city life, the civic realm, and the sense of community. He advocated a return to the preindustrial city of arts and crafts with a strict separation of city from countryside; the elimination of zoning; and the reorganization of the city into a federation of semi-autonomous, aesthetically appealing urban quarters of 15,000 people (limited in territorial size to 87 acres and a comfortable walking distance of no more than 20 minutes). This approach would permit residents to live, work and play in the same urban quarter where everything would be accessible by foot, social segregation by wealth and age would be eliminated, and reduced car use would liberate

Figure 38.1 Aerial view of Seaside and Watercolor, Florida. Source: Google Earth.

time from wasteful commuting and excessive travel for leisure, education and other lofty pursuits. This reconstruction of the city – subsequently endorsed by Prince Charles and the British Urban Village movement – would ostensibly beget more energy-efficient environments, a rebirth of craftsmanship would replace commercial kitsch, and a vital pedestrian city would supersede the boredom and social void of the automobile suburb (Krier 1984; Audirac and Shermyen 1994). Léon Krier's normative vision of the urban quarter adapted to the American context via TND Ordinance similarly offers reclaimed communitas, civitas, social justice, and true democracy by "adapting the conventions which were normal in the US from colonial times until the 1940s" (Foundation for Traditional Neighborhood in Krieger 1991:102). As with Krier's profound anti-metropolitanism – he deemed the metropolitan region theoretically unimportant and inferior to his federation of semi-autonomous quarters[3] – TND's integration into the existing metropolitan fabric and economic structure is largely assumed to happen by TND accretion and seamless street connections between TNDs. However in practice, its regional shopping and office areas are largely oriented to the existing freeway and highway networks as in the regional office centers of Celebration, Florida (Figure 38.2) or in Avalon, Florida, where "extensive parking and department stores are conventionally visible from the expressway" (Krieger 1991: 88).

While Léon Krier's anti-modern-city tirade had a strong influence in New Urbanism's TND, his critique was superseded by Allan Jacobs and Donald Appleyard's livable-places manifesto in the 1970s and even earlier, in the 1950s and 1960s by William H. Whyte's and Jane Jacobs' indictment of suburbia, urban renewal, and zoning. Paradoxically, and

Figure 38.2 Aerial view of Celebration, Florida. Source: Google Earth.

although not so well known, Victor Gruen, America's foremost mall developer also criticized (Gruen 1966) traffic engineering, planning, zoning, city building, and automobile dominance which foreshadowed the urban "growth machine," as argued by Logan and Molotch (1987). A forerunner of New Urbanist censure of the Modern Masters, Gruen viewed with contempt Le Corbusier's built projects as "lonely structures rising out of a dismal sea of asphalt, cement and the tin roofs of moving and stored automobiles" (Gruen 1966: 191) and Wright's Broadacre City's spreading people even further apart than in usual land-wasting suburban patterns" (Gruen 1966: 191). His critique denounced the first twenty years of post-War sprawling, space-consuming, unplanned growth, and the destruction of the traditional urban fabric, extensive land-use segregation negotiable only by car, the car's nefarious wiping out of the pedestrian realm; the dearth of public transit, and the squandering of time in traffic congestion (Gruen 1966: 177–187).[4] Against the centerless city, Gruen's solution – applied in his Fort Worth Plan – was the recentralization of bedroom suburbs around city centers anchored by pedestrian friendly regional shopping centers – the latter conceived as mixed-use urban nuclei (193–194). Gruen's Fort Worth Plan, aimed to domesticate the motorcar, envisioned a metropolitan system of cities consisting of a metrocore (central city) and orbital cellular cities served by metropolitan high-speed rail. Within each city, a nested system of well-defined towns and neighborhoods and industrial centers would be pedestrian-focused and linked by a pattern of pedestrian walks and plazas extending into surrounding green areas. Intercity bus transit would serve each town and city, while neighborhood circulators would be feeders to this system. A scheme of three mixed-mode-freeway loops – with dedicated lanes for express bus transit – would interconnect at the outer loop to the airport and to the national highway system, while internally it would rely on local roads for buses serving downtown and cross city travel. As a transit-cellular metropolis, its designer assumed that people would prefer transit over cars since driving into town would be made intentionally difficult and fast driving would only be possible along the freeway loops supported by huge garden park-and-ride facilities (see Figure 38.3).[5] In the *Exploding Metropolis* (edited by Whyte in 1958), Jane Jacobs complimented the Fort Worth Plan's street treatment, whose public realm was meant to "be more compact, more variegated and busier than before" (Gruen 1966: 200). She noted though, that while the plan had influenced the thinking of many planners, it had met with limited implementation. Thus, despite Gruen's search for alternatives to the centerless city, his built urban legacy of shopping centers could not have been more antithetical to his own ideas. In the 1970s, disillusioned by this legacy, he wrote a manifesto against the modernist Charter of Athens, which together with his book, *Centers for an Urban Environment: Survival of the Cities*, criticized his earlier shopping centers, argued for the need for global sustainability, and refined his cellular metropolis concept extending his metropolitan ideas to the regional and global scales (Hill 1992).

In the sustainable-urban-form debate, David Hill (1992) places at one end the "anarchist" tradition of semiautonomous intimate compact communities such as those proposed by Léon Krier, and at the other end, the "community of interest" tradition with supporters of information-age-megalopolitan urban systems. In this debate, Gruen's cellular transit metropolis occupies a middle position by offering an eclectic multimodal compromise of human scaled pedestrian centers, transit-based diverse and dense urban nuclei, the urban realm and street life of European (upper-class) boulevards, and a system of ring roads with freeways and garden parkways for automotive and bus travel. On the whole, Gruen's integrated transit/auto concept would counter the negative ecological effects of suburbanization and the modern metropolis (Hill 1992).

Transit oriented development

Features of Gruen's urbanist thinking clearly anticipated the TOD, whose regional scope and design synthesis took shape in the San Francisco Bay Area with Sym Van der Ryn and Peter Calthorpe's (1986) work on ecologically-minded redesign of the suburban fabric. However, mass transit, rather than green design, inspired Peter Calthorpe and Douglas Kelbaugh to formulate the "pedestrian pocket,"[6] and later, Peter Calthorpe (1993) to expand the concept to TOD. Inspired by earlier streetcar suburbs, mixed-use centers would be the densest regional TOD, large *urban* TOD would be located on a trunk line, and smaller suburban or *neighborhood* TOD would be sited on feeder bus lines.

To counter the urban sprawling effects inherent in the synergies between transit and automobile systems, TOD is fundamentally a component of a growth management regional strategy (e.g. urban growth boundaries) – the latter was applied with considerable success in Portland, Oregon under the LUTRAQ (Land Use and Transportation and Air Quality Project). The first "on ground" test of Calthorpe's new town TOD idea was Laguna West, an 800-acre-pedestrian-friendly community built around a lake, town center, and future transit station in Sacramento, California. However, its infrequent bus service makes Laguna West incomplete as a TOD and still highly dependent on the automobile. Other TOD such as Orenco Station, on

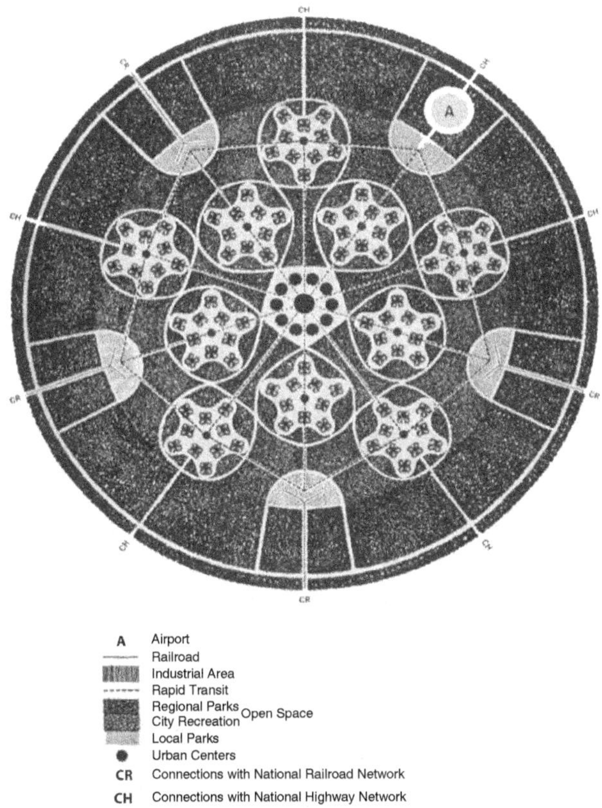

(a)

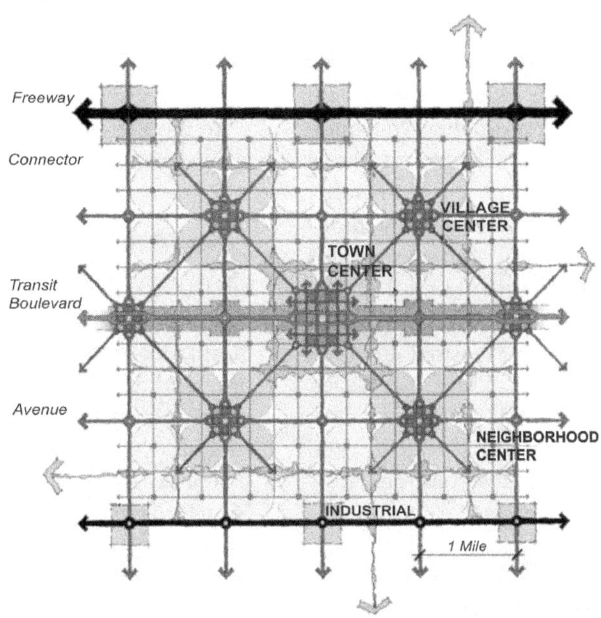

(b)

Figure 38.3 Gruen's Metrocore (a) and Peter Calthorpe's Urban Network (b). Sources: Simon and Schuster, used by permission; and Peter Calthrope – used by permission.

Portland's Westside light-rail line, and the Rosslyn-Ballston Corridor in Arlington VA in the Washington metropolitan area, where nearly 50 percent of corridor residents commute by Metro transit (Virginia Commonwealth University 2008), are considered more successful TOD examples.

All in all, TND and TOD roots clearly point to two different mobility regionalisms. TND, influenced by Krier's precepts dismissive of metropolitan realities, was conceived to produce relatively self-contained communities "bringing within walking distance most of the activities of daily living" (TND Ordinance in Krieger 1991:102). Thus, in practice, TND communities are largely car-based and grafted onto the Old Urbanism's fabric of arterials and freeways.

While not invoked before as part of TOD's lineage, Gruen's cellular transit metropolis, envisioned as remedy to automobile dominance, is clearly a TOD precursor. However, in practice, TOD's mobility regionalism – in addition to energy crises and traffic gridlock, which make driving expensive – is highly dependent on a city's present and future transit service model as well as on strong pro-transit public-private coalitions (e.g. local transit authorities, metropolitan planning organizations, commercial and institutional anchors, developers, and citizens) sufficiently powerful to offset the balkanized nature of public transit and para-transit provision in order to provide the levels of service required to sway drivers to switch from car-based to transit-based lifestyles (Burkhardt *et al.* 2002). Until more such transit systems are built, TOD's regionalism will remain in clear competition with the Old Urbanism's regional system of arterials, freeways and interstates.

New Urbanism trends

A salient trait of New Urbanism is that it is constantly evolving by renewing and adapting ideas from outside its canon, often in response to criticism. In addressing TND's weak regionalism, New Urbanists Duany and Plater-Zyberk proffer the "rural-urban transect" (RUT) and "smart code" for coding the character of places (Duany and Plater-Zyberk 2008). Peter Calthorpe, on the other hand, responds with the Urban Network to the failure of New Urbanist mom and pop retail to survive without the visibility from the freeway, busy arterials and major intersections where retailers want to be (Calthorpe in Langdon 2002). He acknowledges the economic and social realities of contemporary urban systems – regional in scope and intrinsically car and truck oriented – which overwhelm New Urbanist attempts to reorient retail and lifestyles internally toward the neighborhood or town:

> More than ever, regions define our lives. Our job opportunities, cultural interests and social networks are bigger than any neighborhood or town. Even if we double the percent of walkable trips in a neighborhood and triple transit ridership, there still will be massive growth of auto trips – not to mention an exploding quantity of truck miles" (Calthorpe 2001: 2).

The urban network

Similar to Gruen's multimodal, eclectic compromise, Calthorpe proposed the "Urban Network": a multimodal circulation and mobility framework for New Urbanist projects consisting of a system of arterials which bifurcate into one-way roads (couplets) as they reach and brace around New Urbanist walkable-mixed-use-village centers (anchored by 100,000-square-feet retail serving two to four square-mile market areas). As the one-way couplets exit the centers, they merge back into two-way arterials. Within a five-to-ten minute walk from the village centers, quarter-mile radial neighborhoods surround each village perched on an intimate grid of connector streets that provide convenient access to pedestrians, bicycles, and cars, while Transit Boulevards, equipped with dedicated rapid-transit rights of way and frontage roads, connect village centers and town centers (the densest employment and commercial centers) (Figure 38.4). This model would afford a transit-oriented pedestrian friendly streetscape the possibility to blend with high traffic capacities. Furthermore, limited-access throughways supporting truck and auto-oriented warehousing, manufacturing and light industrial areas would be physically segregated, while strip development along these roads would be offset by development opportunities on the boulevards and arterials catering to village and town centers, and by limiting curb cuts (Calthorpe 2001: 3). The Urban Network's street types and circulation corridors would replace the Old Urbanism's functional road classification of arterials, collectors and local streets. It would work with existing freeways but wherever possible, it would replace them with Transit Boulevards and Throughways. Despite its new street hierarchy, intersection design, and land-use types, the Urban Network would rely on the Old Urbanism's bureaucracies and institutions. "Road builders would still lay down asphalt, automakers could build buses, and developers could still build communities" (Calthorpe 2001: 3); however, beyond the neighborhood scale, the Urban Network would be the requisite regional circulation paradigm of the New Urbanism.

The rural–urban transect and urban network

The absence of a regional model has confined New Urbanist practice to the siting of New Urbanist communities as nodes in

Figure 38.4 Village Center One-Way-Couplets in San Elijo, California. Source: Google Earth.

Old Urbanism's transportation networks (Bohl and Plater-Zyberk 2006). The Rural-Urban Transect (RUT) is a regional framework that attempts to solve this issue by expanding the conventional "urban, suburban, and rural" categories into six "ecological zones" (transect-zones) ranging in settlement variety and density from T1 the natural zone (unsuitable for settlement) to T6 the urban core zone (the densest and most varied in uses). This normative typology of ideal regional zones synthesizes previous re-adaptations of Patrick Geddes' Valley Section[7] into a regional framework for regulating the "character of place." Reminiscent of CIAM 8's Valley Section hierarchy and TEAM-10's inclusion of overlapping human associations from the hamlet to the village and the whole city region, the RUT "attempts to distill general physical characteristics of urbanism that have existed for 5,000 years – the hamlet, village, urban neighborhood, town, and city – in relation to each other and the natural world" (Bohl and Plater-Zyberk 2006: 10).

Rather than a true rural-urban transect, or overall theory of human settlement, the RUT is an ingenious and pragmatic sampling method of urban form. As a stylized density gradient, it is essentially a six-category pictographic representation of a romanticized and sanitized urban region – industrial uses and public utilities such as power plants and landfills are zoned out of the RUT into a separate "district" category. New Urbanists use the RUT as a filter for canvassing a city's rural–urban fabric and for selecting from within it prototypical precedents of traditional urban forms that best fit each ideal transect zone (i.e. transect calibration to local character). The outcome, or "synoptic survey," displays the sampled local areas whose physical attributes and metrics at the block level ("quadrat") and public and private frontages ("dissect") serve to generate the urban code regulating the future community's building types, heights, setbacks, open spaces, street character, alleys, streetscape, etc. Both praised as innovative and rebuffed as unauthentic (Moore 2001), the RUT is the generative method for TND's "neotraditional regionalism" and its companion, the smart code, the legal-administrative tool for coding the "DNA" of "good places" and for making such style of form-based coding as compulsory as Euclidean zoning. In contrast to Calthorpe's Urban Network, which adapts to the Old Urbanism's institutions and transportation bureaucracies, Andrés Duany acts as a geneticist of the urban genome, using the transect and smart code to "genetically" alter the Old Urbanism by infiltrating its vast bureaucracies and permeating its regulatory cultures, by penetrating public administrations and design and planning schools, and by winning over the building industry:

> Like the early Christians, our chance is to infiltrate it by using its assets: its secure communication network, its stable currency, its common language and its myths. We will push until the first Emperor of the ULI becomes a New Urbanist (It's close.). There have already been two Governors that are open converts. The vast bureaucracies are everywhere infiltrated, as are a few schools. (Duany in Mehaffy 2004: n.p.)

Andrés Duany's New Urbanist crusade is making inroads as RUT and smart code workshops, conferences, and charrettes organized by an assortment of CNU affiliated institutes and centers have spread the use of transect-form-based codes and certified their apprentices. In 2004, California Governor Schwarzenegger signed a bill that institutionalized form-based coding, a number of jurisdictions throughout the country have adopted form-based codes and the Institute of Transportation Engineers (ITE) in partnership with CNU has used

RUT language to reformulate context-sensitive design of thoroughfares. ITE's intent in the report *Context Sensitive Solutions in Designing Major Thoroughfares for Walkable Communities* (ITE 2006) is to illustrate how established street design guidance such as the American Association of State Highway and Transportation Officials' (AASHTO) *Policy on Geometric Design of Highways and Streets* – considered the bible of transportation planners and engineers – "can be applied to create context sensitive designs in places with the qualities of traditional urbanism" (ITE 2006: 11). Thus, as zealously devised by TND New Urbanists, RUT language has started to seep into the lexicon of traffic engineers and has begun to permeate the bureaucracies that administer the form, structure and mobility options of the Old Urbanism.

Canons of sustainable architecture and urbanism

In spite of early forays into green urbanism by Van der Ryn and Calthorpe (1986) and by Michael and Judy Corbett – the designers of Village Homes – who spearheaded the Ahwahnee principles and became CNU co-founders,[8] green urbanism and New Urbanism, until very recently, have followed parallel paths. For at least a decade after Seaside, in practice "none of the new urbanist developments planned or built addressed ecological concerns beyond the very important problem of reducing dependence on the automobile" (Corbett and Corbett 2000: 16). And although the Ahwahnee declaration blended the formalistic and aesthetic aspects of TND and TOD's regional transportation concerns with the ecological considerations of Village Homes – the prototypical green community built in the 1970s – disagreement over TND gridded streets vis-à-vis cul-de-sacs, used in Village Homes for achieving natural drainage and reducing paved land in favor of common areas and edible gardens, set the two approaches apart (Corbett and Corbett 2000). However, in Civano, the first green New Urbanist community built near Tucson Arizona, Elizabeth Moule and Stefanos Polyzoides with William McDonough and Andrés Duany brought the two approaches together. Perhaps because "the actual cost to develop Civano, with its innovative energy and resource conservation technologies, was $20 million more than for a similarly sized, conventional master-planned community" (Terrain.org 1999), not many similar communities have been built to date. Nonetheless, recent heightened energy and climate change concerns prompted Moule and Polyzoides and other New Urbanists to draft the "Canons of Sustainable Architecture and Urbanism" (Moule *et al.* 2007). Referred to as a companion to the Charter of the New Urbanism and intended to clarify the relationship between New Urbanism and sustainability, the Canons constitute a long overdue "greening" of the Charter and a deliberate effort to harmonize green and form-based codes.

Debates and controversies

Debates and controversies surrounding New Urbanism run the gamut (Southworth 2003). They range from concerns with the environmental determinism (Audirac and Shermyen 1994) of peculiar New Urbanist brand of sociology and romantic assumptions about the inherent sociability, urbanity, participatory citizenship, socio-economic integration, and social equity of the mixed-use walkable neighborhood to New Urbanist communitarianism rooted in a mythical account and nostalgic reconstruction of American small-town society, which was seldom inclusive, healthy, or safe (Harvey 1997 and 2000; Clarke 2005).

Not unlike CIAM's spatial determinism and universalist claims about its modern design constructs, influential members of the CNU harbor a deep conviction that traditional urban forms, obeying scientific principles derived from nature, bear genetic codes that once "cracked" and distilled by designers can be universally applied to cities and urban societies across the globe (Salingaros *et al.* 2006). Stretching this idea, others have argued for the "moral authority" and universality of the transect emanating from the natural law precept of "mixed-use walkable settlements" – "valid for all human beings in all times and places" (Bess 2008: 1). And thus, New Urbanists invoke the ontology of the "mixed-use walkable settlement" as universal cure for the "modern" ills besieging contemporary urban societies regardless of geography or culture. Whether designing new infill and greenfield communities, rebuilding natural-hazards-ravaged cities in the US or reconfiguring slums and shanty towns in Lima, Peru or Kolkata, India, according to New Urbanist dogma, its design precepts are not only scientifically warranted and applicable but also morally binding (Salingaros *et al.* 2006; Bess 2008).

Despite the desirability of the mixed-use walkable neighborhood in American cities, its essentialism, as proffered in TND and RUT design discourses, stems, as here described, partly from Krier's theoretical antipathy for the metropolitan scale in concept and form. This has left New Urbanism vulnerable to criticism regarding ignorance or naïveté about metropolitan structure, agglomeration economies, and the political economy of information-age urban regions, which render anachronistic and chimerical the notion that today's living can be accommodated within a five-to-ten minute walk. However, its scientific and moral universalistic claims have opened up the gate to a flurry of disparaging critique pointing to New Urbanists' "cultural myopia masquerading as universal values" (Knox 2008: 109).

At a more practical level, the mainstreaming of New Urbanist communities is starting to provide a testing field for many professed claims. While indeed, New Urbanist mixed-use developments provide an enhanced public realm along streets and commercial centers and impart a themed or up-scaled character of place absent from conventional subdivisions, critics objecting to the private nature of these public spaces dub them "privatopia." Although New Urbanism deserves praise for envisioning a higher quality physical world that privileges human scale, the spatial reproduction of socioeconomic segregation via New-Urbanist-Hope-VI[9] and regular New Urbanist communities undermines the much touted New Urbanist fight against social inequity and sprawl (Calthorpe and Fulton 2001: 11). The art of place making may beautify and improve the walkability of poor and affluent neighborhoods alike, but in the end, it reproduces the Old Urbanism's socioeconomic and spatial distance that segregates and differentiates poor and affluent "sense of place" on the metropolitan landscape. Furthermore, for the sake of de-concentrating poverty, New-Urbanist-Hope-VI projects replace poor tenants with imported higher income homeowners. This social engineering of mixed-income communities is drawing criticism similar to the one leveled at Modernist urban renewal for being a new form of gentrification, which is not only perpetuating the tradition of social displacement, but also "imposing restrictive architectural and planning strait-jackets onto those who are privileged to remain" (Pyatok 2005: n.p).

CNU merchandizing workshops admit that "developing successful retail centers is one of the more difficult aspects of New Urbanism, where many planned centers fail to attract key retailers or to meet market performance standards" (Gibbs 2008). As TOD proponents have realized, New Urbanist retail will not survive on neighborhood foot traffic alone. It needs to draw customers from the rest of the city or region and compete with established Old Urbanism's retail. Thus, despite vehement pronouncement for social diversity and against automobility and sprawl, paradoxically, successful New Urbanist "lifestyle centers" – the open-air-village concept of upscale boutiques, restaurants, gridded streets and above-store apartments – typically depend on motorists and the Old Urbanism's road network and metropolitan structure to deliver the requisite prosperous clientele from outside the community generating heavy volumes of motorized traffic.[10]

Perhaps the most telling controversy surrounding the New Urbanism comes from within CNU. Although its diverse groups are bound by an oath against sprawl, some deep disagreements reveal internal fissures paralleling CIAM's evolution. One such controversy revolves around Calthorpe's one-way-couplet village centers rejected by some New Urbanists for speeding up traffic flow and jeopardizing walkability and retail viability (Langdon 2002). The second involves the RUT as dogma and guiding framework for all urban development. The insistence of CNU leaders on strict adherence to the transect by the US Green Building Council's (USGBC) green neighborhood certification and rating criteria (LEED-ND) alienated all CNU partners not practicing or speaking the RUT lingo (Wendover 2008). Similarly an impasse over vertical sprawl – among fundamentalists: more than six stories – split opponents and supporters of tall buildings and high density (Wendover 2008). Reminiscent of Modernist hard lines and internal divisions, with time, the parallels between CIAM and CNU only grow more pronounced.

Conclusions

The swift acceptance of New Urbanism in the public imagination was undoubtedly facilitated on the one hand by Neo-traditionalists' nostalgic reinterpretation of pre-motor-age sociability, community utopianism, and revival of the picturesque aesthetic, and on the other hand, by the town of Seaside Florida's incredible real-estate success.[11] In effect, with more than 500 New Urbanist communities built or in construction (Steuteville 2008), the New Urbanism phenomenon has been part and parcel of the building boom of the last twenty years. Economic geographers associate this boom with the end of the last growth cycle – based on homeownership, easy credit, cars, and federal highway construction – that began in 1930 with the Great Depression and the advent of the New Deal programs (Florida 2009).

Paradoxically, New Urbanism's facile equating of sprawl to the last seventy years of American suburbanization and to the root of just about all social evils, has allowed it to declare as practically all urban forms not conforming to New Urbanist principles wholesale anti-urbanism. In this way, postwar conventional sprawl becomes Old Urbanism, or "sprawl urbanism" and the movement extricates itself from the urbanization process that it denounces. By the same token, it is free to repackage a long standing critique of the modern metropolis and to resort to community utopianism, scientific universalism, or moral imperatives to justify the form-based coding of the type of rural-urban region presumed to achieve the good life. Yet few New Urbanists, with notable exceptions, recognize that the reviled contemporary American metropolis (i.e. Old Urbanism) is the built outcome of many such visions from CIAM to the Neighborhood Unit Movement and from Broadacre City to the New Towns Movement, each one containing its own brand of community utopianism, scientific claims, and ethical justification. Their successful and failed attempts to reorganize the metropolitan region were part of the capitalist "spatial fix"[12] afforded by new mobility technologies and massive New Deal programs. As with New Urbanism today, these visions aimed in their time at remedying long-standing problems of poverty, economic decline, and environmental health associated with industrial, rail-based urbanization. In building solutions to these intractable problems – largely inherent in America's evolving capitalism – new problems emerged. Likewise, the fate of previous urban design movements, curtailed by the Great Depression and by everyday and generic urban design and building practices, may presage the future of the movement. However, New Urbanism may prove to be adaptable, innovative and resilient, and may very well become the driving vision of the next long wave of American urbanization. But this will require, as one insider surmises (Kelbaugh 2007), New Urbanism to harness the emerging social and economic forces that can make it inevitable and sustainable, rather than relying too heavily on rigid form-based codes and regulations and its inflexible moral mandates.

Notes

1 Known for ushering the decline of industrial-based central cities and the emergence of contemporary metropolitan regions (of edge cities and waning central cities).

2 Krier claims to have been the originator of the CNU idea as a movement antithetical to CIAM (Congrès Internationaux d'Architecture Moderne) and to have partnered with Andrés Duany and the developer of the Town of Windsor, who financed the first CNU meeting in 1993 (Thompson-Fawcett 1998: 179).

3 "Large scale metropolitan … [areas are] really something [for] which I don't think a theory is needed. … You need rather to theorize the small city … because the large city happens anyway. You need rather theories of how to prevent metropolitanisation" (Krier in Thompson-Fawcett 1998: 173).

4 Victor Gruen charged against the "false friends" of the city, which included: transportation planners ("traffickists"), redevelopment planners and developers ("bulldozerites"), land-use planners ("segregators"), urban designers ("projectites"), and economic development planners ("economizers") whose leading interests were to serve the "well-being of machines (real or political)[…] power, the motorcar, and money" (Gruen 1966: 177).

5 The concept of freeway loops and parking towers acting as the old city walls to protect the city from through traffic was popular in the 1960s. Louis Khan proposed a similar idea for Philadelphia (see Frampton 1983).

6 A residential neighborhood placed within walking distance of transit, jobs, schools, shops, parks, and civic amenities, conceived as a regional design strategy for reducing driving, preserving open space, increasing the supply of affordable housing, and reining in urban sprawl.

7 The rural-urban transect, resurrects Patrick Geddes' ideal urban-regional transect or the "valley section" and its companion "the notation of life" – a diagram that Geddes conceived as a zoning tool specifying concrete architectural proposals for how to realize the ideal city-region along the valley section (Welter 2003).

8 In 1990, at the behest of Peter Katz, Peter Calthorpe, Andrés Duany, Elizabeth Plater-Zyberk, Elizabeth Moule, and Stefanos Polyzoides met for dinner at the Corbetts' house to generate the Ahwahnee principles, which later evolved into the Charter for the New Urbanism (Corbett and Corbett 2000:11–12). The Corbetts belonged to the 15-group of CNU original cofounders.

9 HOPE VI (Housing Opportunities for People Everywhere) is a major public housing redevelopment program administered by the Department of Housing and Urban Development (HUD) that replaces the most distressed public housing projects, typically occupied by the poorest households, with housing redesigned for residents of mixed income. The intent of the program, launched in 1992 and drastically reduced in 2008, has been to "de-homogenize" i.e. to deconcentrate poverty from these projects by bringing in higher income residents and providing housing vouchers to displaced residents. Assessments of the program report evidence that the original residents of HOPEVI projects did not benefit from these redevelopment schemes and that

some ended in similar or more precarious living conditions (Popkin *et al.* 2004). On the other hand, hailed by New Urbanists as their greatest public policy achievement for bringing higher quality housing to the most destructive housing environments (Steuteville 2005), New Urbanist HOPE VI projects have also met with disapproval for bringing about a net loss of low-income housing units.

10 Coincidentally, if inserted in Calthorpe's Urban Network model, New Urbanist lifestyle centers would not only resurrect but also update Gruen's cellular metropolis.

11 In 20 years Seaside property appreciated more than 100-fold their original value. The market validation of walkable, compact mixed-use urban design, rather than New Urbanist loftier social and ecological responsible goals swept middle class Americans awash with easy credit, a segment of the development industry eager to differentiate its residential product line, and Sun Belt localities buoyant largely on real-estate and construction growth.

12 Critical geographer David Harvey (2000) uses the term "spatial fix" to refer to the investments in the geographical landscape (e.g., built environment's physical infrastructure, transportation and communication, land development, territorial organization) that are made during a growing phase of capitalist cycles only to be destroyed and rebuild anew in the next growth cycle in the wake of the next capitalist crisis.

References

Audirac, I. and Shermyen, A.H. (1994). "An Evaluation of Neotraditional Design's Social Prescription: Postmodern Placebo or Remedy for Suburban Malaise?," *Journal of Planning Education and Research*, 13: 161–173

Bess, P. (2008). *The Polis and Natural Law: The Moral Authority of the Urban Transect*. Online. Available HTTP: <http://www.thursdayassociates.net/Texts/The%20Polis%20and%20NL-Center%2015%20rev.pdf> (accessed 15 March 2009).

Bohl, C. and Plater-Zyberk, E. (2006). "Building Community and the Rural-to-Urban Transect," *Places*, 18, 1: 4–17. Online. Available HTTP: <http://repositories.cdlib.org/ced/places/vol18/iss1/Bohl_Plater-Zyberk/> (accessed 15 March 2009).

Burkhardt, J.E., McGavock, A.T., Nelson, C.A. (2002). "Improving Public Transit Options for Older Persons," *TCRP Report 82 (Volume 1: Handbook)*, Washington, DC: Transportation Research Board.

Calthorpe, P. (1993). *The Next American Metropolis*, New York: Princeton Architectural Press.

——(2001) "The Urban Network," San Francisco: Calthorpe Associates. Online. Available HTTP: <www.calthorpe.com/clippings/UrbanNet1216.pdf> (accessed 15 March 2009).

Calthorpe, P. and Fulton, W. (2001). *The Regional City*, Washington, DC: Island Press.

Clarke, P. (2005). "The Ideal of Community and its Counterfeit Construction," *Journal of Architecture Education*, 58: 43–52.

CNU (Congress for the New Urbanism) (1999). *Charter of the New Urbanism*, New York: McGraw-Hill.

Corbett, J. and Corbett, M. (2000). *Designing Sustainable Communities: Learning from Village Homes*, Washington, DC: Island Press.

Duany A. and. Plater-Zyberk, E. (2008). "The Smart Code V.9," viewed March 15, 2009 <http://smartcodecentral.com/>

Duany A., Plater-Zyberk, E. and Speck, J. (2000). *Suburban Nation: The Rise of Sprawl and the Decline of the American Dream*, New York: New Point Press.

Farr, D. (2007). *Sustainable Urbanism: Urban Design With Nature*, New York: Wiley.

Florida, R. (2009). "How the Crash Will Reshape America?," *The Atlantic*, March 2009. Online. Available HTTP: <http://www.theatlantic.com/doc/200903/meltdown-geography> (accessed 15 March 15 2009).

Frampton, K. (1983). "Prospects for a Critical Regionalism," *Perspecta: The Yale Architectural Journal*, 20 (1): 147–162.

Gibbs, B. (2008). "Retail Recipes," *Congress of the New Urbanism*, XVI, Open Source Congress. Online. Available HTTP: <http://www.cnu.org/node/2075> (15 March 2009).

Gruen, V. (1966). "New Forms of Community," in L.B. Holland (Ed.) *Who Designs America? Princeton Studies in American Civilization No. 6*, Garden City, NY: Anchor Books Doubleday.

Harvey, D. (1997). "The New Urbanism and the Communitarian Trap," *Harvard Design Magazine*, Winter/Spring: 68–69.

—— (2000). *Spaces of Hope*, Berkeley, CA: University of California Press.

Hill, D.R. (1992). "Sustainability, Victor Gruen, and the Cellular Metropolis," *Journal of the American Planning Association*, Summer, 58(3): 312–315.

ITE (Institute of Transportation Engineers) (2006). *Context Sensitive Solutions in Designing Major Thoroughfares for Walkable Communities*, Washington, DC: Institute of Transportation Engineers.

Kelbaugh, D. (2007). "Toward an Integrated Paradigm: Further Thoughts on the Three Urbanisms," *Places*, 19(2): 2–19.

Knox, P.L. (2008). *Metroburbia, US*, New Brunswick, NJ: Rutgers University Press.

Krieger, A. (ed.) (1991). *Andrés Duany and Elizabeth Plater-Zyberk: Towns and Town Making Principles*, New York: Rizzoli.

Krier, L. (1984). "Léon Krier Houses, Palaces, Cities," in D. Porphyrios (ed.) *Architectural Design Profile*, London: AD Editions.

Langdon, P. (2002). "Human-Scale Shopping Still Elusive in the Suburbs; Calthorpe and Beyard Propose Solutions," *New Urban News*, July/August. Online. Available HTTP: <http://www.newurbannews.com/Human-scale%20 shopping.html> (accessed 15 March 15 2009).

Logan, J. and Molotch, H. (1987). *Urban Fortunes: The Political Economy of Place*, Los Angeles, CA: University of California Press.

Mehaffy, M. (2004). "An Interview with Andrés Duany," *Katarxis N° 3*, Online. Available HTTP <http://www.katarxis3.com/Duany.htm #Interview%20Text> (accessed 15 March 2009).

Moore, S.A. (2001). "Technology Place and the Nonmodern Thesis," *Journal of Architectural Education*, 54, 3: 130–139.

Moule, E., Dittmar, H. and Polyzoides, S. (2007). "Cannons of Sustainable Architecture and Urbanism a Companion to the Charter of the New Urbanism," *CNU*. Online. Available HTTP: <http://www.cnu.org/canons> (accessed 15 March 2009).

Popkin, S.J., Katz, B., Cunningham, M.K., Brown, K.D., Gustafson, J., and Turner, M.A. (2004). *A Decade of HOPE VI, Research Findings and Policy Challenges*, Washington, D.C.: The Urban Institute. Online. Available HTTP: <http://www.urban.org/publications/411002.html> (accessed 15 March 2009).

Pyatok, M. (2005). "The New Urbanism: To Whom Should We Listen: The Social Policies of Urban Renewal," *New Village Journal*, Issue 2. Online. Available HTTP: <http://www.newvillage.net/Journal/issue2.html> (accessed 15 March 2009).

Salingaros, N.A., Brain, D., Duany, A.M., Mehaffy, M.W. and Philibert-Petit, E. (2006). "Favelas and Social Housing: the Urbanism of Self-Organization," paper presented at the Brazilian and Ibero-American Congress on Social Housing 2006. Online. Available HTTP: <http://zeta.math.utsa.edu/~yxk833/social-housing.pd> (accessed 15 March 2009).

Southworth, M. (2003). "New Urbanism and the American Metropolis," *Built Environment*, 29(3): 210–226.

Steuteville, R. (2005). "Commentary: HOPE VI is dead; long live HOPE VI," *New Urban News*, March. Online. Available HTTP: <http://www.newurbannews.com/Commentary Mar05.html> (accessed 15 March 2009).

—— (2008). *Directory of the New Urbanism*, Ithaca, NY: New Urban News Publications.

Terrain.org (1999). "Community of Civano in Tucson, Arizona: Unsprawl Case Study," *Journal of the Built and Natural Environments*, Number 5, Autumn. Online. Available HTTP: <http://www.terrain.org/unsprawl/5/> (accessed 15 March 2009).

Thompson-Fawcett, M. (1998). "Léon Krier and the Organic Revival within Urban Policy and Practice," *Planning Perspectives*, 13: 167–194.

Van der Ryn, S. and Calthorpe, P. (1986). *Sustainable Communities. A New Design Synthesis for Cities and Towns*, San Francisco: Sierra Club Books.

Virginia Commonwealth University (2008). "Implementation Strategies for Successful Bus TOD Projects," study prepared for the Virginia Transit Association, June 2008, Online. Available HTTP: <http://www.vatransit.com/practices/presentation/r01_VTA_BoilerPlate_Article.pdf> (accessed 15 March 2009).

Welter, W.M. (2003). "Post–war CIAM, Team X, and the Influence of Patrick Geddes Five Annotations," Department of History of Art and Architecture, The University of Reading, England, Online. Available HTTP: <http://www.team10online.org/research/papers/delft1/welter.pdf.> (accessed 15 March 2009).

Wendover, J. (2008). "Dispatch – Congress for the New Urbanism – Green Architecture and Urbanism Council," *Places*, 20, 1: 76–78.

Whyte, W.H. (Ed.), (1958). *The Exploding Metropolis*, Garden City NY: Doubleday.

Further reading

Calthorpe, P. and Fulton, W. (2001). *The Regional City*, Washington, DC: Island Press. The most comprehensive and thoughtful effort at providing a regional framework and foundation to New Urbanism.

CNU (Congress for the New Urbanism) (1999). *Charter of the New Urbanism*, New York: McGraw-Hill. A compendium of New Urbanist thinking and strategies for reshaping the city from the street to the region.

Duany A., Plater-Zyberk, E. and Speck, J. (2000). *Suburban Nation*, New York: New Point Press. A down-to-earth critique of suburbia underpinning TND with practical guidance for TND implementation.

Gruen, V. (1966). "New Forms of Community." In Holland, L.B. (Ed.) *Who designs America? Princeton Studies in American Civilization No. 6.*, Garden City, NY: Anchor Books Doubleday. Gruen's 1960s critique of the modern city predating and foreshadowing New Urbanist critique and transit oriented development solutions.

Krier, L. (2009). *The Architecture of Community*, Washington, DC: Island Press. A collection of theoretical analyses and pronouncement against modern architecture and urbanism together with community design guidance by the urban theoretician behind the New Urbanist TND movement.

39

Form-based codes vs. conventional zoning

Emily Talen

The emergence of form-based codes (FBCs), along with the familiar and near universal rejection of conventional zoning, is a complex story, and more interesting than might first be supposed. The story runs deeper than the conventional wisdom that they were created by a group of architects who wanted to impose stylistic controls sometime in the 1980s. It is instead the culmination of a century-long build-up of frustration over the weak form and poor functioning of American cities.

Looking for a better way to control the built environment, city planning departments across the US have been trying to substitute FBCs for conventional zoning codes. This represents a fundamentally different choice. Unlike conventional, so-called "Euclidean" zoning[1] which regulates building bulk and use and only impacts the design of cities inadvertently, FBCs focus intentionally on urban design: the form, size and location of buildings, streets, and frontages. FBCs provide a mechanism for implementing a prescriptive vision about community form, focusing on pro-active creation of the public realm rather than on rules designed to prohibit incompatibilities in the uses of land.

This chapter has two purposes. First, it recounts the underlying historical trends that gave rise to current interest in form-based codes. The discussion presents two main arguments, one having to do with the failure of zoning, and the other having to do with the failure of planning. Second, the chapter offers a rebuttal to the most common critiques of FBCs. This is warranted because, despite their popularity in city planning offices, FBCs have experienced serious push-back from the beginning.

The trouble with zoning

Although it is widely believed that FBCs represent a heavy-handedness and level of control previously unseen, control of urban form has a long and illustrious history (see Ben-Joseph 2005; Talen 2009; Davis 1999). Regulations specifying the allowable type, height, embellishment, window size and setback of buildings were already in existence in medieval times. Regulations on land use were also common. Islamic codes, for example, are rooted in principles governing the acceptable use of land. The Ancient Romans enacted laws to keep industry out of certain areas, and in the Middle Ages, noxious industries like tanning establishments were kept out of the city center. In seventeenth-century London, shops were not allowed on a main public square, nor the streets leading toward them.

These coding traditions of civilizations past are an important part of the FBC lineage because they speak to the tendency of human beings to want to control the form and function of their cities (Talen 2009). But it is also instructive to focus on the phenomenon of FBCs as a uniquely twentieth-century event – that is, as a reaction to conventional Euclidean zoning. Recounting that history requires a closer look at what transpired over the twentieth century and why there emerged such a backlash to what had become established practice.

Zoning took the country "by storm" in the first decades of the twentieth century (Kimball, cited in Simpson 1985: 126). The aim was to stabilize residential areas – specifically, their property values – and to make industrial zones efficient, functional, and non-harmful. This amounted to the widespread adoption – and unfortunate adaptation – of German zoning philosophy to the American experience. The New York zoning ordinance of 1916 was the first such comprehensive scheme, but the suggestion to separate the city into zones was made earlier at the First National Planning Conference in 1909, where Robert Anderson Pope argued for dispersal of factories to the outskirts and "wider dispersal of the laboring class" (Sies and Silver 1996: 462). By 1926, 400 communities had zoning ordinances, and by 1929, that number had almost doubled to 754 communities (Hubbard and Hubbard 1929).

Opening up the crowded industrial city was a valid humanitarian goal, but the rationale for zoning was not solely to advance public welfare – it was also about the efficient functioning of business and the preservation of land values. The intention of New York City's landmark zoning code of 1916 was to secure property values of merchants along Fifth Avenue, and more generally zoning served as a mechanism to secure land values (Hubbard and Hubbard 1929). Zoning became a mechanism for creating more tax revenue, regardless of its social impact. Designation of slums became commonplace "in the wishful hope that someday someone would buy them up and displace the slums with an apartment or factory" (Lovelace 1992: 92). The result was devastating for many inner city neighborhoods.

European planners were less likely to be swept up by efficiency goals. Most importantly, they were much more adept at applying a legal framework to basic ideals about urban quality. For example, Joseph Stubben, a key framer of German zoning philosophy, was highly attuned to the issue of land use mix. He discussed the need for retailers to "connect their places of business with their dwellings," and that careful planning could situate homes in proximity to businesses and factories. Most importantly, there was an understanding of the need to create a mix of use and building type that would ensure a mix of social classes. Stubben wrote, "the mixing of the wealthy and the poor should be promoted ... [social] grouping ... should never ... be strictly exclusive" (Stubben, in Marsh 1974/1909: 41–42). American planners – spurred on by a segregationist public – lost sight of this essential ideal early on.

By the time of the landmark 1926 zoning case, *Euclid vs. Ambler.*, the segregationist intent of zoning was firmly established. US Supreme Court Justice Sutherland stated: "the apartment house is a mere parasite, constructed in order to take advantage of the open spaces and attractive surroundings created by the residential character of the district. Moreover, the coming of one apartment house [brings] disturbing noises ... depriving children of the privilege of quiet and open spaces for play, enjoyed by those in more favored localities" (US Supreme Court 1926). This decision was lauded by planners. This is the legacy of conventional zoning, the outcome of which has been homogeneous, simplistic, monotonous forms of order. With 20–20 hindsight, we can now see clearly the perverse effects of this approach. Numerous studies have revealed zoning's negative effect on housing costs, market readjustments, spillovers, segregation, environmental quality, and other social and quality of life issues.[2]

The trouble with planning

There is another aspect of the twentieth-century planning experience that has motivated interest in FBCs: the failure of urban planning itself. FBC proponents yearn for the time when the goal of city planning was simpler: promotion of a physical vision. Master plans were top-down, but at least they rendered things in physical terms. Over the course of the twentieth century, urban planning moved away from master planning to an approach that focused on the creation of wordy policy documents. As early as the 1930s, the making of plans had eroded into a set of data gathering and map-making exercises, and Eliel Saarinen (1943: 354) labeled planning methods in 1943 as "the aureole of insipidity." Comprehensive and general plans became devoted to rectifying the problems of urban environments in motherhood-and-apple-pie terms, heavily laden with illusive social goals like "equity" and "opportunities". Thus a combination of reasons, ranging from the democratization of the planning process (Banerjee 2007), to the rise of "nimbyism," to the transformation of planning as "social science," created a vacuum in which the real driver of urban form was, by default, zoning.

In effect, planners had failed to make their new, more scientific approaches applicable to civic art and good urban form. As a result, good design languished underneath the force of zoning codes that paid little attention to the quality of urban form. The problem was that the implications of zoning beyond separation of land uses were only crudely considered by urban planners. Where they tended to rely on architects for the creation of plazas, public housing projects, and other singular arrangements, the codes would be left to dictate large swaths of the urban fabric in intrinsically sterile terms.

The tension between long-range comprehensive planning and short-range zoning control goes back to the beginnings of the profession. Alfred Bettman, the lawyer from Cincinnati who argued the Euclid case before the Supreme Court, faced ridicule for having failed to adequately distinguish the difference between planning and zoning, therefore causing "20 years of confusion," according to renowned planner T.J. Kent (1964/1990; see also Gerckens 1979). On the surface of it, the basic argument separating planning (long-range pattern) and zoning (short-range form) makes sense. Planning is supposed to be about the future visions and long-term aspirations of a community. Zoning, on the other hand, is narrowly focused and piecemeal, dealing directly with immediate building issues that cannot adequately reflect on long-term community goals.

And yet the assumption that zoning and planning must be kept separate is at least partly flawed. The assumption is that if planners are dealing with short-range, physical problems of urban three-dimensional form, they can therefore not be thinking simultaneously in terms of long-range, "big-picture" goals. Or, such long range goals are first set by an all-encompassing plan, only later to be worked out in terms of implementation through zoning. And yet this division between zoning and planning has had very negative effects. It has meant, for one thing, that American cities lack an appropriate definition of space, resulting in an American spatial pattern that is disorganized and often illegible. The narrow application of zoning codes that lacks spatially informed, big picture thinking results in sprawl, car-dependency and social segregation.

The form-based codes "revolution"

Out of this environment of weak, well-intentioned plans and strong, ill-intentioned codes, a revolution of sorts emerged toward the end of the twentieth century. Protagonists of FBC view them as the antidote to two essential problems: conventional, use-based zoning and weak, ineffective plans. FBCs are intended to deliver a predictable, pre-determined effect on urban form[3] by regulating, instead of use land use and setbacks, building form and typology (www.formbasedcodes.org). Figure 39.1 shows an example page from a form-based code, specifying building configuration.

The start of this "revolution" was Duany and Plater-Zyberk's 1982 master-planned community of Seaside, Florida, which specified rules for building form (see Krieger 1991). Subsequent FBCs, promoted in particular by the Congress for the New Urbanism, focused on the creation of visual harmony in the public realm, often by requiring continuous urban frontage as a primary means for ensuring some level of uniformity. FBCs were to be used to instill sensitivity to context, whereby spatial relationships – the importance of where one building sits relative to another – are factored into the code. A transect-based code such as the "SmartCode," is a specific type of form-based code that regulates on the basis of urban intensity, where zones range from being rural to urban (Figure 39.2; see also Duany and Talen 2002).

In addition to their ability to produce a more coherent public realm and an urban fabric that clearly differentiates between public and private space, form-based codes are especially valued for their ability to support social and economic diversity. They do this first by facilitating a mix of housing types, thus reversing the rules by which social segregation has been achieved: allowing multi-family units where previously excluded, and modifying rules that obviated higher density and infill (for example, minimum lot size and setback requirements). The SmartCode approach is also proactive, specifying percentages of housing types required within each zone. In the General Urban Zone, for example, a minimum residential housing mix of three types is required.

Form-based codes are also designed to encourage a mix of neighborhood facilities and services. The degree of mix is likely to vary by zone. The SmartCode, for example, controls use according to levels of intensity, whereby the urban zones allow a variety of lodging, office, retail, and civic uses. In the Sub-Urban zone (T3), mixed use is more controlled, but it does permit corner grocery stores, small scale lodging (such as a bed and breakfast inn), live-work units, and child care centers. This level of mix encouraged in FBCs does not exist in conventional zoning.

Also, form-based codes can incorporate spatial concepts, like centers, edges, and connectivity – all of which would be completely unheard of in a conventional zoning code. Conventional zoning is generally a-spatial – i.e. it does not explicitly consider the meaning and implication of spatial arrangements. An FBC like the SmartCode, by contrast, contains spatial planning concepts: Nested system of sectors, community types, neighborhoods, and pedestrian sheds, all of which are given legal stature. In addition, FBCs might contain specific language about the importance of street connectivity. The SmartCode, for example, requires that "all thoroughfares shall terminate at other thoroughfares," and there are specifications about the size of blocks (which should be kept small) and limitations on cul-de-sacs.

In these ways, FBCs are concerned not only with *what* but with *where*, thus giving

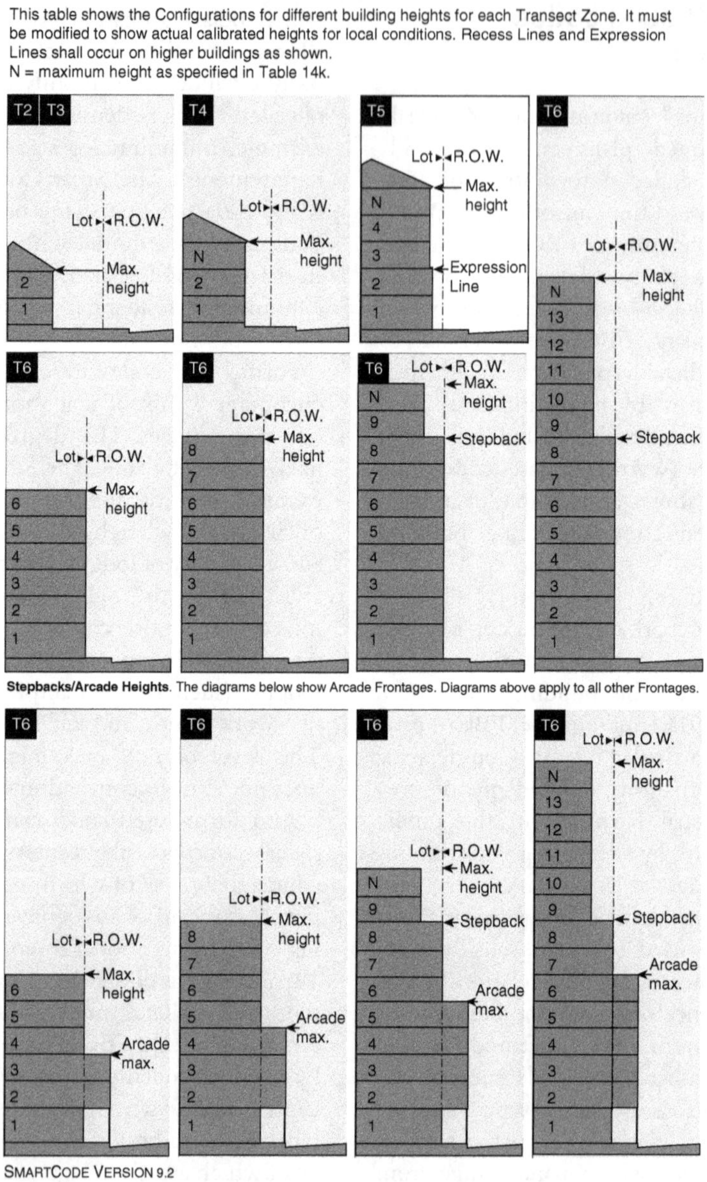

Figure 39.1 Page from *SmartCode*. Source: *SmartCode Version 9.2*, by Duany, Plater-Zyberk and Sorlien – used by permission.
Note: This page shows the graphic way in which building disposition is regulated.

FBCs the ability to reconcile two-dimensional pattern – the locations of neighborhoods, districts and corridors, for example – with three-dimensional form. Neither conventional codes nor conventional comprehensive planning have been able to achieve that on their own. Zoning has an effect on pattern (the land area consumed by segregated housing types, for example) as well as form, but its concern is neither geographically broad nor temporally long-range. Long range comprehensive

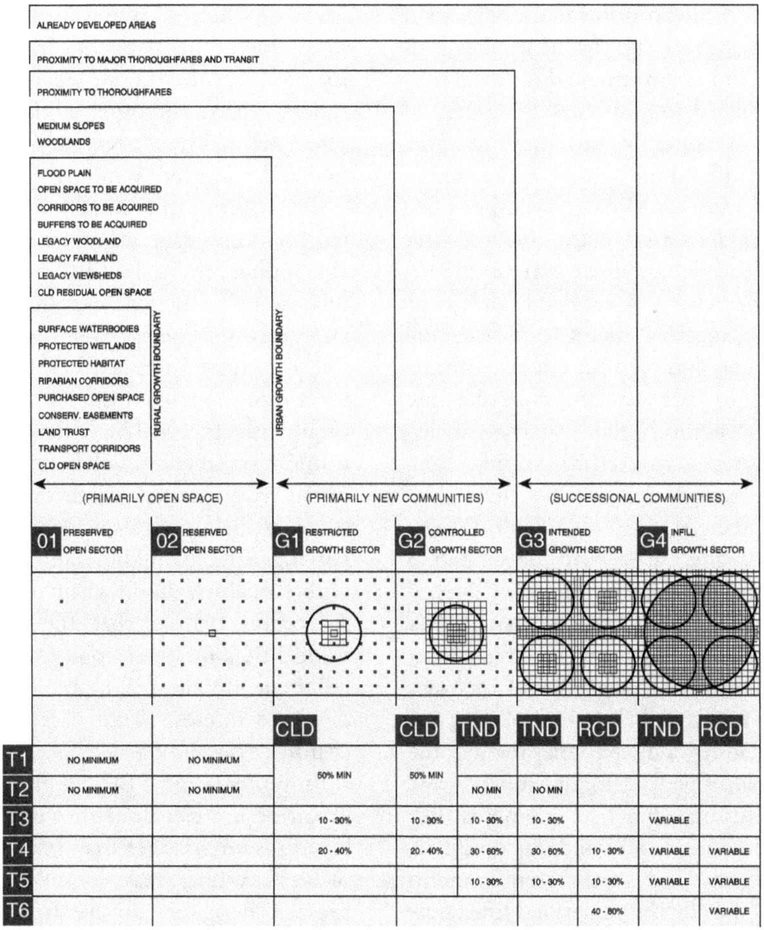

Figure 39.2 Page from *SmartCode.* Source: *SmartCode Version 9.2*, by Duany, Plater-Zyberk and Sorlien – used by permission.

Note: This page determines areas suitable and unsuitable for development and allocates the proportions of Transect Zones within each community type.

planning, on the other hand, seeks to guide the overall urban pattern in a way that considers long-term objectives, but with objectives that are rarely backed by legally enforced codes.

Of course, this level of control and integration comes with significant cost. FBCs have been critiqued from the beginning as an infringement on architectural freedom and an unacceptable level of control over urban form. But an even more significant critique is that FBCs do not go far enough: they treat urban problems superficially, affecting merely the symptoms of deeper causes. A top down focus on codes is therefore unlikely to offer real reform of urban environments. If economic and social systems are the root cause of bad urbanism, as the argument goes, these should be the target of planning reform, not improved physical designs via static codes. In this view, good urbanism starts with building the local jobs base, reconnecting local economic networks and empowering small-scale, independent improvement efforts (Pyatok 2002).

FBC advocates counter that urban form does, in fact, have significant implications for social and environmental justice. From a social welfare point of view, the design of cities affects not only how they function, but also influences who has access to amenities and who doesn't, whether people live qualitatively better or not, and whether the urban environment is safe, well-connected, well-serviced and vital. Codes are necessary because the making of good cities and communities can not be entrusted to private interests that are strictly profit-motivated – private investors cannot be expected to consider long term community benefit on their own accord (Duany *et al.* 2000). With this logic for FBCs, advocates have developed a well-honed set of justifications.

The first justification is that FBCs, unlike conventional codes, are transparent and explicit. FBCs take control of every aspect of regulating the physical city and thus, providing a means of creating the kinds of places people might actually want. In conventional practice, elements like street widths, road and building configurations, setbacks, and signage requirements are reduced to formulas derived from traffic flow counts, public works rights of way requirements, and safety considerations – all enforced through codes that do not always explain why such rules are in place.

The fact that many desirable urban places are the result of explicit rules supports the rationale for making urban form goals explicit in FBCs. Urban historians tell us that beloved urban places like Boston's Back Bay and Edinburgh's New Town were not random accidents, but the result of "a unified control of land and buildings" (Rybczynski 1989). FBC advocates wonder, then, why American cities cannot regain a higher quality urbanism by utilizing the kind of coding approach that shapes the public realm "to invite pedestrian use and social interaction" and produces "walkable, identifiable neighborhoods that provide for daily needs" (www.formbasedcodes.org). One of the main concerns of FBCs is to produce the public realm – streets, squares, plazas, and other public spaces – as often defined by a street wall with a unified and consistent building frontage.

Of course, this kind of consistency requires consensus about what an acceptable public realm is. FBC advocates are relying on a better "win/loss ratio" in architecture (Duany 2003) – a qualitatively improved urban fabric in exchange for a more restricted number of architectural masterpieces. The belief is that some restraint on freedom will be viewed by the public as warranted, when the alternative of no explicit controls on form results in a few memorable buildings amidst an otherwise banal and nondescript urban realm. Advocates believe that the public will value FBCs for their ability to establish a coherent public realm, by coordinating disparate interests, land uses and designs (Parolek *et al.* 2008).

Proponents of FBCs argue also that urban design can flourish within a form-based system. Rather than being the bane of architectural creativity, FBCs can help protect designers from the whims of bureaucrats, "nimbies" and politicians. The logic is this: One might as well leverage the legal authority of codes and put them to good use than allow urban form to evolve by default, subject to the narrow interests of fire marshals, transportation engineers, parking regulations, or land use attorneys – all of whom may impose rules that run counter to what FBC advocates would consider a quality public realm.

Critics have a hard time with this interpretation. The experience of individuality found in the Las Vegas commercial strip, immortalized in Venturi, Scott-Brown and Izenour's *Learning from Las Vegas* (1972), and more recently celebrated in *Everyday Urbanism* (Chase *et al.* 1999), seemingly will be missing in the ordered city that FBCs promise. In the architecture academy, the attempted order of an FBC is likely to be viewed not only as impeding innovation, but also as being out of sync with new, fluid forms of technology, globalization of capital, bottom-up forms of expression, and modern consumption patterns. The creation of urban patterns and architecture, it is believed, should be unfettered and therefore liberating (Kelbaugh 2002).

Apart from the issue of architectural control, there is the question of social control. Detractors may see the order of FBCs as an attempt to sanitize the world through rigid spatial ordering, as seen in Clarence Perry's neighborhood unit formula some years ago (Banerjee and Baer 1984). Order is thus a mechanism for shunning social conflict and controlling the unexpected (Harvey 1997). FBCs fall right into the critique that planning is mostly about imposing "disciplinary order and ceremonial harmony" where humans are organized, but alienated (Boyer 1986: 7).

It is certainly true that urban codes throughout history have had the objective of control, uniformity, and the imposition of order. Motivations have varied from the need to support health, safety or property values on the one hand, and the need for civic order and at times, social control, on the other. Whether or not coded uniformity was good or bad in a qualitative sense is debatable. The Dutch architect Berlage's insistence on the coding of uniform block frontage in Amsterdam around World War I used proportion as the guarantor of permanent value in architecture, creating a "system of definite proportions" which many regard as an especially beautiful urban form (Banham 1960: 142; see also Kostof 1991). But in other instances, such as the nineteenth-century by-law street in England with its standardized terrace housing, the intent was more about reducing costs than achieving visual harmony, and the result was oppressive.

The irony, of course, is that FBCs can play a strong role in accommodating diversity. This is because uniform building frontage can be essential for hiding class distinctions – a deliberate function introduced in the nineteenth century, and later a prominent aspect of Garden City design. What FBC advocates know now is that diverse neighborhoods need codes that specifically address issues related to the mixing of housing types. Diverse areas are prone to wide fluctuations in housing size, type and style, and while this is potentially good for visual interest it is potentially bad for long term stability and neighborhood cohesiveness. Codes are needed to reach beyond the conventional simple floor area ratios and unit sizes, and instead allow flexibility within a framework that promotes a successful integration of housing types (see Talen 2008).

Conclusion

Advocates of FBCs are leery of the critique that their codes are top-down, inflexible, and too reliant on static master plans. In theory, FBC proponents advocate simplicity and the ability to adapt, innovate, and remain culturally distinctive within a coding framework. But the ability to be both adaptive and effective has always been a significant problem. Early twentieth-century planners like Frederick Law Olmsted Jr. wanted codes to be adaptable and responsive, but they became frustrated with zoning's effects (Talen 2005). A major question, then as now, is whether codes are capable of combining what progressive urban designers advocate: "a large measure of performance outcome and a small dose of prescribed rules" (Ben-Joseph 2005; see also Hakim and Zubair 2006, and Jacobs 2002).

The issue is whether it is possible to bank on a system that pits a few simple rules against the historical inertia of technical efficiency, corporate logic, and segregation. It seems that past experience with urban coding made use of, as Witold

Rybczynski (1989) wrote, "a collective wisdom and a shared consensus about what constituted good architectural manners." This may no longer exist. Wolfgang Braunfels, in *Urban Design in Western Europe* (1990: 1), questioned why an ordered urban framework could have previously been achieved by just a few simple laws and ordinances whereas now "the most comprehensive and precise codes no longer suffice to maintain it." He concluded that our predicament could not be blamed on architects and urban designers, but could only be explained by a change in political process. Cities are now too disparate and lacking in collective interest, with too many competing actors and interests.

This is one reason why FBCs are considered to be so important. Conventional zoning codes give no indication of underlying intentions (why must a building be set back 15 feet?), creating a situation where an urban form might very well meet the intent of the law, but will be rejected for failing to meet the letter of it (Davis 1999). To rectify this, FBC advocates understand that they must provide some means of consensus, solidifying as much as possible what Ben-Joseph termed "place-based norms" (Ben-Joseph 2005: 24). Such norms may be dormant. Planners have been accused of relying on conventional zoning codes precisely because design sensibilities and norms about place are missing.

That codes will have to be the substitute, or at least the mechanism through which place-based norms occur, is not a completely new viewpoint. Writing in 1909, Raymond Unwin believed that town planning "and the powers conferred by legislation" were evidence that the "spirit of association" that may have once existed in feudal times was making a comeback. It was a matter of a new order taking the place of an old one, a solution to the problem of the individual in "helpless isolation of his freedom" (Unwin 1909: 375). Some decades later, coding officials in the US had a similar sense of purpose. The 1942 "Dedication of Principles" of the Building Officials Conference of America (BOCA) proclaimed their commitment not only to better methods of construction and to relief "from the uncertainty and confusion of conflicting building laws and regulations," but to "the promotion of civic pride and community well-being" (BOCA International, Inc. 2002: 4).

Code reformers are now trying to help communities uncover the collective purpose they may share when it comes to urban form. The strategy is to ensure that the public is well integrated into the code-making process. This aspect of coding is relatively unique. The new book *Form-Based Codes* (Parolek *et al.* 2008) stresses the importance of public participation emphatically, calling on the community visioning process to stand as a key source for the development of code content.

Thus FBCs aim at imposing limits that are no longer dictated by technological and other constraints; instead they rely on public consensus about urban form. The resulting sophistication of today's codes, stemming from the need to balance use, form, location, safety and public process, is unprecedented. Paradoxically, at the same time that this increased involvement and sophistication is regarded as essential, code reformers are trying to simplify regulation. They are in some ways attempting to reverse the complexity that had been evolving since the onset of modernism and conventional zoning. This is especially evident in the case of zoning controls on land use, where the regulation of what could go where in a city became a complicated set of prohibitions against all imaginable incompatibilities.

Notes

1 The term refers to the conventional practice of zoning primarily by land use, and is named after

the US Supreme court case Euclid v. Ambler (1926), which legalized the practice and resulted in widespread zoning adoption in the US

2 See Dowall (1984) on the effects of land-use regulation; see Booth (1989), McMillen and McDonald (1990), Natoli (1971) on the ineffectiveness of land use zoning; see Babcock (1980) for spatial impacts; see Pogodzinski and Sass (1991) for the effect on the real-estate market, see Talen and Knaap (2003) for counter-urban effects; see Pendall (1999) for the effect on low densities.

3 Three recent publications have documented code reform, especially the emergence of form-based codes: Parolek *et al.*'s (2008) *Form Based Codes*; Steve Tracy's (2004). *Smart Growth Zoning Codes: A Resource Guide*; and the Congress for the New Urbanism's (2004) *Codifying New Urbanism: How to Reform Municipal Land Development Regulations*. See also http://www.formbasedcodes.org/ for information on code reform efforts.

References

Babcock, R.F. (1980). "The Spatial Impact of Land Use and Environmental Controls" In A.P. Solomon (Ed.), *The Prospective City*. Cambridge, MA: MIT Press, pp. 264–287.

Banerjee, T. (2007). The "Public Inc. and the Conscience of Planning." In Verma, N. (Ed.) *Institutions and Planning*. London: Elsevier. pp. 107–127.

Banerjee, T. and Baer, W. (1984). *Beyond the Neighborhood Unit*. New York: Plenum Press.

Banham, R. (1960). *Theory and Design in the First Machine Age*. London: Architectural Press

Ben-Joseph, E. (2005). *The Code of the City: Standards and the Hidden Language of Place-Making*. Cambridge, MA: MIT Press.

BOCA International, Inc. (2002). BOCA at 87: A Retrospective. *The Code Official* September/October.

Booth, P. (1989). "How Effective is Zoning in the Control of Development?" *Environment and Planning B: Planning and Design*, 16: 401–415.

Boyer, C.M. (1986). *Dreaming the Rational City*. Cambridge, MA: MIT Press.

Braunfels, W. (1990). *Urban Design in Western Europe: Regime and Architecture, 900–1900*. Translated by Northcott, K.J. Chicago: University of Chicago Press.

Chase, J., Crawford, M., and Kaliski, J. (1999). *Everyday Urbanism*. New York: The Monacelli Press.

Davis, H. (1999). *The Culture of Building*. New York: Oxford University Press.

Dowall, D.E. (1984). *The Suburban Squeeze*. Berkeley, CA: University of California Press.

Duany, A. (2003). "A Conversation with Dan Solomon and Andres Duany." The Town Paper, Council Report III/IV, April, 2003. http://www.tndtownpaper.com/council/SolomonDuany.htm. (accessed 23 August 2010).

Duany, A. and Talen, E. (2002). "Transect Planning." *Journal of the American Planning Association*, 68(3): 245–266.

Duany, A., Plater-Zyberk, E., and Speck, J. (2000). *Suburban Nation: The Rise of Sprawl and the Decline of the American Dream*. New York: North Point Press.

Gerckens, L.C. (1979). "Historical Development of American City Planning." In So, F., Stollman, I., Beal, F., and Arnold, D. (Eds.) *The Practice of Local Government Planning*, Washington DC: American Planning Association, pp. 21–57.

Hakim, B. and Zubair A. (2006). "Rules for the Built Environment in 19th Century Northern Nigeria." *Journal of Architectural and Planning Research*, 23(1): 1–26.

Harvey, D. (1997). "The New Urbanism and the Communitarian Trap." *Harvard Design Magazine* Winter/Spring, pp. 68–69.

Hubbard, T.K. and Hubbard, H.V. (1929). *Our Cities, Today and Tomorrow: A Study of Planning and Zoning Progress in the United States*. Cambridge, MA: Harvard University Press.

Jacobs, A. (2002). "General Commentary" in Bressi, T.W. (Ed.), *The Seaside Debates: A Critique of the New Urbanism*. New York: Rizzoli International, pp. 136–152.

Kelbaugh, D.S. (2002). *Repairing the American Metropolis*. Seattle, WA: University of Washington Press.

Kent, T.J., Jr. (1964). *The Urban General Plan*. Chicago: Planners Press. Reissued 1990.

Kostof, S. (1991). *The City Shaped*. London: Thames & Hudson, Ltd.

Krieger, A. (1991). *Towns and Town-Making Principles*. Cambridge, MA: Harvard University Graduate School of Design.

Lovelace, E. (1992). *Harland Bartholomew: His Contributions to American Urban Planning*. Urbana, IL: University of Illinois.

McMillen, D.P. and McDonald. J.F. (1990). "A two-limit tobit model of suburban land-use zoning." *Land Economics*, 66: 272–282.

Marsh, B. (1974). *An Introduction to City Planning*. New York: Arno Press. Reprint of 1909 ed., published by the author, New York.

Natoli, S.J. (1971). Zoning and the Development of Urban Land Use Patterns. *Economic Geography*, 47: 169–184.

Parolek, D.G., Parolek, K. and Crawford, P.C. (2008). *Form Based Codes: A Guide for Planners, Urban Designers, Municipalities and Developers*. New York: Wiley.

Pendall, R. (1999). "Do Land Use Controls Cause Sprawl?" *Environment and Planning B: Planning and Design*, 26(4): 555–571.

Pogodzinski, J.M. and Sass, T.R. (1991). "Measuring the Effects of Municipal Zoning Regulations: A Survey." *Urban Studies* 28: 597–621.

Pyatok, M. (2002). "The Narrow Base of the New Urbanists." *Planners Network*, 151: Spring, pp. 1, 4–5.

Rybczynski, W. (1989). "Architects Must Listen to the Melody." *The New York Times*, September 24, 1989.

Saarinen, E. (1943). *The City: Its Growth, Its Decay, Its Future*. New York: Reinhold Publishing Co.

Scott, M. (1969). *American City Planning Since 1890*. Berkeley, CA: University of California Press.

Sies, M.C., and Silver, C. (Eds). (1996). *Planning the Twentieth Century American City*. Baltimore, MD: Johns Hopkins University Press.

Simpson, M. (1985). *Thomas Adams and the Modern Planning Movement: Britain, Canada and the United States, 1900–1940*. London: Alexandrine Press.

Talen, E. (2005). *New Urbanism and American Planning: The Conflict of Cultures*. London: Routledge.

—— (2008). *Design for Diversity: Exploring Socially Mixed Neighborhoods*. London: Elsevier.

—— (2009). "Design by the Rules: The Historical Underpinnings of Form Based Codes." *Journal of the American Planning Association*, 75(2): 1–17.

Talen, E. and Knaap, G. (2003). "Legalizing Smart Growth: An Empirical Study of Land Use Regulation in Illinois." *Journal of Planning Education and Research*, 22(3): 345–359.

Tracy, S. (2004). *Smart Growth Zoning Codes: A Resource Guide*. Sacramento, CA: Local Government Commission.

US Supreme Court (1926). *Village of Euclid, Ohio v. Ambler Realty Co.*, 272 US 365 (1926). 272 US 365.

Unwin, R. (1909). *Town Planning in Practice: An Introduction to the Art of Designing Cities and Suburbs*. London: T. Fisher Unwin.

Venturi, R., Scott-Brown, D. and Izenour, S. (1972). *Learning from Las Vegas*. Cambridge, MA: MIT Press.

Further reading

Ben-Joseph, E. (2005). *The Code of the City: Standards and the Hidden Language of Place-Making*. Cambridge, MA: MIT Press. Provides a historical overview of codes and how they shaped cities. Offers both a detailed critique and a proposal for change.

Braunfels, W. (1990). *Urban Design in Western Europe: Regime and Architecture, 900–1900*. Translated by K.J. Northcott Chicago: University of Chicago Press. A 1,000 year survey of European urbanism. Assesses the link between politics, culture, and urban form, including some discussion of the translation between codes and urban outcomes.

Davis, H. (1999). *The Culture of Building*. New York: Oxford University Press. A focus on the connection between building cultures, institutions and processes and the built environment. One chapter focuses exclusively on the effect of codes.

Parolek, D.G., Parolek, K. and Crawford, P.C. (2008). *Form Based Codes: A Guide for Planners, Urban Designers, Municipalities and Developers*. New York: Wiley. Comprehensive overview of form-based codes, and a practical "how-to" covering everything from the technical details of building frontage to managing an effective public participation process.

Part 8

Global trends

Introduction

The scope of urban design practice will continue to be defined and redefined by the changing circumstances of our society and economy. As it has been hinted in some of the earlier chapters, the rise of the information society, the global economy, and the effects of globalization more broadly defined will no doubt continue to challenge the technical, ethical, and normative competencies of urban design. In this section four authors present four different perspectives on emerging global trends that will continue to engage urban designers to explore a previously uncharted territory of practice.

One of the urban outcomes of globalization (see Banerjee 2008) is that cities are now competing in an "international market place" (Savitch and Kantor 2002). As vendors in the international market place, they are selling their products – i.e. ready and affordable labor, amenities, land, ambience – to the international investment capital, multinational business, and the growing tourist and consumer class. Two decades ago, Frieden and Sagalyn (1989) wrote a book whose title "Downtown Inc." aptly captured the new trend in downtown development, where downtown has become an industry and the cities are functioning as corporate entities that combine both public and private enterprises. In the context of global economy and the international marketplace, we are now beginning to see a new class of "City Inc."

As it is well known in the business world, a brand is important to be successful in the respective businesses. The "City Inc." in the global marketplace therefore must have a brand, and as Jon Lang's chapter points out, many cities attempting to compete in the global marketplace are actively looking for ways to brand the city as a product. In reviewing the literature, and in looking for a conceptual construct, Lang is naturally drawn to Kevin Lynch's seminal work, "The Image of the City," since the concept of "image" is closely linked to the notion of developing a unique brand. But while Lynch's work focused on the imageability of the visual form of the city, the almost parallel work of Anslem Strauss (1961) on "Images of American City," captured the essence of city branding – in the sense of capturing the soul and iconic properties that each city of global renown and significance conveys, whether by design or not.

In the new global economy, urban designers will no doubt be required to think about the challenges of defining a brand for the city. Many cities in the developing world are often leaning on "starchitects" (see Southworth and Ruggeri in Part 7) to create iconic images. Norman Foster's airport building or Rem Koolhaas' CCTV tower in Beijing, or the Burj Dubai and the Palm Islands in Dubai are cases in point. But it is not clear if a single building or a collection of buildings can create the kind of identity cities are looking for. Lang's chapter addresses these issues and points to the inevitable downside of such city branding through redevelopment and gentrification, leading to displacement, social exclusion, and greater inequality (exactly the points also addressed by Madanipour in Part 7).

In his essay Edward Soja addresses another global trend that involves growth and spatial transformation of urbanizing regions. Soja argues that the changes we are seeing today are indeed dramatic and unprecedented in scale. Drawing from his earlier work, he argues that the contemporary trend can be best described as an ongoing "postmetropolitan transition," where earlier patterns of monocentric and polycentric urban forms and density gradients are increasingly displaced by the emergence of what he refers to as "exopolis," that simultaneously involves urbanization of the exurbia and decanting of the inner city jobs and population. In this essay Soja begins by summarizing six overlapping explanations of this ongoing transformation. Here he proposes a new synthesis of his previous arguments on regional urbanization, a topic of current academic and policy interest being defined as the New Regionalism. After discussing the causes and consequences of the emerging social and spatial order in this new pattern of regional urbanization, Soja considers the planning and design implications of this trend. Concluding that the scope of engagement of urban design at this scale has yet to be fully defined, he sees such engagement as necessary and inevitable. While Soja has not explicitly referred to globalization as the force of this new trend of regional urbanization, his examples certainly suggest that this is a global trend. In the next section, chapters by Brenda Scheer, ("Metropolitan Form and Landscape Urbanism") and Anne Whiston Spirn ("Ecological Urbanism") in particular address more directly the challenges outlined by Soja.

The chapter by Clara Irazabal is clearly about one effect of globalization: the phenomenon of "ethnoscape," one of the five "scapes" defined by Arjun Appadurai (1991) as products of globalization. Drawing from Appadurai, Irazabal examines the emerging nature of ethnoscapes in the context of multiculturalism and diversity of American cities, especially the cultural landscape of immigrant communities. While Appadurai's original formulation of "ethnoscape" was not done strictly in spatial or urban form terms, Irazabal uses the construct in the context of urban space and examines how such spaces are produced and used. Although she does not address the social ecological issues of enclavization or ghettoization of the immigrant populations in the metropolitan spaces of the host countries as discussed in the recently edited volume by David Varady (2006), she focuses on the subjective, socio-cultural, spatial, and temporal dimensions of such spaces. More importantly, she also focuses on the implications for practice and pedagogy of urban design and challenges urban designers to think about the future design of multiethnic spaces, especially when they are going to be used by different culturally rooted preferences and behavior patterns. She concludes the essay by raising an intriguing question – whether ethnic urbanism because of its traditional cultural practices, might actually achieve the new urbanist aims of mixed use, compact, and walkable neighborhoods more effectively than the urbanism of mainstream US urban population.

Finally, the chapter by Vinit Mukhija addresses squarely the question that has long haunted design professionals. What does the practice of design have to offer to improve the quality of life and livability of the urban poor – some two billions of them – who live in slums and squatter settlements, if not on pavements, in cities around the world? Mike Davis' "Planet of Slums" (2006) effectively captures the continuing reality of the poor in the developing world. While poverty, slums, and squatter settlements have been around from time immemorial and institutionalized in both land use and language – *bidonville, bustee, chawl, colonia, favela, gecekondu, jhuggie jhompry, paracaidistas,* and so on – their continued growth and densification has been exacerbated by the new global economic order. While globalization has done much good, growing income polarization and poverty continue to symbolize its dark side.

Mukhija's chapter explores what, if any, role might urban designers play in addressing the problems of the poor in urban areas, a phenomenon he defines as "the informal city," drawing from a nuanced analysis of informal economy. He devotes the bulk of his essay to defining the scope of urban design involving the informal city. For example, he argues that the scope of such work might entail small interventions, not clearance and urban redevelopment – where urban designers have typically been involved. These may involve infrastructure upgrading, provision of affordable housing, economic development, and more broadly integrating the informal city with the formal city. He supplements his discussion with examples of successful efforts and innovations.

References

Appadurai, A. (1991). "Global Ethnoscapes: Notes and Queries for a Transnational Anthropology." In Fox, R.G. (Ed.) *Interventions: Anthropologies of the Present.* Santa Fe, NM: School of American Research, 191–210.

Banerjee, T. (2008). "Urban Outcomes of Globalization: Theory, Research, and Practice," [Review Article] *Journal of Urban History*, 34(6): 1044–1054.

Davis, M. (2006). *Planet of Slums.* London, New York: Verso.

Frieden, B. and Sagalyn, L. (1989). *Downtown Inc., How America Rebuilds Cities.* Cambridge, MA: The MIT Press.

Savitch, H. V. and Kantor, P. (2002). *Cities in the International Marketplace: The Political Economy of Urban Development in North America and Western Europe.* Princeton, NJ: Princeton University Press.

Strauss, A. (1961). *Image of the American City.* New York: The Free Press.

Varady, D. (Ed) (2006). *Desegregating the City: Ghettos, Enclaves, and Inequality.* Binghampton, NY: SUNY Press.

40

City branding

Jon Lang

A brand is simply a product with which purchasers have specific associations, positive or negative. Cities can be and are often regarded as products although they have 'more varied "users", "owners", and "governors" than products that one can pull off a supermarket shelf' (Kavaratzis and Ashworth 2005). Specific urban design projects such as Battery Park City in New York or La Défense in Paris or even Singapore can also be regarded as products. The first is a mixed use but primarily residential development, the second a major business district and the third a city. They serve as ideas that can be purchased and reproduced. Clients, particularly municipal authorities, can say 'We'll have one of those', and they have done so. Canary Wharf in London, Lujiazui in Shanghai, and even Naples' new central business districts are modelled on La Défense. A brand gives a building, a city precinct, or even a whole city an identity that differentiates it from others. City branding thus refers to the process of building a clear and positive set of images and associations for a city in the world's mind in order to make it more attractive and, ideally, unique and thus more easily marketable. It is what was once called 'place marketing'. The term 'city branding' for such a process has been used since the early 1990s and has come into common currency since 2000.

There is nothing new about the processes involved in boosting the image of a city to serve a number of purposes. In the nineteenth century the often deceitful description of the assets of a city was used to attract investment. Civic boosterism still exists. Its goal is to bring the world's attention (and often its own citizens' attention) to the positive aspects of a city in order to achieve a number of purposes. Amongst them are the enhancement of the self-esteem of its citizens, the increase in a city's economic standing, and the attraction it offers to tourists. New high-status civic and commercial buildings are often used to bolster a city's attractiveness. Although civic boosterism remains an activity of municipal chambers of commerce, power elites of cities, and political leaders for some time little systematic research exists on city branding (Gold and Ward 1994; Avraham and Ketter 2007). This chapter outlines our present understanding of city branding and the role of urban design in creating a city's brand.

Marketing and branding

The growth of the world economy in the 1990s and the deregulation of markets have resulted in an intense competition amongst cities for attention and influence as they seek to attract foreign direct investment (or FDI) in the form of investments and human skills and talent (Gelder and Allan 2006; Florida 2002). To stay competitive in the world market place cities have to maintain a successful image, restore a past valid image or rebrand themselves after new images. The same observation applies to the competition between precincts of a city: center city versus neighbourhoods, big box stores versus Main Street, or one residential area versus another (Winfield-Pfefferkorn 2005). A growth in entrepreneurial modes of governance has accompanied these changes. Indeed consultants advise cities to be run more like business corporations that are creating and promoting products new to their customers (Kotler *et al.* 1996). City leaders recognize that without investment, cities decline. City branding is thus now regarded as a legitimate and necessary task of municipal governments.

During the last decade of the twentieth century and the first of the twenty-first many cities across the world from Los Angeles to Berlin, and from Dubai to Shanghai have consciously sought to provide themselves with a new image, to re-brand themselves in the eyes of the world, to be in a better position to compete with the major cities of the global economy. Others, such as London, New York and both Cambridge in England and Cambridge in Massachusetts have sought to hang on to already established brand images in the face of international competition. Tempe, Arizona, with the growth of Arizona State University, sees itself as a competitor to both Cambridges, as a centre for research and tertiary education in the twenty-first century (Duhnke 2008). The goal of branding and marketing has thus been and is to attract investment or, in many instances, to not lose investment or status.

Branding sounds like advertising (Ashworth and Voogd 1990; Gold and Ward 1994; Kavaratzis and Ashworth 2005). To many observers advertising and thus branding, has negative connotations associated with attempts to make a product look better than it is in the eyes of consumers. In marketing a city, aspects of it – social, cultural and physical – are selectively appropriated to create a positive image in people's eyes. It has, however, been found that advertising on television and the Internet and sending trade missions around the world to drum up investments yields little positive benefit in promoting a city. Having a clear brand that resonates with some sector of the investment community is more successful (Gelder and Allan 2006). Branding and selling a city does, nevertheless, share many of the characteristics of marketing but involves more than simply selling a city as a product.

Marketing involves differentiating a product from its competitors and promoting its unique qualities. It entails creating a positive image of product – a brand image. A brand is multi-dimensional. It has physical attributes but also many socio-psychological characteristics. Branding thus involves changing people's attitudes towards a city and often the neighbourhoods within it. City-branding can be considered to be the process of positioning a city ahead of its competitors because of its image. It can thus be thought of in terms of a process for changing the personality of a city as seen by its citizens and outsiders. The goal is to attract tourists and business visitors and to retain existing businesses and attract them from elsewhere, promote business expansion and new start-ups, increase investment and/or elevate the socio-economic character of its population (Kotler *et al.* 1993; 1996). Branding is thus similar to marketing but it also involves changing the nature of the image of a city.

City images and city brands

The image that a city has may well be different in the eyes of its citizens and in those of outsiders. It may differ for the various stakeholders amongst its citizenry. The term 'image' itself is ambiguous, if not multivalent. Contemporaneously to Kevin Lynch's well-known book, *The Image of the City* (1960) was a less-heralded book with a similar title: Anslem Strauss's *Images of the American City* (1961). The former dealt with the physical image of cities as represented in a cognitive or three dimensional mental map that people have of them; the latter focused on the meaning – the associations – that different cities have for people. Strauss argued that all cities have what is now called a 'brand' image. Some cities are thought of as industrial cities, others as educational or historic or centres of business or art, and yet others simply as gritty (Proctor and Matuszeki 1978).

Some cities have positive images in both Lynch's and Strauss's terms. Many have prospered over long periods of time. Some cities have boomed over the last three decades and now actively promote themselves particularly in the eyes of younger people by dismissing older cities as antiquated and presenting fewer opportunities for career progress (Marshall 2003). This observation is particularly true of cities in Asia. Others, such as Detroit and other US Rustbelt cities, and a number of European industrial cities, have been in decline for the past four decades. Some cities have negative images because of crime (e.g. Johannesburg), poverty (many African cities) or general decrepitude (e.g. Kolkata). Yet others have negative images not grounded in facts. Some cities are seen as 'gritty' or tough – in fact 'too tough to die' (Tombstone, Arizona). Some cities are certainly seen as boring. W. C. Field's observation about his home town – 'Philadelphia, wonderful town; I spent a week there last night' – still haunts the city. In contrast, some cities benefit from positive images that belie their nature. Can they hang on to them?

The image of a city and thus its brand is not static but evolving. Image formation is a perceptual and cognitive process based on experiences, personal or mediated. An image gets adjusted over time by new experiences or information. The goal of branding and marketing is to disseminate that new information in order to sustain or change a city's image. The conscious imaging or reimaging of a city turns it into a commodity. The objective in identifying or creating a brand for a city is to identify the features of a city that makes it unique or part of a group of cities that share similar attributes. To make it compete effectively with other cities of a class a city's image may well need remaking.

The elements of positive city brands

A city is both a physical and social entity. To have a good image a city must provide attractive employment opportunities, have a strong centre that acts as its heart, be rich in nodes of activities, have attractive and affordable residential neighbourhoods, good schools, good recreational and cultural attractions and, at least, a reasonable public transportation system. Good sporting teams, low crime rates, and a good climate help. Seattle is rainy but that has not stopped it from developing a positive image.

To be successful a city must have the basic attributes that make it operate well on a day-to-day basis (Winfield-Pfefferkorn 2005). As Deng Xiaoping, former Paramount Leader of China, noted, 'It does not matter if a cat is black or white; if it catches mice it is a good cat' (cited in Mahbubani 2007). A city must have a locational advantage. (Locational branding is an aspect of city branding). For instance, historically, cities that prospered were good transportation hubs. Indeed, few significant cities in the world are not located on notable water bodies that were used for transportation. Today too a city must have the qualitative, locational, and topographical characteristics that allow it to be a transportation hub but this mostly means for air travel. In Asia, Bangkok, Hong Kong and Singapore vie to be the premier such hub. Historically a city also had to have a hinterland that served as a service area. One of the reasons for the comparative decline of Kolkata as a major city was the partitioning of Bengal into Indian and East Pakistani (now Bangladesh) components. The city lost more than half its hinterland. Another reason for the city's decline is the continuous silting of the Hooghly River on which it is located.

In addition to sound locational advantages and a clear Lynchian image, a city must possess value-added qualities that make it stand out from its competitors. New York has many such elements, and they are distinctive. To most people New York means Manhattan. Manhattan is highly imageable in Lynch's terms. It has clear streets, a multiplicity of nodes such as Battery Park, Rockefeller Center, Lincoln Center and Times Square, a number of landmarks clearly identified with the city (e.g. the Statue of Liberty), and a distinctive skyline, notable districts, and also clear edges. New York is also highly imageable in Strauss's terms. It has the added value of having a cultural environment that is a world leader in the provision of museums, art galleries, and theatres. It has the association of being cosmopolitan and wealthy. It has Broadway and Wall Street. Manhattan is certainly seen as a centre for the arts, entertainment, business, and, with the presence of the United Nations, international politics. The city, nevertheless, faces significant competition from other centres both within the United States and elsewhere. Much the same can be said for Paris.

Paris may be the city that has the highest brand image in the world (Figure 40.1). It is known for its beauty and style. It possesses a clear structure of boulevards

Figure 40.1 Eiffel Tower and Paris in the background. Source: Wikipedia Commons.

ending in monuments or institutional buildings. It has a wide variety of museums. It is rich in nodes/places and districts. It possesses a great richness in architecture from the Gothic of Notre Dame to the postmodernism of the Pompidou Centre and the Parc de la Villette to more recent works such as the Bibliothèque Nationale. To maintain the city's stature as an international leader in the arts, President François Mitterrand instigated his Grands Travaux (Great Works) programme aimed at both preserving Parisian monuments such as the Grand Louvre and funding such projects as the Cité des Sciences at Parc de la Villette. Now Parisian leaders are considering lifting the height limit of 31 metres on building in Haussmann's Paris in order to compete effectively with the booming and glitzier cities of the world with their new museums designed by renowned architects. Failing to sell itself as the best site for either the 2008 or 2012 Olympic Games still galls.

The acquisition of the 1992 Olympic Games boosted the stature of Barcelona. The city had much going for it already – the medieval city with its plazas and the nineteenth-century city of Idelfonso Corda. With the development of the modern city and the remaking and extending of La Rambla, one of Europe's finest and most imageable streets (Figure 40.2), and the opening up of the Mediterranean waterfront, Barcelona has rapidly created a new brand. It sells itself as a city of innovation. A distinct logo and graphics system adorns the sides of buses and the back of benches promoting the brand. The city has promoted itself with great success. The quality of its football team (European Cup winners in 2009) has helped too.

It is difficult for new towns to have the qualities of older cities which have been built up over time. Company towns have an image associated with the character of the company, and they live and die based on the success of the industry they house.

Figure 40.2 La Rambla, Barcelona. Source: Wikipedia Commons.

Brasília has a clear visual image because of the geometric clarity of its plan and the boldness of the architecture of its central governmental axis. Much the same observation can be made about Chandigarh. They are the most distinctive of the number of new capital cities built around the world during the second half of the twentieth century. Although they were not consciously designed to have a clear brand, they have it because of their architectural character and their association with highly prestigious architects. Other city administrations have set out to either maintain or establish a brand image through architectural and urban design.

Branding strategies and processes

Cities are complex. Creating a brand is not as easy as many city planners and political leaders think. Establishing a good brand needs to be based on what a city offers and on what it can offer. It is a creative act that depends on the ability of those doing the branding to perceive the opportunities that exist to exploit the good features of a city and to enhance them. The goal is to make a city a good place for people to live and to work, and for outsiders to visit. Attracting tourists is important but a city must be more than a good place to visit. It must be a good place to live and work. Tourism creates jobs in hotels, restaurants, and transportation, and it boosts retail sales. How then does one make a city a good place to live, work and visit and have a strong positive image in people's minds?

The process of branding is similar to the process of planning. It consists of six steps: first, identification of the present image, second, asset identification, third, vision development, fourth, project planning, fifth, strategy implementation, and sixth, evaluation of the impact of the whole branding process. The process does not take place in this simple linear fashion but is more likely to occur in a spiralling form. It is an argumentative process during which many conjectures are presented by those involved and tested logically or against empirical information.

All cities have an image. Some, large or small, already have distinctive brand in people's minds. Some are seen as resort cities. The two Cambridges mentioned above are branded as centres of tertiary education excellence. Cities such as Boston and Quebec tend to be considered as historic centres. London is seen as a multi-dimensional financial centre (although its income from design services is higher). San Francisco has a reputation for the quality of its built environment, care in planning, and broad social tolerance. The existing image is not so clear for other cities. Many cities throughout the world have lost a former image and have witnessed radical structural change as a result of competition from other centres. Philadelphia was once one of the largest centres of manufacturing in the world; it now has, for all intents and purposes, no manufacturing industries left. Not much shipbuilding takes place on the Delaware or Clyde or in Belfast. Many major port cities of the past hardly function in that manner today. The list of such cities is extensive.

All cities, however poor their image is in people's minds, have assets. The assets may have to do with physical features such as topography, climate, or specific industrial, artistic or educational niches. Some cities are lucky because of their fine climates or location on fine shorelines or natural harbours. Sydney is such an example. Other cities have a stock of cultural assets because of philanthropic gestures in the past. Some have a pool of entrepreneurial talent looking for opportunities for investment. Will they invest locally or elsewhere where the opportunities are greater? What are the features that are worth exploiting? Perhaps the greatest urban design successes have been in the transformation of the waterfront of cities from environments that had negative meanings associated with declining, smoky industries to the positive image of parks, hotels and modern office complexes (Breen and Rigby 1996). Are the municipal leaders willing to seize on such opportunities and build on them? West Bengal has had a democratically elected and re-elected communist government and is unwilling to develop the Hooghly waterfront or the historical assets of the Kolkata because they are associated with middle class values and are tainted by the city's colonial heritage. Changing the name of the city from Calcutta to Kolkata rebrands it and distances it from its colonial past. That past could be, however, its most immediately exploitable asset.

Vision development, project and implementation planning go hand-in-hand. The vision should have a social, an economic and a physical structural component, but often the last, the urban and architectural characteristics of a place, is given more attention than the former two. It is tangible. What is important is that part of the vision be based on reality. It must also be communicated with confidence that the city 'can get things done' (Berg and Braun 1999). At the same time, it must be noted that successful marketing might make an otherwise implausible vision implementable. San Diego's promoters declared the city to be a 'bio-technology city' before it had any such industries. It now has those (Christian 2007). Vision development draws significantly on precedents. What has been done successfully elsewhere? It makes sense to learn from the successes and failures of other economic, social and physical planning and marketing efforts but questions must be asked about the utility of successful ideas for one city being transported from one location to another.

The implementation phase requires the marketing of the product. It involves developing the organizational mechanism for promoting the vision. Selling the brand of city is something that takes place continuously as does the whole process of branding (Bramezza 2003). Like any other design activity it involves the reiteration of each of the steps mentioned above. This observation also applies to the continuing evaluation of the overall vision and strategy. Circumstances are changing continuously. Cities that can adapt to change and are robust are the ones that prosper in the long-run. Does Bilbao, one city that has recast its image most successfully, have this stamina?

City branding and urban design

Creating a brand requires leadership and a partnership among a city's political leaders, its power elite, and its citizens. City branding involves the effort by governments to create a specifically designed sense of place and to promote it. There are three approaches to enhancing the image of a city: first, changing some activities that take place in the city, second, changing its physical attributes, and third, changing the image of a city as presented in the mass media of television, the Internet, and newspapers and journals. Urban design, landscape architecture, and architecture are very much involved in the first two. The third happens through word-of-mouth and by direct and indirect advertising.

The urban design process involves creating a vision of what a place might be like, and then developing the carrots and sticks that encourage developers to invest in it the way desired and not in another way and certainly not in a different city. The concern is also with providing opportunities for a better quality of residential, work, and recreational life for the city's residents. The focus in urban design is on the public realm – the streets, squares and parks – and how these are formed by the buildings around them. What then has to be done?

Cities range in size considerably, but whatever their size it is almost impossible to achieve a standard of urban design excellence throughout the whole city, although Singapore certainly tries. The focus of urban design is thus on the development of specific precincts and places within a city. The revitalization of central business districts, decaying inner-city neighbourhoods, and abandoned waterfronts have rebranded cities and made them attractive places for investment. Following the examples of La Défense outside Paris and Canary Wharf in London, Shanghai has built a modern central business district of striking buildings in Lujiazui, Pudong east of the Huangpu River. It meets many politicians' and developers' concepts of what a modern city centre should be even though it has been much criticized by others because of its general lifelessness. Naples, because of its reputation as the home of criminal groups, corruption, and ineffectual government has so far failed to rebrand itself through the creation of its distinct new central business area, Il Quartiere Dirigenziale, modelled on La Défense. The goal in all these cases has been to create attractive business districts that brand a city. The goal in residential area design has been similar.

The gentrification of inner city neighbourhoods makes them attractive to the middle-class and improves the image of the cities in which they are located. As these neighbourhoods are the areas of cities that visitors are likely to see, gentrified neighbourhoods reinforce the positive image of a city. There are many examples of gentrification that have completely changed the image of neighbourhoods such as Society Hill in Philadelphia or the North Loop neighbourhood of Minneapolis. Harlem and Bedford-Stuyvesant in New York are African-American neighbourhoods that are being 'middle-classified' and thus transformed into a new brand of neighbourhood. Islington in London and Marais in Paris, where then Mayor Jacques Chirac has encouraged gentrification, are two of many European examples. The processes of gentrification, however, remain controversial but there can be no doubt about gentrification's impact on changing the image of a city and the competitive advantage of one neighbourhood over another.

From Sarawak to Shanghai to Minneapolis-St Paul to Bristol, one of the great urban design success stories in rebranding cities has involved turning decayed and decaying industrial waterfronts into lively destinations for walking, entertainment, and dining (Breen and Rigby 1996). Battery Park City with its esplanade has enhanced the image of lower Manhattan. These redeveloped waterfronts have also acted as catalysts for further development. Baltimore's inner harbour development initiated in 1976 revived that city's image and economic prospects and led to the development of Charles Center. It has been the precedent for the successful development of a number of other abandoned pocket harbours. They include Darling Harbour in Sydney and the Victoria and Alfred Waterfront in Cape Town. What can be learnt from present urban design schemes in forming a brand for a city?

A city must have streets that are full of vitality and where people like to walk. New York has Fifth Avenue; Paris has the Champs Elysées, and London has Oxford and Regent Streets. Nicolae Ceausescu's attempt to make Bucharest a world city with the creation of the Bulevardul Victoria Socialismuli (Avenue of the Victory of Socialism) was, however, not successful (Cavalcanti 1997). Ceausescu wanted a Champs de Elysées but what he got was a boulevard without the defining characteristics and the street life of the Parisian boulevards in terms of the buildings that line it and the activities they house. He also ordered the destruction of the adjacent quarters of the city that gave Bucharest its intellectual character. The city's brand is now a negative one. A city needs to have a variety of districts as well as nodes that make it imageable in Lynch's terms. It also needs to be rich in architecture not only in having landmark buildings by famous architects but also ones that give its precincts an overall texture.

Cities in decline have to be especially clear about what they can offer and how their offerings relate to what is happening in the world. They have to understand the needs of different and potential stakeholders. Any vision that a city has of the future has to be realizable in terms of what it might afford visitors and investors (Avraham and Ketter 2007). Striving to meet high expectations is fine but one also has to be realistic. Glasgow has reinvented itself as a city of culture but in terms of its overall economy these elements are relatively minor but they have acted as a catalyst for other types of investment. Cities such as Glasgow need vast improvements in their infrastructure, population mix, and basic services. The question is: 'Who funds such investments?'

Branding and communication

Promoting a realistic image seems to be fundamental to the utility of branding of cities. The image that people have of cities is obtained through personal experience and/or indirectly through word-of-mouth or through images that appear in newspaper and travelogues on television or the Internet. In much the same way that architects may boost their self-image by writing about themselves in hagiographic terms or getting others to write about them, cities need to be promoted. Much promotion takes place by word-of-mouth, by the vicarious participation in the decisions that others are making, and by what one comes across on a daily basis in journals and on television.

Probably as important as anything else is the first-hand experience that people have when visiting a city. Whom they meet, the attitudes of those people toward them, and what they see and feel about a city are important. Investors need to feel confident that a city has been branded honestly and that it really has something to offer (Kavaratzis and Ashworth 2005). Thus they need to be attracted to visit a city in the first place. If a city has a clear brand image, it will be easier to do so.

Conclusion

Today it seems that a group of cities is pulling ahead of others. Incentives such as tax breaks, tax credits, free land, and low-interest rate loans do attract investments but investors also seek good places in which to place their funds. Such incentives are often difficult to sustain and tend to be the hopeful hallmark of failing economies.

Does city branding work? Conceptually it should work. Singapore has been successfully transformed from a backward colonial entrepôt into arguably the world's most modern city (Mahbubani 2007). It now has a distinct brand as a 'tropical city of excellence'. It has an excellent infrastructure system, good educational and, increasingly, entertainment facilities set in a clean, verdant, and well maintained streetscape. Its recent New-Asia Singapore branding to promote the city as the exemplar of the economic rise of Asia has been less successful (Henderson 2000). Bilbao has certainly been reinvented. Sacramento is promoting itself as an 'eco-friendly city' and seems to be attracting people and activities to support that end.

Berlin is being rebranded from a divided city to the capital not only of Germany but of Europe. Since 1990 the city has been involved in a major marketing effort that has included promoting itself as a progressive modern, informative, and educational city.

It is rich in museums ranging from archaeological to the Beatea Uhse Erotik-Museum, which attract substantial numbers of visitors. Perhaps, the city's unique attribute is that it is not hiding its past. The Holocaust Memorial and the Jewish Museum are reminders of that past. At present, however, the city consists of disparate parts and a series of poorly integrated individual urban design schemes. To obtain a clear cognitive image, it needs to be coordinated to form a whole. It nevertheless is establishing a brand name for itself.

The catalytic or multiplier effect of major investments through the erection of significant buildings or building complexes is difficult to measure. Frank Gehry's design for the Guggenheim Museum in Bilbao has along with other major works by international architects such as Norman Foster and Santiago Calatrava given a previously largely unknown city an international reputation. In the early 2000s the catalytic effect reputedly provided an additional €660 million to the gross domestic product of the city and €117 million to its annual tax base (Vidarte 2002; Lang 2005). Bilbao has clearly been rebranded.

Having a clear and distinct brand may be necessary for a city's prospering and sometimes even survival in the face of competition from other cities. Having a negative image in national and international circles may well be more harmful nowadays than before as cities vie to attract capital for the infrastructure development that makes real estate investment possible (Avraham and Ketter 2007). Good branding can make cities attractive just as poor branding can make them undesirable.

References

Ashworth, G.J. and Voogd, H. (1990). *Selling the City: Marketing Approaches in Public Sector Urban Planning*. London: Belhaven.

Avraham, E. and Ketter, E. (2007). *Media Strategies for Marketing Places in Crisis: Improving the Image of Cities, Countries and Tourist Destinations.* Oxford: Butterworth Heinemann.

Berg, L. van den and Braun, E. (1999). 'Urban Competitiveness, Marketing and the Need for Organising Capacity'. *Urban Studies*, 36(5–6): 987–9.

Bramezza, I. (2003). *Competiveness of the European City and the Role of Management.* West Lafayette, IN: Purdue University Press.

Breen, A. and Rigby, D. (1996). *The New Waterfronts: A Worldwide Success Story*. London: Thames and Hudson.

Cavalcanti, M. de Betânia Uchôa (1997). 'Urban Reconstruction and Autocratic Regimes: Ceausescu's Bucharest in its Historical Context'. *Planning Perspectives*, 12: 71–109.

Christian, S. (2007). *Eco-friendly city branding.* http://www.newsreview.com/Sacremento/Content?oid = 588045 accessed 20th December 2008.

Duhnke, C. (2008). 'Harvesting Success'. *ASU Magazine*, 11(3): 58–67.

Florida, R. (2002). *The Rise of the Creative Class and How It Is Transforming Work, Leisure, Community and Everyday Life*. New York: Basic Books.

Gelder, S. van and Allan, M. (2006). *City Branding: How Cities Compete in the 21st Century.* PlaceBrands

Gold, J.R. and Ward, S.V. (Eds) (1994). *Place Promotion: the Use of Publicity and Marketing to Sell Towns and Regions*. Chichester: John Wiley and Sons.

Henderson, J.C. (2000). 'Selling Places: The New Asia-Singapore Brand'. *The Journal of Tourism Studies*, 11(1): 36–44.

Kavaratzis, M. and Ashworth, J.W. (2005). 'City Branding: An Effective Assertion of Identity or a Transitory Marketing Trick?' *Tijdschrift voor Economische en Sociale Geografie*, 96(5): 506–14.

Kotler, P., Haider, D. and Rein, I. (1993). *Marketing Places: Attracting Investment, Industries and Tourism to Cities, States and Nations.* New York: The Free Press.

Kotler, P., Haider, D. and Rein, I. (1996). *Marketing for Hospitality and Tourism.* Englewood Cliffs. NJ: Prentice Hall.

Lang, J. (2005). *Urban Design: A Typology of Procedures and Products Illustrated With Over 50 Case Studies*. Oxford: Architectural Press.

Lynch, K. (1960). *The Image of the City*. Cambridge, MA: MIT Press.

Mahbubani, K. (2007). 'Creating a Singapore Brand for the International Stage – The Singapore Paradox; The world is not World Class; Singapore is World Class'. http://www/mahbubani.net/speeches/SCII%20presentation.pdf accessed June 18th 2009.

Marshall, R. (2003). *Emerging Urbanity: Global Projects in the Asia Pacific Rim*. London: Spon Press.

Proctor, M. and Matuszeki, B. (1978). *Gritty Cities: a Second Look at Allentown, Bethlehem, Bridgeport, Hoboken, Lancaster, Norwich, Paterson, Reading, Trenton, Troy, Waterbury, Wilmington*. Philadelphia: Temple University Press.

Strauss, A. (1961). *Images of the American City*. New York: The Free Press.

Vidarte, J.I. (2002). 'The Bilbao Guggenheim Museum'. In *Euskal Hiraria*. Victoria-Gasteiz; Central Publishing Services of the Basque Government 153–8.

Winfield-Pfefferkorn, J. (2005). 'The Branding of Cities: Exploring City Branding and the Importance of a Brand Image'. Unpublished Master of Arts in Advertising Design thesis, Graduate School of Syracuse University, New York.

Further reading

Cronin, A. (Ed.) (2008). *Consuming the Entrepreneurial City; Image, Memory, Spectacle.* Hoboken, NJ: Taylor and Francis. (Electronic Resource). The articles in this book present an analysis of the way cities are branded and marketed.

Hemelryk Donald, S., Kofman, E. and Kevin, C. (Eds) (2009). *Branding Cities; Cosmopolitanism, Parochialism and Social Change*, New York: Routledge. Articles highlighting the links between cosmopolitanism and parochialism in city branding and the frequent gaps between branding goals and the life experience of a city's inhabitants, including recent immigrants.

Monclús, J. and Guàrda, M. (Eds.) (2006). *Culture, Urbanism and Planning.* Aldershot: Ashgate. The book comprises papers on urban image making and branding.

Stalnaker, S. (2002). *Hub Culture: the Next Wave of Urban Consumers.* Singapore: John Wiley (Asia). The book includes a comparison of the characteristics of a number of fashionable cities.

Wheeler, A. (2006). *Designing Brand Identity: A Complete Guide to Creating, Building and Maintaining Strong Brands.* Hoboken, NJ: John Wiley. The book presents a guide to sustainable branding for the corporate world, mostly aiming at brand managers and designers of large business organizations.

41

From metropolitan to regional urbanization

Edward W. Soja

The modern metropolis has been experiencing a metamorphosis over the past thirty years. Changes in urban form, function, image, and experience over this time have probably been as great or greater than in any other similar period since the origins of the industrial capitalist city. As these material and imagined changes unfold, they increasingly challenge long established ideas about urban development, leading to a wide variety of interpretive perspectives aimed at making practical and theoretical sense of what has been a significant urban restructuring, short of total transformation but more than minor shifts. (Soureli and Youn 2009) Many different streams of thought have tried to interpret the restructuring of the modern metropolis, each with different implications for the theory and practice of urban design.

One interpretive stream recognizes the enormous magnitude of the changes underway and the serious challenge they pose for the ways we think about, plan, and design our cities. For the followers of this stream, traditional urban theory seems dated as an entirely new era of urbanization and urbanism emerges (Dear 2000; Dear and Flusty 1998). The new era is often seen as chaotic and filled with almost incomprehensible complexity, marking the end of urbanism as we knew it, with little from the past left to help us understand the present or to anticipate the future (Jencks 1995, 2007; Brouwer *et al.* 2002; Talen 2005). There is a widespread belief that we have entered an age of indulgent postmodern urbanism, where almost anything goes and nothing can be depended upon. Although interesting ideas are developed about what is new and different about the contemporary urban condition, knowledge of the past in explaining the present tends to be discarded in a kind of trash-heap of history.

At another extreme, there are others who cling to the past and often quite nimbly see the changes in the modern metropolis as fundamentally more of the same with only minor twists and turns (Duany 2001; Gandelsonas 1999; Jackson 1987). Constancy, with a few perturbations, outweighs change. Traditional ways of thinking about the city and urbanism as a way of life remain as powerful and useful as ever and any claim that a radical disjuncture has occurred is met with suspicion if not derision. Nearly everything that is new today is traceable back to some historical forerunner, making what some see as metamorphosis (or chaos or postmodern urbanism) little more than just another round of evolutionary continuity (Harvey 1989, 2003, 2005).

I have been interpreting the changing patterns of urbanization that have been taking place over the past thirty years in a different way, taking a specifically regional approach that tries to combine the best of the postmodernist and historicist perspectives to make practical and theoretical sense of what has undoubtedly been a profound restructuring of the modern metropolis. In *Postmetropolis: Critical Studies of Cities and Regions* (Soja 2000), I described the metamorphosis as a still ongoing postmetropolitan transition and focused on six distinctive but overlapping discourses, each focusing on particular aspects of the new urbanization processes. Summarizing these discourses, I concentrate here on the major causes and consequences of urban restructuring and argue that these changes reflect the emergence of a new form of urban development that I call *regional urbanization*

It is now widely agreed that the primary causes of urban change over the past three decades have involved the formation of a New Economy of post-Fordist, flexible, and information-intensive industrial capitalism; the accelerated globalization of capital, labor, and culture; and the revolution in information and communication technologies. Arising from these endogenous and exogenous forces of change has been a simultaneously social and spatial reconfiguration. The restructured urban geography, or morphology as some would call it, I depicted as the emergence of "exopolis," a turning of the urban form both inside-out, through an increasing urbanization of peripheral suburbia, and outside-in, a more unpredictable emptying out and refilling of the inner urban core. In this "unbounding" of the metropolis, the once relatively clear division between the urban and the suburban is beginning to disappear.

The reconfiguration of the social order, like the spatial restructuring characterized by an erosion of formerly well defined boundaries, is leading to the formation of what I called the "fractal city," with its widening income gaps, increasing social and cultural polarization, and more complex patterns of fragmentation, especially in the relations between domestic and immigrant populations. How this increasingly volatile postmetropolis has been kept from exploding over its deepening divisions has been the focus of two "post-restructuring" discourses, one richly exemplified by Mike Davis' *City of Quartz* (1990) revolving around the "hard" development of security obsessed urbanism and the "new enclosures" and gated "privatopias" (Mackenzie 1994) of the "carceral city," and the other addressing the softer modes of social control associated with the diverting enchantments of theme-parked hyper-reality and the manipulations of the urban imaginary through scripted and choreographed simulations and simulacra (Sorkin 1992).

In *Postmetropolis*, I argued that these causal, consequential, and reactive discourses needed to be seen together and interwoven to understand and effectively act to improve the contemporary urban condition. Since 2000, however, a new way of looking at the postmetropolitan transition has begun to emerge and consolidate around the concept of regional urbanization.

Regional urbanization: a new synthesis

Embedded in my earlier analysis of the postmetropolitan transition and in nearly all that I have written over the years on urban restructuring is a strong regional perspective. Building on my earlier work and on the resurgence of academic and planning interest in what some now call the New Regionalism (Soja 2002, 2009; MacLeod 2002), I discuss here, with some effort to connect to current debates on urban design, the idea that the restructuring of the modern metropolis has been defined and driven by a new and specifically regional urbanization process.

Over most of the past century, urban growth and development has been seen primarily from the perspective of what can be called *metropolitan urbanization*. In its most simplified version, often associated with Chicago School models, cities grow not just from natural increases in population but from in-migration, in large part into the city center with its peak density of jobs, but then spreading outward into urban and suburban rings of settlement. A sharp and distinct density gradient develops around the defining center, with the core urban agglomeration becoming densely packed while the metropolitan area expands in area and population size through sprawling and low-density suburbanization. In most cases, a clear division emerges between urban and suburban land uses and lifestyles, the urban tending to be more culturally and economically heterogeneous, more dangerous as well as exciting, with a greater concentration of jobs, museums, poverty, gangs, entertainment facilities, and crime.

Metropolitan urbanization—the rise of what is called the modern metropolis – began in the US in the late nineteenth century and became the dominant form in the early twentieth century, so much so that for many it has come to be the only conceivable mode of urban growth and expansion. Cities have been around for perhaps as long as 12,000 years, and suburbs pre-date metropolitan urbanization. It is important to note, however, that the early development of the industrial capitalist city was not yet characterized by metropolitan urbanization. The earliest industrial cities (e.g. Manchester, Chicago) and the industrialization of existing pre-industrial cities (e.g. London, New York) were much more centripetal than centrifugal, attracting huge numbers of workers to dense clusters of factories and stacked housing either in the city center or in districts nearby. Perhaps never before had centralized urban densities (as well as the adversities associated with them) reached such levels as in the highly compact and centralized early industrial capitalist cities.

Characterized by more centrifugal processes of selective decentralization, metropolitan urbanization emerged from the increasing diseconomies and social unrest arising from such dense agglomeration. This involved both a slowing down of growth in the urban core (although many continued to grow well into the twentieth century) and an increasingly expansive suburbanization process. In nearly all metropolitan areas, urban extension through annexation would no longer be the norm, as it was in earlier stages. Beyond the once clearly defined outer edge of the city, where there used to be a distinctive countryside, there grew an expanding aggregation of suburban municipalities. In some cases, new towns were purposely created as "garden cities" to build on the hybridization of city and countryside that was already happening to some extent with metropolitan growth. Although employment opportunities preceded residential growth much more often than most scholars thought (Hise 1999), suburbanization was predominantly residential and classically characterized by single-family detached homes, automobile dependency, daily journeys to work in the city, and all the now familiar attributes of suburbanism as a way of life.

The first step in understanding regional urbanization is to recognize metropolitan urbanization as a distinctive phase and not as an inevitable and immutable end state. What has been happening to the modern metropolis over the past thirty years can then be seen as not just a continuous elaboration or minor inflection of prevailing trends but as the beginnings of a new and different urbanization process, a third phase in the development of the industrial capitalist city. To be clear, the modern and almost always monocentric metropolis, with its pronounced dualism between the urban and suburban, continues to exist and will probably last well into the future. No premature "end of" is being predicted here. But something new and different is being interspersed and embedded within the modern metropolis, demanding new and different modes of analysis and understanding.

What then are the distinctive features of regional urbanization? To begin, regional urbanization is increasingly blurring the dualism and division between urban and suburban areas and ways of life, in some cases to the point that neither category can be identified and studied as it has been in the past. A crude indicator of this blurring is a pronounced *density convergence*. The once characteristic steep gradient declining rapidly around the peak density of the city center is broadly flattening out, with central densities declining to varying degrees while suburban densities increase. This flattening out of the density gradient and the related blurring of the differences between urban and suburban areas helps explain why it can be said that the modern metropolis is being turned inside-out and outside-in at the same time. While the suburbs urbanize, the city becomes more like the suburbs.

Another way of describing what has been happening is a seemingly paradoxical combination of decentralization and recentralization. Decentralization or deconcentration is in itself a continuation of metropolitan suburbanization, but at the same time there has also been a re-centering taking place in the periphery and often in the urban core as well that is reshaping the intra-metropolitan geography. The regional metropolis, for example, is becoming much more polycentric than ever before. Large urban nodes, many with more than 100,000 inhabitants and some with more jobs than dormitory bedrooms, re-punctuate the urban landscape. These changes have triggered many new terms such as "edge cities" and "outer cities," "metroburbs" and "technopoles," to capture the growing urban-suburban hybridization (Muller 1976; Garreau 1991; Knox 2008; Teaford 1996). Some speak now of "postsuburban development," the "urbanization of suburbia," even the "suburbanization of the central city" (Kling *et al.* 1991). Whatever terms are preferred, metropolitan urbanization and suburbanization are no longer what they used to be.

Polycentric regional urbanization increasingly takes the form of an expanding network of cities stretching the old metropolitan boundaries outward to connect with new hinterlands. This has led to the emergence and widespread acceptance of the concept of *city regions* and many other terms that describe the rapidly expanding scope and scale of regional urbanization (Scott 1998, 2001; Calthorpe and Fulton 2001; Ohmae 1993). The United Nations now publishes lists of the world's largest city regions, dropping the older term metropolitan or "greater" metropolitan region. Based on this new categorization, the UN HABITAT office in "The State of the World's Cities" announced that in 2007 the majority of the world's population, more than 3.3 billion people, lived in cities as they are variously defined in different countries, and listed nearly 450 city regions containing more than one million inhabitants, a hundred or so in China. Within the next decade, the number of city regions with over one million residents is expected to increase to well over 500 and to contain within them not just a majority of the world's population but a much greater concentration of the world's wealth and innovative capacity (Soja and Kanai 2008).

The modern metropolis has been bursting out of its former metropolitan boundaries in what can be described as not just a globalization of the urban, creating the most culturally and economically heterogeneous cities ever to exist, but also an urbanization of the globe. This has given rise to such widely used terms as *megacities* to refer to city regions of more than 10 million. Some now use the terms "megacity regions" or "megalopolitan regions" when referring to even larger polycentric urban networks, such as that around Shanghai in the Yangtze Delta, with an estimated population size of 82 million (Florida 2006). If one combines the megacity regions of Tokyo-Yokohama and Osaka-Kobe-Kyoto in Honshu, the total population would be around 100 million. Other megalopolitan regions with populations greater than 50 million include an expanded version of Bos-Wash, Jean Gottmann's (1961) original megalopolis between Boston and Washington D.C., and the Euro-Lowlands, stretching from the prototypical city region of the Randstad in the Netherlands to the Rhine-Ruhr and Luxemburg, now the core of its own cross-border region or Euregio. In this new age of extended regional urbanization, it can even be argued that everywhere on earth is urbanized to some degree, that urban influences and effects extend into the Amazon rainforest, the Siberian tundra, even the shrinking Antarctic icecap.

Inner vs. outer cities

But let us return from this global excursion to what was once called the intra-metropolitan scale, for there is more to be said about the transformations of the inner and outer cities, starting with a brief word about sprawl. With the increasing densification if not urbanization of suburbia, not only must we change our views of suburbanism as a way of life, and how these lifestyles are accommodated and sustained through urban planning and design, it is also necessary to rethink how we understand and respond to the problem of sprawl, as well as how we react to the claims and practices of Neo-Traditional Town Planning or its popular American variant the New Urbanism.

Regional urbanization is characterized by the densification or increasing compactness of suburbia and not as much as in the past by continued outward sprawl devouring agricultural land and greenfield sites. Simply proposing increased density or smartly clustered growth or blanket growth controls is not enough in postmetropolitan suburbia, especially as regional urbanization is throwing the already serious problems of the jobs-housing-transit imbalance ever further out of whack. Increasingly, regional urbanization is demanding specifically regional approaches, whether in dealing with the inefficiencies of sprawl, the fair and equitable delivery of public services, or the provision of affordable housing (Pastor *et al.* 2009).

Perhaps the most advanced form of regional urbanization, with all its problems and opportunities, can be found around the City of Los Angeles. Once the epitome of the automaniacally sprawling low-density metropolis, the "urbanized area" of Los Angeles (another new regional data category based on contiguous census tracts with densities greater than 1000 per square mile) surpassed New York in 1990 as the densest in the US and has been increasing its lead ever since. In 2002, the US census reported that the overall population density of the urbanized area of Los Angeles-Long Beach-Santa Ana CA was 7,068 people/square mile compared with 5,309 people/square mile for New York-Newark-NY-NJ-CT. The closest to Los Angeles was San Francisco-Oakland at 6,130 people/square mile (census.gov/geo/www/ua/ua_natl). Using more detailed data, if one graphed regional census tracts from the most to the least dense, New York (Manhattan) would start out slightly higher than Los Angeles but after fewer than a dozen tracts Los Angeles would pull ahead and remain at higher densities throughout its five county urbanized area, while New York drops precipitously over its 23 county sprawl.

How Los Angeles moved from being the country's least dense major metropolis in 1960 to the densest urbanized area in 1990 and today can be best understood from the perspective of regional urbanization and its impact on both inner and outer cities. The once classical and widely televised suburbia of Los Angeles has experienced an unusually intense urbanization process. Already polycentric to begin with, the region has seen a multiplication of booming municipalities of more than 100,000 inhabitants and the growth of perhaps the country's largest and oldest outer city in Orange County, with its growing cluster of municipalities containing more than 100,000 but less than 500,000 residents. Nearly 2.5 million people live in the amorphous and acephalous (without a dominant center) "postsuburban" city cluster of Orange County (Kling *et al.* 1991). So unclear is its identity that its professional baseball team, the Angels, a name derived from an older minor league team in Los Angeles, has been called California Angels in the past and is now officially described as the Los Angeles Angels of Anaheim.

There is much more to say about the distinctive features of the Outer City of Orange (my title), but the key point is that it can no longer be seen and interpreted as suburban. More people today travel from Los Angeles to Orange County than the other way around. There are probably more jobs than bedrooms in the main urban cluster, and nearly every feature of metropolitan urban life can be found there, from gangs, drugs, and crime to culturally heterogeneous populations, newly arrived migrants, museums, galleries, and concert halls. There are also important differences between the Orange County outer city and other outer cities that have grown around Los Angeles in the San Fernando Valley, the new "suburban" Chinatown in the San Gabriel Valley, and especially the much less economically successful Inland Empire based in San Bernardino and Riverside counties.

Sprinkled throughout the urbanized suburbs of Los Angeles are problem cities of a new sort, not edge cities but more like off-the-edge cities, where availability of affordable housing and developer's promises attracted large numbers of lower middle class and minority families. When job growth did not occur, large percentages of workers were forced to travel more than two hours to their old jobs, generating extraordinary pathologies in what appeared to be brand new suburbs. In such new cities as Moreno Valley and Lancaster, rates of divorce, suicide, spouse and child abuse, and home foreclosures (well before the crash of 2008) rose to exceptionally high levels, while public services declined severely because of the limited tax base (Soja 2000: 259–263). Again, understanding and responding to these new problems needs to be informed by a regional perspective rather than being seen as just the latest degrading of suburbia.

What arises most strikingly from these observations is the need for major new research programs aimed at exploring the extraordinary *differentiation of suburbia* that has been taking place across the US and in all the world's large city regions. We still know very little about where and how suburban municipalities urbanize and become cities in themselves, how race and ethnicity affects and is affected by the transformations of suburbia, how peripheral urbanization has been affected by the vast numbers of highly educated women with children entering the labor market freed from their entrapment in isolated suburbs of the past, and so much more. Seeing the difference between regional and metropolitan urbanization is just a starting point.

The growth of dense outer cities, however, tells only half the story of the densification of Los Angeles and other urbanized regions. New perspectives also need to be developed to understand what has been happening to the urban core. It is more difficult to generalize about the restructuring of the inner city, although here too there has been a wide-ranging differentiation taking place. In the 1970s and early 1980s, at the beginning of the period of crisis-generated urban restructuring, nearly all the larger metropolitan regions of the world experienced some degree of "hollowing out," usually with a loss of jobs (deindustrialization) and reductions of overall density. This led some to speak of the suburbanization of the city while others created a new category for investigation and planning around the category of "shrinking cities, starting in Germany but now spreading to Detroit, Cleveland, and many other US cities"(Oswalt 2004, 2006; Oswalt and Rieniets 2006)

In some cases, such as Detroit, this hollowing out has produced severe stagnation and desolate urban landscapes. In others, such as Osaka, the urban core of what was once the densest of major Japanese cities has been almost emptied of its resident population but is nevertheless vibrant and thriving thanks to its office, retail, civic, and entertainment activities. In still other cases, the urban core has been re-filled with migrant populations, maintaining if not increasing central densities. This was what happened to Los Angeles, where in the last three decades probably as much as 1 million Anglo and Black residents left the inner city in large part driven by factory closures, but around 5 million transnational migrants from almost every country in the world poured in to raise population densities to Manhattan levels. Without major changes in the aging built environment, the mass migration to Los Angeles, one of the largest city-focused migrations in world history, has created the most overcrowded housing in the country as well as contributing to the largest concentration of homeless people anywhere in the US.

Planning and design implications

Clearly then, there is a growing need for more rigorous comparative studies of the differential experiences of inner cities as well as outer cities if we are to deal more effectively as urban planners and designers with the effects of regional urbanization. This task becomes more urgently needed given what has been happening to the city building professions over the past several decades. In almost every city region in at least what was defined as the industrialized world, deindustrialization driven job loss and population movements associated with the shift from Fordist to post-Fordist economies triggered a profound change in urban and regional planning. Neoliberal globalization and related state policies, including some degree of welfare reform as well as deregulation and the privatization of public services, reduced funding from the central state and triggered a move away from anti-poverty efforts and socially progressive forms of planning during a period in which most city regions, with cruel irony, were experiencing widening income gaps and the worst economic and political polarization between the super-rich and the super-poor since the Great Depression.

As the seeds of the present day global economic meltdown were planted, planners were increasingly forced to become entrepreneurs, seeking corporate investment and tourist dollars in the highly competitive global economy. The primary focus of urban and regional planning shifted from helping the most disadvantaged to reviving in any way possible the changing inner cities and old downtowns seemingly threatened by the new urbanization processes. This fiercely competitive mode of entrepreneurial planning stimulated the rise of place-marketing strategies, image-oriented city boosterism, the creation of deregulated enterprise or free-trade zones, state subsidized public-private partnerships and business development areas, and a host of other new tools designed almost entirely to tap external financial sources for local projects, occasionally even ones aimed at alleviating increasing poverty.

As inequalities soared and welfare policies were weakened, the urban and regional planning process around the world became increasingly absorbed in seeking the next Olympic Games or global trade expos or iconic architectural extravaganzas such as the infamously successful Frank Gehry-designed Guggenheim Museum in Bilbao. On a smaller scale, the search for real or imagined place-based comparative advantages contributed to the expansion of New Urbanism projects and a boom for renowned "starchitects" to produce repetitive signature buildings. Sustained by either an "anything goes" postmodern version of chaotic urbanization, in which almost no one knows what is happening, or an arrogant acceptance of neoliberal ideologies of deregulation, privatization, decentralization, and market magic, entrepreneurial forms of urban and regional planning as well as urban design, informed primarily by a metropolitan urbanization perspective, continue to the present in full force.

Fortunately for the future, something else has been happening in the world of urban and regional planning theory and practice. A New Regionalism has been resurgent in recent years, stimulating what has begun to be a shift back to more welfare and equity oriented forms of regional planning and a shift forward in terms of new tools, approaches, and policies that at least in part arise from a regional urbanization perspective as I have been describing it here (Pastor *et al.* 2009; Orfield 1997).

The New Regionalism is not just a matter of re-creating metropolitan government or specialized multi-state regional agencies like the Tennessee Valley Authority or the Appalachian Regional Commission. Nor is it simply a call for more attention to local environments and cultures, as was the case in earlier references to regions and regionalism by architects and designers (Lynch 1976; Frampton 1983). The New Regionalism demands a more far-reaching change in perspective, one that builds on an understanding of the shift from metropolitan to regional urbanization and the growing importance of globalized, polycentric, and networked city regions as generative forces for economic development, technological innovation, and cultural creativity. From this perspective, regions are not just backgrounds or contexts but active and potentially propulsive motors for development and change, operating, some claim, on a level comparable to markets, states, and cultures (Storper 1997).

These "regional worlds" exist at multiple scales, ranging from the human body and an individual building through the local, urban, regional, national, surpranational, to the global or planetary level. Accordingly, they demand multiscalar and comprehensively regional approaches to planning and policy-making. Recognizing this need, the Macarthur Foundation has recently created a research network on Building Resilient Regions aimed at developing responsive regional governance strategies to deal with major national challenges. The multiscalar and networked nature of the New Regionalism has been particularly well understood and responded to in the European Union, where there has been a revival of comprehensive spatial planning (Healey 2006, 2004) and the promotion in all member states of what is called the European Spatial Development Perspective (Faludi and Waterhout 2002).

The specific role of urban design in the New Regionalist approaches has not yet been worked out, but some broad implications are clear. For a start, architects and designers must learn to think regionally, breaking through the scalar trap that often confines their urban imagination to a narrowly defined and localized built environment and to individual projects that are detached from larger scale regional, national, and global perspectives. Accepting the need to design and plan in response to the shift from metropolitan to regional urbanization is also essential, for it leads to a very different view of the differences between urban and suburban environments and the new dynamics shaping inner and outer cities. There is no need to go back to some form of master planning for regions and certainly a strong sense of social responsibility should not be abandoned. But at the same time, all the city building professions must acknowledge in looking toward the future that urbanism and suburbanism as ways of life are no longer what they used to be.

References

Brouwer, J., Brookman, P. and Mulder, A. (2002). *TransUrbanism*, Rotterdam: NAi Publishers.

Calthorpe, P. and Fulton, W.B. (2001). *The Regional City: Planning for the End of Sprawl*, Washington, DC: Island Press.

Davis, M. (1990). *City of Quartz: Excavating the Future in Los Angeles*, New York: Verso.

Dear, M. (2000). *The Postmodern Urban Condition*, Oxford: Blackwell Publishers.

Dear, M. and Flusty, S. (1998). "Postmodern Urbanism," *Annals of the Association of American Geographers*, 88(1): 50–72.

Duany, A. (2001). *Suburban Nation: The Rise of Sprawl and the Decline of the American Dream*, New York: North Point Press.

Faludi, A. and Waterhout, B. (2002). *The Making of the European Spatial Development Perspective*, London: Routledge.

Florida, R. (2006). "The New Megalopolis," *Newsweek*, July 3–10.

Frampton, K. (1983). "Towards a Critical Regionalism: Six Points for an Architecture of Resistance," in Foster, H. (Ed.), *The Anti-Aesthetic: Essays on Postmodern Culture*, Port Townsend: Bay Press.

Gandelsonas, M. (1999). *X-Urbanism: Architecture and the American City*, Princeton, NJ: Princeton Architectural Press.

Garreau, J. (1991). *Edge City: Life on the New Frontier*, New York: Doubleday.

Gottmann, J. (1961), *Megalopolis: The Urbanized Northeastern Seaboard of the Uniteed States*, New York: The Twentieth Century Fund.

Harvey, D. (1989). *The Condition of Postmodernity*, Oxford and Malden, MA: Blackwell Publishers.

—— (2003). *The New Imperialism*, Oxford and New York: Oxford University Press.

—— (2005). *A Brief History of Neoliberalism*, Oxford and New York: Oxford University Press.

Healey, P. (2004). "The Treatment of Space and Place in the New Strategic Spatial Planning in Europe," *International Journal of Urban and Regional Research*, 28(1): 45–67.

—— (2006). *Urban Complexity and Spatial Strategies: a Relational Planning for our Times.* London: Routledge.

Hise, G. (1999). *Magnetic Los Angeles: Planning the Twentieth Century Metropolis*, Baltimore, MD: Johns Hopkins University Press.

Jackson, K. (1987). *Crabgrass Frontier: The Suburbanization of the US*, New York: Oxford University Press.

Jencks, C. (1995). *The Architecture of the Jumping Universe: A Polemic: How Complexity Theory is Changing Architecture and Culture*, London: John Wiley and Sons.

—— (2007). *Critical Modernism: Where is Postmodernism Going?* London: Wiley Academy.

Kling, R., Olin, S. and Poster, M. (1991). *Postsuburban California: The Transformation of Orange County since World War II*, Berkeley and Los Angeles: University of California Press.

Knox, P.L. (2008). *Metroburbia, US*, Piscataway NJ: Rutgers University Press.

Lynch, K. (1976). *Managing the Sense of a Region*, Cambridge, MA: MIT Press.

Mackenzie, E. (1994). *Privatopia: Homeowner Associations and the Rise of Residential Private Government*, New Haven, CT and London: Yale University Press.

MacLeod, G. (2002). "New Regionalism Reconsidered: Globalization and the Remaking of Political Economic Space," *International Journal of Urban and Regional Research*, 25(4): 804–829.

Muller, P.O. (1976). *The Outer City: Geographical Consequences of the Urbanization of Suburbs*, Washington DC: Association of American Geographers Publications.

Ohmae, K. (1993). "The Rise of the Region-State," *Foreign Affairs* (Spring): 78–87.

Orfield, M. (1997). *Metropolitics: A Regional Agenda for Community and Stability*, Washington, DC:

Brookings Institution Press/Lincoln Institute for Land Policy.

Oswalt, P. (Ed.) (2004). *Shrinking Cities, Volume 1: International Research*, Ostfildern: Hatje Cantz Verlag.

Oswalt, P. (2006). *Shrinking Cities, Volume 2: Interventions*, Ostfildern: Hatje Cantz Verlag.

Oswalt, P and Rieniets, T. (20.) (2006) *Atlas of Shrinking Cities*, Ostfildern: Hatje Cantz Verlag.

Pastor, Jr., M., Benner, C. and Matsuoka, M. (2009) *This Could be the Start of Something Big: How Social Movements for Regional Equity are Reshaping Metropolitan America*, Ithaca, NY: Cornell University Press.

Scott, A.J. (1998). *Regions and the World Economy*, Oxford and New York: Oxford University Press.

—— (2001). *Global City-Regions: Trends, Theory, Policy*, New York: Guilford Press.

Soja, E.(2000). *Postmetropolis: Critical Studies of Cities and Regions*, Oxford and Malden: Blackwell Publishers.

—— (2002). "The New Regionalism: A Conversation with Edward Soja," an interview by R. Ehrenfurt, *Critical Planning* 9: 5–12.

—— (2009). "Regional Planning and Development Theories," in Thrift, N. and Kitchin, R. (20.), *The International Encyclopedia of Human Geography*, Amsterdam: Elsevier: 259–270.

Soja, E. and Kanai, J.M. (2008). "The Urbanization of the World" in Burdett, R. and Sudjic, D. (Eds.) *The Endless City*, New York and London: Phaidon, 54–69.

Sorkin, M. (1992). *Variations on a Theme Park*, New York: Hill and Wang-Noonday Press.

Soureli, K. and Youn, E. (2009). "Urban Restructuring and the Crisis: A Symposium with Neil Brenner, John Friedmann, Margit Meyer, Allen J. Scott, and Edward W. Soja," *Critical Planning* 16: 34–60.

Storper, M. (1997). *The Regional World: Territorial Development in a Global Economy*, New York: Guilford Press.

Talen, E. (2005). *New Urbanism and American Planning: The Conflict of Cultures*, New York and London: Routledge.

Teaford, J.C. (1996). *Government and Politics in the Edge Cities*, Baltimore, MD: Johns Hopkins University Press.

Further reading

Knox, P.L. (2008). *Metroburbia, US*, Piscataway, NJ: Rutgers University Press. A major text on the transformation of suburbia with examples from across the US.

Pastor, Jr., M., Benner, C. and Matsuoka, M. (2009). *This Could be the Start of Something Big: How Social Movements for Regional Equity are Reshaping Metropolitan America*, Ithaca, NY: Cornell University Press. An introduction to community-based regionalism and related social movements aimed at regional equity and democracy.

Scott, A.J. (Ed.) (2001). *Global City-Regions: Trends, Theory, Policy*, New York: Guilford Press. A collection of readings on the emerging concept of global city-regions, with contributions from leading urban and regional theorists and practitioners.

Soja, E. (2000). *Postmetropolis: Critical Studies of Cities and Regions*, Oxford: Blackwell Publishers. Comprehensive survey of the literature on urban restructuring and the new urbanization processes transforming the modern metropolis.

Soja, E. and Kanai, J.M. (2008). "The Urbanization of the World" in Burdett, R. and Sudjic, D. (Eds.) *The Endless City*, New York and London: Phaidon, 54–69. Focuses on the extension of regional urbanization to the global scale and the globalization of city regions; part of a well illustrated text on the "new urban age."

Storper, M. (1997). *The Regional World: Territorial Development in a Global Economy*, New York: Guilford Press. A key theoretical work on the importance of regions and regionalism in economic development.

42

Ethnoscapes

Clara Irazábal

Although the urban planning and design literature extensively explores the intersection of race and space, much less work has been done on the study of ethnicity and space, an interrelation I refer to as producing ethnoscapes. This chapter discusses the emergence of ethnoscapes as contemporary spatial typologies, heightened by the new global socio-economic order. The chapter analyzes different processes of ethnicization of space, particularly focusing on Latina/o ethnoscapes in the United States. Latina/os have distinctively used processes of territorializing, regulating, and symbolizing place to sustain distinctive communities flavored by ethnic business and social associations (Arreola 2002; Aguilar San Juan 2005; Irazábal and Gómez-Barris 2007; Irazábal and Farhat 2008).

The chapter is composed of four sections. The first three sections discuss critical dimensions of ethnoscapes, including socio-cultural and subjective, spatial and temporal, and political dimensions. The last section discusses emerging challenges and opportunities for urban design and development presented by the increased saliency of ethnicity in the urban world.

Socio-cultural and subjective dimensions of ethnoscapes

The new disjunctures between economy, culture, and politics brought about by globalization, require new theories. Arjun Appadurai (1990; 1991) proposed a framework for exploring the relationship between five dimensions of global cultural flows: ethnoscapes, mediascapes, technoscapes, financescapes, and ideoscapes. This chapter takes Appadurai's (1991: 192) concept of ethnoscape as a point of departure. According to Appadurai, ethnoscape means,

> [T]he landscape of persons who constitute the shifting world in which we live: tourists, immigrants, refugees, exiles, guestworkers, and other moving groups and persons who constitute an essential feature of the world, and appear to affect the politics of and between nations to a hitherto unprecedented degree.

He further explains,

> This is not to say that there are not anywhere relatively stable communities and networks ... But that is not to say that the warp of these stabilities is everywhere shot through with the woof of human motion, as more persons and groups deal with the realities of having to move, or the fantasies of wanting to move. What is more, both these realities as well as these fantasies now function on larger scales ... And as international capital shifts its needs, as production

> and technology generate different needs, as nation-states shift their policies on refugee populations, these moving groups can never afford to let their imaginations rest too long, even if they wished to.

There are several dimensions worth highlighting in this definition: human motion, physically and imaginatively, is a sign of the times that disrupts place- and community-based stabilities. Moving groups (tourists, immigrants, refuges, exiles, guest workers) constitute a leading force transforming politics of and between nations. In the process, their cultural-political economy (i.e. traits that constitute their ethnos) denotes their strategies to survive, cope, and prosper in the midst of a fluxing and harsh world.

Ethnoscape thus evokes an intricate and dynamic relation between people (ethnos) and place (scape). Cultural identity (ethnicity) has become a prominent way of building individual and collective subjectivities and constructing urban lifestyles. The ideal of public spaces – open, accessible, inclusive, and capable of supporting encounters of difference – makes them privileged sites in this quest (Makowski 2003). Public spaces are also sites for the negotiation of values, rights, duties, and rules of sociability in a community. Identity politics – issues of legal status, gender, sexuality, race, and ethnicity – are increasingly played out in public spaces, thus creating ethnoscapes (Irazábal 2008). The notion goes beyond, but builds upon, more stable and traditional conceptions of ethnic landscape or urbanism.

Ethnoscapes are constituted as historical palimpsests of fluctuating layers and degrees of mediascapes, technoscapes, finanscapes, and ideoscapes, and can be examined as such. Throughout history, ethnoscapes can be understood as material expressions of culture. Their physicality shows signs of change that accompany and at times reflect or propel transformations of cultural practices in the societies they belong to, beyond those explicitly related to the practices of placemaking. Given their materiality, ethnoscapes have a resilient capacity of preserving the physical imprints of time. Hence, we can read ethnoscapes as texts in which to decipher the historical trajectories and contemporary conditions of communities *in* and *through* place.

Planners and designers can deepen their appreciation and analyses of ethnoscapes by focusing on the sensuous, vital, embodied, and affective practices through which they are performed. To ignore those practices "is to sideline both a key aspect of these spaces themselves, and a key element in the circulation of [ethnic] discourses and the identities that produce and are produced through them" (Holloway 2006: 182). Ethnic practices can bind people together in ways that other institutions cannot, and in the process they create particular spatialities (Miller *et al.* 2002: 120; Irazábal and Dyrness forthcoming 2010).

Given the palimpsestic and situated nature of ethnoscapes and the mobility of people that populate and traverse them, processes of ethnicizations of space are ongoing. These phenomena can be examined in different historical or geographical contexts. In the case of the Americas, for instance, we can appreciate the dual and uneven phenomenon of the Latinization of the United States and the Americanization of Latin America. In effect, the production of urban space in the two regions is closely intertwined, as places, culture, and social practices are influenced by the movement of capital, labor, and urban planning/design ideas. This Americanization of Latin American cities and Latinization of US cities is not limited to a benign, fluid urbanism. Rather, those processes include resistance, homogenization, colonization, and power struggles. They, however, defy generalizations. The specificities of the recent resurgence of Mexican Phoenix, among other examples of Latinization in the US, prompted Oberle's and Arreola's (2008: 171) call "for greater understanding of the internal heterogeneity of Mexicans [and other Latin American and Latina/o groups] in the United States and how this can inform our geographical interpretations of the growing Latinization of American cities."

Appadurai's notion of ethnoscape, in some instances, may mean migrantscape or diasporicscape. In the contemporary context of the United States and countries of the European Union, different immigrant experiences and institutional regimes underscore the fact that the marginalization of immigrants is derived not only by *who* they are (i.e. their ethnorace and country of origin), but also by *what* they are (i.e. their legal status), and *where* they are (i.e. in foreign land). In this context, enactments of ethnic culture often become a mechanism for both *coping* with current conditions and *hoping* for a better future. Ethnic practices perform as a connecting tissue between time and space for immigrants, i.e. between the different places they emotionally or physically inhabit and between the past and the present (Jasper 1997). In the Latina/o experience, migrantscapes not only de/reterritorialize cultural practices from country to country, but also frequently from rural to urban areas (Smith 2002). They also disrupt traditional models of immigrant settlement in central city neighborhoods, recently occupying maturing suburbs and developing suburban enclaves, such as in Charlotte, North Carolina (Smith and Furuseth 2004).

The rise of conventional and diasporic tourism (expatriates and their descendents visiting the homelands) is fueling the invention of traditions (Hobsbawm and Ranger 1983) and the manufacturing of ethnoscapes. Recently, many cities have (re)created urban traditional forms that fall "out of context in their nostalgic references to (an imagined) social and economic order of the past" (Holston 1989: 317) with the goal of transforming the past into a commodity for mass consumption and profit-making (Hobsbawm and Ranger 1983). Such representations may invoke a past from which traces of exploitative social relations have been expunged (Davis 1987). Nostalgia and imagination could be seen as manipulated "to stimulate our acts of consumption, by the spectacle of history made false" (Boyer 1992: 204). Such romanticized invocations of tradition may also create images without real referents, a "simulacrum or pastiche" (Harvey 1990: 303).

Early on, spatial thematization of an idealized Spanish and Mexican past became popular in some Anglo communities in the US. In San Antonio, TX, for example, Workers Progress Administration funds were used to create Spanish and Mexican quarters such as Paseo del Rio, La Villita, and the Latin Quarter (Arreola 2004; Camarillo 1979). Other examples are the thematization of Santa Barbara, CA, and Placita Olvera in LA. Although criticized for being "Anglo" visions of Mexican architectural styles and detrimental to Mexican American interests at the time, these projects survived to become emblems of the Mexican American heritage and generators of economic activity (Irazábal and Farhat 2008).

A contemporary and more complex ethnoscape, the shopping mall Plaza Mexico in Southern California, embodies a case of invention and commodification of traditions for immigrants and US citizens of Mexican descent (Figures 42.1 and 42.2). Conceived and owned by Korean investors, the Plaza is a unique architectural recreation of Mexican regional and national icons that make its patrons feel "as if you were in Mexico." Plaza Mexico produces a space of diasporic, bounded tourism, whereby venture capitalists opportunistically reinvent tradition within a structural context of constrained immigrant mobility. While most of the contemporary

Figure 42.1 Plaza Mexico in Lynwood, California. Source: Clara Irazábal.

Figure 42.2 Plaza Mexico in Lynwood, California. Source: Clara Irazábal.

theory of tourism, travel, and place emphasize the erosion of national boundaries and the fluidity of territories, the case of Plaza Mexico brings us to appreciate this phenomenon and its opposite as well – the strengthening of national borders and their impact on the (in)mobility of millions of individuals (Irazábal and Gómez-Barris 2007).

The arts are fertile sites for the production of ethnoscapes, and artistic expressions of the ethnic find their way into cityscapes through a myriad of media. Neighborhoods such as Boyle Heights in LA and The Mission in San Francisco, and parks and plazas such as Placita Olvera in LA and Logan Park in San Diego, derive a potent character of place from their artistic murals, sculptures, pavings, landscaping, etc. Many of these ethnic places become attractive places for the community to congregate and celebrate ethnic festivals and rituals – performative arts of music, dance, poetry, etc. Nonetheless, the attractiveness of some Latina/o places, rich in vernacular architecture, community festivals, and public art has hastened gentrification in some cities, as was the case in San Francisco's Mission District (Godfrey 2004; Irazábal and Farhat 2008).

Spatial and temporal dimensions of ethnoscapes

There are several disciplines and practices that contribute to the spatial dimensions of ethnoscapes. The disciplines with the most direct impact are those associated with the design of the public realm – urban planning, urban design, architecture, landscape architecture, and public arts. However, it is people in their everyday life practices that constitute the most recurrent and meaningful spatial markers of ethnicity in and through place. The spatial markers of ethnoscapes can have different temporal dimensions. They can be relatively permanent, as those that compose the physicality of a place such as its architecture; semi-permanent or cyclical, as those that have predictable and repetitive occurrences such as farmers markets; or impermanent, such as political demonstrations or unique artistic events. Transient or impromptu ethnoscapes include drum circles, hip-hop circles, skateboard sites, dance and sport circles, etc. Due to their celebratory or rebellious nature, some of these ethnoscapes become sites of multiple ethnic fusions and crossovers.

Time also impacts ethnoscapes in a different way. As social constructs, ethnic places have received different appreciation throughout history, which has led to distinct policy measures. Modernist planners subjected them to slum clearance, because they were viewed as areas of overcrowding, social pathology, and obstacles to modernization. More recently, however, a renewed appreciation of ethnic culture has been mobilized by preservationist activists, place entrepreneurs, and planners to stimulate neighborhood revitalization (Lin 1995).

Political dimensions of ethnoscapes

Often ethnoscapes are sites of struggle for the recognition and/or expansion of citizenship rights. The notion of citizenship is being redefined to include the right to be different from the dominant national community, and citizenship is understood as fluid and dynamic with rights and values constructed through practices and discourses (Winocur 2003). These conditions make ethnoscapes akin to Miraftab's (2004: 1) notions of invented and invited spaces of citizenship, respectively:

> "Invited" spaces are defined as the ones occupied by those grassroots and their allied non-governmental organizations that are legitimized by

> donors and government interventions. "Invented" spaces are those, also occupied by the grassroots and claimed by their collective action, but directly confronting the authorities and the status quo .... the latter challenges the status quo in the hope of larger societal change and resistance to the dominant power relations.

Ethnoscapes can be invited – legitimized by donors and government interventions. However, they have the potential, under certain circumstances, to expand invented spaces of citizenship. Invented ethnoscapes are created, used, and appropriated by people when recourse to an invited ethnoscape is ineffective. Frequently, these invented ethnoscapes subvert or expand the invited public sphere and create new spaces and practices (Isin 1999; Rose 2000). Ethnoscapes can contribute to "the significance of both invited and invented spaces of citizen participation in the formation of inclusive cities and citizenship" (Miraftab and Wills 2005: 212).

Arlene Dávila's study of the marketing of Latina/o space in El Barrio (Spanish Harlem in New York) revealed that the politics of marketing ethnicity promoted gentrification and Latinization. The transformation of rental housing in condominiums, school privatization, and the enactment of Empowerment Zone legislation in El Barrio were presented as empowerment strategies, but ended up displacing long-time residents. These struggles confronted developers, politicians, long-time residents, newcomers, Puerto Ricans, Mexicans, and African Americans over space, gentrification, and cultural representation. They also revealed paradoxes of development, where the commodification of neighborhood cultures for the promotion of gentrification can also provide an obstacle to it (Dávila 2004). Thus, the reinvited ethnoscape of El Barrio made by planning institutions and commercial corporations collided with the reinvented ethnoscape of the residents' resistance.

Invented ethnoscapes can have a lasting impact and a transformational effect on cities and nations, and thus can constitute spaces of insurgent citizenship. Holston and Appadurai (1996: 50) define spaces of insurgent citizenship as "situations which engage, in practice, the problematic nature of belonging to society." Insurgent citizenship problematizes the normative basis of citizenship in capitalist societies, in which citizenship rights are taken away through the privatization of open space, the creation of gated communities and edge cities, the criminalization of the homeless and immigrants, and disciplinary actions against insurgent groups (MacLeod nd; Irazábal 2008).

For communities across the world, struggles between the defense of ethnic identity and ways of life vis-à-vis consumerism and homogenization have increased (Leclerc *et al.* 1999; Villa 2000; Dávila 2004; Irazábal and Farhat 2008). Marketing strategies are used to select and edit urban images to distort reality to different degrees (Del Rio 1992; Irazábal 2005: 99). The image of the city and the citizens' identity can be molded with an orchestrated marketing of urban planning ideas with multi-media tools and showy architectural interventions. Thus, aggressive image making and city marketing campaigns can distort social practices of appreciation/appropriation of the city by different groups of inhabitants, favoring urban plans and interventions that cater to the needs of the middle- and high-income classes (Sánchez 1996; Irazábal 2005).

In the post-World War II era, the Chicano movement was an impetus for the celebration of spatial and cultural rights in the face of calls for wholesale assimilation (Frausto 1999). In this resurgence of pride, the barrio was reconstituted, physically and discursively, as a site of resistance and celebration (Arreola 2004). The physical and social appropriation of private, semi-private, and public space furthered the assertion of cultural pride. In the growing Latina/o barrios, the social use of front yards, the symbolic appropriation of public space in the muralization movement, and the cultural appropriation of the built environment through vernacular architecture proliferated (Gonzalez 1999; Rojas 1999; Arreola 2002). This period also witnessed a revival of native cultural traditions with the organization of fiestas and celebrations in public spaces (e.g. in Placita Olvera and Mariachi Plaza, LA).

A more recent example of an invented ethnoscape of resistance was the South Central Farm in Los Angeles – a 14-acre urban farm in one of the highest concentrations of impoverished residents in the county – which was destroyed in 2006 after a long legal and socio-cultural battle for its preservation (Irazábal and Punja 2009). In addition to relieving food insecurity, the farm offered a much needed community-development venue for the surrounding Latina/o ethnoscape. The farm provided a medium to preserve and recreate community traditions of agriculture and heirloom seeds, survival strategies of indigenous cultures, as well as the farmers' ability to pass on their living traditions to their children (Radford and Santos 2006). For the youth, it was an alternative from gangs and drugs, and a place where the elderly could contribute in a substantial way to the community. Furthermore, the community's decision-making structure, planning, political outreach, and presence at City Hall and courts turned the farm into a "democracy workshop" (Kuipers 2005), offering critical tools to participants that even now continue to be applied in the promotion of larger environmental justice goals in the region (Irazábal and Punja 2009: 10–11).

Challenges and opportunities for urban design practice and pedagogy

The notion of ethnoscape underscores the significance of ethnic-based reformulations of urban practices and living preferences in cities and the potential these have for the transformation of policy making, planning, and development. In order to make urbanism more responsive to these challenges, urban design, planning, and development practices and teaching have to pay increased and more sophisticated attention to issues of ethnicity, identity, and culturally-based, urban lifestyle practices. This should both expand urban scholarship and be applied to the policy-making effort of drafting effective and culturally-sensitive solutions to the challenges of multiethnic cities. Below, three critical areas of exploration are discussed: design strategies for the creation of culturally-sensitive and adaptable public spaces, the retrofitting of cities, and the progression toward spatial justice.

Multiethnic design of public spaces

The rise of the importance and prevalence of ethnicity as a marker of people and place identity, prompts planners and designers to rethink the way they understand and interact with communities and partake in the process of placemaking with/for them. Contemporary urban spaces and regulations need to be more versatile and flexible to fulfill a host of different cultural, educational, and environmental requirements for multiethnic societies (Loukaitou-Sideris and Stieglitz 2002). The appropriate and versatile design of public spaces is paramount in this quest.

As ethnic communities proliferate, planners are looking back to historic plaza-centered villages for inspiration (e.g. in

Las Vegas, NM and San Diego, TX), which are notable for their socially engaged life-style centered on ethnic traditions and events (Arreola 2002; Smith 2004). The study of Los Angeles parks, however, reveals the increasing difficulty faced by governments to provide and maintain public spaces, the inequitable distribution of them in the city, and the challenges of addressing different and competing public space needs for an increasingly heterogeneous public (Loukaitou-Sideris and Stieglitz 2002; Loukaitou-Sideris 2006). Although these conditions are particularly dire in Los Angeles, they are present in many cities around the world facing increased population and cultural diversity, and hence, more complex processes of creating ethnoscapes. Private developers are also emulating ethnic markers for the creation of invented places in the United States, such as Plaza Mexico in greater LA, where Latina/os can buy ethnic commodities, enact their cultural traditions, and even celebrate their religious rituals as if they were in Mexico. This may be a perfectible model for public-private and community partnership that produces both economic and community development gains (Irazábal and Gómez-Barris 2007).

Through continuous community-specific analysis, planners and designers can take advantage of spaces in neighborhoods to host recreational opportunities and displays of ethnic culture. Underutilized or empty lots in neighborhoods and along freeways, railway lines, riverfronts, waterfronts and transportation corridors can become redesigned for these purposes. Linkages to other land uses, such as housing, schools, sports and cultural facilities should be pursued (Loukaitou-Sideris 2006). Planners and designers also need to accommodate less conventional uses in public space and promote single-to multi-use space conversions. Cultural programs, urban gardening, group sports (most saliently soccer and volleyball in the case of Latina/os), entrepreneurial activities, and volunteerism can infuse liveliness to public spaces and benefit ethnic communities. Professionals, city agents, and community members can also seek the development of the educational and environmental potential of public space for instilling appreciation for ecology, traditions, sociability, and multiculturalism. Community-based design efforts need to be mindful of both processes and outcomes to democratically develop multiethnic public spaces (Main 2008). Urban design and planning pedagogy should incorporate theoretical and practical exercises for students to get educated, sensitized, and skilled in the participatory analysis and production of ethnoscapes.

The retrofitting of our cities: is new urbanism ethnic urbanism?

The retrofitting of cities in an era of global climate change, economic crisis, and rapid urbanization would require a substantial transformation of design, planning, and development practices along the lines of promoting more compact living, mixed uses, lively use of public spaces, and heavy reliance on walking, cycling, and transit for mobility.

For almost three decades, New Urbanism advocates have aimed to create more compact cities with an array of different housing choices, mixed-use development, improved public transportation, and pedestrian-friendly environments. The purpose of such conditions is to attain greater environmental sustainability, curb sprawl, enhance mobility, and accommodate growing urban populations in spatial arrangements that improve opportunities for socialization and a greater sense of place (see chapter by Audirac). But the New Urbanism advocates and practitioners have largely ignored a dynamic central to understanding how cities are shaped and transformed by ethnic groups. "That the movement claims to remedy complex social and economic issues without serious consideration of non-mainstream populations amounts to a willful disengagement from issues of race, ethnicity and poverty" (Hall 1998: 56; Vanderbeek and Irazábal 2007). If these development projects do not attend to the specific ways in which ethnic groups create and appropriate urban spaces, their effectiveness and scope will be limited.

Latino New Urbanism (LNU) emerged in response to this lacuna by explicitly examining the ways in which ethnic identity and traditions shape the city-making practices and urban preferences and lifestyles of Latina/os. The assumptions behind the notion are that Latina/os have a culturally-driven preference for living more compactly, in multigenerational households, relying on public transportation, and socializing in public spaces – thus transforming the uses and meanings of front yards, sidewalks, streets, and parks. Rojas (1999), Méndez (2005), and Myers (2001), among others, claim that by tapping into the greater propensity of compact living and transit use of Latina/os, LNU may present a sustainable development alternative to the suburban, low-density development patterns that have been prevalent in the US for decades. This presents a window of opportunity for the preservation and expansion of "smart growth" communities in the United States at a moment when concerns regarding oil depletion, sustainability, and health are growing (see chapter by Inam). Planners and policymakers would have to be mindful, however, that this window of opportunity decreases with the expansion of the length of stay of immigrants and with newer generations (Irazábal and Farhat 2008).

Beyond the labels, it is important that urban planning and design pedagogy furthers the training of students' analytical and propositional skills to identify and advance the dimensions of ethnoscapes that contribute to the creation of more sustainable places.

Progressing toward spatial and environmental justice

Multiethnic leadership and mobilization around ethnoscape formation can prove to be crucial political strategies for the advancement of environmental justice. By combining environmental concerns and social and racial justice in a single, visible project – e.g. an urban farm – leaders may have greater mobilizing and interethnic, interclass coalitional potential than Smart Growth, New Urbanist, or New Regionalist advocates, whose concerns have tended to be more middle-class and environmentally driven. More critically, while white middle-class suburbanites should expand inclusion of non-white lower-income urbanites in smart growth or regionalist coalitions, they should also understand and respond to the latter's own initiated calls for solidarity and leadership toward different conceptions of fairer regional growth models. Notably, in the case of the South Central Farm, it was the SC Farmers, representing non-white lower-income urbanites, who were heading the regional, and even international, call for environmental justice. The case signaled the growth of ethnic – in this case Latina/o – environmental justice activism, a movement identified as a "greening" of ethnic/ Latina/o politics, or inversely a "browning" of environmentalism (George 2006; Irazábal and Punja 2009).

As these processes *take place* in and through space, the greening of ethnoscapes *and* the ethnoscaping of environmentalism constitute frontiers of development for urban design and planning in the ongoing quest for the just city. Along with their specific technical training, students of urban planning and design should be given persistent theoretical and practical opportunities to develop a sense of ethical professional responsibility to safeguard and promote spatial and environmental justice for all ethnoracial groups.

These critical challenges and opportunities for urban design, planning, and development arise from the increased saliency of ethnicity as a determinant marker in the urban world. Tackling these challenges should prompt us to probe the politics and policy of how we think about the practice and pedagogy of urban design, including effective design strategies for the creation of culturally-sensitive and adaptable public spaces, the retrofitting of our cities, and the progression toward spatial justice.

References

Aguilar San Juan, K. (2005). "Staying Vietnamese: Community and Place in Orange County and Boston," *City and Community*, 4: 37–66.

Appadurai, A. (1990). "Disjuncture and Difference in the Global Cultural Economy," *Public Culture*, 2(2), pp. 1–24.

—— (1991). "Global Ethnoscapes: Notes and Queries for a Transnational Anthropology," in Fox, R.G. (Ed.) *Interventions: Anthropologies of the Present*, Santa Fe, NM: School of American Research, 191–210.

Arreola, D. (2002). *Tejano South Texas: A Mexican American Cultural Province*, Austin: University of Texas Press.

Arreola, D. (Ed.) (2004). *Hispanic Spaces, Latino Places: Community and Cultural Diversity in Contemporary America*, Austin, TX: University of Texas Press.

Boyer, C. (1992). "Cities for Sale: Merchandising History at South Street Seaport," in Sorkin, M. (Ed.) *Variations on a Theme Park: The New American City and the End of Public Space*, New York: Hill and Wang.

Camarillo, A. (1979). *Chicanos in a Changing Society: From Mexican Pueblos to American Barrios in Santa Barbara and Southern California*, Cambridge, MA: Harvard University Press.

Dávila, A. (2004). "Empowered Culture? New York City's Empowerment Zone and the Selling of El Barrio," *Annals of the American Academy of Political and Social Science*, 594: 49–64.

Davis, D. (1987). "Late Postmodern: The End of Style," *Art in America*, 21.

Del Rio, V. (1992). "Urban Design and Conflicting City Images of Brazil," *Special Series on Urban Design*. Cities. Butterworth-Heinemann Ltd. November.

Frausto, T.Y. (1999). "El Movimiento: The Chicano Social Project since 1960." In Leclerc, G., Villa, R. and Dear, M. (Eds.) *La Vida Latina en LA: Urban Latino Cultures*, Thousand Oaks, CA: Sage, 23–34.

George, E. (2006). "Browning the Green Movement," *Los Angeles Alternative*, September 15.

Godfrey, B. (2004). "Barrio under Siege: Latino Sense of Place in San Francisco, California," In Arreola, D. (Ed.) *Hispanic Spaces, Latino Places: Community and Cultural Diversity in Contemporary America*, Austin, TX: University of Texas Press, 79–102,

Gonzalez, R. (1999). "Learning from East LA," In Leclerc, G., Villa, R. and Dear, M. (Eds.) *La Vida Latina en LA: Urban Latino Cultures*, Thousand Oaks: Sage, 185–198.

Hall, D. (1998). "Community in the New Urbanism: Design Vision and Symbolic Crusade," *Traditional Dwellings and Settlements Review*, 9: 2

Harvey, D. (1990). *The Condition of Postmodernity: An Enquiry into the Origins of Cultural Change*, Cambridge, MA: Blackwell.

Hobsbawm, E. and Ranger, T. (Eds.) (1983). *The Invention of Tradition*, New York, Cambridge University Press.

Holloway, J. (2006). "Enchanted Spaces: The Séance, Affect, and Geographies of Religion," *Annals of the Association of American Geographers*, 96(1), pp. 182–187.

Holston, J. (1989). *The Modernist City: An Anthropological Critique of Brasilia*, Chicago, University of Chicago.

Holston, J. and Appadurai, A. (1996)."Cities and Citizenship," *Public Culture*, 8: 187–204.

Irazábal, C. (2005). *City Making and Urban Governance in the Americas: Curitiba and Portland*, Aldershot: Ashgate.

Irazábal, C. (Ed.) (2008). *Ordinary Places, Extraordinary Events: Citizenship, Democracy, and Public Space in Latin America*, London: Routledge.

Irazábal, C. and Gómez-Barris, M. (2007). "Bounded Tourism: Immigrant Politics, Consumption, and Traditions at Plaza Mexico," *Journal of Tourism and Cultural Change*, 5(3): 186–213.

Irazábal, C. and Farhat, R. (2008). "Latino Communities in the United States: Place-Making in the Pre-World War II, Post-World War II, and Contemporary City," *Journal of Planning Literature*, 22(3): 207–228.

Irazábal, C. and Punja, A. (2009) "Cultivating Just Planning and Legal Institutions: A Critical Assessment of the South Central Farm Struggle in Los Angeles." *Journal of Urban Affairs*, 31(1): 1–23.

Irazábal, C. and Dyrness, G.R. (forthcoming 2010). "Promised Land? Immigration, Religiosity, and Space in Southern California." *Space & Culture*, (issue to be decided).

Isin, E.F. (1999). "Cities and Citizenship," *Citizenship Studies*, 3(2): 165–172.

Jasper, J. (1997). *The Art of Moral Protest: Culture, Biography and Creativity in Social Movements*, Chicago: University of Chicago Press.

Kuipers, D. (2005). "Trouble in the Garden," *LA City Beat*, Available at www.lacitybeat.com. (accessed 15 December 2007).

Leclerc, G., Villa, R. and Dear, M. (Eds.). (1999). *La Vida Latina en LA: Urban Latino Cultures*, Thousand Oaks, CA: Sage.

Lin, J. (1995). "Ethnic Places, Postmodernism, and Urban Change in Houston," *The Sociological Quarterly*, 36, No. 4, 629–647.

—— (Ed.) (2002) "Studying and Understanding Ethnic Landscapes." *Journal of Architectural and Planning Research*, 19: 4.

Loukaitou-Sideris, A. (2006). "Urban Parks," In *Southern California Environmental Report Card*. Los Angeles: UCLA Institute of the Environment, http://www.ioe.ucla.edu/reportcard/article.asp?parentid = 1455. (accessed 22 June 2008).

Loukaitou-Sideris, A. and Stieglitz, O. (2002). "Children in Los Angeles Parks: A Study of Equity, Quality, and Children Satisfaction with Neighborhood Parks," *Town Planning Review*, 73(4): 467–488.

MacLeod, G. (nd). "Citizenship in the 'post-justice' city," in Geography of Leisure and Tourism Research Group, *Geographies of Citizenship*. Royal Geographical Society (with the Institute of British Geographers). http://www.exeter.ac.uk/geography/tourism/gltrg/Events/London/citizenship.html. (accessed 21 June 2007).

Main, K. (2008). "Playing Out Democracy in MacArthur Park: Spatial Struggles in the Everyday Use of Public Space," *Progressive Planning*, http://www.plannersnetwork.org/publications/2008_summer/main.html. (accessed 23 June 2007).

Makowski, S. (2003). "Alteridad, Exclusión y Ciudadanía: Notas para una reescritura del espacio public," In P. Ramírez Kuri (Ed.) *Espacio Público y Reconstrucción de Ciudadanía*, Mexico: Flacso and Miguel Angels Porrúa Grupor Editorial.

Méndez, M. (2005). "Latino New Urbanism: Building on Cultural Preferences," *Opolis*, 1(1): 33–48.

Miller, D., Miller, J. and Dyrness, G.R. (2002). "Religious Dimensions of the Immigrant Experience in Southern California." In E.J. Heikkila and R. Pizarro (Eds.) *Southern California and the World*, Westport, CT: Praeger Press, 101–131.

Miraftab, F. (2004). "Invented and Invited Spaces of Participation: Neoliberal Citizenship and Feminists' Expanded Notion of Politics," *Journal of Transnational Women's and Gender Studies*, 1(1). http://web.cortland.edu/wagadu/vol1–1toc.html (accessed December 2005).

Miraftab, F. and Wills, S. (2005). "Insurgency and Spaces of Active Citizenship: The Story of Western Cape Anti-eviction Campaign in South Africa," *Journal of Planning Education and Research*, 25(2): 200–217.

Myers, D. (2001). "Demographic Futures and a Guide to Planning. California's Latinos and the Compact City." *APA Journal*, 67(4): 383–397.

Oberle, A.P. and Arreola, D.D. (2008). "Resurgent Mexican Phoenix." *Geographical Review*, New York: 98(2): 171–197.

Radford, L. and Santos, J. (2006). "Race, Class and the Battle for South Central Farm: Seeds of Hope, Seeds of War," *Counterpunch*, Available at http://www.counterpunch.org/radford07132006.html, accessed August 13, 2006.

Rojas, J. (1999). "The Latino Use of Urban Spaces in East Los Angeles." In Leclerc, G., Villa, R. and Dear, M. (Eds.) *La Vida Latina en LA: Urban Latino Cultures*, Thousand Oaks: Sage, 131–138.

Rose, N. (2000). "Governing Cities, Governing Citizens," in Isin E.F. (Ed.) *Democracy, Citizenship and the Global City*, London: Routledge.

Sánchez, F. (1996). "O City Marketing de Curitiba: Cultura e Comunicação na Construção da Imagem Urbana," in Del Rio, V. and de Oliveira, L. (Eds.) *Percepção Ambiental: A Experiencia Brasileira*, São Paulo: Studio Nobel/Editora da UFSCar.

Smith, H.A. and Furuseth, O.J. (2004). "Housing, Hispanics, and Transitioning Geographies in Charlotte," *North Carolina Southeastern Geographer*, 44(2): 216–235.

Smith, J.S. (2002). "Rural Place Attachment in Hispano Urban Centers," *Geographical Review*, 92(3): 432–451.

—— (2004). "The Plaza in Las Vegas, New Mexico: A Community Gathering Place," In. Arreola, D. (Ed.) *Hispanic Spaces, Latino Places: Community and Cultural Diversity in Contemporary America*, Austin, TX: University of Texas Press, 39–53.

Vanderbeek, M. and Irazábal, C. (2007). "Urban Design as a Catalyst for Social Change: A Comparative Look at Modernism and New Urbanism," *Traditional Dwellings and Settlements Review*, 19(1): 41–57.

Villa, R.H. (2000). *Barrio-Logos: Space and Place in Urban Chicano Literature and Culture,* Austin: University of Texas Press.

Winocur, R. (2003). "La Invención Mediática de la Ciudadanía," in Ramírez Kuri, P. (Ed.) *Espacio Público y Reconstrucción de Ciudadanía.* Mexico: Flacso and Miguel Angels Porrúa Grupor Editorial.

Further reading

Abrahamson, M. (1995). *Urban Enclaves: Identity and Place in the World*, New York: Worth Publishers. Discusses the formation of urban enclaves in many cities in the US and around the world, critically examining processes of assimilation, transnationalism, inclusiveness, and border-making.

Appadurai, A. (1991). "Global Ethnoscapes: Notes and Queries for a Transnational Anthropology." In R.T.G. Fox (Ed.) *Interventions: Anthropologies of the Present*, Santa Fe: School of American Research, 191–210. Calls for a transdisciplinary approach to the study of diasporic public spheres and the new global, cultural processes where diacritic identities are constantly mobilized to reflect shifting and negotiated identities.

Martín Alcoff, L. and Mendieta, E. (Eds.) (2003). *Identities: Race, Class, Gender, and Nationality*, Malden, MA: Wiley-Blackwell. A reader in the theoretical sources of contemporary thinking about identity, including explorations of race, class, gender, nationality, difference, power, and construction of the individual and social self.

Sandercock, L. (Ed.) (1998). *Making the Invisible Visible: A Multicultural Planning History*, Berkeley and Los Angeles: University of California Press. A scrutiny of the class, race, gender, ethnic, or other biased agendas previously hidden in planning histories and proposals for new planning paradigms for multicultural cities.

43

Urban design for a planet of informal cities

Vinit Mukhija

For the first time in human history we live in a primarily urban world. More than half of us live in urban areas, and commentators have begun referring to the dawn of the *Century of the City* (Pierce and Johnson 2008). While urban designers have also celebrated this epochal transformation (see for example Brown *et al.* 2009), they may not have noticed that this urban revolution has come at an unprecedented human and social cost. Close to a third of all urban residents – almost one billion people – live in slums today (UN-Habitat 2003). I refer to these often forgotten neighborhoods as the Informal City.

Urban designers, I suspect, are likely to have missed the dramatic and unprecedented rise in the growth of slums because the formal city remains urban design's essential domain of practice. The formal city is the city of wealth, power, and grandeur. It is also the city of boulevards, downtowns, public plazas, waterfronts, office parks, retail centers, transit-oriented developments, historic neighborhoods, museums, cultural centers, and theme parks. As a number of chapters in this book suggest, urban design's traditional horizons and narrow concerns have broadened, and continue to expand, to include ordinary, low-income and minority neighborhoods in many affluent countries. Urban designers are also becoming increasingly engaged in affordable housing debates and endeavors. Nonetheless, the Informal City and its neighborhoods still fail to consistently attract the attention and skills of urban designers.

In this chapter, I argue that urban designers need to pay more attention to the Informal City. If urban design aims to move beyond mere elitist concerns and be a progressive practice, it cannot afford to ignore neighborhoods that house the majority of urban residents in many parts of the world. The Informal City has numerous disadvantages and structural barriers. Urban design is neither a panacea nor a solution for such momentous problems. But it can play a small, positive role in the struggles of these neighborhoods and their residents. The Informal City can benefit from the engagement of urban designers, and urban design can renew itself as a vital and engaged social practice through such a commitment. In the next pages, I elaborate on this and discuss the opportunities and challenges for urban design. I also explain my preference for the term Informal City, and share examples of practice from around the world to illustrate the rich and varied possibilities of involvement.

The informal city

New research has emphasized the exponential growth in informality and slums around the world, and noted its coincidence with the recent decades of globally expansive, neoliberal economic policies (UN-Habitat 2003; Verma 2003). Mike Davis' (2006) trenchant and penetrating critique of neoliberalism as the defining cause for the rapid growth of slums in *Planet of Slums* is probably one of the best known works on such topic. The United Nations estimates that unless present circumstances change dramatically, either through a transformation in the current urbanization trends or through a dramatic increase in the rate of supply of affordable housing in the formal sector (both of which seem unrealistic and unlikely), there could be almost *two billion* slum-residents by about 2030–2035 (UN-Habitat 2003: xxv).

The name slum, however, is controversial and problematic. It is prejudiced and evokes popular descriptions of a disorderly, disorganized, and crime-ridden place. Moreover, there is no consensus on what constitutes a slum, and this makes the UN's data and projections somewhat suspect. The UN, like many other public agencies, assumes that insecure tenure, substandard housing – poor structural quality and overcrowding – and poor infrastructure conditions – inadequate access to safe water and sanitation – define a slum (UN-Habitat 2003). It is, however, difficult to operationalize these attributes in a consistent manner for cities across the world. But it is also clear that the housing and living conditions of an increasing number and proportion of urban residents in developing countries need upgrading. These housing conditions span a wide range of situations, and include central city tenements, squatter settlements, and illegal subdivisions on the urban periphery.

There is no consensus regarding the ideal lexicon for these disparate housing conditions, but many urban planning scholars and practitioners try to avoid using the terminology of slums (Gilbert 2007; Neuwirth 2005). They consider the label of slums pejorative and worry that its use encourages the policy of slum clearance (Gans 1962). I share such concerns and in the context of urban design, prefer the term Informal City. The name Informal City, like other alternatives to the terminology of slums, is problematic too but offers some useful advantages also. Its problems include the fuzziness or difficulty in defining informality. Keith Hart (1973) is credited with coining the term "informal sector" and distinguishing it from the formal, rational, and modern economy that followed state-mandated regulations and procedures. Most of the early literature approaches informality through a lens of duality and defines economic activities, and spatial settlements, as formal or informal based on whether or not they follow state regulations and laws.

Subsequent and contrarian literature challenges the duality and suggests that there is a continuum, overlap, and linkage between the formal and informal spheres (Laguerre 1994; Peattie 1987; Roy 2005). For urban design, the interpretation that informality is not a distinct sector but is socially constructed and has its own organizing rules and norms is more promising. The presence of underlying rules in the Informal City suggests that they can be leveraged for design interventions. Moreover, the moniker City indicates a permanence and resilience that urban designers appreciate, and applies to most slums. For example, the favelas in Rio de Janiero, Brazil, like the *Morro da Providência*, have been around since the late nineteenth century (Fabricus 2008). Research also indicates that in socio-demographic terms slums are strikingly similar to the rest of the city (2001 Census of India data reported in Sivaramakrishnan *et al.* 2005: 158). Finally, as Lisa Peattie (1987) notes, recognizing informality gives a standing to previously ignored activities. In contrast, urban designers have the shameful legacy of ignoring the informal economy while designing new cities like Brasilia (Holston 1989) and Chandigarh (Sarin 1982).

Urban design and the poor

Urban designers' less than successful past forays to improve the housing and living conditions of the poor may account for some of their reluctance to get actively engaged in the Informal City. The interest of modernist urban designers and architects in improving poor housing conditions by eliminating urban slums is well documented. Their controversial perspectives on housing problems and their solutions have been clearly stated and celebrated in various manifestos. Their endeavors and projects have also been critically analyzed, often as naïve and reckless failures (Frampton 1985). The lifeless landscapes of many American cities remind us poignantly of the disasters of modernism-inspired, urban renewal policies. Such misguided, large-scale redevelopment projects were championed and implemented by architects and urban designers across the world. Although real-estate greed, motivated by the central city location of many urban slums, played a pivotal role in urban renewal and slum clearance, these projects were also driven by the modernist credo in the great power of design and physical determinism (Gans 1962; Jacobs 1961).

Their victims were more than the built environment of cities. Slums despite all their disadvantages provide the poor with an affordable housing option (Peattie 1994). We now also know that ambitious slum clearance projects almost invariably reduced the total stock of available housing, not just the supply of affordable housing. It is far easier to clear and demolish substandard housing than to build new, decent, and appropriately located units. Consequently, urban renewal and slum clearance inevitably displaced the former slum-dwellers and made their living conditions worse. In the few instances where affordable housing was built, say in the form of public housing projects, critics charged that they were misconceived and poorly designed (Turner 1967; 1977). The apparent failure of public housing seems to be captured in the moving images of the dynamiting of Pruitt Igoe in St. Louis, Missouri – a former public housing project completed in 1955 and demolished in 1972. Although the story of Pruitt Igoe and public housing is infinitely more complicated (Bristol 2004), the project's demolition helped shatter the optimism and faith in the promises of urban design and architecture to improve the lives of the poor. As Peter Hall (1995: 234) devastatingly noted, "What emerges ... is that it might actually have been better to leave the poor alone."

To some extent, as a result of the failed grandiose experiments of urban renewal, slum clearance, and public housing, the orthodox policy response to slums and informal settlements is either a hands-off strategy of "do no damage," or a property rights-based, tenure legalization approach based on minimal government intervention. The conventional wisdom of tenure legalization as the preferred slum upgrading policy owes its intellectual legacy to the pioneering work of the anarchist architect John Turner in the 1960s and 1970s (Turner 1972), and a more recent campaigner, the free-market economist Hernando de Soto (1989; 2000). Urban design, architecture, or urban planning have almost no role in their visions of improving the Informal City. But the private property rights-based strategy has also been criticized for being simplistic (Payne 2001). Its limited success in upgrading indicates that legalization is an inadequate response, and suggests that nuanced and deliberate design and policy responses are necessary.

In spite of the tremendous shortcomings, modernism's rich legacy also suggests some potential directions for future urban design and planning endeavors. Two examples are particularly illustrative. First, is the seminal work of N. John Habraken (1972), who emphasized the design and provision of shared infrastructure. He drew from the modernist belief in public but pluralist infrastructure and proposed an open infrastructure approach as a flexible matrix in which users have opportunities to adapt and build according to their specific needs. Second, is the scholarly work of Gwendolyn Wright (1994), who has helped recover and reaffirm some of the less acknowledged strands within early modernism's design interventions for housing. These include the willingness to pursue collaboration and compromise, the emphasis on communities rather than private dwellings, and the readiness to forego authoritative excesses. Such approaches and qualities are likely to serve twenty-first-century urban designers well in the Informal City.

Urban design and the informal city

Although most of the development planning discourse focuses on the institutional, legal, and financial aspects of slum upgrading, a closer reading of the improvement efforts around the world indicates that urban design skills, if not urban designers per se, are being regularly employed and demanded in the Informal City. A majority of these engagements are on the fringes of urban design practice, and are rarely discussed or fêted in design forums. Together, however, they are a potent depiction of the practical role of urban design. For the ease of illustrating, I have divided these engagements into five separate but interrelated categories.

Small interventions

The value of simple, place-based interventions like pocket parks, mini-plazas with safe seating areas, and landscaped flowerbeds in enhancing the quality of life is often ignored or underestimated. Yet a walk through any informal settlement is likely to show that individual residents are constantly investing time and resources in making small improvements to their private dwellings. These modest enhancements are often as simple as a painted doorway, a decorated threshold, or a single flowerpot next to a window. They contribute to the beauty and identity of both the private and public realms. Given the relative permanence of the Informal City, there is also a need for similar concerted improvements to the everyday life of its public realm. These improvements could include design interventions to create small communal facilities and amenities. Such humble contributions might be the perfect place for urban designers to start getting acquainted with the Informal City. A noteworthy example of this approach is the work of the Kounkuey Design Initiative (KDI) in Kibera, Nairobi, one of Africa's densest settlements (Gendall 2008). KDI is working with the community to create small pocket parks, envisioned as productive public spaces that can also be used for income generation activities, like compost farms.

KDI's designs, and other similar strategies, recognize that informal settlements often form along creeks, streams, riverbeds, etc. Their proposals try to carve out usable spaces along the embankments (Figures 43.1 and 43.2). In addition to creating a public amenity, these design strategies also contribute to safer and healthier communities by better articulating the interface between communities and adjacent water resources. Perhaps the most impressive scaling up of this strategy is the Favela Bairro – slum-to-neighborhood – program in Rio de Janeiro, Brazil (Petersen 2008). This is a holistic upgrading program, which has as one of its key objectives the creation of modest, shared spaces. Projects often include wide sidewalks, seating alcoves, mini-plazas, and undersupplied, communal amenities like day care centers.

Figure 43.1 Diagram of open space proposal in Kibera, Nairobi, Kenya. Source: Kounkuey Design Initiative – used by permission.

Figure 43.2 Usable open space in Kibera, Nairobi, Kenya. Source: Kounkuey Design Initiative – used by permission.

The Informal City can be extremely dense and making room for even small interventions can be a big design challenge. It requires using leftover spaces more efficiently and creatively. It is also likely to require places and amenities that are considerably smaller than what we are conventionally accustomed to. Low maintenance costs – for example, using more hardscapes than softscapes – are imperative. Additional small interventions can help augment the infrastructure of the Informal City, and the next subsection elaborates on such possibilities.

Infrastructure upgrading

Infrastructure is inadequate and needed in the Informal City. The existing high density of many neighborhoods can make this challenging. Thus, it is necessary that urban designers creatively make the public infrastructure more than a mere utilitarian use. For example, communities struggle to create the room for shared water tanks, like the Awami (people's) water tanks in Orangi, Karachi, Pakistan (Ahmed and Sohail 2003). These tanks and the spaces around them can be used to create public gathering areas. Similarly, many informal settlements in Mumbai lack adequate toilet facilities. Non-profit and civil society groups like the Society for Promotion of Area Resource Centres (SPARC) and the National Slum Dwellers Federation (NSDF) have been working with local communities to create toilet-blocks that also include community facilities (Burra *et al.* 2003).

Often providing infrastructure in the Informal City involves "reblocking," or rearranging the street layout from an irregular pattern to a more regular layout. But the importance of this exercise in infrastructure upgrading is rarely given its due in the development planning literature. Readers of this chapter would recognize rearrangement of the street network, and the existing houses and businesses, as quintessentially an urban design exercise. Re-blocking the streets, however, can be a slippery slope. After all, Baron Haussmann's dramatic re-blocking of Paris tore apart the social fabric and physical structure of existing neighborhoods. In contrast to following the Baron, urban designers should pay heed to Patrick Geddes' call for "conservative surgery" (Hall 1995; Tyrwhitt 1947). Geddes, while working in India about a century ago, recognized the need to tread lightly in improving the infrastructure of dense, fine-grained, often irregular, old cities. He argued that engineers had to change their infrastructure standards. For instance, access for cars may not be the priority in many dense neighborhoods. The appropriate standards are likely to vary significantly from neighborhood to neighborhood, and determining what is feasible or desirable will inevitably require urban designers to work closely with communities.

Urban designers' ability to comprehend the physical morphology of settlements, nonetheless, is an essential skill in slum upgrading (Mukhija 2001). For example, the renowned Slum Networking Project in the Indian cities of Ahmedabad, Baroda, and Indore leverages the undulating terrain and floodplains of settlements to structure a low cost, gravity-based, sewer network (Parikh and Parikh 2009). New GIS technologies can also be used to better understand the layout and topography of Informal City neighborhoods (Joshi *et al.* 2002).

Integration of the informal and formal city

Infrastructure provision allows for the gradual integration of informal and formal neighborhoods in a city (Durand-Lasserve and Clerc 1996). Some of the more successful slum upgrading programs, for example Indonesia's Kampung Improvement Program – KIP (Kessides 1997), are noteworthy for their ambitious infrastructure investments and attempts to facilitate relatively seamless movement across different neighborhoods. More audacious are the design proposals from Caracas, Venezuela, to build funiculars or cable cars to the tops of hillsides, where vehicular access is impossible because of steep slopes and densely built housing (Brillembourg *et al.* 2005).

But ultimately, the regularization and integration of the Informal City with other urban neighborhoods needs more than shared infrastructure. Urban design interventions at the edges can also help. Practice suggests at least two intriguing directions. First, the edges can house shared amenities as catalyst projects. In Medellin, Colombia, public libraries are being built at the interfaces of formal and informal neighborhoods (Romero 2007). Urban designers need to be cautious, though, that the Informal City is not merely used as a space for placing the institutions of the formal city. Second, design interventions might also succeed in making the edges more permeable. Through Rio's Favela Bairro program some informal neighborhoods have built gateways, which serve as identity markers helping to both define and integrate the Informal City with the formal neighborhoods (Petersen 2008).

Jobs and economic development

Although not all the urban poor live in informal neighborhoods, and not everyone who lives there is poor, economic development and more jobs are needed in the Informal City. Most of the examples discussed in the previous subsections also have the advantage of creating employment opportunities. Urban designers, nonetheless, have to find creative ways to overcome the orthodox planning belief in the undesirability of mixing land uses, particularly residential with commercial or industrial uses. Mixed land uses tend to be the norm in the Informal City, and are a necessity for economic survival (Hart 1973). We also know that women-owned businesses are more likely to be home-based enterprises (Tipple 2005). But it is a formidable design challenge to organize the mixing of land uses while minimizing their nuisances.

This is likely to be even more daunting in the dense fabric of the Informal City. Density, however, can also be an asset. Solomon Benjamin (1991) documented the clustering and opportunities to integrate trades in the dense informal neighborhoods of East Delhi and suggested the idea of "neighborhood-as-factory." John Loomis (1997) has similarly argued for a closer integration between work and housing, and suggested the concept of "manufacturing communities." Sophisticated urban design strategies can help, but ultimately our sensibilities about separating land uses need to evolve and a new kind of urbanism must emerge (Pyatok 2000).

Affordable housing

Housing developments, both inside and outside the Informal City, provide opportunities for urban designers to get involved. Affordable housing projects outside the informal neighborhoods help reduce some of the demand for growth in the Informal City. Well known, though only partially realized, projects include Hassan Fathy's New Gourna village (Fathy 1976) and Balkrishna Doshi's Aranya Township (Curtis 1988). More substantial in scale are the affordable housing projects being developed through the MetroVivienda program in Bogotá, Colombia (Gilbert 2009). It is not easy to design affordable housing and pleasant communities. For example, urban economics suggests that housing is more affordable with longer blocks (Dowall 1992), but streets in such blocks are less walkable (Jacobs 1961; Jacobs 1995). Balancing such contrasting imperatives is a classic urban design challenge.

Numerous formal housing development projects constantly take place within the Informal City as well, and integrating them with the existing urban fabric can be a formidable design problem. Many projects include the demolition and redevelopment of substandard housing into higher-standard, replacement housing for the original residents. Two of the best known programs of this approach are in Bangkok (Angel *et al.* 1983) and Mumbai (Mukhija 2003). Their designs, however, have rarely been noteworthy. A recent on-site resettlement program that has gained much positive attention is the Quinta Monroy housing project in Chile (Architecture for Humanity 2006), designed by the Elemental Design Team as a "parallel building" arrangement. Lots are arranged around communal courtyards, and 2 families share each lot but have individual units that can expand both horizontally and vertically. It is an innovative design that maintains a community orientation and allows for both high density and incremental expansion of the dwellings.

A frequent challenge for housing projects in the Informal City is adding more density to already dense neighborhoods. Conventional sensibilities towards density need to change, but we also need urban design innovations that can accommodate higher density with a superior quality of life. In contrast to most new neighborhoods of formal cities, densities are almost invariably increasing in the Informal City. Take the case of Cairo, where around two million additional residents live on the rooftops of the existing dense urban fabric (Shaath and Kamel 2004). Or consider Mumbai, where controversial proposals for redeveloping Dharavi – one of the densest places in the world – at a higher density are being floated. Most civil society groups opposing the proposals are not completely against additional density. They want some limits to the extra density, but are demanding urban design guidelines too (Patel *et al.* 2009). They also insist that Dharavi's residents and workers participate in the development of the design guidelines. Urban designers should be up to such challenges.

Conclusion

Our world is dramatically changing. The global share of the urban population has increased from almost 13 percent in 1900 to around 50 percent today. Population trends indicate that the world's rural population has more or less leveled-off, and we can even expect some small declines in the near future. The world, however, will continue to rapidly urbanize in the twenty-first century. Some estimates suggest that a greater proportion of this growth, more than a half, will take place in informal settlements and slums, likely characterized by substandard housing and living conditions with inadequate infrastructure (UN-Habitat 2003). Nonetheless, the Informal City will continue to be extremely diverse and vital. It will provide a rich variety of challenges and opportunities for urban design interventions.

Urban designers can play a substantive role in making informal neighborhoods livable, improving their infrastructure, integrating them with other parts of the city, and creating better work and housing conditions. To achieve successful outcomes, urban designers will have to develop new skills and attitudes. They will also have to be adept in participatory design, where urban designers do not design but educate, engage, and offer technical assistance. Getting actively involved in the endeavors of the Informal City and experientially learning-by-doing may be the best and only way to build up the necessary knowledge and craft of the discipline.

Urban designers must be cautious. They need to be cognizant of past failures and aware of the perils of imprudently overreaching. Urban design history clearly indicates that design cannot be the panacea. The complexity of the problems in the Informal City cannot be trivialized. It is worth reiterating that substantive progress in the growth and conditions of informal settlements and slums will require dramatic improvements in economic development and poverty alleviation approaches, radical transformations in the markets for finance and land, and far reaching institutional innovations. It is likely that conventional institutions and expectations will have to evolve and transform to match and respond to the needs of the Informal City, and not the other way around. The same is likely to be true for orthodox urban design ideas and practices.

The challenges that urban designers are likely to face in the Informal City are overwhelming and foreboding. Nonetheless, there is no credible excuse to stand on the sidelines and not get involved. If this involvement can be modest but meaningful, the practice and ethics of urban design will be dramatically transformed. It is likely to emerge as a more progressive, participatory, democratic and socially-vital practice. Urban designers will learn and gain as much from their association with these communities, as they will contribute.

References

Ahmed, N. and Sohail, M. (2003). "Alternate Water Supply Arrangements in Peri-Urban Localities: Awami (people's) Tanks in Orangi Townships, Karachi," *Environment and Urbanization*, 15(2): 33–42.

Angel, S., Archer, R., Tauphiphat, S. and Wegelin, E. (1983). *Land for Housing the Poor*, Bangkok: Select Books.

Architecture for Humanity (2006). *Design Like you Give a Damn: Architectural Responses to Humanitarian Crises*, New York: Metropolis Books.

Benjamin, S. (1991). *Jobs, Land and Urban Development*, Cambridge, MA: Lincoln Institute of Land Policy.

Brillembourg, A., Feireiss, K. and Klumpner, H. (2005). *Informal City: Caracas Case*, Munich; New York: Prestel.

Bristol, K.G. (2004). "The Pruitt-Igoe Myth," in K. Eggener (ed.) *American Architectural History: A Contemporary Reader*, pp. 352–364, New York: Routledge.

Brown, L.J., Dixon D. and Gillham O. (2009). *Urban Design for an Urban Century: Placemaking for People*, Hoboken, NJ: Wiley.

Burra, S., Patel, S. and Kerr, T. (2003). "Community-Designed, Built and Managed Toilet Blocks in Indian Cities," *Environment and Urbanization*, 15(2): 11–32.

Curtis, W. (1988). *Balkrishna Doshi: An Architecture for India*, New York: Rizzoli.

Davis, M. (2006). *Planet of Slums*, New York: Verso.

De Soto, H. (1989). *The Other Path: The Invisible Revolution in the Third World*, New York: Harper and Row.

—— (2000). *The Mystery of Capital: Why Capitalism Triumphs in the West and Fails Everywhere Else*, New York: Basic Books.

Dowall, D. (1992). "The Benefits of Minimal Land Development Regulation," *Habitat International*, 16(4): 15–26.

Durand-Lasserve, A. and Clerc, V. (1996). *Regularization and Integration of Irregular Settlements: Lessons from Experience* (Working Paper No. 6), Nairobi: Urban Management Programme.

Fabricus, D. (2008). "Resisting Representation: The Informal Geographies of Rio de Janeiro," *Harvard Design Magazine*, 28(1): 4–17.

Fathy, H. (1976). *Architecture for the Poor*, Chicago: University of Chicago Press.

Frampton, K. (1985). *Modern Architecture: A Critical History*, London: Thames and Hudson.

Gans, H. (1962). *The Urban Villagers: Group and Class in the life of Italian-Americans*, New York: Free Press of Glencoe.

Gendall, J. (2008). "Kibera Public Space Project by Kounkey Design Initiative: Co-Designing Productive Parks with the Poorest of Kibera, Kenya," 28(1): 67–69.

Gilbert, A. (2007). "The Return of the Slum: Does Language Matter?" *International Journal of Urban and Regional Research*, 31(4): 697–713.

—— (2009). "The Rise (and Fall?) of a State Land Bank," *Habitat International*, 33(4): 425–435.

Habraken, N.J. (1972). *Supports: An Alternative to Mass Housing*, London: Architectural Press.

Hall, P. (1995). *Cities of Tomorrow: An Intellectual History of Urban Planning and Design in the Twentieth Century*, Oxford: Blackwell.

Hart, K. (1973). "Informal Income Opportunities and Urban Employment in Ghana," *Journal of Modern African Studies*, 11(3): 61–89.

Holston, J. (1989). *The Modernist City: An Anthropological Critique of Brasilia*, Chicago: University of Chicago Press.

Jacobs, A. (1995). *Great Streets*, Cambridge, MA: MIT Press.

Jacobs, J. (1961). *The Death and Life of Great American Cities*, New York: Random House.

Joshi, P., Sen, S. and Hobson, J. (2002). "Experiences with Surveying and Mapping Pune and Sangli Slums on a Geographical Information System (GIS)," *Environment and Urbanization*, 14(2): 225–240.

Kessides, C. (1997). *World Bank Experience with the Provision of Infrastructure Services for the Urban Poor*, Washington, D.C.: World Bank.

Laguerre, M. (1994). *The Informal City*, New York: St. Martin's Press.

Loomis, J. (1997). "Manufacturing Communities," *Places*, 10(1): 48–57.

Mukhija, V. (2001). "Upgrading Housing Settlements in Developing Countries: The Impact of Existing Physical Conditions," *Cities*, 18(4): 213–222.

—— (2003). *Squatters as Developers? Slum Redevelopment in Mumbai*. Burlington, VT: Ashgate.

Neuwirth, R. (2005). *Shadow Cities: A Billion Squatters: A New Urban World*, New York: Routledge.

Parikh, H. and Parikh, P. (2009). "Slum Networking? A Paradigm Shift to Transcend Poverty with Water, Environmental Sanitation and Hidden Resources," in Feyen, J., Shannon, K. and Neville, M. (Eds.) *Water and Urban Development Paradigms: Towards an Integration of Engineering, Design and Management Approaches*, Boca Raton, FL; London: CRC Press, 357–370.

Patel, S., Arputham, J. Burra, S. and Savchuk K. (2009). "Getting the Information Base for Dharavi's Redevelopment," *Environment and Urbanization*, 21(1): 241–251.

Payne, G. (2001). "Urban Land Tenure Policy Options: Titles or Rights?" *Habitat International*, 25(3): 415–429.

Peattie, L. (1987). "An Idea in Good Currency and How it Grew: The Informal Sector," *World Development*, 15(7): 851–860.

—— (1994). "An Argument for Slums," *Journal of Planning Education and Research*, 13(2): 136–143.

Petersen, M.L. (2008). "Interventions for the Socio-Urban Integration of the Favelas of Rio de Janeiro," *Harvard Design Magazine*, 28(1): 50–57.

Pierce, N. and Johnston, C.W. (2008). *Century of the City: No Time to Lose*, New York: Rockefeller Foundation.

Pyatok, M. (2000). "Martha Stewart vs. Studs Terkel," *Places*, 13(1): 40–43.

Romero, S. (2007). "Medellin's Nonconformist Mayor Turns Blight into Beauty," *New York Times*, July 15.

Roy, A. (2005). "Urban Informality: Toward an Epistemology of Planning," *Journal of the American Planning Association*, 71(2): 147–158.

Sarin, M. (1982). *Urban Planning in the Third World: The Chandigarh Experience*, London: Mansell.

Shaath, R. and Kamel, N. (2004). *Randa Shaath: Under the Same Sky, Cairo*, Rotterdam: Witte de With; Barcelona: Fundació Antoni Tapies.

Sivaramakrishnan, K.C., Kundu, A. and Singh, B.N. (2005). *Handbook of Urbanization in India: An Analysis of Trends and Processes*, New Delhi: Oxford University Press.

Tipple, G. (2005). "The Place of Home-Based Enterprises in the Informal Sector: Evidence from Cochabamba, New Delhi, Surabaya and Pretoria," *Urban Studies*, 42(4): 611–632.

Turner, J. (1967). "Barriers and Channels for Housing Development in Modernizing Countries," *Journal of the American Institute of Planners*, 34(6): 354–363.

—— (1972). "Housing as a Verb," in J. Turner and R. Fichter (eds.) *Freedom to Build: Dweller Control of the Housing Process*, pp. 148–175, New York: Macmillan Company.

—— (1977). *Housing by People: Towards Autonomy in Building Environments*, New York: Pantheon Books.

Tyrwhitt, J. (1947). *Patrick Geddes in India*, London: Lund Humphries.

UN-Habitat (2003). *The Challenge of Slums: Global Report on Human Settlements 2003*, Earthscan (for United Nations Human Settlements Programme): London and Sterling, VA.

Verma, G.D. (2003). *Slumming India: A Chronicle of Slums and their Saviours*, Penguin: New Delhi.

Wright, G. (1994). "Inventions and Interventions: American Urban Design in the Twentieth Century," in Ferguson, R. (Ed.) *Urban Revisions: Current Projects for the Public Realm*, pp. 26–37, Los Angeles: Museum of Contemporary Art.

Further reading

Serageldin, I. (1997). *The Architecture of Empowerment: People, Shelter and Livable Cities*, London: Academy Edition. Essays and case studies on how architects can enable the poor.

Turner, J. and Fichter, R. (1972). *Freedom to Build: Dweller Control of the Housing Process*, New York: Macmillan. Papers on the importance of flexible and self-help approaches in squatter settlements and illegal subdivisions.

UN Millennium Project (2005). *A Home in the City*, Task Force on Improving the Lives of Slum Dwellers, London and Sterling, VA: Earthscan. A comprehensive discussion of planning interventions for providing adequate shelter and improved infrastructure in slums.

Part 9

New directions

Introduction

In this section we will consider the questions: Which are the important new directions that are influencing urban design research and practice? How has urban design responded to the contemporary challenges of climate change, depletion of energy resources, population growth and expanding urbanization, need for safety and security, and protection from natural and human disasters? The different directions and trends discussed in Part 9 are generated by particular understandings about the nature of problems attended by urban design, and the ways they should be addressed. They are also responses to specific challenges faced by society in the early twenty-first century.

For one, the chapters that follow reveal an increasing realization that urban design problems cannot be addressed in isolation of their social, economic, and physical consequences. The complex – some have called "wicked" – problems that urban designers are facing require a synergy of actions, utilization of knowledge from different disciplines, and consideration of issues at different scales. This leads to calls and efforts for an "integrated" approach to urban design.

Second, there is an increasing understanding of the growing imperative to incorporate ecological thinking into design action, decrease the negative externalities generated by urban environments, deal with pollution, reduce the depletion of natural resources, and minimize the urban footprint by reducing sprawl. Ultimately, the calls for ecological urbanism, landscape urbanism, sustainable urbanism or smart urbanism are calls for design "with" instead of "against" nature.

Third, the trends discussed in the chapters that follow range from small-scale improvements of the "ground-floor" of cities to calls for metropolitan design. There is, however, an interrelationship of urban environments of different scales and a need for urban design interventions which are sensitive to more than one scale. Urban design is asked to give attention to the interdependencies of the lot, the block, the neighborhood, the city, the metropolitan, and even the regional context. The house and the lot represent the smallest units of design. They are nested within the block, which in turn joins with similar units to compose the neighborhood. Cities represent assemblies of neighborhoods, and regions contain constellations of cities. Design actions that take place at one scale should be cognizant of their impacts on other scales.

Fourth, some new directions are also defined by the understanding that urban design should privilege the users and their needs. The notion that urban design should be community-oriented and place-driven situates the discipline in a physical and social context, where interventions are guided by the particularities of the place and the desires of the community.

Finally, some of the trends are the response of urban design to particular problems that have emerged or become accentuated in recent times. The unfettered urbanization of the last century has often resulted in loss of habitat, congestion, pollution, and climate change. The increasing emphasis placed on sustainable urban design and smart growth is in direct response to such issues. Cities and their citizens have also been challenged by natural and human-made disasters, and some groups have emerged as more vulnerable than others. Urban design efforts for the creation of "safe" or "resilient" cities suggest possibilities and ways to respond to such threats.

In the first chapter of this section, Nan Ellin argues that a variety of design initiatives in the last decade are restoring the well-being of contemporary cities by demonstrating the qualities of hybridity, connectivity, porosity, authenticity, and vulnerability. While such initiatives vary, they are loosely classified under the rubric of "Integral Urbanism." Ellin views this as a new model which seeks to integrate buildings with nature, center with periphery, local contexts with global forces, and draws from different disciplines to address diverse and multi-cultural social contexts.

The next three chapters integrate the theories of landscape urbanism, ecological urbanism, and sustainability into urban design. For example, Anne Whiston Spirn introduces the notions that cities are first, parts of the natural world instead of antithetical to it; second, habitats for both human and animal species; and third, parts of dynamic and connected ecosystems. She articulates specific urban design implications that such notions entail. Brenda Scheer focuses urban design's attention to the metropolitan scale. The different urban districts, the systems of movement and infrastructure, the network of open spaces are all components of the metropolitan landscape. Nevertheless, Scheer argues that traditional urban design, which views urban form as architectural, is not well equipped to intervene at the metropolitan scale. In contrast, urban design can draw from landscape ecology to emphasize and recover existing natural systems and privilege them as the "shapers of the metropolitan image." Randy Hester and Marcia McNally describe two generations of urban design thought in the context of sustainability. They argue that urban designers have long advocated for principles that contribute to sustainable cities such as legibility of urban form, livable density, walkability, mixing of land uses, adaptive reuse of buildings, etc. More recently, ecological thinking has introduced new mandates to the practice of urban design: the responsibility to build and rebuild cities in ecological niches, to consider the footprint of urban interventions, and to reduce dependence on nonrenewable sources of energy.

The next two chapters discuss two urban design strategies – smart growth and transit-oriented development – that aspire to contribute to more compact and less automobile-dependent development patterns. Aseem Inam explains that smart growth is both a political discourse and a set of urban design strategies, which may have different types of effects and different degrees of effectiveness, depending on the mix of measures, and the particular designs, policies, and contexts. For Stefanos Polyzoides, concentrating development around transit is a basic premise of sustainable urbanism. However, different prerequisites need to be in place for transit-oriented development to be successful: a physical vision, shared with the neighborhood, which includes catalytic projects, a sound development strategy, form-based codes that establish building and open space standards, and an implementation framework which delineates the responsibilities of the public and private sectors.

In the next chapter, Kathy Madden calls attention to the small public spaces of everyday life and the elements that make them successful. Drawing from and extending theories first developed by William H. Whyte, she articulates *placemaking* as a design approach which is place- and user-oriented and treats a neighborhood's stakeholders as the experts. In this approach, urban designers provide technical support and information to communities helping them to reach a common vision for the improvement of their public spaces.

The importance of the resident as an expert in achieving safer environments through design and policy is also stressed by Carolyn Whitzman in her chapter about Secure Cities. She juxtaposes crime prevention through environmental design techniques, that at times have resulted to privatization and exclusivity, with safety audits and design guidelines that include consultation with users, particularly the most vulnerable ones, and recognition of their needs and perspectives. She also urges for a holistic understanding of secure cities, in which good public space design is one component but other strategies including economic development and empowerment should also be in place.

In light of recent major catastrophes that some cities have experienced because of natural and human-made disasters, design for resiliency has emerged as an important concern. In the last chapter of this section Mahyar Arefi broadens the concept of resiliency beyond the context of hazard mitigation, arguing that resilient urban form is comprised of components which can adapt to new conditions. Urban design can help by identifying liabilities, transforming them into assets, and creating flexible city forms, functions, and flows.

44

Postmodern and integral urbanism

Nan Ellin

From tightly-woven hubs of activity, much of our urban fabric frayed over the last century in the US, separating homes from workplaces, commerce, leisure and recreational opportunities, and so forth. The result was urban fragmentation, environmental degradation, social isolation, and the widely-decried loss of a "sense of place" and "sense of community." While numerous efforts to address these issues during the 1970s and 1980s only exacerbated them, the tide turned over the last two decades with impressive efforts to enhance urban vitality. I have grouped these efforts under the rubric "integral urbanism."

Postmodern urbanism: form follows fiction, finesse, finance, and fear

During the 1970s and 1980s, a prevalent response to rapid change and globalization was a backlash: nostalgia for the clarity of the older boundary markers and efforts to resurrect them. This was apparent in a desire to "re-tribalize," or to assert cultural distinctions. It was apparent in the search for "roots" through the tracing of family lineages, the call to return to traditional values and institutions, resurrecting old customs, and even inventing "new" traditions.

In architecture and urban planning, the nostalgic reflex was apparent through ubiquitous references to past cities. The threat to previously clear boundaries incited an anxious effort to produce places that appeared to have grown spontaneously over time without planning. There was also a tendency to mask what is going on behind facades and escape into fantasy worlds, apparent in a proliferation of theme parks and megastructures devoted to leisure and recreational activities. This impulse to drag and drop forms from other places and other times into the present might be described as *Form follows Fiction* (Ellin 1999).

Another defense mechanism for coping with change and uncertainty during this period was irony. With the challenge posed to beliefs in progress and moral clarity, there was a lack of consensus and loss of innocence. Ultra-relativity reigned: the view that all options are equally good or bad, or equally constructed, because there are no absolute truths. The ironic response acknowledged that one's choices are just an arbitrary selection from things that have been done before; it was manifest through tone of voice, wink of eye, tongue in cheek. There was an emphasis on surface rather than substance; heroes had been replaced by celebrities; camp (self-conscious sentimentalizing) had become kitsch (bad taste).

But irony was a cop out. It was a way to hide and not take responsibility for improving the world. Irony precluded deep commitments, convictions, or passions. It was too sophisticated to laugh aloud, or find something truly funny. And it ultimately led to complacency and detachment. All that remained was images and texts, representations and discourses referring to each other. The ironic attitude said: "Nothing I do really matters. We can only live in and create fictions. So we may as well just distract ourselves with bread and circuses – food and entertainment – rather than take care of our environment, others, and ourselves." The void allowed by this attitude was too often filled by the self-serving agenda of market processes.

Sometimes, it was filled by the designer. I've described architects who shirk from taking a stand or strive for betterment, striving instead to please themselves and impress colleagues, as falling into the category *Form follows Finesse* (Ellin 1999). Striving to produce "starchitecture" profiled in the architectural press, their emphasis is on formalism and self-gratification, in search of the three Ps: prestige, power, and profits. For "Finessers," architecture is primarily a personal expression or means to an end (three Ps), rather than a social art with the goal of improving life quality for all.

While garnering the lion's share of media attention, starchitecture actually accounts for just a tiny fraction of what is built. Conversely, the vast majority of buildings, those produced by the private sector and motivated principally by the bottom line, receive the least attention. Such *Form follows Finance* (Ellin 2006) is manifest, for instance, in the sprawling suburbs and transnational business operations housed in cookie-cutter forms repeated around the globe. Although divergent in their agendas, *Form follows Finance* and *Form follows Finesse* tend to share a cynicism, or at minimum a resignation, regarding the potential for improving the world through urban design.

The fourth response to rapid change, under which the other three might ultimately be subsumed, is *Form Follows Fear* (Ellin 1997, 1999, 2008). Along with historicism, nostalgia for traditional boundary markers was also apparent in a tendency to cluster with one's own kind. Segregated urbanism is most blatant in the growth of age-restricted (55 and older) communities, such as Sun City in Arizona, but metropolitan areas also became strongly segregated along ethnic and social class lines.

The impulse to retreat was epitomized by the proliferation of gated communities for all age and income levels during these decades, despite findings that gating communities fail to effectively diminish crime and may even elevate it (Blakely 1999; Blakely and Snyder 1997; Ellin 1997; Flusty 1997; Low 2004). Outside gated communities, individually gated homes also grew rapidly along with the building of "safe rooms." Popularized by the movie "Panic Room" in 2000, starring Jody Foster, these security rooms may be converted closets or more elaborately concealed in the house plan and accessed by sliding panels and secret doors.

This period also witnessed exponential growth in homeowners associations, privately-managed groups that exercise a good deal of power, regulating house colors and renovations, pets, basketball nets, lawn care, and more (McKenzie 1996). Although these "shadow governments" are not consensually supported, people who choose to live in homes governed by them submit to their rules in an effort to protect property values and/or be with others like themselves.

The mentality of fear among homeowners of all kinds led to a pronounced anti-growth movement. People who did not want development to occur near them became known as NIMBYs (not in my back yard). Those opposed to growth of any kind were referred to as BANANAs (build absolutely nothing anywhere near anything).

The popularity of the four-wheel drive sports utility vehicle (SUV), especially in cities, during this period suggested a widespread desire to defend oneself. Although equipped for off-road driving, very few of these vehicles actually leave the roads. Their appeal was epitomized by the vogue for the Humvee (the human military vehicle or high-mobility vehicle) which was released in a civilian edition called the Hummer, available for $65,000 and up, along with an exorbitant car insurance rate. While the Hummer may have been "the ultimate in body armor" (Rugoff 1995), the safety of all cars became a major selling point, including a wide range of options from alarms to car phones, built-in car seats for children, air bags, bulletproof glass, and more.

The escapist nature of these design and development trends – behind gates, away from downtowns; or transported into the past, other places, or fantasy worlds – emitted signals that the present was indeed unsavory. The rising tide of fear led people to stay at home more, as activities that once occurred in the public realm were increasingly satisfied now in the private one via television or computer (Pawley 1973; Sennett 1974). Venturing out became increasingly restricted to the controlled settings of the shopping mall, theme park, or sports arena. Partaking in the unplanned and unpredictable public pageant of the city, a characterizing feature of urban life, became increasingly rare.

As a result, private spaces gained importance while public spaces diminished in quantity and quality. The public spaces that endured often conveyed the message, "Go away," or, "Don't linger long," since they were largely stripped of rest rooms, telephones, and even water fountains.

All four of these tendencies – Form Follows Fiction, Finance, Finesse, and Fear – are reactive attempts to cope with the anxiety wrought by rapid change through escapist and self-serving means. They are ultimately not sustainable.

Integral urbanism: urban thresholds and the five qualities

While the downward spiral of the four postmodern tendencies continues, it has been offset in recent years by a marked upward spiral. Indeed, a quiet revolution has been underway, aiming to heal the wounds inflicted upon the landscape over the last century. Practicing "integral urbanism," these restorative efforts exemplify five qualities: *hybridity, connectivity, porosity, authenticity*, and *vulnerability*.

Hybridity and *connectivity* bring activities and people together at all scales. While modern urbanism espoused the separation of functions, integral urbanism reaffirms their symbiotic nature by combining and linking them. These various integrations can be accomplished through cross-programming buildings and regional plans – spatially (plan and section) as well as temporally. Examples of cross-programming include the office building with basketball court and daycare center, the community center and library (Figure 44.1), the intergenerational community building (combining day care, teenage community center, adult education, and seniors center), the public school/community center, the integrated parking structure (into office, residential, and office buildings), the movie theater/restaurant, and the urban plaza by day/movie theater at night.

Transposing this concept onto the larger scale can increase density of activity without necessarily increasing building density, translating into reduced commuting, greater convenience, preservation of the natural environment, an increase in quality public space, and richer opportunities for social interaction. The outcome is new hybrid typologies and morphologies that pool human and natural resources to the benefit of all, conserving energy, time, money, water, fuel, building materials, and other resources. This approach activates

Figure 44.1 Palo Verde Library and Maryvale Community Center in Phoenix, AZ, 2006, designed by Gould Evans & Wendell Burnette Architects. Source: Bill Timmerman – used by permission.

places by creating thresholds, or places of intensity, where diversity thrives. By increasing density of activity, and perhaps building mass as well, these thresholds weave connections between places, people, and experiences (Figure 44.2).

Porosity preserves the integrity of that which is brought together, while allowing mutual access through permeable membranes, in contrast to the modernist attempt to dismantle boundaries or postmodernist fortification (for examples, see Ellin 2006: 62–94). *Authenticity* involves actively engaging and drawing inspiration from actual social and physical conditions with an ethic of care, respect, and honesty. Like all healthy organisms, the "authenti-City" is always growing and evolving according to new needs that arise, thanks to a self-adjusting feedback loop that measures and monitors success and failure (see more on the Authenti-City below.) Finally, *Vulnerability* calls upon us to relinquish control, listen deeply, value process as well as product, and re-integrate space with time (for examples, see Ellin 2006: 118–132).

In sum, integral urbanism emphasizes re-integration (functional, social, disciplinary and professional), permeable membranes, and design with movement in mind, both movement through space (circulation) and through time (dynamism and flexibility). Bringing together the functions that the twentieth-century city separated (living, working, circulating, learning, creating, and recreating), integral urbanism offers a new model that additionally integrates buildings with nature, center with periphery, local character with global forces, the various professions involved with urban growth and development, and people of different ethnicities, incomes, ages, and abilities. While not forming a "school" of thought, since the interventions vary widely, these efforts share an emphasis on first, drawing from the best aspects of pre-modern, modern,

Figure 44.2 The Grove at Arizona State University. Source: Bill Timmerman – used by permission.
Note: The Grove, a series of shade structures designed by Studio Ma for ASU, converts a nondescript space between buildings into a vibrant threshold.

and postmodern urban design; second, incorporating new technologies in a humane way; and third, respecting physical, historical, social, and environmental contexts. The result is a reorientation in urban design theory and practice, ranging from small-scale interventions to regional plans, that is enhancing the health and well-being of the contemporary landscape.

Practicing integral urbanism

Integral urbanism involves an approach as well as an outcome. Just as a good manager builds on existing strengths of an organization, so good urbanism builds upon given assets of a place. In contrast to the modern ethos that started from a clean slate (or *tabula rasa*), integral urbanism begins by identifying what we already value and assuring its preservation, be it buildings, neighborhoods, businesses, cultural institutions, natural landscapes, or creative and intellectual capital. Integral urbanism similarly recognizes exemplary practices from which we can learn and upon which to build. Recognizing existing assets and capacities inflects the process, invariably leading to a consideration of what we might value more with minor adjustments. After protecting what is valued and enhancing what may be underperforming, this approach addresses what is missing and should be added. Rather than neglect, abandon, or erase our urban heritage, integral urbanism *preserves* buildings, neighborhoods, and natural landscapes that we value; *rehabilitates, reclaims, restores, or renovates* what is underperforming; and *adds* what we do not have yet but would like, as informed by effective community involvement.

When this process is applied, transformations are inspired by the "DNA" of a place, allowing for unique and meaningful expressions to unfold. Skillfully inserted, these interventions into urban fabrics can perform "urban acupuncture" (Frampton 1999; de Solà Morales 1999, 2004; Lerner 2009), clearing blockages and liberating energy to fuel additional positive growth and change. Integral urbanism thereby engages in strategic interventions that may have a tentacular effect, catalyzing other interventions in an ongoing dynamic process.

This process contrasts with master planning which, in its focus on controlling everything, has tended ironically to generate fragmented cities without soul or character. The integral urbanism approach additionally integrates professions and academic disciplines that grew increasingly specialized and balkanized over the years: architecture, planning, landscape architecture, engineering, geographical sciences, interior design, industrial design, graphic design, sculpture, and more.

Authenti-city: system meets serendipity

In stark contrast to the excesses of irony, cynicism, and escapism characterizing postmodern urbanism, integral urbanism expresses a widespread and broad-based yearning for authenticity. In architecture and urbanism, clarion critiques of the collapse of reality (e.g Huxtable 1999; Leach 1999) have abounded along with propositions for bringing it back, such as Rem Koolhaas's advocacy of "Bigness" to "resurrect the Real" (Koolhaas 1997). A symposium on the topic stated: "For contemporary architectural critics, authenticity has replaced the Vitruvian triad of firmness, commodity, and delight as the primary standard of judgment" (Savannah College of Art Design 2001).

But how can we avoid the ersatz environments and achieve this highly sought-after authentic urbanism? Should we step aside and allow the city to grow and change without any guidance whatsoever, allowing market forces to drive development? No, because markets are only designed to allocate resources in the short-term, without regard for what may not have obvious financial value like the purity of our air and water or the quality of our communities. As Paul Hawken, Amory Lovins, and L. Hunter Lovins eloquently caution in *Natural Capitalism*, "Markets were never meant to achieve community or integrity, beauty or justice, sustainability or sacredness – and by themselves, they don't" (Hawken *et al*. 2000).

In contrast, an authenti-city draws from a combination of large-scale and small-scale interventions, both systematic and the serendipitous. How it happens is just as important – and goes hand in hand – with what happens. An authenti-city is responsive to community needs and tastes, which have to do with local climate, topography, history, and culture. The best urban plans contain both urban design *and* policy frameworks upon which a city can grow and change in a never-ending dynamic process. Like a good parent, a good plan nurtures healthy growth and change without being "over-involved," without determining everything, allowing the city to blossom and define itself. While providing some overall defining guidelines, these frameworks should not prescribe every land use and every architectural detail.

Like all healthy organisms, an authenti-city is always evolving according to new needs that arise, thanks to a self-adjusting feedback loop that measures and monitors success and failure. When people hatch an idea for improving the city such as a network of linear parks, a public market, better crime prevention and educational opportunities, or the development of small business incubators, an authenti-city has the ability to implement these. In contrast to postmodern escapist tendencies – that may deny unpleasant social and urban conditions or retreat into formalism, nostalgia, fantasy, or cyberspace – this integral urbanism engages contemporary realities by honoring the local community and landscape as the greatest source of inspiration, rather than hindrances to overcome or obstacles to surmount.

Urban designer and critic Mark Hinshaw has described these places as "True Urbanist" communities:

> Not the product of a singular vision, they emerge from the collective decisions of many organizations, associations, corporations and government bodies. They value the results of democracy – however messy, unpredictable, and uneven they may be ... They are constantly evolving, infilling, and re-developing, with a broad mixture of architectural styles and sensibilities. ... They have a gritty urbanity that values variety over uniformity. Rarely are they subject to a highly prescriptive set of design standards; rather, they revel in the idea that everything need not fit an ideal. They may be subject to design guidelines and a design review process, but those techniques encourage creativity over conformity (Hinshaw 2007).

The International Making Cities Livable Movement promotes True Urbanism, advocating generative design guidelines based upon the DNA of places. This DNA, the website states,

> is expressed in those architectural and spatial characteristics best loved by the city's inhabitants. These may consist of certain building materials and colors, a typical arrangement of scale and architectural forms, building lot size, rooflines, scale of public and semi-public spaces. In order to fit into the context, new buildings have respected this "genetic code," reflecting at least some existing patterns, or interpreting them in a contemporary idiom (IMCLM http://www.livablecities.org/).

The quest for authenticity among urban designers has taken various directions. A "dirty realist"[1] transgressive approach implicitly critiques the manifestations of economic/social disparities in the landscape. "Everyday urbanism" implicitly critiques the high/mass/popular culture divides (Harris and Berke 1997; Chase *et al.* 1999; Kelbaugh 2005). The New Urbanism claims to produce an "authentic urbanism" by learning from urban wisdom passed down through the ages (Kelbaugh 2005). Integral urbanism extends the everyday urbanist respect for spontaneous expressions of popular culture and the New Urbanist respect for urban traditions, while infusing these with local knowledge gained through appreciative listening (AI Commons http://appreciativeinquiry.case.edu/). While acknowledging an important place for the de-familiarizing tactics of the dirty realists, integral urbanism does not consider these appropriate for large-scale interventions.

For urban integrity to flourish at the larger scale of districts or cities, there must be infinite opportunities for the "unofficial plans," developed by many different people with a wide range of ideas, as described by Jane Jacobs (Jacobs 1961). These can only be effective, as Jacobs also pointed out, if certain tools are made available by the public sector. Redevelopment agencies, such as San Diego's Centre City Development Corporation (formed in 1975), and Tax Increment Financing[2] are essential to oversee and coordinate revitalization efforts that include important infrastructural improvements (especially transit) and to preserve social diversity. Initial public incentives to bring private development into targeted areas are also important for "priming the pump." Supporting local independent retail is critical for places to be distinctive and to retain sales dollars. Arts districts, as legislated in Maryland and in Providence (RI), are extraordinarily effective catalysts toward urban revitalization.

It is also important to have programs in place to ensure affordable housing (such as San Diego's SRO program and Seattle's taxpayer-approved low-income housing levy) and to assure the preservation of buildings and neighborhoods that are valued by the community. Finally, regulatory practices can support urbanism by requiring "build-tos" rather than setbacks, pedestrian-friendly uses on the ground level, specifying maximum rather than minimum parking spaces, and other means.

Speaking across the fissures: a new vocabulary

The modern era divided the world and our thinking about it into fragments. Our landscape followed in step, and we have been suffering the results. To bring it back together, we need to overcome the divisions in our thinking, so we can envision and implement the integration.

A decade and a half ago, Herbert Muschamp described the "Urban Revisions" exhibition at the Los Angeles Museum of Contemporary Art as "a sprawling mess of undigested ideas," reflecting "the field of urbanism that it sets out to survey." He concluded, "If nothing else, the show exposes the need for a new vocabulary of urbanism – a language capable of bridging the differences among those who shape the public realm." Muschamp maintained, "if designers want to reinforce the connective tissue of cities, they will have to speak across the fissures that have opened up among themselves" (Muschamp 1994).

Introducing the term "integral urbanism" is an effort to speak across these fissures. Rather than pose yet another contender for the war-banism, integral urbanism draws from the most compelling aspects of all contemporary trends from dirty realism to everyday urbanism, New Urbanism, authentic urbanism, true urbanism, incremental urbanism, "her-banism" (feminist planning theory), re-urbanism,[3] posturbanism,[4] market urbanism, and more. To clearly convey the keys to contemporary best practices, it distills this synthesis into their constituent five qualities (described above). These five qualities offer a point of departure, like the basic chord structure in jazz from which musicians improvise, or any set of technical skills (artistic, technological, business, sports, culinary, etc.), essential for generating something of value.

Coming full spiral: the integral project

The goal of integral urbanism is to create adjacencies of uses and people, allowing relationships among them to develop and flourish. Rather than separate and control – the guiding ethos of modern urbanism – this approach aims for integration, inclusion, and dynamism. The strength and resilience of relationships and communities relies upon trust, but urban fragmentation during the second half of the twentieth century diminished trust, allowing an "architecture of fear" to occupy the void (Ellin 1997; 2008). Integral urbanism re-builds community, along with a high quality public realm for the twenty-first century, by cultivating relationships through a process that engages and builds mutually supportive networks of people (Alexander 1987). The trust upon which relationships and communities rely ensues.

Although there remain certain social and professional obstacles impeding this integration, we are nonetheless passing through an opportune historical moment when urban design theory is fortuitously aligned with political, economic, and social trends. In some cases, these trends have been initiated by urban designers while, in other instances, integral urbanism is occurring without their input at all.

Powerful trends lending toward integral urbanism include the sustainability, environmental, smart city, creative city, historic preservation, community garden, and land trust movements (Ellin 2006). Numerous grassroots organizations, some global in reach, have become powerful advocates for creating livable places.[5] Consequently, much new development and many older suburbs and urban cores have been striving to create mixed-use hubs of activities, along with park networks, and other strategies for enhancing livability and sense of community.

The marked growth of these activities suggests a reorientation toward restoring the connections that were severed over the last century between body and soul, people and nature, and amongst people. This reorientation may be characterized as a shift from acceleration, accumulation, irony, and escapism towards slowness, simplicity, sincerity, spirituality, and sustainability. With regards to places, this upward spiral is apparent at all scales, from the wastebasket to the watershed.

Indeed, we have been coming full circle or, more accurately, full spiral by infusing the inherent wisdom of nature and cities of the past with contemporary sensibilities arising from new technologies, expectations, and sensibilities. Rather than choosing to continue or abandon the modern project, our reliance upon information technologies along with the simultaneous revalorization of process, relationships, and complementarity has been enabling us to do both simultaneously. In the process, each provides feedback for and adjusts the other accordingly, holding potential for achieving integration at another level.

The modern project is thus revised, or supplanted, by an integral project. The modern project sought liberation through scientifically and creatively controlling nature and the irrational. The integral project cultivates liberation (from oppression, inequality, ignorance, pain, and discomfort) by understanding our place in nature, including the irrational, and drawing upon science, technology, creativity, and our own deep empathy or greater intelligence in pursuit of the common good, personal fulfillment, and global cooperation.

Although integral urbanism pertains specifically to urban design, its five qualities (hybridity, connectivity, porosity, authenticity, and vulnerability) might effectively apply to governance, homeland security, management, business, education, mediation, technology, the arts, and other realms. Applied generally, these qualities translate into regarding organizations as dynamic networks with built-in feedback mechanisms; acknowledging the primacy of relationships and process over products; bringing people and other resources together to achieve efficiencies (optimization); and maintaining an ethic of care and respect for self, others, and the environment. Incorporating these qualities brings a profound shift from competition to synergism, the kind of collaboration that yields outcomes larger than the sum of its parts, not the lowest common denominator.

Conclusion: the re-generation

As our connections to the environment and other people grow increasingly tenuous – a condition commonly described as a "breakdown" in community and family along with an ecological "crisis" – efforts to re-think urban design have been seeking to reconnect or provide places allowing connections to occur. Rather than respond to specific problems with piecemeal solutions that only exacerbate the problems or push them elsewhere (reactive solutions), integral urbanism emphasizes holism and forging connections at another level. Without shifting into reverse, integral urbanism seeks to put a brake on the continual fragmentation of our landscapes and our lives through proactive design solutions. Resolutely refusing to idealize the past or to escape the present, it seeks to mend seams in the urban and social fabrics by acknowledging contemporary challenges and formulating inspired alternatives for an enriched future.

If our places are to sustain us, they must of course offer clean air and water along with other essentials for survival. But if that is all they offer, we will only survive. Applying the five qualities of integral urbanism has been offering the soul food essential for our cities and communities to blossom and truly thrive.

The 1960s produced the "We" generation emphasizing peace and love, the 1970s the "Me" generation emphasizing self-awareness and self-actualization, the 1980s the "Whee" generation emphasizing materialism and escapism, and the 1990s the "Whoa" generation, placing a self-imposed brake upon the rapid changes that were wreaking havoc upon our landscapes and our well-being. This new millennium has been spawning the "Re-generation," with a clear-eyed vision and the courage to Re-build our towns and cities, Re-vitalize our communities, Re-store what has been taken from the earth, and Re-align design with the goal of supporting humanity.

Notes

1 This term was applied by Liane Lefebvre (1989) who notes similarities between certain late 1980s architects and the school of literature that charts the "dirty realities" of late twentieth-century life rather than flee from them into escapism and narcissism as postmodern literature had. In literature and architecture alike, dirty realists engage in "de-familiarization," seeking to make people aware of ordinary conditions in a new way.

2 When municipalities create Tax Increment Financing Districts, they can retain a portion of property and/or sales tax from new development within that district for a predetermined number of years and use this revenue for new development in the district.

3 Reurbanism is a broad category covering everything from high-end examples of "positive redevelopment and revitalization of American cities that is now happening piecemeal" to local architecture with its default urbanism" (Kelbaugh 2005, v. III, 8–10).

4 Post-urbanism is avant-garde and "driven by aesthetics." Michael Speaks suggests calling it "Not Urban" (Kelbaugh 2005 v. I, 35).

5 See for example: Active Living by Design – www.activelivingbydesign.org
Project for Public Spaces – www.pps.org
Walkable Communities – www.walkablecommunities.org
Well community Association and Foundation – www.wellcommunity.org
Project for Livable Communities – www.projectlivablecommunities.org
Slow Cities Movement – www.cittaslow.net

References

AI Commons, http://appreciativeinquiry.case.edu/. (accessed 16 August 2010).

Alexander, C. (1987). *A New Theory of Urban Design*, New York: Oxford University Press.

Blakely, E. (1999). *Fortress America*, Washington D.C.: Brookings Institution Press.

Blakely, E.J. and Snyder, M.G. (1997) "Divided We Fall: Gated and Walled Communities in the United States." In Ellin, N. (Ed.), *Architecture of Fear*, NY: Princeton Architectural Press, 85–100.

Chase, J., Crawford, M. and Kaliski, J. (Eds.) (1999). *Everyday Urbanism*, NY: Monacelli.

De Solà-Morales, M. (1999). "Progettare citta/Designing Cities," Lotus Quaderni Documents, no. 23, Mirko Zardini, ed., Milan: Electa.

De Solà-Morales, M. (2004). "The Strategy of Urban Acupuncture" at Structure Fabric and Topography Conference, Nanjing University.

Ellin, N. (1997). "Shelter from the Storm or Form Follows Fear and Vice Versa." In Ellin, N. (Ed.), *Architecture of Fear*, NY: Princeton Architectural Press, 13–46.

Ellin, N. (1999). *Postmodern Urbanism*, Revised Edition, NY: Princeton Architectural Press.

Ellin, N. (2006). *Integral Urbanism*, NY: Routledge.

Ellin, N. (2008). "Life Support: Nacirema Revisited," *Journal of Urbanism: International Journal of Placemaking and Urban Sustainability*, (March), 47–55.

Flusty, S. (1997). "Building Paranoia," in Ellin, N. (Ed.) *Architecture of Fear*, NY: Princeton Architectural Press, 47–60.

Frampton, K. (1999). "Seven Points for the New Millennium: An Untimely Manifesto," *Architectural Record*, August, November, 76–80.

Harris, S. and Berke, D. (1997). *Architecture of the Everyday*, NY: Princeton Architectural Press.

Hawken, P., Lovins, A. and Lovins, L.H. (2000). *Natural Capitalism*, Boston: Back Bay Books.

Hinshaw, M. (2007). *True Urbanism*, Chicago: American Planning Association.

Huxtable, A.L. (1999). *The Unreal American: Architecture and Illusion*, New York: New Press.

IMCLM, http://www.livablecities.org/. (accessed 16 August 2010).

Jacobs, J. (1961). *The Death and Life of Great American Cities*, New York: Vantage.

Kelbaugh, D., (Ed.) (2005). *Michigan Debates on Urbanism I, II, and III*, Ann Arbor: University of Michigan Press.

Koolhaas, R. (1997). "Bigness: Or the Problem of Large," in Rem Koolhaas/OMA and Bruce Mau *S M L XL*, Taschen: Cologne, p. 513.

Leach, N. (1999). *The Anesthetics of Architecture*, Cambridge, MA: MIT Press.

Lefebvre, L. (1989). "Dirty Realism in European Architecture Today," *Design Book Review*, 17, Winter, 17–20.

Lerner, J. (2009). *Urban Acupuncture*, Barcelona: Institute for Advanced Architecture of Catalonia.

Low, S. (2004). *Behind the Gates*, New York: Routledge.

McKenzie, E. (1996). *Privatopia: Homeowner Associations and the Rise of Residential Private Government*, New Haven, CT: Yale University Press.

Muschamp, H. (1994). "The Polyglot Metropolis and Its Discontents," *New York Times*, July 3, 1994.

Pawley, M. (1973). *The Private Future*, London: Pan Books.

Rugoff, R. (1995). "L.A.'s New Car-tography," *LA Weekly*, October 6, 35.

Savannah College of Art and Design (2001). *Second Savannah Symposium: Authenticity in Architecture*, Savannah, Georgia.

Sennett, R. (1974). *The Fall of Public Man*, New York: Random House.

Further reading

Beatley, T. (2004). *Native to Nowhere: Sustaining Home and Community in a Global Age*, New York: Island Press. Overview of a wide range of efforts to recover sense of place in the US and Western Europe.

Calthorpe, P., Fulton, W. and Fishman, R. (2001). *The Regional City: Planning for the End of Sprawl*, New York: Island Press. On the importance of regional coordination for sustainable urban growth and development.

Goldsmith, S. and Elizabeth, L. (Eds.), (2010). *What We See: Advancing the Investigations of Jane Jacobs: Contemporary Thinkers Observe Our World*, Oakland, CA: New Village Press. Original essays inspired by the work of Jacobs.

Hallsmith, G. (2003). *The Key to Sustainable Cities*, New Society Publishers: Gabriola Island, BC, Canada. Why and how to apply system dynamics to city planning.

Kelbaugh, D. and McCullough, K.K. (2008). *Writing Urbanism*, New York: Routledge. Presents 40 articles about current issues in urban design.

Kunstler, J.H. (1993). *The Geography of Nowhere*, New York: Simon & Schuster. A synopsis of American urbanization over the last three centuries by an astute and acerbic observer.

Landry, C. (2003) *The Creative City: A Toolkit for Urban Innovators*, London: Earthscan; and *The Art of City Building* (2007), London: Earthscan. On the centrality of human creativity and innovation for twenty-first century urban vitality, including guidelines for achieving it.

Larice, M. and Macdonald, E. (Eds.) (2007). *The Urban Design Reader*, London: Taylor & Francis. Presents 43 classic and contemporary selections on urban design theory and practice.

Leinberger, C. (2008). *The Option of Urbanism*, Washington, DC: Island Press. How financial and real estate professionals can contribute to retrofit cities and suburbs to enhance quality of life.

45

Ecological urbanism

Anne Whiston Spirn

Human survival depends upon adapting ourselves and our settlements in life-sustaining ways, designing places that reflect the interconnections of air, earth, water, life, and culture, that help us feel and understand these connections, places that are functional, sustainable, meaningful, and artful (Spirn 1998: 26). Ecological urbanism aims to advance this goal.

Ecological urbanism weds the theory and practice of urban design and planning, as a means of adaptation, with the insights of ecology – the study of the relationships between living organisms and their environment – and other environmental disciplines, such as climatology, hydrology, geography, psychology, and history. Ecological urbanism has an aesthetic dimension, but it is not a matter of style; the works of its practitioners may be radically different in appearance.

Ecological urbanism is not a new idea. Its roots are ancient, and it is grounded in a tradition of key concepts and principles. Ecological urbanism is critical to the future of urban design: it provides a framework for addressing challenges that threaten humanity (global warming, rising sea level, declining oil reserves, rising energy demands, and environmental justice) while fulfilling human needs for health, safety, and welfare, meaning, and delight.

Ecological urbanism: historic roots and current trends

The theory and practice of ecological urbanism has a long history, a foundation of knowledge to support it, and projects that demonstrate its benefits. The roots of this tradition in Western culture are deep: from Hippocrates' treatise *Airs, Waters, Places* of the fifth century B.C. to Ian McHarg's *Design with Nature* of 1969 and Kevin Lynch's *Good City Form* of 1981 to contemporary authors (see Spirn 1985 for a review of this tradition).

More than 2,000 years ago, Hippocrates described the effects of "airs, waters, and places" on the health of individuals and communities. Vitruvius (*c.* first century B.C.) described how the layout of streets and the orientation and arrangement of buildings should respond to seasonal patterns of sun and wind. Leon Battista Alberti in 1485 expanded on these recommendations, advocating that the siting of cities and the design of streets, squares, and buildings should be adapted to the character of their environment so that cities might promote health, safety, convenience, dignity, and pleasure. "We ought never to undertake any Thing that is not exactly agreeable to Nature," Alberti warned, "for nature, if you force or wrest her out of her

Way, whatever Strength you may do it with, will yet in the End overcome and break thro' all Opposition and Hindrance ... forced to yield to her daily and continual Perseverance" (Alberti 1485). Alberti underscored this warning by cataloguing the disasters suffered by cities that had disregarded the power of nature, a warning issued several centuries later by George Perkins Marsh, who predicted that "human improvidence" was reducing the earth "to such a condition of impoverished productiveness, of shattered surface, of climatic excess" as to threaten the "extinction of the species" (Marsh 1864). Marsh proposed that "in reclaiming and reoccupying lands laid waste by human improvidence or malice ... the task ... is to become a co-worker with nature in the reconstruction of the damaged fabric." This was an approach embraced by Marsh's contemporary, landscape architect Frederick Law Olmsted, who designed urban parks, parkways, and neighborhoods as part of a broad program to promote the health, safety, and welfare of urban residents. In designs for landscape infrastructure of parkways, streetcar lines, rivers, and sewers, Olmsted sought to "hasten the process already begun" by nature, thereby achieving more than the "unassisted processes of nature" (Olmsted and Vaux 1887: 19, 8).

By the beginning of the twentieth century, some disagreed over whether the task was to rebuild existing cities or to build new "garden cities" in the countryside, such as those advocated by Ebenezer Howard (Howard 1902). Patrick Geddes opposed Howard's approach: "Here or nowhere" is our Utopia, he argued (Geddes 1915, p. 2). Geddes, who was educated as a biologist, viewed each city and its surrounding countryside as an evolving organic whole whose future plan should be based on an understanding of its natural and cultural history and its "life-processes in the present" (Geddes 1915, p. 2). To attain such understanding, Geddes advocated "regional surveys" of "things as they are and as they change ... towards things as they may be," which would serve as the vehicle for town plans and city designs tailored to the particular "character and spirit" of each city. (Geddes 1915: 138–139). Lewis Mumford (1968: 164) like his mentor Geddes, promoted an integrative approach to cities and their regions:

> Once a more organic understanding is achieved of the complex interrelations of the city and its region, the urban and the rural aspects of environment, the small-scale unit and the large-scale unit, a new sense of form will spread through both architecture and city design.

To Mumford, this new urban form "must include the form-shaping contributions of nature, of river, bay, hill, forest, vegetation, climate, as well as those of human history and culture, with the complex interplay of groups, corporations, organizations, institutions, personalities" (Mumford 1968: 164). Mumford influenced Kevin Lynch and Ian McHarg, who shared the conviction that cities must be viewed in their regional context and that the natural environment has a social value to be cultivated in urban design. From that common ground, they diverged.

For Lynch, the city is first and foremost a human habitat, and he judged "good city form" by how well it sustains human life (Lynch 1981). Lynch stressed the importance of how people perceive the city, proceeding from human perception to understanding the sense of place. He explored the role that natural features play in enhancing the identity, legibility, coherence, and immediacy of urban form from the scale of the street to that of the region, for "the mental sense of connection with nature ... is a basic human satisfaction, the most profound aspect of sensibility. ... The movement of sun and tides, the cycles of weeds and insects and men, can also be celebrated along the city pavements" (Lynch 1981: 257). His last book, *Wasting Away*, takes an ecological approach to managing resources and waste (Lynch 1990).

McHarg's point of departure was the natural environment: "Let us accept the proposition that nature is process, that it is interacting, that it responds to laws, representing values and opportunities for human use with certain limitations and even prohibitions" (McHarg 1969: 7). Like Geddes, McHarg asserted that "any place can only be understood through its physical evolution" (McHarg 1967: 105). As a prerequisite for planning and design, he advocated a survey, the "ecological inventory." It was always the same list (climate, geology, hydrology, soils, vegetation, wildlife, and ecosystems) no matter what the site's location, scale, or land use. McHarg's ecological inventory is a checklist of interrelated systems, useful not only to understand how a place came to be, but also as a diagnostic tool with which to identify problems and opportunities that might otherwise be missed (Spirn 2000). For McHarg, design was an evolutionary strategy, a means of adaptation. His approach is valuable for urban design, even though he viewed the city as a pathological environment (McHarg 1969).

For Jane Jacobs, as for McHarg and Lynch, "human beings are ... part of nature" as are cities (Jacobs 1961: 446). "Nature, sentimentalized and considered as the antithesis of cities, is apparently assumed to consist of grass, fresh air and little else," Jane Jacobs scorned, "and this ludicrous disrespect results in the devastation of nature." Like Lynch, she focused on the city as a human habitat and saw urban design as a way to support and fulfill human needs. Jacobs advocated an ecological approach to designing and managing cities, arguing that cities are problems of organized complexity, akin to living organisms, and that there are lessons for urban design from the study of systems where "half-dozen or even several dozen quantities are all varying simultaneously and *in subtly interconnected ways*" (Jacobs 1961: 433). Jacobs urged urban designers and planners to think in terms of processes and to "work inductively, reasoning from particulars to the general, rather than the reverse," from grand theories to specific proposals (Jacobs 1961: 440).

Many researchers, practitioners, and critics have contributed to the theory and practice of ecological urbanism since Jacobs, McHarg, and Lynch, far too many to treat adequately in the context of this brief summary. Ecological urbanism is a broad approach to urban design and planning; related to it are aspects of several contemporary movements: ecological design (Spirn 1984; Hough 1995; Van der Ryn and Cowan 1996; Thompson and Steiner 1997; Johnson and Hill 2002; Berger 2009), sustainable design (Calthorpe and Van der Ryn 1986; Lyle 1994; Hester 2006), green architecture (Wines 2000; Fromonot 2003), green infrastructure (Wenk 2002; Benedict and McMahon 2006), landscape urbanism (Mostafavi and Najle 2003; Waldheim 2006; Almy 2007), and industrial ecology (Graedel and Allenby 2003). Not all the works – written, drawn, or built – produced under these rubrics, however, qualify as ecological urbanism; they belong to the extent that they embody key concepts and principles.

Ecological urbanism: key concepts and principles for urban design

Key concepts of ecological urbanism include: cities are part of the natural world; every city has a deep, enduring context; cities are habitats; cities are ecosystems; urban ecosystems are connected and dynamic; urban design is a tool of human adaptation. These fundamental propositions are the foundation from which principles for urban design and planning derive (those listed here are illustrative not exhaustive).

Cities are part of the natural world

Natural processes do not stop operating at city limits: paving and building stone, for example, affect heat gain and water runoff just as exposed rock surfaces do anywhere. The urban environment is the consequence of a complex interaction between the many purposes and activities of human beings and the natural processes that govern the transfer of energy, the movement of air, the erosion of the earth, and the hydrologic cycle. Despite their differences, all cities transform their natural environment in similar ways. Human activities interact with natural processes to create a typical urban climate, urban soils, urban hydrology, urban plant and animal communities, and characteristic urban ecosystems (Spirn 1984: 4–5). Cities are part of the natural world. Recognition of this basic fact has powerful implications for how the city is designed, built, and maintained, and for the health, safety, and welfare of every resident.

Despite overwhelming evidence to the contrary, the belief that the city is an entity apart from nature and even antithetical to it has dominated the way in which the city is perceived and continues to effect how it is designed, built, and maintained. This attitude has aggravated and even created many environmental problems, both local and global: poisoned air and water; depleted or destroyed resources; more frequent and more destructive floods, more damage from geological hazards; increased energy demands and higher construction costs (Spirn 1984: 5). At the root of this failure to recognize the city as part of nature is the notion that nature is a place (wilderness and countryside, but not city) or a thing (mountain, river and tree, but no thing made by humans). But nature is an idea, not a place or a thing. The idea of nature as consisting of the biological, physical, and chemical processes that create and sustain life, the earth, and the universe is fundamental to ecological urbanism. If one embraces this idea, then the false oppositions between city and nature, the given and the built, fall away.

Recognize cities as part of the natural world and design them accordingly

Many authors have described how this might be accomplished. See, for example, *The Granite Garden: Urban Nature and Human Design*, which is organized into sections on air, earth, water, life, and ecosystems, with successful cases from the scale of the house and garden to those of the neighborhood, city and region (Spirn 1984). The key is to think in terms of the ways that human activities and urban form interact with natural processes of air (heat transfer and air flow), earth (geology and soils), water (water flow), life (reproduction, growth, and behavior), and ecosystems (flows of energy, information, and materials, succession of plant species and behavior of plants and animals). Note, this is not a just a matter of imitating or echoing the forms of natural features or of using indigenous materials, but, as Marsh put it more than a century ago, "the task ... is to become a co-worker with nature" and thus to achieve more than the "unassisted processes of nature" (Marsh 1864; Olmsted and Vaux 1887). By focusing on the natural processes that shape and structure the environment, urban designers can accommodate dynamic change in the natural environment, make connections among seemingly unrelated elements and issues, recognize that not all traditional settlement patterns or building forms should be repeated, and realize opportunities.

Plot the interplay of natural and social processes that shape and structure the city

Ian McHarg overlaid maps of diverse natural and social factors in order to better understand this interaction (McHarg 1969). Such overlays can reveal surprising relationships among seemingly unrelated phenomena, such as the correlation between buried floodplains and vacant land in low-income inner-city neighborhoods, but they portray these relationships as static (Spirn 2005). Anu Mathur and Dilip da Cunha, among others, have developed mapping techniques that enable designers to visualize how processes operate in space and time (Mathur and da Cunha 2001; Berger 2006).

Every city has a deep, enduring context

While urbanization radically changes the surface of the urban landscape, there is a more enduring context, with distinctive rhythms, which is the product of climatic, geomorphic, and biological processes operating and interacting across millennia. The enduring context of a city is expressed in its climate (hot, cold, or temperate; humid or arid; and seasonal variations), geology (rock type and structure, seismic and volcanic activity), physiography (plain, basin, foot hills, or mountain), and biome (tundra, forest, prairie, or desert). In the history of a place this "deep" context is a constant that successive human generations must re-address. Traditions, values, and policies may change, but deep context remains key to the history and future of a place – why it was settled, its initial location, its transportation routes, its economic development and population distribution, the character of its buildings, streets, and parks, and the health and safety of its residents (Spirn 1998: 158). When urban form obscures or opposes a city's enduring context (by planting trees and lawn in a desert; by burying a river in a sewer; and filling in its floodplain), it will require additional energy and materials to sustain.

Adapt the physical shape and structure of a city – the infrastructure of roads and sewers, the buildings and parks – to its deep context

Urban form that reveals and responds to deep context is likely to be more functional, more economical, and more resilient than design that disregards it (Spirn 1998). This is especially important for the design of the infrastructure (water, sewer, power, transportation) that supports the city, whether at the scale of building, neighborhood, city, or region. Such design may also afford an aesthetic experience of unity with the processes which shape the landscape and which sustain life (Spirn 1988).

Anticipate and exploit catastrophic events

Every city is prone to specific natural hazards whose precise timing is unknown. San Francisco will experience a major earthquake; Las Vegas and Phoenix, severe drought; St. Louis and Pittsburgh, major floods. It was inevitable that a major hurricane would strike New Orleans. After a catastrophe, there is a will to rebuild and to "do things right," but that window of opportunity is small. Urban designers should plan ahead for redesigning and rebuilding in order to seize the opportunity when catastrophe strikes.

Cities are habitats

As habitats, cities must provide settings for the biological and social needs of the organisms – humans and other species – who dwell there. There must be places for reproduction and growth, movement and exchange, communication, making and building, teaching and learning, work and play, reflection and worship. What could be more obvious? And yet, cities are full of places that are ill-adapted to the needs of their inhabitants. They are dysfunctional, contaminated, and vulnerable to natural hazards, exposing residents to discomfort, inconvenience, and even to danger.

Cities provide habitats for many nonhuman species (from microbes to trees, from insects to fish, birds, and mammals). Some are indigenous, others are typical urban species, some are central to human health and prosperity, a few are hostile. Ironically, most urban pests were imported by humans, deliberately or inadvertently, and are well-adapted to the habitats that humans create. Urban development tends to reduce biodiversity, with far-reaching adverse effects (Johnson and Hill 2002).

Design the city as a life-sustaining and life-enhancing habitat

Every urban design project should enhance the quality of the urban habitat for humans and other species, even if that goal is not an explicit part of the designer's brief. Kevin Lynch provides measures of "good city form" in terms of how well urban form sustains life ("vitality"), by how clearly it is perceived in space and time ("sense"), how well environment and behavior "fit," and by whether these standards are provided in a manner that provides "access," "control," "efficiency," and "justice" (Lynch 1981). Many others, too numerous to cite here, have elaborated on how this might be accomplished (e.g. Alexander *et al.* 1977; Hester 2006; Spirn 1984, 1987, and 1998).

Celebrate the natural processes that shape the urban habitat and that sustain life, make them tangible and understandable

Pleasure and meaning are fundamental human needs, and "the mental sense of connection with nature is a basic human satisfaction, the most profound aspect of sensibility" (Lynch 1981, p. 257). Urban design that fosters and intensifies the experience of the natural processes that sustain life fulfills this need (Spirn 1988 and 1998). Aesthetic experience of such places has the potential for "recentering human consciousness from an egocentric to a more bio-centric perspective" (Meyer 2008, p. 6).

Design and manage the urban habitat for nonhuman species

Like humans, each species has specific needs, and the most effective way to enhance their survival or establish control is often through the design and management of their habitat.

Cities are ecosystems

The urban ecosystem, like any ecosystem, consists of all the organisms that dwell within it (including humans) and their interactions with each other and with their physical environment, which comprises built artifacts like buildings, roads, and sewers, as well as water, soil, and plants. The urban ecosystem is an open system: energy, material, and information flow through it as resources are imported, transformed, and consumed, then exported as wastes. The less efficiently resources are used, the more wastes are produced and contamination is increased. The urban ecosystem encompasses all the processes which flow within and through the city: cultural processes as well as natural processes, flows of capital, people, and goods, as well as flows of water, air, nutrients, and pollutants. The city as a whole, itself an ecosystem, is composed of many smaller ecosystems, such as ponds and river corridors, parks, and buildings.

Design the city and its rural periphery, as well as every park, building, and district within that larger whole, as ecosystems that require minimal inputs of energy and resources to build and sustain

The design of an urban ecosystem entails not just the composition of its structure, shape, and materials, but should include as well the means by which it will be built and maintained over time. The city, and every building, park, and infrastructure system within it, should be designed as much as possible as "closed" ecological systems, systems that import and consume fewer resources, produce fewer wastes, and whose wastes are recycled as resources. This goal is most easily understood and achieved at the scale of a park or a building and its immediate surroundings, and there are successful models (Lyle 1994; Wines 2000; Fromonot 2003). At the district scale, increasing the density of urban development can make energy-conserving strategies such as shared transportation systems and district heating more feasible. At all scales from house to metropolitan region, wastes – the byproducts of one activity – may be a resource for another. The home composting of kitchen waste to produce garden amendments is analogous to the regional project of combining leaf litter and sewage sludge to produce new soil (Spirn 1984). Industrial ecology brings together industries whose waste and resource streams are symbiotic (Lynch 1990; Graedel and Allenby 2003).

Urban ecosystems are connected

The many ecosystems that comprise the larger urban ecosystem are linked by the physical space they share and by the channels through which energy, material, and information flow. There are ecosystems within ecosystems. A pond ecosystem, for example, exists within the larger ecosystem of its watershed (and there are watersheds within watersheds, from that of a small stream to a continental river basin); a building is an ecosystem within a neighborhood. Given this connectivity, changes to one ecosystem may produce repercussions in many others, and an ecosystem may be externally regulated (Pickett *et al.* 2004). Problems felt in one place may be caused by activities that take place elsewhere: strong winds at the base of a tall building aggravated by conditions upwind; floods and pollution by discharge upstream; vulnerability to hurricane-driven waves by erosion of marshes and swamps. Environmental and social problems in low-income neighborhoods are often created or aggravated by flows of capital and wastes to and from suburban communities (Spirn 2005). In these and many other cases, local intervention alone is doomed to failure. Solving social and environmental problems may require taking action in a different location than where the problem is felt.

Address social and environmental challenges within appropriate boundaries at the appropriate spatial and temporal scales

Designers should identify the systems to which their project site is connected and track the flows of energy, materials (food, water, and waste water), information, and capital that move in and out. No matter how small or large the project, the designer's responsibility is to address the impact on the ecosystems to which it is connected. Design proposals should not be limited to the area enclosed by the client's boundaries, but should be expanded to include that area necessary to effectively address the challenges posed by site, program, and context.

Define multi-purpose solutions to comprehensively defined problems

Urban designers and planners should seek integrated solutions to social, economic, cultural, and environmental problems. One strategy is to start with a city's most pressing problem, one for which there is widespread public support, and then find ways to address other concerns as well. Air pollution, water pollution, or flooding may be the central, organizing issue within which social, economic, and other environmental problems are addressed. Or, alternatively, a social or economic problem, such as unemployment or extensive abandonment of property, may serve as the focus, and suggest ways to incorporate solutions to environmental problems. The integration of open urban land into a green infrastructure, for example, could extend the aesthetic and recreational value of parks and parkways to a crucial role in health, safety and welfare. Parks and plazas, rivers, streams, and floodplains, steep hillsides, and even parking lots and highway corridors could be part of a cohesive system to improve air quality and climate, to reduce flooding and improve water quality, to limit the impact of geological hazards such as earthquakes, subsidence, and landslides, to provide a diverse community of plants and animals within the city, to conserve energy, water and mineral resources, and to enhance the safe assimilation of the city's wastes (Spirn 1984; Wenk 2002).

Urban ecosystems are dynamic

Urban design is an art of time as well as space; it is a projection into the future, which is complicated by the fact that the urban ecosystem is constantly changing. Studying environmental change over time fosters an understanding of urban landscapes as dynamic, how natural processes shape human settlements and how social and cultural processes shape urban ecosystems, in turn (Spirn 2005). Knowing how a place has been shaped over time is key to understanding its present and possible future.

Take account of history

The urban designer must ask: What is this place in the process of becoming? Which of its features are clues to ongoing processes that continue to exert a decisive influence, and which are merely artifacts of the past that assert little influence now? Which features are amenable to change and which are resistant? It is difficult to answer such questions without understanding how a place evolved, through what processes and actions, when, and which of its features have had a sustained impact on their surroundings over time. The environmental history of a place provides a window into the ways natural and social processes interact through time, and how planners have intervened, for good or bad (Cronon 1991; White 1996; Klingle 2007). Readers may note that this has nothing to do with imitating or adapting historic built form. Taking account of history means more than preserving historical structures and more than using history as a source of formal precedent. History is a way of extending human memory beyond the human life span.

Urban design is a powerful tool of adaptation

Through culture, technology, and the design of physical habitats, the human species has spread across the earth, from warm savannahs to cool forests to the cold Arctic tundra, and has continued to evolve. Most humans now live in cities, and urban design is a powerful tool of adaptation. No matter how well one understands a city's history, its ecosystems, and its enduring context, no matter how carefully one tries to anticipate the future, there will always be unforeseen circumstances to which a city must adapt.

Design resilient cities

An ecosystem's resilience is a measure of its ability to adapt or adjust to change, whether caused by internal or external processes (Pickett *et al.* 2004). Resilience is a useful concept for urban designers in thinking about cities that are adaptable to changing conditions and needs (as opposed to the concept of sustainability, for example, which implies maintenance of a stable state). In *The Resilient City*, Vale and Campanella (2005) brought together cases of how twelve cities recovered from a variety of historic disasters (war, fire, earthquakes, floods) to draw lessons for how cities might better prepare for and respond to catastrophe, whether predicted or not. Unlike earthquakes, hurricanes, floods, and drought, whose risks in a particular place are known, phenomena like economic and cultural shifts, new technology, and changes in the global environment are less predictable.

Kevin Lynch describes a range of additional physical design strategies for enhancing a city's ability to adapt to future change: avoid urban form that is too narrowly specialized such as districts that consist entirely of a single, specialized land-use; encourage a diversity of buildings and neighborhoods; adopt an additive structure, one that can accommodate growth or decline at the periphery without major change to the overall structure at the center of a neighborhood or city (a grid, is an example of additive structure); employ temporary structures or uses, when appropriate, especially true for uses in which technology is changing rapidly; utilize communication systems to accommodate changing needs rather than radical alteration of the city's physical structure (Lynch 1958). Urban form that is congruent with the enduring context of a city's natural environment will also be more resilient.

Act comprehensively and incrementally

Major challenges like climate change and recovery from economic recession may require a comprehensive and rapid response, but it is dangerous to implement a single model for change. Massive large-scale interventions often produced unforeseen effects, which may be devastating, such as those precipitated by urban renewal of the 1950s and 1960s. Diverse approaches, implemented incrementally, provide the opportunity to learn from failure and success and to respond; such solutions should fit local conditions, tailored to the needs of specific people in particular places. But incremental projects should be undertaken as part of a comprehensive framework for large-scale investment that addresses regional needs. The local view gives an intimate view of the habitat of individuals and small groups; an overview gives a broader perspective of larger systems.

Ecological urbanism and the future of urban design

Much is known about the urban natural and social environment, and there exist many successful models of ecological urbanism. Yet most of these examples are not known to the public, to natural and social scientists, or even to urban designers and planners. Ignorant of existing knowledge and precedents, researchers and practitioners repeatedly reinvent the wheel. What is needed is a series of literature reviews on ecological urbanism and its subfields, which provide a critical, comprehensive overview of what is known: the principal themes and threads of inquiry; the keys works and contributions in each area; regions of agreement and the disputed territories; gaps in knowledge; potentially fertile areas of inquiry; and models of practice that deserve to be replicated.

Much is still not known about the urban natural environment and the processes that shape it, and there is great opportunity for future research. Particularly promising are recent collaborations between urban designers and experts in other disciplines, such as ecology, economics, engineering, and art. Landscape architect Alex Felson and ecologist Steward Pickett, for example, describe design projects that are also ecological experiments (Felson and Pickett 2005).

The reasons for embracing and promoting ecological urbanism are compelling. At stake is the future of humanity and the human habitat, and whether we can adapt our behavior and settlements to meet the challenges we face (those posed by climate change and environmental contamination, for example, and by inequities in exposure to the hazards they represent) and whether we can do so in ways that are life-enhancing and life-expanding. Urban designers have an essential role, not merely in producing safer and healthier urban habitats, but in making legible and tangible the systems that support life, and in changing the perception of what is possible.

References

Alberti, L.B. ((1485) 1966). *Ten Books on Architecture*, Rykwert, J. (Ed.), New York: Transatlantic Arts.

Alexander, C., Ishikawa, S. and Silverstein, M. (1977). *A Pattern Language: Towns, Buildings, Construction*, New York: Oxford University Press.

Almy, D.J. (Ed.) (2007). *On Landscape Urbanism*, Austin, TX: Center for American Architecture and Design, University of Texas.

Benedict, M.A. and McMahon, E.T. (2006). *Green Infrastructure: Linking Landscapes and Communities*, Washington, DC: Island Press.

Berger, A. (2006). *Drosscape: Wasting Land in Urban America*, New York: Princeton Architectural Press.

—— (2009). "An Interview with Alan Berger," *Abitare*, posted February 10, 2009, http://abitare.it/featured/an-interview-with-alan-berger. (accessed 12 February 2009).

Calthorpe, P. and Van der Ryn, S. (1986). *Sustainable Communities: A New Design Synthesis for Cities, Suburbs and Towns*, San Francisco: Sierra Club.

Cronon, W. (1991). *Nature's Metropolis: Chicago and the Great West*, New York: W.W. Norton.

Felson, A. and Pickett, S.T.A. (2005). "Designed Experiments: New Approaches to Studying Urban Ecosystems," *Frontiers in Ecology and Environment*, 3: 549–556.

Fromonot, F. (2003). *Glenn Murcutt: Buildings and Projects, 1962–2003*, London: Thames and Hudson.

Geddes, P. (1915), *Cities in Evolution*, rev. ed. (1949). London: Williams and Norgate.

Graedel, T.E. and Allenby, B.R. (2003). *Industrial Ecology*, Upper Saddle River, NJ: Prentice-Hall.

Hester R. (2006). *Design for Ecological Democracy*, Cambridge: MIT Press.

Hippocrates (*c.* fifth century B.C.), "Airs, Waters, Places," in *Hippocrates*, vol 1, The Loeb Classical Library, T. E. Page, ed., Cambridge, MA: Harvard University Press, 1962.

Hough, M. (1995). *Cities and Natural Process*, London: Routledge.

Howard, E. (1902). *Garden Cities of To-Morrow*, Cambridge, MA: MIT Press, 1965.

Jacobs, J. (1961). *The Death and Life of Great American Cities*, New York: Vintage.

Johnson, B.R. and Hill, K. (Eds.) (2002). *Ecology and Design: Frameworks for Learning*, Washington: Island Press.

Klingle, M. (2007). *Emerald City: An Environmental History of Seattle*, New Haven, CT: Yale University Press.

Lyle, J.T. (1994). *Regenerative Design for Sustainable Development*, New York: John Wiley.

Lynch, K. (1958). "Environmental Adaptability," *Journal of the American Institute of Planners*, 24(1): 16–24.

—— (1981). *A Theory of Good City Form*, Cambridge: MIT Press.

—— (1990). *Wasting Away*, San Francisco: Sierra Club.

Marsh, George Perkins (1864). *Man and Nature*, New York: Charles Scribner.

Mathur, A. and da Cunha, D. (2001). *Mississippi Floods: Designing a Shifting Landscape*, New Haven, CT: Yale University Press.

McHarg, I.L. (1967). "An Ecological Method for Landscape Architecture," *Landscape Architecture*, 57 (January): 105–107.

—— (1969). *Design with Nature*, Garden City: Natural History Press.

Meyer, E. (2008). "Sustaining Beauty," *Journal of Landscape Architecture*, 1: 6–23.

Mostafavi, M. and Najle, C. (Eds.) (2003). *Landscape Urbanism*, London: Architectural Association.

Mumford, L. (1968). *The Urban Prospect*, New York: Harcourt Brace Jovanovich.

Olmsted, F.L. and Vaux, C. (1887). *General Plan for the Improvement of the Niagara Reservation*, New York: Niagara Falls.

Pickett, S.T.A., Cadenasso, M.L. and Grove, J.M. (2004). "Resilient Cities: Meaning, Models, and Metaphor for Integrating the Ecological, Socio-economic, and Planning Realms," *Landscape and Urban Planning*, 69(4): 369–384.

Spirn, A.W. (1984). *The Granite Garden: Urban Nature and Human Design*, New York: Basic Books.

—— (1985). "Urban Nature and Human Design: Renewing the Great Tradition," *Journal of Planning Education and Research*, 5(1): 39–51. Reprinted in J.M. Stein, (Ed.) (1995). *Classic Readings in Urban Planning*, New York: McGraw-Hill.

—— (1987). "Air Quality at Street Level: Strategies for Urban Design," In Vernez Moudon, A. (Ed.) *Public Streets for Public Use*, New York: Van Nostrand Reinhold. Abridged, see full report at: http://www.annewhistonspirn.com/pdf/Air-Quality_1986.pdf. (accessed 10 August 2010).

—— (1988). "The Poetics of City and Nature: Toward a New Aesthetic for Urban Design," *Landscape Journal*, 2(2): 108–126.

—— (1998). *The Language of Landscape*, New Haven, CT: Yale University Press.

—— (2000). "Ian McHarg, Landscape Architecture, and Environmentalism: Ideas and Methods in Context." In Conan, M. (Ed.) *Environmentalism and Landscape Architecture*, Washington, DC: Dumbarton Oaks.

—— (2005). "Restoring Mill Creek: Landscape Literacy, Environmental Justice, and City Planning and Design," *Landscape Research*, 30: 359–377.

Thompson, G. and Steiner, F. (Eds.) (1997). *Ecological Design and Planning*, New York: Wiley.

Vale, L.J. and Campanella, T.J. (Eds.) (2005). *The Resilient City: How Modern Cities Recover from Disaster*, New York: Oxford University Press.

Van der Ryn, S. and Cowan, S. (1996). *Ecological Design*, Washington: Island Press.

Waldheim, C. (Ed.) (2006). *The Landscape Urbanism Reader*, New York: Princeton Architectural Press.

Wenk, W. (2002). "Toward an Inclusive Concept of Infrastructure." In Johnson, B.R. and Hill, K. (Eds.), *Ecology and Design: Frameworks for Learning*, Washington, DC: Island Press.

White, R. (1996). *The Organic Machine*, New York: Hill and Wang.

Wines, J. (2000). *Green Architecture*, New York: Taschen.

Further reading

Hester, R. (2006). *Design for Ecological Democracy*, Cambridge, MA: MIT Press. Integrates ecological and social dimensions of urban design.

Johnson, B.R. and Hill, K. (Eds.) (2002). *Ecology and Design: Frameworks for Learning*, Washington, DC: Island Press. A compendium of concepts and applications of ecological design.

Spirn, A.W. (1984). *The Granite Garden: Urban Nature and Human Design*, New York: Basic Books. A comprehensive overview of ecological urbanism.

46

Metropolitan form and landscape urbanism

Brenda Scheer

Do we know how to design a metropolitan region, the now-ubiquitous urbanized territory sprawling fifty or one hundred miles without a break? Can we even conceive of it as a place with its own identity? Even if we can imagine ways to conceptualize design ideas at the metropolitan scale, can we imagine a level of control that still corresponds to our traditional idea of "design"?

Much of the contemporary urban landscape is a loose, flat, agglomerated field, interspersed with natural landscape, large industrial uses, airports, shopping malls, high schools with enormous sports facilities, stadiums, office parks, subdivisions and a vast, flattened landscape devoted to parking. Most commentators decry it as formless sprawl: without structure and too amorphous to have identity.

Even describing this landscape is difficult. Although the notions of concentric rings of "center, suburb, and periphery" are clearly obsolete, urban designers have not coalesced around a conceptual framework of metropolitan form that embraces both its scale and its physical diversity.

Robert Lang (2003) postulates two formal conceptions. One is the idea that the metropolis is (or could be) multi-centered, with the "ur-center" of the historic downtown, and a distributed set of mini-downtowns. These are imagined as mixed use centers with higher density than the usual suburban development, preferably connected by transportation networks. The second conception is that of a non-centered metropolis, or, as Lang puts it, "edgeless" city, where business land uses (for example) do not coalesce in significant centers, and do not coincide with higher density housing or with mixed uses, since this is not a necessary condition in an auto-centered metropolis (Lang 2003: 10).

Drawing on the first conception, a frequently suggested metropolitan design strategy is to propose more, higher density urban centers (Ewing *et al.* 2008) to absorb growth and offer greater potential for sustainability. Dunham-Jones and Williamson (2008) note an increasing suburban trend to redevelop large malls and other derelict sites into mixed use housing and retail, which they consider a significant first step in creating dispersed centers.

But even those who firmly support the multi-centric strategy concede that the metropolitan landscape cannot be substantially reconfigured into something resembling a traditional urban setting. Even if we stopped adding territory to metropolitan areas tomorrow (which is unlikely), what has already been built is difficult to reshape. Highways, low-density housing, and the corresponding vast extent of the metropolis will remain the dominant urban form in the US for many decades.

In fact, after twenty years of promotion, compact mixed use projects still constitute less than half of 1 percent of the urbanized land area – trailer parks are more prolific (Wheeler 2008: 406–407).

Complicating our ability to conceptualize the metropolitan landscape is the significant change in how we inhabit and understand this kind of city. In traditional cities, the center was a necessary place of shared economic, cultural and social experiences. The central city's key monuments and public spaces were inhabited and understood by all residents. Today, the distributed form and uses of the metropolis make it unnecessary to inhabit or even visit the center of a large metropolis. Robert Fishman suggests that our idea of "urban" – a place of common understanding and coming together, simply does not apply anymore. He suggests that a reordering of our perceptions has already occurred: the "center" of a metropolis is now the individual household, not a shared place (Fishman 1990). Each household develops a distinct perception of the urban landscape, circumscribed by its daily trips and choices. My Starbucks, my job, my movie theater, my daycare – these tend to be located in a limited orbit, which may be a substantially different orbit than my neighbors', and is likely to have very little overlap with a person living five or twenty miles from me.

Urban design has traditionally involved shaping the public realm as a series of outdoor rooms or axial spaces defined by built form and cultivated landscape. Urban designers cannot apply these concepts to the metropolitan scale, with its characteristic lack of central focus and low density. The urban designer's obsession with pedestrian scale also loses meaning in a city where speed and vastness are characteristic. Problematic, too, is the pervasive idea of urban design as designing a "product" – a large project conceived and built as a whole, which is impractical at the scale of the extended metropolis. Are there other ways to think of urban design that could have more impact on the metropolitan landscape?

Ecological urbanism

Charles Waldheim (2006) has written, "Landscape Urbanism describes disciplinary realignment currently underway in which landscape replaces architecture as the basic building block of urbanism." Although it goes by many names (urban ecology, landscape urbanism, landscape ecology), this reinvigorated movement is potentially a very powerful response to the problems created by metropolitan form (see also chapter by Spirn). Waldheim (2006) calls upon the groundbreaking work of landscape architect James Corner (Corner and MacLean 2000), as well as drawing on much earlier principles of landscape ecology developed under traditional urban configurations.

In 1984, Michael Hough proposed that ecological processes be used as a principle and model of urban design. Hough was only the latest in a series of important landscape architects and planners to foreground the natural setting as a key component of urban form. For centuries, the dominant conception of urban form was architectural – the ideal city consisted of buildings, streets and civic spaces, and the countryside was its treasured opposite: a place of natural repose or bucolic productivity. When Patrick Geddes first set about defining modern planning in the nineteenth century, he specifically turned to biological conceptions and analogies to articulate the relationship between a city, its inhabitants, and its corresponding countryside (Welter 2002).

In the mid-twentieth century, Ian McHarg reinvigorated the notion that urban design and planning should account for the natural environment. In his highly influential, *Design with Nature* (1969), he proposed to selectively limit urban development by directing it away from fragile, beautiful, or critically important natural ecologies, especially in areas that were in the path of urban expansion. Natural areas thus preserved could serve as an outlet for city dwellers. McHarg's invention of the layered mapping system of analysis led directly to today's computerized mapping GIS tools.

Hough's ideas took him in a different direction. He explicitly rejected the conceptual separation of nature and city, insisting that the city exists within an important natural landscape and has reciprocal and critical effects on it. He particularly disdained the high-energy cultivated urban landscape (lawns and streetscape) for its unnecessary lack of ecological diversity and productivity. He imagined a city that was designed to mimic natural processes by waste re-use, species diversification, water collection and recharge, food production, and wildlife support. He also firmly supported an enlightenment ideal, popularized by Frederick Law Olmsted, that contact with the natural environment was a necessary, civilizing force for society.

In recent years, urban ecology has once again been invoked as a potential design approach. The global warming crisis is certainly one provocation, but the extensive loss of the countryside to development has effectively distanced all city dwellers from the natural landscape.

Landscape urbanism specifically references the metropolitan sprawl that now physically characterizes the city (Figure 46.1). In this design conception, landscape is both an analogue of the city and its description. The analogue suggests how the city has become like a landscape, an endless and

Figure 46.1 Aerial image of Texas Stadium. Source: Alan Berger, *Drosscape*, Princeton Architectural Press, pp. 162–163. Used by permission.
Note: Landscape urbanists use the aerial photograph as an analytical tool. In this image of Texas Stadium, Alan Berger documents the substantial land area devoted to what he terms, "*drosscape.*"

boundless territory of diverse fields and flows, both natural and human-made. This conceptualization sees the city as, necessarily, an ecosystem, but one that has dependencies on imported energy and human-made intervention that can overwhelm natural systems. The urban landscape contains surfaces, areas and systems that overlap, collide, and shift. It is characterized by a wide variety of urban typologies, analogous to plant communities. Some are named and well described, like office parks and subdivisions, freeway intersections and airports, but some are nearly invisible or lack identity, like vehicle storage lots, utility corridors and edgeless corridors of single office buildings.

The "city as landscape" analogy suggests that the city can have common ground with nature: it invokes ideas of evolution, rapid and incremental change, interdependency of parts (ecology), and the productive reuse of waste.

Another conceptualization of the "city as landscape" is the nature of the physical situation of the city itself: broad and without boundaries, the city lies within a natural landscape and is defined and limited by it in ways that have not been important in a hundred years. Rejecting the dichotomous concept of "city" as a place of vertical density opposed by the "country," a relatively natural setting, the urban landscape is neither. Instead, it is everywhere both at once, ideally using the framework of the regional landscape as an important urban design element and motivator of change. For this expanded role, the term "landscape" must escape the confines of green formal lawns, gardens or parks and regain McHarg's concept as the space of potential and realized urban development, with the resultant dependencies and intermingling of natural and human-made systems and architecture.

Landscape urbanism's most pervasive design idea is to emphasize the natural systems that already exist in the metropolis, recovering them and foregrounding them as shapers of metropolitan image. Topographic changes, waterways, and natural landscapes are interpreted and expressed as a way of regional differentiation. The geography of the place is not only an aesthetic component. It is intimately tied to the historic and economic foundation of all places and remains a powerful determinant of urban form, shaping culture and identity. Living in concert with the landscape, while broadly and widely inhabiting it, is different from thinking of landscape as an element of design in contrast to architecture.

The natural systems also become a stepping off point for imitating natural processes. The ideal is to model the city as a self-sustaining dynamic system: recycling its own waste, producing its own energy, and otherwise balancing inputs and outputs. To even begin this task requires looking holistically at urban processes and accepting the idea that waste, for example, might become a resource (Figure 46.2) (Berger 2006). The city also contains reciprocal and responsive conditions, which are rarely accounted for in urban design (Lerup 1995). For example, disordered strip centers are the necessary resultant and the support system of the orderly subdivisions behind them. Outside the boundaries of exclusively residential neighborhoods are the gas stations, storage lockers and big box theaters that serve the residents of these neighborhoods, but are not allowed in (Scheer 2007). Every shop lining an urban street generates multiple shipping containers stored in a rail yard or loaded on a truck.

In all the ideas of urban ecology, the metropolitan landscape is not considered a static object, but a living and growing system. Like a forest, it is complicated and has elements that change on many different time scales. The current form of the city is a palimpsest of modern functionalist buildings and parking, superimposed upon and

Figure 46.2 High Line Park in New York City. Source: Yuka Yoneda, courtesy of Inhabitat http://www.inhabitat.com

Note: The High Line Park in New York City is an example of a project where a "wasted" piece of land (an abandoned railway line) is reclaimed for landscape and recreation. The High Line design is led by James Corner Field Operations, with Diller Scofidio + Renfro.

securely bounded by the property lines of former farms and small towns, nestled in ancient valleys that are fed by streams that are captured and controlled over generations (Scheer 2001). This is a solid representation of the time and scale in the shaping of a metropolis: from ancient landform to tomorrow's new construction.

Like any evolving system, the urban landscape requires flexibility and elasticity to accommodate change. Kevin Lynch (1981) proposed that the ability to change was essential to the definition of good city form, but despite this early warning, the static "master plan" is still the *sine qua non* of traditional urban design.

By contrast, landscape urbanism takes explicit account of change and has developed several strategies to accommodate continuous evolution. The first is to design and privilege open systems of physical infrastructure, rather than a full and specific architectural plan. The city's infrastructure defines important systems of order for designers. Infrastructure includes streets, transit, highway interchanges, but also water distribution and importantly, energy networks. Infrastructure can also include air terminals and routes, interstate trading networks, and communications networks. "Infrastructure" can also refer to ownership and political subdivisions that structure land and limit its uses.

Importantly, infrastructure systems are resistant to rapid large-scale change, unlike buildings or land uses which are relatively impermanent and short-lived (Scheer 2001). The potentials and limitations of the infrastructure are thus critical tools for the urban designer, easily as important as individual buildings or the codes that shape them, and with greater influence over longer periods of time. Location and design of infrastructure, which is the relatively static component of the city, provides a rigid framework that allows land use, architecture, and landscape to remain flexible but orderly and defined.

Another strategy for dealing with change is the planned obsolescence of particular uses or forms. A temporary use, including a landscape or building, can be cycled out in phases. Landscape has particular potential for short-term healing of abused places, or as a placeholder for the next planned cycle of more intense use. Designed landscapes or natural areas thus become a healing mechanism, especially in concert with built form. For example, devastated inner cities can be revived as landscape temporarily replacing vacant lots, as in proposals for Detroit (Shane 2004) or Brooklyn (Brown and Morrish 1994).

Because of the fluid and dynamic nature of the metropolitan form, urban design as landscape urbanism requires a critical balance between control and flexibility. Limited control of the field of design distinguishes landscape urbanism ideas from the "big architecture" camp of urban design – plans for large scale projects that describe every building and every open space and require large scale ownership or heavy-handed political control.

Individuals actively working in this vein are commonly some combination of ecologist, landscape architect, politician, urban designer, planner, scientist, engineer, or architect. Designers, broadly defined, may or not may not work for a "client" in the traditional sense of having a discrete task (master plan or building design), a site, a time scale, and a contract. Frequently, the designer instigates the work or advocates for it or simply carries it out and leads a change in direction (Berger 2006). Organizations like *Envision Utah*, which identifies and funds its own design projects, and then markets the recommendations to constituents and agencies, provide a template for this kind of design. In the absence of regional government, civic and advocacy groups may provide the only possible method of implementation (Yaro 2000).

In these roles designers act more as researchers or activists, seeking support for propositions and experiments, testing ideas and theories. This alternative approach and cross-disciplinary participation yields ideas and plans which are fragmented, incomplete, suggestive, loose, and yet distinctive (see examples of projects in Czerniak and Hargreaves 2007). Partial completion is often the norm, since the "design" may not be much more than setting up a series of strong frameworks (including natural systems) and effective processes for managing transformation. It may be necessary to imagine and design a cross-boundary "authority" to carry out the plan. A metropolitan landscape strategy may also require public relations, branding and promotion of the central idea so that the "summoning up" of the metropolitan perception has life outside specific designs for "projects" (Healey 2007).

Metropolitan scale and urban design

What would be a successful metropolitan design? Our goal as urban designers is always to improve the daily life and sensibility of the inhabitants and visitors, to bring greater access and opportunity to all, to create places for people to come together, and of course, to assist with the great project of making a more sustainable world. In addition to these traditions, metropolitan design would need to account for all typologies of place, not just traditional centers. It would need to distinguish and create places within the metropolitan landscape. It would recognize speed and movement and the variable daily circuits of household life. It would recognize the need for flexibility and different rates of change. It would celebrate the diversity of the metropolitan landscape and conserve its resources. Finally, it would need to operate within the values of democracy, entrepreneurship, local control and individualism that shape the fabric of this kind of city.

The struggle to design at the regional scale began as early as the late nineteenth century with Ebenezer Howard's ideas of a central city surrounded by reserved open spaces and smaller satellite settlements. Early twentieth-century planning advocates like Lewis Mumford, Benton MacKaye, and Clarence Stein moved expeditiously to import this regionalism to the fast growing cities of the east coast, by proposing dispersed centers or corridors and associated green belts. These ideas, which separated nature and settlements, were frustrated by the lack of a regional governing mechanism and the low-density sprawl that subsequently consumed the countryside (Fishman 2000).

These same frustrations exist today, but the problem is compounded by actual artifacts on the ground – existing networks, sprawling subdivisions, suburban typologies – and the urgent need to conserve resources. At the scale of the region, it is tempting to work on technical solutions (transit, drainage, air pollution, land use, governance) without taking account of the regional, aesthetic "sensibility" issues identified by Lynch (1976).

At the metropolitan scale, our sense of the city is not immediate and graspable in a pictorial way, like the common picture of a downtown street or a riverfront park, which a person or a group can literally grasp in its entirety by being there. As we have seen, a metropolitan sense is shaped by a series of experiences so that the metropolitan form is created as an abstract in the mind of each individual.

Creating a collective metropolitan sense would seem to be one important order of business for designers. This collective sense could aid in the perception of the region's unique character, its accessibility and diversity, and in the protection and enhancement of valued places. If the metropolitan form continues to be seen as hopelessly disordered, there may be a tendency to overlook the potential for large-scale design in favor of small-scale interventions that leave most of the urban landscape without guidance of any kind.

The first step in recognizing the scale and scope of the metropolitan design problem is a reordering of design priorities, which is well underway. It is not too difficult to imagine a time soon when interpreting, reviving, and integrating natural systems is the very first order of business for the urban designer. These systems are all-encompassing, historically significant, uniquely beautiful, and critical to the ecological functioning of the region. Landscape urbanism, with its emphasis on large and small natural systems, a multi-layered physical infrastructure, cradle-to cradle ideals, and a flexible level of development control, offers a way of managing urban design at a metropolitan scale.

References

Berger, A. (2006). *Drosscape: Wasting Land in Urban America*, New York: Princeton Architectural Press.

Brown, C. and Morrish, B. (1994). *The Productive Park: New Waterworks as Neighborhood Resources*, New York: The Architectural League of New York and Princeton Architectural Press.

Corner, J. and MacLean, A. (2000). *Taking Measures across the American Landscape*, New Haven, CT: Yale University Press.

Czerniak, J. and Hargreaves, G. (Ed.) (2007). *Large Parks*, New York: Princeton Architectural Press.

Dunham-Jones, E. and Williamson, J. (2008). "Retrofitting Suburbs: Instant Cities, Instant Architecture and Incremental Metropolitanism," *Harvard Design Magazine*, Spring/Summer.

Ewing, R., Bartholomew, K., Winkelman, S., Walters, J. and Chen, D. (2008). *Growing Cooler: The Evidence on Urban Development and Climate Change*, Washington, DC: The Urban Land Institute.

Fishman, R. (1990). "Megalopolis Unbound," *Wilson Quarterly*, 14(1): 25–47.

—— (2000). "The Death and Life of American Regional Planning," in Katz, B. (Ed.) *Reflections on Regionalism*, Washington, DC: Brookings Institution Press. 107–123.

Healey, P. (2007). *Spatial Complexity and Territorial Governance*, London: Routledge.

Hough, M. (1984). *City Form and Natural Process*, London: Elsevier Science.

Lang, R. (2003). *Edgeless Cities: Exploring the Elusive Metropolis*. Washington, DC: Brookings Institution Press.

Lerup, L. (1995). "Stim and Dross: Rethinking the Metropolis," *Assemblage 25*, Cambridge: MIT Press.

Lynch, K. (1976). *Managing the Sense of a Region*, Cambridge, MA: MIT Press.

—— (1981). *A Theory of Good City Form*, Cambridge: MIT Press.

McHarg, I. (1969). *Design with Nature*, American Natural Museum of History.

Scheer, B.C. (2001). "The Anatomy of Sprawl," *Places: A Forum of Environmental Design*, Fall, 14(2): 26–37.

—— (2007). "The Shape of the City: The Future of Master Plans," *Planning*, American Planning Association, July, 30–33.

Shane, G. (2004). "The Emergence of Landscape Urbanism: Reflections on Stalking Detroit," *Harvard Design Magazine*, Fall/Winter.

Waldheim, C. (2006). "A Reference Manifesto," in Waldheim, C. (Ed.) *Landscape Urbanism Reader*, New York: Princeton Architectural Press, pp. 15–19.

Wheeler, S.M. (2008). "Built Landscapes in Metropolitan Regions," *Journal of Planning Education and Research*, 27: 400–416.

Welter, V. (2002). *Biopolis: Patrick Geddes and the City of Life*, Cambridge, MA: MIT Press.

Yaro, R.D. (2000). "Growing and Governing Smart: A Case Study of the New York Region," In Katz, B. (Ed.) *Reflections on Regionalism*, Washington, DC: Brookings Institution Press, pp. 43–77.

Further reading

Lynch, K. (1976). *Managing the Sense of a Region*, Cambridge, MA: MIT Press. One of the first urban design texts to consider design at the regional scale.

McHarg, I. (1969). *Design with Nature*, Washington, DC: American Natural Museum of History. Classic treatise on how to design cities and urban areas privileging the natural ecology and regional landscape.

Thompson, G. and Steiner, F. (Eds.) (1997). *Ecological Design and Planning*, New York: John Wiley and Sons, Inc. Collection of essays focusing on the design and planning towards an ecological landscape.

47

Intertwist and intertwine

Sustainability, meet urban design

Randolph T. Hester and Marcia J. McNally

Today's widespread popularity and increasing applicability of sustainability is intertwined with the practice and conceptual underpinnings of urban design, most notably the city as democratic setting, as ecological antagonist, and as aesthetic delight. Although the sustainability argument now embraces deep and ongoing participation of community members in city building and urbanity made impelling by social and sensory engagement, it originally addressed a worldwide ecological crisis. Problems of ecosystem collapse, species extinction, resource scarcity, and suburban sprawl – the results of profit-driven decision making for short-term results – demanded that almost every aspect of American life, including city design, be rethought. The idea of sustainable development meant that the health of future generations as well as our own was to be considered. Thus it is through this crisis that sustainability has most dramatically influenced urban design. This chapter discusses two generations of thinking and best practices; first, those early design principles that have contributed to the concept of sustainability and second, those aspects of city form that sustainability is most reshaping.

Best practices, undeniable mandates

The 1987 Bruntland Report and the 1992 Earth Summit spawned a generation of thinking about sustainability and created a foundation for new thinkers who see the world and its systems in an even more dense and at-risk web than their predecessors. Interdisciplinary problem solving is now the norm, not the exception. One must think globally, act locally, and oh, but yes, act globally, too.

The 1990s saw the proliferation of definitions, principles, and frameworks for sustainability. Fortunately there are now plans that are implemented, projects on the ground, theory and science put to use. As a result, sustainable design practices provide guidelines that are simultaneously high-minded, imageable, and pragmatic; demonstrating how sustainable cities can 1) have centers of social life; 2) be interconnected between virtually everything across every scale; 3) promote transparent fairness in the process and production of place; 4) display sensible and innocuous expressions of aspiration and status; 5) enable satisfaction of our most deeply held and noble values; 6) employ form particular to a region and climate in the broadest sense; 7) encourage cultural and biological diversity; 8) favor density and smallness; 9) define a limited extent to the city such that it provides most of what it requires from within the region; 10) engage adaptable management practices to guarantee that no crisis is ever wasted in the pursuit of sustainability; 11) satisfy everyday life patterns even as the city radically reshapes; 12) ensure ready access to nature; 13) allow citizens to inhabit science by reducing ecological illiteracy and making natural processes part of daily life; 14) expect reciprocal stewardship among people and between people and place; and 15) choreograph a pace of life that encourages thoughtful decision making (Hester 2006).

But what is the origin of this prescription? In considering the most appropriate rules of engagement for sustainability, it is important to look to the past before looking forward. This chapter reviews the various strains of urban design that are foundational to the first generation's ideas about sustainable design along with new principles that have recently emerged to address the crises of global warming, habitat loss, and public health epidemics.

Tracing the three Es

While one does not typically attribute the "three Es" – Environment, Equity, and Economy – to urban design thought, urban designers were some of the earliest integrated thinkers in the physical interventionist professions, working across the space spectrum and in concert with a wide-ranging set of disciplines. By the 1960s and 1970s community designers were translating the research of anthropologists and geographers to develop spatial patterns for social engagement. Planners took on citizens as clients and advocated their positions as if lawyers. Environmental psychologists and planners together created simulation methodologies to anticipate and test the impact of building heights on pedestrians. During this same time the application of the underlying ecological thinking was also rapidly emerging. Michael Hough (1984), for example, had begun reshaping Toronto to maintain watershed ecologies and create "natural" social spaces in the city's center. By the mid-1960s, ecology became a requirement in landscape architectural curricula. Some of this thought percolated and mainstreamed into urban design practice. A tacitly-agreed upon set of urban design best practices was born out of these collaborations and professional adaptations, which provided a powerful starting point for sustainable design thinking. For the first time, the mandate to plan for future generations demanded that design of the built environment consider the entire life cycle of the city and the ecological footprint of consumption.

The first "E": walkable, livable density

What were some of the starting points? Urban designers think about cities as people experience them in their everyday lives; employing "the principles of classic urbanism: walkable streets, human-scaled buildings, an active public realm, and meaningful and context-relevant places" (Larice and Macdonald 2008). An enduring example is Kevin Lynch's (1960) *The Image of the City* in which he pointed out that people have to be able to "read" the city in order to make sense of it, move around, and enjoy dwelling in it. Concluding that a city's "legibility" is essential not just for way finding but also for developing a sense of orientation and worldview, Lynch developed a simple vocabulary that consisted of nodes, landmarks, districts, edges, and paths. This vocabulary combined environmental psychology with city form making and became one of the most enduring urban design tools. Detailed studies of territoriality, sociopedality, and personal space (Hall 1966; Sommer 1969) and broader issues of how the environment influences our psyche (Proshansky *et al.* 1976) strengthened Lynch's hypotheses and significantly informed urban design practice (see also chapter by Nasar in this volume).

As twentieth-century urbanity came to be dominated by the automobile, pedestrian environments were undistinguished and neglected. Urban designers resisted this trend. Most notably Donald Appleyard (1981) documented the negative impacts of car traffic on pedestrianism and neighborhood life. This led to a movement to make streets livable, curb the car, and improve public transportation. Today, in addition to the social interaction and neighboring concerns that motivated Appleyard's work, a variety of sustainability concerns – some old and some new – are in the forefront. These include walking to prevent heart disease, diabetes, and obesity; reduction of air and water pollution; and curbing greenhouse gas emissions.

Flawed experiments with high rise public housing in the 1940s and 1950s stigmatized density in the American mind for decades. Partly because of the poor living conditions, and partly because of the large-scale clearance that such high-rise development required, urbanists, sociologists, and journalists decried this grave misstep (Gans 1962). For the last half century, urban designers have sought to rescue density's honor. Inspired by the writings of Jane Jacobs (1961), and emboldened by considerable research, professional practice, and political experience that followed, designers searched for parameters of livable density. In the 1980s, Appleyard and Allan Jacobs (1987) joined forces on an urban design manifesto that set goals for urban life and defined a "fabric" which would encourage it. Usable metrics appeared from studies of urban housing types (Lozano 1990; Cooper Marcus and Sarkissian 1986). Their conclusions suggested that densities of 15–100 units per acre could be delightfully livable depending upon life cycle stage and social class.

In their writings and projects architects such as Leon Krier and Donlyn Lyndon have shown how density can be urbane and beautiful (Duany *et al.* 2003; Lyndon and Moore 1996). Others have researched the qualitative characteristics of density, concluding that it is not actual density (net or gross) but perceived and affective density that matters to people. One study discovered that whereas nearly one-third of American homebuyers still disapproved of higher density in any form, 20 percent desired it, and almost another 50 percent might be enticed to live in it, provided it was accompanied by certain added benefits such as better access to shopping, parks, and transit (American LIVES, Inc. 1995; Cervero and Bosselmann 1994).

Density today is seen as creating resilient cities in a multitude of ways. Reasonable, concentrated development affords exurban biological diversity and slows the extinction of rare and endangered species, provides access to nearby nature, enhances centeredness and innovation, reduces vehicle miles traveled and health costs, and tempers global warming (Hester 2006). Vancouver's EcoDensity project, for example, will accommodate a doubling of the city's population without inducing sprawl, thereby reducing the per capita carbon footprint. The project concentrates development within one square mile of the downtown. It is predicted that in excess of 75 percent of the city's newcomers will live within walking distance of work (Condon 2008).

The second "E": social diversity, civic engagement, and justice

Some modernist urban designers were among the primary architects of post World War II urban renewal and freeway construction that, in retrospect, dislocated poor people of color, discouraged public participation, and created widespread social inequities in access and distribution of resources. In contrast, others joined disenfranchised citizens to fight these projects. In parallel and in concert with civil rights struggles, this gave rise in the mid-1960s to the advocacy and community design approaches to city making that challenged the objectivity of professional experts, questioned whose goals and values were shaping the city, and sought to help those who bore the disproportionate burden of these decisions (Davidoff 1965; Sanoff 2000; Hartman 2002).

Design principles that emerged from these protests foretell by at least a decade some of the basic tenets of sustainability, including the need for widespread public participation in city building, freedom of information, and the recognition of different needs of different social groups in space. Changing attitudes towards grassroots democracy, fairness, and cultural relativism were redirected to build a new sense of community at the same time that the physical city was being remade. Barn raising projects (Linn 2007), co-housing neighborhoods (McCamant and Durrett 1994), and design for tribal units (Alexander *et al.* 1977) became more widespread as urban design interventions. Today this line of inquiry focuses on everyday urbanism, globalization of culture, and far more complex immigration and integration issues than those of the civil rights era (Crawford 1999; Sandercock 2000).

The recent American experience of embracing multicultural participation and place making has been lumpy. One success can be found in the redesign of the Sixteenth Street Mission BART Station in San Francisco. Located in a neighborhood of lower-income immigrant families, the station is part of a lively commercial district with its share of social problems which caused residents and transit patrons to pass through it "quickly and uneasily." In 1996, a series of community "dialogues" was launched in hopes of addressing eroding neighborhood conditions and improving the limited public spaces while keeping a watchful eye on gentrification. Participants developed the idea to allow artists and vendors onto the plaza space, making it into an animated, outdoor exhibit and market area. Contested and unresolved claims on this postage-stamp-sized site were resolved through years of debate, design proposal, adaptation, and open-mindedness (Rios 2008).

The third "E": reuse, multipurpose actions, and new economies

Charmed by the architectural virtues of old buildings, designers sought ways to reuse them often with mixed rather than single uses. Only later did green architects realize the resource savings of such actions. A reused older building is almost always more sustainable than new construction; designing for multiple functions simultaneously conserves resources. Recently these ideas have evolved into the guiding principles for creating more sustainable economies.

As historic preservationists challenged the demolition of traditional buildings to make way for new construction, urban designers extended these practices to entire neighborhoods, districts, and towns. This often required finding ways to retrofit old typologies of built form for new functions, one of the earliest examples of large-scale reuse and recycling. One of the most dramatic cases was Michael Southworth's 1970s Urban National Park for Lowell, Massachusetts in which his team sought to recycle the derelict textile city into a new economy by reusing historic factories and canals. His Discovery Network introduced the idea of "the educative city" whereby historic functions were made transparent (Southworth and Southworth 1974).

Urban reuse exemplifies the more recent sustainability principle of adaptive management. One such makeover can be found in the reinvention of Durham, North Carolina, once the cigarette capital of the world, now the City of Medicine. This transformation has entailed more than replacing the economy of Liggett & Meyers Tobacco Company with the medical industrial complex of Duke Hospital, however. Literally the factories once used to manufacture tobacco products are being recycled for high tech research space, housing, and commercial uses.

By the middle of the last century, American city building relied not only on clearing the old to make way for the new, it was also predicated on specialization and segregation of functions. Municipal departments operated in the vacuum of their own disciplines of traffic, water, sewer, public housing, parks, recreation, economic development, public health, and so on. City planning departments zoned single land uses, while developers specialized in single project types. Together these decisions created a pattern of single-purpose uses with little consideration of how one affected the other. Urban designers were among the first to realize the many undesirable impacts of such practices on livability and community and argued that these systems had to be considered simultaneously to revive urbanity.

Until very recently this idea met with extraordinary resistance, but this is changing as multifunctional infrastructure, for example, is being found to be economically efficient as well as ecologically preferable. Today the principle of holistic systems thinking is fundamental to sustainable development. Basic concepts of adjacency, mutualism, chains, webs, flows, networks, cycles, resource footprints, and conservation biology have been better conceptualized by proponents of sustainable development. Urban designers have played an important role, finding urban venues for mixing aesthetics, livability, and the environment with other functions. The living systems of the Todds (1994), John Lyle (1985), and Michael Hough (1990); the mixed use zoning of the new urbanists (Calthorpe 1993), and the multiple-use boulevards of Alan Jacobs and Elizabeth Macdonald (Jacobs *et al.* 2002) are notable examples.

Sustainability reel two

While contributions of urban design to sustainability have been substantial, they were largely socially and aesthetically motivated and less ecologically informed. Likewise the early environmental movement defined ecology primarily in terms of natural and life sciences, excluding humankind and development. People and cities were viewed as detrimental to healthy ecosystems. But as we argued previously, this has changed in recent years as systems thinkers, regardless of discipline, have embraced urban ecology (Urban Ecology 1996). The next section looks at ways that more ecologically-derived sustainability is influencing, and in some cases, dramatically changing the practice of urban design. Sustainability poses at least four challenges to design including: first, incorporating ecological thinking across every design scale to encourage green living; second, making nature a framework for city design in order to provide biological diversity and healing benefits; third, organizing urban areas to be self-sufficient in meeting their basic needs; and fourth, reshaping cities to avoid, accommodate, or mitigate natural hazards.

Ecological dwelling

City legibility and delight were not major considerations of the early ecological movement. Yet sustainable cities must be readable and impelling. To be impelling a city must be comprehensible and sensually delightful. The sensual experience, or the phenomenology of place, has long been a focus of urban design. Practical concerns about physical and psychological comfort, like the provision of sun pockets for hanging out in a cold climate (Hester 1984), more theoretical constructs of place and placelessness (Relph 1976; Lynch 1976; Seamon 1996), and ideas about place making as expressed in the journal *Places* emerged in the 1970s and 1980s. Now we see a more direct move to correlate the experience of urban landscapes to include daily access to nature. Experiencing nature helps keep us healthy, heals us when we are sick, shortens the time we are in the hospital, and diminishes the recurrence of illness (Ulrich 1984; Kaplan and Kaplan 1989). Being in nature reduces stress, combats mental fatigue, and makes us less fearful (Louv 2005). It stimulates our creativity, our primal and civic selves, and our commingling with ecological forces (Hester 2006).

Ecological design thinking, which is not linear but rather cyclical, messy, patchy, flowing, branching, webbing, and looping, can inspire ecological living as people live smaller in their housing square footages and other "footprints," and integrate disciplined conservation into their daily routines. This is not a new idea. As Louise Mozingo pointed out, we have been "insinuating ecological landscapes" into the city for some time (1997). Thus the leap from creek daylighting to manipulating flows and closing loops in stormwater systems may not be as radical as we think. How cities manage urban runoff in the future to prevent nonpoint source pollution will increasingly shape urban form in addition to meeting regulatory demands (France 2002). Detention and treatment will require re-engineering the entire stormwater system and in some cases the sanitary sewer system as well. Best practices are evolving, as early attempts to capture and detain runoff and filter pollutants were counterproductive to other sustainability goals. Suburbia provided demonstrations of on-site stormwater ponds, rainwater swales, and even residential landscaping but required low densities and sometimes undesirable materials to incorporate these single-purpose mechanisms. More successful cases – from Chicago to Portland to San Francisco – have incorporated new low-impact design treatments to manage stormwater runoff at the site level and are bringing cities into Clean Water Act compliance while at the same time creating exciting, useable public amenities without sacrificing city densities or livability.

Cultural and biological diversity

In order to sustain a healthy human population, city design must reincorporate nature. This is the phenomenological partner of ecological science. First and foremost it must be accessible from home in the immediate neighborhood but also as a structure of the overall city. The design of nature must provide calm and restful settings, ephemeral distractions, and spontaneous interactions. Urban nature must be elemental and simple, wild for some and tame for others. It should provide passive perspective and active engagement for a variety of family types and cultural groups. August Hawkins Nature Park in South Central Los Angeles provides one such example. Reusing a city storage yard for old pipes and culverts, the site was transformed to an 8.5 acre urban wilderness at the corner of Compton and Slauson Avenues (Figure 47.1). Today a visitor is greeted by exuberant neighborhood kids who want to show off the park's functioning arroyo fed by a wind-powered fountain (Figure 47.2 and Figure 47.3) (Hester 2006).

Concerns about biological diversity must shape urban form as much as cultural

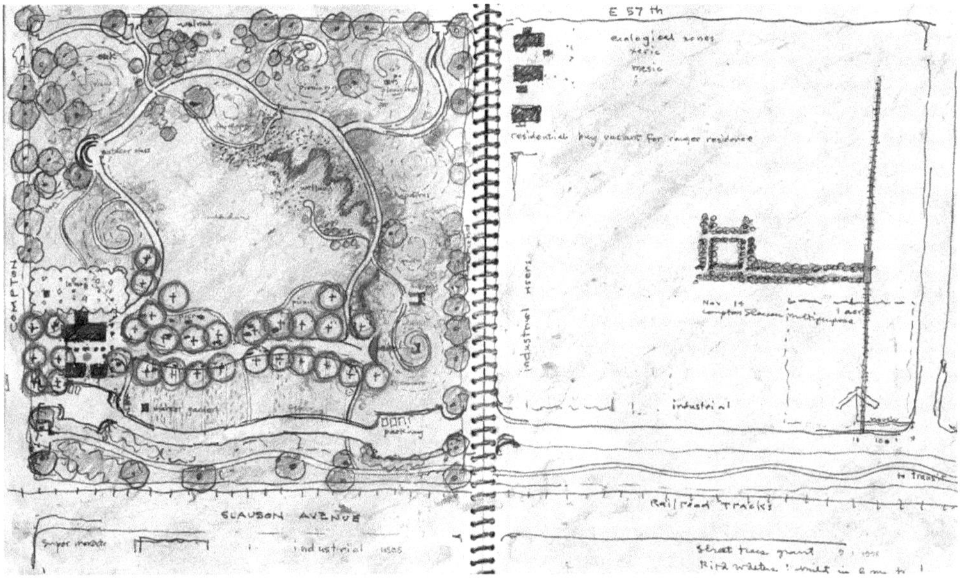

Figure 47.1 Authors' sketches for Augustus Hawkins Park in Los Angeles. Source: Randy Hester and Marcia McNally.

Figure 47.2 Windmill at Augustus Hawkins Park. Source: Randy Hester and Marcia McNally.

Figure 47.3 Children at Augustus Hawkins Park. Source: Randy Hester and Marcia McNally.

diversity in the future. Scientists believe that about 50,000 species are becoming extinct each year, and the cause is most attributed to habitat loss from urbanization. Conservation biology applied to terrestrial and aquatic ecosystems calls for protection and creation of core habitat and connecting corridors that support top predators and metapopulations. Actions to protect endangered species and ecosystems require nuanced design at the urban edge but also in ribbons and patches throughout the city that connect to intercontinental habitat and flyways as envisioned by the Wildlands Project (Dramstad *et al.* 1996). Among the first prescriptions are limiting the extent of urbanization and densification within a growth limit to create a greenbelt such as was pioneered in Portland. Other examples can be found in unlikely places such as Los Angeles, where the Santa Monica Mountains Conservancy and its partners have nearly completed an interconnected Big Wild that surrounds the city stretching from Hollywood Boulevard to the Los Padres National Forest. This greenway provides more than the 640,000 acres needed for a sustainable population of mountain lions and the entire ecosystem – urban biodiversity writ large (Hester 2006). New projects on the Los Angeles River recreate habitat for the 444 species of birds found in the region, provide new open space in some of the most underserved neighborhoods of the city, and reduce peak flows during storm events (Community Development by Design 2005).

Regional supply

Climate change will exacerbate an already epidemic urban problem. Resources that cities depend on like food, water, energy and most consumer goods are produced far from the cities that need them. The average carrot travels thousands of miles before it is eaten (Urban Ecology 1996). The multiple, deadly effects of long-haul transportation and environmental degradation in producer regions are now joined with emerging concerns about food safety and security and ecological illiteracy to offer a primary sustainability dictum: produce more of what the city needs within its region. For design, sustainable or urban, the challenges are reforming the city to incorporate these uses, particularly food and fiber production and processing, drinking water supply, and building materials. Where should they go? From what should they be isolated or be adjacent to? Additionally, these uses compete for limited space. What will they displace? With what uses can they be combined?

Natural hazards

Since Ian McHarg popularized hazard overlay mapping forty years ago (1969), landscape architects have sought to employ these methods in city design. Because the overlays of natural systems determined where to develop and where not to, this approach served primarily new town and green field development rather than influencing the form of existing cities. Still, avoiding flood-prone and landslide areas, liquefaction zones, aquifer recharge areas, and agricultural lands was useful to urban design decision making, albeit primarily at the policy and regulatory level. Unfortunately urbanity has increasingly concentrated in areas of high natural risk, denial being the unconscious strategy when the resource is out of sight, distant, and/or abstracted from daily life. It is not likely that many cities will be relocated to avoid the risks of floods, hurricanes, earthquakes, and rising sea level; rather it is expected that most cities will seek to accommodate or mitigate natural hazards through ecologically-informed zoning and building codes. But a few cities have relocated and more will, presenting unique urban design opportunities to restructure city form based on levels of hazard danger to become more resilient and catastrophe-immune.

The aquifer is a good example – it is underground and the recharge area may be hundreds of miles from the water customer who depends on the supply of clean water. Aquifer hydrology is not easily comprehended by common metaphors or everyday urban experience, as we are more likely to take responsibility for matters that are observable, comprehensible, and unabstracted. As a result, the protection of recharge areas is usually neglected until the aquifer is contaminated or intruded upon. Even surface water supply may be so removed from the urban experience that its role is denied in shaping the city. This has been the case of Southern California, which gets drinking water from as far away as Wyoming. In cities where limited water supply is revealed, such as Kyoto, water significantly reshapes urban form and public awareness. In Kyoto, water use is expressed through a diverse palette: the city's rivers and creeks are managed jointly for flood control and recreation, Beaux Arts aqueducts and channels openly send water from Lake Biwa to supply neighborhoods, while recycling fountains delightfully call attention to its arrival from the other side of the ridge (Hester 2006).

The dangers of aquifer breakdown are only the beginning. The ecological threats described by Al Gore as "inconvenient truths" will have devastating effects on cities unless they are restructured to adapt to climate change (Gore 2006). Sea level rise is the most immediately terrifying challenge to urban sustainability. This is in no small part because of simulations appearing in the popular media which allow us to envision thousands of coastal cities disappearing, including most of the east coast metropolises, from New York to Miami (Hertsgaard 2006). Some cities lacking political power will be left to their own devices. Most of the powerful ones will be fortified. Almost all cities will have to completely reconstruct primary infrastructure that supports transportation, stormwater management, sewer discharge, and energy distribution. The total cost of such adaptations will be astronomical.

Kristina Hill and Jonathan Barnett offer a sobering overview of our limited choices (2007). The most common retrofit approach has been to protect cities with sea walls, pumps, and other mechanical means that create a vertical barrier between city and sea menace, but the location and design of these barriers has been blunt as they stop ecological flows and divide the public from the day-to-day natural pleasures and ecoliteracy benefits associated with an urban waterfront. Unfortunately this approach to flood protection will likely dominate over more resilient ecological design alternatives for some time.

Other interventions are more challenging. As Hill and Barnett (2007) note, moving away from the coast is the last resort for larger cities. This may be reasonable in small towns with modest infrastructure, but will likely take many disasters before residents can be convinced to relocate. For example, portions of Cherokee, Iowa, a Little Sioux River community, reluctantly moved only after decades of flooding wore them down (Wagner 1998). It is possible that inhabited floor levels can be raised above anticipated sea level rise and storm surge using semi floating structures, stilt buildings, and/or regrading to raise the entire ground level. Most of these solutions have precedent in vernacular village architecture such as is found in the indigo farmsteads of Japan's Yoshino River valley. Here, the floor areas that must stay dry, such as those used for sleeping or crop storage, were traditionally built high to protect against flooding. Each farm stored a wooden life boat in the rafters which could easily be lowered in high water. The roofs of farm buildings were made to detach in the most serious flood events. Despite dam building, all of these techniques are still found in the valley today (Hester 2006). The transferability of these small-scale solutions to meet the necessities of large-scale topographical and hydrologic regrading, new building engineering, and civic typologies will admittedly require considerable experimentation.

Hill and Barnett (2007) believe that the use of restored and newly created wetlands, if done in concert with mechanical interventions, has the potential to offer significant protection to coastal cities. These "horizontal solutions" are based on the ecological principle that barrier islands, wetlands, and shallow water reduce the impact of storm surge. Such approaches entail filling shallow ocean fronts to build artificial islands and reefs with solid waste on a much larger scale, much like shipwrecks do now. A variation would be to float the barriers, as is done on Lalu Island in Taiwan (Hester 2006). In addition to surge reduction, these solutions provide the nursery habitat for almost all coastal ecosystems including fisheries, in contrast to the vertical barriers which, as mentioned above, exacerbate the loss of wetlands by starving them from land while the rising ocean drowns them from the sea.

In total, sea level rise and other climate change forces offer challenges at a scale urban design has never faced simultaneously before. And urban design can only meaningfully assist within a framework more forcefully structured by ecological principles than the one that has guided city making in the past. In fact, this crisis may be an opportunity for urban design to reshape cities to be more sustainable in multiple ways which few have considered. Design, however, will only be effective if approached through site-specific experimentation and by engaging an array of partners including coastal hydrologists, ecologists, and environmental engineers.

Sustainable and urban design practices must both be operative.

Final thought

Where does this leave us? Viable sustainability has required Americans to disassociate from modernist environments and to embrace ecological urbanity. Urban designers have been prime enablers of our realignment, impelling landscapes the siren. As the rules are expressed in mundane infrastructure and economically beautiful, public venues, hopefully we will be lured into eco living through place literacy (Orr 1992, Mozingo 1997). But make no mistake. Clever metaphor is not enough. To rewire our cities to be livable, safe from natural disaster, and refunctioning with the ecosystem requires considerable intestinal fortitude on our parts. Green principles, designers, scientists, and engineers will go nowhere without supportive leadership. Active civic engagement is more urgently required than ever before.

References

Alexander, C., Ishikawa, S. and Silverstein, M. (1977). *A Pattern Language*, New York: Oxford University Press.

American LIVES, Inc. (1995). "1995 New Urbanism Study: Revitalizing Suburban Communities," Paper presented at the Urban Land Institute Seminar on Master Planned Communities 2000 and Beyond.

Appleyard, D. (1981). *Livable Streets*, Berkeley, CA: University of California Press.

Appleyard, D. and Jacobs, A. (1987). "Toward an Urban Design Manifesto," *Journal of the American Planning Association*, 53(1): 112–120.

Calthorpe, P. (1993). *The Next American Metropolis: Ecology, Community and the American Dream*, New York: Princeton Architectural Press.

Cervero, R. and Bosselmann, P. (1994). *An Evaluation of the Market Potential for Transit-Oriented Development Using Visual Simulation Techniques*, Berkeley, CA: Institute of Urban and Regional Development, University of California.

Community Development by Design (2005). *The Los Angeles River Urban Wildlife Refuge: A Vision for Parks, Habitat, and Urban Runoff*, Berkeley, CA: CDbyD.

Condon, P. (2008). "Planning for Climate Change," *Land Lines*, 20(1): 2–7.

Cooper Marcus, C. and Sarkissian, W. (1986). *Housing As If People Mattered: Site Design Guidelines for Medium Density Family Housing*, Berkeley, CA: University of California Press.

Crawford, M. (1999). "Blurring the Boundaries: Public Space and Private Life," in Chase, J., Kaliski, J. and Crawford, M. (Eds.) *Everyday Urbanism*, New York: The Monacelli Press.

Davidoff, P. (1965). "Advocacy and Pluralism in Planning," *Journal of the American Institute of Planners*, 31(4): 331–338.

Dramstad, W., Olson, J.D. and Forman, R.T.T. (1996). *Landscape Ecology Principles in Landscape Architecture and Land-Use Planning*, Washington DC: Island Press.

Duany, A., Plater-Zyberk, E. and Alimana, R. (2003). *The New Civic Art: Elements of Town Planning*, New York: Rizzoli.

France, R. (2002). *Handbook of Water Sensitive Planning and Design*, Boca Raton, LA: Lewis Publishers/CRC Press.

Gans, H. (1962). *The Urban Villagers: Group and Class in the Life of Italian-Americans*, New York: Free Press.

Gore, A. (2006). *An Inconvenient Truth: The Planetary Emergency of Global Warming and What We Can Do About It*, Emmaus, PA: Rodale Books.

Hall, E. (1966). *The Hidden Dimension*, Garden City, NY: Doubleday.

Hartman, C. (2002). *City for Sale: The Transformation of San Francisco*, Berkeley, CA: University of California Press.

Hertsgaard, M. (2006). "While Washington Slept," *Vanity Fair* (May 2006). http://www.vanityfair.com/politics/features/2006/05/warming200605. Accessed November 28, 2009.

Hester, R. (1984). *Planning Neighborhood Space with People*, New York: Van Nostrand Reinhold.

Hester, R. (2006). *Design for Ecological Democracy*, Cambridge, MA: MIT Press.

Hill, K. and Barnett, J. (2007). "Design for Rising Sea Levels," *Harvard Design Magazine*, 27 (Fall 2007/Winter 2008).

Hough, M. (1984). *City Form and Natural Process*, New York: Van Nostrand Reinhold.
—— (1990). *Out of Place: Restoring Identity to the Regional Landscape*, New Haven, CT: Yale University Press.
Jacobs, A., Macdonald, E. and Rofe, Y. (2002). *The Boulevard Book: History, Evolution, Design of Multiway Boulevards*, Cambridge: MIT Press.
Jacobs, J. (1961). *The Death and Life of Great American Cities*. New York: Random House.
Kaplan, R. and Kaplan, S. (1989). *The Experience of Nature: A Psychological Perspective*, Cambridge: Cambridge University Press.
Larice, M. and Macdonald, E. (2008). *The Urban Design Reader*, London: Routledge.
Linn, K. (2007). *Building Commons and Community*, Oakland, CA: New Village Press.
Louv, R. (2005). *Last Child in the Woods: Saving Our Children from Nature-Deficit Disorder*, Chapel Hill, NC: Algonquin Books.
Lozano, E. (1990). *Community Design and the Culture of Cities*, Cambridge, MA: Cambridge University Press.
Lyle, J. (1985). *Design for Human Ecosystems: Landscape, Land Use, and Natural Resources*, New York: Van Nostrand Reinhold.
Lynch, K. (1960). *The Image of the City*, Cambridge: MIT Press.
—— (1976). *Managing a Sense of a Region*, Cambridge: MIT Press.
Lyndon, D. and Moore, C. (1996). *Chambers for a Memory Palace*, Cambridge, MA: MIT Press.
McCamant, K.M. and Durrett, C. (1994). *CoHousing: A Contemporary Approach to Housing Ourselves*, Berkeley, CA: Ten Speed Press.
McHarg, I. (1969). *Design with Nature*, Garden City, NJ: Natural History Press.
Mozingo, M. (1997). "The Aesthetics of Ecological Design: Seeing Science as Culture," *Landscape Journal*, 16(1): 46–59.
Orr, D. (1992). *Ecological Literacy: Education and the Transition to a Postmodern World*, Albany: State University of New York Press.
Proshansky, H.M., Ittelson, W.H. and Rivlin, L.G. (1976). *Environmental Psychology*, New York: Holt, Rinehart and Winston.
Relph, E. (1976). *Place and Placelessness*, London: Pion.
Rios, M. (2008). "Envisioning Citizenship: Toward a Polity Approach in Urban Design," *Journal of Urban Design*, 13(2): 213–229.
Sandercock, L. (2000). "Cities of (in)difference and the challenge for planning," *Documents and Information on Swiss Local, Regional, and State Planning (DISP)*, 140: 7–15.
Sanoff, H. (2000). *Community Participation in Design and Planning*, Hoboken, NJ: John Wiley and Sons.
Seamon, D. (1996) "A Singular Impact," *Environmental and Architectural Phenomenology Newsletter*, 7(3): 5–8.
Sommer, R. (1969). *Personal Space: The Behavioral Basis of Design*, Englewood Cliffs, NJ: Prentice Hall.
Southworth, M. and Southworth (1974). "The Educative City," in Coats, G. (Ed.) *Alternative Learning Environment*, Stroudsburg: Dowden, Hutchinson & Ross.
Todd, N.J. and Todd, J. (1994). *From Eco-cities to Living Machines: Principles of Ecological Design*, Berkeley, CA: North Atlantic.
Ulrich, R. (1984). "View through a Window May Influence Recovery from Surgery," *Science*, 224: 420–421.
Urban Ecology (1996). *Blueprint for a Sustainable Bay Area*, Oakland, CA: Urban Ecology.
Wagner, M. (1998). *Acquisition-Relocation of Urban Flood Damaged Landscapes: Decision Making and Local Acceptance of Floodplain Resource Enhancement*, unpublished master's thesis, Iowa State University.

Further reading

Hester, R. (2006). *Design for Ecological Democracy*, Cambridge, MA: MIT Press. Integrates ecological principles with democratic practices informed by many disciplines from sociology to conservation biology. Ideas presented as fifteen principles of city making with many case studies to elaborate.

Hough, M. (1984). *City Form and Natural Process*, New York: Van Nostrand Reinhold. An early writing that explained how city form could be retrofitted to function better using ecological principles such as stormwater management, urban agriculture, and wind patterns.

Kaplan, R. and Kaplan, S. (1989). *The Experience of Nature: A Psychological Perspective*, Cambridge: Cambridge University Press. Integrates years of research, both theirs and others. Presents the health benefits of the experience of nature and how to apply the research findings to planning and design.

Lyle, J. (1985). *Design for Human Ecosystems: Landscape, Land Use, and Natural Resources*, New York: Van Nostrand Reinhold. Offers projects at the site scale that create natural living systems out of water purification, energy flows, etc. that would traditionally be engineered as hard infrastructure.

48

Smart growth

A critical review of the state of the art

Aseem Inam

What exactly is smart growth, and what is its relationship to urban design? The term smart growth appeared in public in the mid-1990s in the United States and was initially associated with urban growth management initiatives of the Governor of the State of Maryland, Parris Glendenning (Levy 2006). In 1996, the US Environmental Protection Agency joined with several non-profit and government organizations to form the Smart Growth Network. The network (Smart Growth Network 2006: 1) defines smart growth as regional and urban development that:

- mixes land uses;
- takes advantage of compact building design;
- creates a range of housing opportunities and choices;
- creates walkable communities;
- fosters distinctive, attractive communities with a strong sense of place;
- preserves open space, farmland, natural beauty, and critical environmental areas;
- strengthens and directs development toward existing communities;
- provides a variety of transportation choices;
- makes development decisions predictable, fair, and cost-effective, and;
- encourages community and stakeholder collaboration in development decisions.

The most successful cities take a multi-pronged approach that incorporates many of these principles, including several urban design strategies at the local and regional level. This chapter will discuss the complex picture of smart growth practices and point to successful efforts but also to continuing challenges in its implementation.

About a dozen years after the introduction of smart growth into the lexicon of planning and design practices, we can observe the following. First, the most effective smart growth measures tend to be regional in scale (such as at the county or even state level), and certain states (such as Oregon, Maryland, and Florida) are much more active in this regard than others. Second, policies labeled as smart growth have different types of effects and different degrees of effectiveness. Third, smart growth has its share of critics, such as libertarian think tanks (e.g. Cato Institute, Reason Foundation), who claim that its policies increase the cost of land and development and hinder the operation of a free market at the local level. Fourth, in order for smart growth to become more than a vacuous platitude, there remains much critical analysis and on-the-ground implementation to be done. Effective implementation of smart growth includes its deployment as a tool for mobilizing communities, creating political capital for elected officials, suggesting creative partnerships between developers and planners, and offering concrete design guidelines. Only then could smart growth become, over time, an effective strategy for the design of cities.

The chapter begins with a discussion of the rationale for smart growth, followed by an overview of major smart growth initiatives in the United States, a discussion of key elements of a smart growth strategy such as New Urbanism, transit oriented development, and affordable housing, an enumeration of the obstacles and critiques faced by smart growth, an explanation of the broad political appeal of the movement, and concludes with directions for the implementation of smart growth.

The rationale for smart growth

Smart growth is a reaction to the pattern of urban growth popularly known as "sprawl," defined as "a form of urbanization distinguished by leapfrog patterns of development, commercial strips, low density, separated land uses, automobile dominances, and a minimum of public open space" (Gillham 2002: 4). In the United States, such dispersed patterns of urban growth in many metropolitan regions in the late twentieth century have created real challenges for cities suffering from chronic traffic congestion and attendant increases in air pollution from vehicular emissions. In addition, concerns ranging from inadequate supply of land for housing, loss of agricultural land and open space, increasing economic inequality and social fragmentation, and even the growing epidemic of obesity have been blamed on increasingly suburban and ex-urban lifestyles within dispersed, low-density, land-use-segregated, and automobile-oriented metropolitan landscapes (Chavan *et al.* 2007). Furthermore, a study by the US Census bureau in 2004 estimated a 50 percent growth in population by the year 2050, an addition of nearly 140 million people (Barnett 2007), which suggests that many of the concerns associated with urban growth are likely to be exacerbated in the coming decades.

As a reaction to these concerns, a variety of regulatory and incentive-driven policy tools have emerged in the United States, including urban containment, growth management, and smart growth (Bae 2007). Urban containment, most famously seen in Portland, Oregon, attempts to impose a defined boundary around a city beyond which development will be prohibited to simultaneously prevent sprawl outside the boundary and promote higher density inside it (Figure 48.1). Growth management policies encompass a wide array of policy instruments intended to slow growth within a specific jurisdiction and achieve economic development, ensuring quality of life and environmental quality. Smart growth emphasizes incentives and disincentives rather than direct regulation, and attempts to encourage different stakeholders such as developers, homeowner associations, and environmentalists to reach some consensus about the direction and growth of future development.

Smart growth initiatives

One of the first serious attempts at smart growth was the State of Maryland's Smart Growth program, which consisted of five pieces of legislation (Frece 2008: 82):

- Smart Growth and Neighborhood Conservation – Smart Growth Areas
- Smart Growth and Neighborhood Conservation – Rural Legacy Programs

Figure 48.1 Street in Portland. Source: Aseem Inam.
Note: This is an example of a higher-density compact urban form with walkable streets and streetcars associated with smart growth strategies in Portland, Oregon.

- Brownfields – Voluntary Cleanup and Revitalization Programs
- Job Creation Tax Credit Act of 1997
- Maryland Right to Farm.

In addition, a sixth component was introduced as a budget program: a pilot program called Live Near Your Work through which state and local governments and participating employers would provide stipends to home buyers who purchased homes in certain designated revitalization areas. A key insight from Maryland's Smart Growth program is that even though it was formally adopted in 1997, the program was in fact built over the long term, starting with precedents such as the creation of the Maryland State Planning Commission in 1927.

An evaluation demonstrated that Maryland's Smart Growth Area Act and Rural Legacy Act have been generally successful in achieving their policy objectives by reinforcing a pattern of relatively concentrated development (Shen and Zhang 2007). The evaluators' models of land conversion showed that smart growth policies reinforced relatively compact patterns of urban growth in counties that had a tradition of managing growth, and had drastically changed the spatial distribution of land conversion in some counties previously characterized by highly scattered developments.

While eschewing the term smart growth, other states in the US also redirected resources to encourage more compact growth, promote housing near employment centers, reuse existing urban land, and allow local governments to cooperate regionally (Flint 2006). Initiatives have included the appointment of a senior official coordinating housing, environment, transportation, and energy efforts in the Commonwealth of Massachusetts, and programs encouraging brownfield redevelopment and local planning in the state of New Jersey. For example, starting in 2003, Massachusetts began a smart growth initiative that was more about economic development and less about sustainability (Krueger and Gibbs 2008). At the time, Governor Mitt Romney's administration focused its efforts along smart growth policies such as the Commonwealth Capital program, transit oriented development (TOD), and affordable housing. The Commonwealth Capital program attempted to direct the state's $500 million in development funds to cities and towns that were using their planning and zoning powers to promote smart growth. The TOD program provided cities and towns with technical assistance and a $30 million fund earmarked for investment in pedestrian, bicycle and housing projects that occur within ¼ mile of a transit station. A third feature included reforms to the state planning guidelines to encourage affordable housing in which local governments were eligible for funding when they chose to create smart growth districts in town centers, downtowns, near transit stations, and on vacant industrial land.

An evaluation of smart growth policies at the local and county levels in different states of the US examined their effects in Montgomery County in Maryland, Orange County in Florida, and the Portland metropolitan area in Oregon (Song 2005). The study found that while neighborhoods in these regions were better connected internally through street networks, they had less external linkages to other areas. Similarly, while densities of development have increased, a mix of land uses within residential neighborhoods was absent in the study areas. The evaluation suggests that only a synergy over time of the different components of a smart growth strategy would have a significant impact: street connectivity, density, mix of land uses, accessibility, and pedestrian walkability.

Elements of a smart growth strategy

A smart growth strategy includes a number of elements, and the urban design movement known as New Urbanism is an important development cognate to the smart growth movement. New Urbanism draws its inspiration from patterns of urban growth that are pre-World War II. Both smart growth and New Urbanism work through the markets (Flint 2006), but while smart growth consists of a larger framework of policy measures, funding mechanisms, and incentives for developers, New Urbanism focuses more on design and building at the neighborhood scale. While these two movements overlap considerably in their advocacy of similar principles in patterns of urban growth, New Urbanists focus particularly on the following principles (Frug 1999: 150–152):

- Multiuse environments that reintegrate commercial and residential spaces, work with home, and incorporate schools, parks, public squares, and civic buildings into these multiuse neighborhoods;
- Grid systems of streets that facilitate intra-neighborhood connections, create multiple ways of getting from one destination to another, thus relieving congestion on collector and arterial streets;
- Pedestrian-oriented streets with limited speeds for cars, interconnected paths and sidewalks, trees and plantings, and parked cars to protect pedestrians from traffic;
- Mass transit to provide greater choice of transportation modes and reinforce walking to and from transit stops;
- Public spaces such as squares, parks, and civic amenities to introduce areas of landscape on a regular basis in the city and foster social interaction;

- Centers and edges for neighborhoods and districts to create focal points and boundaries for urban space, making cities more legible and memorable for residents and visitors.

New Urbanism provides not only an overlap with the smart growth principles mentioned at the beginning of this chapter, but perhaps more significantly, a tangible tool for its three-dimensional implementation at the local level. In this manner, urban design can be a critical implementation tool in the smart growth tool kit, an issue we will return to towards the end of this chapter.

Transit oriented development (TOD) is becoming a common form of urban growth in the US for three reasons (Cervero 2007). First, and this is most relevant for our purposes, TOD is a visible, cogent form of smart growth (Figure 48.2). Citizens, politicians and even those of different ideological persuasions can appreciate the logic of concentrating urban growth around transit stations. Second, demographic and lifestyle trends favor TOD as part of an overall smart growth strategy because it appeals to young single professionals, childless couples, and retirees, all of whom value convenient access to urban amenities in walkable areas. Third, TOD is market-based urbanism in the sense that if the true social costs of automobile oriented development were fully accounted for, the market would privilege such alternatives as mixed-use developments around major transit nodes over the conventional approaches.

The relationship between smart growth and affordable housing is ambiguous (Connerly 2007). On the one hand, smart growth emphasizes density, which can lower the cost of housing. On the other hand, it has the propensity to draw urban growth boundaries, and limit housing to relatively compact areas, thus constraining the supply of land for housing development

Figure 48.2 Rio Vista West transit-oriented development in San Diego. Source: Aseem Inam.
Note: The project contains compact and affordable forms of residential development within walking distance of the light rail station.

and increasing the cost of developable land. At the same time, given the potentially broad appeal of smart growth in achieving compact cities with mixed land uses and enhanced public transit, affordable housing as an integrated component of this approach may gain a new acceptance in the US There are multiple tools available for state and local governments to link smart growth to affordable housing: flexible land use regulations, inclusionary zoning, mixed income redevelopment, community land trusts, housing trust funds, and financing tools such as location-efficient and energy-efficient mortgages.

Research on the state of Florida (Connerly 2007) suggests that even when a state establishes a basic framework to merge smart growth and affordable housing goals, unless that state is prepared to enforce its affordable housing objectives with specific performance expectations, local communities will lack the incentives to integrate affordable housing with smart growth. Instead, they will be influenced by the NIMBY (Not In My Back Yard) attitudes of local residents as well as fiscal concerns. Because the state plays a key role in the US in determining the rules which local governments must follow in planning their communities, affordable housing advocates could target the state as the critical actor towards encouraging local governments to embrace affordable housing as part of an overall smart growth strategy. In Florida, housing advocates were successful in building a broad coalition to pressure state government to create the State Housing Initiatives Partnership trust fund (i.e. to provide direct financial assistance to local governments for affordable housing), but did not succeed in employing this coalition to push for other critical reforms such as flexible land use regulations (e.g. for higher density housing) and inclusionary zoning (e.g. requiring new residential developments to set aside a fixed percentage of dwelling units for affordable housing).

Obstacles and critiques of smart growth

A number of obstacles impede the implementation of smart growth principles (Willmer 2006). Local land use regulations in many parts of the US either encourage or require patterns of urban growth consisting of larger single-family detached houses and retail strip centers, and often prohibit alternatives. Larger scale mixed use or infill projects create resistance from existing residents who fear the impact of such projects, whether in the perception of lowered property values, increased traffic, or additional burdens on existing infrastructure. Technical standards such as minimum parking requirements or minimal street widths also foster automobile oriented development rather than the compact, walkable designs that smart growth promotes.

In addition to obstacles at the local level, there exist other forms of resistance to smart growth, including political and policy differences. Critics of smart growth, particularly those who favor so-called free market and libertarian ideologies, label smart growth policies as draconian (Cox 2007). They claim that in the United States and Europe, most urban destinations are reasonably accessible only by automobile, and that mass transit can be an effective alternative to the automobile only in the dense core areas, such as the largest downtowns. Furthermore, they point out that in cities with smart growth initiatives such as Portland, Oregon, housing affordability has declined, making it difficult for low-income and many minority citizens to purchase their own homes. Critics who define the so-called American dream in terms of mobility and homeownership (O'Toole 2007) consider smart growth to be coercive land-use planning aimed at compact cities, often combined with expensive and ineffective rail transit. In this view, automobiles have helped make

Americans wealthy by giving people access to jobs, lower-cost consumer goods, better housing, and higher mobility levels that cannot be reached by mass transit, by bike, or by foot. The critics propose alternatives to smart growth, including highway tolls, air pollution emission fees, vouchers to subsidize transit-dependent people, and devolving zoning power to individual neighborhoods.

Critics such as the conservative Heritage Foundation have gone so far as to claim that smart growth exacerbated the international financial crisis that began in 2007 (Cox 2008). The argument suggests that strict land use regulations forced higher prices than would have been the case if the previous more traditional yet "environmentally sound" regulation had been retained. Thus, California, Florida, and the northeastern and northwestern US implemented excessive land use controls and were unable to accommodate the stronger demand created in the more profligate lending environment. Further, because state and local restrictions on housing supplies sent prices soaring, families who ordinarily would have qualified for prime loans were compelled to borrow at subprime rates (O'Toole 2008).

On the one hand, smart growth proponents would be wise to seriously consider these obstacles and critiques in order to promote more effective policies and increase the probability of their implementation. Ironically, however, some of these critiques – especially the ideological ones – mirror the overly simplistic and overly-broad critiques of sprawl, which blame most social ills and even obesity on low-density, segregated land uses, and automobile-oriented patterns of urban growth. These ideologically-driven critics of smart growth and other growth management efforts choose to ignore market factors such as job growth and subsequent demand for housing in particular metropolitan regions as being at least partially responsible for the rise in housing costs. They also seem to ignore the benefits of smart growth regulations in terms of the conservation of the natural environment.

Broad appeal of smart growth

Since smart growth lacks a precise definition and has a broad appeal, the term means different things to different people and hence can represent a large political movement including not only professionals such as urban designers and city planners, but also elected officials, public administrators, neighborhood associations, environmental groups, and forward-thinking real estate investors and private developers. In addition, smart growth policies may be pursued at multiple geographic scales (Levy 2006) – state, county, or local jurisdictions – as well as at the regional scale as a coalition of multiple jurisdictions.

Although the term smart growth is uniquely associated with the American context, the ideas behind the concept are often subsumed under the broader banner of sustainability in other parts of the world (Ruth 2006). For example, by 2002, there were over 6,000 municipal and local governments in 113 countries that had either made a formal commitment to sustainability principles or were actively undertaking the process, although many are facing difficulties in implementing the principles because of competing short-term economic interests or because negative attitudes about living in higher-density environments persist. American urban designers and planners have looked to Europe and its sustainable cities movement as models (e.g. Beatley 1999). However, the most surprising insights may yet emerge from studying those cities of the developing regions of the world that make highly efficient use of land and space through innovative mixes of land uses and higher densities, find ways of recycling buildings and construction materials on an ongoing basis, create extremely vibrant urban centers that are accessible to all, and contain many different modes of transportation, including low-energy ones such as walking, bicycles, and rickshaws.

Smart growth claims to unite the interests of environmentalists, developers, and civic boosters by promoting development that serves the environment, the community, and the economy (Krueger and Gibbs 2008). Furthermore, like sustainable development, smart growth is hard to pin down in its practical application. For many municipal governments, smart growth provides a way of organizing disparate elements of land-use planning goals (e.g. open space preservation, regeneration, housing choice, and economic development) and approaches (e.g. comprehensive planning, business improvement districts, the use of existing infrastructure). The movement has different adherents with different agendas under the larger umbrella of the smart growth moniker. For example, for the National Association of Home Builders, the core area of interest in smart growth is access to housing and home ownership, both on infill sites as well as greenfield land. The US Department of Housing and Urban Development also emphasizes access to housing, although with a focus on compact development. In contrast, the Trust for Public Land and the Sierra Club focus their efforts on open space preservation, such as productive agricultural lands, wetlands, and wildlife habitats. Thus, while smart growth may be an overly-broad and ambiguous term, it has the political potential of drawing together stakeholders with different interests in sometimes tenuous but ultimately beneficial coalitions in order to craft alternatives to current patterns of urban growth. For example, there was a successful partnership between environmental groups and labor unions in Contra Costa County, California to restrict development of environmentally sensitive land with tighter growth controls.

Part of the impetus for this partnership arose from the concerns of construction workers, as voiced by Bill Nack, the head of San Mateo's Building and Construction Trade Council (cited in Goodno 2002): "Smart Growth isn't smart if it doesn't consider the people who are building the project."

Thus, much of the strength and potential of smart growth as a policy term derives from its ambiguity and ability to provide an umbrella large enough to shelter stereotypically antagonistic groups such as residential builders and environmentalists, who could see mutual benefits within a unifying metaphor and marketable urban design. The "large umbrella" metaphor for building broad smart growth coalitions is reinforced through institutional initiatives such as the US Environmental Protection Agency Smart Growth Awards (which represents a group of government and non-profit organizations) and the Urban Land Institute Smart Growth Awards (which represents the interests of real estate developers). The range of efforts which have won these awards include regional ones such as the Livable Centers Initiative by the Atlanta Regional Commission, and site-specific ones such as Solara in San Diego, California, the first apartment complex in California to be fully powered by the sun.

Conclusion: implementing smart growth

While sound policy making and planning is important, a sound political approach is just as important. A plan without an accompanying political strategy is probably a plan that will never be implemented (Frece 2008). Thus, while the rhetoric of smart growth may be lofty, the challenge is in its implementation and effectiveness. A study exploring the social roots and policy choices of the smart growth movement with data from 202 US cities (O'Connell 2008) offers clues to probabilities of likely implementation. The study found that the number of smart growth policies adopted in a city was significantly and positively associated with the city's percentage of college graduates and the degree of environmental activism. Thus, urban designers, planners, and policy makers may attempt to form political coalitions with environmental activists and educated professionals to push through land use reform. Furthermore, a number of tools already exist to help address different challenges, including easy-to-read toolkits of practical implementation strategies and financial suggestions (Smart Growth Network 2003). *The Getting to Smart Growth* booklet contains 100 such strategies (e.g. use transportation enhancement funds to create places of distinction), as well as a list of funding institutions and examples of implemented projects.

Policy makers and urban designers can also reflect upon and learn from the effectiveness of implementing smart growth policies in many American cities at the local and neighborhood scales. One of the best known efforts in Portland, Oregon is the Pearl District, a former warehouse area in the center of the city that was transformed into apartments and condominiums, often built over stores, galleries, and other retail businesses, as well as offices, grocery stores, and parks (Knickerbocker and Wood 2006). Around 1,000 affordable housing units were built, with a local nonprofit organization owning the land for these units in order to preserve affordability. However, even in this success story, there remain challenges: to provide housing that is suitable for children, to find financially viable ways to extend public transit into more neighborhoods, to reduce traffic in downtown, and to deal effectively with aging infrastructure. This suggests a multipronged approach over the long term and applying the full range of smart growth principles, as suggested in the introduction of this chapter. Another example is the Clarendon Corridor in Arlington, Virginia, which contains four Metro subway stops, a mix of residential and commercial development, and the diversion of major commuting roads around the city center (Goffman 2006). The planning board of Arlington has long supported development around transit stations, and in 1984, the revised master plan incorporated high-rise buildings with parks, while future plans mandated a variety of housing types, including low-income housing, high-rise condominiums, garden-style apartments, and single-family houses. A development scenario that would have occupied 14 square miles under conventional patterns of suburban growth was contained within 2 square miles, thus improving efficiencies of land use and creating walkable neighborhoods.

There is one coalition of individuals and organizations that is particularly experienced and politically savvy in the implementation of the alternative patterns of urban growth, the New Urbanists. New Urbanist practitioners – architects, urban designers, planners, real estate developers – understand that current patterns of low-density, land-use-segregated, and automobile-oriented urban form (Calthorpe and Fulton 2001). Conventional developers, builders, engineers, and contractors tend to repeat past successes despite changing times and unforeseen consequences, and local governments look for further growth and an expanded tax base without regard for cumulative development quality or regional implications. Similarly, neighborhood groups and homeowner associations attempt to enhance property values through exclusionary practices, while even environmental groups sometimes promote such patterns of urban growth by encouraging low-density development or no development at all in the localities where they operate.

The promise of New Urbanism in offering a viable alternative to such inefficient and destructive patterns of urban growth in the face of considerable resistance is multifold (Urban Design Associates 2003). First, the movement can implement smart growth principles at the local or neighborhood scale through tangible design and attractive visuals to generate community dialogue. For example, the National Resources Defense Council hosted a website in 2009 entitled "Picturing Smart Growth: Visions for Sustainable Communities Across America" (http://www.nrdc.org/smartGrowth/visions/), which consists of a series of powerful before and after images of conventional development transformed into attractive destinations through New Urbanist design interventions. Second, New Urbanism operates at multiple scales, from regional plans (e.g. Envision Utah by Calthorpe and Associates) to city master plans (e.g. Miami 21 by Duany Plater-Zyberk) and neighborhood scale projects (e.g. Del Mar Station transit oriented development, Pasadena, California by Moule and Polyzoides), thus offering a range of smart growth prototypes. Third, New Urbanism embraces multiple facets of smart growth, from public policy to development codes, to innovative financing and community based design charrettes. New Urbanists are able to integrate multiple physical and non-physical strategies into fairly cohesive approaches. Fourth and finally, New Urbanism is a healthy movement – with internal debates and evolving strategies – that can buttress the goals of smart growth through overlaps and complementarities of purpose.

In order for smart growth to evolve, adapt, and be effective, it needs to be part of a larger and more visible movement for the redesign of our cities. Rather than expecting a set of ideal political conditions, smart growth advocates have adopted an approach of a policy tool kit that contains many different strategies, from regional coordination to the design of prototypical neighborhoods. The tool kit approach offers the advantage of being flexible and adaptive to the needs of a region during different economic growth conditions, including extremely challenging times. For example, what are the prospects for efforts such as smart growth during periods of economic decline, such as the recession in the United States? A study conducted by the Urban Land Institute (ULI 2008), while forecasting a difficult period for commercial real estate markets, notes that higher-density residential projects with retail components and core urban markets with mass transportation alternatives will be highly favorable in the next round of development following the recession. The potential of smart growth and similar urban design strategies to make more efficient use of land, manage our resources in wiser ways, and design urban areas to be more humane, continues to be promising.

References

Bae, C. (2007). "Containing Sprawl." In Knaap, G. *et al.* (Eds.) *Incentives, Regulations and Plans: The Role of States and Nation-states in Smart-Growth Planning*. Cheltenham: Edward Elgar Publishing Limited, 36–56.

Barnett, J. (Ed.) (2007). *Smart Growth in a Changing World*. Chicago: American Planning Association.

Beatley, T. (1999). *Green Urbanism: Learning from European Cities*. Washington, DC: Island Press.

Bullard, R. (Ed.) (2007). *Growing Smarter: Achieving Livable Communities, Environmental Justice, and Regional Equity*. Cambridge MA: MIT Press.

Calthorpe, P. and Fulton, W. (2001). *The Regional City: Planning for the End of Sprawl*. Washington, DC: Island Press.

Cervero, R. (2007) "Transit-Oriented Development in the US: Contemporary Practices, Impacts and Policy Directions." In Knaap, G. *et al.* (Eds.) *Incentives, Regulations and Plans: The Role of States and Nation-states in Smart-Growth Planning*. Cheltenham: Edward Elgar Publishing Limited, 149–167.

Chavan, A., Peralta, C. and Steins, C. (Eds.) (2007). *Planetizen Contemporary Debates in Urban Planning*. Washington, DC: Island Press.

Connerly, C. (2007). "Smart Growth: Opportunity or Threat to Affordable Housing." In Knaap. G. *et al.* (Eds.) *Incentives, Regulations and Plans: The Role of States and Nation-states in Smart-Growth Planning*. Cheltenham, UK: Edward Elgar Publishing Limited, 103–123.

Cox, W. (2007). "The Argument against Smart Growth." In Chavan, A. *et al.* (Eds.) *Planetizen Contemporary Debates in Urban Planning*. Washington, DC: Island Press. 10–17.

—— (2008). *How Smart Growth Exacerbated the International Financial Crisis*, web memo no. 1906 published April 29, Washington, DC: The Heritage Foundation.

Flint, A. (2006). *This Land: The Battle Over Sprawl and the Future of America*. Baltimore, MD: The Johns Hopkins University Press.

Frece, J. (2008). *Sprawl and Politics: The Inside Story of Smart Growth in Maryland*. Albany, NY: State University of New York Press.

Frug, G. (1999). *City Making: Building Communities Without Building Walls*. Princeton, NJ: Princeton University Press.

Gillham, O. (2002). *The Limitless City: A Primer on the Urban Sprawl Debate*. Washington, DC: Island Press.

Goffinan, E. (2006). "Smart Growth: A Tale of Two Cities in Maryland and Virginia." *E Magazine*. November/December, 21–24.

Goodno, J. (2002). "Labor Embraces Smart Growth." AFL-CIO Housing Investment Trust. *http://www.aflcio-hit.com/frame-templates/print_template.cfm* Accessed October 3, 2008.

Knaap, G., Haccou, H. Clifton, K. and Frece, J. (Eds.) (2007). *Incentives, Regulations and Plans: The Role of States and Nation-states in Smart-Growth Planning*. Cheltenham. UK: Edward Elgar Publishing Limited.

Knickerbocker, B. and D. Wood (2006)." Gilbert, Ariz. and Portland, Ore. Have Approached Expansion in Two Very Different Ways." *The Christian Science Monitor*, October 3 http://www.csmonitor.com/2006/1003/p01s02-ussc.htm. Accessed March 5, 2009.

Krueger, R. and Gibbs, D. (2008). "Third Wave Sustainability? Smart Growth and Regional Development in the US." *Regional Studies*, 49(9): 263–1274.

Levy, J. (2006). *Contemporary Urban Planning*, 7th edition. Upper Saddle River, NJ: Pearson Prentice Hall.

O'Connell, L. (2008). "Exploring the Social Roots of Smart Growth Policy Adoption by Cities." *Social Science Quarterly*, 89(5): 1356–1272.

O'Toole, R. (2007) "Preserving the American Dream by Cost, Not Coercion." In Chavan, A. *et al.* (Eds.) *Planetizen Contemporary Debates in Urban Planning*. Washington, DC: Island Press. 34–38.

—— (2008). "Big Burdens from Growth Management." Washington, DC: Cato Institute. http://www.cato.org/pub_display.php?pub_id =9701. Accessed March 10 2009.

Ruth, M. (Ed.) (2006). *Smart Growth and Climate Change: Regional Development, Infrastructure, and Adaptation*. Cheltenham, UK: Edward Elgar Publishing Limited.

Shen, Q. and Zhang, F. (2007). "Land Use Changes in a Pro-Smart-Growth State: Maryland, US." *Environment and Planning A*, 39: 1457–1477.

Smart Growth Network (2003). *Getting to Smart Growth II: 100 More Policies for Implementation*. Washington, DC: Smart Growth Network and International City/County Management Association.

—— (2006). *This is Smart Growth*. Washington, DC: Smart Growth Network.

Song, Y. (2005). "Smart Growth and Urban Development Pattern: A Comparative Study." *International Regional Science Review*. 28(2): 239–265.

Soule, D. (Ed.) (2006). *Urban Sprawl: A Comprehensive Reference Guide*. Westport, CT: Greenwood Press.

ULI (2008). *Emerging Trends in Real Estate 2009*. Washington, DC: Urban Land Institute.

Urban Design Associates (2003). *The Urban Design Handbook: Techniques and Working Methods*. New York: W.W. Norton and Company, Inc.

Willmer, R. (2006). "Planning Framework: A Planning Framework for Managing Sprawl." In Soule, D. (Ed.) *Urban Sprawl: A Comprehensive Reference Guide*. Westport CT: Greenwood Press. 61–78.

Further reading

Brown, M. and Southworth, F. (2008). "Managing Climate Change Through Green Buildings and Smart Growth" *Environment and Planning A*, 40: 653–675. Suggests that by the middle of the twenty-first century the combination of green buildings and smart growth could deliver the deeper reductions that are needed to mitigate climate change, especially with the implementation of measures such as retrofitting existing buildings, constructing net-zero-energy buildings, and improving the locational efficiency of urban systems and neighborhoods.

Carruthers, J. and Ulfarsson, G. (2008). "Does Smart Growth Matter to Public Finance," *Urban Studies*, 45(9): 1791–1823. Based on a number of spatial econometric models, the article argues that conventional low-density development in the US nearly always raises the per capita spending on services and that the reasoning behind fiscally motivated smart growth policy frameworks is sound.

DeGrove, J. (2005). *Planning Policy and Politics: Smart Growth and the States*, Cambridge, MA: Lincoln Institute of Land Policy. Recounts the story of nine states in the US that have pursued smart growth and related programs to different degrees of success, and highlights the roles played by key actors such as governors, legislators, agency heads, home builders, farmers, and environmentalists.

Ingram, G., Carbonell, A., Hong, Y.H. and Flint, A. (Eds.) (2009). *Smart Growth Policies: An Evaluation of Programs and Outcomes*, Cambridge, MA: Lincoln Institute of Land Policy. An evaluation of smart growth policies in the United States. It finds that the most successful states use a variety of regulatory controls, market incentives, and institutional policies to achieve their objectives, including supporting local government actions to pursue effective land use planning.

49

Notes on transit-oriented development

Stefanos Polyzoides

Throughout human history, people have settled on the land based on two fundamental desires: to be both in motion and in place. Cities, as we understand them today, represent both stasis and flux. Thus they came into being when the necessary movements of people and vehicles converged in space, leading to the design of distinct rights of way. In response, distinct blocks, lots, and buildings of various kinds appeared to enable stasis and habitat, human comfort and security, pleasure and production. The physical character and livability of human settlements has always existed at the intersection of a dual system of urbanism: blocks defining streets for mobility, and streets defining blocks for habitation.

The particular design of these two key ingredients of urban form has changed radically in pattern and scale over time, adapting to evolving human needs and new technologies. But their interdependence has not. We can experience it in all kinds of beloved traditional places that still exist today: in villages designed for hand cart service, in towns organized around the movement of animal-drawn carriages, in cities built to the scale of motorized vehicles, and in regions supported by transit. This chapter will offer some thoughts on the development of particular districts in the city, which serve both as paths enhancing mobility as well as nodes accommodating a mix of residential and commercial land uses: the neighborhoods adjacent to transit systems.

Density and mobility: the link between development and transit

Development related to transit is a nineteenth-century phenomenon whose benefits are again becoming apparent. The new, expansive scale of mass production of the Industrial Revolution demanded greater concentrations of workers living in cities in order to service the labor needs of factories and expanding businesses. The combined environmental effects of factory pollution and poorly designed housing, associated with the densification of cities, fueled an exodus of residential development to the suburbs. As cities expanded vertically and horizontally, their scale was forever transformed both by intensity and distance.

Mass transit was modern society's first response to the twin challenge of moving more people further than ever imagined or attempted before. The systems of choice were the streetcar and the railroad. The resultant urban forms were the central business district (CBD) connected to the rail commuter suburbs via the streetcar

(Warner 1962). The proliferation of the automobile, especially after World War II, dramatically altered this pattern, even though many such "streetcar suburbs" are still thriving today with admirable efficiency, architectural distinction, and high quality living. Most of them are to be found in the metropolitan regions of New York, Philadelphia, Boston, and Chicago.

The recent rediscovery of transit as a catalyst for development and redevelopment is more than anything else, a response to the ravages of auto-based sprawl that developed in the latter half of the twentieth century. Reclaiming cities from automobiles, selectively densifying them, remixing their uses, re-charging their economic ingredients, and recasting their environmental performance, depends on introducing mobility strategies beyond the automobile. The choice of walking, biking, or taking transit to work or shopping, allows people to forego the remarkable cost of owning and maintaining a car, and eliminating the catastrophic environmental effects associated with operating cars and car-generated growth.

TOD and DOT at the core of sustainable urbanism

There are two basic strategies for using transit as an agent of urbanization: Transit Oriented Development (TOD) and Development Oriented Transit (DOT). The first strategy, TOD, was brilliantly utilized in the United States in the late nineteenth century and up to the 1930s for the purpose of opening up greenfield sites for development. Streetcar lines were constructed to gain access to the periphery of cities, and transit-accessible suburbs were built in almost every major city in the United States. By the beginning of sprawl-based suburbanization in the 1950s, almost every city in this country was organized around streetcar-centered arterial roads. Indeed, the definition of arterial road included a streetcar corridor in its center. These streetcar systems were almost entirely demolished by 1965, as all growth became strictly and exclusively oriented towards the automobile.

The second strategy, DOT, is currently being practiced to redevelop underutilized urban land or to intensify the building fabric in underperforming or underserviced parts of cities. The catalytic effect of inserting transit into such areas can be substantial. Station sites and the neighborhoods, and the district and corridors surrounding them become accessible to metropolitan populations as living, working, shopping, and entertainment destinations. As a result their economic potential and physical forms are radically transformed.

TODs and DOTs are an important model of an alternative sustainable urbanism, a key tool in the reconstruction of the American metropolis. Both strategies provide a development option that reverses auto-centric sprawl composed of single-family subdivisions and commercial strips, which are isolated by immense arterial roads and disconnected megastructures surrounded by a sea of cars.

Choosing a route

The choice of a transit route is essential in determining both the viability of a transit system and the evolving character of the places it affects by the location of its stops. At a regional scale, the introduction of transit systems has the potential to strengthen the economic primacy of the city centers as their terminus. If there is a good coordination between land use and transportation, the various sub-centers linked through the transit system will also benefit, as well as the areas around the various transit stops along its way (Transit Cooperative Research

Program 2004). For this to happen, the choice of a route and stops for greenfield development must enable opportunities for mixed use housing. A terminus must provide the maximum employment potential, supported by large concentrations of housing and jobs, so as to attract people to commute from the metropolitan edge to its center or even live there.

A transit route should connect potential origin and destination sites, established or planned places of intense urban activity, whether rich in just housing, or any combination of housing, employment, retail, entertainment or special uses. Each of these places around stops must have the potential to be further intensified and enriched by an appropriate mix of uses that encourage pedestrian activity and transit ridership. An adequate street network and space for future parking facilities must be available to enable redevelopment.

Station types by mode: heavy rail, light rail, streetcar, and bus-based

All transit-related development is organized around a quarter-mile pedestrian shed, and a half-mile bicycle riding shed around stations. Such development is possible around the stop of all modes of transit, heavy rail, light rail, streetcar, and bus. The design of stations should be sensitive to the technical requirements of each mode, as well as to its surrounding urban context.

Within a development intensity of transects from rural to urban (see Duany and Talen 2002),[1] stations assume a particular design character depending on which transect zone they are located within (Figure 49.1). For example, stations in suburban locations should be designed as free standing platforms, corresponding and reflecting the order of their urban settings. In contrast, stations in dense metropolitan

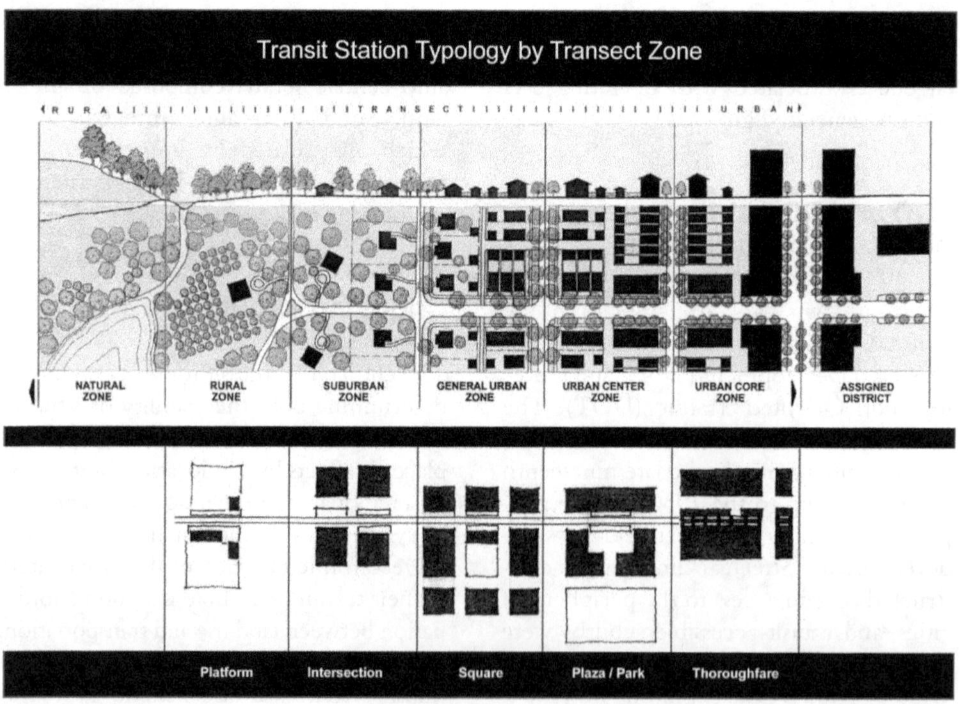

Figure 49.1 Diagram of transit station types by transect zone. Source: Moule & Polyzoides, 2006 – used by permission.

downtown locations should be designed under buildings in subway configurations.

As stations are typically a vital part of the public realm, they should be prominently located and highly visible to the people trying to access them, or should be accessed through a coherent way-finding system. They should be so designed as to support the idea that other buildings and their uses, including housing, can be located as physically close to them as possible.

Station planning: access and parking

The most crucial dimension of station planning is the means by which people access a station to either board transit or to get to a destination after having traveled by transit. There are five ways to access a station: walking, using other forms of transit, by taxi or jitney, being dropped off or picked up in a private vehicle, and by a private vehicle to or from parking. All of these have to be accommodated in the design of the right of way surrounding a station.

Essential to the performance of a station as an urban place is the design of the public realm surrounding it in a manner that favors pedestrian convenience and safety over vehicular speed. The larger and more prominent the station, the more it should be framed by appropriately sized plazas or parks that enhance the ease of pedestrian access.

Cars must be always removed from view in this public space surrounding stations. When parking is accommodated in garages or lots, retail or work-live linear buildings should screen them. When parking is provided in the underground garages of mixed-use buildings, these should be entered peripherally (Figure 49.2). Pedestrian portals off parking garages or lots should be

Figure 49.2 Mission/ Meridian Village, South Pasadena, CA. Source: Moule & Polyzoides, 2003 – used by permission.

located more remotely, up to 750 feet away from the station entrance. The less intense the development around a station, however, the more proximate the location of parking should be, in part to enhance a sense of security for transit riders during off peak travel.

Retail and commercial activities can transform stations into destination draws, useful beyond their purpose as arrival or departure points. Pedestrians walking from their parking spaces to the station platforms can boost the performance of retail businesses located along the way.

Kinds of TOD/DOT by place character, T1 to T6

As in station design, buildings and places around them should correspond to a transect of development intensity from rural to urban. Buildings, open space, landscape and infrastructure should assume a particular character, depending on which transect zones they are located within (Duany and Talen 2002). Small, low rise, detached, horizontal mixed-use buildings should be expected in lower intensity urban settings. Attached, mid-rise, mixed use buildings and large, high-rise, vertical mixed-use buildings should be common in middle and high intensity urban locations respectively (Figure 49.3).

Within an urban fabric, the density of buildings should follow a gradient according to their proximity to the transit station – thus the closer the station, the denser the buildings. Readers will note that beyond the immediate surroundings of a station, and as its direct physical presence and influence wanes, urban and architectural design challenges become the same as the design or redesign criteria of any sustainable neighborhood, district or corridor. These include: inter-connectedness, public realm definition, compactness, diversity, efficiency. Such criteria, in turn, lead to rights of way that are interconnected and multimodal; buildings

Figure 49.3 Del Mar Station, Pasadena, CA. Source: Moule & Polyzoides, 2006 – used by permission.
Note: The fabric of buildings surrounding a transit station should be configured in block-appropriate massing that supports a pedestrian-friendly public realm.

that frame the public realm of thoroughfares, parks and plazas by containing and hiding parking; projects and their uses that enjoy good pedestrian access and are accommodated within a broad range of places; buildings and unit types, and infrastructure that is affordable, effective, and green.

Neighborhood, district, and corridor visioning

Planning should encompass the entire neighborhood, district or corridor surrounding a station. The physical changes brought about by the construction of new transit systems, and the intensity of expected private development around them can be unsettling to the public. The transformation of existing conditions is immediate and can be intrusive. The benefits generated by this kind of intense change may seem distant. For this reason, it is very advantageous if the design of transit stations and their adjacent development is linked through a community process to a visioning urban design charrette (see chapter by Kelbaugh in this volume). The community of neighbors living in proximity to stations can then discern the mobility, economic development, and physical design benefits that a station design may produce for them. Such a process should be transparent, participatory, and engaging all constituencies involved, including a mix of public and private interests. Most typically, the product of such an effort should be a Special Area Master Plan or a Specific Plan.

Master Plans for Transit Oriented Development must include at a minimum aspects of a development strategy that incorporates economic goals, a physical vision including catalytic or priority projects, a public infrastructure analysis and projection, and an implementation framework that outlines public and private responsibilities for seeing the Master Plan through to fruition.

Neighborhood, district, and corridor coding

Transit stations, particularly fixed rail stations, provide unique opportunities for durable urban development. Their permanent locations guarantee that large numbers of riders will access them every day in perpetuity. It is not surprising, therefore, that they become places where housing and other uses can be assembled to serious economic and fiscal advantage. Transit Oriented Development Master Plans typically establish the vision for such transit-centered neighborhoods, districts and corridors. They are best implemented through the actions of many developers, their projects often including many and diverse buildings and public spaces.

Conventional urban growth is highly regulated. Yet, the current, dominant mode of zoning is vague when it comes to managing the form of the city. It induces the kind of unpredictability that deters developers from investing in projects that promote collateral development in their vicinity. The tendency to produce oversized, metropolitan-scale projects follows. Isolated from and unresponsive to context, these are reflexive attempts by developers to gain some modicum of control over the quality and character of their projects, by sheer size accompanied by physical and economic isolation.

Form-based codes are the indispensable tool for seeding the alternative to mega projects (see also chapter by Talen): incrementally assembled ensembles of smaller buildings and the human scaled places between them. This method of coding offers predictability by establishing the building, open space, landscape, and right-of-way standards that deliver an orderly urban form, serving many development interests over time. They also regulate uses in a flexible manner that allows a fast response to changing economic patterns and space needs. As form-based codes support phased, incremental design actions, they become ultimately supportive of both private and community interests. They maximize the performance of private projects, while building up a stable and permanent public realm.

Architectural design must follow upon neighborhood or district Master Plans that include a vision and code for the station area. It is not politically viable to design in the context of TODs and have to fight planning and architectural battles simultaneously. The gross size and scale of expected projects must be clearly established in advance of any architectural design.

A transit proximate architecture

Contrary to current common understandings of architecture, style is not the essential design ingredient of Transit Oriented Development. A "space-first" strategy is. New buildings that embrace station and train right-of-ways and frame them into a coherent realm of defined public space are the appropriate response to the design of sites adjacent to transit. This is particularly true in the case of light rail, where buildings should not shy away from being located as close to moving trains as possible (Figure 49.4).

Supporting a high pedestrian connectivity shapes transit-proximate buildings in a variety of ways. The proper configuration of buildings and public space in the vicinity of stations depends on reciprocal, people-friendly thoroughfare design. Street parking, drop-off lanes, and slow-moving traffic generate large volumes of pedestrian traffic and affect both the character of buildings and the experience of living in them.

Most importantly, ground floors are essential ingredients of TOD project design. Buildings in a transit station context should be designed to accommodate

Figure 49.4 Del Mar Station, Pasadena, CA. Source: Moule & Polyzoides, 2006 – used by permissison.

a variety of uses over time. Their ground floors should be activated continuously, with commercial frontages predominating. If the buildings are residential, ground floors should be open and accessible every twenty or so feet. Their parking should be placed behind and under them, and car entrances should be located discreetly, to also interfere as little as possible with the pedestrian-oriented public space.

TOD projects should be fitted into existing contexts in a manner that validates the historical continuity of towns and cities. New buildings should be designed in conjunction with adjacent existing ones to generate thoroughfares and public space of distinct character. This can be most often accomplished by designing available entitlement programs into building types of various densities and combining them into site plans that accomplish both internal project coherence and a better fit into the collective form of a neighborhood or district as a whole.

Most often, buildings of diverse types can also be designed to incorporate a variety of dwelling units by type and size, and resulting in an assortment of vernacular and contemporary styles. It is this complexity and variety, this attention to both the measured definition of new projects and the completion of existing street and city block form that generates an authentic sense of place. The broad consumer choice inherent in projects so designed is also a key ingredient to their financial success.

Notwithstanding issues of density, the design of residential and mixed-use buildings should express their residential character. Their gross form should speak of human habitation. The fabric of such buildings should be made of materials and building components and assemblies that reflect a human scale and invite people to use them and experience them close up.

Designing for sustainability at the building scale should capitalize on issues that are common practice in residential design, cross ventilation, natural lighting, highly insulated shells, locally available construction materials and low-cost technologies. This emphasis on passive modes of environmental control reflects the importance of permanence and durability in green design. Architecture dependent on exotic construction techniques and expensive environmental control devices may not be a preferred option for new housing projects, since buildings so endowed offer little incentive for developers to risk higher project construction costs.

Phasing and implementation

The difference between the commodified production of single family housing characteristic of sprawl versus mixed use urban housing is the rich mix of amenities that is typically associated with mature urban neighborhoods. Housing developed around transit creates competitive advantages for the dwellings and supporting commercial uses that are proximate to it. This increased accessibility overcomes opportunity costs associated with congestion. In this manner, transit becomes an amenity that can catalyze rich development opportunities in the vicinity of stations, and can produce significant competitive economic premiums in the housing market.

Transit Oriented Development is a process that should recognize the role of the station in the overall context of the regions' economic structure. There is no standard recipe for transit-supportive development. As projects are defined, their programming should distinguish between departure versus destination uses in order to generate places, and by extension building and development strategies, that are unique to the special economic profile of each station.

Joint development opportunities with transit or redevelopment agencies should be actively pursued. These may maximize public investment by offering land write-downs to private developers in order to stimulate development. Other opportunities to cooperate should also be followed through with the local authorities that have control over the entitlement and development processes. Changes in development standards, such as parking-to-building ratios or the establishment of a "Park Once" district, and increasing densities in the vicinity of stations, are prime examples of the benefits of public/private partnership on transit oriented projects.

Despite their clear dependence on public sector cooperation, these kinds of developments need to be solidly anchored in the logic of the market. Their economics should be based on a realistic assessment of prevailing conditions in their setting at the time of their design and construction. In a mixed-use strategy, each use should be economically feasible and able to be financed in its own right. Together, transit and a mix of uses should produce increasing returns on investment stemming from their synergy and the resulting physical and economic integration of the project as a whole.

Transit oriented projects are typically large and complex enough to demand multiple phases in the development process. Such projects should be deployed on sites as large as possible, so as to avoid isolated and incomplete improvements and to produce the economic benefits expected of sequential project development.

Conclusion: toward a successional urbanism

Urban design can contribute to a city's social and economic betterment and also be the means by which change is incrementally managed in the city today. Currently, potent forces operating on both our natural and urban environment become visible in two distinctly different forms: first, as a chaotic process of assembling mobility and utility infrastructures, buildings, open space, and landscape elements into uncoordinated and fragmented subdivisions; or second, as highly planned, highly-controlled, and well-managed places, incorporating the same set of urban design ingredients into homogeneous neighborhoods, districts, and corridors.

Both of these twenty-first century models lack the prospect of growing their foundation urbanism into a more mature settlement. On the one hand, endless sprawl locks in the urban configurations of separate use zones, anemic choices of building types, enormous city blocks, and inadequately connected thoroughfares. On the other hand, highly planned and controlled places are typically frozen into a legal and fiscal framework that leads to architectural fixity and aesthetic predictability. The design pattern under which both models were first developed becomes permanent. This is what we experience almost everywhere in the world as arbitrary and random (as opposed to intentional) change.

A true urbanism is one that encourages and delivers successional stages of urban development by horizontal extension or by vertical transformation, or both. There are many key factors that encourage this kind of growth from one transect intensity to the next. As population densities increase, as use mixes become more varied, as institutions proliferate, and employment concentrations intensify, as the need for all kinds of services multiplies, the demand for convenient mobility skyrockets. Transit and transit-related development become the key way to maintain many city functions intact, while their character is transformed from one stage of maturity and service to their inhabitants to the next.

Note

1 The term transect has been developed by New Urbanists to refer to the varieties of land use from an urban core to a rural boundary. Transect classifications (from lowest to highest density),

include: T1: natural, T2: rural, T3: suburban, T4: general urban; T5: Urban center; T6: Urban core.

References

Duany, A. and Talen, E. (2002). "Transect Planning." *Journal of the American Planning Association*, 68(3): 245–266.

Transit Cooperative Research Program (2004). *Transit-Oriented Development in the United States: Experiences, Challenges, and Prospects*, Washington, DC: TCRP Report 102.

Warner, Bass S. (1962). *Streetcar Suburbs: The Process of Growth in Boston, 1870–1900*, Cambridge, MA: Harvard University Press.

Further reading

Dittmar, H. and Ohland, G. (Eds.) (2004). *The New Transit Town*. Washington DC: Island Press. Case studies of the first generation of TOD projects and discussion of lessons learned.

Dunphy, R., Cervero, R., Dock, F., McAvey, M. and Porter, D. (2005). *Developing around Transit*. Washington DC: Urban Land Institute. The book discusses different ways to tap the prospects of undeveloped and underdeveloped areas around transit stations, whether large scale or small scale, downtown or suburban.

50

Placemaking in urban design

Kathy Madden

> Placemaking is both an overarching idea and a hands-on tool for improving a neighborhood, city or region. It has the potential to be one of the most transformative ideas of this century.
>
> (PPS 2008:1)

After decades of urban planning policies geared towards facilitating the movement of automobiles and imposing order and development from the top down, a broader bottom-up approach to community-building is taking hold globally. This approach, which we refer to as *Placemaking*, is geared toward the "ground floor" of a city – streets, sidewalks, parks, buildings and other public spaces. Simply, *Placemaking* aims to create places in cities that can invite greater interaction among people, while fostering healthier and more economically viable communities.

An alternative to the approach that has shaped the built environment during the past fifty years, *Placemaking* offers a new direction for the field of urban design. Rather than requiring professionals to define the parameters of a project, this new approach is based on the community's vision and employs the skills of professionals (e.g. civil engineering, architecture, urban planning and community development) as resources in implementing this vision.

Placemaking is a multi-faceted approach to improving public spaces, involves looking at, listening to, and asking questions of the people who live, work and otherwise use a particular public space to discover how they use the space, their perceptions about it, and how they think it can be improved. This information is the basis of a common vision for that place and can evolve quickly into an implementation strategy, beginning with small-scale, incremental improvements that can immediately bring benefits to public spaces. An important component of this approach is becoming familiar with research about how a place or similar places are used to avoid repeating mistakes and creating spaces that are not used.

Thus, *Placemaking* focuses on the creation of the public places of everyday life: the street corners, bus stops, and parks (see Figures 50.1 and 50.2). They provide the setting for people to engage in a variety of activities at different times of the day, and consequently, draw people to use them again and again. Good "places" are busy because they have many reasons for people to use them, and they differ from "spaces," which do not provide reasons for people to be there and use them. In a sense, "spaces" are primarily physical settings that have yet to be turned into "places." People may notice them but rarely stop,

Figure 50.1 Los Angeles bus stop before improvements. Source: Project for Public Spaces – used by permission.

Figure 50.2 Los Angeles bus stop after improvements. Source: Project for Public Spaces – used by permission.

and if they stop they do not linger. Simply, "spaces" become "places" when they begin to develop a multitude of reasons for people to go there.

The foundation of placemaking

It is hard to design a place that will not attract people; what is remarkable is how often it has been accomplished.

(William Holly Whyte)

The above statement about the state of public space design was made in the 1990s by Holly Whyte and grew out of his research of many years about how people interact with buildings and their surrounding public spaces. Whyte was reacting to the fact that many designed public spaces, especially those of the last 50 years, have never become completely successful. Although he did not use the term *Placemaking*, Whyte examined the micro-characteristics of places and elements that together encourage the types of activity that result in a successful or well-used place. Whyte's philosophy was best expressed in his book, *The Social Life of Small Urban Spaces* (1980), which examines why some city spaces work for people while others do not, and what practical lessons could be learned.

In 1970, Whyte started the Street Life Project to research urban spaces using direct observation, which entailed, among other methods, interviews with people using public spaces, mapping of their behavior and levels of activity, time lapse photography and studies of density and other patterns of use. Whyte felt that direct observation had not been used to any great extent in cities in the United States to examine issues of concern to city planners and others, such as urban crowding. According to him, "most of the research on the issue was done somewhere other than where [crowding] supposedly occurred." In contrast, he felt that the direct observation of spaces could yield unique and vital insights into the success and failure of particular spaces.

With the goal of studying urban crowding, the Street Life Project began its work examining the use of public spaces in parks and playgrounds in New York City. Whyte and his researchers soon found something that surprised them. Most of these spaces, rather than being crowded, were characterized by a lack of crowding even in very dense neighborhoods. The simple conclusion was that the space alone was not enough to attract users. They also found that while most playgrounds were empty, the informal areas on city blocks – mainly the streets, were full of children playing. They questioned the common assumption that children play in streets because they lack playground space in their neighborhood, and instead hypothesized that many children play in the streets simply because they find them exciting.

This initial research yielded the additional conclusion that most crowding resulted from "choke points," such as at entrances to small parks, intersections or transit facilities that required people to move through spaces that were constricted in some way. Consequently, crowding was occurring frequently, but only for a short time, yet this experience had an impact on people's perceptions of the city. The perception that the city was a "negative" or crowded place was disproportionate to the amount of time that people actually spent in places where the discomfort occurred (Whyte 1980).

In the following years, Whyte and the Street Life Project researchers conducted further studies of the use of plazas through observations and time-lapse filming in order to learn more about the characteristics of spaces that were used versus those that were empty, and why people chose to use some spaces over others. The researchers concluded that well used places have several common characteristics including a mix of people (more couples, people in groups, more people meeting other people or saying "good bye," and generally more social interaction). They also have a larger proportion of women than men (women are more particular about the quality of the spaces they choose to use), and a higher presence of children and seniors (because they can often be in a place when others are working). Physical elements such as seating, water, food, and shade (especially movable seating, water that people can touch, food, and a choice of sun or shaded places) were found to be critical to the use of a place. In addition, other qualities were important such as the combination and location of physical elements such as seating, the relationship between a space and its edges, and the relationship between elements in a space such as a bus shelter, a waste basket, and a bench or ledge (Whyte 1980).

Why don't we have better public spaces? The need for a new approach

Many public spaces never evolve into being "places" because they were never conceived that way. For example, many of the plazas adjacent to office buildings built in New York City in the 1970s were lifeless and devoid of activity even though they represented the state of the art of public space planning and design at that time. Ironically, such places mushroomed in the ensuing decades all over the United States and even abroad.

An example is the sunken plaza that was built in front of the RCA Building at Rockefeller Center (now the General Electric Building). Originally designed as an entrance to the underground shopping concourse and subway, it failed to entice people to descend from the street. Later, efforts were made by the Rockefeller Center to improve the plaza by first adding a roller skating rink and later an ice skating rink in the winter. Many years later, outdoor dining was added to the sunken plaza in the summer, as well as temporary exhibits and events such as Orchid Shows, etc. These uses drew people down into what was an unsuccessful public space, mainly because it was located below grade. However, although the space did not function in its initial form, it was imitated all over the world.

The adjacent street level Channel Gardens (named as such because it is the walkway between the British and French buildings) evolved in a similar way. In the mid-1970s, the Rockefeller Center Inc. (RCI) management was concerned because people were sitting on the ledges of the planters, damaging the yew trees planted behind them. In determining a solution to this "problem" they asked Project for Public Spaces[1] (PPS) what kind of spikes to place on the ledges to discourage sitting. After studying how the space was used, PPS found out that a wide variety of people were using the ledges simply because there was nowhere else to sit. Rather than preventing people from using the space, PPS recommended adding benches to legitimize sitting and encourage people to stay in the plaza. The experiment was a great success – people sat on the benches and the use of the space increased. Seeing the benefit of attracting people who would sit and also perhaps shop, RCI removed the yews and replaced them with horticultural and art displays that change eleven times per year. The entire area soon became a destination, and the ground floor retail spaces facing Channel Gardens – previously occupied by banks and travel agencies – were filled by retail tenants such as the Metropolitan Museum Gift Shop. As time progressed, other ground floor spaces also filled with new tenants including NBC's Today Show, the auction house Christies, and Dean and Deluca grocers.

By starting with the simple idea of providing benches next to the planters in Channel Gardens, which created demand for other amenities, Rockefeller Center has become one of the top destinations in New York City, and indeed, the entire United States. This success is the result of the management's ability to respond in a creative manner to the issues at hand, demonstrating how a simple process based on observing how a space is used can yield unique and vital insights into the success and failure of particular spaces.

Project/discipline driven approach

Broadly speaking, the process that leads to failed spaces tends to be "project driven" and is initiated in response to a predetermined problem (Figure 50.3). This process starts out with fairly narrow goals, such as the need to widen a street. As described in the diagram below, input is provided by the "community" or stakeholders after the project has been defined, which is generally late into the process. Sometimes, but not always, modifications to the plans are made.

The problem with this approach is that it does not begin with what the community has defined as an issue. It leaves the community or stakeholders no opportunity to raise issues they are concerned about after the project has been initiated, and as a result, important questions are left unaddressed. This process is used in many government funded projects such as building roads or improving streets.

Place/community driven approach

A different approach grows out of the experience and vision that the community has for a particular place in a neighborhood and is essentially opposite from the approach described above. At its simplest, a place oriented approach (or *Placemaking*)

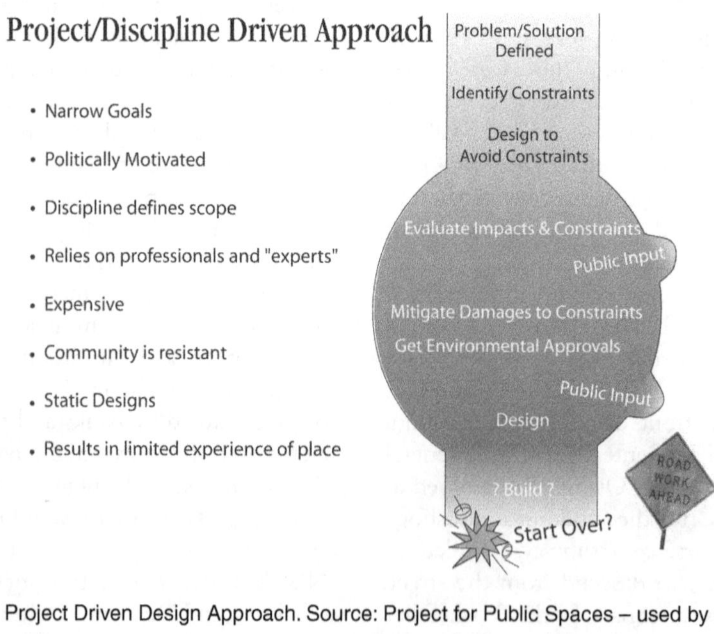

Figure 50.3 Project Driven Design Approach. Source: Project for Public Spaces – used by permission.

starts with the community or stakeholders who evaluate a particular place and develop a vision for it. Professionals then function as implementers of their vision (Figure 50.4).

Generally speaking, this approach has several benefits: it results in more flexible solutions; evolves over time rather than being built all at once; leads to a stronger sense of ownership of the place and of the improvements; ensures community support and leverages stakeholder time; and often generates financial support to make the improvements. For example, if a main street is the project being considered, local retailers might be inclined to make improvements to their façades because the vision is the result of their own evaluation, which consequently allows them to see the direct benefit.

The idea and relevance of "the power of ten" to urban design

A great city requires many different places, with a multitude of activities or reasons to use each one, in order to thrive. By extension, a great city should have at least ten successful public spaces. The term the "Power of Ten," first coined by Charles and Ray Eames in their seminal film of the same name, provides an excellent framework for thinking about the city as a whole and for evolving a community engagement process into a larger public space plan for that city.

When a city is fortunate to have ten good places, it is likely that each of these places offers at least ten things to do or ten reasons to be there. Each place is also characterized by its *accessibility*, the range of

Figure 50.4 Place/Community Driven Design Approach. Source: Project for Public Spaces – used by permission.

activities that people can engage in, the *level of social interaction* such as talking, shaking hands or kissing, its *image*, and *comfort*. Comfort is essential and having comfortable places to sit, art to touch, water to play in, food to purchase, etc. will encourage people to use the space. It follows that if there are ten of these great places with many things to do in each of them in a neighborhood, then a broader, more successful district and a great city would ensue.

Having a number of dynamic and interesting destinations within a city can define the public's experience of a city because the result is that people would keep coming back, and the district and city would keep evolving. In terms of real actions that can change the direction of urban design, the "Power of 10" is a simple way for citizens and urban designers to understand the potential of their city. Not only is it a common sense approach for communities to think about their neighborhoods, but it is also an effective tool for leaders to communicate with communities and professionals (Project for Public Spaces 2000).

An example of this concept in action can be found in downtown Houston at the newly created Discovery Green. Project for Public Spaces facilitated a place-oriented process to develop a program for a new park, which was conceived as "Houston's backyard." Building on the idea of the Power of 10, the local community outlined a number of irresistible destinations or "places" in the future park. The ten places identified included places for people to meet each other, places to eat, areas for children to play (Figure 50.5), and places to showcase the assets of the community. As built, Discovery Green includes a family destination on a small "lake" with a café, playground and interactive fountain, library branch, and a stage for performances. There is also a restaurant with a

Figure 50.5 Discovery Green, Houston. Source: Jim LaCombe.

treehouse terrace and an oak shaded promenade for markets, fairs, exhibits and other events. In winter, there is an ice rink which has become a major attraction combined with movie showings, holiday markets, and concerts. By embracing the Power of 10, Discovery Green has become a major destination in downtown Houston and has already attracted new investment surrounding the park, including plans for a retail center, office building and convention hall.

Implications of a placemaking approach for the field of urban design

Placemaking is turning a neighborhood, town or city from a place you can't wait to get through to one you never want to leave.

(PPS Survey Respondent)

The aforementioned principles of Placemaking suggest a new and broader role for urban designers that centers around using design skills to build places and destinations based on a community's vision. This new role requires turning the typical planning process upside down and allows the community to take on a new role as the "expert" or the people who know the place best. Urban designers and other professions then act as resources to respond to the community's unique vision for itself. A program that is grounded in the community's vision challenges professionals, encourages them to be more creative and in many ways, allows them more freedom. It also results in greater public benefits and creates more demand and a more positive role for the skills of urban designers.

Neither people in communities nor professionals, however, have much experience developing programs for public spaces. Although common for architectural and interior design projects, it is not as common to develop a program for the activities that occur in public spaces. Yet it is the program that ultimately defines how a space or street is to be used, and is the essential component for developing both design and management solutions that result in successful public spaces. In the future, well-trained representatives of community based organizations can be responsible for facilitating a *Placemaking* process and developing a program for many different types of public spaces. If urban designers take on a new role, they could have a great impact on how cities evolve in the future.

Conclusion

Placemaking is a dynamic human function: it is an act of liberation, of staking claim, and of beautification; it is true human empowerment.

(PPS Survey Respondent)

There are several key elements that, if taken into consideration, could fundamentally change how urban design is carried out. First is an agreed upon definition of a successful "place" as one that is well-used. Second is consensus that a different process is necessary where the professional is the facilitator and implementer of a community's vision rather than the one who defines the vision. Third is the acknowledgement that post-construction evaluation and other research into models of successful urban spaces can legitimately inform decisions about that place. Finally, it is the acceptance that both design and management are key ingredients in creating a successful urban space.

If urban designers were to adapt this more holistic view and learn to become generalists in *Placemaking*, they may be able to create an entire agenda around urban places that is transformative in affecting how people live in cites in the future. The result can be new and more meaningful opportunities for professionals and a better and more livable public realm for people who live in communities.

Acknowledgment

Craig Raphael assisted in the writing of this chapter.

Note

1 Project for Public Spaces (PPS) is a nonprofit organization dedicated to creating and sustaining public places that build communities. Established in 1975, PPS has built on the techniques of William H. Whyte's Street Life Project, and elaborated the principles of placemaking to help communities envision and build successful public spaces. PPS has written several publications describing a variety of tools for analyzing spaces including "User Analysis in Park Planning and Management", "Main Street – a Look at How they Work" "Film in User Analysis" "Achieving Great Federal Spaces," "Placemaking in Chicago: A Guide to Neighborhood Placemaking" and, most recently three books about how citizens can get involved in transportation related issues in their communities.

References

Project for Public Spaces (2000). *How to Turn a Place Around*. New York: Project for Public Spaces, Inc.

Project for Public Spaces (PPS) (2008). Metropolitan Planning Council (MPC). *A Guide to Neighborhood Placemaking in Chicago*. Chicago: PPS/MPC.

Whyte, William H. ([1980]; 2001). *The Social Life of Small Urban Spaces*. New York: Project for Public Spaces.

Further reading

Appleyard, D. (1981). *Livable Streets*. Berkeley, CA: University of California Press. An eloquent account of the negative effects of vehicular traffic on the fabric of neighborhoods.

Jacobs, A. (1985). *Looking at Cities*. Cambridge, MA: Harvard University Press. A thorough account of how one can gain rich information and visual cues about a neighborhood through observation of its physical context.

Kretzmann, J.P. and McKnight, J. (1993). *Building Communities from the Inside Out*. Chicago, IL: ACTA Publications. Describes an asset-based approach to community building and development that focuses on the positive elements of neighborhoods rather than their problems.

Marcus Cooper, C. and Francis, C. (1997). *People Places: Design Guidelines for Urban Open Space*. New York: John Wiley and Sons, Inc. Useful guidelines for different types of urban open spaces.

Whyte, W.H. (1988). *City: Rediscovering the Center*. New York: Doubleday. This follow-up book to Whyte's classic text. *The Social Life of Small Urban Spaces* is an examination of human interactions and design elements which make cities lively.

51

Secure cities

Carolyn Whitzman

This chapter will trace two streams of thought about urban design for safe and secure cities. The first stream emerged from a late nineteenth- and early twentieth-century concern with the criminogenic properties of low-income housing in the central city. From the late 1940s to the 1970s, densely populated urban housing, with children playing on the street instead of in supervised parks and recreation centers, was torn down in many cities around the world for supposedly more scientific and modern public housing. From the work of Jane Jacobs (1992 – originally published 1961) onwards, this public housing was derided by scholars, who stressed the importance of formal and informal controls over public space, particularly in low-income areas. Writers like Oscar Newman (1972) and Alice Coleman (1985) are associated with Crime Prevention Though Environmental Design (CPTED), which focuses on how the design and maintenance of built environments can control crime; for instance, by controlling access to particular locations or improving surveillance of particular spaces. Later scholars, such as Ron Clarke, broadened their concerns with opportunities for crime by focusing on issues such as "target hardening" of public spaces, and developing zero tolerance policies for minor incivilities such as graffiti, begging, and sleeping rough (Kelling and Coles 1996; Clarke 1992). Critics of these theories complained, however, that when translated into policies they promoted exclusionary public spaces, thus exacerbating increasing socio-economic polarization in cities (Mitchell 2003; Kohn 2004).

A second stream of thought on building secure cities can also be traced back to Jane Jacobs. This stream of thought emphasized land use and social mix in both low-income and higher income areas, and importance of the resident as expert in developing local solutions. Many of Jacobs' ideas were taken up by a generation of feminist writers on urban space, who developed tools such as the Women's Safety Audit (METRAC 1989), which is intended to identify unsafe spaces and suggest improvements based on the experiences of female users of the space. This approach began by emphasizing political empowerment and inclusion of women. Now it has matured to include other marginalized or excluded social groups. Particularly innovative examples of this second perspective on secure cities are occurring in the slums of low-income nations, where women's safety audits and children's participatory planning tools have been modified to raise concerns about basic infrastructure, such as roads and footpaths, toilets, public transport, water, and social services. This chapter will trace tensions between these two schools of thought in terms of protection versus empowerment, exclusion versus risk, and design versus community development.

Secure cities for whom?

Early fortified cities kept invaders and trouble-makers at bay, and were a safe haven not only for their inhabitants but for the people in the surrounding vicinity. It was the suburbs outside the wall that were unsafe. Six hundred years ago, in *The Canterbury Tales*, an alchemist's servant could describe his "sly" and "crafty" master living in "the suburbes of a toun," "lurkinge in hernes [hedges] and lanes blinde/ Whereas thise robbours and thise thieves by kinde/ Holden hir pryvee fereful residence" (Chaucer 1974 [1387]: 90). Almost fifty years ago, Jane Jacobs contended that the first and foremost use of sidewalks, those hallmarks of urban life, is safety: "if a city's streets are safe from barbarism and fear, the city is thereby tolerably safe from barbarism and fear" (Jacobs 1992 [1961]: 30). In this understanding of secure cities, design allows people to carry out their business in public spaces by protecting them from threats from the dangerous "other." This formulation ignores violence within families (Whitzman 2007) as well as institutional violence (Moser 2008). Further, it fails to anticipate the impact of recent terrorist attacks in New York City, London, and Mumbai, which confound traditional notions of how to fortify cities and citizens against attack.

In contrast to an emphasis on protection from dangerous others is an inclusive viewpoint that "city life is an openness to unassimilated otherness" (Young 1990: 227). Some see public space as a forum for diversity and productive conflict (Fincher and Iveson 2008), and safer urban design as an outcome of mechanisms that will encourage people of all ages and backgrounds to linger, meet, mingle, and enjoy "life between buildings" (Gehl 2006 [1971]: title). Many contend that the main danger to secure cities is the abandonment of public space and public expression rights within public space (Kohn 2004). It is no longer necessary to use public spaces like sidewalks in order to travel between home, work, shops, and recreation. Now the affluent can shut themselves up in cars and gated communities, shop in malls, and limit their interaction with other citizens in public space. Abandonment of the public realm is exacerbated by increasing car dependence, and resultant urban sprawl. The costs of neglecting public realm interactions include reduced everyday physical activity leading to increased obesity, increased social isolation leading to increased risk of depression, and decreased social capital (Frumkin *et al.* 2004). The challenge of creating livable, healthy, and convivial cities is to encourage optional or leisure-related use of public space by all people, as well as encourage relatively sociable walking, cycling, and public transport as alternatives to cars. In this second understanding, safer urban design improves individual and collective health and equity outcomes. Instead of an emphasis on protecting people from robbers, thieves, and barbarians, the emphasis instead is on encouraging encounters with other people, recognition of different needs of people in public space, and redressing inequalities of access (Fincher and Iveson 2008).

The abandonment of public space has particularly inequitable results for certain groups, such as children. The proportion of primary school aged children allowed to walk or cycle to school in the UK decreased from 80 percent in 1971 to 9 percent in 1990 (Hillman *et al.* 1990), with similar decreases reported in most English-speaking developed countries (Kingham and Ussher 2007). The primary reason for this decline is parental concern about traffic and personal safety (Valentine 1997). Not only does the abandonment of public space by unaccompanied children lead to negative physical and mental health outcomes (Tranter and Pawson 2001), it also places additional strain on "helicopter parents" who are expected to ferry their children to a series of social activities that they used to be able to accomplish on their own: recreation, amusement, meetings with friends (Malone 2007).

Use of public spaces and amenities is also unevenly distributed along gender lines. Recent mapping exercises of women and men's use of public space in Mumbai have found that adult men in public space overwhelmingly outnumber adult women and children at all times and in most places. For instance, in Nariman Point, a business district in South Mumbai, head counts taken between noon and 2 p.m. on a weekday, found 88 percent of pedestrians were adult men, 10 percent women, and only 2 percent children. Men tended to eat outdoors or at least linger and talk with one another whilst buying lunch, while women quickly bought lunch and re-entered their workplaces to eat at their desk, at least partly because of fear and discomfort in public space (Ranade 2007).

According to these scholars, there is a tyranny of purpose in assumptions about the use of public space by women, children, and other subject groups. Women have been excluded from lingering in and enjoying public space not only by busy lives, but from a sense that they should not court risk of harassment or assault by overstaying their welcome in public space (Phadke 2007). Young children must be protected from dangerous strangers by constant supervision by a parent in public space, while older children, particularly minority male youth, are construed as a threat in public space, particularly when they linger around shops or in parks (Valentine 1997). Elizabeth Wilson (1991) has spoken of how public space, particularly in the central city, became seen as both masculine and dangerous in the nineteenth century, and how women and children needed to be excluded from independent exploration of these dangerous spaces for their own good. Apparently, this attitude lingers in the twenty-first century.

CPTED and its discontents

The foundational works of CPTED, Jane Jacobs' *Death and Life of Great American Cities* (1992 [1961]) and Oscar Newman's *Defensible Space* (1972), continue to exert considerable influence today (Sutton *et al.* 2008). Jacobs argued that the primary safety mechanism of any street, or public space for that matter, was not the police, but informal social control, what she called "eyes on the street" (Jacobs 1992: 35). Well-used city streets are likely to be safe streets, particularly if people are walking along the sidewalks day and evening, and there are also shop windows or residences overlooking the space (ibid. 34–35). It is for that reason that Jacobs favored a fine-grained urban fabric of small lots, small blocks, and mixed uses, concerns that were later taken up by New Urbanists (Duany *et al.* 2000).

Jacobs provides a third element of safe streets, a sense of symbolic ownership created by a clear demarcation between public space and private space (Jacobs 1992: 35). A decade later, Oscar Newman (1972) focused on this particular element in his work in high-crime public housing in the US. Newman used detailed crime records provided by the public housing authorities, along with surveys of residents, to pinpoint particular problem areas. Places which had the least direct visibility – lobbies not visible from the street, fire stairs, corridors that took sharp turns, circuitous outdoor paths framed by high bushes, laundry rooms in isolated basements – were found to be the most dangerous and the most feared. Large tracts of park space "unassigned" to any groups of buildings were also avoided. Newman suggested greater territorial definition of space by breaking up large public spaces into small zones of influence, making them the symbolic property of individual or small groups of households. In fact, Newman believed that the breakdown of social control mechanisms was more likely in public housing developments, where home ownership would not act as a spur to neighborhood self-protection.

In the UK, Alice Coleman blamed "modern problem estates" (public housing) for "lapses in civilized behavior" such as "litter-dropping, graffiti-scrawling, vandalism, pollution by excrement, and family breakdown leading to children taken into care," as well as "crime, fear, anxiety, marital breakdown, and physical and mental disorders that would largely be avoidable in more socially stabilizing environments." In fact, most people could "cope perfectly well with life in more traditional houses" (Coleman 1985: 2–3). Her solution was that "no more [public housing] should be built," that people should be "allowed to find their own houses" according to the "age-old principle of natural selection" (ibid. 171). As interim measures for existing estates, she proposed fencing off communal green space to create private gardens, limiting the number of dwellings per entrance by partitioning larger buildings, walling off apartment blocks from one another, and integrating projects with the surrounding streets and neighborhoods (ibid. 170–177).

Newman and Coleman's work fed into the neoliberal urban agenda that was on the ascendant by the 1980s. Increasingly, residents of public housing became socially, economically, and politically marginalized. An attack on the admittedly brutalist design of many estates became an environmentally determinist excuse to build less public housing, to blame bad design for increasing social polarization, and to tear down estates without necessarily constructing affordable alternative housing (Murie 1997). Other prominent CPTED proponents went even further in supporting a right wing social agenda than Newman and Coleman. Barry Poyner attempted to systematize the growing research on CPTED in *Design Against Crime* (1983). After studying Newman's work on privatizing streets in St. Louis, Poyner recommended that "access on foot and by car to residential streets or groups of streets be limited," that deed restrictions be allowed to limit homes to single family residential occupation, and that cul-de-sacs should be encouraged (1983: 23–25). In order to limit burglaries, Poyner recommended that "areas of middle-class/ middle income housing be separated as far as possible from poorer housing" (ibid. 36), and that "houses not face on main through roads," but be separated from these roads by walls or hedges (ibid. 42). Poyner thus provided the intellectual justification for the development of gated communities, which have become prominent features of US cities, and are an increasing phenomenon internationally (Blakely and Snyder 1995; Schneider and Kitchen 2007). Ironically, CPTED's encouragement of protective hedges and cul-de-sacs replicate those design elements that Geoffrey Chaucer had associated with unsafe suburbs 600 years earlier.

In the UK, "Secured by Design" guidelines, developed and administered by Police Architectural Liaison Officers, provide official recognition to developers and builders who abide by their principles (Schneider and Kitchen 2007). These guidelines have inevitably found themselves in collision with New Urbanist proponents who also claim Jane Jacobs as their intellectual godmother. A journal article in the *Police Review* (Knowles 2003: 23; see also Schneider and Kitchen 2007: 114–124) explained New Urbanist principles of creating "neighborhoods that are accessible and easy to move through," encouraging "the development and use of public transport in terms of provision and by generating a sufficient density of people to make public transport a feasible proposition," limiting "the environmental impact of cars," and emphasizing "the importance of sustainable approaches to environmental design." These are deemed to be in conflict with Secured by Design principles of restricting "the amount of public space," restricting "the number of escape routes available to criminals," and restricting "the number of crime generators, for instance ... 'honeypots' (such as fast food take-aways) that encourage people to concentrate in an area." The article estimates that the financial implications of building a development of 4,500 homes along New Urbanist principles instead of Secured by Design guidelines to be 17 additional police officers, a cost of almost 1 million USD per annum, although if "under-reporting and the true economic cost of crime is considered," the cost might rise to an additional 3 million USD annually. This contention is made in defiance of the fact that CPTED evaluations have not shown particular efficiency or effectiveness in decreasing crime, in the UK and elsewhere (Shaftoe and Read 2005), although others would contend that it is a promising, if not proven, crime prevention technique (Eck 2002).

A further attack on public space and the use of public space by particular populations such as poor people and young men is found in Situational Crime Prevention, a set of theories associated with CPTED, which seeks "to understand how people perceive opportunities for crime, and to remove these opportunities, or at least ensure that they are less easily recognized" (Sutton *et al.* 2008: 51; see also Clarke 1992). Perhaps the most famous theory associated with Situational Crime Prevention is George Kelling's "broken windows" hypothesis, which contends that if minor signs of disorder, including graffiti, vandalism, and perhaps most contentiously, begging, are not controlled, then more serious crimes will occur in that space (Kelling and Coles 1996). This popular theory has led to thousands of people, mostly young men, in the UK being served with Anti-Social Behavior Orders for "offenses" such as loud music, verbal abuse, and fouling by dogs (Brown 2004). Anti-begging ordinances and other mechanisms are used to exclude particular groups, such as visibly homeless or mentally ill people, from public space (Mitchell 2003).

In the 1990s, an attempt was made to insert a humanizing element in CPTED discourses through so-called second generation CPTED. For instance, Greg Saville and Gerry Cleveland (1998) discuss Dutch Secure by Design principles which attempt to integrate smaller scale public housing developments, provide meeting places such as public squares and youth clubs within low income neighborhoods, and encourage residents' participation through organized activities. Schneider and Kitchen (2007: 25) argue against simplistic situational crime prevention solutions, and endorse Saville and Cleveland's view that the social aspect of home and neighborhood is as important as "bricks and mortar." Nevertheless, it can be concluded that CPTED and its associated theory of situational crime prevention have excluded social inequities from their analysis and have relied on a small group of "experts" to evaluate the safety and security of public space. In turn, the application of these theories have had destructive and exclusionary impacts on the use of public space, particularly by certain groups labeled as "the problem," such as young men and homeless people.

Women's safety audits and child-friendly cities

While CPTED principles and practices were being adopted, particularly in

English-language high-income countries, another group of researchers in those countries were developing design guidelines that emphasized consultation with users, and recognition of the needs and perspectives of women and children, particularly those with low incomes. Books such as *Housing as if People Mattered: Site Design Guidelines for Medium-density Family Housing* (Marcus and Sarkissian 1986), *People Places: Design Guidelines for Urban Open Space* (Marcus and Francis 1990), and *Safe Cities: Guidelines for Planning, Design, and Management* (Wekerle and Whitzman 1995) were dedicated to extending the use and enjoyment of public space by all citizens. During the same time that such design guidelines were developed in the 1980s and early 1990s, the Women's Safety Audit was created by the Toronto group METRAC (Metro Toronto Action Committee on Violence Against Women and Children). The *Women's Safety Audit Guide* (METRAC 1989) allows community members as well as local government and other authorities to identify places which they feel unsafe and recommend means of improvement. Unlike CPTED or Situational Crime Prevention, the Women's Safety Audit takes the perspective of potential victims, or at least people who feel unsafe, rather than looking at places through the eyes of potential offenders. Although focusing on many similar elements as CPTED – lighting, informal surveillance, land use mix, vandalized or poorly maintained spaces – the tool also recognizes that women want to be autonomous actors, not dependent potential victims. Thus, it also emphasizes good signage, empowerment of users of a space through participatory planning mechanisms, and increasing awareness of the issue of violence and how it can be successfully addressed at the local level. In other words, it treats particular users of a public space as "experts of local experience," rather than helpless and uninformed potential victims, who need to rely on professional experts like police or criminologists to protect them (Whitzman 2008: 111–112).

A recent report summarizes 69 sources – websites, reports, and academic articles – that have described the gender-specific use of safety audits, and a further 18 organizations answered a survey on how they have used women's safety audits (WICI 2008). The Women's Design Service, a non-profit consultancy and advocacy group in the UK, developed a safety audit resource guide that allowed residents to influence regeneration projects in London, Bristol and Manchester (Cavanaugh 1998, Berglund 2007). They worked with a variety of groups, from young people with intellectual disabilities in the London Borough of Hammersmith, to minority women living on housing estates in Longsight, Manchester. The safety audits led to concrete changes like paths being re-routed in a high-crime park, but also to a more generalized sense of empowerment, probably because decision makers participated in the audits as active listeners, and there was funding available for immediate improvements (WICI 2008: 29). The audits also informed workshops for planners and community groups called Planning Safer Places, which in turn fed into the new London Plan (Berglund 2007: 56).

In Dar es Salaam Tanzania, safety audits that involved 100 women in the low income neighbourhoods of Midizini and Mnazi Mmoja in 2000 and again in 2002 brought out a range of secure design issues that are also commonly identified in high income cities: lack of accessibility of streets to emergency vehicles, absence of street lights and signs, lack of cleanliness and maintenance. The audits also brought out some concerns less typically noted in high-income city audits, such as unemployment of young people and female heads of household, and related "grog" selling and prostitution. Some of the recommendations did not require external funding, such as the suggestion that households buy a light bulb for their front and back doors to create street lighting; public education led to the implementation of this recommendation. Some did require local government assistance, such as unblocking roads and footpaths, and developing job creation programs. The safety audit findings were used to justify funding received from the World Bank in 2006 for neighborhood upgrading, which has resulted in tarmac streets, municipal street lighting, and an enhanced job creation program. Seed capital was provided to the women who were brewing grog and acting as prostitutes, to enable them to establish less risky income-generating activities such as food kiosks, second-hand cloth stalls and charcoal stalls instead. While this initiative started as a municipal project, it soon expanded into a national program (Mtani 2007; Whitzman 2008: 128–131).

Another African example of participatory planning recommendations to address violence in both public and private spheres from a gendered perspective comes from the Khayelitsha Project in Cape Town, South Africa (Moser 2008). Violence in domestic spaces, such as assault, rape, and emotional abuse, are recommended to be addressed through developing new houses of refuge, counseling, and conflict resolution facilities as part of slum upgrading. Police facilities also need to be equipped with trauma facilities and female officers. Violence in public space such as open fields, narrow lanes, and empty stalls, needs to be addressed through improvements to street lighting and getting a telephone system, but also through the provision of rape relief centers providing self-defense training, safe walkways, and an improved public transport system. Rape at or near sanitary facilities (also a concern in the Indian example described below) could be addressed by sewers being installed and outside toilets being phased out, and supervision of communal sanitary facilities. Drug and alcohol related violence in and near shebeens (establishments where alcohol is illegally brewed and sold), could be reduced if these are re-located to where social and police control is more efficient. Physical violence and group rape around schools could be prevented by installing better fencing and guards; guarded schools might then also serve as safe playgrounds after hours. Assault and sexual harassment on roads and transport could be addressed by jobs and services being brought closer to slum residents, thus reducing transport needs. These recommendations show an attention to synergies between design and community development approaches.

Similar examples of children identifying unsafe or unused spaces, and suggesting improvements can be found in the growing literature on children's participation and Child Friendly Cities. An example of space that balances the needs of young men, on the one hand, and women with young children, on the other hand, is Dufferin Grove, a park in west central Toronto. There, shared space has been negotiated in a skating rink and club house, as well as in community gardens next to a basketball court (Whitzman 2008: 235–238). In Mexico, neighborhood safe havens have been built for children to escape sometimes violent or disruptive home lives, and to provide a range of social and recreational activities for young people whose opportunities might otherwise be seriously limited (Bartlett *et al.* 1999). In Karnataka India, 149 children, aged 9 to 18, tracked their typical routes to school and work, and identified hazards such as streams that flooded in the rainy season, and informal waste disposal dumps near playgrounds filled with glass and syringes. Their work led to building or fixing foot bridges and filling in holes used for waste disposal (Lolichen 2007).

These examples suggest a process of developing secure cities that explicitly addresses poverty reduction, violence in both public and private space, and the identification of key needs and resources by the ultimate users of a space. These efforts work towards safer cities in three ways. The first and most self-evident aspect is making public spaces safer and more accessible to groups previously ignored. The second aspect involves developing "safe spaces" such as neighborhood centers, refuges, job creation programs, and other spaces where people can come together to discuss issues of common concern, and improve their autonomy and resilience. The third aspect of "safe space" involves bringing previously hidden issues like family violence (both intimate partner violence and child abuse) out into the public sphere, and thus making them public issues (Whitzman 2007).

As pointed out in a recent paper by urban development theorist Caroline Moser (2008), there remain two large challenges within this work. First, to what extent is women or children's safety a separate issue, and to what extent can these needs be mainstreamed into broader "Safe Cities" research, policy and practice? Second, to what extent should issues of poverty and exclusion, as well as elements of identity and agency, be considered as determining factors in urban safety issues along with gender and age? There are no simple answers to these questions, but given that the majority of urbanization is now occurring in low-income nations, and that there are increasing numbers of women-led households in those cities (UN-Habitat 2001), these questions deserve to be at the forefront of debate around secure cities.

Conclusion

The debate over the extent to which good design determines good behavior has raged for well over a century. Certainly, design encourages or enables certain behaviors, while constraining others (Dovey 2000, citing Anthony Giddens). The quality and quantity of safe and accessible public space varies widely within and between cities. Design, combined with differing cultural norms, can enable some users to feel comfortable and valued within public space, while constraining the access of others (Gehl 2006).

A holistic understanding of secure cities must include recognition that violence occurs in both the public and private realms, and is committed both by individuals and by institutions. Good design may not have a direct impact on domestic or family violence, but safe havens can be provided, and public space discourses identified, that can bring this hidden violence out into open discussion. Providing basic water, sewer, and public transport infrastructure and developing new jobs through local service provision in low-income communities can redress health and income inequalities as well as improve individual and collective security (Moser 2008; Whitzman 2008).

There are inevitable tensions between groups over the use of public space, and tensions between individual and collective desires for "enclosure" or safety on the one hand, and "encounter" or risk on the other. There are a variety of approaches to secure cities that range on a continuum between reliance on police, security, and other formal controls to enforce norms and protect subject groups such as women and children on one end, and promoting autonomy, empowerment, and even risk-taking at the other end. As Phadke (2007) and others argue, women and children should have the right to risk and the right to freely explore in sometimes unsettling or unsafe cities. The risk of assault needs to be weighed against the physical and mental health risks of not having access to education, employment, and recreation opportunities because of fear of crime. Users of a space, particularly those often discouraged from exploration "for their own good" such as children, should have a much greater voice in designing secure cities. Making cities safer for these excluded groups may result in safer cities for everyone.

There are at least three rights associated with public space: the right to encounter; the right to recognition as a distinct group with needs and resources; and the right to redistribution, or greater access to communal resources (Fincher and Iveson 2008). A fourth right, to expression, dissent, and even conflict within public space, may also be under threat because of the desire to make cities more secure (Kohn 2004; Davis 1992). For cities to be liberatory – as in the medieval adage "city air makes men [sic] free" – as well as secure, safer urban design must attend to these linked rights to public space, as well as encourage tools that promote greater autonomy of excluded groups such as women, children, and poor people.

References

Bartlett, S., Hart, R., Satterthwaite, D., de la Barra, X., Missair, A. (1999). *Cities for Children: Children's Rights, Poverty and Urban Management*. London: Earthscan.

Berglund, E. (2007). *Doing Things Differently: Women's Design Service at 20*. London: Women's Design Service.

Blakely, E. and Snyder, M.G. (1995). *Fortress America: Gated and Walled Communities in the United States*. Cambridge, MA: Lincoln Institute of Land Policy.

Brown, A. (2004). "Anti-Social Behaviour, Crime Control, and Social Control." *The Howard Journal of Criminal Justice*, 43(2): 203–211.

Chaucer, G. (1974) [originally published c. 1387]. *The General Prologue and the Canon's Yeoman's Prologue and Tale*. London: University of London Press.

Cavanaugh, S. (1998). *Making Safer Places: A Resource Book for Neighbourhood Safety Audits*. London: Women's Design Service.

Clarke, R. (Ed.) (1992). *Situational Crime Prevention: Successful Case Studies*. New York: Harrow and Heston.

Coleman, A. (1985). *Utopia on Trial: Vision and Reality in Planned Housing*. London: Hilary Shipman.

Davis, M. (1992). *City of Quartz: Excavating the Future in Los Angeles*. New York: Vintage.

Dovey, K. (2000). "Redistributing Danger: Enclosure and Encounter in Urban Design," *Australian Planner*, 37(1): 10–13.

Duany, A., Plater-Zyberk, E., Speck, J. (2000). *Suburban Nation: The Rise of Sprawl and the Decline of the American Dream*. New York: North Point Press.

Eck, J. (2002). "Preventing Crime at Places," In Sherman, L., Farrington, D., Welsh, B., and MacKenzie, D. (Eds.), *Evidence-based Crime Prevention*. New York: Routledge. 241–294.

Fincher, R. and Iveson, K. (2008). *Planning and Diversity in the City: Redistribution, Recognition, and Encounter*. Houndsmills: Palgrave Macmillan.

Frumkin, H., Frank, L., Jackson, R. (2004). *Urban Sprawl and Public Health: Designing, Planning, and Building for Healthy Communities*. Washington, DC: Island Press.

Gehl, J. (2006) [originally published 1971]. *Life Between Buildings: Using Public Space*. Skive: Danish Architectural Press.

Hillman, M., Adams, J., Whitelegg, J. (1990). *One False Move: A Study of Children's Independent Mobility*. London: Policy Studies Institute.

Jacobs, J. (1992) [originally published 1961] *The Death and Life of Great American Cities*. New York: Vintage Books.

Kelling, G. and Coles, C. (1996). *Fixing Broken Windows: Restoring Order and Reducing Crime in our Communities*. New York, Martin Kessler Books.

Kingham, S. and Ussher, S. (2007). "An Assessment of the Benefits of Walking School Bus in Christchurch, New Zealand," *Transportation Research Part A*, 41: 502–510.

Knowles, P. (2003). "Designs on Crime," *Police Review*, 18 July 2003.

Kohn, M. (2004). *Brave New Neighbourhoods: The Privatization of Public Space*. New York and London: Routledge.

Lolichen, P. (2007). "Children in the Drivers' Seat: Children Conducting a Study of Their Transport and Mobility Problems," *Children, Youth and Environments*, 17(1): 238–256.

Malone, K. (2007). "The Bubble-Wrap Generation: Children Growing Up in Walled Gardens," *Environmental Education Research*, 13(4): 513–527.

Marcus, C. and Francis, C. (1990). *People Places: Design Guidelines for Urban Open Space*. New York: Van Nostrand Reinhold.

Marcus, C. and Sarkissian, W. (1986). *Housing as if People Mattered: Site Guidelines for Medium-Density Housing*. Berkeley, CA: University of California Press.

METRAC [Metro Toronto Action Committee on Public Violence Against Women and Children] (1989). *Women's Safety Audit Guide*. Toronto: METRAC.

Mitchell, D. (2003). *The Right to the City: Social Justice and the Fight for Public Space*. New York, Guilford Press.

Moser, C. (2008). "Safety, Gender Mainstreaming, and Gender-based programmes," keynote address to International Centre for the Prevention of Crime Annual Colloquium: Focus on women's safety, Queretaro, Mexico, November 12, 2008.

Mtani, A. (2007). "Local Innovations for Crime Prevention: The Case of Safer Cities Dar es Salaam." In Shaw, M. and Travers, K. (Eds.), *Strategies and Best Practices in Relation to Urban Areas and Youth at Risk*. Montreal: International Centre for the Prevention of Crime, 69–79.

Murie, A. (1997). "The Social Rented Sector, Housing, and the Welfare State," *Housing Studies*, 12(4): 437–461.

Newman, O. (1972). *Defensible Space*. New York: Macmillan.

Phadke, S. (2007). "Dangerous Liaisons: Women and Men, Risk and Reputation in Mumbai," *Economic and Political Weekly*, April 28, 2007, 1510–1518.

Poyner, B. (1983). *Design Against Crime: Beyond Defensible Space*. London: Butterworth.

Ranade, S. (2007). "The Way She Moves: Mapping the Everyday Production of Gender-Space," *Economic and Political Weekly*, April 28, 2007, 1519–1526.

Saville, G. and Cleveland, G. (1998). "2nd Generation CPTED: An Antidote to the Social Y2K Virus of Urban Design," Paper presented at the Third Annual International CPTED conference, Washington, December 1998.

Schneider, R. and Kitchen, T. (2007). *Crime Prevention and the Built Environment*. London: Routledge.

Shaftoe, H. and Read, R. (2005). "Planning out Crime: The Appliance of Science or an Act of Faith?" In Tilley, N. (Ed.), *Handbook of Crime Prevention and Community Safety*. Cullompton, Devon: Willan Publishing, 245–265.

Sutton, A., Cherney, A. and White, R. (2008). *Crime Prevention: Principles, Perspectives, and Practices*. Port Melbourne: Cambridge University Press.

Tranter, P. and Pawson, E. (2001). "Children's Access to Local Environments: A Case Study of Christchurch, New Zealand," *Local Environments*, 6: 27–48.

UN-Habitat [United Nations Commission on Human Settlements] (2001) *Women and Urban Governance*. Nairobi: UN-Habitat Policy Dialogues Series no. 1.

Valentine, G. (1997). "'Oh Yes I Can' 'Oh No You Can't': Children and Parents' Understandings of Kids' Competence to Negotiate Public Space Safely," *Antipode*, 29(1): 65–89.

Wekerle, G. and Whitzman, C. (1995). *Safe Cities: Guidelines for Planning, Design and Management*. New York, Van Nostrand Reinhold.

Whitzman, C. (2007). "Stuck at the Front Door: Gender, Fear of Crime, and the Challenge of Creating Safer Space," *Environment and Planning A*, 39: 2715–2732.

—— (2008). *The Handbook of Community Safety, Gender, and Violence Prevention: Practical Planning Tools*. London: Earthscan.

WICI [Women in Cities International] (2008). *Women's Safety Audits: What Works and Where?* Nairobi: UN-Habitat Safer Cities Programme.

Wilson, E. (1991). *The Sphinx in the City: Urban Life, the Control of Disorder, and Women*. London, Virago Press.

Young, I.M. (1990). *Justice and the Politics of Difference*. Princeton, NY: Princeton University Press.

Further reading

Jacobs, J. (1992) (originally published 1961). *The Death and Life of Great American Cities* (New York: Vintage Books). Still cited by most researchers on secure cities, Jacobs describes how "eyes on the street" (informal surveillance by strangers who are willing to intervene in an unsafe situation) is the most important determinant of secure public space, and how mixed use and street-oriented design might assist in improving public space.

Gehl, J. (2006) (originally published 1971). *Life Between Buildings: Using Public Space* (Skive, Denmark, Danish Architectural Press). Another writer who does not focus solely on secure cities, but whose ideas have been enormously influential. Gehl emphasizes the importance of optional and social activities (as opposed to necessary activities) in encouraging eyes on the street and symbolic ownership of public space.

Moser, C. (2005). "Urban Violence and Insecurity: An Introductory Roadmap," *Environments and Urbanization*, 16(3), pp. 3–16: The strongest current conceptualization of urban violence in both the public and private spheres, in both developing and developed countries, and how this violence has differential impacts based on age, income, and gender.

WICI (Women in Cities International) (2008). *Women's Safety Audits: What Works and Where?* (Nairobi: UN-Habitat Safer Cities Programme). An examination of the use of one practical tool to improve public space, and how it has had an impact on cities in both developing and developed countries.

52

Design for resilient cities

Reflections from a studio

Mahyar Arefi

Resiliency is an emerging concept in urban design which fosters new thinking about designing less vulnerable and more flexible cities. From Chicago and San Francisco which recovered from big fires in the nineteenth century, Berlin and Beirut which survived wars in the twentieth century, and New Orleans which – in the face of its geologic and hydrologic limitations – is re-emerging from the Hurricane Katrina in the twenty-first century (Kates *et al.* 2007), to vibrant pockets of everyday urbanism observed in Istanbul, Mumbai and New York, resilient cities manifest the saga of survival, governance, sustainability, adaptability and flexibility.

Rooted in ecology, resiliency incorporates environmental considerations into urban design. Both disciplines find resiliency a potent metaphor for understanding ecosystems and cities (Pickett 1999). Metaphors such as resiliency, tree, or organism stress the nature of the city as a "living thing" (Kostof 1991: 15). Ecologists typically study both what causes organisms to survive extreme environmental conditions, and what causes them to fail or perish.

Just as ecologists think about the persistence of resilient organisms and ecosystems against environmental threats, urban designers think about the benefits of resilient cities captured by new models including "the Photosynthetic City," "the Renewable Energy City," "the Eco-efficient City," "the Carbon Neutral City," and "the Place-Based City" (Newman 2009). In these models resiliency ranges from increasing efficiency by producing energy from waste, wind, and sun; decreasing reliance on oil, consumption of non-renewable resources, and carbon emissions; decentralizing water and power grids; and water recycling to localizing economic development initiatives and promoting green jobs.

The degree to which urban designers can draw inspirations from ecology depends on the two key elements cities and organisms have in common: recovery from "disaster" (Vale and Campanella 2005) or "illness" and "absorbing change" (Hester 2006: 139). Recovery from disaster or illness in a city or an organism conjures up two options: reverting back to a *previous* status or pursuing a *preferred* option. The former represents an "equilibrium" model associated with capacity building and reaching normalcy. Exposure to natural and human-made disasters (i.e. floods, earthquakes, fires, wars, and terrorist attacks such as 9/11) has prompted the need to reduce vulnerability by increasing safety and adaptability in the city. Post-disaster recovery reconstruction efforts tend to mitigate risk and reduce vulnerability (Musacchio and Wu 2002) through capacity building (i.e. by increasing the city's infrastructure capacity or reducing carbon emissions and streamlining mobilization efforts when disaster strikes).

In a "non-equilibrium" model (Pickett *et al.* 2004), however, the goal is to promote "flexibility" rather than reduce "vulnerability." This model offers broader applications for urban design by focusing on three aspects of the built environment: "form," "function," and "flow."

"Forms" define buildings. Just as resilient organisms adapt to their habitats, certain building types (i.e. modules and lattice-like structures) can increase the urban form's adaptability to change. Modules represent "standardized parts of one of more sizes," and lattices illustrate "a repeating plane or solid regular grid of dimensions, within which parts must fit" (Lynch, 1990: 384). The flexibility of modules and lattices occurs within fixed, standardized pieces of replicable structures.

"Functions" reflect purpose in urban form (and organisms). Architects and planners have examined whether "non-specialized" forms better adapt to new purposes and functions compared to "specialized" forms. The question is whether the design for "special purposes" or "functions" renders urban form "inflexible?" The debate on the nexus between specialized versus non-specialized forms and flexibility shows that cities need both "narrowly specialized" and "unspecialized units," which should be kept apart so that adaptability in one would not disturb the functions and growth in the other (Lynch 1990: 382).

Urban form facilitates the "flows" of information, movement, services, and people that form separate but interconnected webs of critical relationships in the long-term vitality of the city. Infrastructure handles the "flows" of services (i.e. traffic, gas, water, and sewer) whereas public spaces (i.e. squares, plazas, and what some call "loose spaces" or "nooks and crannies") facilitate social interaction and flows of people. In an ecological context, flows facilitate the ways in which organisms interact with the surrounding environment across different scales from small catchment areas to regional ecosystems and watersheds.

These three attributes conceptualize a continuum from fixed or rigid (i.e. infrastructure) to semi-fixed (i.e. public space), to flexible, adaptable, or fluid (i.e. loose spaces) urban form. Forms, functions, and flows characterize the spatial and non-spatial (i.e. social and cultural) attributes which vary widely in scale and size, and can cover a metropolitan area, a neighborhood, or a public square, park, or plaza.

The main premise here is that within a non-equilibrium paradigm resilient urban forms comprise of components which can adapt to new conditions. For example, a gentrifying neighborhood may not necessarily bounce back to its previous condition (in an equilibrium model), may even lose part of its original population, and may find a new status, a new identity, with a new population. Such a neighborhood could adapt to new forms, functions, and flows which may either result from planned intervention, or from bottom-up grassroots' efforts.

Fulton (2005) considers Los Angeles a resilient city not just because of bouncing back from earthquakes or fire, but because it consists of resilient, adaptable areas. Fulton finds the resiliency of areas such as Watts or South Central in the energy which flows from "Angelinos" more so than government investments in an aging infrastructure inherited from an earlier era. The resiliency in South Central Los Angeles and Watts, hence, has less to do with (generally long-term) implemented policies, but has more to do with people's (typically mid- to short-term) efforts to reenergize and transform urban liabilities into new forms, functions, and flows of assets. This is important in understanding resiliency in urban design. Three important points in developing urban design schemes are: first, identifying and examining liabilities (forms, functions, flows) within the city; second, transforming liabilities into assets; and third, exploring adaptable, flexible forms, functions, and flows. These prerequisites do not necessarily require the same amount of time and effort for change. Investments in infrastructure (i.e. freeways) are typically government-led and long-term compared to investments in loose spaces (i.e. spaces which vendors occupy and use) are often grassroots and short-term.

Against this conceptual backdrop, in the spring of 2008, a group of University of Cincinnati planning students explored resiliency in downtown Cincinnati. Cincinnati is not particularly prone to severe natural disasters such as earthquakes or wildfires – although floods constitute problems from time to time. It also exhibits the problems of an average-sized typical American city (i.e. rising suburban sprawl and struggling inner-city neighborhoods such as Over-The-Rhine and the West End). Familiarity with forms, functions and flows in downtown Cincinnati helps better understand the periods of boom and bust against the broader backdrop of its long-term vitality.

Based on the form, function, flow distinction the design elements in downtown Cincinnati were arrayed in a continuum-like fashion with fixed infrastructure covering "specialized forms" for a modular mixed-use complex located at one end and highly fluid "unspecialized forms" for spontaneous activities at the other end. In between these poles, lie various intermediate forms of possibilities, which are neither "specialized" nor "unspecialized."

This research discerns three areas in downtown Cincinnati for adapting to new conditions. These areas grow, thrive, and develop over time based on their internal logic. Resiliency here transcends its typical post-disaster recovery normalcy observed in an equilibrium model, and represents forms that can adapt to short-range, mid-range, or long-range change within a non-equilibrium model. Infrastructure (i.e. roads, sewer and electricity lines or even urban districts) includes the areas subject to long-range change, whereas areas subject to mid-range change include public spaces, while areas subject to short-term change consist of temporary urban spaces (or nooks-and-crannies). These three areas embody the physical and social aspects of flexibility described above.

Project outline

The study area lies at the northwest of downtown Cincinnati near the City Hall. Large, underutilized surface parking lots are located directly to the north and east of City Hall. These lots account for almost 20 percent of the total land use in the area. The author selected three concepts which aim to capture the three aforementioned types of resiliency: the "fixed city" concept focuses on infrastructure with specialized, often long-range and less flexible forms, where every piece has a unique design and purpose. To explore urban spontaneity and multiple "temporalities," the "kinetic city" concept stresses less specialized more temporary fluid forms, which cannot be captured by permanent forms and single uses. In between these poles lies the "good city" concept which represents semi-specialized forms in public spaces for strengthening social solidarity. Three types of resiliency emerged along this continuum: "opportunity," "flexibility," and "spontaneity." Opportunity addresses long-term resiliency, whereas flexibility and spontaneity capture mid-range and short-range resiliency respectively. Long-term opportunities involve investments in infrastructure (building new highways, investing on superhighways and fiber optics, or rearranging and reconfiguring the vacant parking lots in the city). Mid-range investments can more or less capture other decisions cities face like repairing, redesigning, and restoring public spaces. Less strategic decisions including modifying the nooks and crannies of the city for people's day to day activities belong to the realm of short-range decisions.

Three concepts

Fixed city

One group of students explored the fixed elements of the city (i.e. sewers, power feeds, and other utilities that guide development), which often outlive the buildings' effective life span and people. The group examined the transformation of these elements in the study area over time, and focused on vacant buildings and parking lots. Students questioned the vast swathes of surface parking lots close to major freeways since the automotive boom of the 1950s, 1960s, and 1970s, and explored options such as the mixed-use and green architecture, which are currently in vogue.

The underlying premise of the group was that buildings change, governments come and go, but the infrastructure persists over time. Whatever is plugged into the infrastructure also remains for a long time. Two interrelated concepts were examined: "modularity" (Figure 52.1) and "the ages of space" (Figure 52.2). While the former serves as a response to the fixed infrastructure, the latter represents the periods of boom and bust, changes in technology, and building innovation over time. Modularity provides the "plug and play" system that interacts with infrastructure whereas the "ages of space" showcase the degrees of "spatial death" (University of Cincinnati School of Planning 2008) or spaces ranging from robust (i.e. vivid public spaces) to dead (i.e. underutilized parking lots where form, function and flow no longer reinforce one another). To show modularity and interchangeability of "forms," and plugging them to the "functions" and "flows" of infrastructure, students found the Rubik's Cube an appropriate analogy. They overlaid the Rubik's Cube concept of modularity with the spatial degrees of death to define how the areas which need intervention could change. As Figures 52.1 and 52.2 show new forms, functions and flows in the area become possible even though infrastructure (i.e. roads and utility lines) remains fixed.

Functional symbolism reflects an additional interpretation of the fixed city. These symbols or objects may or may not necessarily remain fixed while keeping their symbolic content, value, and meaning. For example, while Fountain Square in Cincinnati has moved several times over the last decades, its purpose as the city's symbolic center has remained constant.

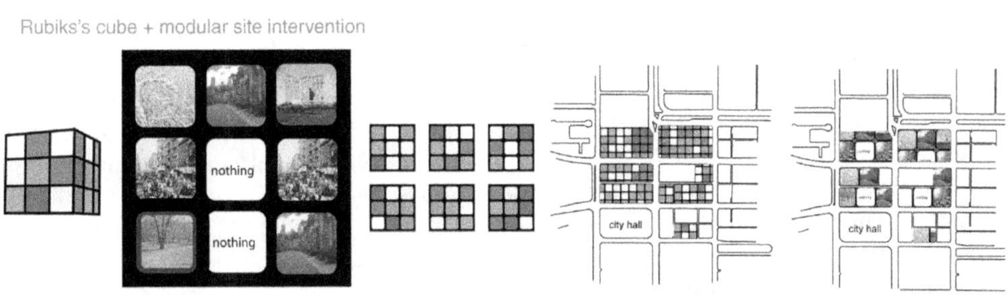

Figure 52.1 Rubik's Cube overlaid on study area. Source: University of Cincinnati, Planning Studio Report 2008.

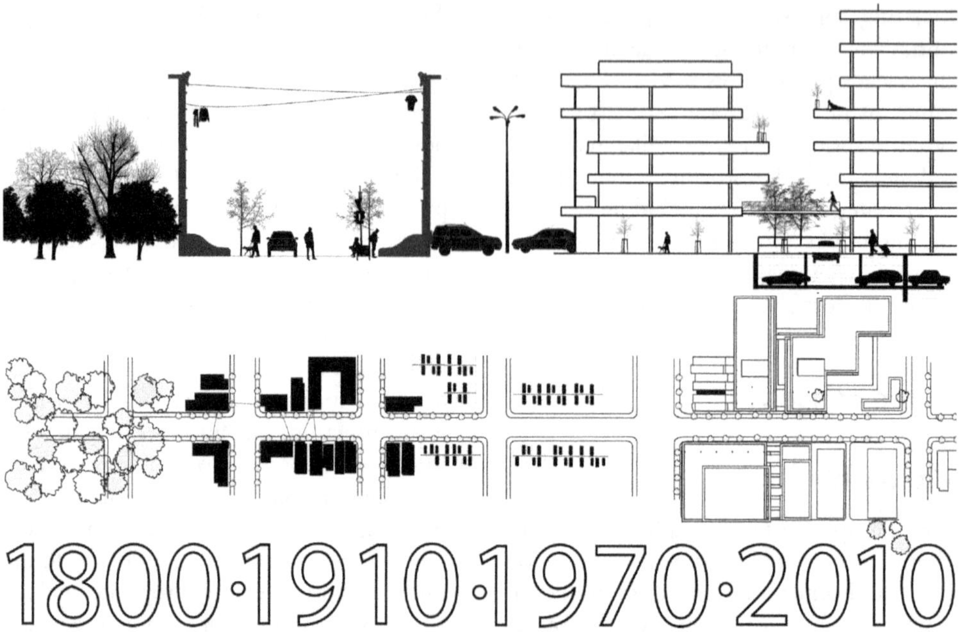

Figure 52.2 Schematic diagram from urban design studio. Source: University of Cincinnati, Planning Studio Report 2008.

This fixed identity sparked the idea of a new district with symbolic permanence. This new services district could accommodate public services including the courthouse, the Police precinct, the public library, the Federal Building etc. Many of these already existing elements can delimit the boundaries of such a district within a fixed city paradigm.

Good city

The good city group examined the role and significance of the public space in social life. The second group used urban geographer Amin's "Good City" theory as a framework for mobilizing social relations and promoting governance (Amin 2006). The good city model sought to incorporate social values into planning and design. If an area needed reconstruction, better to rebuild to be conducive to social diversity, harmony, and unity. The group conceptualized the "four Rs" of "rights," "repair,""relatedness,"and"re-enchantment" to create vibrant public spaces. Citizens have rights. The right of presence in the public space drives the good city theory. Restricting access and use and preventing people from carrying out desired activities in public places contradict the notion of rights in a good city. Repair is the city's ability to maintain and replenish its capacities and capabilities. The premise behind relatedness is that people and places are socially and spatially related (Lynch 1981). The group explored social diversity and the idea that changes to the study area would change the whole area in some manner. Rediscovering a place over and over in a good city is crucial. Re-enchantment is vital if people aim to further these ideas. Through rediscovery, people appreciate what they know about different parts of the city.

Kinetic city

The kinetic city concept emerges from the literature on Everyday Urbanism

(Crawford 2005) along with the contemporary nature of multiple, fragmented, and often competing publics. Everyday Urbanism celebrates the importance of daily life and its social, cultural and political dimensions in designing the city by reclaiming it. In its orientation towards the lived experience, kinetic city contrasts the conventional approaches to urban design and planning. A rationally-ordered system of permanent divisions has dominated the design of cities for several decades. The city as an ephemeral experience, however, draws from the energy that users bring and augment in its public spaces.

This group identified several key concepts such as *temporality, multiplicity, spontaneity*, and *user-centricity* pertinent to their theme. Temporality (time of day, seasonal, cultural holidays and civic occasions) recognizes and encourages multiple functions, uses, and activities by taking advantage of differences in how space shapes activities. The spatial structure of the city allows design to accommodate a diverse set of social experiences based on a temporal logic. Multiplicity reveals the inclusive nature of the citizenry regardless of economic, social or physical differences while acknowledging, accepting and accommodating differences in opinion, taste, values, and purpose. Increasingly, design and management of cities support homogeneous urban spaces. Finding ways to reinsert the cacophony of people, views, uses, aesthetics, and other aspects of urban life immensely benefits the city. Not only does it provide a sense of vitality, but it allows for social interactions that would not happen otherwise and breed tolerance and respect for fellow citizens.

Spontaneity allows for unpredictable and unplanned needs to be met as long as it does not promote criminal activity against persons or property. The students consulted Sennett's (1989: 84) short piece the *Civitas of Seeing* and viewed spontaneity as a venue for "learning to talk and learning from people who are unlike oneself." They also noted that conventional planning and design methods do not particularly lend themselves to creating or promoting spontaneity in typical urban spaces. Planning and design should allow for decisions to be made "on the fly," for unanticipated encounters and a "forum for discourse" to derive our "ethics," as Sennett argues. In accomplishing this, design moves from the conceptual to the practical realm and becomes part of the everyday experience of urban space. User-centricity enables residents to participate and define forms, functions, and movement according to their needs, rights and pleasures, and to engage in a range of physical, social and economic activities. This interaction with the built environment helps people to mobilize and make decisions based on their learned experiences. The students defined areas with potentials for "kinetic" as opposed to areas for "static" activities with their respective publics. Bureaucratic, legal, political and other types of boundaries exclude potential users (i.e. homeless, vendors, and other marginalized people) from falling into multiple categories.

Broadening the concept of resilience

This chapter seeks to broaden the applicability of the concept of resiliency beyond the "equilibrium" model based on risk mitigation and post-disaster recovery. Risk mitigation and physical, social, economic, and political stability play key roles in this model. In a "non-equilibrium" model however, the goal is not merely to revert to "previous," pre-disaster capacities, but to reach a "preferred" condition. Fulton's (2005) account of South Central Los Angeles discusses resiliency as a process different from reaching normalcy and stability by federal or local government intervention. The preferred condition, which

Table 52.1 Concepts, themes, principles, and types of urban resiliency.

	Fixed City	*Good City*	*Kinetic City*
Main theme	Infrastructure (vacant and parking lots)	Public space	Nooks and crannies (loose space)
Resiliency type	Opportunity	Solidarity/flexibility	Spontaneity
Principles	• Interchangeability of forms • Modularity • Ages of space	• Repair • Rights • Relatedness • Re-enchantment	• Multiplicity • Multiple temporalities • Spontaneity • User experience

in this case, involves government incentives for mobilizing grassroots efforts for managing change requires flexibility in identifying and leveraging resources – especially by transforming liabilities (i.e. abandoned buildings or vacant lands) to assets (new functions, forms and flows).

Three interpretations of resiliency emerged from the planning studio project. Examining "modular," "adaptable" or "flexible," and "ephemeral" forms, functions, and flows enabled the students to explore resiliency in the study area. The "Fixed City," "Good City," and "Kinetic City" capture these three attitudes toward the area's transformation respectively (Table 52.1). Each group explored the characteristics and principles of each attitude. The project focused on the linkages between functions, forms and flows within the study area. Three types of changes were identified: long-range (i.e. the infrastructure); mid-range (i.e. public spaces); and short-range (i.e. nooks and crannies, loose or invisible spaces). Both specialized and non-specialized forms associated with infrastructure, forms of public space that induce social interaction and facilitate connectivity between multiple areas, and their compatibility with people's daily and temporary activities were examined.

The Fixed City group recognized street right-of-ways, utilities, and especially parking lots as the elements associated with infrastructure, and proposed two projects: a mixed-use complex consisting of building modules and an urban services district. In both cases, long-term needs and opportunities for change justify new forms, functions, and flows (Table 52.2). In the case of an urban services district, the partial infrastructure of public services and

Table 52.2 Resiliency concepts by form, function, and flow.

	Form	*Function*	*Flow*
Fixed City	• Specialized • Unspecialized • Modules	• Urban services district • Modules • Infrastructure	• People • Services • Information • (Long-range)
Good City	• Plazas • Squares • Open spaces	• Public space • Social interaction • Connectivity • Forma/ceremonial	• People • Formal • Flexible • (Mid-range)
Kinetic City	• Freeway off ramps • Spaces between buildings • Nooks and crannies	• Informal relations • Temporary • Spontaneity	• People • Spontaneous/temporary • (Short-range)

institutions (i.e. the courthouse, the public library, the police precinct, the federal building, and the municipality) already exists. The rest of this area can materialize over the long-term. The idea behind the mixed-use project emerged from the ages of space and modularity. Similar to the concept of "form-based codes," which is gaining national popularity (Plater-Zyberk 2008) modularity and ages of space provide long-term functional flexibility within fixed urban form.

Based on the idea of the ages of space and modularity, building forms can outlive building functions. The principles of repair, rights, relatedness, and re-enchantment allowed the Good City group to propose how to re-energize the public spaces in the study area. The Kinetic City group used the principles of multiplicity, multiple temporalities, spontaneity, and user experience to tap into the area's potential for spontaneity. Urban resiliency presents a powerful pedagogical tool for exploring the limitations and potentials of conceptualizing and communicating the city as a fixed object, a flexible phenomenon, and a venue for spontaneous activities. The following matrix summarizes the three aspects of urban resiliency used in this study:

Exploring the relationship between "resiliency" and "rigidity" sparked curiosity among the students. Cities with seemingly rigid infrastructure still embody fairly resilient identifiable areas. The infrastructure in the fixed city concept with long-term capital costs remains fairly rigid whereas formal public spaces (i.e. the good city concept) and loose or "invisible" spaces (i.e. the ones identified by the kinetic city group) lend themselves to mid-range and short-range changes. After collecting information about the study area students explored three visions of the city as parts of a larger puzzle of resiliency. The emerging principles from each vision paved the way for the intervention stage. With respect to the city's infrastructure, students realized that since it "locks us into patterns of behavior for years to come," (Muller 2007: 100) it also provides long-term development opportunities. The students recognized the importance of this opportunity and explored the concepts of ages of space and modularity.

The idea of ages of space, and the Rubik's Cube which possesses some flexibility within its rigid structure, initiated this proposal. The proposed design consisted of a latticework of modular spaces for housing, retail and commercial, parking space, and also public and semi-public spaces that are well-connected to the surrounding context. Just like the Rubik's cube allows various possibilities based on the arrangement of colors, the proposed design provides a flexible latticework of possibilities in the place of the vacant parking lots in the study area. This type of flexibility creates long-term resiliency.

In their second proposal the City Hall, the Courthouse, the Public Library, and the Federal Building make up part of the downtown's public infrastructure. These important landmarks also provide opportunities for a new district which can accommodate major public services for the city of Cincinnati. Interestingly, Piatt Park, a linear green space, is located midway between these important local markers and can act as a connecting tissue. Identifying these services as part of the existing infrastructure, in turn, carries the seeds of long-term urban resiliency based on the idea of a service district which concentrates more on the other public services such as the fire department, a promenade, and the Police Precinct to name a few.

The Good City group, used geographer Amin's (2006) four registers (repair, relatedness, rights, and re-enchantment) as their design principles. Starting out with the non-spatial dimensions of resiliency, this group focused on the role public space plays in facilitating social resiliency. The proposed changes by this group include multi-use public spaces and areas that facilitate dialogs among multiple publics. These dialogs serve as conduits for materializing social solidarity, hope, and happiness. By restoring public venues and promoting an "architecture of engagement" (Bothwell *et al.* 1998) students aimed to increase social solidarity and social capital within the study area. Changes made to the interior courtyard of the City Hall (i.e. increasing connectivity and people's visual and physical access) will also serve as a forum for celebrating people's rights to the city. The other aspect of this proposal is building new facilities for consolidating the existing city government services and departments that work closely together. Furthermore, the City Hall's interior courtyard forms the base for a pedestrian corridor which connects downtown Cincinnati with two other important neighborhoods – the West End and Over-The-Rhine. Other projects included creating open spaces that allowed the public to exercise more freedom and liberty on the one hand, and facilitate communication between the public and the city administration on the other.

The Kinetic City concept examined the nooks and crannies of the study area, and developed four relevant principles: multiple temporalities, multiplicity of uses, spontaneity, and user-centricity. Each of these principles portrays conditions with which kinetic spaces can thrive.

The three groups developed plans based on different aspects of the study area and using the theory that they had adopted. The Fixed City dealt with the massive parking fields and the infrastructure still running through them. The design focused on modularity as a response to the long-term nature of buildings, especially, the infrastructure. The Good City group designed the public spaces around the city hall and mid-range resiliency. The Kinetic City group identified and designed the loose, underutilized spaces, or the "nooks and crannies" or short-term resiliency. Ideas like beer gardens, centralized information structures, and the redesign of alleys were central in the design. These three sets of ideas complemented each other along a continuum ranging from short-range to long-range urban design resiliency.

Conclusion

The non-equilibrium paradigm of resiliency shows more utility in urban design than the equilibrium paradigm, which ultimately seeks stability and balance in an ecological or urban system. The literature on the former is well-documented – especially where resiliency implies reaching pre-disaster capacities and capabilities. Where resiliency focuses on persistence and flexibility rather than stability, short-range, mid-range, and long-range planning become relevant. The non-equilibrium model focuses more on the process rather than the product. The utility of modularity, for example, lies in establishing a framework for future change as opposed to reaching equilibrium based on a predictable end result. The group's second proposal for creating an urban services district also manifests the long-term utility of the non-equilibrium model of resiliency.

Observations from this study suggest that resiliency in a non-equilibrium model has a temporal dimension. By discovering nooks and crannies, the kinetic city concept illustrates short-term urban resiliency, promotes "spontaneity," and captures the essence of Everyday Urbanism. Such activities occur in unplanned, "loose" or "invisible" as opposed to planned spaces, and "escape the restrictions of more visible sites" (Franck and Stevens 2007: 231). Cities across America and elsewhere are experiencing activities and types of public spaces such as the "urban leftovers, grass carpet, street advertising the flexible way, take your home with you) as the emerging "Pop-up City" (Schwartz 2008). The Fixed City concentrates on long-term changes arising from the gradual transformation of the infrastructure. For example, it takes years if not decades for public services to form an urban district in close proximity from each other. This constitutes one way to identify and leverage the local assets. Furthermore, the modularity of the Rubik's Cube concept provides long-term spatial flexibility and resiliency. These modules provide flexible forms, flows and functions that sustain long-term changes. The Good City concept lies along a continuum between the short-term resiliency of the Kinetic City and the long-term resiliency of the Fixed City. Good cities promote social solidarity and governance by energizing public spaces, strengthening people's access to resources, and celebrating their citizenship rights.

Urban designers and policymakers can operationalize resiliency by: tapping into opportunities; pursuing flexibility and adaptability; and fostering spontaneity (Figure 52.3). Identifying the underutilized parking lots and planning a new district by capitalizing on the synergistic relationship of existing public services

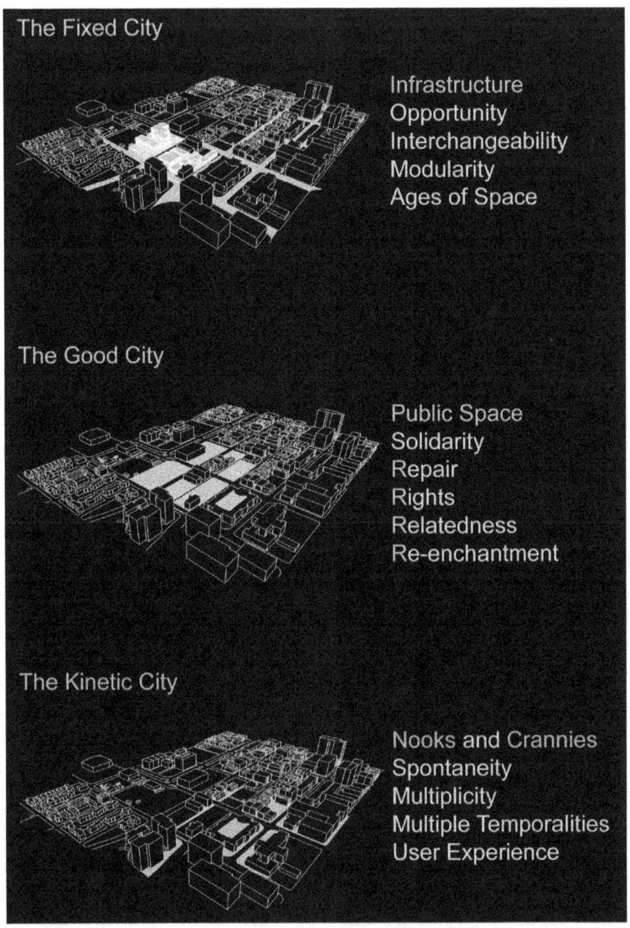

Figure 52.3 Elements of three concepts of resiliency. Source: University of Cincinnati, Planning Studio Report 2008.

and infrastructure provide long-term resiliency. However, Amin's four Rs provide middle-range resiliency. These "registers," specifically repair, promote governance and diminish the proliferation of junkspaces – especially, if we sympathize with Koolhaas's (2002: 180) doom and gloom observation: "a department store, a night club, ... turns into a slum overnight without warning: wattage diminishes imperceptibly, letters drop out of signs, air-conditioning units start dripping, cracks appear as if from otherwise unregistered earthquakes, sections rot." To enhance spontaneity on the other hand, public officials can relax restrictions in using loose spaces, de-emphasize architecture as the only spectacle of the city by promoting temporary spectacles and social activities that "leave no memory" (quoting Mehrotra in Mack 2008) once they are over (i.e. street closures during weekends). The emergent themes from the studio provide a forum for more discussion on the importance and variations of resiliency in urban life and design.

References

Amin, A. (2006). "The Good City," *Urban Studies*, 43(5): 1009–1023.

Bothwell, S., Gindroz, R. and Lang, R. (1998). "Restoring Community through Traditional Neighborhood Design: A Case Study of Diggs Town Public Housing," *Housing Policy Debate*, 9(1): 89–114.

Crawford, M. (2005). "Everyday Urbanism." In Mehrotra, R. (Ed.) *Everyday Urbanism: Margaret Crawford vs. Michael Speaks*, Ann Arbor, MI: The University of Michigan, 17–32.

Franck, K. and Stevens, Q. (2007). *Loose Space: Possibility and Diversity in Urban Life*, New York: Routledge.

Fulton, W. (2005). "After the Unrest: Ten Years of Rebuilding Los Angeles following the Trauma of 1992." In Vale, L.J. and Campanella, T.J. (Eds.) *The Resilient City: How Modern Cities Recover from Disaster*, Oxford: Oxford University Press, 299–312.

Hester, R. (2006). *Design for Ecological Democracy*, Cambridge: MIT Press.

Kates, R., Colten, C., Laska, S. and Leatherman, S. (2007). "Reconstruction of New Orleans after Hurricane Katrina: A Research Perspective," *Cityscape: A Journal of Policy Development and Research*, 9(3): 5–22.

Koolhaas, R. (2002). "Junkspace," *October*, 100: 175–190.

Kostof, S. (1991). *The City Shaped: Urban Patterns and Meanings Through History*. Boston, MA: Little, Brown, and Co.

Lynch, K. (1981). *Good City Form*, Cambridge: MIT Press.

—— (1990). "Environmental Adaptability." In Banerjee, T. and Southworth, M. (Eds.) *City Sense and City Design: Writings and Projects of Kevin Lynch*. Cambridge: MIT Press, 379–395.

Mack, M. (2008). *The "Kinetic City": Key to Development in India's Urban Boom*. Online. Available HTTP: <http://web.mit.edu/cre/education/kinetic-city_prof-mehrotra_alum-wknd-08.html>. 18 September 2009.

Muller, M. (2007). "Adapting to Climate Change: Water Management for Urban Resilience," *Environment and Urbanization*, 19(1): 99–113.

Musacchio, L. and Wu, J. (2002). "Cities of Resilience: Four Themes of the Symposium." In ESA 2002 Annual Meeting *Understanding and Restoring Ecosystems: Integrating Ecology into Urban Planning, Design, Policy and Management*. Online. Available HTTP: <http://abstracts.co.allenpress.com/pweb/esa2002/document/4722> (accessed 19 September 2009).

Newman, P. (2009). *The Renewable Energy City*, Eco-Compass Blog. Online posting. Available HTTP: <http://blog.islandpress.org/322/peter-newman-the-place-based-city> (accessed 9 March 2009).

Pickett, S. (1999). "The Culture of Synthesis: Habits of Mind in Novel Ecological Integration," *Oikos*, 87: 479–487.

Pickett, S., Cadenasso, M. and Grove, M. (2004). "Resilient Cities: Meaning, Models, and Metaphor for Integrating the Ecological, Socio-Economic, and Planning Realms," *Landscape and Urban Planning*, 69: 369–384.

Plater-Zyberk, E. (2008). "Foreword: An Optimistic Moment." In Parolek, D.G., Parolek, K.

and Crawford, P.C. (Eds.) *Form-Based Codes: A Guide for Planners, Urban Designers, Municipalities, and Developers*. Hoboken, NJ: John Wiley & Sons: ix-xii.

Schwartz, T. (2008). *The Pop-Up City*. Available HTTP: <http://popupcity.net/>.

Sennett, R. (1989). "The Civitas of Seeing," *Places*, 5(4): 82–84.

University of Cincinnati School of Planning (2008). *Exploring Urban Resiliency: Three Theories*, Spring.

Vale, L.J. and Campanella, T.J. (2005). *The Resilient City: How Modern Cities Recover from Disaster*. Oxford: Oxford University Press.

Further reading

Adger, W.N. (2000). "Social and Ecological Resilience: Are they Related?" *Progress in Human Geography*, 24 (3): 347–364. Draws parallels in resiliency between social and ecological systems: first, based on the dependency of institutions as social systems on the environment; second, based on the resilience of such institutions.

Bosselmann, P. (2008). *Urban Transformation: Understanding City Design and Form*. Washington: Island Press. Makes a distinction between the urban designer as an agent for the status quo or an agent for change who understands the adaptability and resiliency of physical, natural and social processes.

Brand, F.S. and Jax, K. (2007). "Focusing the Meaning(s) of Resilience: Resilience as a Descriptive Concept and a Boundary Object," *Ecology and Society*, 12 (1): 1–16. Offers a typology of resilience ranging from descriptive (the ability of a system to absorb change) to hybrid (capacity of a system to maintain its original behavior) to normative (flexibility of a system over long term).

Dubbins, M. (2009) *Urban Design and People*, Hoboken, NJ: Wiley and Sons. Sees change as an outcome of opposing situations or dualities. A desirable resolution may emerge when agents of change opt for a "both-and" instead of an "either-or" option where the former illustrates a more resilient option as opposed to the other in which one or more of the parties involved in the process of change are eliminated.

Epilogue

Coming to the end of this volume on urban design, it is time to take stock and reflect on what was accomplished and what was missed. In undertaking this project we knew that we did not want to produce another handbook on urban design, although we are aware that increasingly this term is being used in assembling foundational anthologies on various professional topics. Nevertheless we disliked the term, and the concept more broadly, because a handbook implies a "how to" cookbook or a compendium of best practice. In our view that was not the purpose of this project, for several reasons. First, there are already many such publications in related fields such as architecture, landscape architecture and civil engineering. Some of these manuals and handbooks actually contain standards, principles and guidelines for the practice of urban design, involving large scale design decisions. Site planning standards such as turning radius, parking layout, etc. are routinely included in architectural and engineering handbooks. *Planning the Neighborhood*, a well-known document of standards published by the Committee of Healthful Hygiene of the American Public Health Association (1950) in the middle of the last century or the more recent publication entitled *Planning Design Criteria* authored by DeChiara and Koppelman (1969), are examples of such technical sourcebooks within the field of planning that also address aspects of the practice of urban design. Second, the aim of our project was not to specify a set of prescriptive and definitive, if not deterministic, normative mores for the design of large scale built environments, as suggested in various occasions by authors such as Camillo Sitte in *The Art of Building Cities* (transl. 1945), Werner Hegemann and Elbert Peets (1922) in *The American Vitruvius*, or in more recent times by Rob Krier (1979) in *Urban Space*, Christopher Alexander *et al.* (1977) in *A Pattern Language*, Donald Appleyard and Allan Jacobs (1982) in *Toward an Urban Design Manifesto*, Kevin Lynch (1981) in *A Theory of Good City Form*, and the most recent flurry of New Urbanist guidelines discussed in some of the chapters here. We were more interested in assembling critical ideas and ruminations about the past, present, and future applications of urban design, rather than various exhortations and specifications that are familiar and a bit tiring perhaps. Third, and more importantly, we saw this project as one of defining the contours of urban design as a field, rather than as a profession or as a domain of practice. We wanted to make the claim that while the professional identity of urban design may have remained stunted and diffused from the variable claims and practices made by other allied professions, the very absence of a professional identity or branding may have actually helped it to draw from other fields and professions, and capture a larger terrain of intellectual contributions from the humanities and social sciences. Finally, we wanted to convey the sense of the rather remarkable scope of growing interests and reflections from artists, humanists, philosophers, social scientists, lawyers, as well as environmental designers on the human experiences and consequences of the built environment and its design in a rapidly globalizing world.

This expanding scope of interests has produced a remarkable array of new subjects and voices that are pertinent for the field. For this reason, we did not want to produce another Reader of preassembled classical texts on urban design. We strongly believed that there is enough new material that could be generated from scholars and practitioners of urban design, and that the time has come to collect, present, and make sense of new texts, which one day may be indeed perceived as "classical."

As we continued to search for an appropriate term for what we aspired to achieve, the notion of *Companion*, as suggested by our editor Andrew Mould, hit the mark for us. The term captured the sense of what we had in mind: a collection of independent and original essays that would address various aspects of the field, from its roots and influences to critical review of contemporary outcomes and trends in the built environment, and the future challenges of emerging scales, technologies, and sustainability.

Although we wrote the initial script for what we believe is the first ever *Companion* in urban design and allied areas like architecture, planning, and landscape architecture, the final product is very much a collective work of the invited contributors. Soliciting these specific contributions, we asked authors to place their work within the overall spirit and scope of the *Companion*, which we defined in the broadest terms: to provide an authoritative sourcebook that will complement the various available Readers on urban design; to provide critical reviews and appraisals of the current state of the urban design field, including conceptual foundations, disciplinary influences, and trends affecting urban design pedagogy, scholarship, and praxis; to engage with and reflect upon the politics and policy of how we think about and practice urban design; to identify the unresolved issues, emerging challenges, and opportunities for urban design and urban development presented by the new global economic order.

We were impressed by the contributions, which exceeded our initial expectations of how the contours of the field would be defined. We were struck by the creative and critical interpretations of past developments and current outcomes, and how they imagined the future scope, scale, and challenges for the field and the practice of urban design. If we use the metaphor of symphony to describe this project, in the end we were neither the composers nor the conductors. The symphony (or should we call it a jazz production?) was composed by our authors, with remarkable coherence and some distinct leitmotifs and reprises that we did not consciously anticipate.

Of course we had some disappointments, as some of the experts in the field we initially approached could not oblige us for various reasons. Because of this and other reasons, we are aware of some omissions in the *Companion*, which we want to briefly address here, and hope to include them if the publishers choose to undertake a second edition in the future.

One of the lacunae we are aware of is the absence of a comprehensive global perspective, although we have devoted a section to this issue and invited several contributors to address them. Indeed, the global context and the global trends emerged as one of the leitmotifs we referred to previously. Yet, and more specifically, we find the global comparative perspective presented in this book rather limited in several ways. First, the collection has a primarily North American focus. Even though we invited contributors from all over the world, the North American perspective dominates. Second, there is relatively little that represents the experience of the developing world generally, and the emerging economies more specifically, where urbanization and urban development are occurring at an unprecedented scale. Third, a related bias inheres in the English language's dominance of the relevant literature, thus limiting the breadth of possible contributions and insights.

A second lacuna involves several substantive areas of practice and application of urban design. These include specific coverage of such important areas as historic preservation and urban conservation; the nexus between urban design and economic development; urban design and environmental justice, and the like. It is not that these important topics have not been addressed at all, but only peripherally and without a specific focus.

A third lacuna involves the issue of best practices. We had hoped to address the question of best practice – commonly cited – from a conceptual, theoretical, and critical perspective. This proved to be difficult and may require a more nuanced coverage than a single chapter can provide. At the core of this inquiry should be an elaboration of what constitutes a "best practice," for whom, and why; as well as how the notion of "best practice" may vary in different geographical and cultural contexts.

Finally, in a time of post-Copenhagen conference on global warming and climate change, the role of urban design in shaping the built environments of a hot, crowded, and endangered planet remains to be addressed directly, although several chapters address some of the derivative and common urban problems.

In the end, we believe that the *Companion* has outlined the contours of the field of urban design and established the richness and diversity of ideas, criticisms, and controversies that define its content. We also hope that in this *Companion*, we have been able to emphasize the wide ranging applications and intellectual roots of the field. Indeed the collection represents a growing and evolving field, which is seeking to make sense of the human experience of the built environment and the processes and values that will shape it in the future.

References

Alexander, C., Ishikawa, S. and Silverstein, M. (1977). *A Pattern Language: Towns, Buildings, Construction*. New York: Oxford University Press.

Appleyard, D. and Jacobs, A. (1982). *Toward an Urban Design Manifesto*. University of California, Berkeley: Institute of Urban and Regional Development.

American Public Health Association, Committee on the Hygiene of Housing (1950). *Planning the Neighborhood*. Washington, DC.

DeChiara, J. and Koppelman, L. (1969). *Planning Design Criteria*. New York: Van Nostrand Reinhold.

Hegemann, W. and Peets, E. (1922) in *The American Vitruvius: An Architect's Handbook of Civic Art*. New York: The Architectural Book Publishing Company.

Krier, R. (1979). *Urban Space*. New York: Rizzoli International Publications.

Lynch, K. (1981). *A Theory of Good City Form*. Cambridge, MA: MIT Press.

Sitte, C. (Stewart, C. translator) (1945). *The Art of Building Cities*. New York: Reinhold Publishing Corp.

Index

Page numbers in *italics* denotes a table/illustration

AARON 270
Abernethy, James 232
abstract plan 250–1
Accra (Ghana) 139, 143
Active Living movement 152, 388
Adams, Thomas 357, 358
adaptation: urban design as powerful tool of 607–8
adaptation level theory 164
Adorno, Theodor and Horkheimer, Max: *Dialectic of Enlightenment* 99
advertising 542
advocacy design approach 622
advocacy organizations 332
advocacy planning 25, 339
aerovilles 396
aesthetics 57, 177, 199, 373–4
aesthetics plus doctrine 177
affordances 164–5, *164*, 170
AIAS (American Institute of Architecture Students) 226–7
air quality 607
airports 396, 500
Alberti, Leon Battista 600–1
Alexander, Christopher 19, 24, 54, 62, 133, 285; "A City is Not a Tree" 24
AlSayyad, N. 214
American Association of State Highway and Transportation Officials (AASHTO) 429, 519
American Institute of Architects (AIA) 224, 226–7, 324
American Planning Association 467
American Public Health Association (APHA) 384
American Society of Planning Officials 364
American Sociological Association 125
Americans with Disabilities Act (1990) (ADA) 201–2, 203
Amman 143
Amsterdam 35, 122, 132, 426, 533
Anthony, Kathryn H. 47, 219–20, 223–35
anthropology 110, 137–46; critique of urban design practices 144–6; ethnic identity and urban sites 140–2; global cities and cultural hybridity 142–4; urban poor studies 138–40
Anti-Social Behavior Orders 667
Appadurai, Arjun 52, 143, 538, 564, 567
Appleyard, Donald 24, 54, 208, 210, 499–500, 507, 512, 621
aquifers 627
arcades, shopping 393–4, *394*
Archer, John 341, 342, 356–66
architectural competitions 305
architecture 1, 41, 58, 59, 85; branded 92–3; and city structure 117–21; community design movement 330; development of profession of 224; and studio model 223–8, 219, 235; and transit oriented development 650–1; and urban design 7, 85, 447
Architecture for Humanity 337
archives 220, 241–2, 244
Arefi, Mahyar 587, 674–84
Arizona State University *593*
Arlington (Virginia): Clarendon Corridor 640
Arnstein, S.R. 332
Arreola, D.D. 564
art: collective public 453–4, 456; representation of cities in 121–2
Artificial Intelligence 70, 71, 270
Artificial Life 70
arts districts 595
Asea Brown Boveri (ABB) 312, 314, 315
Asian cities 449
asset-based approach: and citizen participation 336

Association of Collegiate Schools of Architecture (ACSA) 224, 229, 230
asthma: and housing 203
atelier culture 223–4
Atget, Eugène 239, 240
Atlanta 145, 469, 474; BeltLine redevelopment 338
Audirac, Ivonne 465, 510–22
Augmented Reality (AR) 262, 268
Augmented Round Table for Architecture and Urban Planning 268
aural documentation 245, 246
Australia 288
Australian Institute of Planners (AIP) 84
Austrian Research Institute for Artificial Intelligence 270
authenticity: and integral urbanism 592, 594–6; and place identity 500–1, 502; versus placelessness 120
auto-centric urbanization: critique of 510, 511, 513
automobiles 38, 372, 465, 510, 621, 637–8; impact of on transit 645; safety of 591; social cost of use of 472–3; and suburbs 362–3; and women 154
autopoietic systems theory 77–8
Avenida Juárez (Mexico City) *193*

Bacon, Edmund 16, 19, 24–5; *Design of Cities* 19, 447
Badiou, Alain 97, 106
Baer, William C. 276, 277–85
Bakema, Jaap 22
Baldwin Park (California) *478*, 479
Baltimore 21, *119*, 350, 548
Banerjee, Tridib 345–54
Banfield, E.C. 133
Bangkok 581
Baptistery San Giovanni (Florence) 249
Barcelona 145; brand image 545, *546*
Barnett, Jonathan 21, 25, 628
barrios 567–8
Bartholomew, Keith 463, 464, 467–79
Battery Park City (New York) 19, 30, 31, 32, 57, 541, 548
Baudrillard, Jean 91, 449
Bauer, Catherine 18; "Good Neighbourhoods" 18
Bauhaus school 224
beaches 441
Beal, E. 143
Beaux Arts tradition 32, 59, 304, 346
see also École de Beaux-Arts
Becher, Hilla and Bernd 239, 240
behavior: in the environment 126–8, 168–70
Beijing 143, *400*, 401
Ben-Joseph, Eran 221, 261–72, 292
BenchWorks 268
Benjamin, Solomon 580
Benjamin, Walter 101
Bennett, Edward 32, 33, 447
Berkeley, UC 43, 253, 497, *498*
Berlin 549–50; Das Schloss shopping arcade 394, *394*; Potsdamer Platz 276, 309, 310–14, *311*, *313*, *314*, 453
Berlin, Isaiah 63
Berman v. Parker (1954) 177
Better Philadelphia Exhibition (1947) 16
Bettman, Alfred 528
Bickford, Susan 193
Bilbao 547, 549, 550
Bilbao effect 343
billboards 176, 449, 461
biological diversity 605, 624–6
Birch, Eugénie 7, 9–26
black box/glass box metaphors:, and design process 275
Blackpool funfair 210
Blakely, E. 385
block-by-block inventories 201
blueprint planning 204
Boarnet, Marlon G. 111, 198–204
Boeri, Stefano 54, 71, 78–81, 81
Bollier, David: *Silent Theft* 187
Booher, D.E. 333, 336–7
Bosselmann, Peter 220–1, 249–59
Boston 20, 22, 130, 474, 497, 546; Big Dig project 351; Quincy Market 349
Boston Museum of Fine Arts 406
Boston Redevelopment Authority 267
Boston Southwest Corridor 334
boulevards: airport 396; as consumption space 393; introduced to Latin American cities 115; multiway 423–4, *424*; Parisian 32–3
bourgeois public sphere 190–2
Boyce, D.E. 282
Boyer, Christine M. 54, 70–82
Boyer, Ernest and Mitgang, Lee: *Building Community* 228
brainstorming: and charrettes 319–21
branded architecture 92–3
branded chain hotels 93
branding, city 91, 92–3, 537–8, 541–50; and communication 549; definition 541; elements of positive 543–6; goal of 542, 543; and image of city 542–3, 549; and marketing 541–2; strategies and processes 546–7; and urban design 547–9; and vision development 547
Brasilia 138, 145, 307, 373, 374, 546
Braunfels, Wolfgang: *Urban Design in Western Europe* 534
Brazil 138, 145, 426
Brecht, George 449
Brennan, William J. 177, 178, 181
Bridgman, Percy 59
Brill, Michael 436

Britain 484; and design guidance 288–303; new towns 376; postwar housing programs 22; town planning scheme 183; urban design 64
"broken windows" hypothesis 167, 667
Brooklyn: Army Plaza 352
Brown, Denise Scott 18
Bruegman, Robert 464
Brunelleschi, Filippo 249, 250
Brunswick (Germany) 398, *399*
Bruntland Report (1987) 25, 619
Bubinas, K. 143
Bucharest 548–9
Building Information Modeling (BIM) 265
Building Officials Conference of America (BOCA) 534
Building Resilient Regions 559
bureaucratization: and citizen participation 332–3
Burgess, Ernest 116
Burnham, Daniel 32, 33, 447

café 190
Cage, John 449
Cairo 581
Calatrava, Santiago 550
Caldeira, T. 144
calmness 166, *166*, 168
Calthorpe, Peter 25, 38, 514, *515*, 516, 519, 5221; "The Traditional American Town" 362–3
Campanella, T.J. 608
Cannon Design 234
Canter, David 61
capital web concept 21
capitalism 54, 97; relationship with urbanism 98–9, 103
Caracas (Venezuela) 580
carbon dioxide emissions 476
Carmona, Matthew 64, 276, 288–303
Carr, Stephen 63, 64–5
cars *see* automobiles
cartography: representation of cities in 121–2
case study 133
Castells, Manuel 88–9, 90, 465, 495; *The Urban Question* 103–4
Castro Street (Mountain View) 427
catalytic planning 408–9
Caudill, Bill 325
Ceausescu, Nicolae 548
Celebration (Florida) 130, 512, *513*
cellular transit metropolis 513–14, 516
central business district (CBD) 345–6, 644–5
Central Park (Davis, California) 438, *439*, *442*
centrality 104
Centre Pompidou (Paris) 409–10, *410*, 416
certification 184
Cervero, R. 474
Chandigarh 546
change: forces of urban 553; physical design strategies for enhancing city's ability to adapt to future 608; responses to rapid 589–90; simulating magnitude, rate and nature of 254–9
Chapin, Jr, F. Stuart 132
Charlottesville Community Chalkboard 461
charrettes 220, 224, 276, 306, 317–27; academic value of 327; benefits and contributions 325–7, 327; brainstorming 319–21; briefings 319; definition and types of 317, 318–19; distilling the options 321; non-academic 324–5; origin of term 317; participants 319; pitfalls and challenges 326; presentation to the public 321, 323, 326; production and design 321; site selection 323; sponsors and funders 323–4
Charte d'Athènes, La 25
Charter of the New Urbanism (CNU) 386, 510, 511
Chicago 122, 129, 133, 352; downtown 350, 351; Millennium Park 33, 453, 501, 504; neighborhoods 380; parks 33; Plan of (1909) 32, 33, 199, 447; Tribune Tower competition (1922) 307
Chicago School 65, 87, 114, 126, 127, 128, 202, 554
Chicano movement 567
Child Friendly Cities 669
children: abandonment of public space by 664–5; response to environment 170–1
Chile 581
China 7, 118–19, 377
China, ancient 379
Chinatowns 140, *141*, 395
Chipperfield, David 396
choke points 656
Chomsky, Noam 71
Chow, Renée 366
Christmas decorations 450
Christo 453
Chtcheglov, Ivan 101
Churchill, Robert 122
CIAM (*Congrès Internationaux d'Architecture Moderne*) 7, 12–13, 14, 15, 16, 23, 25, 145, 465, 510, 518, 521
Cincinnati: and resiliency 676–9, 681–2
cinematic arts 111, 208–14, 238; and cognitive mapping 210–11; directions for further development 212–13; as interpretive media to study cities 211–12, 214; as tools for urban designers and design educators 210–11, 213–14; and urbanization 208–9
CIRA Center (Philadelphia) 453
cities: changes and transformations in 484–7; as corporate entities 537; design properties of within modernism and postmodern globalization *90*; form of 92–3; as generator of surprise and learning 64–5; geographers and models of

structure of 114–17; growth in populations 484; redevelopment of 486, 490
Cities of the World 115
citizen design 276, 329, 338 *see also* citizen participation
citizen experts 334
citizen participation 276, 329–38; and advocacy organizations 332; and bureaucratization 332–3; challenges to 332; contribution to remaking of urban spaces 337–8; and design aesthetic 334; extending 335–7; growth of 329–30; institutionalization of 330, 332; and legal mandates 331; limitations and challenges 332–4; and multiculturalism 333; and new technology 334–5; and non-profit practice 337–8; and privatization 333–4; reviewing and commenting on projects 331; self-interest advancement 333; and visioning exercises 331
citizenship: and ethnoscapes 566–7
City Beautiful Movement 32, 194, 199, 346, 347, 357–8, 406, 447
city centers *see* downtowns
City Functional 199, 347, 510
city as landscape analogy 613–14
city regions 555
City of San Gabriel (California) *141*
Ciudad Guyana (Venezuela) 139
Civano (Tuscon, Arizona) 519
civic architecture 342
civic boosterism 541
Clarke, Ron 663
classical economic theory 192
Cleveland, Gerry 667
climate change 354, 467, 475–6, 608, 626
Cloud Computing 270
clustering 364
CNU (Congress for New Urbanism) 7, 22, 26, 36, 38, 363, 387, 520, 521, 529
CoBrA (Copenhagen-Brussels-Amsterdam group of radical artists) 101
coding 526–7, 533–4; form-based *see* FBCs; *see also* design codes
cognition 163; and digital technology 265–70; environmental 165
cognitive mapping 75, 132; and cinema 210–11
Coleman, Alice 663, 666
Colins, John F. 20
collaboration 336–7; and design codes 300; and digital technology 262–5
collective public art 453–4, 456
Colombia 213
colonial cities 345
colonialism 137
Columbian Exposition (1893) 447
Columbine High School massacre (1999) 432
Comcast Centre (Philadelphia) 453
commodity fetishism 91, 92
Common Interest Developments (CIDs) 188, 189–90, 192, 381, 382
commons 187–90
communities of practice 47
community 186; and City Beautiful movement 357–8; and neighborhood 382
Community Action Program 330
community art projects 332
Community Concepts 232
community design centers/studios 44, 219, 229–31, 232–3, 235, 330, 622
community gardening 332, 441
compact city 375, 463, 464, 465; and consumer preferences 471–2; and density 467–8; and energy efficiency 476; impact of on physical activity and obesity 476–7; and vehicle miles travelled 474–5; versus sprawl 467–79
competitions: architectural 305; design *see* design competitions
"Complete Streets" 441
complexity 168
Computational Creativity 270
Computer Aided Drafting (CAD) 254, 264–5
computer animation 247
computer-based modeling 253–4
Conant, James 15
concentric ring theory 352, 611
conferences, urban design 19
congestion 31
congestion pricing 477–8
Congress for New Urbanism *see* CNU
conservation, land 364
conservation biology 626
Constitution, US 178–9
constructed situation 101
consumption spaces 342, 343, 391–403, 485; architectural styles 399; cultural heritage issues 398; design dimensions 397; ethnic consumer islands 395; and flagship design 398, 399; hierarchy of 393; international versus local styles in design 401–2; location 392–3; theming 401–2; types and varieties 392–7; and urban image 401; and urban renaissance 398–9
contrived communities 127
Cooper, Alexander 21
Cooper, M. 145
Copenhagen 170
Corbett, Michael and Judy 519
Corner, James 612
corpus of media 239, 241–2, 247
corridor street 35
CPTED (Crime Prevention Through Environmental Design) 663, 665–7, 668

Craik, Kenneth 61
Cranbrook Academy 16
Crane, David 19, 20–1
Crawford, M. 337
Creative Class Struggle 106
creativity: and digital technology 270–1
Cret, Paul 447
Crewe 334
crime: factors associated with reduction in 170; fear of 144, 167; and public housing 133, 170, 665–6; rates 492n; women's fear of 155; *see also* secure cities
Crime Prevention Through Environmental Design *see* CPTED
criteria 280–1, *281*, 284–5
critical urbanism 54–5, 97–106
cross-programming 591
crowding 164, 656; and women 153
Cullen, Gordon 18, 24, 37, 208, 210; *Townscape* 24, 496
cultural buildings 352–3
cultural complexes 343
cultural diversity 624–6
cultural heritage issues: and consumption spaces 398
cultural hybridity: and global cities 142–4
cultural identity 563
cultural institutions 342, 405–16; Centre Pompidou 409–10, *410*, 416; contemporary trends 406–7; emphasis on corporate display and marketing 407; emphasis on iconic buildings and star architects 407–8; functions 416; Guggenheim Bilbao 408, 410–11 *412*, 416; Harley-Davidson Museum (HDM) 411–13, *413*, 416; Lincoln Center redevelopment 406, 407, 408, 413–14, *415*, 416; origins and development 406; revenue-producing activities 407; shift towards a new openness rooted in consumer culture 408; and urban revitalization 405, 407, 408–9, 416
culture industry 91
culture of poverty theory 138, 139, 142
Cumbernauld (Scotland) 373, 374–5, *375*
Cunha, Dilip da 604
Cuthbert, Alexander 54, 65, 84–94
cybernetics 54, 70–82; first order 70, 71–6, 78, 81; second order 70, 76–81

da Vinci, Leonardo 249, 250
Daimler-Benz 312, 314–15
Dar es Salaam: women's safety audits 668–9
data analysis 132–3
Davidoff, Paul 25, 330
Dávila, Arlene 567
Davis (California) 441; Central Park 438, *439*, *442*; Village Homes 441, 579
Davis, Mike 106, 575; *City of Quartz* 553; *Planet of Slums* 539, 575
Day, Jacob 227
Day, Kristen 110, 150–7
Dearborn, Lynne 232–3
Debord, Guy 55, 99, 100, 102, 103, 106; *The Society of the Spectacle* 98
decentralization 32, 555
defensible space 133, 487
Delucchi, Mark 473
density 621; and compact city and sprawl 467–9; and informal city 581; livable 621
density convergence 555
Denver: Civic Center Park 352–3
department stores 393–4
design: and planning 60; tension between science and 58–9, 61, 67
design charrette *see* charrettes
design codes 276, 289, 291–303, 363; arguments against 302; benefits and strengths 292, 294–5, 300, 302; building upon a spatial vision 300, *301–2*; and collaboration 300; fundamental factors for the success of coding projects 299–302; and leadership 300, 302; phases in successful implementation 295–6, *297*; reasons for choosing 292–5; role of in development process 295–6, *297*; roles and motivations of stakeholders 296, *298–9*; setting quality thresholds 299–300; and up-front investment 300
design competitions 48, 275, 276, 304–16, 401; and attraction of public attention 304; benefits 276, 305, 308, 309–10, 316; and decision-making processes 309; ideas 306; importance of the program 308, 309; invited 307–8; limitations 315–16; limited 307; media attention and press coverage 310, 315; open 306–7, 310; and Potsdamer Platz 310–14, *311*, *313*, *314*, 315; project 306, 310; purpose of 304, 305; and spectacularization of building process 304; types of 305–8; versus direct hiring 308–10
design education *see* education, design
design guidance 288–303; characteristics and nature of 289–91; definition 288; and design codes *see* design codes; purpose and goals of 290; types of 288–9
design juries 224–5, 226, 234
design review commissions 176, 182
design review laws 182
design standards *see* standards, design
design studios 47, 219–20, 223–35, *228*; advantages 226, 227; community 44, 219, 229–31, 232–3, 235, 330, 622; and design juries 224–5, 226, 234; and desk critique 225–6; disadvantages and call for improvement 226, 227–8, East St Louis Action Research Project (ESLARP) 219, 223,

225, 231–3; educational benefits of community 232–3; evolution and history 223–8; and Hurricane Katrina aftermath 230–1; impact of new technology and internet on 233–5; as a vehicle for teaching urban design 228–31; and videoconferencing 234; *see also* charrettes
desk critique 220, 225–6
determinism 126, 128
Detroit 350, 558
Detroit Collaborative Design Center 230
Developers 9, *10*, 11, 16–25, 26
Development Oriented Transit (DOT) 645
Dewey, John 229, 384
Diana, Princess 450–1
Diderot, Denis 239; *Déscriptions des Arts et Métiers* 238; *Encyclopédie* 239
digital immersion tools 266–7
digital technology 220, 221, 261–72; and archival 244; and cognition 265–70; and collaboration 262–5; and creativity 270–1; limitations and considerations 271–2
Dillon's Rule state 178
direct action: and citizens 332
direct hiring: versus design competitions 308–10
disability: and accessibility 201–2, 203
discretion-versus-rule debate 183–4
Disney Corporation 401, 447
Disneylands 401, 402
distance education 48
districts 165
DNA of places 595
documentation 220, 245–6; archival 241; and indexing 245; and storytelling 247–8
Doges Palace (Venice) 451, *452*
Doo Dah Parade 142
Doorn Manifesto 22
Douglas, William O. 177
downtowns 342, 345–54, 537; building of sports stadiums and entertainment complexes 350; and City Beautiful movement 346–7; and City Functional model 347; cultural and civic buildings 352–3; cultural projects pursued 350; decline and factors contributing to 353; failure of large public housing projects 353; festival marketplaces 349–50; future of design 353–4; and globalization 353–4; and infrastructure investments 351; and land economics 351–2; in late twentieth/early twenty-first century 349–51; logic of urban design 351–3; and market-driven urbanism 348; origin and development of 345–7; pedestrian improvements 349; reliance on private investment and impact of 348; renewal of and problems 20, 347–8, 353; residential projects 350–1; transit malls 349; transportation and development of 346
Drucker, Peter: *Adventures of a Bystander* 57
Duany, Andrés 22, 38, 362, 511, 516, 518, 529
Dubai 116, 538
Dublin: regeneration project 487, *488*
Dunham-Jones, Ellen 435, 611
Durham (North Carolina) 623
dwelling 496; ecological 623–4; suburbs and design of 365, 366

Eagle's Nest of St. Clair County 232
Eames, Charles and Ray 659
Earth Summit (1992) 619
EarthCam 265
East St. Louis Action Research Project (ESLARP) 219, 223, *225*, 231–3
Eclectic Atlases 80
École de Beaux-Arts 220, 223–4, 224, 317
ecological approach to perception 164–5
ecological design 507
ecological dwelling 623–4
ecological inventory 602
ecological issues 585; and new towns 375; and New Urbanism 519; and streets 427–8; and suburbs 365
ecological psychology 168–9
ecological urbanism 600–9, 612–16; cities as ecosystems 605–7, 613–14; cities as habitats 604–5; cities as part of the natural world 603–4; deep, enduring context of city 604; and future of urban design 608–9; historic roots 600–2; key concepts and principles 602–8; reasons for promoting 609; and resilient cities 607–8
economic crisis (2008) 485
economic development: and informal city 580; and new towns 574–5
Economic Opportunity Act (1964) 330
ecosystems: cities as 605–7, 613–14; connection of 606–7; protecting of 626
edge-city phenomenon 385
edgeless city 611
education, urban design 8, 41–9; and cinematic arts 211, 213–14; and design studio *see* design studios; and globalization 43; and new technologies 48; pedagogic techniques employed in active teaching 47–8; promoting dialectic process with communities and sites 43–4; survey of programs in universities 44–7, *45–6*; and *téchne/poiesis* dualism 49
educative city 622
Eesteren, Cornilis van 12
Eisenstein, Sergei 210
El Barrio (Spanish Harlem in New York) 567
Ellin, Nan 586, 589–98
"Emerald Necklace" 33
enclosure movement 188

enclosure paradigm 31, 36–8
encoding/decoding reception theory 213
energetic public spaces 442
energy efficiency: and enclosure paradigm 38–9; and neighborhoods 388; sprawl vs compact city 476
Engel, F.: *The Civil War in France* 100
England: design codes 291–303; *see also* Britain
entertainment districts 35
entropy 71, 81
environment: and behavior 126–8, 168–70; impact of 126–8; and social structure 128–9; variety in human response to 170–1; *see also* ecological urbanism
Environment and Behavior (journal) 127
environmental cognition 165
Environmental Design Research Association (EDRA) 127
environmental evaluations 165–7
environmental justice: and ethnoscapes 570–1
environmental perception 164–5
environmental psychology 61–2, 110, 127, 162–71, 620; basis for environmental response 162–3, *163*; features 162
Envision Utah 616
Epstein, D. 138–9
Erikson, Erik 498
ethnic city 140
ethnic complexity 116–17
ethnic consumer islands 395
ethnic enclaves 140, 142, 194
ethnic urbanism 538
ethnicity 140
ethnography 110, 137–46
ethnoscapes 143, 538, 562–71; and the arts 566; and citizenship 566–7; definition and nature of 562–3; and environmental justice 570–1; multiethnic design of public spaces 568–70; and New Urbanism 569–70; political dimensions 566–8; socio-cultural and subjective dimensions of 562–6; spatial and temporal dimensions 566; and urban design practice 568
Euclid vs. Ambler (1926) 527, 528
European Commission 403
European space 78–81
European Union 559
European urban village movement 374
evaluations, environmental 165–8, *166*
evaluative image 110
everyday life 102–3, 105, 365–6, 500
everyday urbanism 337, 595, 622, 678–9, 682
Ewing, Reid 463, 464, 467–79
exciting environments 166, *166*, 168
exclusion *see* social exclusion
exopolis 538, 553
facial recognition software 271
Faga, B. 334
Faneuil Hall Marketplace (Boston) 22
farmers' markets 440
Favela Bairro program (Rio de Janeiro) 577, 580
FBCs (form-based codes) 466, 518–19, 526, 526–34; abilities of 529–30; advocation and justifications for 532, 533; criticism of 531, 532–3; example *530*; incorporation of spatial concepts 529; and problems with zoning 526–7, 529; and public participation 534; reasons for importance of 534; and Smartcode 529, *530*, *531*; supporting of social and economic diversity 529, 533; and transit-oriented development 649–50; and trouble with planning 528, 529
fear 590–1; of cities 487; of crime 144, 167; men and 155–6; women and 154–5
Federal-Aid Highway Act (1956) 19; (1969) 330
Felson, Alex 609
feminist approaches 110, 150–7
Ferguson, J. 142
Ferris, Hugh 13
festival marketplaces 349–50
field theory 126
film *see* cinematic arts
filmic montage 208
Finitude 268, *269*
first order cybernetics 70, 71–6, 78, 81
Fisher, Thomas 227
Fishman, Robert 7, 30–9, 612
fixed city 676, 677–8, *677*, 680–1, *680*, 682, 683, *683*
flagship design 398, 399, 408
flexible streets 426–7
Flint, Anthony 251
flooding 628
Florida 637
Florida, Richard 399
flows 675
flux, urban 344, 446–61; difficulty in regulating 460; ethical and legal issues 458–60; as a force for civic improvement 454–8; implications for urban design 460–1; possibilities of 449–54; transformation of neglected neighborhoods 456; versus permanence 446–9
Fluxus movement 449
Foerster, Heinz von 70, 81
Foglesong, Richard E. 357
Ford Foundation 19
Ford, Larry R. 110, 113–23
form-based codes *see* FBCs
formal city 574; integration with informal city 379–80
forms 675
Forsyth, Ann 342, 369–77

Fort Worth Plan 513–14
fortress architecture 144
Foster, Norman 538, 550
Founders 9, *10*, 12–14, 26
fractal city 553
France 288; and new towns 376
Francis, Mark 332–3, 343, 432–43
Frank, L. 71
French Revolution 406
Frieden, B. 537
Friedman, Avi 366
Fromm, Erich: *The Sane Society* 360
Fukuyama, Francis 97
Fulton, W. 675, 679
Functional Classification of streets 420–1
functional theory 53, 99, 199
functions 675
Future Systems Architects 399

G-Speak 271
galleries 393
Galster, G. 469
Gandhi Marg (Chicago) 143
Gans, Herbert 130; *The Urban Villagers* 24
Garde, Ajay 341, 343, 379–89
garden cities 12, 14, 25, 36–7, 357, 358, 362, 369, 370, 371–2, 486, 533, 601
garden suburbs 36, 37, 38
Garvin, Alexander 2, 21
gated communities 144, 188, 193, 361, 384–5, 590, 666
gating 487–8
Geddes, Patrick 12, 22–3, *23*, 25, 518, 579, 601, 612
Gehl, Jan 130–1, 432, 439
Gehry, Frank 93, 265, 343, 410, 449, 550
general systems theory 88
gentrification 119–20, 329, 334, 392, 398, 464–5, 489–92, 520, 548; changing image of neighborhoods 548; and displacement 464–5, 490–1; factors contributing to 490; and government policy 490; and urban change 490
geo-referencing 250
Geographic Information Science *see* GIS
Geographical Review 113, 114
geography 110, 113–23; architecture and city structure 117–21; emphasis on micro elements in landscape 114; models of city structure 114–17, 122; monitoring of changes in urban form 113–14; and representation of cities in cartography and art 121–2; study of housing 119–20
Germany 183, 288, 306
Getty Center (Los Angeles) 407
ghettos 488
Gibberd, Frederick 37
Gibson, J.J. 252
Giedion, Sigfried 14–15
GIS (Geographic Information Systems) 122, 200, 254, 613
Glasgow 459
Glass, Ruth 18
global cities: and cultural hybridity 142–4
Global Positional System (GPS) 267
global warming 354, 387, 476, 613; planting street trees to combat 427–8
globalization 43, 89–91, *90*, 116, 353, 359, 401, 485, 537, 538
Golden, Jane 456
good city 676, 678, *678*, 681–2, 682, 683
Google 234
Google Earth 233, 263
Google *Maps* 265
Google Street View 241, 245, 247, 263
Goonewardena, Kanishka 54–5, 97–106
Gordon, N.J. 379
Gordon, Peter 464, 467–8, 474, 475, 477
Gore, Al 627
Gosling, David 41, 44, 210
Gottdiener, M. 129
Gottmann, Jean 556
government power 176–8
gradient approach 465
Graduate School of Design (GSD) (Harvard) 14, 15, 19, 230
graffiti 449, 460
Graphical User Interfaces (GUIs) 271
Grèber, Jacques 447
Greece, ancient 379
Green Pix Zero Energy Media Wall (Beijing) 453, *454*
green streets 427–8
green urbanism 519 *see also* ecological urbanism
Greenbelt towns 362
greenhouse gases 476
greenways 440
Greenwich Street (New York) 21
Greenwich Village 251
Greenwich Village Halloween Parade 142
Gregory, S. 142
Grodach, Carl 342, 343, 405–16
Gropius, Walter 12, 15, 224
ground water supplies 427
"growth machine" 128–9, 513
growth management policies 633
Gruen, Victor 18, 20, 435, 513–14
Guggenheim Museum (Bilbao) 93, 343, 408, 410–11, *412*, 416, 500, 550, 559
Gulf Coast Community Design Studio (GCCDS) 230
Gupta, A. 142
Gurian, E.H. 416

Gustafson, Kathryn 504
Gutkind, E.A. 19

Habermas, Jurgen 54, 190–1; *The Structural Transformation of the Public Sphere* 190–1
habitats: cities as 604–5
Habraken, N. John 576–7
Hack, Gary 342, 343–4, 446–61
Hadacheck v. Sebastian (1915) 179–80
Hall, Peter 576
Hall, Stuart 213
Halperin, Lawrence 20
Hampstead Garden Suburb 37
Hampton, Keith 130
Hannigan, John 129
Hardin, Garrett: "The Tragedy of the Commons" 187
Harley-Davidson Museum (HDM) 411–13, *413*, 416
Harris, C.D. 352
Hart, Keith 575
Harvard 15, 42, 43 *see also* Graduate School of Design
Harvey, David 103, 104
Haussmann, Baron Eugene 32, 239, 393, 579
Hawken, P. 594
Hayden, Dolores 361; *Redesigning the American Dream* 152
health 109 *see also* obesity; public health
heat islands 204; and health 203; minimizing of effect of on streets 427
Hegemann, Werner: *The American Vitruvius* 13–14
Heidegger, Martin 102, 495–6
Heritage Foundation 638
Herzfeld, M. 144, 146
Hester, Randolph T. 333, 334, 335, 436, 437–8, 500, 586, 619–29
high theory 58, 65, 65–7, *66*
highway urbanism 31
Hill, David 514
Hill, Kristina 628
Hinshaw, Mark 595
Hippocrates 600
Hiss, Tony 258
historic preservation commissions 176
Historical Materialism 87
historical significance 168
historically preserved neighborhoods 144
history: taking account of 607
Hockney, David 245
Holden, Meg 67
holistic systems thinking 623
Hollywood 209, 210
Holston, J. 145–6, 567
Holtzer, Jenny 449
home rule state 178
homeless individuals 192
homeowners associations 189, 371–2, 381, 590
Hong Kong 139, 449, *450*
Hood, Walter 434
HOPEVI scheme 490, 520
Hou, Jeffrey 276, 329–38
Hough, Michael 507, 612, 613, 620
housing 119–20, 472, *472*; and asthma 203; demand versus supply 472, *472*; and informal city 580–1; Latin American cities 116; need for upgrading of conditions 575; and the poor 576–7; smart growth and affordable 636–7; *see also* public housing
Housing and Slum Clearance Act (1949) 19
Houston: Discovery Green 660–1, *660*
Howard, Ebenezer 36–7, 369, 370, 371–2, 601, 617; *Garden Cities of To-morrow* 358, 371
Huber, N. 211
Hudnut, Joseph 14, 15
Hudson-Smith, A. 270
Human Interface Technology Laboratory New Zealand (HIT Lab NZ) 268
Human-Computer Interactions (HCI) 262
Hummer 591
"100 percent corner" 352
Hurricane Katrina 230, 325, 674
Hutchison, R. 129
Hybridity: and integral urbanism 591–2
hypermarkets 396

iconic architecture 92–3, 407–8, 538, 559
identity 401; and non-places 498–9; place *see* place identity
Illuminating Clay 268, *269*, 271
image 91, 401, 537; and city branding 542–3, 549
image/object recognition 271
imageability 165, 496, 500, 502, 537
imaginative draughtsmanship 243, 244
immigrant communities 538, 564
Imola: da Vinci's map 249, 250
Inam, Aseem 586, 632–41
incentive zoning 21
incremental projects 608
India 7, 212
Indian-Americans 143
Indonesia 580
industrial cities, early 554
inequality 485, 488, 492
informal city 539, 574–82; and density 581; and housing 580–1; infrastructure upgrading 579; integration with the formal city 579–80; jobs and economic development 580; small interventions 577–9; and urban design 576–7; *see also* slums
information society, rise of 495

infrastructure: and ecological urbanism 604; and landscape urbanism 615–16; upgrading of in informal city 579
inner cities: vs outer cities 556–8
Innes, J.E. 333, 336–7
installation art 461
Institute of Transportation Engineers (ITE) 421, 518–19
integral urbanism 586, 591–8; and authenticity 592, 594–6; goal of 596; hybridity and connectivity 591–2; and porosity 592; practicing 593–4; and preservation 593; qualities 591–3, 597; speaking across fissures 596; trends lending toward 596–7; and vulnerability 592
integrated approach 585
International Making Cities Livable Movement 595
International Movement for an Imaginist Bauhaus (IMIB) 101
International Sociological Association 125
Internet 261; and citizen participation 334–5; impact on design studios 233–4; shopping on 397; use of 261
Irazábal, Clara 438, 562–71
irony 589–90
Irvine (California) 502–4, *503*
Isaacs, Reginald 15–16, 18
Islamic city 116
Islamic codes 526
Italy 156, 253
Izenour, Steven 18

Jacobs, Allan 22, 25, 512; "Toward an Urban Design Manifesto" 24
Jacobs, Jane 18, 19, 24, 31, 35, 170, 251, 409, 436–7, 511, 512, 595, 602, 621, 663, 664; *The Death and Life of Great American Cities* 19–20, 35–6, 330, 437, 665; *Exploding Metropolis* 514
Jahn, Helmut 313
James, William 58, 67
Jameson, Fredric 98
Japan 253
Jencks, Charles 434
Jorn, Asger 101
Journal of the American Planning Association 467
Journal of Environmental Psychology 61–2
Journal of Urban Design 41
Just Compensation clause 179, 181

Kahn, Louis 16
Kaliski, J. 334
Kampung Improvement Program (KIP) (Indonesia) 580
Kant, Immanuel 62, 66, 99
Kantor, P. 354, 465
Karnataka (India) 669
Katz, Peter 64
Kayden, Jerold S. 110, 175–85
Keats, John 360
Kelbaugh, Douglas 276, 317–27, 514
Keller, Susanne 131, 381
Kelling, George 667
Kelo v. City of New London (2005) 181–2
Kennedy Center (Washington, DC) 406
Kenny, J. 212
Kent, T.J. 528
Kepes, Gyorgy 54, 71–3, 75, 81; *Arts of the Environment* 75–6; *Language of Vision* 73–4; *The New Landscape in Art and Science* 74
Khayelit sha Project (Cape Town) 669
Kibera (Nairobi) 577, *578*
Kim, Joongsub 232
kinetic city 676, 678–9, *680*, 681, 682, *683*
Kitchen, T. 667
knowledge-based economy 485
Kohler Act 180
Kohn, Margaret 110–11, 186–95
Kolkata 544
Koolhaas, Rem 35, 54, 71, 76–7, 78, 81, 312, 449, 538, 594
Kopp, Anatole: *Town and Revolution* 102
Kotkin, Joel 364–5
Kotler, M. 380
Kounkuey Design Initiative (KDI) 577, *578*
Kreditor, Alan 64
Krieger, Martin H. 220, 238–48
Krier, Leon 51, 447, 465, 511–12, 516, 520, 621
Kuhn, Thomas 58
Kumic, I. 92
Kunstler, James Howard: *Geography* 361
Kunzmann, Klaus 342, 343, 391–403
Kyoto: water use 627

LabelMe 271
Lagos 78
Laguna West (Sacramento) 514
land economics: and downtowns 351–2
land uses: feminist critique of the separation of 151–2
landmarks 165, 171
landscape architecture 1, 41, 85, 330
landscape ecology 586
landscape urbanism 41; and metropolitan form 611–17
Lang, Jon 537, 541–50
Lang, Robert 611
language 71
Las Vegas 92, 401–2, 447, 500, 532
Latin America: Americanization of 563–4; model of city structure 115–16; and murals 121; slums and squatter settlements 138–9

Latina/o ethnoscape 562, 564, 566, 567, 568, 570
Latino New Urbanism (LNU) 570
law, urban design 109, 110, 175–85; author's principles of 184–5; and government power 176–8; and individual rights 178–83; and local governments 177–8; and private governance regimes 182–3; and private property rights 179–81; rule versus discretion debate 183–4; and zoning 177, 178
Lawrence-Zúñiga, Denise 110, 137–46
Laws of the Indies (1573) 115
Le Corbusier 12–13, 16, 18, 513; and open paradigm 34–5; Plan Voisin for Paris 12, 30, 34; Radiant City concept 12, 34; and tower-in-the-park model 7, 12, 30, 34–5, 504; *La Ville Contemporaine* 12, 347
Le Prince, Louis 208
leadership: and design codes 300, 302
Leadership in Energy and Environmental Design (LEED) 387
"learning by surprise" vision 64–5
LED arrays 453
LEED-ND rating system 343, 387–8, 389, 521
Lefebvre, Henri 55, 88, 98, 100, 102–5, 106; concept of totality 104–5; *Critique of Everyday Life* 102–3, 105; *De l'État* 105; *The Production of Space* 103, 104; *The Urban Revolution* 103, 104–5, 106
legibility 409, 620; elements affecting 165
Lehrer, Ute 276, 304–16
Leigh, N.G. 212
Lessig, Lawrence 187, 188
Letchworth Garden City 37, 357, *359*, 374
Lettrist International (LI) 101
Levine, Jonathan: *Zoned Out* 473, 474
Levittown (New Jersey) 130
Lewis, Mumford: *The City* 360
Lewis, Oscar 138
Libeskind, Daniel 316, 399, 449
Librino (Sicily) 504–6, *505*, *506*
light-rail systems 38
likeability 166
Lima (Peru) 138
Lin, Maya 309
Lincoln Center (New York) 406, 407, 408, 413–14, *415*, 416
Lindsay, Mayor John V. 21
Litman, Todd 473
livable street project 170
live mapping 263–4
Liverpool University 42
Llewellyn Park (New Jersey) 361
Lobo, S. 139
local governments: and law 177–8
locality: building of, by immigrants 143
locational branding 543
locational socialization 500
Lofland, Lyn 361, 382
loft living 350
Logan, John 128, 129, 513
Logue, Edward 20
London 258, 490, 546; Bangladeshi neighborhood in East *117*; congestion charge 262; virtual model 267, *267*
Loomis, John 680
Los Angeles 116, 144, 337, *468*, 556–7; Augustus Hawkins Park 624, *625*, *626*; bus stop *655*; and compactness 468, 469; density of urbanized area 556–7; downtown 350, 351, *351*, 353; The Grove 402; mass migration to 558; Melrose Avenue *459*; neighborhoods 380; and Orange County 557; parks 569; as a resilient city 675–6, 679; South Central Farm 568, 570; surrounding of by Big Wild 626; urbanization of suburbia 557; virtual model of 266–7
Los Angeles County Performing Arts Center 406
Loukaitou-Sideris, Anastasia 345–54
Louvre 406, 408
Low, S.D. 140, 144, 145
low theory 58, 65–7, *66*
Lowell (Massachusetts): Urban National Park 622
Lukács, Georg 98
Lund (Sweden) 132
LUTRAQ (Land Use and Transportation and Air Quality Project) 514
Lynch, Kevin 18, 24, 43, 53, 54, 55, 63, 64–5, 71, 75, 81, 109, 208, 210, 289, 352, 409, 436, 463, 496, 497, 601–2, 608, 615; *Good City Form* 24, 99, 600, 601, 605; *The Image of the City* 19, 64, 75, 330, 496, 537, 543, 620; *Wasting Away* 602
Lyndon, Donlyn 621

Macarthur Foundation 559
Macdonald, Elizabeth 342, 343, 419–30
McHarg, Ian 23–4, 25, 37, 502, 602, 604, 627; *Design with Nature* 23, 27, 600, 612–13
Maciunas, George 449
McLoughlin, Brian: *Centre or Periphery* 88
McLuhan, Marshall 92
McNally, Marcia J. 586, 619–29
McPartland, John 360
Madanipour, Ali 464, 484–92
Madden, Kathy 587
Madeleine Church (Paris) 451, *452*
malls *see* shopping malls
Mangin, William 138
Manhattan 257
Manheim, Marvin 54
maps *see* cartography
market versus planning debate 464
market-driven urbanism: and downtowns 348

marketing: and city branding 541–2
markets 394–5
Marsh, George Perkins 601
Marville, Charles 220, 239, 241
Marx, Karl 100, 103, 104, 105, 128; *Capital* 87, 98
Marxism 98, 102, 104
Maryland: Smart Growth program 633–4
mass production: and suburbs 359–61
Massachusetts: and smart growth initiative 634–5
master plans/planning 361, 362, 363, 366, 528, 594, 615; and suburbs 361; and transit oriented development 649
Mathur, Anu 604
Maturana, Humberto 70
Maya 247
Mayor's Institute on City Design 325
meanings: conveying of by places 165–6
Medellin (Colombia) 580
media tools (for urban design) 238–48
megacities 555–6
megalopolitan regions 555–6
memorials, temporary 450–1
Memphis: sanitary survey (1879-1980) 201
METRAC (Toronto) 156, 668
metrocore 513–14, *515*
metropolitan urbanization 554–5
metropolitan form 586, 611–17
Metropolitan Museum of Art 406
Mexican Americans 564
Mexico 121, 669
Mexico City *195*
Meyerson, Martin 16, 17–18, 19, 133
Miami 140
Michael Reese Hospital (Chicago) 16, *17*
Michelson, William 110, 125–33
Middlesborough 18
migrantscapes 564
migration 485
Milan 258
Milan Polytechnic 225
Miles, Malcolm 408
Milgram, Stanley 460
Millennium Park (Chicago) 33, 453, 501, 504
Milligan, Melinda 500
Minneapolis: Nicollet Transit Mall 349, *349*; Public Library 453
minority women 151, 152
Miraftab, F. 566–7
Mirrored Cities 267
Mitchell, Don 189
Mitchell, Robert B. 16
Mitgang, Lee 228
Mitterand, François 545
mixed communities 489, 491
mixed-life places 343, 432–43; definition 436; as democratic space 438–9; and nature 438; limitations of 343, 433, 434–6; precedents for 436–7; principles of *438*; qualities needed to create 437–8; role of form and urban design 442–3; typology of 439–41, *440*; users of 439
mobile phones 261; tracking use of 263
Mobilizy 268
Model Cities program (1966) 330
models/modeling: 3-D 263, 265, 266; city structure and geographic 114–16; computer-based 253–4; physical 253
modern: versus tradition aesthetics 373–4
Moholy-Nagy, L. 71
Molotch, Harvey 128, 129, 513
mono-centric model 352
montage 210, 238
Monterey Park (California) 140
Montgomery, Roger 20, 25
Moore, Charles 325
Moos, R. 132
Morgan, Julia 224
Morocco 145
Moser, Caroline 670
Moses, Robert 18, 251, 359, 407
Moudon, Anne V. 41
Moule, Elizabeth 51
Mozingo, Louise 624
Mukhija, Vinit 538–9, 574–82
multi-centric strategy 611
multi-modal streets 424–6, *425*
multiculturalism: and citizen participation 333
multiethnic spaces 538
multinational franchise architecture 93
Multiplicity 78, 679
multiway boulevards 423–4, *424*
Mumbai 579, 581, 665
Mumford, Lewis 14, 15, 22, 39, 360, 372, 380, 601
Murakami, Takashi 407
murals 121, 456, *457*, 459, 461
Muschamp, Herbert 596
museums 406, 407, 500
mystery 167

Nairn, Ian 37
Naples 548
Napoleon III 32
Nasar, Jack L. 110, 162–71
National Architectural Accrediting Board (NAAB) 227
National Association of Home Builders 639
National Council on Disability 202
National Endowment for the Arts (NEA) 324
National Resources Defense Council 641

National Science Foundation 252
National Slum Dwellers Federation (NSDF) 579
National Survey on Communities 471
Native Americans 142
natural disasters/hazards 587, 604, 608, 627–9, 674
Natural Resources Defense Council (NRDC) 387
naturalness 167
nature/natural world 22, 586, 602; benefits of access to 438, 624; and cities 603–4, 613; and mixed-life spaces 438; and new towns 375; reincorporating 624; *see also* ecological urbanism
Négri, Tony 106
Neighborhood Matching Fund (Seattle) 336, *336*
neighborhood unit 14, 15, 18, 26, 31, 199, 343, 372, 382–3, *383*, 388–9, 521, 533; critics of 84; essential components of 382; influence of on design of residential environments 384; layout 383–4, *383*; location of school in center 382–3, 384; and New Urbanism 386–7, 388; size of 382–3
neighborhood(s) 341, 343, 379–89, 488–9, 501–2; adoption and changing of by ethnic groups 458–9; as basis for social integration 488–9; characteristics 380; and common interest developments 381, 382; and community 382; defining 379–82, 388; as economic entities 381–2; health disparities 202–3; and mixed-life places 441; New Urbanist development strategies for 385–7; sustainable 387–8; types of 171; use of flux in transforming neglected 456
Nelson, Arthur C. 463, 464, 467–79
neocorporatism 92
neoliberalism 192, 575, 666
Neo-Traditional Town Planning 556
Netherlands: postwar housing 22
network society 78
New Babylon 101
New Orleans 496, 497, 674
New Regionalism 538, 553, 559
New School and Cooper Union 15
"new town blues" 374
new towns 22, 37, 342, 358, 369–77, 486; definition 369–70; ecological issues 375; and economic development 574–5; and garden city tradition 370, 371–2, 376; and nature 375; physical design and aesthetics 373–4; planning and design traditions 370–3; planning from scratch 375–6; size issue 371; social aims 374; transportation issues 374–5
New Urban Design: difference between Mainstream Urban Design 90
New Urban Sociology 128–9
New Urbanism 14, 19, 64, 152, 186, 204, 325, 342, 358, 385–7, 447, 510–22, 556, 559; and authenticity 595; benefits of 385; and compact city 465; in conflict with Secured By Design guidelines 666–7; debates and controversies 519–21; and ecological issues 519; and ethnoscapes 569–70; features of 363, 510; influence on public policy 387; Latino (LNU) 570; neighborhood development strategies 385–7; and place identity 496; principles and features of 363, 385, 510, 635–6; roots and influences 385, 511; and rural-urban transect (RUT) 516, 517–19; and smart growth 635–6, 640–1; and suburbs 363; and sustainability 519; and Traditional Neighborhood Development (TND) 510, 511; and transit-oriented development (TOD) 511, 514–16; and urban network 516–17
New York/New York City 181, 335, 350, 460; Battery Park City 19, 30, 31, 32, 37, 541, 548; brand image 544; and certification 184; and complete streets 424; emergence of downtown 345–6; and ethnic enclaves 140, 142; failure to adopt congestion pricing program 477–8; gentrification of Lower East Side of 465; Highline project 338, *615*; Lincoln Center 406, 407, 408, 413–14, *415*, 416; Penn Central Transportation case (1978) 181; plazas 657; population growth 485; Project for Public Spaces 332, 657, 660; Rockefeller Center 117–18, 657–8; Soho neighborhood 35; Stuyvesant Town 18; Times Square 350, 454–5, *455*, *456*, 485–6, *486*; Urban Design Group 21–2; Washington Square Park 35; zoning in 21–2, 177, 527
New York Department of Transportation: *Street Design Manual* 22
Newhall *294*
Newman, Oscar 170, 487, 663, 665–6; *Defensible Space* 133, 665
NIMBY (not in my back yard) 590, 637
nodes 165, 167
Nolen, John 14, 358, 362
Nolli, Giambattista: *Pianta Grande di Roma* 19
non-places 463–4, 465, 498–9
normative theory 53, 67, 99, 463, 464
nostalgic reflex 589

Oakland (California) 337–8, *338*
Obama, Barack 235
Oberle, A.P. 564
obesity 421, 467, 476–7, 633
observation 110, 130–1, 169
Office of Technology Assessment 472
Oglethorpe, James 419
oil dependence: and climate change 475–6
Oldenberg, Ray 436
Olmstead, Frederick Law 14, 33, 352, 533, 601, 613
one-way-couplet village centers 521
open paradigm 30, 31–5
order 168

O'Reilly, Tim 263
Orenco Station (Portland) 514, 516
Orlando 479
Osaka 558
Osborne, Peter 103
Oslo (Norway) 156
Ostrom, Elinor 187
Ouro Preto (Brazil) 145
outer cities: vs inner cities 556–8

Pachube 266
Palace of the Soviets competition 307
Palazzo, Danilo 8, 41–9
Pareto Superiority 464
Paris 121, 239; Avenue de l'Imperatrie 33; Avenue Jean Jaurès 426; Boulevard Magenta 424, 426, *426*; boulevards 32–3; brand image 544–5, *544*; Hausmann's transformation of 32–3, 220, 239, 393, 579; La Défense 541, 548; Le Corbusier's plan for 34; Madeleine Church 451, *452*; Marville's images 220, 239, 241; Rue des Petits Carreaux 423
Paris Commune 100, 106
Park, Robert 380, 383
Parker, Richard Barry 37, 358
parks, urban 32, 33, 189, 441
parkway 32
participation, citizen *see* citizen participation
Pasadena: Del Mar Station *648*, *650*
Pasadena Rose Parade 142
paths 165, 171; curved 167
patriarchy 150
pattern language 62
peak-oil 354
Peattie, Lisa 139, 145, 146, 575
pedagogical traditions 8, 41–9
pedestrian crossings 429
"pedestrian pocket" 514
Pedestrian Pocket Book, The 325
pedestrianized shopping streets 399, 401
pedestrianized urban centers 15, 25
peer review 63–4
Peets, Elbert 13–14
Pei, I.M. 18, 20
Peirce, Charles 66
Pellow, Deborah 139, 143
Penn Central Transportation Company v. New York City (1978) 177, 181
Pennsylvania Coal Co. v. Mahon (1922) 180, 181
perception 162–5, 167–8; ecological theory of 252; probalistic theory of 252
performance art 449
performance standards 278, 280, 282
performing arts centers 405, 406, 407
Perkins, G. Holmes 15, 42
Perlman, J. 138–9
Permanence: versus flux 446–9
Perry, Clarence 14, 199, 343, 372, 382–4
Peru 139
Phadke, S. 670
phenomenology of images 242–4
Philadelphia 16, 543; Center City 350; image 546; Mural Arts Program 456, *457*; The Village of Arts and Humanities 332, 456–8, *458*
Philadelphia Planning Commission 16
Phoenix *592*
physical activity 200–1, 203, 204; and built environment 200–1; Piano, Renzo 409
piazzas, public 399
Pickett, Steward 609
Picturesque 256, 342
Pioneers 9, *10*, 14–16, 26
Pizarro, Rafael E. 111, 208–14
place(s) 63, 122, 138, 151, 252, 463–4, 465; and design codes 299–300; DNA of 595
place attachment 382, 495
place/community driven approach 658–9, *659*
place identity 465, 495–508; and authenticity 500–1, 502; high design versus the vernacular 500; and imageability 496; importance of 495, 506–7; and individual experience 496–8; Irvine (California) 502–4, *503*; Librino (Sicily) 504–6, *505*; Millennium Park (Chicago) 504; as a multi-faceted gradient 501–2, 507; and shared meanings 500; and urban design in the future metropolis 506–8
placelessness 500, 501, 624; versus authenticity 120
placemaking 587, 624, 654–61; aim of 654; foundation of 656–7; implications for urban design field 661
planned communities 369–70 *see also* new towns
planned unit development (PUD) 381
planning 262, 530–1; advocacy 24, 339; and design 60; problems with 528; and public health 199; and spatial political economy 88; and zoning 528
Planning Action 106
planning theory 53, 99
Plater-Zyberk, Elizabeth 38, 362, 511, 516, 529
Plaza Mexico (Lynwood, California) 564, *565*
plazas 18, 21, 25, 36, 170, 195, *195*, 393, 657
pleasantness 166, *166*, 167, 168
Plensa, Jaume 453
police power 176–7
political economy 87, 90, 91, 104 *see also* spatial political economy
political theory 109, 110–11, 186–95
politics: and simulations 253–4
polycentric city model 352
Polyzoides, Stefanos 519, 586–7, 644–52

Pompidou Centre (Paris) 409–10, *410*
poor, urban 138–40, 539; and gentrification 464–5; and urban design 576–7
pop-up city 683
Pope, Robert Anderson 527
porosity, and integral urbanism 592
Portland (Oregon) 335, 337, 469, *470*; Oregon's Urban Growth Boundary 364; Orenco Station 514, 516; Pearl District 435, *436*, 640; and smart growth 640; and urban containment 633, *634*
possibilism 127
postmetropolitan transition 538
postmodern urbanism 589–91, 594
postwar reconstruction 22
Potlatch (journal) 101
Potsdamer Platz (Berlin) 276, 309, 310–14, *311*, *313*, *314*, 453
Poundbury 447, *448*
poverty 485, 670 *see also* poor, urban
"Power of Ten": and public spaces 659–60
Poyner, Barry 666; *Design Against Crime* 666
pragmatism 58
Prairie Crossing (Grayslake) 364, *364*
privacy, rise of 191
private property rights 179–82
privately managed communities 182–3
privatization: and citizen participation 333–4; critique of 187, 188; and downtown renewal 348
proactive planning: as solution to sprawl 478
probabilism 126–7
probalistic theory of perception 252
Proctor, Robert 62
"production of space" 42, 54, 55, 79, 98, 101, 104
professionalism 26
Project for Public Spaces (PPS) 332, 657, 660
project/discipline driven approach 658, *658*
Proshansky, Harold 498
Pruitt Igoe (St Louis) 35, 576
psychogeography 100–1
psychology 252
Public Architecture in San Francisco 337
public art 453–4, 456, 459
public health 111, 198–204; accessibility and disability 201–2; and birth of urban planning 199; and built environment 198, 202–3; disparities at the neighborhood level 202–3, 203–4; and heat islands 203; housing and asthma 203; and physical activity 200–1, 203, 204; and urban design 109
public housing 353, 576, 621, 663, 665–6
public participation 61, 329; direct action 332; and FBCs 534; and project design 331–2; *see also* citizen design
public space ethnography 144–6
public space(s) 24, 25, 110–11, 341–2, 343, 432–4, 563, 654–61; agonistic character of 184; approach to making better 657–8; characteristics of well-used 657; children and use of 665; and commons 189; contemporary theories of 192–4; definition 432; and democratic theorists 193–4; diminishing and abandonment of 191, 591, 664–5; energetic 442; and FBCs 532; growth in research and literature on 432–3; lack of 432; and mixed-life spaces *see* mixed-life spaces; and mixed-use projects 434–6; multiethnic design of 568–70; place/community driven approach 658–9, *659*; "power of ten" 659–60; project/discipline driven approach 658, *658* ; regulation of 189; rights associated with 671; roads as 192–3; and secure cities 670; and segregation 193–4; and Street Life Project 656; and streets *see* streets; successful ingredients of 433; and urban design 433–4, 442; women's use of 152–3, 665
public sphere, bourgeois 190–2
public subsidies 472–3
public transportation: and suburbs 363; and women 154
public/private partnerships 490; transit oriented development 651–2
Punter, John 64
Pushkarev, Boris 18

quality of life 109, 341
Quebec 546
Quinta Monroy housing project (Chile) 581

Rabinow, P. 145
Radburn (New Jersey) 14, 372–3
Radburn planning 370, 372, 373–4
radiant city 31, 34
Radović, Darko 44
Raleigh-Durham 469, *470*
Rapoport, Amos: *House Form and Culture* 122
rational city 347
"Re-imagining Cities" conference (2008) 25
real-time city 264, *264*
realism 63
recentralization 555
redevelopment authorities (RDAs) 20
reflected appraisal mechanism 498
reflection-in-action 226
Regional Planning Association of America (RPAA) 372
regional supply: and sustainability 626–7
regional urbanization 538, 552–60; features of 555; planning and design implications 558–60
Regional/Urban Design Assistance Teams 324
regionalism 25, 510

regulations 277, 278 *see also* standards
Relph, Edward 500–1; *Place and Placelessness* 122
remote sensing 265–6
res communes 188
res nullis 188
res publicae 188–9
res universitatis 189
Residential Partnerships 171
resilient cities 587, 607–8, 674–84; equilibrium model 674–5, 679, 682; key concepts 679; non-equilibrium model 675, 679, 682; prequisites for developing urban design and 676; *see also* fixed city; good city; kinetic city
Reston City Center (Virginia) 19
retail centers 520
Rethymnon (Crete) 144
Retreats 171
reuse, urban 622–3
Richardson, Harry 464, 467–8, 474, 475, 477
Right to the City Alliance 106
Right to the City, The 491
rights: law and individual 178–83
Rio de Janeiro 575, 577
Rios, M. 336
Ripstein, Arthur: "Roads to Freedom" 192
Rittel, Horst 57, 60
ritualized events 142
roads: as public spaces 192
Robert Wood Johnson Foundation 476
Robertston, Jacqueline 21
Robin Hood 209
Robin Hood Gardens 24
Robinson, Charles Mulford 357, 358
Rockefeller Urban Design studies program 19
Rockfeller Center (New York) 117–18, 657–8
Rodwin, Lloyd 19
Roetenberg, R. 145
Rogers, Richard 312, 409
Roman law 188–9
Romans 526
Ross, Kristin: *Fast Cars, Clean Bodies* 106
Rouse, James 22
Rowe, Colin 18, 24
Rubik's Cube 677, *677*, 681, 683
Ruggeri, Deni 463, 465, 495–508
rule-versus-discretion debate 183–4
rules 278 *see also* standards
runoff 624
Rural-Urban Transect (RUT) 516, 517–19
rus in urbe 356
Ruscha, E. 242
Rutheiser, C. 145
Rybeczynski, Witold 533–4
Rykwert, Joseph 460
Saarinen, Eliel 16, 528
Sacramento 549
Sacramento Region Blueprint Plan 204
Sadler, Simon: *The Situationist City* 100
safe rooms 590
"safe spaces" 670
safety: and women 153, 154–6
safety audits: women's 663, 667–9
Sagalyn, L. 537
San Antonio 564
San Diego 547, 595, 596; Hillcrest 380; Horton Plaza 121; Rio Vista West TOD development *636*
San Francisco 35, 262, 350, 422; block structure of 257–8; Civic Center Plaza 352; downtown transformation 351; image of 546; Octavia Boulevard 423–4, *424*; simulation and development of downtown 221, 254–7, *255*, *256*, *257*; Sixteenth Street Mission BART Station 622
San Francisco Bay Area 332, 354, 514
San José (Costa Rica) 145
Sander, August 239, 240
sanitary movement 199
Santa Fe 189
Santa Monica: Third Street Promenade 20
São Paulo 144
Sasaki, Hideo 16, 502
satellite cities 14
Satoh, Shigeru 258
Savage, Susan 43
Savannah (Georgia) 419
Saville, Greg 667
Savitch, H.V. 354, 465
scenario building 25
Scheer, Brenda 538, 586, 611–17
Schnedier, R. 667
Schön, Donald 54, 220, 225–6, 233
Schumaker, S.A. 500
Schuman, Anthony 229
science: tension between design and 58–9, 61, 67
Science City (Tokyo) *376*
SCRUB (Society Created to Reduce Urban Blight) 459
sea levels, rise in 627–8
Seaside (Florida) 511, *512*, 521, 529
Seattle 350, 543, 596; and citizen participation 338; Neighborhood Matching Fund 336, *336*; Northgate Mall 435–6; Pike Place Market 331, 497, *497*
Second Life 267–8, 270
second order cybernetics 70, 76–81
secure cities 587, 663–71; challenges 670; and children 669; and CPTED 665–7, 668;

holistic understanding of 670; protection versus empowerment issue 664, 670; and public housing 665–6; and public space 664–5, 670; and Secured by Design principles 666–7; and Situational Crime Prevention 667; and streets 665; and Women's Safety Audit 663, 667–9
Secured by Design guidelines 666
segregation 193–4, 465, 485, 487–9, 491, 492, 590
Semantic Web 262
Sennett, Richard 59, 126, 382, 436; *Civitas of Seeing* 679
sense of place 252, 259, 497, 589
sensory overload 460
serial vision 208, 210, 213
Sert, Josep Luís 13, 15, 19, 22; *Can Our Cities Survive?* 13, *13*, 15; *The Heart of the City* 15; "The Human Scale in City Planning" 15
service economy 485
set pieces: creation of 343
setting records 169
Shanghai 258, 377, 548, 556
Shelley v Kraemer (1948) 18
shopping arcades 393–4, *394*
shopping malls 20, 120–1 190, 348, 353, 399, 432; cenetral city 393–4; *see also* suburban shopping malls
shopping streets 393, 399–400
shrinking cities 558
sidewalks 33, 167, 176, 203, 347, 399, 427, 664
Sieber, T. 145
Sierra Club 639
significant buildings 550
signage: laws 176; temporary 458–9; Times Square (New York) 455
signature architecture 343, 348, 354, 391, 559
SimCity 132, 487
Simon, Herbert 67, 279
simulations 110, 132, 220–1, 249–59, 265, 266–8; conceptual 249; and development of downtown San Francisco 221, 254–7, *255*, *256*, *257*; functions 251; historic examples 249; laboratories 253; and magnitude of change 254–5; modelling future experiences 251–2; and nature of change 258–9; perceptual 249, 250, 252; politics of 253–4; and rate of change 255–8; and real world 250, 252; as tool of manipulation 250; validation experiment 252
Singapore 541, 549
Sites, W. 353, 464–5
Sitte, Camillo 19, 36; *City Planning According to Artistic Principles* 36
Situational Crime Prevention 667
Situationists 100–1, 101–2
Sixth Sense 268–70
Sixtus V 447
Sjoberg, Gideon 115
skate parks 441
Skladanowsky Brothers 208
Sklair, Leslie 92–3
skyscrapers 12, 13, 117–19, *118*, 176; Asian 118; and Le Corbusier 34; role in shaping city structure 117–18; worldwide distribution of 118
slum clearance 12, 13, 19, 20, 330, 576
Slum Networking Project (India) 579
slums 106, 138–9, 539; designation of 527; growth of 574–5; label of 575; numbers living in 574; policy towards 576; *see also* informal city
Small Towns 171
Smart, Alan 139
smart code 516, 518
smart growth 152, 204, 387, 467, 471, 586, 632–41; and affordable housing 636–7; broad appeal and benefits of 638–9; definition 632; elements of a strategy of 635–7; factors for effective implementation 633; impact of 635; implementing of 639–41; initiatives 633–5; and Maryland 633–4; and New Urbanism 635–6, 640–1; obstacles and critiques of 632, 637–8; origin of term 632; rationale for 633; and sustainability 638–9; tool kit approach 641; and transit-oriented development 636
Smart Growth Network 632
Smartcode 529, *530*, *531*
smartphones 245
Smith, Adam: *The Wealth of Nations* 87
Smithson, Peter and Alison 22, 24
social capital 23
social distances 169–70
social exclusion 464–5, 485, 487–9
social inequality 485, 488, 492
social mixing 374, 489, 491, 492
social sciences 61–2; and urban design 59, 109, 110
social structure: and environment 128–9
Society for Promotion of Area Resource Centres (SPARC) 579
sociology 9, 110, 125–33; challenges facing connection between urban design and 129; diversity of 125–6; environment and behavior 126–8; methodologies 129–33; social structure and environment 128–9
Soja, Edward W. 538, 552–60; *Postmetropolis* 553
Sony 313, 315
Soto, Hernando de 576
Southern California Association of Governments (2009) COMPASS plan 204
Southworth, Michael 463, 465, 495–508
Soviet Union 102; and new towns 376
space syntax 19
spatial political economy 54, 84–94; and human geography 87–8; and planning 88; and urban design 88–9

special districts 21
specimen records 169
speed limits, streets 428–9
speed-bumps 422
Spirn, Anne Whiston 438, 586, 600–9
spontaneity 679
sports stadiums 350
sprawl 38, 364–5, 465, 467–79, 511, 521, 556, 633; causes of 469–71; and consumer preferences 471–2; costs of 474–7; cures for 477–9; definitions and characteristics of 467–8, 469; and density 468–9; and energy efficiency 476; impact of on physical activity and obesity 476–7; and land use regulation 473–4; measuring 468–9; and public subsidies 472–3; and regional urbanization 556; suburban 364–5, 370; and vehicle miles travelled 474–5; versus compact city 463, 464, 467–79; and zoning 473
squatter settlements 116, 138–9, 539
Staehli, Lynn 189
standards, design 276, 277–85, 292; apprehending the world 279–80; characteristics of 279–83; and criteria 280–2, *281*, 284–5; negative vs. positive 279; performance 278, 280, 282, 284; prescriptive 278, 280, 282, 284; product vs. process 279–80, 283; professional mind-set surrounding 278; specificity of 280–3; and streets 428
starchitects/starchitecture 343, 348, 354, 500, 537–8, 559, 590
stations, transit 646–50; access and parking at 647–9; building design 650–1; neighborhood, district and corridor coding 649–50; neighborhood, district and corridor visioning 649; types of by mode 646–7, *646*
Stein, Clarence 14, 18, 42, 372
Sternberg, E. 409
Stevenage 384
Stimmann, Hans 312
Stockholm 373, 396
Stokols, D. 500
Stonorov, Oskar 16, 18
stormwater systems 624
storytelling 238, 239, 247–8; digital 262
Strachan, D.P. 203
strangers, welcoming of 489
Strauss, Anslem: *Images of the American City* 57, 543
Street Life Project 656
street vending 395
streetcar 644–5
streets 18, 25, 35, 36, 343, 419–30, 519; flexible 426–7; Functional Classification of 420–1; green 427–8; intersection designs 429; lack of pedestrian crossings 429; and livable street project 170; as mixed-life places 441; movement functions 420–1; multi-modal 424–6; and multiway boulevards 423–4, *424*; as places for social exchange 420; as public space opportunities 421–2; re-blocking 579; and safety 665; shopping 393, 399–400; spatial challenge 428; speed limits and wide travel lanes 428–9; standards challenge 428; and traffic 420–1; and traffic calming 422–3, 441; tree restrictions 429–30; widths 419, 428; and woonerfs 170
"Streets as Places" 441
Strickland 212
Stubben, Joseph 527
studio process *see* design studios
suburban shopping malls 348, 353, 393, 394, 396
suburban sprawl 364–5, 370
suburbanization 19, 486, 554
suburbs/suburbia 14, 32, 37, 341, 342, 356–66, 370; and automobiles 362–3; and City Beautiful movement 357–8; critical discourse on 359–60, 361; differentiation of 557; dwelling design 365, 366; early 356; and ecological issues 365; efforts to maintain tradition and enhance community 362–3; establishing control 361–2; features 356; flattening out of density gradient between urban areas and 555; and Garden City movement 357, 358; growth of 360, 362; and mass production 359–60; and master planning 361; need for substantial changes in design 365; and New Urbanism 363; and new-town model 362; and Picturesque 356; primary objectives of 359; and public transportation 363; recasting design 365–6; reformist agenda 362; and *rus in urbe* 356; and single-family house 357; urbanization of 555, 556, 557; and women 150, 152
Summerson, John 34
Sunstein, Cass 57
surveillance 220
surveillance videos 242
surveys 110, 131–2; cross-sectional 131; longitudinal 131; time-use 131–2
sustainability 7, 24, 25, 26, 154, 299, 341, 441, 510, 586, 619–29; basic tenets of 622; building design and transit-oriented development 651; challenges to design 623; cultural and biological diversity 624–6; and design codes 299; ecological dwelling 623–4; and holistic systems thinking 623; and LEED-ND rating system 387–8, 389; and livable density 620–1; and multifunctional infrastructure 623; and natural hazards 627–9; and New Urbanism 519; principles for 619–20; and regional supply 626–7; and reuse 622–3; and rise in sea levels challenge 627–8; and smart growth 638–9
sustainable cities: prescription for 619–20
Sustainable Design Assessment Team (SDAT) 324
sustainable neighborhoods 387–8, 389

Suttles, Gerald 127, 129, 380
SUVs (sports utility vehicles) 591
Swindon *293*
Sydney 92
symbol-making 72–3
Synder, N.G. 385

Tafuri, Manfredo 89
Takahashi, Lois M. 111, 198–204
Talen, Emily 285, 465–6, 526–34
Tange, Kenzo 504–5
Tangible Infoscapes 266
tangible user interfaces (TUIs) 268, 271
Tate Modern (London) 408
Taut, Bruno 210
Tax Increment Financing 595
teaching *see* education, design
Team 10: 22–3, *23*, 25
téchne/poiesis dualism 49
television 209
temporality 679
temporary elements: on construction sites 451–3
temporary signage 458–9
Terminal Complex (Cleveland, Ohio) 117
territorial spaces 169
terrorism 664
Terrytown 230
Texas Stadium *613*
Thailand 146
Thames Valley 356
theme parks 391, 589
theming 92, 129, 343; and consumption design 401–2
theories-in-action 53, 54
theory: and urban design 65, 85–6, *86*
Thief of Baghdad, The 209
Thiel, Philip 54, 208, 210
Thompson, Ben 22
Tijuana (Mexico) 121
time lapse photograph 170
time-use survey 131–2
Times Square (New York) 350, 454–6, *455*, *456*, 485–6, *486*
Tokyo 258, 450, *451*; Ginza district 258; Rappongi Hills complex 490, *491*
tomography 240, 242, 245, 248
Toronto 130; Dufferin Grove 669
Toronto School of Criticism and Innovation (TSCI) 106
totality: Lefebvre's concept of 104–5
tourist circuits 395
tower-in-the-park model 7, 12, 20, 30, 31, 34–5, 504
Town and County Planning Act (1947) (Britain) 183
town planning 534
Town Planning Associates 15
Town of Tioga 186, 188, 195
townscape movement 36, 37
Townsend, Anthony 264
Traditional Neighborhood Development (TND) 465, 510, 511, 511–14, 516, 519
traffic calming 170, 422–3, 441
traffic congestion 513
traffic management: and digital technology 261–2
training *see* education, design
transaction costs 192
Transect 288, 466
transit: and development 644–5
transit malls 349, *349*
transit-oriented development (TOD) 38, 325, 363, 465, 510, 511, 514–16, 519, 586–7, 635, 644–52; access and parking at stations 647–9; and building design 650–1; choosing a route 6456; and form-based codes 649–50; and master plans 649; neighborhood, district and corridor visioning 649; neighborhood, district and corridor coding 649–50; phasing and implementation 651–2; and place character 648–9; prerequisites for successful 586–7; public/private partnership 651–2; and smart growth 636; station types by mode 646–7, *646*
transit priority streets 426
transit route, choosing 645–6
transportation: and development of downtowns 346; digital technology and planning 261–2; new towns 374–5; and women 153–4; *see also* public transportation
trees: restrictions of on streets 429–30
Trotsky, Leon 102
True Urbanism 595
Truman Show, The 210
Trust for Public Land 639
truth: correspondence theory of 62, 66–7; and distinction between high and low theory 66–7
Tuan, Yi-Fu 122, 497
Tung, Antony 38
Tunnard, Christopher 18, 19; *Manmade America* 19
Turner, John C. 138, 576
Twin Rivers (New Jersey) 131
Tyrwhitt, Jacqueline 17

Ullman, E.L. 352
UNESCO: "Paths of Thought" program 62
unitary urbanism 100, 101
United Nations 575
universities: internationalization of 47–8; survey of urban design programs 44–7, *45–6*
University of Michigan (UM) 317, 318, 327
University of Pennsylvania: Civic Design Program 42–3

University of Washington (UW): charrettes 317–27; "PK (Post Katrina) studio" 230
Unwin, Raymond 14, 36, 36–7, 37, 358, 362, 534
upkeep 167
urban complexity 76–7
urban containment 633
urban cultural landscape 117–21
urban design: functions 341; and independence 84, 85; interdisciplinary nature 85; roots and thinkers of modern 9–26; tension between science and 58–9, 61, 67; ultimate goal of 109
urban design graduate programs 19
Urban Design Group (New York) 20–1
Urban Design International 41
urban flux *see* flux, urban
Urban Land Institute (ULI) 324, 472, 641; Advisory Services 324–5
urban morphology 18, 25
Urban Network 516–17, *517*, 518
urban problems: nature of 60–1
urban redevelopment/renewal projects 19, 20, 21, 347–8, 353, 408–9, 486, 490 576
urban renaissance 398–9
Urban Renewal Act (1954) 330
Urban Renewal Agency (US) 19
"Urban Revisions" exhibition 596
urban sensorium 220, 239, 240–1, 246–7
Urban Simulation Team (University of California) 266
urbanism: and capitalism 98, 103; old versus new 465, 510–22
urbanization 9, 25, 103, 383, 406, 586; and cinematic arts 208–9
U.S. Green Building Council (USGBC) 387–8, 521
user-centricity 679

Vale, L.J. and Campanella, T.J.: *The Resilient City* 608
Valley Section 22–3, *23*, 25, 518
van der Rohe, Mies 17
Van der Ryn, Sym 514, 519
Vancouver: EcoDensity project 621
Varela, Francisco 70, 77
Vastu Shilpa principles 7
Vaux, Calvert 33, 352
vehicle miles of travel (VMT) 474–5, *475*
Venturi, Robert 18; *Learning from Las Vegas* 447
Venusfort (Tokyo) 02, *402*
Verma, Niraj 53–4, 57–67
vernacular 342
video games 213
videoconferencing: and design studios 234
videos 241–2
Vienna Ringstrasse 36, 307
Viennese gardens 145
Vietnam memorial competition (1981) 309
Village of Euclid v. Ambler Reality Co. 180–1
Virolio, Paul 209
Virtual Cities 266–7
virtual reality 266–70
virtual shopping 396–7
virtual simulation 266–7
vision: Kepes' language of 71–4
visioning exercises, and citizen participation 331
visual documentation 220
Visual Rights Act 459
Vitruvius 600
vulnerability: and integral urbanism 592

Waldheim, Charles 612
Waldron, Jenny 192
walkability 7, 32, 37, 38, 60–1, 66
Walker, Peter 502
walking: and built environment 200–1
Wallace, David 24
walls 487, 487–8
walls of communication 453–4
Walzer, Michael 434
Washington, D.C. 132, 144, 251; Rosslyn-Ballston Corridor 516
Washington Monument 451, *453*
water quality/supply 607, 627
waterfront redevelopment 145, 547, 548
Web 2.0 262, 262–3, 270
Web based Geographic Information Systems (WGIS) 263
Webber, Melvin 57, 60, 382, 464, 498–9
webcams 265–6
Webster, C. 381
Wiener, Norbert 70, 71, 81 Weinstein, Richard 20, 21
Wellman, Barry 130
Wells, H.G. 31
West Bengal 547
West Indian Labor Day Parade (Brooklyn) 142
whirlpool cities 31, 32
Whitzman, Carolyn 587, 663–71
Whyte, William H. 18, 22, 25, 131, 170, 266, 360, 433, 436, 512, 587, 656; *The Social Life of Small Urban Space* 330, 656
"wicked problems" 57, 60, 585
WikiCity 263–4
Wildlands Project 626
wildlife 605
Williams, B. 144
Williams, J. 611
Williamson, June 435
Willmott, P. 24, 384
Wilman, Gil 101
Wilson, Elizabeth 665
wireless sensor networking (WSN) 262
Wirth, Louis 126

women 110, 150–7; constraints on use of environments and mobility 151, 153; and distinction between public and private roles 151–2; fear in city experiences 154–5; leadership roles in shaping of cities and communities 156–7; and patriarchy 150; programs to improve gender equity in urban design 156; and safety in urban environments 153, 154–6; and separation of land uses 151–2, 153–4; and suburbs 150, 152; and transportation 153–4; use of public spaces 152–3, 665
Women's Design Service 668
Women's Safety Audit 663, 667–8
woonerfs 170, 422
World Trade Center Memorial Competition 306
World Trade Center (New York) 316
Wright, Gwendolyn 577
Wright, Henry 14, 372

Xerox Star workstation 271

Yahkleef, A. 93
Yahoo's TagMaps 265
Yaro, Robert: *Dealing with Change in the Connecticut River Valley* 25
Yeh, Lilly 45
Young, Iris Marion 193, 194
Young, Michael 24
Youngstown (Ohio) 331

Zizek, Slavoj 106
zoning/zones 21, 115–16, 176, 366, 466, 526–34, 649; and downtowns 347; and *Euclid vs. Ambler* (1926) 527, 528; failure of 526–8; form-based 466; history 527; incentive 21; Latin American cities 115–16; and law 177, 178; rationale for 527; special districts 21; and sprawl 473; tension between planning and 528; vs form-based codes 526–34
zoning ordinances 277
Zukin, Sharon 129

www.ingramcontent.com/pod-product-compliance
Lightning Source LLC
LaVergne TN
LVHW080758011026
849160LV00015B/1485
* 9 7 8 1 1 3 8 7 7 6 5 4 8 *